STUDENT'S SOLUTIONS MANUAL

EMILY KEATON

INTERMEDIATE ALGEBRA

THIRD EDITION

Tom Carson

Bill Jordan

Seminole State College of Florida

Addison-Wesley
is an imprint of

PEARSON

Reproduced by Pearson Addison-Wesley from electronic files supplied by the author.

Publishing as Pearson Addison-Wesley, 75 Arlington Street, Boston, MA 02116.

ISBN-13: 978-0-321-62694-3
ISBN-10: 0-321-62694-X

1 2 3 4 5 6 BB 14 13 12 11 10

Addison-Wesley
is an imprint of

PEARSON

www.pearsonhighered.com

CONTENTS

Chapter 1

Real Numbers and Expressions

Exercise Set 1.1

1. A collection of objects.

3. If every element of a set B is an element of a set A, then B is a subset of A.

5. Rational numbers can be expressed as a ratio of two integers; irrational numbers cannot.

7. {Saturday, Sunday}

9. {January, June, July}

11. {North Carolina, North Dakota}

13. $\{0, 1, 2, 3, 4\}$

15. $\{3, 6, 9, \ldots\}$

17. $\{-1, 0, 1\}$

19. $\{\ \}$ or $\varnothing$

21. $\{x|x \text{ is an integer}\}$

23. $\{x|x \text{ is a letter of the English alphabet}\}$

25. $\{x|x \text{ is a day of the week}\}$

27. $\{x|x \text{ is a natural-number multiple of } 5\}$

29.

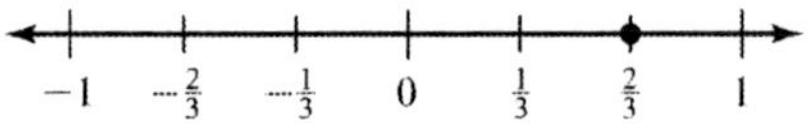

31.

33.

35.

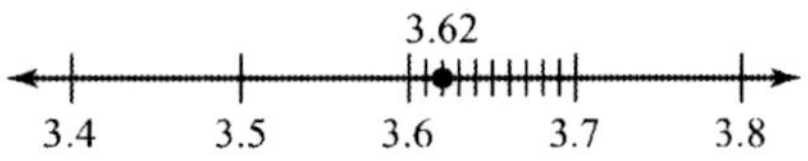

37. False because n is not a vowel.

39. True because "go" is a verb.

41. True because 4 is a rational number.

43. True because "James" is not the last name of any U.S. president.

45. False because -0.6 is a real number.

47. True because each of the vowels is a letter of the English alphabet

49. True because each number listed is an integer.

51. True because the set of integers is a subset of the rational numbers.

53. True because the set of irrational numbers is a subset of the real numbers.

55. False because a real number can be either rational or irrational but not both.

57. True because the whole numbers consists of the natural numbers plus 0.

59. True because the definition of rational number is "a real number that can be expressed in the form $\frac{a}{b}$, where a and b are integers and $b \neq 0$."

61. True because each natural number can be expressed in the form $\frac{a}{1}$, where a and 1 are integers and $1 \neq 0$, which is a rational number.

63. $|2.6| = 2.6$

65. $\left|-1\frac{2}{5}\right| = 1\frac{2}{5}$

67. $|-1| = 1$

69. $|-8.75| = 8.75$

71. $-6 > -8$ because -6 is farther right on a number line than -8.

73. $0 > -1.8$ because 0 is farther right on a number line than -1.8.

75. $3\frac{4}{5} > 3\frac{3}{4}$ because $3\frac{4}{5}$ is farther right on a number line than $3\frac{3}{4}$.

77. $|-3| = |3|$ because the absolute value of -3 is 3, which is the same as the absolute value of 3.

79. $6.7 = |6.7|$ because the absolute value of $|6.7|$ is the same as 6.7.

81. $\left|-\frac{2}{3}\right| < \left|-\frac{4}{3}\right|$ because $\left|-\frac{4}{3}\right| = \frac{4}{3}$, which is farther right on a number line than $\left|-\frac{2}{3}\right| = \frac{2}{3}$.

83. $-0.6, -0.44, 0, |-0.02|, 0.4, \left|1\frac{2}{3}\right|, 3\frac{1}{4}$

85. $-12.6, -9.6, 1, |-1.3|, \left|-2\frac{3}{4}\right|, 2.9$

87. $\{2001, 2006, 2008\}$

89. $\{2001, 2006, 2008\}$

91. {Comedy, Action/Adventure, Family}

93. {Action/Adventure, Comedy}

95. $\{885, 560, 544\}$

97. { } or ∅

Exercise Set 1.2

1. The order of the addends is changed with the commutative property of addition, whereas the grouping is changed with the associative property of addition.

3. Their sum is zero.

5. To add two numbers that have the same sign, add their absolute values and keep the same sign.

7. To write a subtraction statement as an equivalent addition statement, change the operation symbol from a minus sign to a plus sign and change the subtrahend to its additive inverse.

9. Additive inverse

11. Additive identity

13. Multiplicative identity

15. Multiplicative inverse

17. Multiplicative identity

19. Additive inverse

21. The additive inverse of 8 is -8 because $-8+8=0$.

    The multiplicative inverse of 8 is $\frac{1}{8}$ because $8\cdot\frac{1}{8}=1$.

23. The additive inverse of -7 is 7 because $-7+7=0$.

    The multiplicative inverse of -7 is $-\frac{1}{7}$ because $-7\cdot\left(-\frac{1}{7}\right)=1$.

25. The additive inverse of $-\frac{5}{8}$ is $\frac{5}{8}$ because $-\frac{5}{8}+\frac{5}{8}=0$.

    The multiplicative inverse of $-\frac{5}{8}$ is $-\frac{8}{5}$ because $-\frac{5}{8}\cdot\left(-\frac{8}{5}\right)=1$.

27. The additive inverse of 0.3 is -0.3 because $-0.3+0.3=0$.

    The multiplicative inverse of $0.3=\frac{3}{10}$ is $\frac{10}{3}$ because $0.3\cdot\frac{10}{3}=1$.

29. Commutative property of addition

31. Distributive property

33. Associative property of multiplication

35. Associative property of addition

37. Commutative property of addition

39. Commutative property of multiplication

41. $9+(-16)=-7$

43. $-27+(-13)=-40$

45. $-15+9=-6$

47. $14+(-19)=-5$

49. $$\begin{aligned}-\frac{3}{4}+\frac{1}{6}&=-\frac{3(3)}{4(3)}+\frac{1(2)}{6(2)}\\&=-\frac{9}{12}+\frac{2}{12}\\&=\frac{-9+2}{12}\\&=-\frac{7}{12}\end{aligned}$$

51. $$\begin{aligned}-\frac{1}{8}+\left(-\frac{2}{3}\right)&=-\frac{1(3)}{8(3)}+\left(-\frac{2(8)}{3(8)}\right)\\&=-\frac{3}{24}+\left(-\frac{16}{24}\right)\\&=\frac{-3+(-16)}{24}\\&=-\frac{19}{24}\end{aligned}$$

53. $-0.18+6.7=6.52$

55. $-3.28+(-4.1)=-7.38$

57. $-(-7)=7$

59. $-(-(-2.7))=-(2.7)=-2.7$

61. $-|12|=-12$

63. $-\left|-\frac{3}{4}\right|=-\left(\frac{3}{4}\right)=-\frac{3}{4}$

65. $2-(-3)=2+3=5$

67. $10-(-2)=10+2=12$

69. $7-11=7+(-11)=-4$

71. $8-3=8+(-3)=5$

73. $\frac{7}{10}-\left(-\frac{3}{5}\right)=\frac{7}{10}+\frac{3}{5}=\frac{7}{10}+\frac{6}{10}=\frac{13}{10}$

75. $-\frac{1}{5}-\left(-\frac{1}{5}\right)=-\frac{1}{5}+\frac{1}{5}=0$

77. $6.2-3.65=6.2+(-3.65)=2.55$

79. $-6.1-(-4.5)=-6.1+4.5=-1.6$

81. $4(-3)=-12$

83. $(-2)(-1)=2$

85. $-2\cdot\frac{1}{4}=-\frac{2}{4}=-\frac{1}{2}$

87. $-25\div(-5)=5$

89. $-\frac{1}{4}\div\frac{2}{3}=-\frac{1}{4}\cdot\frac{3}{2}=-\frac{3}{8}$

91. $-12\div0.3=-40$

93. $-1(-2)(-3)=-6$

95. $2.7(-0.1)(-2)=0.54$

97. $-2(5)(-2)(-3)=-60$

99. $(-1)(-3)(-5)(2)(3)=-90$

101. $2261.18-(-19.65)=2261.18+19.65=2280.83$

103. $-1475.84+1200+(-124.75)+(-12.50)$
$+(-225.65)+(-175.92)=-\$814.66$

105. $560.5+(-2402.5)+560.5=-1281.5$ N

107. $4.8(-0.035)=-0.168$
$4.8+(-0.168)=\$4.632$ million

109. a) Earth: $-32.2(5.5)=-177.1$ lb.
The Moon: $-5.5(5.5)=-30.25$ lb.
Mars: $-12.3(5.5)=-67.65$ lb.
b) Earth

111. $-6.4(8)=-51.2$ V

Review Exercises

1. {Washington, Adams, Jefferson, Madison}
2. No, because an element, 6, is in *B* but not in *A*.
3. infinite
4. -6 can be written as $-\frac{6}{1}$
5. $|-25|=25$
6. $-\frac{5}{6}=-\frac{15}{18}$ because $-\frac{15}{18}=-\frac{\cancel{3}\cdot5}{\cancel{3}\cdot6}=-\frac{5}{6}$

Exercise Set 1.3

1. To evaluate an exponential form raised to a natural-number exponent, write the base as a factor the number of times indicated by the exponent and then multiply.

3. The square of every real number is positive (or 0) because of the sign rules for multiplying two numbers.

5. -8^6 is negative because this is the additive inverse of 8^6, which is positive.

7. $(-4)^3$
base = -4; exponent = 3; negative four to the third power, or negative four cubed.

9. -1^7
base = 1; exponent = 7; the additive inverse of one raised to the seventh power

11. $5^4=5\cdot5\cdot5\cdot5=625$

13. $(-3)^4 = (-3)\cdot(-3)\cdot(-3)\cdot(-3) = 81$

15. $(-4)^3 = (-4)\cdot(-4)\cdot(-4) = -64$

17. $-6^2 = -6\cdot 6 = -36$

19. $(-1)^{10}$
$= (-1)(-1)(-1)(-1)(-1)(-1)(-1)(-1)(-1)(-1)$
$= 1$

21. $\left(-\frac{3}{8}\right)^2 = \left(-\frac{3}{8}\right)\left(-\frac{3}{8}\right) = \frac{9}{64}$

23. $\left(-\frac{5}{6}\right)^3 = \left(-\frac{5}{6}\right)\left(-\frac{5}{6}\right)\left(-\frac{5}{6}\right) = -\frac{125}{216}$

25. $(0.4)^3 = (0.4)(0.4)(0.4) = 0.064$

27. $(-2.1)^3 = (-2.1)(-2.1)(-2.1) = -9.261$

29. ± 15

31. ± 16

33. $\pm\frac{2}{3}$

35. No real-number roots exist

37. $\sqrt{25} = 5$

39. $\sqrt[5]{32} = 2$

41. $\sqrt{0.36} = 0.6$

43. $\sqrt[3]{-27} = -3$

45. $\sqrt[3]{\frac{8}{27}} = \frac{2}{3}$

47. $\sqrt{-36}$ is not a real number

49. $\sqrt[3]{\frac{24}{3}} = \sqrt[3]{8} = 2$

51. $-4 + 3(-1)^4 + 18 \div 3 = -4 + 3(1) + 18 \div 3$
$= -4 + 3 + 6$
$= -1 + 6$
$= 5$

53. $-8^2 + 36 \div (9-5) = -64 + 36 \div 4$
$= -64 + 9$
$= -55$

55. $12^2 \div \sqrt{45-9} - 8 = 12^2 \div \sqrt{36} - 8$
$= 144 \div 6 - 8$
$= 24 - 8$
$= 16$

57. $-2|-8-2| \div (-5)(2) = -2|-10| \div (-5)(2)$
$= -2\cdot 10 \div (-5)(2)$
$= -20 \div (-5)(2)$
$= 4(2)$
$= 8$

59. $-24 \div (-6)(2) + \sqrt{169-25}$
$= -24 \div (-6)(2) + \sqrt{144}$
$= -24 \div (-6)(2) + 12$
$= 4(2) + 12$
$= 8 + 12$
$= 20$

61. $-18\cdot\frac{2}{9} \div (-2) + |9 - 5(-2)|$
$= -18\cdot\frac{2}{9} \div (-2) + |9 + 10|$
$= -18\cdot\frac{2}{9} \div (-2) + |19|$
$= -4 \div (-2) + 19$
$= 2 + 19$
$= 21$

63. $13.02 \div (-3.1) + 6^2 - \sqrt{25}$
$= 13.02 \div (-3.1) + 36 - 5$
$= -4.2 + 36 - 5$
$= 26.8$

65. $(1-0.8)^2 + 2.4 \div (0.3)(-0.5)$
$= (0.2)^2 + 2.4 \div (0.3)(-0.5)$
$= 0.04 + 2.4 \div (0.3)(-0.5)$
$= 0.04 + 8(-0.5)$
$= 0.04 - 4$
$= -3.96$

67. $\frac{4}{5} \div \left(-\frac{1}{10}\right) \cdot (-2) + \sqrt[5]{16+16}$
$= \frac{4}{5} \div \left(-\frac{1}{10}\right) \cdot (-2) + \sqrt[5]{32}$
$= \frac{4}{5} \cdot \left(-\frac{10}{1}\right) \cdot (-2) + 2$
$= 16 + 2$
$= 18$

69. $\frac{9}{8} \cdot \left(-\frac{2}{3}\right) + \left(\frac{1}{5} - \frac{2}{3}\right) \div \sqrt{\frac{125}{5}}$
$= \frac{9}{8} \cdot \left(-\frac{2}{3}\right) - \frac{7}{15} \div \sqrt{25}$
$= -\frac{3}{4} - \frac{7}{15} \div 5$
$= -\frac{3}{4} - \frac{7}{75}$
$= -\frac{253}{300}$

71. $\frac{12-2^3}{5-3\cdot 2} = \frac{12-8}{5-6}$
$= \frac{4}{-1}$
$= -4$

73. $\frac{6^2 - 3(4+2^5)}{5+20-(2+3)^2} = \frac{6^2 - 3(4+32)}{5+20-5^2}$
$= \frac{6^2 - 3(36)}{5+20-25}$
$= \frac{36 - 3(36)}{0}$
$=$ undefined

75. Associative property of multiplication is used to multiply $3 \cdot 3$ instead of multiplying $1 \cdot 3$ from left to right.

77. Distributive property was applied instead of adding $-1+36$ in the parentheses.

79. Mistake: Multiplied before division.
Correct: $12 \div 4 \cdot 3 - 11 = 3 \cdot 3 - 11$
$= 9 - 11$
$= -2$

81. Mistake: Found the square root of the addends 16 and 9 instead of their sum.
Correct: $30 \div 2 + \sqrt{16+9} = 30 \div 2 + \sqrt{25}$
$= 30 \div 2 + 5$
$= 15 + 5$
$= 20$

83. a) $\frac{(76.5+74.5+71.4+69.2)}{4} = \frac{291.6}{4} = 72.9$ min.
b) No, her average is greater than 72 minutes.

85. $\frac{157{,}000+159{,}190+163{,}048+165{,}288+168{,}648}{5}$
$= \frac{813{,}174}{5}$
$= 162{,}634.8$

87. $\frac{(1349.88+1352.99+1341.13+1325.76+1315.22)}{5}$
$= \frac{6684.98}{5}$
$= 1336.996$

89. $\frac{4(3)+4(4)+3(2)+3(1)}{14} = \frac{37}{14} \approx 2.64$

91. $35.5 + 0.10(658-500) + 0.12(45)$
$= 35.5 + 0.10(158) + 0.12(45)$
$= 35.5 + 15.8 + 5.4$
$= \$56.70$

93. $0.35(814) + 54.50 + 3(89.90) + 112.45$
$= 284.9 + 54.50 + 269.70 + 112.45$
$= \$721.55$

95. $[2(8.95) + 2(10.95) + 6.95 + 2(1.45)] \div 5$
$= (17.90 + 21.90 + 6.95 + 2.90) \div 5$
$= 49.65 \div 5$
$= \$9.93$

97. $1200 + 349(3)(12) + 0.15(52{,}000 - 45{,}000)$
$= 1200 + 349(3)(12) + 0.15(7000)$
$= 1200 + 12{,}564 + 1050$
$= \$14{,}814$

99. $8^4 = 4096$

101. $2^7 = 128$

103. $3^3 \cdot 10^4 = 27 \cdot 10{,}000 = 270{,}000$

Review Exercises

1. $\{0, 1, 2, 3, 4, 5, 6, 7, 8\}$
2. $19-(-54)=19+54=73$
3. $-2(-5)(-6)=-60$
4. $\frac{-38}{2}=-19$
5. It is an expression because it has no equal sign.
6. Commutative property of multiplication

Exercise Set 1.4

1. Addition is commutative.
3. To evaluate a variable expression, (1) replace each variable with its corresponding given value and (2) simplify the resulting numerical expression.
5. Like terms are variable terms that have the same variable(s) raised to the same exponents, or constant terms.

7. $5n$ 9. $3n-2$ 11. $5-p$

13. $\frac{n^4}{8}$ 15. $2n-20$` 17. x^4y^2

19. $\frac{p}{q}-\frac{1}{2}$ 21. $m-3(n+5)$

23. $(4-t)^5$ 25. $\frac{6}{7}x+7$

27. $(m-n)-(x+y)$

29. Mistake: incorrect order
Correct: $y-6$

31. Mistake: *sum* means use parentheses
Correct: $4(r+7)$

33. $5w$ 35. $2-3w$ 37. $2r$

39. $17-n$ 41. $\left(t+\frac{1}{2}\right)$ hr. 43. πd

45. $\frac{1}{2}h(a+b)$ 47. $\frac{1}{3}\pi r^2h$ 49. $\frac{1}{2}mv^2$

51. $\frac{Mm}{d^2}$ 53. $\sqrt{1-\frac{v^2}{c^2}}$

55. $$\begin{aligned}-0.4(x+2)-5&=-0.4(3+2)-5\\&=-0.4(5)-5\\&=-2-5\\&=-7\end{aligned}$$

57. $$\begin{aligned}-3m^2+5m+1&=-3\left(-\frac{2}{3}\right)^2+5\left(-\frac{2}{3}\right)+1\\&=-3\left(\frac{4}{9}\right)+5\left(-\frac{2}{3}\right)+1\\&=-\frac{4}{3}-\frac{10}{3}+\frac{3}{3}\\&=-\frac{11}{3}\end{aligned}$$

59. $$\begin{aligned}\left|2x^2-3y\right|+x&=\left|2(5)^2-3(-1)\right|+5\\&=\left|2(25)+3\right|+5\\&=\left|50+3\right|+5\\&=\left|53\right|+5\\&=53+5\\&=58\end{aligned}$$

61. $$\begin{aligned}\sqrt[3]{c}-2ab^2&=\sqrt[3]{8}-2(-1)(-2)^2\\&=2-2(-1)(4)\\&=2+8\\&=10\end{aligned}$$

63. a) $ad-bc=5(7)-(0.2)(-3)=35+0.6=35.6$

b) $ad-bc=-8\left(-\frac{5}{6}\right)-\left(\frac{2}{3}\right)(2)=\frac{20}{3}-\frac{4}{3}=\frac{16}{3}$

65. a) $\frac{y_2-y_1}{x_2-x_1}=\frac{-7-(-1)}{5-3}=\frac{-7+1}{2}=\frac{-6}{2}=-3$

b) $\frac{y_2-y_1}{x_2-x_1}=\frac{-2-(-1)}{-1-3}=\frac{-2+1}{-4}=\frac{-1}{-4}=\frac{1}{4}$

67. If $x = 0$, this expression is undefined because the denominator is 0.

69. $$\begin{aligned}y-6&=0\\y&=6\end{aligned}$$
If $y = 6$, this expression is undefined because the denominator is 0.

71. $y+5=0$ $y-1=0$
$y=-5$ $y=1$
If $y = -5$ or $y = 1$, this expression is undefined because the denominator is 0.

73. $4x+1=0$

$4x=-1$

$x=-\frac{1}{4}$

If $x=-\frac{1}{4}$, this expression is undefined because the denominator is 0.

75. $9(3x-5)=9\cdot 3x-9\cdot 5=27x-45$

77. $-5(m+2)=-5\cdot m-5\cdot 2=-5m-10$

79. $\frac{3}{8}\left(\frac{2}{9}x-24\right)=\frac{3}{8}\cdot\frac{2}{9}x-\frac{3}{8}\cdot 24=\frac{1}{12}x-9$

81. $-2.1(3x+2.4)=-2.1\cdot 3x-2.1(2.4)=-6.3x-5.04$

83. $2x-13x=-11x$

85. $\frac{6}{7}b^2-\frac{8}{7}b^2=-\frac{2}{7}b^2$

87. $4x-9y-12+y+3x=4x+3x-9y+y-12$
$=7x-8y-12$

89. $1.5x+y-2.8x+0.3-y-0.7$
$=1.5x-2.8x+y-y+0.3-0.7$
$=-1.3x-0.4$

91. $2.6h^2+\frac{5}{3}h-\frac{2}{5}h^2+h+7$
$=2.6h^2-\frac{2}{5}h^2+\frac{5}{3}h+h+7$
$=2.2h^2+\frac{8}{3}h+7$

93. $2(5n-6)+4(n+1)-8=10n-12+4n+4-8$
$=10n+4n-12+4-8$
$=14n-16$

95. $7a+5b-3(4a+2b)-12+8b$
$=7a+5b-12a-6b-12+8b$
$=7a-12a+5b-6b+8b-12$
$=-5a+7b-12$

97. a) $14+(6x-8x)$ b) $14-2x$
c) $14-2(-3)=14+6=20$

Review Exercises

1. $\{x|x \text{ is an integer and } x\ge -2\}$

2. $-5^2+3(6-8)-\sqrt{16}=-25+3(-2)-4$
$=-25-6-4$
$=-35$

3. $(20-24)^3-4|3-8|=(-4)^3-4|-5|$
$=-64-4(5)$
$=-64-20$
$=-84$

4. $-7^2=-49$

5. Commutative property of addition

6. Distributive property

Chapter 1 Review Exercises

1. True
2. False; when using the order of operations, multiplication and division are performed from left to right, in the order as they occur.
3. True
4. True
5. True
6. True
7. change or vary
8. positive, negative
9. positive
10. replace
11. {Alaska, Hawaii}
12. {…,–3, –1, 1, 3, 5, …}
13. {5, 10, 15, …}
14. {s, i, m, p, l, f, y}
15. $\{x|x \text{ is a natural-number multiple of } 3\}$
16. $\{x|x \text{ is a whole number}\}$
17. $\{x|x \text{ is a prime number}\}$
18. $\{x|x \text{ is a day of the week}\}$
19. True, because March is not one of the days of the week.
20. False, because the numbers in the set *A* are not all contained in the set *B*.

21. False, because $\frac{2}{3}$ is not an irrational number.

22. True, because $\sqrt{2}$ is not an integer.

23. $-(-5)=|-5|$

24. $\left|-3\frac{1}{4}\right|>-3\frac{1}{4}$

25. $0.5=\frac{1}{2}$

26. $-|-3|>-4$

27. Additive inverse

28. Multiplicative inverse

29. Additive identity

30. Multiplicative identity

31. Distributive property

32. Associative property of multiplication

33. Commutative property of addition

34. Commutative property of multiplication

35. Associative property of addition

36. Distributive property

37. $6+(-7)=-1$

38. $-4+9=5$

39. $-15+(-2)=-17$

40. $-2+(-5)=-7$

41. $7-9=7+(-9)=-2$

42. $-2-8=-2+(-8)=-10$

43. $15-(-2)=15+2=17$

44. $-8-(-1)=-8+1=-7$

45. $-2(4)=-8$

46. $-3(-5)=15$

47. $7(-8)=-56$

48. $25\div(-5)=-5$

49. $-10\div(-5)=2$

50. $-8\div 4=-2$

51. $-1(-2)(-3)=-6$

52. $-50\div[4\div(-2)]=-50\div(-2)$
$=25$

53. -2^7
base = 2; exponent = 7; the additive inverse of two raised to the seventh power; –128

54. $(-1)^4$
base = –1; exponent = 4; negative one raised to the fourth power; 1

55. $-3^2=-(3\cdot 3)=-9$

56. $-2^3=-(2\cdot 2\cdot 2)=-8$

57. $(-4)^2=(-4)(-4)=16$

58. $\left(-\frac{2}{5}\right)^3=\left(-\frac{2}{5}\right)\left(-\frac{2}{5}\right)\left(-\frac{2}{5}\right)=-\frac{8}{125}$

59. $\sqrt{121}=11$

60. $\sqrt[4]{81}=3$

61. $\sqrt[3]{27}=3$

62. $\sqrt[8]{1}=1$

63. $-7\left|8-2^4\right|+7-3^2=-7|8-16|+7-3^2$
$=-7|-8|+7-3^2$
$=-7\cdot 8+7-9$
$=-56+7-9$
$=-58$

64. $8(1-3^2)+\sqrt{16-7}=8(1-9)+\sqrt{9}$
$=8(-8)+3$
$=-64+3$
$=-61$

65. $\sqrt[3]{8}-7\cdot 3^2=2-7\cdot 3^2$
$=2-7\cdot 9$
$=2-63$
$=-61$

66. $\sqrt{16}+\sqrt{9}-3(2-7)=4+3-3(-5)$
$=4+3+15$
$=22$

67. $5^2(3-8)^2 = 5^2(-5)^2 = 25\cdot 25 = 625$

68. $\dfrac{3^4-5[3-4(-2)]}{16\div 8(-5)} = \dfrac{3^4-5[3+8]}{2(-5)}$
$= \dfrac{3^4-5\cdot 11}{-10}$
$= \dfrac{81-55}{-10}$
$= \dfrac{26}{-10} = -2.6$

69. $14-8n$

70. $2(n-2)$

71. $n+\dfrac{1}{3}(n-4)$

72. $\dfrac{m}{\sqrt{n}}$

73. $\dfrac{1}{2}(n-8)-16$

74. $(n+5)+20$

75. $2w$

76. $\left(\dfrac{1}{3}+t\right)$ hr.

77. $3a^2-4a+2 = 3(-1)^2-4(-1)+2$
$= 3(1)+4+2$
$= 3+4+2$
$= 9$

78. $-4|-b+2ac| = -4|-(-2)+2(1)(-1)|$
$= -4|2-2|$
$= -4|0|$
$= -4(0)$
$= 0$

79. $15(x+y)^2+x^2 = 15(1+(-2))^2+1^2$
$= 15(-1)^2+1^2$
$= 15\cdot 1+1$
$= 15+1$
$= 16$

80. $\sqrt{a-b}+3^0-a = \sqrt{25-16}+3^0-25$
$= \sqrt{9}+3^0-25$
$= 3+1-25$
$= -21$

81. $x-3=0$
$x=3$
If $x=3$, this expression is undefined because the denominator is 0.

82. $2x-1=0$
$2x=1$
$x=\dfrac{1}{2}$
If $x=\dfrac{1}{2}$, this expression is undefined because the denominator is 0.

83. $-2(5x+1) = -2\cdot 5x-2\cdot 1 = -10x-2$

84. $4(2a+3b-4) = 4\cdot 2a+4\cdot 3b+4\cdot(-4)$
$= 8a+12b-16$

85. $3x+2x^2-4x-x-3x^2$
$= 2x^2-3x^2+3x-4x-x$
$= -x^2-2x$

86. $5m^5+3mn-2mn^2-mn-4m^5$
$= 5m^5-4m^5-2mn^2+3mn-mn$
$= m^5-2mn^2+2mn$

87. $-7ab+3ab^2+2ab+3a-7a^2-8$
$= -7a^2+3ab^2-7ab+2ab+3a-8$
$= -7a^2+3ab^2-5ab+3a-8$

88. $6r-3-r-2r-7$
$= 6r-r-2r-3-7$
$= 3r-10$

89. $-245.85+125.00+(-72.34)+(-12.50)$
$+(-14.75) = -\$220.44$

90. $\dfrac{2(3)+4(4)+3(3)+3(1)}{12} = \dfrac{6+16+9+3}{12}$
$= \dfrac{34}{12}$
≈ 2.83

91. $59+0.15(37) = 59+5.55 = \$64.55$

92. $5^3 \cdot 10^5 = 125 \cdot 100{,}000 = 12{,}500{,}000$ codes

Chapter 1 Practice Test

1. $|8.1| = 8.1$

2. $-\left|-\frac{11}{4}\right| = -\frac{11}{4}$

3. $\sqrt{169} = 13$

4. $\sqrt[3]{125} = 5$

5. $\sqrt[5]{\frac{1}{32}} = \frac{1}{2}$

6. Commutative property of addition because the order of the addends is changed.

7. Associative property of multiplication because the grouping of the factors is changed.

8. $9 + (-1) = 8$

9. $$\begin{aligned}\frac{2}{3} - \left(-\frac{1}{4}\right) &= \frac{2}{3} + \frac{1}{4}\\ &= \frac{2(4)}{3(4)} + \frac{1(3)}{4(3)}\\ &= \frac{8}{12} + \frac{3}{12}\\ &= \frac{11}{12}\end{aligned}$$

10. $(-3)(2.5) = -7.5$

11. $(-5)^2 = (-5)(-5) = 25$

12. $-\frac{2}{5} \div \frac{5}{2} = -\frac{2}{5} \cdot \frac{2}{5} = -\frac{4}{25}$

13. $\sqrt[4]{16} = 2$

14. $8 \div 4 \cdot 2 = 2 \cdot 2 = 4$

15. $$\begin{aligned}-3^2 + 7 - 2(5-1) &= -3^2 + 7 - 2(4)\\ &= -9 + 7 - 2(4)\\ &= -9 + 7 - 8\\ &= -10\end{aligned}$$

16. $$\begin{aligned}\sqrt[4]{16} + 8 - (3+1)^2 &= \sqrt[4]{16} + 8 - 4^2\\ &= 2 + 8 - 16\\ &= -6\end{aligned}$$

17. $$\begin{aligned}4 \div |8-6| + 2^4 &= 4 \div |2| + 2^4\\ &= 4 \div 2 + 16\\ &= 2 + 16\\ &= 18\end{aligned}$$

18. $$\begin{aligned}(5-4)^5 + (2-3)^3 &= 1^5 + (-1)^3\\ &= 1 - 1\\ &= 0\end{aligned}$$

19. $$\begin{aligned}&\sqrt{9+16} + \left[-2^2 + 3(2-5)\right]\\ &= \sqrt{25} + \left[-2^2 + 3(-3)\right]\\ &= 5 + (-4-9)\\ &= 5 + (-13)\\ &= -8\end{aligned}$$

20. $-423.75 + (-84.50) + (-24.80) + 500$
$+(-356.45) = -\$389.50$

21. $\frac{23.1+6.2+5.9+3.9+3.1+2.7+2.2+2.1+1.4+1.3}{10}$
$= 5.19$ million

22. Let $x = 2$ and $y = -3$.
$$\begin{aligned}-2\left|3 - 4xy^2\right| &= -2\left|3 - 4(2)(-3)^2\right|\\ &= -2|3 - 4(2)(9)|\\ &= -2|3 - 72|\\ &= -2|-69|\\ &= -2(69)\\ &= -138\end{aligned}$$

23. Let $a = 16$ and $b = 4$.
$$\begin{aligned}\frac{a}{b} - \sqrt{a} + \sqrt{b} &= \frac{16}{4} - \sqrt{16} + \sqrt{4}\\ &= 4 - 4 + 2\\ &= 2\end{aligned}$$

24. $$\begin{aligned}-7(3x+5) &= -7 \cdot 3x + (-7) \cdot 5\\ &= -21x + (-35)\\ &= -21x - 35\end{aligned}$$

25. $\frac{2}{5}x+\frac{3}{4}y-5x+6y+2.7$

$=\frac{2}{5}x-5x+\frac{3}{4}y+6y+2.7$

$=\frac{2}{5}x-\frac{25}{5}x+\frac{3}{4}y+\frac{24}{4}y+2.7$

$=-\frac{23}{5}x+\frac{27}{4}y+2.7$

Chapter 2

Linear Equations and Inequalities in One Variables

Exercise Set 2.1

1. A solution for an equation is a number that makes the equation true when it replaces the variable in the equation.

3. The solution set for an identity contains every real number for which the equation is defined.

5. Divide both sides of the equation by r.

7.
$$\begin{aligned} 2x-2&=10\\ 2x-2+2&=10+2\\ 2x&=12\\ \frac{2x}{2}&=\frac{12}{2}\\ x&=6 \end{aligned}$$

Check: $2(6)-2\overset{?}{=}10$

$$12-2\overset{?}{=}10$$
$$10=10$$

9.
$$\begin{aligned} 9-3m&=12\\ 9-9-3m&=12-9\\ -3m&=3\\ \frac{-3m}{-3}&=\frac{3}{-3}\\ m&=-1 \end{aligned}$$

Check: $9-3(-1)\overset{?}{=}12$

$$9-(-3)\overset{?}{=}12$$
$$9+3\overset{?}{=}12$$
$$12=12$$

11.
$$\begin{aligned} 2x+2&=x+3\\ 2x-x+2&=x-x+3\\ x+2&=3\\ x+2-2&=3-2\\ x&=1 \end{aligned}$$

Check: $2(1)+2\overset{?}{=}1+3$

$$2+2\overset{?}{=}4$$
$$4=4$$

13.
$$\begin{aligned} 4t+3&=2t+9\\ 4t-2t+3&=2t-2t+9\\ 2t+3&=9\\ 2t+3-3&=9-3\\ 2t&=6\\ \frac{2t}{2}&=\frac{6}{2}\\ t&=3 \end{aligned}$$

Check: $4(3)+3\overset{?}{=}2(3)+9$

$$12+3\overset{?}{=}6+9$$
$$15=15$$

15.
$$\begin{aligned} 2(3z+1)&=8\\ 6z+2&=8\\ 6z+2-2&=8-2\\ 6z&=6\\ \frac{6z}{6}&=\frac{6}{6}\\ z&=1 \end{aligned}$$

Check: $2(3(1)+1)\overset{?}{=}8$

$$2(3+1)\overset{?}{=}8$$
$$2(4)\overset{?}{=}8$$
$$8=8$$

17. $n+2(3n+1)=9$
$$n+6n+2=9$$
$$7n+2=9$$
$$7n+2-2=9-2$$
$$7n=7$$
$$\frac{7n}{7}=\frac{7}{7}$$
$$n=1$$
Check: $1+2(3(1)+1)\stackrel{?}{=}9$
$$1+2(3+1)\stackrel{?}{=}9$$
$$1+2(4)\stackrel{?}{=}9$$
$$1+8\stackrel{?}{=}9$$
$$9=9$$

19. $6u-17=4(u+3)-3$
$$6u-17=4u+12-3$$
$$6u-17=4u+9$$
$$6u-4u-17=4u-4u+9$$
$$2u-17=9$$
$$2u-17+17=9+17$$
$$2u=26$$
$$\frac{2u}{2}=\frac{26}{2}$$
$$u=13$$
Check: $6(13)-17\stackrel{?}{=}4(13+3)-3$
$$78-17\stackrel{?}{=}4(16)-3$$
$$61\stackrel{?}{=}64-3$$
$$61=61$$

21. $\frac{1}{2}z-1=4z-3-3z$
$$\frac{1}{2}z-1=z-3$$
$$2\left(\frac{1}{2}z-1\right)=2(z-3)$$
$$z-2=2z-6$$
$$z-z-2=2z-z-6$$
$$-2=z-6$$
$$-2+6=z-6+6$$
$$4=z$$
Check: $\frac{1}{2}(4)-1\stackrel{?}{=}4(4)-3-3(4)$
$$2-1\stackrel{?}{=}16-3-12$$
$$1=1$$

23. Multiply both sides by the LCD, 12.
$$\frac{1}{4}x+\frac{1}{3}x+\frac{5}{6}=\frac{15}{4}$$
$$12\left(\frac{1}{4}x+\frac{1}{3}x+\frac{5}{6}\right)=12\left(\frac{15}{4}\right)$$
$$12\cdot\frac{1}{4}x+12\cdot\frac{1}{3}x+12\cdot\frac{5}{6}=12\cdot\frac{15}{4}$$
$$3x+4x+10=45$$
$$7x+10=45$$
$$7x+10-10=45-10$$
$$7x=35$$
$$\frac{7x}{7}=\frac{35}{7}$$
$$x=5$$
Check:
$$\frac{1}{4}x+\frac{1}{3}x+\frac{5}{6}\stackrel{?}{=}\frac{15}{4}$$
$$\frac{1}{4}(5)+\frac{1}{3}(5)+\frac{5}{6}\stackrel{?}{=}\frac{15}{4}$$
$$\frac{5}{4}+\frac{5}{3}+\frac{5}{6}\stackrel{?}{=}\frac{15}{4}$$
$$\frac{15}{12}+\frac{20}{12}+\frac{10}{12}\stackrel{?}{=}\frac{15}{4}$$
$$\frac{45}{12}\stackrel{?}{=}\frac{15}{4}$$
$$\frac{15}{4}=\frac{15}{4}$$

25. Multiply both sides by 100 to clear decimals.

$$0.5x+0.95=0.2x-1$$
$$100(0.5x+0.95)=100(0.2x-1)$$
$$100\cdot 0.5x+100\cdot 0.95=100\cdot 0.2x-100\cdot 1$$
$$50x+95=20x-100$$
$$50x-20x+95=20x-20x-100$$
$$30x+95=-100$$
$$30x+95-95=-100-95$$
$$30x=-195$$
$$\frac{30x}{30}=\frac{-195}{30}$$
$$x=-6.5$$

Check:

$$0.5(-6.5)+0.95\stackrel{?}{=}0.2(-6.5)-1$$
$$-3.25+0.95\stackrel{?}{=}-1.3-1$$
$$-2.3=-2.3$$

27. Multiply both sides by 10 to clear decimals.

$$1.6x-18+0.9x=0.1x+6$$
$$2.5x-18=0.1x+6$$
$$10(2.5x-18)=10(0.1x+6)$$
$$25x-180=x+60$$
$$25x-x-180=x-x+60$$
$$24x-180=60$$
$$24x-180+180=60+180$$
$$24x=240$$
$$\frac{24x}{24}=\frac{240}{24}$$
$$x=10$$

Check:

$$1.6(10)-18+0.9(10)\stackrel{?}{=}0.1(10)+6$$
$$16-18+9\stackrel{?}{=}1+6$$
$$7=7$$

29. $17a-5-7a=2a+19$

$$10a-5=2a+19$$
$$10a-2a-5=2a-2a+19$$
$$8a-5=19$$
$$8a-5+5=19+5$$
$$8a=24$$
$$\frac{8a}{8}=\frac{24}{8}$$
$$a=3$$

Check:

$$17(3)-5-7(3)\stackrel{?}{=}2(3)+19$$
$$51-5-21\stackrel{?}{=}6+19$$
$$25=25$$

31. $8r-2(r+5)=5(2r-6)-8$

$$8r-2r-10=10r-30-8$$
$$6r-10=10r-38$$
$$6r-6r-10=10r-6r-38$$
$$-10=4r-38$$
$$-10+38=4r-38+38$$
$$28=4r$$
$$\frac{28}{4}=\frac{4r}{4}$$
$$7=r$$

Check:

$$8(7)-2(7+5)\stackrel{?}{=}5(2(7)-6)-8$$
$$56-2(12)\stackrel{?}{=}5(14-6)-8$$
$$56-24\stackrel{?}{=}5(8)-8$$
$$32\stackrel{?}{=}40-8$$
$$32=32$$

33. $5-5(3x-2)=38-2(x-8)$
$$5-15x+10=38-2x+16$$
$$15-15x=54-2x$$
$$15-15x+2x=54-2x+2x$$
$$15-13x=54$$
$$15-15-13x=54-15$$
$$-13x=39$$
$$\frac{-13x}{-13}=\frac{39}{-13}$$
$$x=-3$$
Check:
$$5-5(3(-3)-2)\stackrel{?}{=}38-2(-3-8)$$
$$5-5(-9-2)\stackrel{?}{=}38-2(-11)$$
$$5-5(-11)\stackrel{?}{=}38-(-22)$$
$$5-(-55)\stackrel{?}{=}38+22$$
$$5+55\stackrel{?}{=}60$$
$$60=60$$

35. $8h-27+5(2h-3)=62-(3h+8)$
$$8h-27+10h-15=62-3h-8$$
$$18h-42=54-3h$$
$$18h+3h-42=54-3h+3h$$
$$21h-42=54$$
$$21h-42+42=54+42$$
$$21h=96$$
$$\frac{21h}{21}=\frac{96}{21}$$
$$h=\frac{32}{7}$$
Check:
$$8\left(\frac{32}{7}\right)-27+5\left(2\left(\frac{32}{7}\right)-3\right)\stackrel{?}{=}62-\left(3\left(\frac{32}{7}\right)+8\right)$$
$$\frac{256}{7}-27+5\left(\frac{64}{7}-3\right)\stackrel{?}{=}62-\left(\frac{96}{7}+8\right)$$
$$\frac{256}{7}-27+5\left(\frac{43}{7}\right)\stackrel{?}{=}62-\frac{152}{7}$$
$$\frac{256}{7}-27+\frac{215}{7}\stackrel{?}{=}62-\frac{152}{7}$$
$$\frac{282}{7}=\frac{282}{7}$$

37. Multiply both sides by the LCD, 15.
$$\frac{2}{3}q-4=\frac{1}{5}q+10$$
$$15\left(\frac{2}{3}q-4\right)=15\left(\frac{1}{5}q+10\right)$$
$$10q-60=3q+150$$
$$10q-3q-60=3q-3q+150$$
$$7q-60=150$$
$$7q-60+60=150+60$$
$$7q=210$$
$$\frac{7q}{7}=\frac{210}{7}$$
$$q=30$$
Check:
$$\frac{2}{3}(30)-4\stackrel{?}{=}\frac{1}{5}(30)+10$$
$$20-4\stackrel{?}{=}6+10$$
$$16=16$$

39. Multiply both sides by the LCD, 35.
$$\frac{2}{7}(x-5)=-2+\frac{2}{5}x$$
$$35\left(\frac{2}{7}(x-5)\right)=35\left(-2+\frac{2}{5}x\right)$$
$$10(x-5)=-70+14x$$
$$10x-50=-70+14x$$
$$10x-14x-50=-70+14x-14x$$
$$-4x-50=-70$$
$$-4x-50+50=-70+50$$
$$-4x=-20$$
$$\frac{-4x}{-4}=\frac{-20}{-4}$$
$$x=5$$
Check:
$$\frac{2}{7}(5-5)\stackrel{?}{=}-2+\frac{2}{5}(5)$$
$$\frac{2}{7}(0)\stackrel{?}{=}-2+2$$
$$0=0$$

41. Multiply both sides by 100 to clear decimals.

$$0.5(x-2)+1.76=0.3x+0.8$$
$$0.5x-1+1.76=0.3x+0.8$$
$$100(0.5x+0.76)=100(0.3x+0.8)$$
$$50x+76=30x+80$$
$$50x-30x+76=30x-30x+80$$
$$20x+76=80$$
$$20x+76-76=80-76$$
$$20x=4$$
$$\frac{20x}{20}=\frac{4}{20}$$
$$x=0.2$$

Check:

$$0.5(0.2-2)+1.76\stackrel{?}{=}0.3(0.2)+0.8$$
$$0.5(-1.8)+1.76\stackrel{?}{=}0.06+0.8$$
$$-0.9+1.76\stackrel{?}{=}0.06+0.8$$
$$0.86=0.86$$

43. Multiply both sides by 100 to clear decimals.

$$2x-3.24+2.4x=6.2+0.08x+3.52$$
$$4.4x-3.24=0.08x+9.72$$
$$100(4.4x-3.24)=100(0.08x+9.72)$$
$$440x-324=8x+972$$
$$440x-8x-324=8x-8x+972$$
$$432x-324=972$$
$$432x-324+324=972+324$$
$$432x=1296$$
$$\frac{432x}{432}=\frac{1296}{432}$$
$$x=3$$

Check:

$$2(3)-3.24+2.4(3)\stackrel{?}{=}6.2+0.08(3)+3.52$$
$$6-3.24+7.2\stackrel{?}{=}6.2+0.24+3.52$$
$$9.96=9.96$$

45. $6(m+3)-5+2m=3(3m+1)-m$

$$6m+18-5+2m=9m+3-m$$
$$8m+13=8m+3$$
$$8m-8m+13=8m-8m+3$$
$$13\neq 3$$

No solution

47. $7(n+2)-3n=4+4n+10$

$$7n+14-3n=14+4n$$
$$4n+14=14+4n$$
$$4n-4n+14=14+4n-4n$$
$$14=14$$

All real numbers

49. Mistake: The distributive property was not used correctly.

Correct:

$$2(x+3)=7x-1$$
$$2x+6=7x-1$$
$$2x-2x+6=7x-2x-1$$
$$6=5x-1$$
$$6+1=5x-1+1$$
$$7=5x$$
$$\frac{7}{5}=\frac{5x}{5}$$
$$\frac{7}{5}=x$$

51. Mistake: Subtracted before distributing into the parentheses.

Correct: $4-5(x+1)+4x=7-11$

$$4-5x-5+4x=-4$$
$$-1-x=-4$$
$$-1+1-x=-4+1$$
$$-x=-3$$
$$x=3$$

53. Mistake: Did not distribute the − with the 2.

Correct:

$$-2(x+4)+6x=2x+6$$
$$-2x-8+6x=2x+6$$
$$4x-8=2x+6$$
$$4x-2x-8=2x-2x+6$$
$$2x-8=6$$
$$2x-8+8=6+8$$
$$2x=14$$
$$\frac{2x}{2}=\frac{14}{2}$$
$$x=7$$

55.
$$\begin{aligned} P &= R - C \\ P + C &= R - C + C \\ P + C &= R \\ P - P + C &= R - P \\ C &= R - P \end{aligned}$$

57.
$$\begin{aligned} A &= bh \\ \frac{A}{h} &= \frac{bh}{h} \\ \frac{A}{h} &= b \end{aligned}$$

59.
$$\begin{aligned} A &= 2\pi pw \\ \frac{A}{2\pi w} &= \frac{2\pi pw}{2\pi w} \\ \frac{A}{2\pi w} &= p \end{aligned}$$

61.
$$\begin{aligned} A &= \frac{1}{2}\theta r^2 \\ 2 \cdot A &= 2 \cdot \frac{1}{2}\theta r^2 \\ 2A &= \theta r^2 \\ \frac{2A}{r^2} &= \frac{\theta r^2}{r^2} \\ \frac{2A}{r^2} &= \theta \end{aligned}$$

63.
$$\begin{aligned} F &= \frac{kMm}{d^2} \\ d^2 \cdot F &= d^2 \cdot \frac{kMm}{d^2} \\ Fd^2 &= kMm \\ \frac{Fd^2}{km} &= \frac{kMm}{km} \\ \frac{Fd^2}{km} &= M \end{aligned}$$

65.
$$\begin{aligned} A &= \pi s(R + r) \\ \frac{A}{\pi(R + r)} &= \frac{\pi s(R + r)}{\pi(R + r)} \\ \frac{A}{\pi(R + r)} &= s \end{aligned}$$

67.
$$\begin{aligned} P &= 2l + 2w \\ P - 2w &= 2l + 2w - 2w \\ P - 2w &= 2l \\ \frac{P - 2w}{2} &= \frac{2l}{2} \\ \frac{P - 2w}{2} &= l \end{aligned}$$

69.
$$\begin{aligned} 3x + 2y &= 6 \\ 3x - 3x + 2y &= 6 - 3x \\ 2y &= 6 - 3x \\ \frac{2y}{2} &= \frac{6 - 3x}{2} \\ y &= \frac{6 - 3x}{2} \end{aligned}$$

71.
$$\begin{aligned} F &= \frac{9}{5}C + 32 \\ F - 32 &= \frac{9}{5}C + 32 - 32 \\ F - 32 &= \frac{9}{5}C \\ \frac{5}{9}(F - 32) &= \frac{5}{9} \cdot \frac{9}{5}C \\ \frac{5}{9}(F - 32) &= C \end{aligned}$$

73.
$$\begin{aligned} x &= vt + \frac{1}{2}at^2 \\ x - vt &= vt - vt + \frac{1}{2}at^2 \\ x - vt &= \frac{1}{2}at^2 \\ 2(x - vt) &= 2 \cdot \frac{1}{2}at^2 \\ 2(x - vt) &= at^2 \\ \frac{2(x - vt)}{t^2} &= \frac{at^2}{t^2} \\ \frac{2(x - vt)}{t^2} &= a \end{aligned}$$

75. Mistake: Subtracted lw instead of dividing lw.

Correct:
$$\begin{aligned} V &= lwh \\ \frac{V}{lw} &= \frac{lwh}{lw} \\ \frac{V}{lw} &= h \end{aligned}$$

77. Mistake: Subtracted the coefficient of l instead of dividing by the coefficient.

Correct:
$$\begin{aligned} P &= 2l + 2w \\ P - 2w &= 2l + 2w - 2w \\ P - 2w &= 2l \\ \frac{P - 2w}{2} &= \frac{2l}{2} \\ \frac{P - 2w}{2} &= l \end{aligned}$$

Review Exercises

1. {1, 3, 5, 7, 9, 11, 13}

2. $14-12\left[6-8\left(3+2^5\right)\right]+\sqrt{9\cdot 16}$
$=14-12\left[6-8(3+32)\right]+\sqrt{144}$
$=14-12[6-8\cdot 35]+12$
$=14-12[6-280]+12$
$=14-12[-274]+12$
$=14+3288+12$
$=3314$

3. $7n-9$

4. $-3(n+8)$

5. $-2x-6y+5$

6. $-54m+24$

Exercise Set 2.2

1. (1) Understand the problem. (2) Devise a plan. (3) Execute the plan. (4) Check results.

3. Choose a variable for one of the unknowns. Use one of the relationships to describe the other unknown in terms of the chosen variable. Translate the second relationship to an equation.

5. Two angles are supplementary if the sum of their measures is 180º.

7.
$$F=\frac{9}{5}C+32$$
$$-109.3=\frac{9}{5}C+32$$
$$-141.3=\frac{9}{5}C$$
$$\frac{5}{9}(-141.3)=\frac{5}{9}\cdot\frac{9}{5}C$$
$$-78.5^\circ=C$$

9. Find the area of the wall: $A=10\cdot 22$
$A=220\text{ ft.}^2$

Find the area of one window: $A=5\cdot 4$
$A=20\text{ ft.}^2$

Subtract the area of the 2 windows from the area of the wall to find the area covered by paint.

$220-2(20)=220-40=180\text{ ft.}^2$

11.
$$B=P+Prt$$
$$3330=P+P(0.055)(2)$$
$$3330=P+0.11P$$
$$3330=1.11P$$
$$\$3000=P$$

13. Let x be the number. Translate to an equation and solve for x.
$$5x-3=37$$
$$5x-3+3=37+3$$
$$5x=40$$
$$\frac{5x}{5}=\frac{40}{5}$$
$$x=8$$

15. Let x be the number. Translate to an equation and solve for x.
$$3(x+4)=-6$$
$$3x+12=-6$$
$$3x+12-12=-6-12$$
$$3x=-18$$
$$\frac{3x}{3}=\frac{-18}{3}$$
$$x=-6$$

17. Let x be the number. Translate to an equation and solve for x.
$$5(x-2)-8=2x$$
$$5x-10-8=2x$$
$$5x-18=2x$$
$$5x-5x-18=2x-5x$$
$$-18=-3x$$
$$\frac{-18}{-3}=\frac{-3x}{-3}$$
$$6=x$$

19. Let p represent the wholesale price. Because \$135.68 is the retail price, it is the result of adding 25% of the wholesale price to the wholesale price. Translate to an equation and solve for p.
$$0.25p+p=135.68$$
$$1.25p=135.68$$
$$\frac{1.25p}{1.25}=\frac{135.68}{1.25}$$
$$p\approx 108.54$$
The wholesale price is approximately \$108.54.

21. Let p represent the original price. Because \$699.95 is the discounted price, it is the result of subtracting 30% of the original price from the original price. Translate to an equation and solve for p.

$$p - 0.30p = 699.95$$
$$0.70p = 699.95$$
$$\frac{0.70p}{0.70} = \frac{699.95}{0.70}$$
$$p \approx 999.93$$

The original price is approximately \$999.93.

23. Let x represent the number of units sold in the previous year. Because the 45,000 units sold this year is a 20% increase over the previous year, it is the result of adding 20% of x to x. Translate to an equation and solve.

$$0.20x + x = 45,000$$
$$1.20x = 45,000$$
$$\frac{1.20x}{1.20} = \frac{45,000}{1.20}$$
$$x = 37,500$$

The number of units sold the previous year was 37,500 units.

25. The consecutive integers are x and $x + 1$.

$$x + (x+1) = 191$$
$$2x + 1 = 191$$
$$2x + 1 - 1 = 191 - 1$$
$$2x = 190$$
$$\frac{2x}{2} = \frac{190}{2}$$
$$x = 95$$

The integers are 95 and 95 + 1 = 96.

27. The consecutive even integers are x, $x + 2$, and $x + 4$.

$$x + (x+2) + (x+4) = 78$$
$$3x + 6 = 78$$
$$3x + 6 - 6 = 78 - 6$$
$$3x = 72$$
$$\frac{3x}{3} = \frac{72}{3}$$
$$x = 24$$

The integers are 24, 24 + 2 = 26, and 24 + 4 = 28.

29. If two angles are complementary, their sum is 90 degrees. Let one of the angles be x and the other be $2x + 15$.

$$x + (2x+15) = 90$$
$$3x + 15 = 90$$
$$3x + 15 - 15 = 90 - 15$$
$$3x = 75$$
$$\frac{3x}{3} = \frac{75}{3}$$
$$x = 25$$

The angles are 25° and 2(25) + 15 = 65°.

31. If two angles are supplementary their sum is 180 degrees. Let one of the angles be x and the other angle be $2x + 6$.

$$x + 2x + 6 = 180$$
$$3x + 6 = 180$$
$$3x + 6 - 6 = 180 - 6$$
$$3x = 174$$
$$\frac{3x}{3} = \frac{174}{3}$$
$$x = 58$$

The angles are 58° and $2(58) + 6 = 122°$.

33.

Categories	Value	Number	Amount
12 oz.	1.00	$3600 - x$	$3600 - x$
16 oz.	1.25	x	$1.25x$

35.

Categories	Rate	Time	Distance
Freight	50	t	$50t$
Passenger	75	$t - 3$	$75(t-3)$

37.

Solutions	Concen-trate	Vol. of Solution	Vol. of HCl
40%	0.40	x	$0.40x$
20%	0.20	2000	$0.20(2000)$
35%	0.35	$x + 2000$	$0.35(x+2000)$

39. Translate the information from the table in Exercise 33 to an equation, and solve.

$$3600 - x + 1.25x = 4100$$
$$3600 + 0.25x = 4100$$
$$3600 - 3600 + 0.25x = 4100 - 3600$$
$$0.25x = 500$$
$$\frac{0.25x}{0.25} = \frac{500}{0.25}$$
$$x = 2000$$

2000 16-oz. and 3600 – 2000 = 1600 12-oz. drinks were sold.

41. Translate the information from the table in Exercise 35 to an equation, and solve.

$$50t = 75(t-3)$$
$$50t = 75t - 225$$
$$50t - 75t = 75t - 75t - 225$$
$$-25t = -225$$
$$\frac{-25t}{-25} = \frac{-225}{-25}$$
$$t = 9$$

It will take the passenger train 9 – 3 = 6 hr. to overtake the freight train. The trains will be 6(75) = 450 mi. from the station.

43. Translate the information from the table in Exercise 37 to an equation, and solve.

$$0.40x + 0.20(2000) = 0.35(x+2000)$$
$$0.40x + 400 = 0.35x + 700$$
$$0.40x - 0.35x + 400 = 0.35x - 0.35x + 700$$
$$0.05x + 400 = 700$$
$$0.05x + 400 - 400 = 700 - 400$$
$$0.05x = 300$$
$$\frac{0.05x}{0.05} = \frac{300}{0.05}$$
$$x = 6000$$

6000 L of the 40% solution are added.

45. Create a table.

Categories	how many	value	total amount
iMac	16	x	$16x$
MacBook	6	$x+600$	$6(x+600)$

Translate the information in the table to an equation and solve.

$$16x + 6(x+600) = 29{,}978$$
$$16x + 6x + 3600 = 29{,}978$$
$$22x + 3600 = 29{,}978$$
$$22x + 3600 - 3600 = 29{,}978 - 3600$$
$$22x = 26{,}378$$
$$\frac{22x}{22} = \frac{26{,}378}{22}$$
$$x = 1199$$

iMacs are $1199 each and MacBook Pros are $1199 + 600 = $1799 each.

47. Create a table.

Categories	how many	value	total amount
Motorola	x	74.95	$74.95x$
Nokia	$2x$	49.95	$49.95(2x)$

Translate the information in the table to an equation and solve.

$$74.95x + 49.95(2x) = 1398.80$$
$$74.95x + 99.90x = 1398.80$$
$$174.85x = 1398.80$$
$$\frac{174.85x}{174.85} = \frac{1398.80}{174.85}$$
$$x = 8$$

He ordered 8 Motorola phones and 16 Nokia phones.

49. Create a table.

Categories	rate	time	distance
car 1	65	t	$65t$
car 2	50	t	$50t$

Translate the information in the table to an equation and solve.

$$65t + 50t = 230$$
$$115t = 230$$
$$\frac{115t}{115} = \frac{230}{115}$$
$$t = 2$$

The cars will meet in 2 hours.

51. Create a table.

	rate	time	distance
Andrea	55	x	$55x$
Paul	65	$x-0.5$	$65(x-0.5)$

Translate the information in the table to an equation and solve.

$$55x = 65(x-0.5)$$
$$55x = 65x - 32.5$$
$$55x - 65x = 65x - 65x - 32.5$$
$$-10x = -32.5$$
$$\frac{-10x}{-10} = \frac{-32.5}{-10}$$
$$x = 3.25$$

Paul will catch up in 3.25 – 0.5 = 2.75 hours.

53. Create a table.

	concentrate	vol. of solution	vol. of antifreeze
20%	0.20	24	$0.20(24)$
10%	0.10	x	$0.10x$
16%	0.16	$x+24$	$0.16(x+24)$

Translate the information in the table to an equation and solve.

$$0.20(24)+0.10x=0.16(x+24)$$
$$4.8+0.10x=0.16x+3.84$$
$$4.8+0.10x-0.10x=0.16x-0.10x+3.84$$
$$4.8=0.06x+3.84$$
$$4.8-3.84=0.06x+3.84-3.84$$
$$0.96=0.06x$$
$$\frac{0.96}{0.06}=\frac{0.06x}{0.06}$$
$$16=x$$

He should add 16 oz. of the 10% solution.

55. Create a table.

	concentrate	vol. of sol'n	vol. of pesticide
10%	0.10	20	$0.10(20)$
40%	0.40	x	$0.40x$
20%	0.20	$x+20$	$0.20(x+20)$

Translate the information in the table to an equation and solve.

$$0.10(20)+0.40x=0.20(x+20)$$
$$2+0.40x=0.20x+4$$
$$2+0.40x-0.20x=0.20x-0.20x+4$$
$$2+0.20x=4$$
$$2-2+0.20x=4-2$$
$$0.20x=2$$
$$\frac{0.20x}{0.20}=\frac{2}{0.20}$$
$$x=10$$

He should use 10 gallons of 40% pesticide.

Review Exercises

1. $-5\frac{3}{8},-\frac{1}{6},0.02,4.5\%,\sqrt{48},|-15.8|$

2. 2 is the smallest prime number.

3. $-|5.8|=-|-5.8|$

4. $-(-6)>-(-(-8))$

5. $$6x-19=4x-31$$
$$6x-4x-19=4x-4x-31$$
$$2x-19=-31$$
$$2x-19+19=-31+19$$
$$2x=-12$$
$$\frac{2x}{2}=\frac{-12}{2}$$
$$x=-6$$

6. $$-\frac{4}{9}n=\frac{5}{6}$$
$$-\frac{9}{4}\cdot-\frac{4}{9}n=-\frac{9}{4}\cdot\frac{5}{6}$$
$$n=-\frac{15}{8}$$

Exercise Set 2.3

1. Any value that makes the inequality true.

3. The graph shows that every real number greater than -1, but not including -1, is in the solution set.

5. The set of all values of x such that x is greater than or equal to 2. This means that every real number greater than 2 and including 2 is in the solution set for the variable x.

7. a) $\{x|x\ge 5\}$ b) $[5,\infty)$

c) 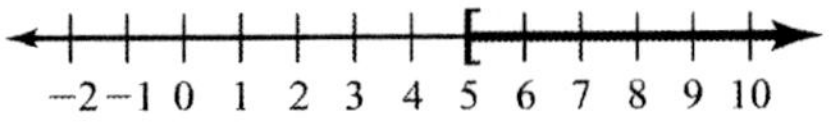

9. a) $\{q|q<-1\}$ b) $(-\infty,-1)$

c) 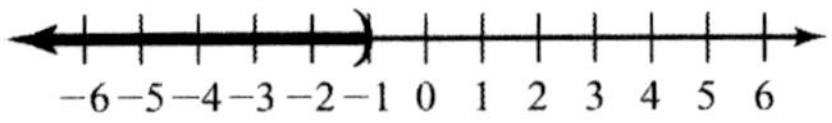

11. a) $\left\{p\middle|p<\frac{1}{5}\right\}$ b) $\left(-\infty,\frac{1}{5}\right)$

c) 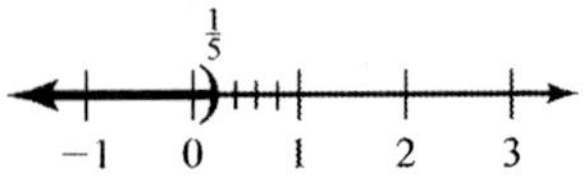

13. a) $\{r|r\le 1.9\}$ b) $(-\infty,1.9]$

c) 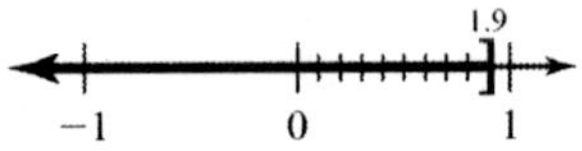

15. $$r-6<-12$$
$$r-6+6<-12+6$$
$$r<-6$$

a) $\{r|r<-6\}$ b) $(-\infty,-6)$

c) 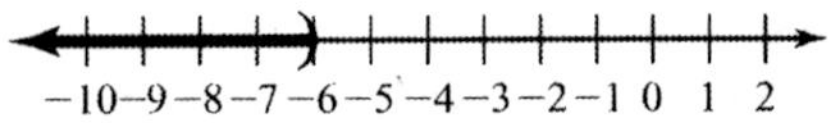

17. $-4y \le -16$

$\frac{-4y}{-4} \ge \frac{-16}{-4}$

$y \ge 4$

a) $\{y | y \ge 4\}$ b) $[4, \infty)$

c) 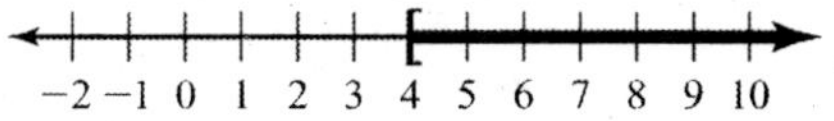

19. $3p + 9 < 21$

$3p + 9 - 9 < 21 - 9$

$3p < 12$

$\frac{3p}{3} < \frac{12}{3}$

$p < 4$

a) $\{p | p < 4\}$ b) $(-\infty, 4)$

c) 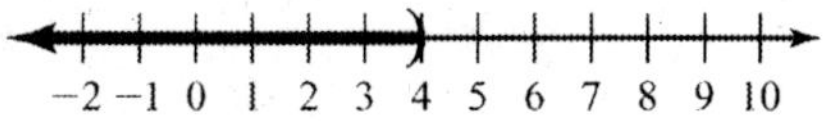

21. $6 - 5x > 29$

$6 - 6 - 5x > 29 - 6$

$-5x > 23$

$\frac{-5x}{-5} < \frac{23}{-5}$

$x < -\frac{23}{5}$

a) $\left\{x \middle| x < -\frac{23}{5}\right\}$ b) $\left(-\infty, -\frac{23}{5}\right)$

c) 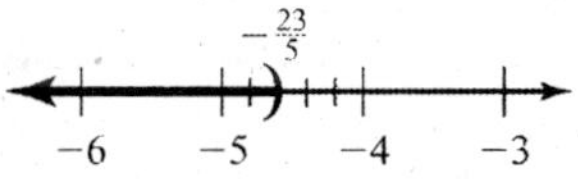

23. $\frac{a}{8} + 1 < \frac{3}{8}$

$8 \cdot \left(\frac{a}{8} + 1\right) < 8 \cdot \left(\frac{3}{8}\right)$

$a + 8 < 3$

$a + 8 - 8 < 3 - 8$

$a < -5$

a) $\{a | a < -5\}$ b) $(-\infty, -5)$

c) 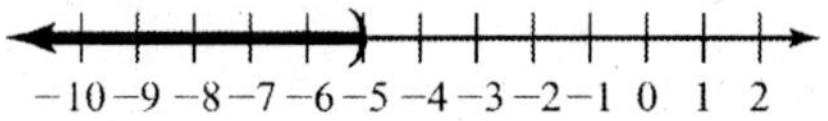

25. $7 + 2x < -2 + x$

$7 + 2x - x < -2 + x - x$

$7 + x < -2$

$7 - 7 + x < -2 - 7$

$x < -9$

a) $\{x | x < -9\}$ b) $(-\infty, -9)$

c) 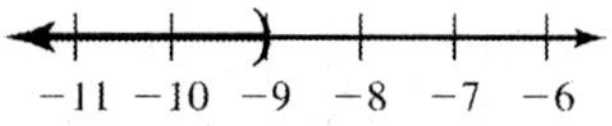

27. $-11k - 8 > -16 - 9k$

$-11k + 9k - 8 > -16 - 9k + 9k$

$-2k - 8 > -16$

$-2k - 8 + 8 > -16 + 8$

$-2k > -8$

$\frac{-2k}{-2} < \frac{-8}{-2}$

$k < 4$

a) $\{k | k < 4\}$ b) $(-\infty, 4)$

c) 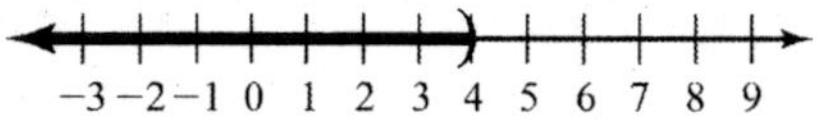

29. $2(3w - 4) - 5 \le 17$

$6w - 8 - 5 \le 17$

$6w - 13 \le 17$

$6w - 13 + 13 \le 17 + 13$

$6w \le 30$

$\frac{6w}{6} \le \frac{30}{6}$

$w \le 5$

a) $\{w | w \le 5\}$ b) $(-\infty, 5]$

c) 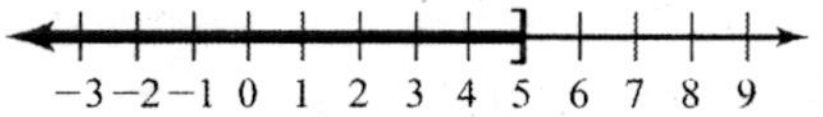

31.
$$4(y+2)+5>3(2y-1)$$
$$4y+8+5>6y-3$$
$$4y+13>6y-3$$
$$4y-6y+13>6y-6y-3$$
$$-2y+13>-3$$
$$-2y+13-13>-3-13$$
$$-2y>-16$$
$$\frac{-2y}{-2}<\frac{-16}{-2}$$
$$y<8$$

a) $\{y|y<8\}$ b) $(-\infty,8)$

c) 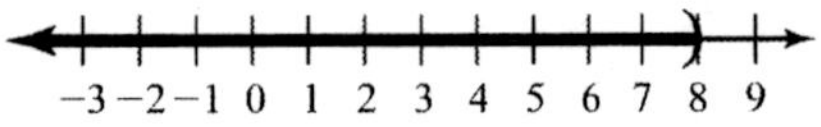

33.
$$\frac{1}{6}(5x+1)<-\frac{1}{6}x-\frac{5}{3}$$
$$6\cdot\frac{1}{6}(5x+1)<6\cdot-\frac{1}{6}x-6\cdot\frac{5}{3}$$
$$5x+1<-x-10$$
$$5x+x+1<-x+x-10$$
$$6x+1<-10$$
$$6x+1-1<-10-1$$
$$6x<-11$$
$$\frac{6x}{6}<\frac{-11}{6}$$
$$x<-\frac{11}{6}$$

a) $\left\{x\middle|x<-\frac{11}{6}\right\}$ b) $\left(-\infty,-\frac{11}{6}\right)$

c) 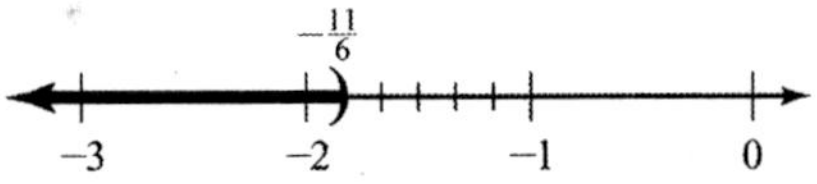

35.
$$\frac{1}{6}(7m+2)-\frac{1}{6}(11m-7)\geq 0$$
$$6\cdot\frac{1}{6}(7m+2)-6\cdot\frac{1}{6}(11m-7)\geq 6\cdot 0$$
$$7m+2-(11m-7)\geq 0$$
$$7m+2-11m+7\geq 0$$
$$-4m+9\geq 0$$
$$-4m+9-9\geq 0-9$$
$$-4m\geq -9$$
$$\frac{-4m}{-4}\leq\frac{-9}{-4}$$
$$m\leq\frac{9}{4}$$

a) $\left\{m\middle|m\leq\frac{9}{4}\right\}$ b) $\left(-\infty,\frac{9}{4}\right]$

c) 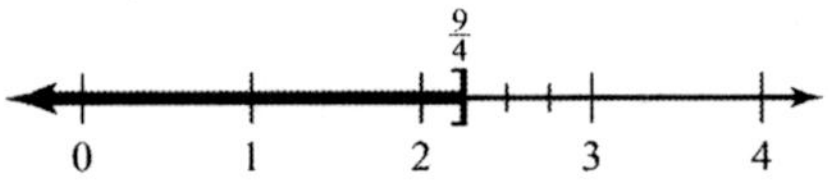

37.
$$0.7x-0.3\leq 0.8x+0.7$$
$$0.7x-0.8x-0.3\leq 0.8x-0.8x+0.7$$
$$-0.1x-0.3\leq 0.7$$
$$-0.1x-0.3+0.3\leq 0.7+0.3$$
$$-0.1x\leq 1$$
$$\frac{-0.1x}{-0.1}\geq\frac{1}{-0.1}$$
$$x\geq -10$$

a) $\{x|x\geq -10\}$ b) $[-10,\infty)$

c) 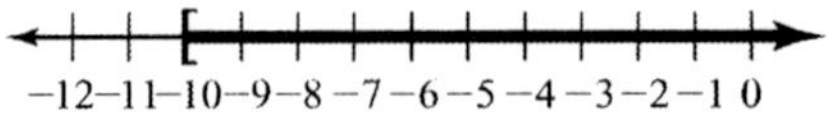

39.
$$0.09z + 20.34 < 3(1.4z - 1.5) + 2.1z$$
$$0.09z + 20.34 < 4.2z - 4.5 + 2.1z$$
$$0.09z + 20.34 < 6.3z - 4.5$$
$$0.09z - 6.3z + 20.34 < 6.3z - 6.3z - 4.5$$
$$-6.21z + 20.34 < -4.5$$
$$-6.21z + 20.34 - 20.34 < -4.5 - 20.34$$
$$-6.21z < -24.84$$
$$\frac{-6.21z}{-6.21} > \frac{-24.84}{-6.21}$$
$$z > 4$$

a) $\{z \mid z > 4\}$ b) $(4, \infty)$

c) 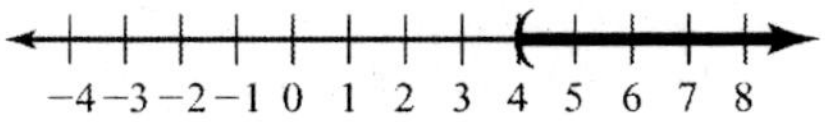

41. $3(x+2) - 4 < x + 5 + 2x$
$$3x + 6 - 4 < 3x + 5$$
$$3x + 2 < 3x + 5$$
$$3x - 3x + 2 < 3x - 3x + 5$$
$$2 < 5$$

Because there are no variable terms left in the inequality and the inequality is true, every real number is a solution for the original inequality.

a) $\{x \mid x \text{ is a real number}\}$

b) $(-\infty, \infty)$

c) 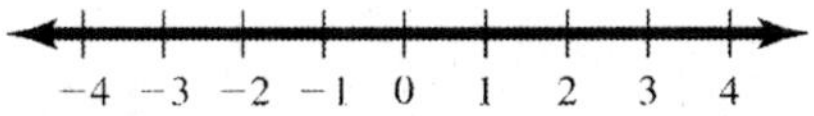

43. $4b - 3(b+5) \ge 2(2b+3) - 3(b+5)$
$$4b - 3b - 15 \ge 4b + 6 - 3b - 15$$
$$b - 15 \ge b - 9$$
$$b - b - 15 \ge b - b - 9$$
$$-15 \ge -9$$

Because there are no variable terms left in the inequality and the inequality is false, there are no solutions for the original inequality.

a) $\{\ \}$ or $\varnothing$

b) No interval notation

c) 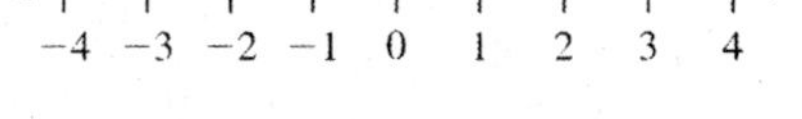

45.
$$\frac{3}{4}x < -6$$
$$\frac{4}{3} \cdot \frac{3}{4}x < \frac{4}{3} \cdot (-6)$$
$$x < -8$$

47.
$$5x - 1 > 14$$
$$5x - 1 + 1 > 14 + 1$$
$$5x > 15$$
$$\frac{5x}{5} > \frac{15}{5}$$
$$x > 3$$

49.
$$1 - 4x \le 25$$
$$1 - 1 - 4x \le 25 - 1$$
$$-4x \le 24$$
$$\frac{-4x}{-4} \ge \frac{24}{-4}$$
$$x \ge -6$$

51. $6 + 2(x - 5) \le 12$
$$6 + 2x - 10 \le 12$$
$$2x - 4 \le 12$$
$$2x - 4 + 4 \le 12 + 4$$
$$2x \le 16$$
$$\frac{2x}{2} \le \frac{16}{2}$$
$$x \le 8$$

53. Let x represent Yvonne's score on the fifth test.
$$\frac{92 + 100 + 83 + 81 + x}{5} \ge 90$$
$$\frac{356 + x}{5} \ge 90$$
$$356 + x \ge 450$$
$$356 - 356 + x \ge 450 - 356$$
$$x \ge 94$$

Yvonne must score 94 or higher.

55. Let x represent Blake's score on his fourth round.
$$\frac{86 + 82 + 88 + x}{4} \le 84$$
$$\frac{256 + x}{4} \le 84$$
$$256 + x \le 336$$
$$256 - 256 + x \le 336 - 256$$
$$x \le 80$$

Blake must score 80 or less.

57. Let w represent the width of the bedroom.
$$18w \le 234$$
$$\frac{18w}{18} \le \frac{234}{18}$$
$$w \le 13$$

The width must be 13 ft. or less.

59. Let r represent the radius of the pot.

$$150 \ge 2\pi r$$
$$150 \ge 2(3.14)r$$
$$150 \ge 6.28r$$
$$\frac{150}{6.28} \ge \frac{6.28r}{6.28}$$
$$23.9 \ge r$$

The radius must be approximately 23.9 in. or less.

61. Let r represent Juan's average rate.

$$180 \le r(3)$$
$$\frac{180}{3} \le \frac{3r}{3}$$
$$60 \le r$$

He must drive at least 60 mph.

63. To find profit, we subtract costs from revenue.

$$(12.5n + 3000) - (8n + 21{,}000) \ge 0$$
$$12.5n + 3000 - 8n - 21{,}000 \ge 0$$
$$4.5n - 18{,}000 \ge 0$$
$$4.5n - 18{,}000 + 18{,}000 \ge 18{,}000$$
$$\frac{4.5n}{4.5} \ge \frac{18{,}000}{4.5}$$
$$n \ge 4000$$

4000 or more lamps must be sold to break even or make a profit.

65. $F = \frac{9}{5}C + 32$

$$F = \frac{9}{5}(1064.58) + 32$$
$$F = 1948.244$$

a) $t < 1948.244°\text{F}$ b) $t \ge 1948.244°\text{F}$

67. current × resistance = voltage

$$x \cdot 6 \le 15$$
$$\frac{6x}{6} \le \frac{15}{6}$$
$$x \le 2.5$$

The current can be 2.5 amps or less

Review Exercises

1. Yes, because each element of A is a member of B.

2. Commutative property of addition

3. $0.5r^2 - t = 0.5(-6)^2 - 8$

$$= 0.5(36) - 8$$
$$= 18 - 8$$
$$= 10$$

4. $|x^3 - y| = |(-4)^3 - (-19)|$

$$= |-64 + 19|$$
$$= |-45|$$
$$= 45$$

5.
$$6x + 9 = 13 + 4(3x + 2)$$
$$6x + 9 = 13 + 12x + 8$$
$$6x + 9 = 21 + 12x$$
$$6x + 9 - 21 = 21 - 21 + 12x$$
$$6x - 12 = 12x$$
$$6x - 6x - 12 = 12x - 6x$$
$$-12 = 6x$$
$$\frac{-12}{6} = \frac{6x}{6}$$
$$-2 = x$$

6.
$$\frac{3}{4}n - 1 = \frac{1}{5}(2n + 3)$$
$$20\left(\frac{3}{4}n - 1\right) = 20 \cdot \frac{1}{5}(2n + 3)$$
$$15n - 20 = 4(2n + 3)$$
$$15n - 20 = 8n + 12$$
$$15n - 8n - 20 = 8n - 8n + 12$$
$$7n - 20 = 12$$
$$7n - 20 + 20 = 12 + 20$$
$$7n = 32$$
$$n = \frac{32}{7}$$

Exercise Set 2.4

1. Two inequalities joined by either *and* or *or*.

3. For two sets A and B, the intersection of A and B, symbolized by $A \cap B$, is a set containing only elements that are in both A and B.

5. We graph the region of overlap of the two inequalities.

7. a. $\{1, 3, 5\}$
   b. $\{1, 3, 5, 7, 9\}$

9. a. $\{7, 8\}$
   b. $\{5, 6, 7, 8, 9\}$

11. a. $\varnothing$
    b. $\{a, c, d, g, o, t\}$

13. a. $\{x, y, z\}$
    b. $\{w, x, y, z\}$

15. $-4 < x < 5$

17. $0 < y \le 2$

19. $-7 < w < 3$

21. $0 \le u \le 2$

23. $2 < x < 7$

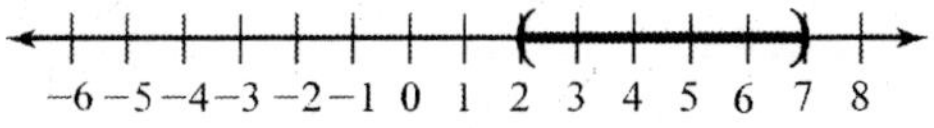

25. $-1 < x \le 5$

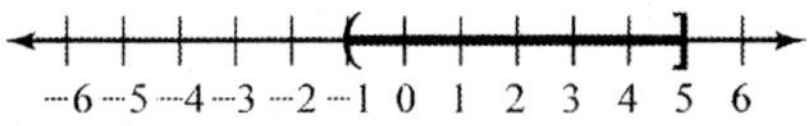

27. $1 \le x \le 10$

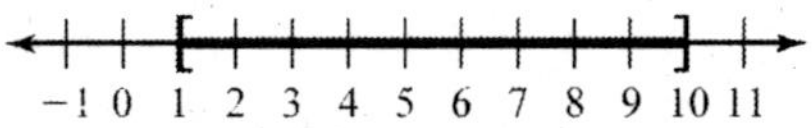

29. $-3 \le x < 4$

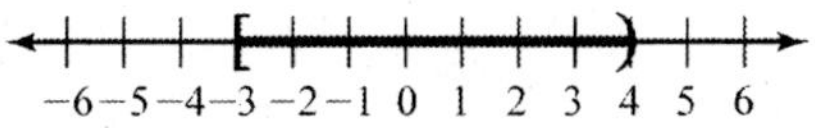

31. $-3 < x < -1$

a)

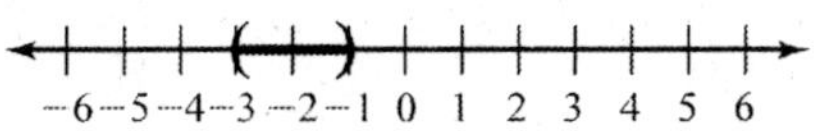

b) $\{x \mid -3 < x < -1\}$ c) $(-3,-1)$

33. $x+2 > 5$ and $x-4 \le 2$

$x > 3$ $x \le 6$

a)

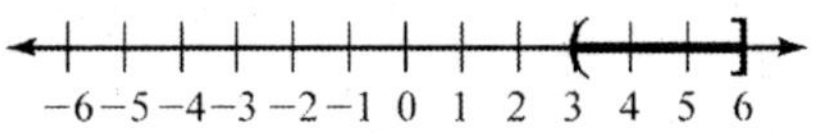

b) $\{x \mid 3 < x \le 6\}$ c) $(3,6]$

35. $-x > 1$ and $-2x \le -10$

$x < -1$ $x \ge 5$

a)

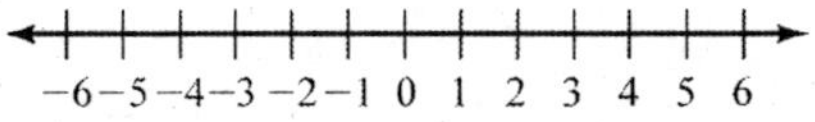

b) Ø c) no interval notation

37. $3x+8 \ge -7$ and $4x-7 < 5$

$3x \ge -15$ $4x < 12$

$x \ge -5$ $x < 3$

a)

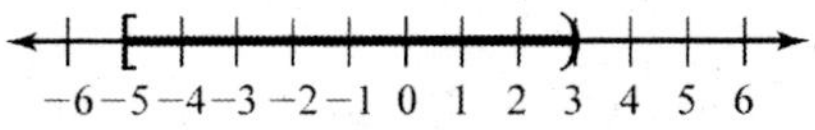

b) $\{x \mid -5 \le x < 3\}$ c) $[-5,3)$

39. $-3x-8 > 1$ and $-4x+5 \le -3$

$-3x > 9$ $-4x \le -8$

$x < -3$ $x \ge 2$

a)

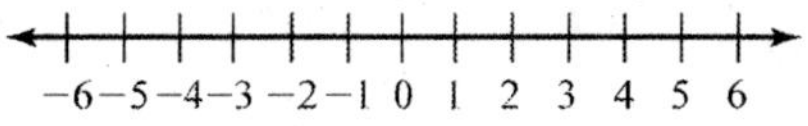

b) Ø c) no interval notation

41. $-3 < x+4 < 1$

$-7 < x < -3$

a)

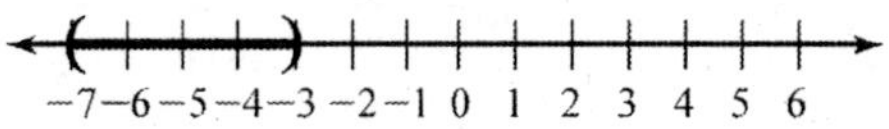

b) $\{x \mid -7 < x < -3\}$ c) $(-7,-3)$

43. $-7 \le 4x-3 \le 5$

$-4 \le 4x \le 8$

$-1 \le x \le 2$

a)

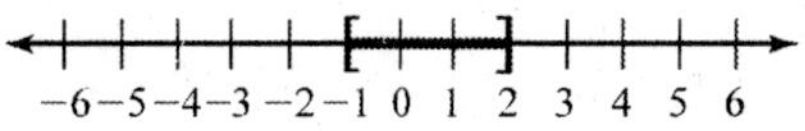

b) $\{x \mid -1 \le x \le 2\}$ c) $[-1,2]$

45. $0 \le 2+3x < 8$

$-2 \le 3x < 6$

$-\dfrac{2}{3} \le x < 2$

a)

b) $\left\{x \mid -\dfrac{2}{3} \le x < 2\right\}$ c) $\left[-\dfrac{2}{3},2\right)$

47. $-1 < -2x+5 \le 5$

$-6 < -2x \le 0$

$3 > x \ge 0$

a)

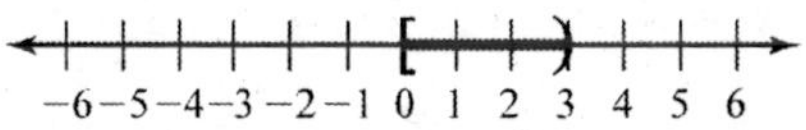

b) $\{x \mid 0 \le x < 3\}$ c) $[0,3)$

49. $3 \le 6-x \le 6$

$-3 \le -x \le 0$

$3 \ge x \ge 0$

a)

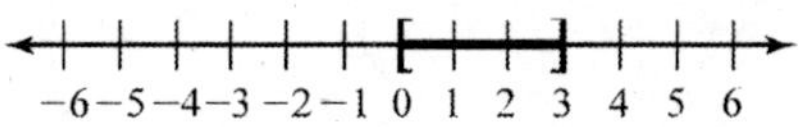

b) $\{x \mid 0 \le x \le 3\}$ c) $[0,3]$

51. $x < -2$ or $x > 6$

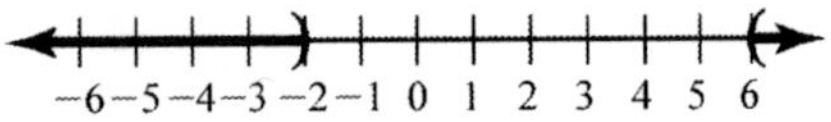

53. $y < -3$ or $y \geq 0$

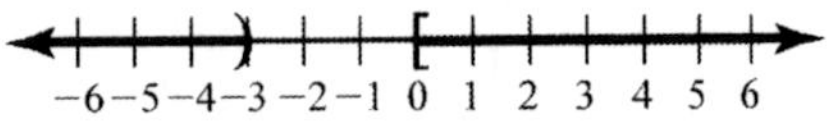

55. $w > -3$ or $w > 2$

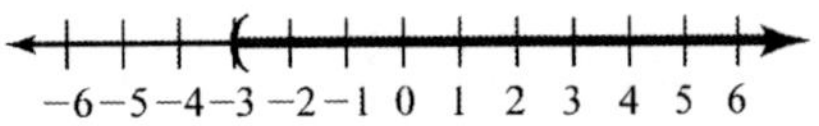

57. $u \geq 0$ or $u \leq 2$

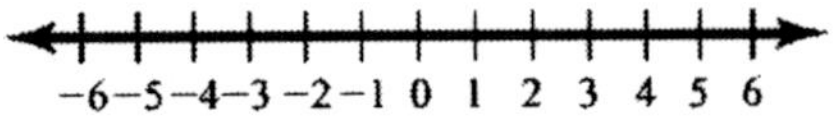

59. $y + 2 < -7$ or $y + 2 > 7$

$y < -9$ $\quad$ $y > 5$

a)

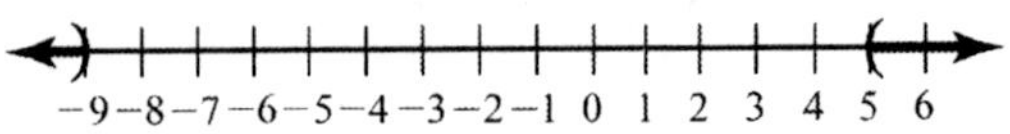

b) $\{y | y < -9 \text{ or } y > 5\}$

c) $(-\infty, -9) \cup (5, \infty)$

61. $4r - 3 < -11$ or $2r - 3 > -1$

$4r < -8$ $\quad$ $2r > 2$

$r < -2$ $\quad$ $r > 1$

a)

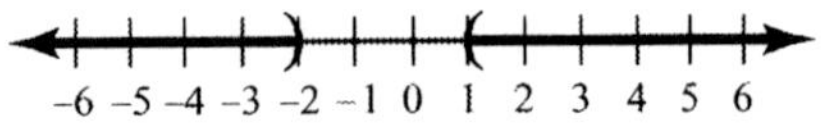

b) $\{r | r < -2 \text{ or } r > 1\}$

c) $(-\infty, -2) \cup (1, \infty)$

63. $-w + 2 \leq -5$ or $-w + 2 \geq 3$

$-w \leq -7$ $\quad$ $-w \geq 1$

$w \geq 7$ $\quad$ $w \leq -1$

a)

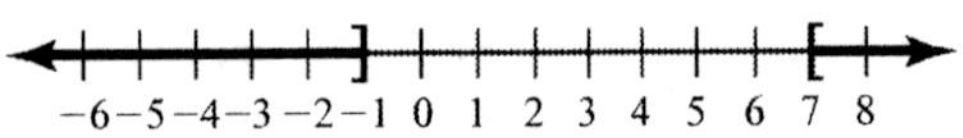

b) $\{w | w \leq -1 \text{ or } w \geq 7\}$

c) $(-\infty, -1] \cup [7, \infty)$

65. $7 - 4k \leq -5$ or $6 - 2k \geq 2$

$-4k \leq -12$ $\quad$ $-2k \geq -4$

$k \geq 3$ $\quad$ $k \leq 2$

a)

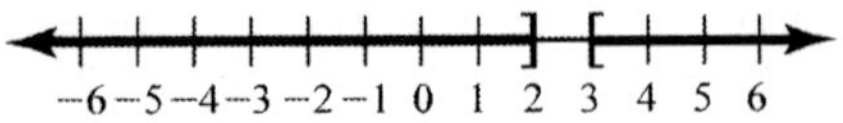

b) $\{k | k \leq 2 \text{ or } k \geq 3\}$

c) $(-\infty, 2] \cup [3, \infty)$

67. $\frac{2}{3}x - 4 \geq -2$ or $\frac{2}{5}x - 2 \leq -4$

$\frac{2}{3}x \geq 2$ $\quad$ $\frac{2}{5}x \leq -2$

$2x \geq 6$ $\quad$ $2x \leq -10$

$x \geq 3$ $\quad$ $x \leq -5$

a)

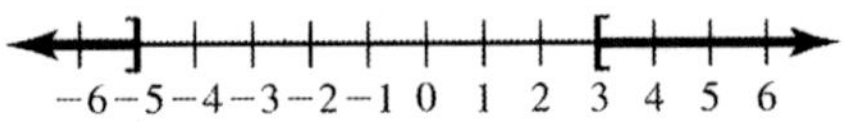

b) $\{x | x \leq -5 \text{ or } x \geq 3\}$

c) $(-\infty, -5] \cup [3, \infty)$

69. $-2(c-1) < -4$ or $-2(c-2) < -6$

$-2c + 2 < -4$ $\quad$ $-2c + 4 < -6$

$-2c < -6$ $\quad$ $-2c < -10$

$c > 3$ $\quad$ $c > 5$

a)

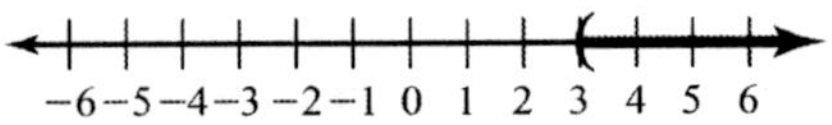

b) $\{c | c > 3\}$ $\quad$ c) $(3, \infty)$

71. $2(x+3) + 3 \leq 1$ or $2(x+1) + 7 \leq -3$

$2x + 6 + 3 \leq 1$ $\quad$ $2x + 2 + 7 \leq -3$

$2x + 9 \leq 1$ $\quad$ $2x + 9 \leq -3$

$2x \leq -8$ $\quad$ $2x \leq -12$

$x \leq -4$ $\quad$ $x \leq -6$

a)

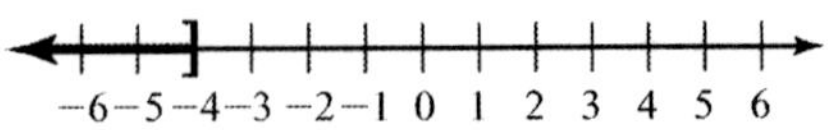

b) $\{x | x \leq -4\}$

c) $(-\infty, -4]$

73. $-3(x-1)-1 \le -1$ or $-3(x+1)+5 \ge -2$

$-3x+3-1 \le -1$ $\quad$ $-3x-3+5 \ge -2$

$-3x+2 \le -1$ $\quad$ $-3x+2 \ge -2$

$-3x \le -3$ $\quad$ $-3x \ge -4$

$x \ge 1$ $\quad$ $x \le \frac{4}{3}$

a) 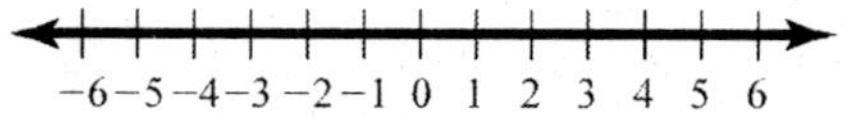

b) $\{x | x \text{ is a real number}\}$ or $\mathbb{R}$

c) $(-\infty, \infty)$

75. $-3 < x+4$ and $x+4 < 7$

$-7 < x$ $\quad$ $x < 3$

a) 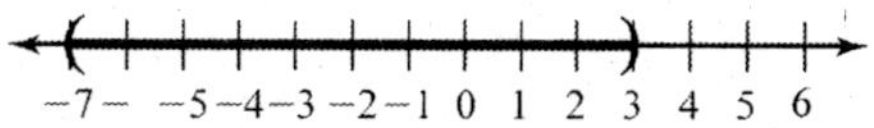

b) $\{x | -7 < x < 3\}$ c) $(-7, 3)$

77. $4x+3 < -5$ or $3x+2 > 8$

$4x < -8$ $\quad$ $3x > 6$

$x < -2$ $\quad$ $x > 2$

a) 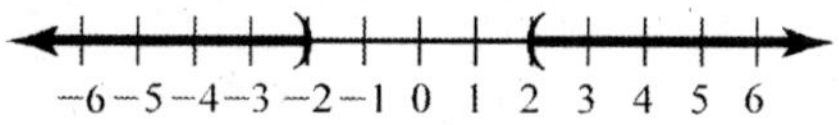

b) $\{x | x < -2 \text{ or } x > 2\}$

c) $(-\infty, -2) \cup (2, \infty)$

79. $4 < -2x < 6$

$-2 > x > -3$

a) 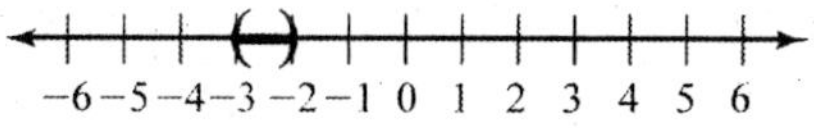

b) $\{x | -3 < x < -2\}$ c) $(-3, -2)$

81. $7 \le 4x - 5 \le 19$

$12 \le 4x \le 24$

$3 \le x \le 6$

a) 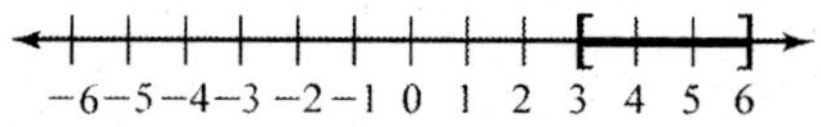

b) $\{x | 3 \le x \le 6\}$ c) $[3, 6]$

83. $-5 < 1 - 2x < -1$

$-6 < -2x < -2$

$3 > x > 1$

a) 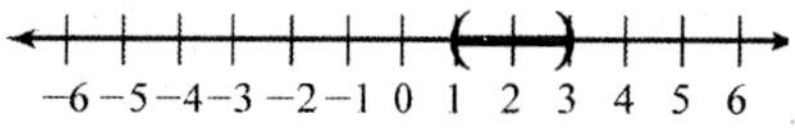

b) $\{x | 1 < x < 3\}$ c) $(1, 3)$

85. $x - 3 < 2x + 1$ and $2x + 1 < 3x$

$-x < 4$ $\quad$ $-x < -1$

$x > -4$ $\quad$ $x > 1$

a) 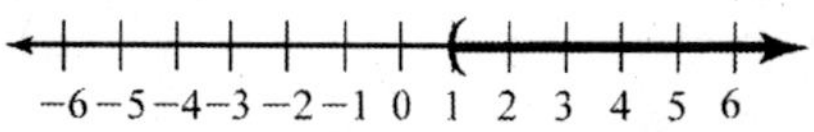

b) $\{x | x > 1\}$ c) $(1, \infty)$

87. $36 \le 0.09x \le 45$

$400 \le x \le 500$

a) 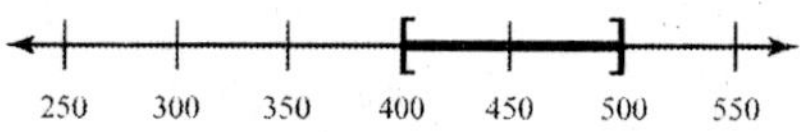

b) $\{x | 400 \le x \le 500\}$

c) $[400, 500]$

89. $80 \le \frac{95+80+82+88+x}{5} < 90$

$400 \le 345 + x < 450$

$55 \le x < 105$

a) 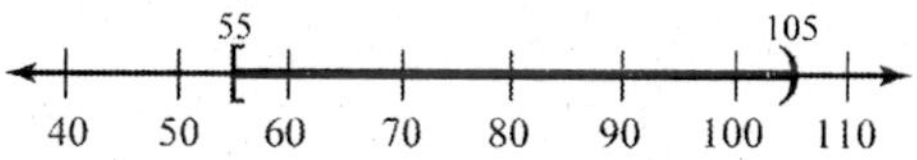

b) $\{x | 55 \le x < 105\}$ c) $[55, 105)$

91. a) 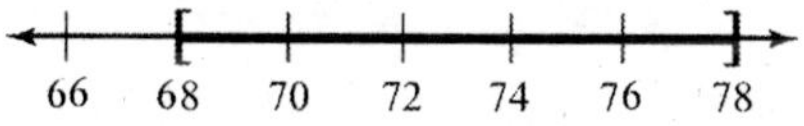

b) $\{x | 68° \le x \le 78°\}$ c) $[68, 78]$

93. a) 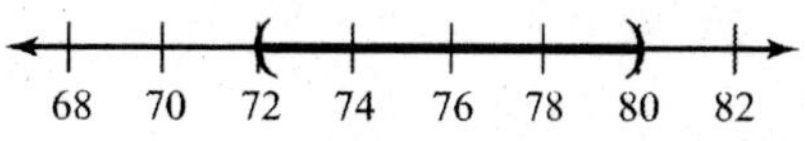

b) $\{x | 72° < x < 80°\}$ c) $(72, 80)$

95. $6000 \le \frac{1}{2} \cdot 50(60 + x) \le 8000$

$$6000 \le 25(60 + x) \le 8000$$
$$6000 \le 1500 + 25x \le 8000$$
$$4500 \le 25x \le 6500$$
$$180 \le x \le 260$$

a) 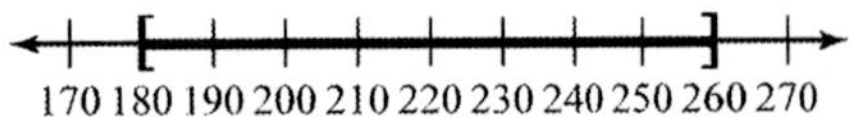

b) $\{x \mid 180 \le x \le 260\}$ c) $[180, 260]$

Review Exercises

1. No, the absolute value of zero is zero and zero is neither positive nor negative.

2. $-|2-3^2| - |-4| = -|2-9| - |-4|$
$$= -|-7| - 4$$
$$= -7 - 4$$
$$= -11$$

3. $-|-|-16|| = -|-16| = -16$

4. $3x - 14x = 17 - 12x - 11$
$$-11x = 6 - 12x$$
$$x = 6$$

5. $\frac{3}{5}(25 - 5x) = 15 - \frac{3}{5}$
$$15 - 3x = \frac{72}{5}$$
$$5 \cdot 15 - 5 \cdot 3x = 5 \cdot \frac{72}{5}$$
$$75 - 15x = 72$$
$$-15x = -3$$
$$x = \frac{1}{5}$$

6. $\frac{1}{2}(x + 2) = 0$
$$2 \cdot \frac{1}{2}(x + 2) = 2 \cdot 0$$
$$x + 2 = 0$$
$$x = -2$$

Exercise Set 2.5

1. The distance a number is from zero.

3. If $|n| = a$ where n is a variable or an expression and $a \ge 0$, then $n = a$ or $n = -a$.

5. Separate the absolute value equation into two equations: $ax + b = cx + d$ and $ax + b = -(cx + d)$.

7. $x = 2$ or $x = -2$

9. $|a| = -4$

Because the absolute value of every real number is a positive number or zero, this equation has no solution.

11. $x + 3 = 8$ or $x + 3 = -8$
$$x = 5 \qquad x = -11$$

13. $2m - 5 = 1$ or $2m - 5 = -1$
$$2m = 6 \qquad 2m = 4$$
$$m = 3 \qquad m = 2$$

15. $6 - 5x = 1$ or $6 - 5x = -1$
$$-5x = -5 \qquad -5x = -7$$
$$x = 1 \qquad x = \frac{7}{5}$$

17. $4 - 3w = -6$ or $4 - 3w = 6$
$$w = \frac{10}{3} \qquad w = -\frac{2}{3}$$

19. $|4m - 2| = -5$

Because the absolute value of every real number is a positive number or zero, this equation has no solution.

21. $4w - 3 = 0$
$$4w = 3$$
$$w = \frac{3}{4}$$

23. $|2y| - 3 = 5$
$$|2y| = 8$$
$$2y = 8 \text{ or } 2y = -8$$
$$y = 4 \qquad y = -4$$

25. $|y + 1| + 2 = 4$
$$|y + 1| = 2$$
$$y + 1 = 2 \text{ or } y + 1 = -2$$
$$y = 1 \qquad y = -3$$

27. $|b-4|-6=2$
$|b-4|=8$
$b-4=8$ or $b-4=-8$
$b=12$ $b=-4$

29. $3+|5x-1|=7$
$|5x-1|=4$
$5x-1=4$ or $5x-1=-4$
$5x=5$ $5x=-3$
$x=1$ $x=-\frac{3}{5}$

31. $1-|2k+3|=-4$
$-|2k+3|=-5$
$|2k+3|=5$
$2k+3=5$ or $2k+3=-5$
$2k=2$ $2k=-8$
$k=1$ $k=-4$

33. $4-3|z-2|=-8$
$-3|z-2|=-12$
$|z-2|=4$
$z-2=4$ or $z-2=-4$
$z=6$ $z=-2$

35. $6-2|3-2w|=-18$
$-2|3-2w|=-24$
$|3-2w|=12$
$3-2w=12$ or $3-2w=-12$
$-2w=9$ $-2w=-15$
$w=-\frac{9}{2}$ $w=\frac{15}{2}$

37. $|3x-2(x+5)|=10$
$|3x-2x-10|=10$
$|x-10|=10$
$x-10=10$ or $x-10=-10$
$x=20$ $x=0$

39. $2x+1=x+5$ or $2x+1=-(x+5)$
$x=4$ $2x+1=-x-5$
$3x=-6$
$x=-2$

41. $x+3=2x-4$ or $x+3=-(2x-4)$
$7=x$ $x+3=-2x+4$
$3x=1$
$x=\frac{1}{3}$

43. $3v+4=1-2v$ or $3v+4=-(1-2v)$
$5v=-3$ $3v+4=-1+2v$
$v=-\frac{3}{5}$ $v=-5$

45. $2n+3=3+2n$ or $2n+3=-(3+2n)$
$3=3$ $2n+3=-3-2n$
$4n=-6$
$n=-\frac{3}{2}$

One equation leads to a solution with no variables yet is true. The solution to this absolute value equation is all real numbers.

47. $2k+1=2k-5$ or $2k+1=-(2k-5)$
$0=-6$ $2k+1=-2k+5$
no solution $4k=4$
$k=1$

This absolute value equation has only one solution, 1.

49. $|10-(5-h)|=8$
$|10-5+h|=8$
$|5+h|=8$
$5+h=8$ or $5+h=-8$
$h=3$ $h=-13$

51. $\frac{b}{2}-1=4$ or $\frac{b}{2}-1=-4$
$\frac{b}{2}=5$ $\frac{b}{2}=-3$
$b=10$ $b=-6$

53. $\frac{4-3x}{2}=\frac{3}{4}$ or $\frac{4-3x}{2}=-\frac{3}{4}$
$4(4-3x)=2\cdot 3$ $4(4-3x)=-3\cdot 2$
$16-12x=6$ $16-12x=-6$
$-12x=-10$ $-12x=-22$
$x=\frac{5}{6}$ $x=\frac{11}{6}$

55. $\left|2y+\frac{3}{2}\right|-2=5$

$\left|2y+\frac{3}{2}\right|=7$

$2y+\frac{3}{2}=7$ or $2y+\frac{3}{2}=-7$

$2y=\frac{11}{2}$ $\quad$ $2y=-\frac{17}{2}$

$y=\frac{11}{4}$ $\quad$ $y=-\frac{17}{4}$

Review Exercises

1. $(-2,0]$

2. 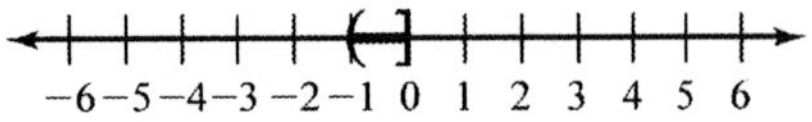

3. $6-4x>-5x-1$

   $x>-7$

4. $-4x\le 12$

   $x\ge -3$

5. $-1<\frac{3x+2}{4}<4$

   $-4<3x+2<16$

   $-6<3x<14$

   $-2<x<\frac{14}{3}$

6. $n-2<5$

   $n-2+2<5+2$

   $n<7$

Exercise Set 2.6

1. $x\ge -a$ and $x\le a$, or $-a\le x\le a$

3. Shade between the values of $-a$ and a.

5. when $a<0$

7. a) 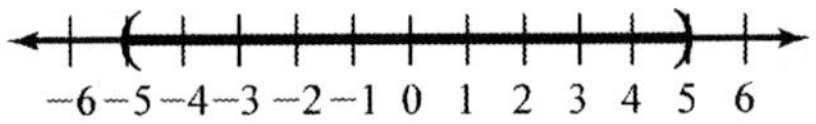

   b) $\{x|-5<x<5\}$ c) $(-5,5)$

9. $-7\le x+3\le 7$

   $-10\le x\le 4$

   a) 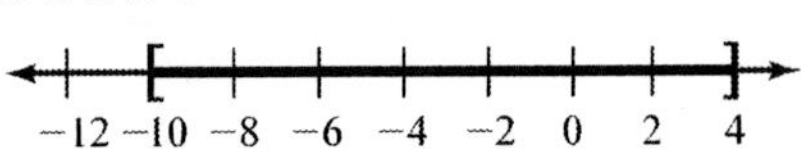

   b) $\{x|-10\le x\le 4\}$ c) $[-10,4]$

11. $|s+3|+6<9$

   $|s+3|<3$

   $-3<s+3<3$

   $-6<s<0$

   a) 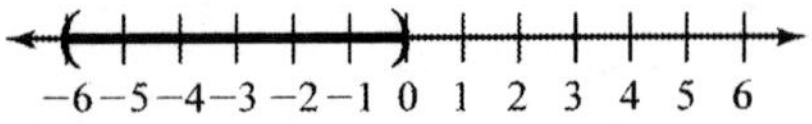

   b) $\{s|-6<s<0\}$ c) $(-6,0)$

13. $|2m-5|-3<6$

   $|2m-5|<9$

   $-9<2m-5<9$

   $-4<2m<14$

   $-2<m<7$

   a) 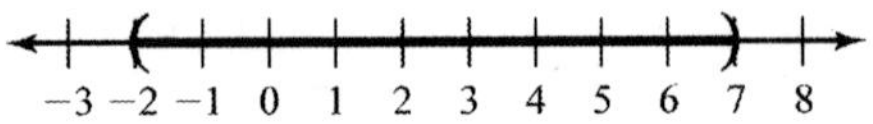

   b) $\{m|-2<m<7\}$ c) $(-2,7)$

15. $|-3k+5|+7\le 8$

   $|-3k+5|\le 1$

   $-1\le -3k+5\le 1$

   $-6\le -3k\le -4$

   $2\ge k\ge \frac{4}{3}$

   a) 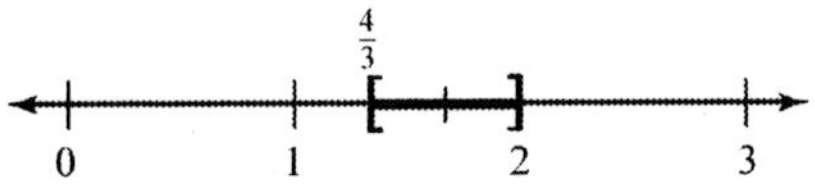

   b) $\left\{k\middle|\frac{4}{3}\le k\le 2\right\}$ c) $\left[\frac{4}{3},2\right]$

17. $2|x|+7\le 3$

   $2|x|\le -4$

   $|x|\le -2$

   Because absolute values cannot be negative, this inequality has no solution.

   a) 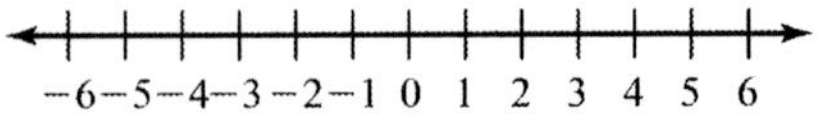

   b) $\{\ \}$ or $\varnothing$ c) no interval notation

19. $2|w-3|+4<10$

$2|w-3|<6$

$|w-3|<3$

$-3<w-3<3$

$0<w<6$

a) 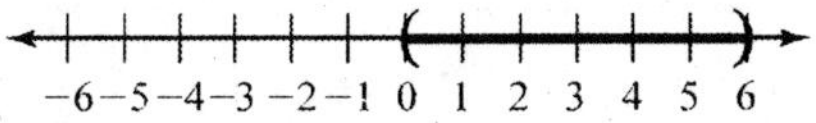

b) $\{w|0<w<6\}$ c) $(0,6)$

21. a) 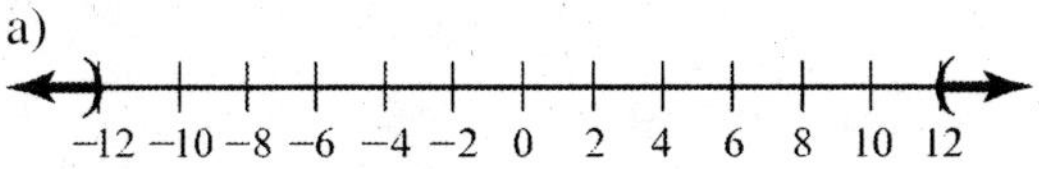

b) $\{c|c<-12 \text{ or } c>12\}$

c) $(-\infty,-12)\cup(12,\infty)$

23. $y+2\le-7$ or $y+2\ge7$

$y\le-9$ $y\ge5$

a) 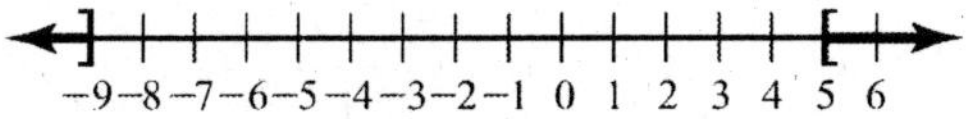

b) $\{y|y\le-9 \text{ or } y\ge5\}$

c) $(-\infty,-9]\cup[5,\infty)$

25. $|p-6|-3>5$

$|p-6|>8$

$p-6<-8$ or $p-6>8$

$p<-2$ $p>14$

a) 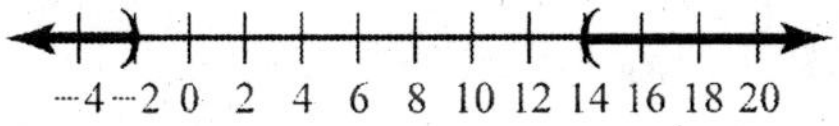

b) $\{p|p<-2 \text{ or } p>14\}$

c) $(-\infty,-2)\cup(14,\infty)$

27. $|3x+6|-3\ge9$

$|3x+6|\ge12$

$3x+6\le-12$ or $3x+6\ge12$

$3x\le-18$ $3x\ge6$

$x\le-6$ $x\ge2$

a) 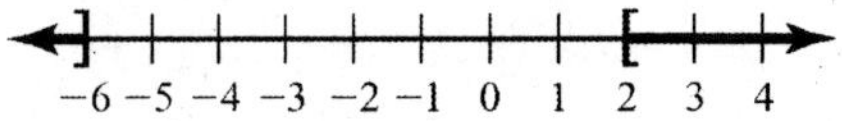

b) $\{x|x\le-6 \text{ or } x\ge2\}$

c) $(-\infty,-6]\cup[2,\infty)$

29. $|-4n-5|+3>8$

$|-4n-5|>5$

$-4n-5<-5$ or $-4n-5>5$

$-4n<0$ $-4n>10$

$n>0$ $n<-\frac{5}{2}$

a) 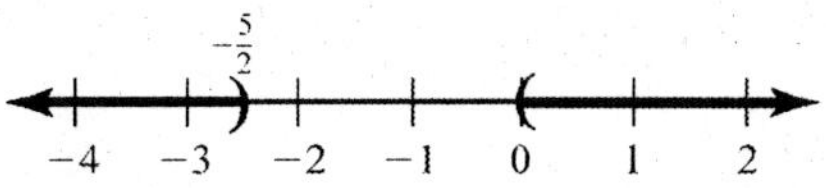

b) $\left\{n\middle|n<-\frac{5}{2} \text{ or } n>0\right\}$

c) $\left(-\infty,-\frac{5}{2}\right)\cup(0,\infty)$

31. $4|v|+3\ge7$

$4|v|\ge4$

$|v|\ge1$

$v\le-1$ or $v\ge1$

a) 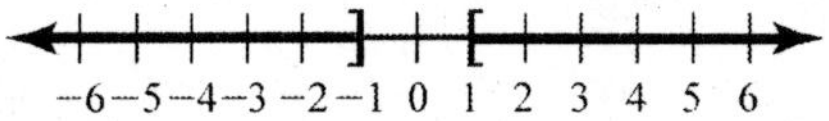

b) $\{v|v\le-1 \text{ or } v\ge1\}$

c) $(-\infty,-1]\cup[1,\infty)$

33. $4|y+2|-1>3$

$4|y+2|>4$

$|y+2|>1$

$y+2<-1$ or $y+2>1$

$y<-3$ $y>-1$

a) 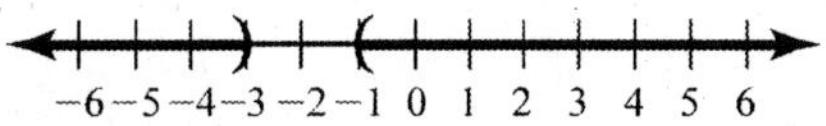

b) $\{y|y<-3 \text{ or } y>-1\}$

c) $(-\infty,-3)\cup(-1,\infty)$

35. $|4m+8|-2>10$

$|4m+8|>12$

$4m+8<-12$ or $4m+8>12$

$4m<-20$ $4m>4$

$m<-5$ $m>1$

a) 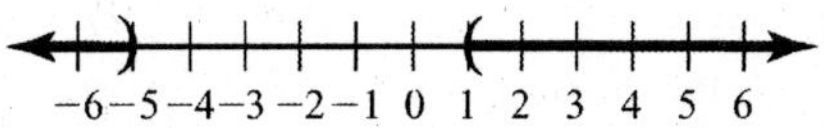

b) $\{m|m<-5 \text{ or } m>1\}$ c) $(-\infty,-5)\cup(1,\infty)$

37. $-6 < -3x + 6 < 6$

$-12 < -3x < 0$

$4 > x > 0$

$0 < x < 4$

a) 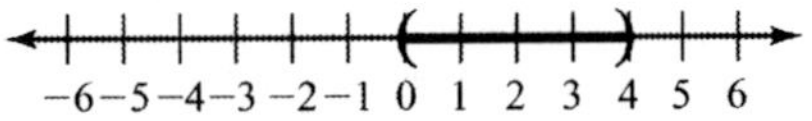

b) $\{x | 0 < x < 4\}$ c) $(0, 4)$

39. $|2r - 3| > -3$

This inequality indicates that the absolute value is greater than a negative number. Because the absolute value of every real number is either positive or 0, the solution set is $\mathbb{R}$.

a) 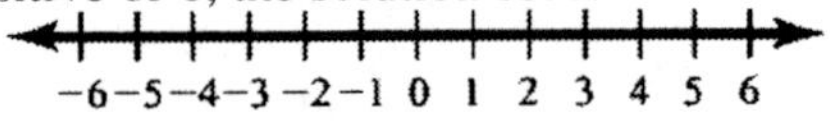

b) $\{r | r \text{ is a real number}\}$

c) $(-\infty, \infty)$

41. $4 - 2|x + 3| > 2$

$-2|x + 3| > -2$

$|x + 3| < 1$

$-1 < x + 3 < 1$

$-4 < x < -2$

a) 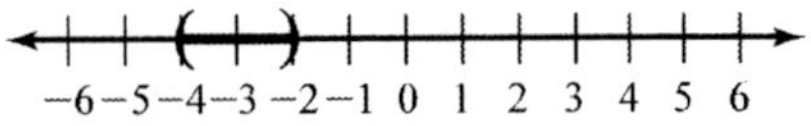

b) $\{x | -4 < x < -2\}$ c) $(-4, -2)$

43. $6|2x - 1| - 3 < 3$

$6|2x - 1| < 6$

$|2x - 1| < 1$

$-1 < 2x - 1 < 1$

$0 < 2x < 2$

$0 < x < 1$

a) 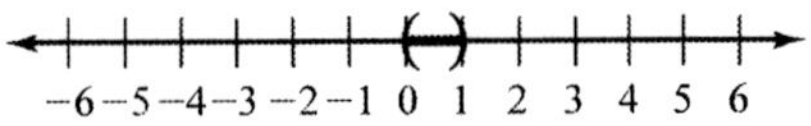

b) $\{x | 0 < x < 1\}$ c) $(0, 1)$

45. $5 - |w + 4| > 10$

$-|w + 4| > 5$

$|w + 4| < -5$

Because absolute values cannot be negative, this inequality has no solution.

a) 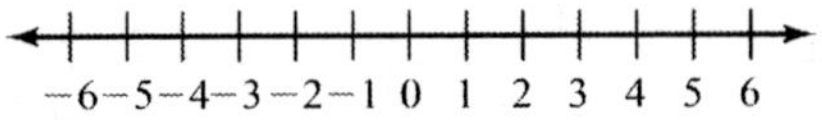

b) $\varnothing$ c) no interval notation

47. $-5 \le 2 - \frac{3}{2}k \le 5$

$-10 \le 4 - 3k \le 10$

$-14 \le -3k \le 6$

$\frac{14}{3} \ge k \ge -2$

a) 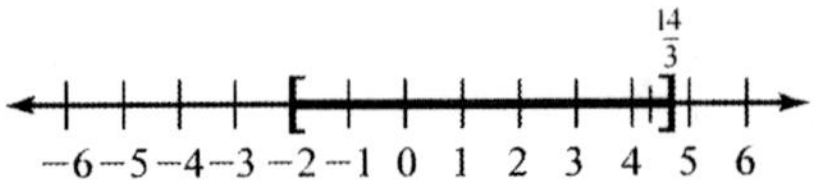

b) $\left\{k \middle| -2 \le k \le \frac{14}{3}\right\}$ c) $\left[-2, \frac{14}{3}\right]$

49. $|0.25x - 3| + 2 > 4$

$|0.25x - 3| > 2$

$0.25x - 3 < -2$ or $0.25x - 3 > 2$

$0.25x < 1$ $\qquad$ $0.25x > 5$

$x < 4$ $\qquad$ $x > 20$

a) 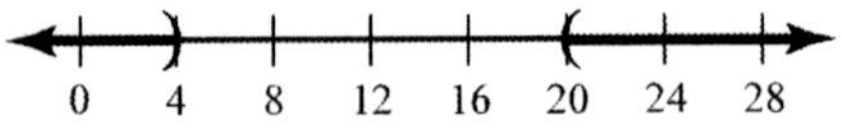

b) $\{x | x < 4 \text{ or } x > 20\}$

c) $(-\infty, 4) \cup (20, \infty)$

51. $-7.2 \le 2.4 - \frac{3}{4}y \le 7.2$

$-9.6 \le -\frac{3}{4}y \le 4.8$

$12.8 \ge y \ge -6.4$

a) 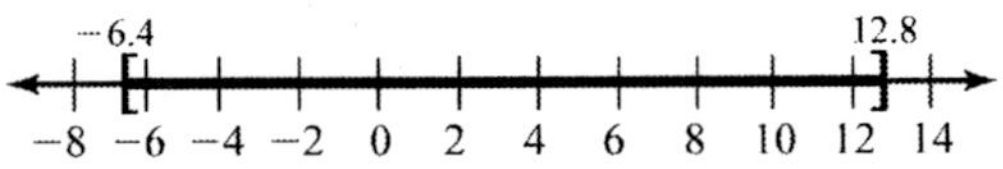

b) $\{y | -6.4 \le y \le 12.8\}$ c) $[-6.4, 12.8]$

53. $|2p-8|+5>1$

$|2p-8|>-4$

This inequality indicates that the absolute value is greater than a negative number. Because the absolute value of every real number is either positive or 0, the solution set is $\mathbb{R}$.

a) 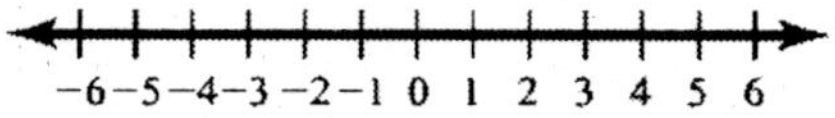

b) $\{p \mid p \text{ is a real number}\}$ c) $(-\infty, \infty)$

55. $|x|<3$

57. $|x|\geq 4$

59. $|x+1|>1$

61. $|x-3|\leq 2$

63. $|x|>$ any negative number

Review Exercises

1. true 2. true

3. $23-5(2-3x)=12x-(4x+1)$

$23-10+15x=12x-4x-1$

$13+15x=8x-1$

$13+7x=-1$

$7x=-14$

$x=-2$

4. $\frac{5}{6}x-\frac{3}{4}=\frac{2}{3}x+\frac{1}{2}$

$\frac{1}{6}x-\frac{3}{4}=\frac{1}{2}$

$\frac{1}{6}x=\frac{5}{4}$

$6\cdot\frac{1}{6}x=6\cdot\frac{5}{4}$

$x=\frac{15}{2}$

5. $Ax+By=C$

$By=C-Ax$

$\frac{By}{B}=\frac{C-Ax}{B}$

$y=\frac{C-Ax}{B}$

6. Let the first integer be x, the next be $(x+1)$ and the third be $(x+2)$.

$x+(x+1)+(x+2)=141$

$3x+3=141$

$3x=138$

$x=46$

The integers are 46, 47, and 48.

Chapter 2 Review Exercises

1. False; of these, only $\{\ \}$ indicates an empty set.
2. False; this equation has no solution.
3. True
4. True
5. False; absolute value equations may have two, one, or zero solutions.
6. True
7. contradiction 8. 90°
9. $ax+b=c; ax+b=-c$
10. $x>-a; x<a$
11. $3x-9=12$

$3x=21$

$x=7$

Check:

$3(7)-9\stackrel{?}{=}12$

$21-9\stackrel{?}{=}12$

$12=12$

12. $7m-3=4m+9$

$3m-3=9$

$3m=12$

$m=4$

Check:

$7(4)-3\stackrel{?}{=}4(4)+9$

$28-3\stackrel{?}{=}16+9$

$25=25$

13. $6(3a-4)+(3a+4)=5(a-2)-10$

$$18a-24+3a+4=5a-10-10$$
$$21a-20=5a-20$$
$$16a-20=-20$$
$$16a=0$$
$$a=0$$

Check:

$$6(3\cdot 0-4)+(3\cdot 0+4)\stackrel{?}{=}5(0-2)-10$$
$$6(0-4)+(0+4)\stackrel{?}{=}5(-2)-10$$
$$6(-4)+(4)\stackrel{?}{=}5(-2)-10$$
$$-24+4\stackrel{?}{=}-10-10$$
$$-20=-20$$

14. $2-2y-6(y+3)=5-7y$

$$2-2y-6y-18=5-7y$$
$$-16-8y=5-7y$$
$$-16-y=5$$
$$-y=21$$
$$y=-21$$

Check:

$$2-2(-21)-6(-21+3)\stackrel{?}{=}5-7(-21)$$
$$2-(-42)-6(-18)\stackrel{?}{=}5-(-147)$$
$$2+42+108\stackrel{?}{=}5+147$$
$$152=152$$

15. $\frac{1}{5}d-2=-3+d$

$$-\frac{4}{5}d-2=-3$$
$$-\frac{4}{5}d=-1$$
$$-5\left(-\frac{4}{5}d\right)=-5(-1)$$
$$4d=5$$
$$d=\frac{5}{4}$$

Check:

$$\frac{1}{5}\left(\frac{5}{4}\right)-2\stackrel{?}{=}-3+\frac{5}{4}$$
$$\frac{1}{4}-2\stackrel{?}{=}-\frac{7}{4}$$
$$-\frac{7}{4}=-\frac{7}{4}$$

16. $\frac{1}{2}k+\frac{5}{8}=\frac{1}{4}$

$$8\left(\frac{1}{2}k+\frac{5}{8}\right)=8\left(\frac{1}{4}\right)$$
$$4k+5=2$$
$$4k=-3$$
$$k=-\frac{3}{4}$$

Check:

$$\frac{1}{2}\left(-\frac{3}{4}\right)+\frac{5}{8}\stackrel{?}{=}\frac{1}{4}$$
$$-\frac{3}{8}+\frac{5}{8}\stackrel{?}{=}\frac{1}{4}$$
$$\frac{2}{8}\stackrel{?}{=}\frac{1}{4}$$
$$\frac{1}{4}=\frac{1}{4}$$

17. $2w+2(7-w)=2(5w+10)+4$

$$2w+14-2w=10w+20+4$$
$$14=10w+24$$
$$-10=10w$$
$$-1=w$$

Check:

$$2(-1)+2(7-(-1))\stackrel{?}{=}2(5(-1)+10)+4$$
$$-2+2(7+1)\stackrel{?}{=}2(-5+10)+4$$
$$-2+2(8)\stackrel{?}{=}2(5)+4$$
$$-2+16\stackrel{?}{=}10+4$$
$$14=14$$

18. $5r-2(5r-3)=6-5r$

$$5r-10r+6=6-5r$$
$$-5r+6=6-5r$$
$$0+6=6$$
$$6=6$$

All real numbers

19. $0.53-0.2z=0.2(2z-13)-0.47$

$0.53-0.2z=0.4z-2.6-0.47$

$0.53-0.2z=0.4z-3.07$

$0.53-0.6z=-3.07$

$-0.6z=-3.6$

$z=6$

Check:

$0.53-0.2(6)\stackrel{?}{=}0.2(2(6)-13)-0.47$

$0.53-1.2\stackrel{?}{=}0.2(12-13)-0.47$

$0.53-1.2\stackrel{?}{=}0.2(-1)-0.47$

$0.53-1.2\stackrel{?}{=}-0.2-0.47$

$-0.67=-0.67$

20. $18(v+2)-6v=30+12v-2$

$18v+36-6v=28+12v$

$12v+36=28+12v$

$36\neq 28$

No solution

21. $\frac{2}{5}p-\frac{3}{20}=\frac{13}{20}-\frac{3}{10}(6-2p)$

$20\left(\frac{2}{5}p-\frac{3}{20}\right)=20\left(\frac{13}{20}-\frac{3}{10}(6-2p)\right)$

$8p-3=13-6(6-2p)$

$8p-3=13-36+12p$

$8p-3=-23+12p$

$-4p-3=-23$

$-4p=-20$

$p=5$

Check:

$\frac{2}{5}(5)-\frac{3}{20}\stackrel{?}{=}\frac{13}{20}-\frac{3}{10}(6-2(5))$

$2-\frac{3}{20}\stackrel{?}{=}\frac{13}{20}-\frac{3}{10}(6-10)$

$2-\frac{3}{20}\stackrel{?}{=}\frac{13}{20}-\frac{3}{10}(-4)$

$2-\frac{3}{20}\stackrel{?}{=}\frac{13}{20}+\frac{12}{10}$

$\frac{37}{20}=\frac{37}{20}$

22. $0.5(l+4)=0.3(3l-1)-0.9$

$0.5l+2=0.9l-0.3-0.9$

$0.5l+2=0.9l-1.2$

$2=0.4l-1.2$

$3.2=0.4l$

$8=l$

Check:

$0.5(8+4)\stackrel{?}{=}0.3(3(8)-1)-0.9$

$0.5(12)\stackrel{?}{=}0.3(24-1)-0.9$

$6\stackrel{?}{=}0.3(23)-0.9$

$6\stackrel{?}{=}6.9-0.9$

$6=6$

23. $I=Prt$

$\frac{I}{Pr}=\frac{Prt}{Pr}$

$\frac{I}{Pr}=t$

24. $P=2l+2w$

$P-2l=2w$

$\frac{P-2l}{2}=\frac{2w}{2}$

$\frac{P-2l}{2}=w$

25. $A=\frac{1}{2}bh$

$2A=bh$

$\frac{2A}{b}=h$

26. $P=a+b+c$

$P-b-c=a$

27. $-8n\geq -32$

$n\leq 4$

a) $\{n|n\leq 4\}$ b) $(-\infty,4]$

c) 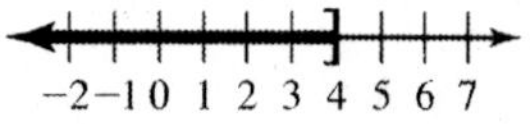

28. $3x+5>-1$

$3x>-6$

$x>-2$

a) $\{x|x>-2\}$ b) $(-2,\infty)$

c) 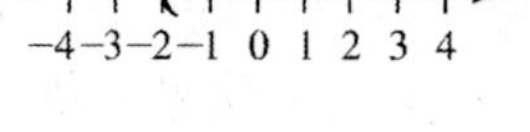

29. $9-2m>15$

$-2m>6$

$m<-3$

a) $\{m|m<-3\}$ b) $(-\infty,-3)$

c) 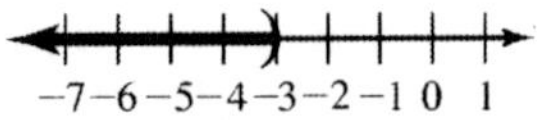

30. $5h-11\ge 9h+9$

$-4h-11\ge 9$

$-4h\ge 20$

$h\le -5$

a) $\{h|h\le -5\}$ b) $(-\infty,-5]$

c) 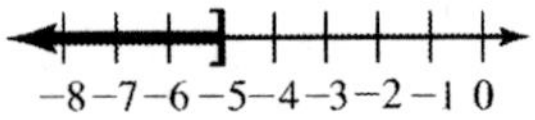

31. $\frac{2}{3}t-1\le\frac{1}{4}t+\frac{1}{2}$

$12\left(\frac{2}{3}t-1\right)\le 12\left(\frac{1}{4}t+\frac{1}{2}\right)$

$8t-12\le 3t+6$

$5t-12\le 6$

$5t\le 18$

$t\le\frac{18}{5}$

a) $\left\{t\middle|t\le\frac{18}{5}\right\}$ b) $\left(-\infty,\frac{18}{5}\right]$

c) 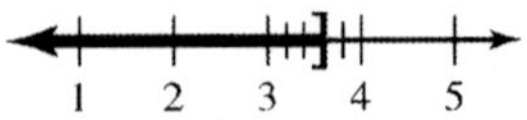

32. $12-(u+7)<3(u-1)$

$12-u-7<3u-3$

$-u+5<3u-3$

$-4u+5<-3$

$-4u<-8$

$u>2$

a) $\{u|u>2\}$ b) $(2,\infty)$

c) 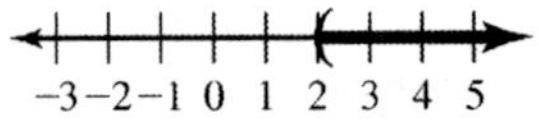

33. $A\cap B=\{2,4\}$

$A\cup B=\{1,2,3,4,5,6\}$

34. $A\cap B=\{a\}$

$A\cup B=\{a,b,c,d,e,i,o,u\}$

35. $-6<2x<-4$

$-3<x<-2$

a) 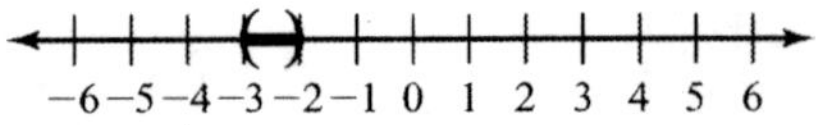

b) $\{x|-3<x<-2\}$ c) $(-3,-2)$

36. $9\le -3x\le 15$

$-3\ge x\ge -5$

a) 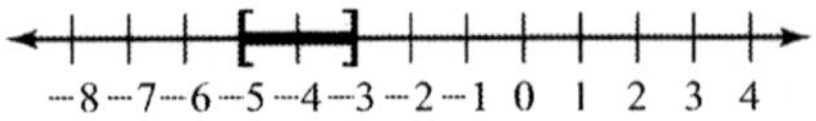

b) $\{x|-5\le x\le -3\}$ c) $[-5,-3]$

37. $-3<x+4<7$

$-7<x<3$

a) 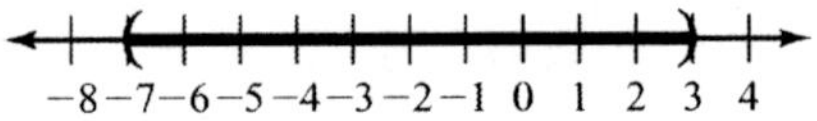

b) $\{x|-7<x<3\}$ c) $(-7,3)$

38. $0<x-1\le 3$

$1<x\le 4$

a) 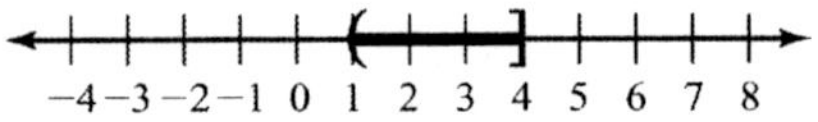

b) $\{x|1<x\le 4\}$ c) $(1,4]$

39. $-5\le 2x-1<-1$

$-4\le 2x<0$

$-2\le x<0$

a) 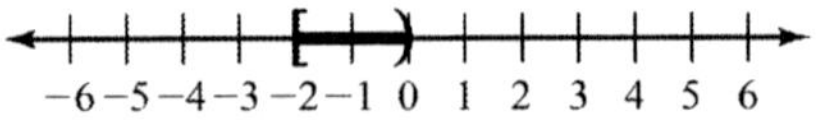

b) $\{x|-2\le x<0\}$ c) $[-2,0)$

40. $-5<3x-2<7$

$-3<3x<9$

$-1<x<3$

a) 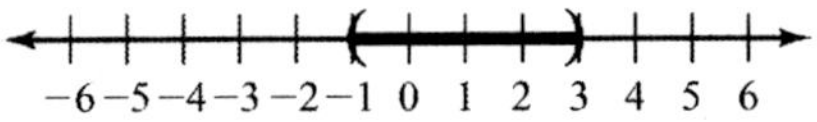

b) $\{x|-1<x<3\}$ c) $(-1,3)$

41. $w+4 \le -2$ or $w+4 \ge 2$

$w \le -6$ $\quad$ $w \ge -2$

a) 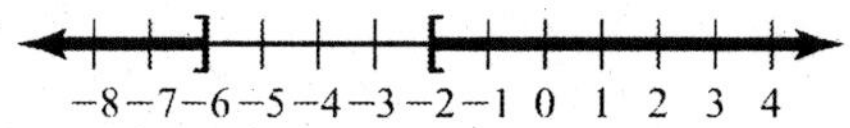

b) $\{w \mid w \le -6 \text{ or } w \ge -2\}$

c) $(-\infty,-6] \cup [-2,\infty)$

42. $4w-3<1$ or $4w-3>0$

$4w<4$ $\quad$ $4w>3$

$w<1$ $\quad$ $w>\frac{3}{4}$

a) 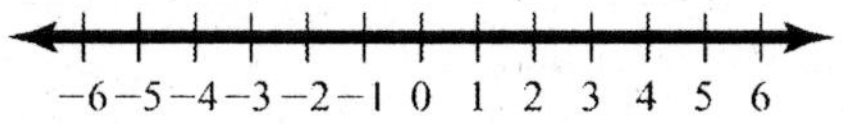

b) $\{w \mid w \text{ is a real number}\}$

c) $(-\infty,\infty)$

43. $2m-5<-1$ or $2m-5>5$

$2m<4$ $\quad$ $2m>10$

$m<2$ $\quad$ $m>5$

a) 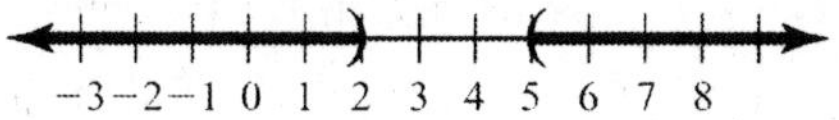

b) $\{m \mid m<2 \text{ or } m>5\}$

c) $(-\infty,2) \cup (5,\infty)$

44. $3x+2 \le -2$ or $3x+2 \ge 8$

$3x \le -4$ $\quad$ $3x \ge 6$

$x \le -\frac{4}{3}$ $\quad$ $x \ge 2$

a) 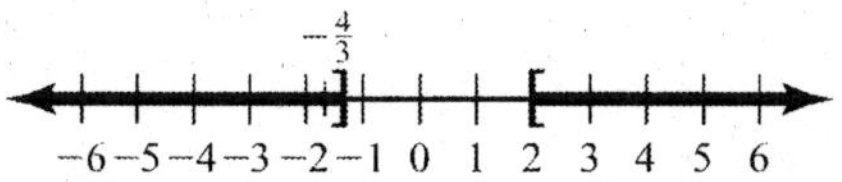

b)

$\left\{x \mid x \le -\frac{4}{3} \text{ or } x \ge 2\right\}$

c) $\left(-\infty,-\frac{4}{3}\right] \cup [2,\infty)$

45. $-x-6 \le -2$ or $-x-6 \ge 3$

$-x \le 4$ $\quad$ $-x \ge 9$

$x \ge -4$ $\quad$ $x \le -9$

a) 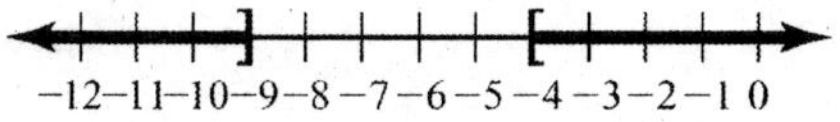

b) $\{x \mid x \le -9 \text{ or } x \ge -4\}$

c) $(-\infty,-9] \cup [-4,\infty)$

46. $-4w+1 \le -3$ or $-4w+1 \ge 5$

$-4w \le -4$ $\quad$ $-4w \ge 4$

$w \ge 1$ $\quad$ $w \le -1$

a) 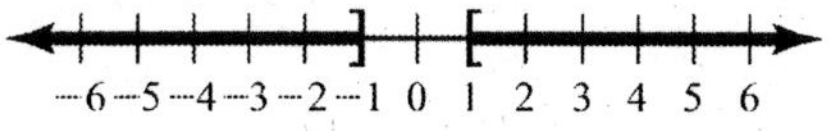

b) $\{w \mid w \le -1 \text{ or } w \ge 1\}$

c) $(-\infty,-1] \cup [1,\infty)$

47. $-4, 4$

48. $x-4=-7$ or $x-4=7$

$x=-3$ $\quad$ $x=11$

49. $2w-1=-3$ or $2w-1=3$

$2w=-2$ $\quad$ $2w=4$

$w=-1$ $\quad$ $w=2$

50. $3u-4=-2$ or $3u-4=2$

$3u=2$ $\quad$ $3u=6$

$u=\frac{2}{3}$ $\quad$ $u=2$

51. $|5r+8|=-3$

Because the absolute value of every real number is a positive number or zero, this equation has no solution.

52. $|q-4|+3=14$

$|q-4|=11$

$q-4=-11$ or $q-4=11$

$q=-7$ $\quad$ $q=15$

53. $|5w|-2=13$

$|5w|=15$

$5w=-15$ or $5w=15$

$w=-3$ $\quad$ $w=3$

54. $2|3x-4|=8$

$|3x-4|=4$

$3x-4=-4$ or $3x-4=4$

$3x=0$ $\quad$ $3x=8$

$x=0$ $\quad$ $x=\frac{8}{3}$

55. $1-|2x+3|=-3$

$-|2x+3|=-4$

$|2x+3|=4$

$2x+3=-4$ or $2x+3=4$

$2x=-7$ $\quad$ $2x=1$

$x=-\frac{7}{2}$ $\quad$ $x=\frac{1}{2}$

56. $4-2|r-5|=-8$

$-2|r-5|=-12$

$|r-5|=6$

$r-5=-6$ or $r-5=6$

$r=-1$ $\quad$ $r=11$

57. a) 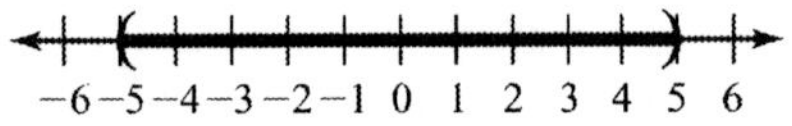

b) $\{x|-5<x<5\}$ c) $(-5,5)$

58. a) 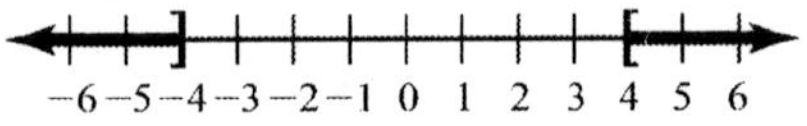

b) $\{p|p\le -4 \text{ or } p\ge 4\}$

c) $(-\infty,-4]\cup[4,\infty)$

59. $x-3<-7$ or $x-3>7$

$x<-4$ $\quad$ $x>10$

a) 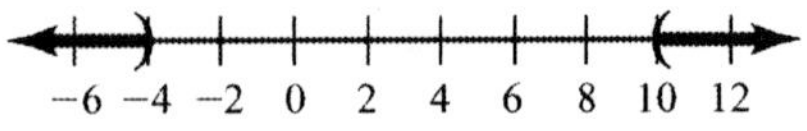

b) $\{x|x<-4 \text{ or } x>10\}$

c) $(-\infty,-4)\cup(10,\infty)$

60. $-4<2m+6<4$

$-10<2m<-2$

$-5<m<-1$

a) 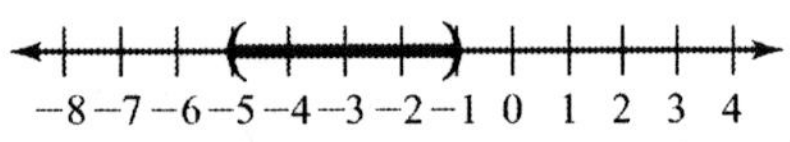

b) $\{m|-5<m<-1\}$ c) $(-5,-1)$

61. $|3s-1|\ge -2$

This inequality indicates that the absolute value is greater than a negative number. Because the absolute value of every real number is either positive or 0, the solution set is $\mathbb{R}$.

a) 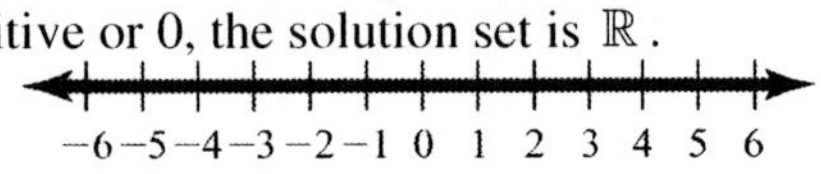

b) $\{s|s \text{ is a real number}\}$

c) $(-\infty,\infty)$

62. $5|b|+8<3$

$5|b|<-5$

$|b|<-1$

Because absolute values cannot be negative, this inequality has no solution.

a) 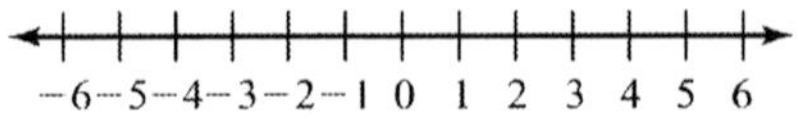

b) $\{\ \}$ or $\varnothing$

c) no interval notation

63. $7|m+3|\le 21$

$|m+3|\le 3$

$-3\le m+3\le 3$

$-6\le m\le 0$

a) 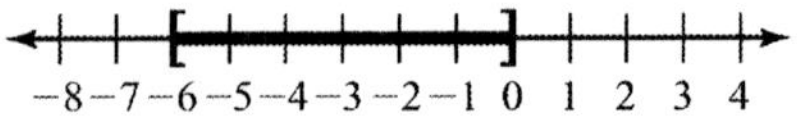

b) $\{m|-6\le m\le 0\}$ c) $[-6,0]$

64. $-2|t-5|<-10$

$|t-5|>5$

$t-5<-5$ or $t-5>5$

$t<0$ $\quad$ $t>10$

a) 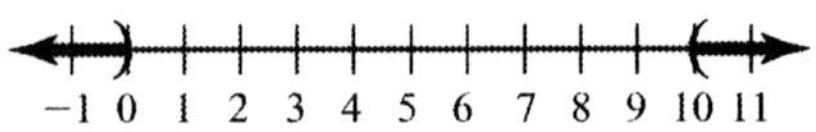

b) $\{t|t<0 \text{ or } t>10\}$

c) $(-\infty,0)\cup(10,\infty)$

65. $5-2|2k-3|\le -15$

$-2|2k-3|\le -20$

$|2k-3|\ge 10$

$2k-3\le -10$ or $2k-3\ge 10$

$2k\le -7$ $\quad$ $2k\ge 13$

$k\le -\frac{7}{2}$ $\quad$ $k\ge \frac{13}{2}$

a) 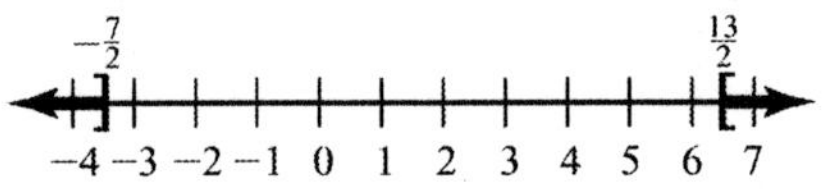

b) $\left\{k \middle| k \le -\frac{7}{2} \text{ or } k \ge \frac{13}{2}\right\}$

c) $\left(-\infty, -\frac{7}{2}\right] \cup \left[\frac{13}{2}, \infty\right)$

66. $3-7|2p+4| \le 24$

$-7|2p+4| \le 21$

$|2p+4| \ge -3$

a) 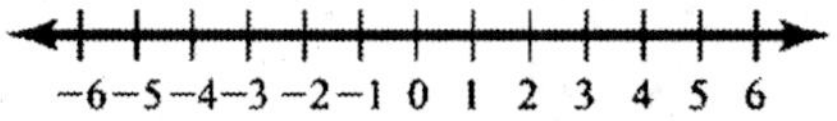

b) $\{p \mid p \text{ is a real number}\}$ c) $(-\infty, \infty)$

67. $C = \frac{5}{9}(F-32)$

$-196 = \frac{5}{9}(F-32)$

$-\frac{1764}{5} = F - 32$

$-352.8 = F - 32$

$-320.8° = F$

68. $B = P + Prt$

$4368 = P + P(0.04)(1)$

$4368 = P + 0.04P$

$4368 = 1.04P$

$\$4200 = P$

69. Let n represent the number.

$\frac{2}{3}n = 2 + \frac{1}{2}n$

$\frac{1}{6}n = 2$

$n = 12$

70. Let n represent the number.

$2(n-7) = -6$

$2n - 14 = -6$

$2n = 8$

$n = 4$

71. Let p represent the original price. Because \$44.94 is the discounted price, it is the result of subtracting 30% of the original price from the original price. Translate to an equation and solve for p.

$p - 0.30p = 44.94$

$0.70p = 44.94$

$\frac{0.70p}{0.70} = \frac{44.94}{0.70}$

$p = 64.20$

The original price is \$64.20.

72. Let x represent the number of units sold in the previous year. Because the 37,791 units sold this year is a 10.5% increase over the previous year, it is the result of adding 10.5% of x to x. Translate to an equation and solve.

$0.105x + x = 37{,}791$

$1.105x = 37{,}791$

$\frac{1.105x}{1.105} = \frac{37{,}791}{1.105}$

$x = 34{,}200$

The number of units sold the previous year was 34,200 units.

73. Let one integer be x and the other be $x + 1$.

$x + (x+1) = 125$

$2x + 1 = 125$

$2x = 124$

$x = 62$

The integers are 62 and 63.

74. Let the integers be x, $x + 2$, and $x + 4$.

$x + (x+2) + (x+4) = 108$

$3x + 6 = 108$

$3x = 102$

$x = 34$

The integers are 34, 36, and 38.

75. If two angles are complementary, their sum is 90°. Let the first angle be a and the other be $3a + 6$.

$a + (3a+6) = 90$

$4a + 6 = 90$

$4a = 84$

$a = 21$

The angles are 21° and $3(21) + 6 = 69°$.

76. If two angles are supplementary, their sum is 180°. Let the first angle be a and the other be $2a - 15$.

$a + (2a-15) = 180$

$3a - 15 = 180$

$3a = 195$

$a = 65$

The angles are 65° and $2(65) - 15 = 115°$.

77. Create a table.

	how many	value	total amount
CD 1	x	10.50	$10.5x$
CD 2	$400-x$	12.00	$12(400-x)$

Translate into an equation and solve.

$$10.5x+12(400-x)=4320$$
$$10.5x+4800-12x=4320$$
$$-1.5x=-480$$
$$x=320$$

She purchased 320 CDs at \$10.50 and 80 CDs at \$12.00.

78. Create a table.

	rate	time	distance
car 1	50	x	$50x$
car 2	55	x	$55x$

Translate into an equation and solve.

$$50x+55x=315$$
$$105x=315$$
$$x=3$$

They will meet in 3 hr.

79. Create a table.

	concentrate	vol. of solution	vol. of HCl solution
15%	0.15	45	$0.15(45)$
35%	0.35	x	$0.35x$
25%	0.25	$x+45$	$0.25(x+45)$

Translate into an equation and solve.

$$0.15(45)+0.35x=0.25(x+45)$$
$$6.75+0.35x=0.25x+11.25$$
$$6.75+0.1x=11.25$$
$$0.1x=4.5$$
$$x=45$$

He must add 45 ml of 35% HCl solution.

80. Let l be the length of the pool.

$$12l\le 300$$
$$l\le 25$$

The length of the pool can be 25 ft. or less.

81. Let x represent the score on the fourth test.

$$80\le\frac{70+89+83+x}{4}<90$$
$$320\le 70+89+83+x<360$$
$$320\le 242+x<360$$
$$78\le x<118$$

The student could receive scores in the range $78\le x<118$ to receive a B.

82. a) $32<t<212$ b) $t\le 32$ or $t\ge 212$

c) $C=\frac{5}{9}(32-32)$ $\quad C=\frac{5}{9}(212-32)$

$C=\frac{5}{9}(0)$ $\quad C=\frac{5}{9}(180)$

$C=0°$ $\quad C=100°$

$$0<t<100$$

d) $t\le 0$ or $t\ge 100$

Chapter 2 Practice Test

1. Let $A=\{h,o,m,e\}$ and $B=\{h,o,u,s,e\}$

$A\cap B=\{e,h,o\}$ because e, h, and o are the only elements in both A and B.

$A\cup B=\{e,h,m,o,s,u\}$ because these are the elements that are either in A or B or both.

2. $2(w+2)-3=3(w-1)$

$$2w+4-3=3w-3$$
$$2w+1=3w-3$$
$$2w-2w+1=3w-2w-3$$
$$1=w-3$$
$$1+3=w-3+3$$
$$4=w$$

3. Multiply both sides by the LCD, 15.

$$\frac{2}{3}q-4=\frac{1}{5}q+10$$
$$15\left(\frac{2}{3}q-4\right)=15\left(\frac{1}{5}q+10\right)$$
$$10q-60=3q+150$$
$$10q-3q-60=3q-3q+150$$
$$7q-60=150$$
$$7q-60+60=150+60$$
$$7q=210$$
$$\frac{7q}{7}=\frac{210}{7}$$
$$q=30$$

4. $|x+3|=5$

$x+3=-5$ or $x+3=5$

$x=-8$ $\quad x=2$

5. $3-|2x-3|=-6$

$-|2x-3|=-9$

$|2x-3|=9$

$2x-3=-9$ or $2x-3=9$

$2x=-6$ $\quad$ $2x=12$

$x=-3$ $\quad$ $x=6$

6. $|2x+3|=|x-5|$

$2x+3=x-5$ or $2x+3=-(x-5)$

$x+3=-5$ $\quad$ $2x+3=-x+5$

$x=-8$ $\quad$ $3x+3=5$

$3x=2$

$x=\frac{2}{3}$

7. $|5x-4|=-3$

This equation has the absolute value equal to a negative number. Because the absolute value of every real number is a positive number or zero, this equation has no solution.

8. $A=\frac{1}{2}h(b+B)$

$2A=h(b+B)$

$\frac{2A}{h}=b+B$

$\frac{2A}{h}-B=b$

9. $-3<x+4\le 7$

$-7<x\le 3$

a) 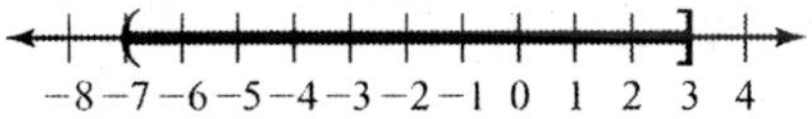

b) $\{x|-7<x\le 3\}$ c. $(-7,3]$

10. $4<-2x\le 6$

$-2>x\ge -3$

a) 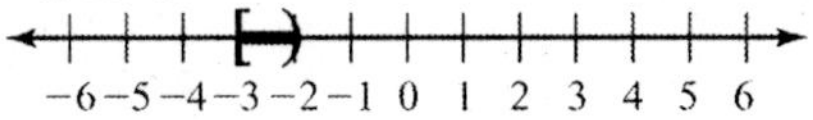

b) $\{x|-3\le x<-2\}$ c. $[-3,-2)$

11. $|x+4|<9$

$-9<x+4<9$

$-13<x<5$

a) 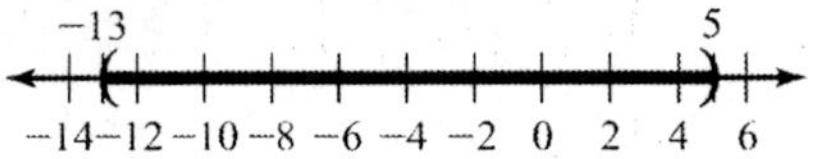

b) $\{x|-13<x<5\}$ c. $(-13,5)$

12. $2|x-1|>4$

$|x-1|>2$

$x-1<-2$ or $x-1>2$

$x<-1$ $\quad$ $x>3$

a) 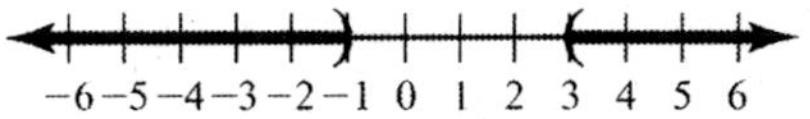

b) $\{x|x<-1 \text{ or } x>3\}$ c. $(-\infty,-1)\cup(3,\infty)$

13. $3-2|x+4|>-3$

$-2|x+4|>-6$

$|x+4|<3$

$-3<x+4<3$

$-7<x<-1$

a) 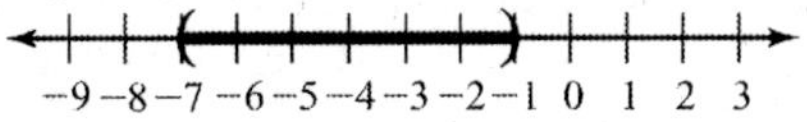

b) $\{x|-7<x<-1\}$ c. $(-7,-1)$

14. $|3y-2|<-2$

Because the absolute value of every real number is a positive number or zero, this inequality has no solution.

a) 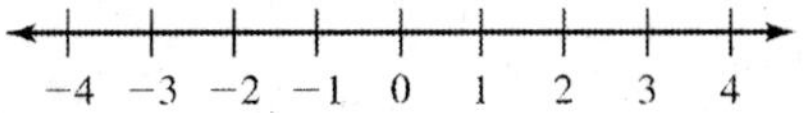

b) $\{\ \}$ or Ø c. no interval notation

15. $|8t+4|\ge -12$

Because the absolute value of every real number is greater than or equal to -12, this inequality is true for all real numbers.

a) 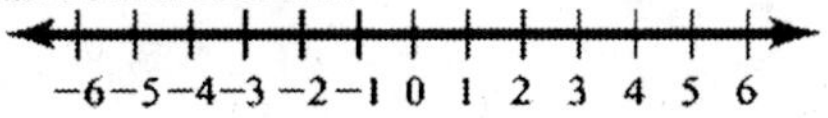

b) $\{t|t \text{ is a real number}\}$ c. $(-\infty,\infty)$

16. $2|3x-4|\le 10$

$|3x-4|\le 5$

$-5\le 3x-4\le 5$

$-1\le 3x\le 9$

$-\frac{1}{3}\le x\le 3$

a) 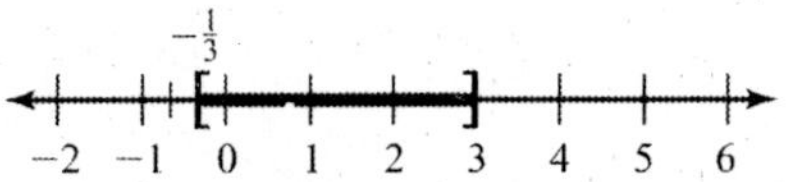

b) $\left\{x\middle|-\frac{1}{3}\le x\le 3\right\}$ c. $\left[-\frac{1}{3},3\right]$

17. Let n represent the unknown number. We are given that five decreased by three times a number is forty-one. Translate to an equation and solve.

$$5-3n=41$$
$$-3n=36$$
$$n=-12$$

The number is -12.

18. Let x represent the original price. Translate to an equation and solve.

$$31.36=x-0.30x$$
$$31.36=0.70x$$
$$\frac{31.36}{0.70}=\frac{0.70x}{0.70}$$
$$44.8=x$$

The original price of the dress is \$44.80.

19. Let x represent the number of boxes sold to client 1. We are given that a total of 500 boxes were sold, and that the combined amount of the sale was \$120,000. Complete a table.

	# of boxes	price	total amount of sale
client 1	x	250	$250x$
client 2	$500-x$	225	$225(500-x)$

$$250x+225(500-x)=120,000$$
$$250x+112,500-225x=120,000$$
$$25x=7500$$
$$x=300$$

300 boxes were sold at \$250 and 200 boxes were sold at \$225.

20. Let x represent the number of books Johnson orders.

$$40<8x<50$$
$$5<x<6.25$$

Since he would have to order between 5 and 6.25 books, he needs to order 6 books.

Chapters 1 and 2 Cumulative Review

1. False; $\frac{3}{2}$ is a rational number, but not an integer.
2. False; subtraction is not commutative.
3. False; $-a<x<a$ is not the same as $x>a$ and $x<-a$.
4. identity
5. real number
6. reverse the inequality symbol
7. $-\left|-\frac{5}{9}\right|=-\frac{5}{9}$
8. $\sqrt[3]{-125}=-5$
9. $-16+9=-7$
10. $(-4)(-5)(4)=80$
11. $-9^2=-(9)\cdot(9)=-81$
12. $14-4(-2)^2=14-4(4)$
$=14-16$
$=-2$
13. $|7-3(8-2)\div 6|+\left[4(2-4^2)\div 2\right]$
$=|7-3(6)\div 6|+\left[4(2-16)\div 2\right]$
$=|7-18\div 6|+\left[4(-14)\div 2\right]$
$=|7-3|+(-56\div 2)$
$=|4|+(-28)$
$=4+(-28)$
$=-24$
14. $\left[-3^2-2(-6+2)\right]+\sqrt{25+144}$
$=\left[-3^2-2(-4)\right]+\sqrt{169}$
$=\left[-9-2(-4)\right]+13$
$=(-9+8)+13$
$=-1+13$
$=12$
15. $\frac{4y-5x^2}{3y^3-z^2}=\frac{4(2)-5(-3)^2}{3(2)^3-(-4)^2}$
$=\frac{4(2)-5(9)}{3(8)-(16)}$
$=\frac{8-45}{24-16}$
$=\frac{-37}{8}$
16. $-3(2x^2-4x+3)=-3\cdot 2x^2-(-3)\cdot 4x+(-3)\cdot 3$
$=-6x^2+12x-9$
17. $\frac{x-5}{x+4}$ is undefined when the denominator is equal to zero, $x+4=0$. This occurs when $x=-4$.

18. $4d^2 - 4 + 4d + 4d^2 - 9 - 3d$
$= 4d^2 + 4d^2 + 4d - 3d - 4 - 9$
$= 8d^2 + d - 13$

19. $4(2y-5) - 2(3y-6) + 6y$
$= 8y - 20 - 6y + 12 + 6y$
$= 8y - 8$

20. $12 - 4(5-x) = 12 - 20 + 4x = 4x - 8$

21. $8(2z-3) - z = 5z - 34$
$16z - 24 - z = 5z - 34$
$15z - 24 = 5z - 34$
$10z - 24 = -34$
$10z = -10$
$z = -1$

22. $1.4u - 0.5(44 - 6u) = 66 + 2.4u$
$10(1.4u - 0.5(44 - 6u)) = 10(66 + 2.4u)$
$14u - 5(44 - 6u) = 660 + 24u$
$14u - 220 + 30u = 660 + 24u$
$44u - 220 = 660 + 24u$
$20u - 220 = 660$
$20u = 880$
$u = 44$

23. $\frac{3x}{4} - \frac{2}{5} = \frac{3x}{10} + \frac{1}{2}$
$20\left(\frac{3x}{4} - \frac{2}{5}\right) = 20\left(\frac{3x}{10} + \frac{1}{2}\right)$
$15x - 8 = 6x + 10$
$9x - 8 = 10$
$9x = 18$
$x = 2$

24. $7u - 10(2u+3) \geq 22$
$7u - 20u - 30 \geq 22$
$-13u - 30 \geq 22$
$-13u \geq 52$
$u \leq -4$

a. $\{u \mid u \leq -4\}$

b. $(-\infty, -4]$

c. 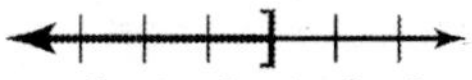

25. $|2x-5| - 1 \leq 4$
$|2x-5| \leq 5$
$-5 \leq 2x - 5 \leq 5$
$0 \leq 2x \leq 10$
$0 \leq x \leq 5$

a. $\{x \mid 0 \leq x \leq 5\}$

b. $[0, 5]$

c. 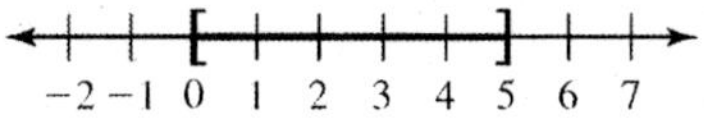

26. $3 < -2y + 5 \leq 7$
$-2 < -2y \leq 2$
$1 > y \geq -1$
$-1 \leq y < 1$

a. $\{y \mid -1 \leq y < 1\}$

b. $[-1, 1)$

c. 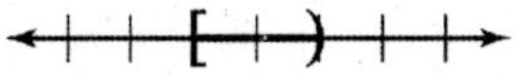

27. $|3x-9| + 6 = 12$
$|3x-9| = 6$
$3x - 9 = -6$ or $3x - 9 = 6$
$3x = 3$ $\quad$ $3x = 15$
$x = 1$ $\quad$ $x = 5$

28. Let n represent the number.
$3 - 2(n-6) = 3n - 5$
$3 - 2n + 12 = 3n - 5$
$15 - 2n = 3n - 5$
$15 = 5n - 5$
$20 = 5n$
$4 = n$
The number is 4.

29. Create a table.

Categories	rate	time	distance
passenger jet	560	t	$560t$
private jet	440	t	$440t$

Translate the information in the table to an equation and solve.
$560t + 440t = 3000$
$1000t = 3000$
$t = 3$
The jets will meet in 3 hours.

30. Create a table.

	concentrate	vol. of solution	vol. of baking soda
2%	0.02	40	$0.02(40)$
10%	0.10	x	$0.10x$
5%	0.05	$x+40$	$0.05(x+40)$

Translate the information in the table to an equation and solve.

$$0.02(40)+0.10x=0.05(x+40)$$
$$100(0.02(40)+0.10x)=100(0.05(x+40))$$
$$2(40)+10x=5(x+40)$$
$$80+10x=5x+200$$
$$80+5x=200$$
$$5x=120$$
$$x=24$$

24 oz. of the 10% solution should be added.

Chapter 3

Equations and Inequalities in Two Variables and Functions

Exercise Set 3.1

1. Beginning at the origin, move to the left 4 units along the x-axis, then up 3 units.

3. Replace the variables in the equation with the corresponding coordinates from the ordered pair. If the resulting equation is true, the ordered pair is a solution.

5. The graph of an equation represents the equation's solution set.

7. Plot the solutions as points in the rectangular coordinate system; then draw a line through the points to form a straight line. Put arrowheads on both ends of the line.

9. A: (2, 4), B: (–2, 1), C: (0, –3), D: (4, –5)

11. a. I b. IV c. III d. y-axis

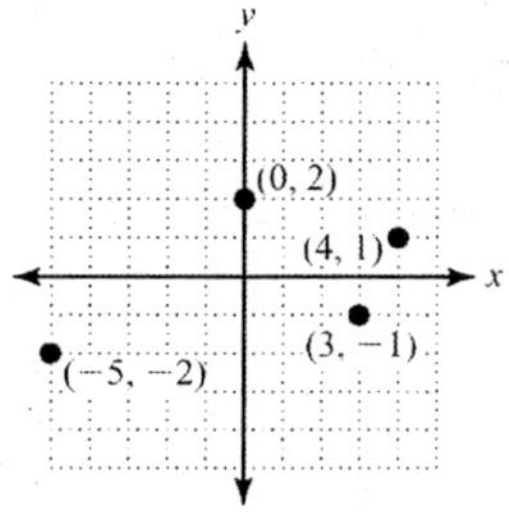

13. a. x-axis b. IV c. III d. I

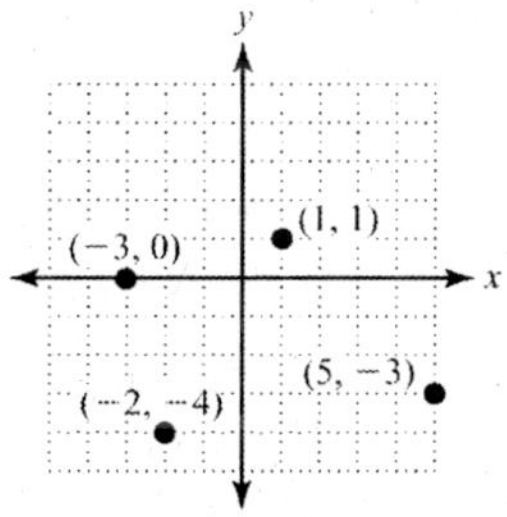

15. a. The figure is shown below.

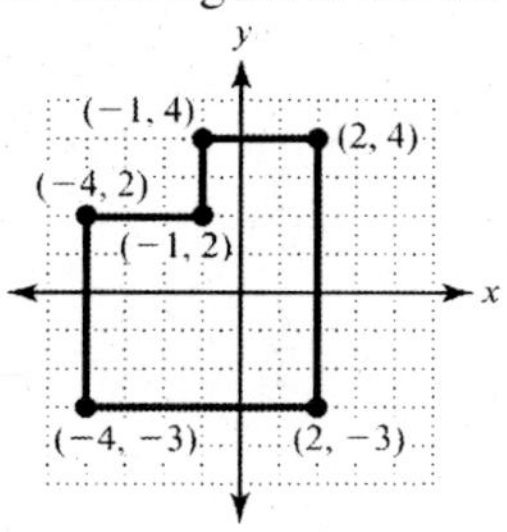

b. The perimeter of the figure is found by adding up the lengths of each side: 3 + 7 + 6 + 5 + 3 + 2 = 26 units.

c. The area of the figure is found by counting the grid squares included within the figure: 36 square units.

17. Replace x with 4 and y with 2 and see if the equation is true.

$$\begin{aligned} 2x+3y &= 14 \\ 2(4)+3(2) &\stackrel{?}{=} 14 \\ 8+6 &\stackrel{?}{=} 14 \\ 14 &= 14 \end{aligned}$$

Yes, $(4,2)$ is a solution.

19. Replace x with $-\frac{1}{3}$ and y with $\frac{2}{3}$ and see if the equation is true.

$$\begin{aligned} x-y &= -1 \\ -\frac{1}{3}-\frac{2}{3} &\stackrel{?}{=} -1 \\ -\frac{3}{3} &\stackrel{?}{=} -1 \\ -1 &= -1 \end{aligned}$$

Yes, $\left(-\frac{1}{3},\frac{2}{3}\right)$ is a solution.

21. Replace x with 5 and y with -1 and see if the equation is true.

$$\begin{aligned} \frac{2}{3}x-y &= 8 \\ \frac{2}{3}(5)-(-1) &\stackrel{?}{=} 8 \\ \frac{10}{3}+1 &\stackrel{?}{=} 8 \\ \frac{13}{3} &\neq 8 \end{aligned}$$

No, $(5,-1)$ is not a solution.

23. Replace x with 2.5 and y with -2.1 and see if the equation is true.

$$\begin{aligned} 5.2x &= 6.7-3y \\ 5.2(2.5) &\stackrel{?}{=} 6.7-3(-2.1) \\ 13 &\stackrel{?}{=} 6.7+6.3 \\ 13 &= 13 \end{aligned}$$

Yes, $(2.5,-2.1)$ is a solution.

25. $x - y = 4$

x	y	Ordered Pair
0	−4	$(0,-4)$
4	0	$(4,0)$
2	−2	$(2,-2)$

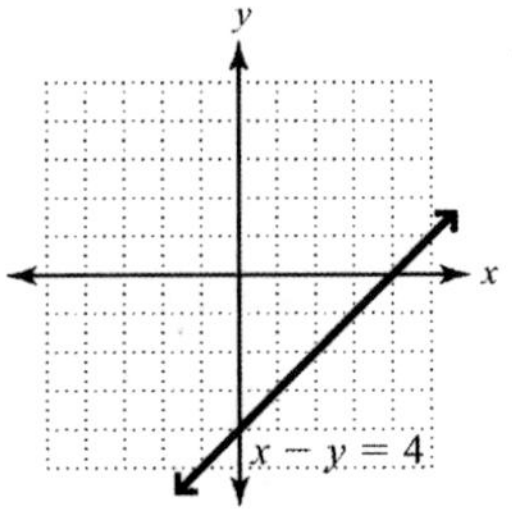

27. $2x + y = 6$

x	y	Ordered Pair
0	6	$(0,6)$
3	0	$(3,0)$
2	2	$(2,2)$

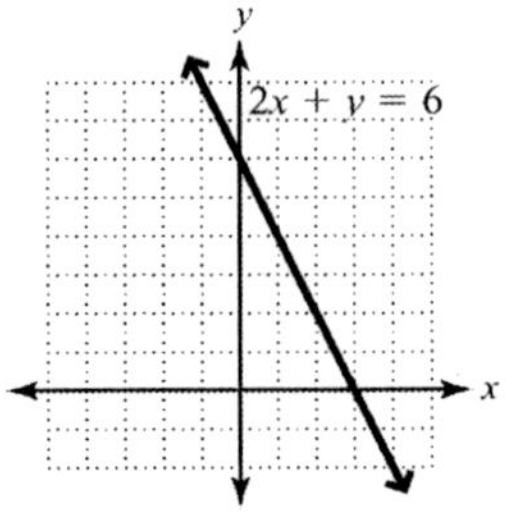

29. $x - 2y = 10$

x	y	Ordered Pair
0	−5	$(0,-5)$
10	0	$(10,0)$
2	−4	$(2,-4)$

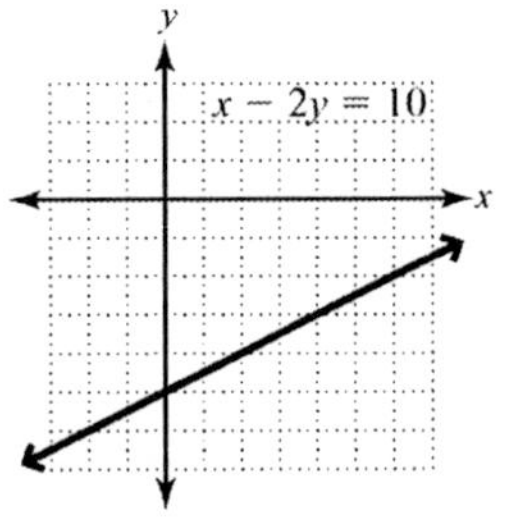

31. $2x - 5y = 10$

x	y	Ordered Pair
0	−2	$(0,-2)$
5	0	$(5,0)$
−5	−4	$(-5,-4)$

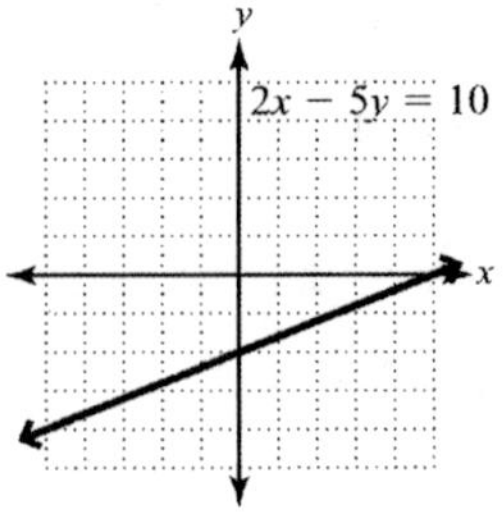

33. $y = x$

x	y	Ordered Pair
0	0	$(0,0)$
3	3	$(3,3)$
−2	−2	$(-2,-2)$

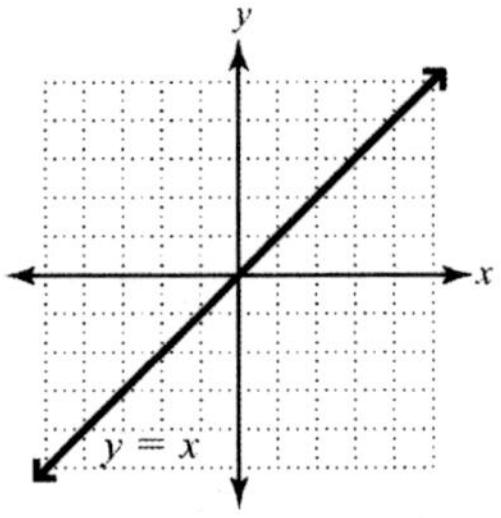

35. $y = 3x$

x	y	Ordered Pair
0	0	$(0,0)$
1	3	$(1,3)$
2	6	$(2,6)$

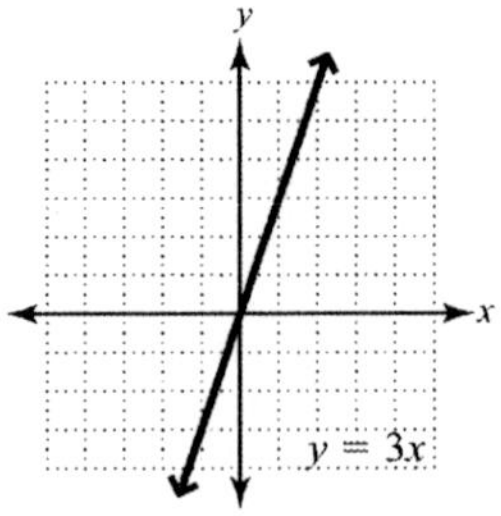

37. $y=-x$

x	y	Ordered Pair
0	0	$(0,0)$
−2	2	$(-2,2)$
3	−3	$(3,-3)$

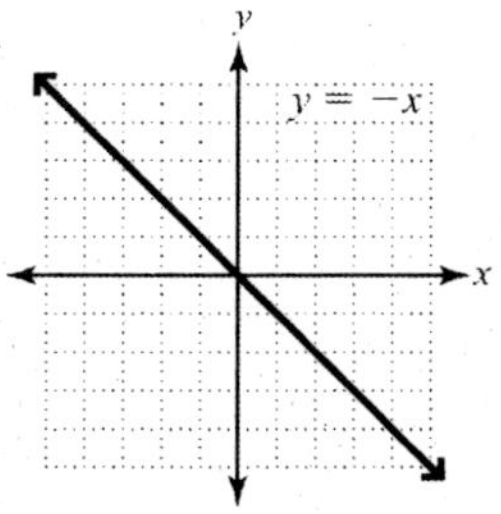

39. $y=-3x$

x	y	Ordered Pair
0	0	$(0,0)$
−1	3	$(-1,3)$
1	−3	$(1,-3)$

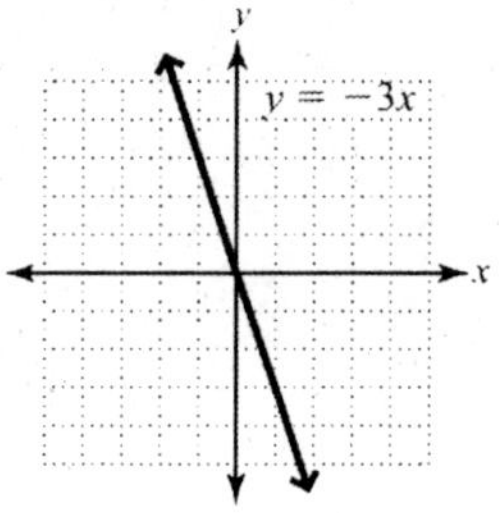

41. $y=\frac{1}{3}x$

x	y	Ordered Pair
0	0	$(0,0)$
3	1	$(3,1)$
−3	−1	$(-3,-1)$

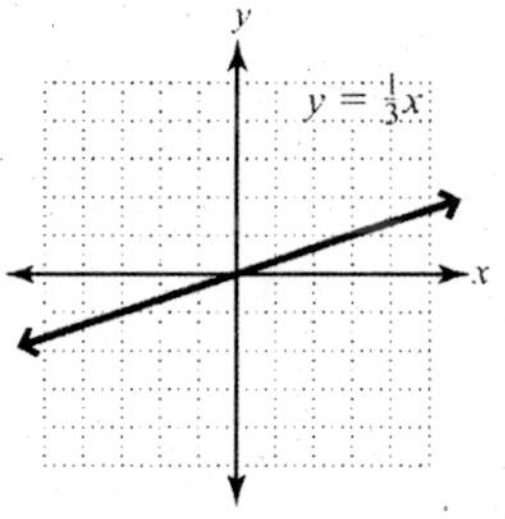

43. $y=-\frac{2}{3}x$

x	y	Ordered Pair
0	0	$(0,0)$
−3	2	$(-3,2)$
3	−2	$(3,-2)$

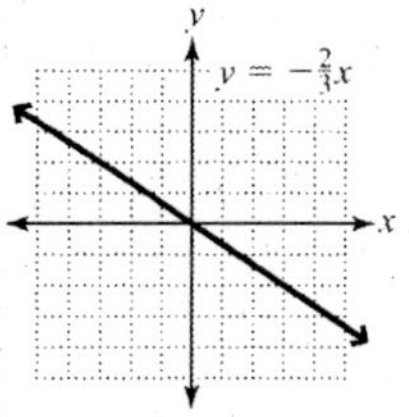

45. $y=3x+4$

x	y	Ordered Pair
0	4	$(0,4)$
−1	1	$(-1,1)$
−2	−2	$(-2,-2)$

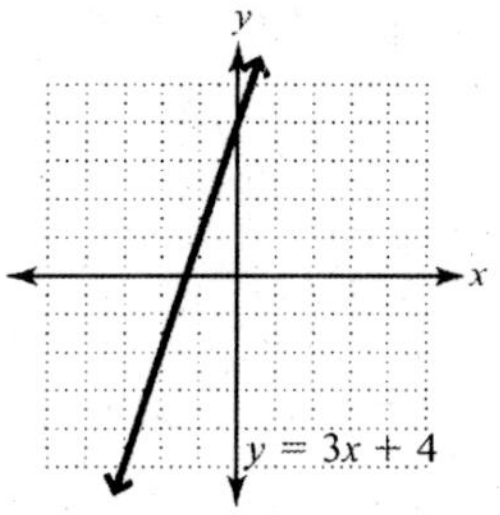

47. $y=-2x+1$

x	y	Ordered Pair
0	1	$(0,1)$
−1	3	$(-1,3)$
1	−1	$(1,-1)$

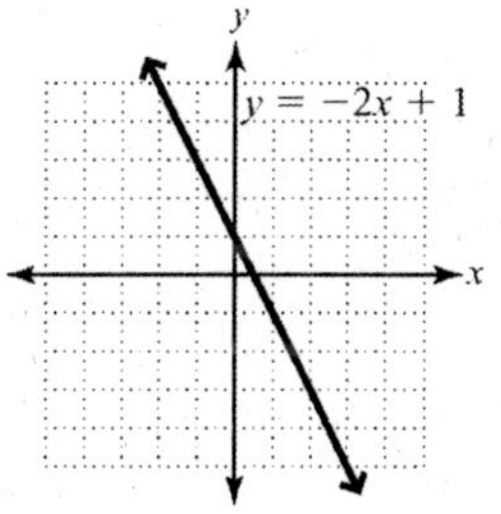

49. $y = \frac{3}{4}x + 2$

x	y	Ordered Pair
0	2	$(0,2)$
4	5	$(4,5)$
−4	−1	$(-4,-1)$

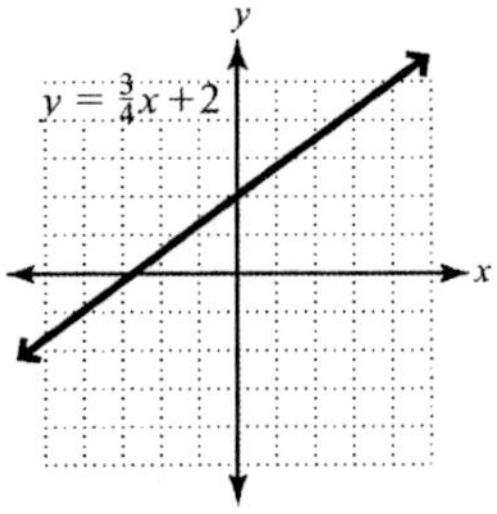

51. $y = -\frac{1}{3}x - 1$

x	y	Ordered Pair
0	−1	$(0,-1)$
3	−2	$(3,-2)$
−3	0	$(-3,0)$

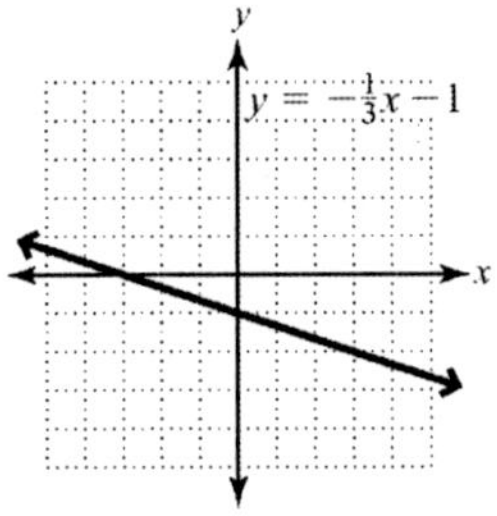

53. $y = 4$

x	y	Ordered Pair
0	4	$(0,4)$
2	4	$(2,4)$
−2	4	$(-2,4)$

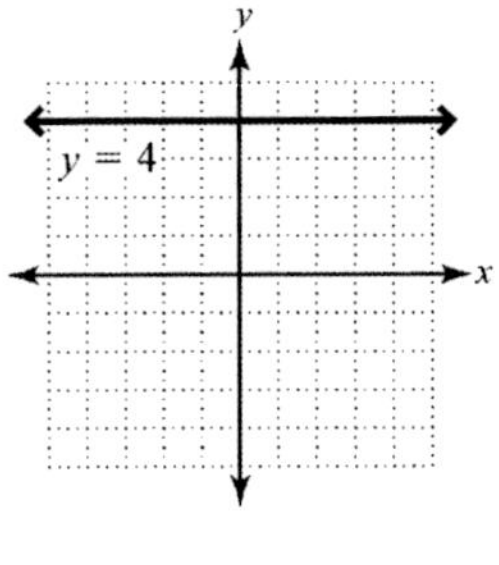

55. $x = 2$

x	y	Ordered Pair
2	0	$(2,0)$
2	4	$(2,4)$
2	−1	$(2,-1)$

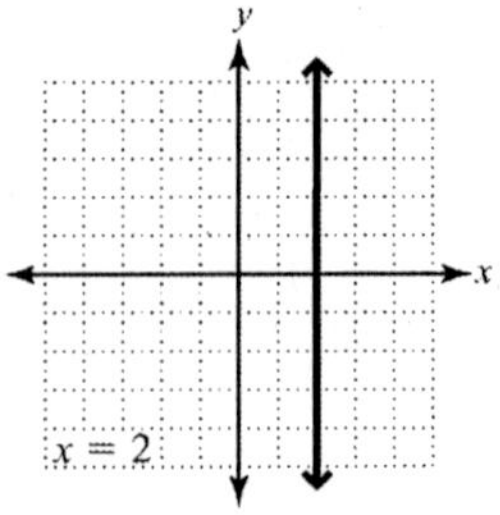

57. $2x + y = 4$

x-intercept	y-intercept
$y = 0:\ 2x + y = 4$	$x = 0:\ 2x + y = 4$
$2x + 0 = 4$	$2(0) + y = 4$
$2x = 4$	$0 + y = 4$
$x = 2$	$y = 4$
$(2,0)$	$(0,4)$

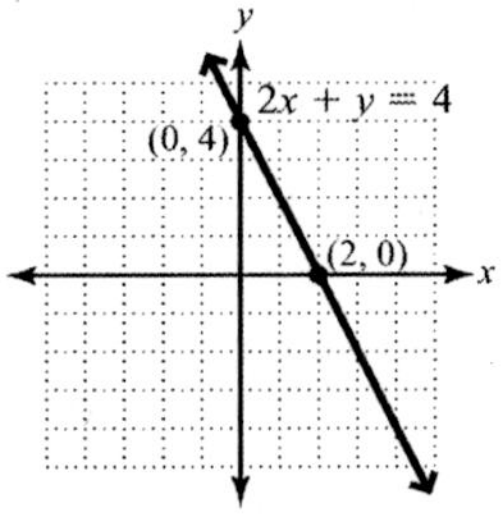

59. $3x + 2y = 6$

x-intercept	y-intercept
$y = 0:\ 3x + 2y = 6$	$x = 0:\ 3x + 2y = 6$
$3x + 2(0) = 6$	$3(0) + 2y = 6$
$3x = 6$	$2y = 6$
$x = 2$	$y = 3$
$(2,0)$	$(0,3)$

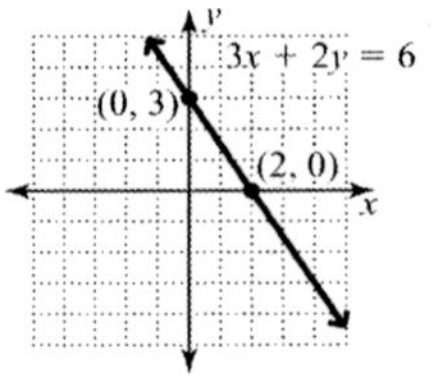

61. $2x-3y=-6$

x-intercept	y-intercept
$y=0:\ 2x-3y=-6$	$x=0:\ 2x-3y=-6$
$2x-3(0)=-6$	$2(0)-3y=-6$
$2x=-6$	$-3y=-6$
$x=-3$	$y=2$
$(-3,0)$	$(0,2)$

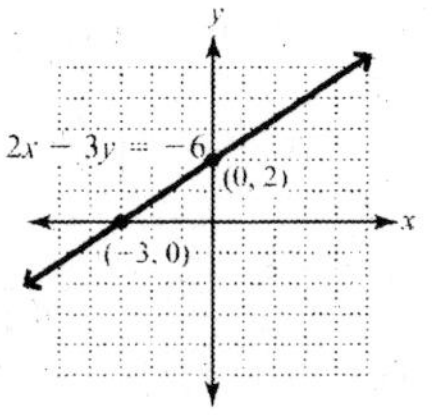

63. $y=x-3$

x-intercept	y-intercept
$y=0:\ y=x-3$	$x=0:\ y=x-3$
$0=x-3$	$y=0-3$
$3=x$	$y=-3$
$(3,0)$	$(0,-3)$

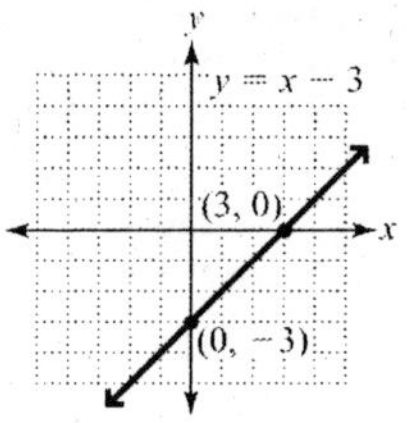

65. $y=-3x$

x-intercept	y-intercept
$y=0:\ y=-3x$	$x=0:\ y=-3x$
$0=-3x$	$y=-3(0)$
$0=x$	$y=0$
$(0,0)$	$(0,0)$

Extra points:

$x=-1:\ y=-3x$	$x=1:\ y=-3x$
$y=-3(-1)$	$y=-3(1)$
$y=3$	$y=-3$
$(-1,3)$	$(1,-3)$

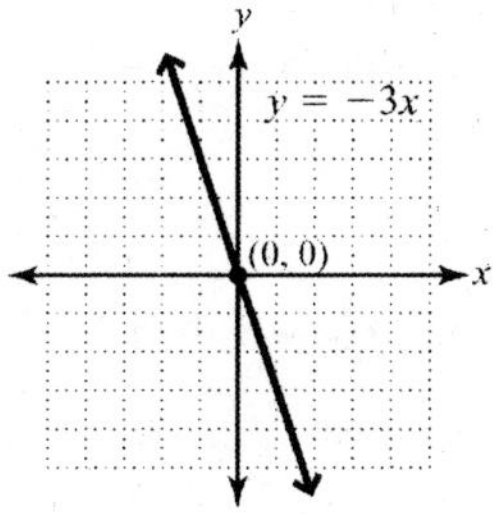

67. $y=-\frac{3}{4}x+1$

x-intercept	y-intercept
$y=0:$	$x=0:$
$y=-\frac{3}{4}x+1$	$y=-\frac{3}{4}x+1$
$0=-\frac{3}{4}x+1$	$y=-\frac{3}{4}(0)+1$
$\frac{4}{3}\cdot\frac{3}{4}x=\frac{4}{3}\cdot 1$	$y=1$
$x=\frac{4}{3}$	$(0,1)$
$\left(\frac{4}{3},0\right)$	

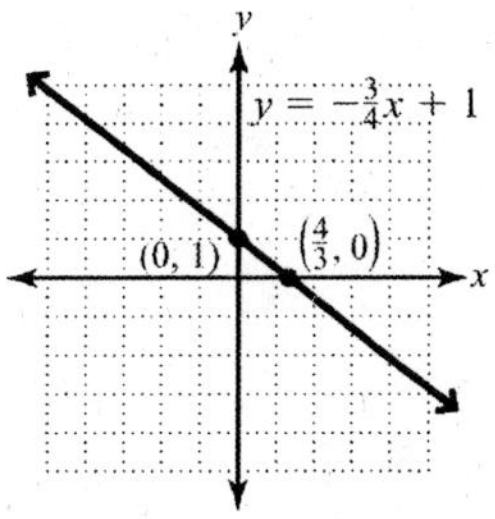

69. The graph gets steeper.

71. Adding b (where $b>0$) shifts the graph up the y-axis.

73. a) $(-3, 1), (0, 1), (-2, -3), (1, -3)$
b) $(1, 2,), (4, 2), (2, -2), (5, -2)$
c) $[(x+4),(y+1)]$

75. a) $c=5n+60$
$c=5(7)+60$
$c=35+60$
$c=\$95$

b) $c=5n+60$
$90=5n+60$
$30=5n$
$6\text{ hr.}=n$

c) The graph is shown below.

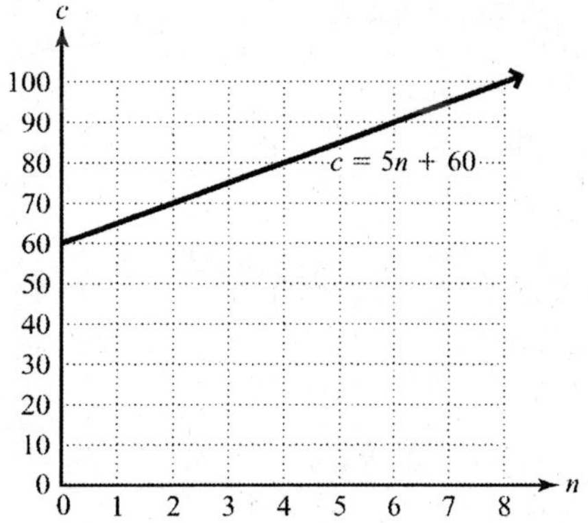

77. a) The number of minutes used over 450 was $600 - 450 = 150$.

$c = 0.05n + 40$

$c = 0.05(150) + 40$

$c = 7.5 + 40$

$c = \$47.50$

b) $c = 0.05n + 40$

$41.75 = 0.05n + 40$

$1.75 = 0.05n$

$35 \text{ min.} = n$

c) The graph is shown below.

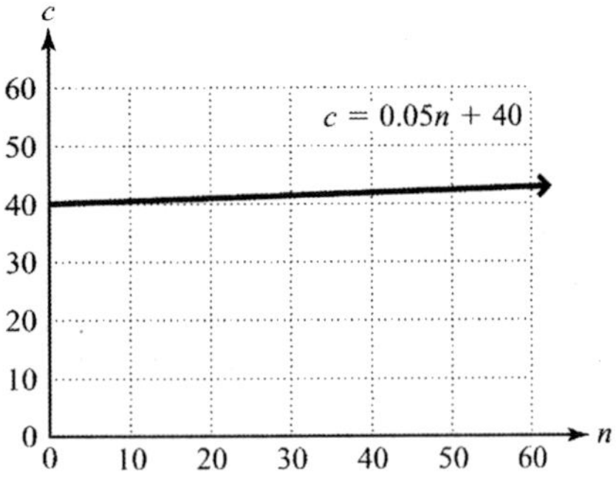

79. a) The number of copies beyond 200 was $300 - 200 = 100$.

$c = 0.03n + 5$

$c = 0.03(100) + 5$

$c = 3 + 5$

$c = \$8$

b) $c = 0.03n + 5$

$11 = 0.03n + 5$

$6 = 0.03n$

$200 = n$

Since the 200 copies is then added to the "first 200 copies," the total number of copies is 400.

c) The graph is shown below.

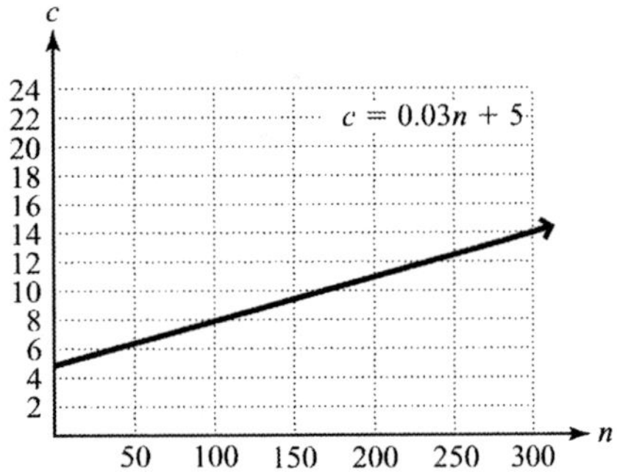

Review Exercises

1. Commutative property of addition

2. $\dfrac{3(5-8)+21}{8-3(6)} = \dfrac{3(-3)+21}{8-18} = \dfrac{-9+21}{-10} = \dfrac{12}{-10} = -\dfrac{6}{5}$

3. $\dfrac{y_2 - y_1}{x_2 - x_1} = \dfrac{-2-(-8)}{1-4} = \dfrac{-2+8}{-3} = \dfrac{6}{-3} = -2$

4. $\dfrac{y_2 - y_1}{x_2 - x_1} = \dfrac{5-13}{2-(-10)} = \dfrac{-8}{2+10} = \dfrac{-8}{12} = -\dfrac{2}{3}$

5. $12 - 3(n-2) = 6n - (2n+1)$

$12 - 3n + 6 = 6n - 2n - 1$

$18 - 3n = 4n - 1$

$19 = 7n$

$\dfrac{19}{7} = n$

6. $7x - 5 > 4x + 13$ $\quad \{x | x > 6\}$ $\quad (6, \infty)$

$3x - 5 > 13$

$3x > 18$

$x > 6$

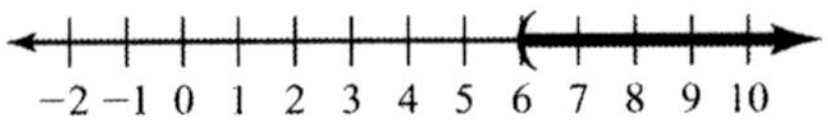

Exercise Set 3.2

1. Slope is the incline of a line.

3. If $m > 0$, the graph is a line that slants upward from left to right.

5. It is a horizontal line because all the y-coordinates are the same.

7. The graph of $y = 3x + 1$ is steeper than the graph of $y = 2x - 7$ because a slope of 3 is greater than a slope of 2.

9. The graph of $y = x - 4$ is steeper than the graph of $y = 0.2x + 3$ because a slope of 1 is greater than a slope of 0.2.

11. The graph of $y = 3x + 2$ has an upward incline from left to right because the slope is positive ($m = 3$).

13. The graph of $y = -\dfrac{4}{5}x - 3$ has a downward incline from left to right because the slope is negative $\left(m = -\dfrac{4}{5}\right)$.

15. From the graph, we can see that the line rises vertically 2 units and then runs horizontally 3 units for a slope of $m = \frac{2}{3}$. The y-intercept is $b = (0,1)$.

17. From the graph, we can see that the line is vertical, so it has an undefined slope. It does not cross the y-axis, so it has no y-intercept.

19. e

21. g

23. h

25. d

27. $y = \frac{2}{3}x + 5$

$m = \frac{2}{3}$; $y -$ intercept: $(0,5)$

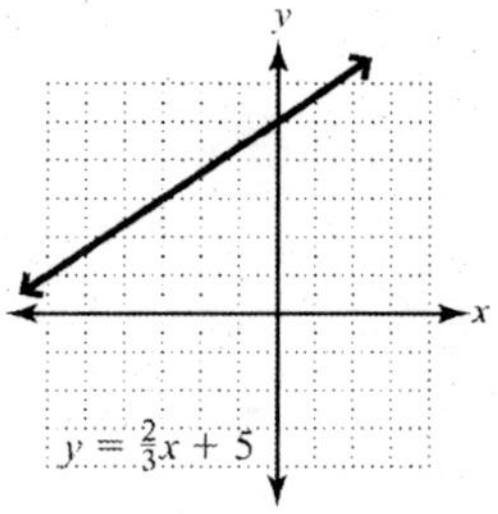

29. $y = -\frac{3}{4}x - 2$

$m = -\frac{3}{4}$; $y -$ intercept: $(0,-2)$

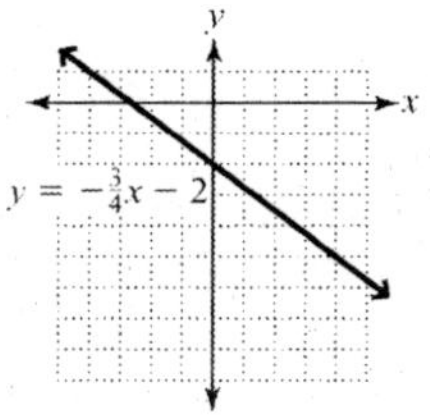

31. $y = x + 4$

$m = 1$; $y -$ intercept: $(0,4)$

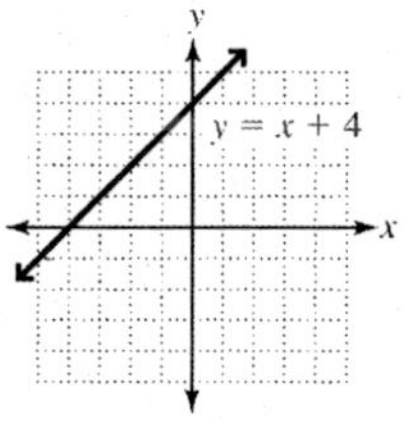

33. $y = -5x + \frac{2}{3}$

$m = -5$; $y -$ intercept: $\left(0, \frac{2}{3}\right)$

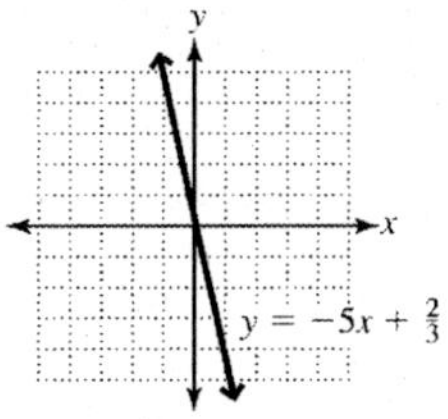

35. $2x + 3y = 6$

$$3y = -2x + 6$$

$$y = -\frac{2}{3}x + 2$$

$m = -\frac{2}{3}$; $y -$ intercept: $(0,2)$

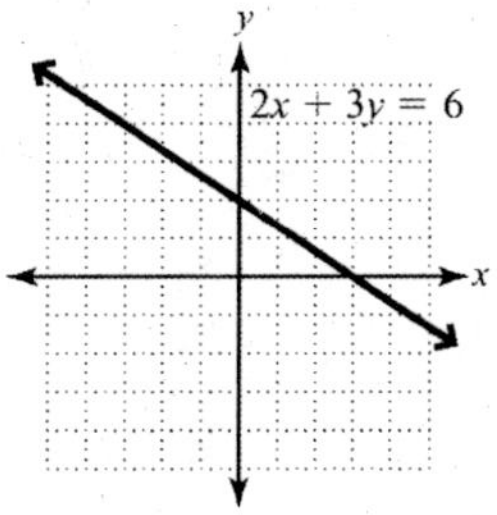

37. $x - 2y = -7$

$$-2y = -x - 7$$

$$y = \frac{1}{2}x + \frac{7}{2}$$

$m = \frac{1}{2}$; $y -$ intercept: $\left(0, \frac{7}{2}\right)$

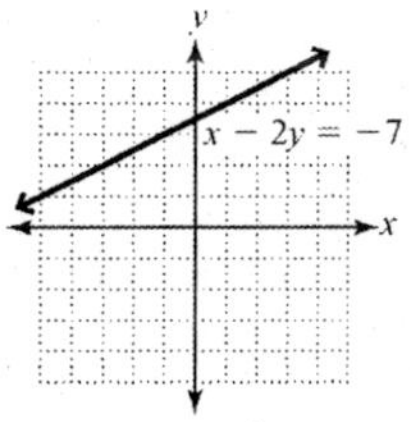

39. $2x-7y=8$

$$-7y=-2x+8$$

$$y=\frac{2}{7}x-\frac{8}{7}$$

$$m=\frac{2}{7};\ y-\text{intercept:}\left(0,-\frac{8}{7}\right)$$

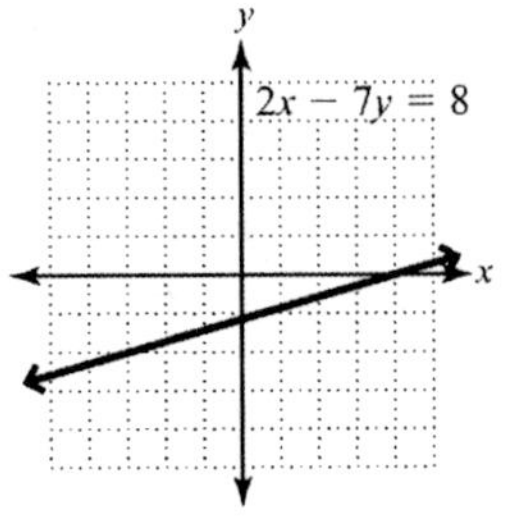

41. $-x+y=0$

$$y=x$$

$$m=1;\ y-\text{intercept:}(0,0)$$

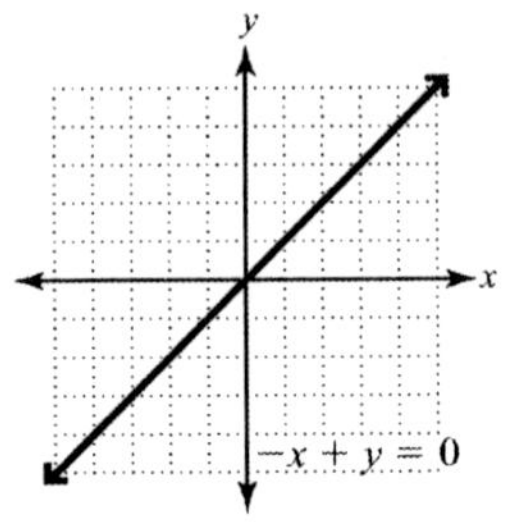

43. $(4,2),(2,6)$

$$m=\frac{y_2-y_1}{x_2-x_1}=\frac{6-2}{2-4}=-\frac{4}{2}=-2$$

45. $(-1,3),(4,-6)$

$$m=\frac{y_2-y_1}{x_2-x_1}=\frac{-6-3}{4-(-1)}=\frac{-9}{4+1}=-\frac{9}{5}$$

47. $(1,4),(-3,10)$

$$m=\frac{y_2-y_1}{x_2-x_1}=\frac{10-4}{-3-1}=\frac{6}{-4}=-\frac{3}{2}$$

49. $(8,2),(8,-5)$

$$m=\frac{y_2-y_1}{x_2-x_1}=\frac{-5-2}{8-8}=-\frac{7}{0}\text{ is undefined}$$

51. $(5,12),(-1,12)$

$$m=\frac{y_2-y_1}{x_2-x_1}=\frac{12-12}{-1-5}=\frac{0}{-6}=0$$

53. $(0,0),(10,-8)$

$$m=\frac{y_2-y_1}{x_2-x_1}=\frac{-8-0}{10-0}=\frac{-8}{10}=-\frac{4}{5}$$

55. a) The parallelogram is shown below.

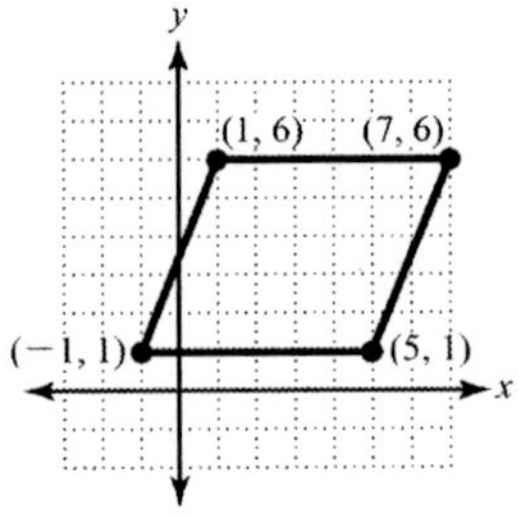

b) Right: $m=\frac{6-1}{7-5}=\frac{5}{2}$

Left: $m=\frac{6-1}{1-(-1)}=\frac{5}{1+1}=\frac{5}{2}$

Top: $m=\frac{6-6}{7-1}=\frac{0}{6}=0$

Bottom: $m=\frac{1-1}{5-(-1)}=\frac{0}{5+1}=\frac{0}{6}=0$

c) Slopes of parallel sides are equal.

57. $m=\frac{29}{348}=\frac{1}{12}$

59. $m=\frac{215-210}{20-0}=\frac{5}{20}=0.25$

61. a) $C(50, 15)$, $D(100, 80)$

b) $m=\frac{y_2-y_1}{x_2-x_1}=\frac{80-15}{100-50}=\frac{65}{50}=\frac{13}{10}$

c) The coaster climbs 13 ft. vertically for every 10 ft. that it moves horizontally.

63. a)

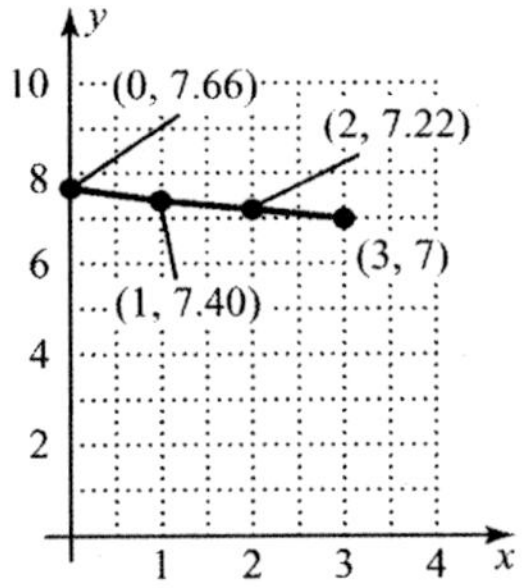

b) $m=\frac{y_2-y_1}{x_2-x_1}=\frac{7-7.66}{3-0}=-\frac{0.66}{3}=-0.22$

c) $7-0.22=6.78=\$6.78$

65. a)

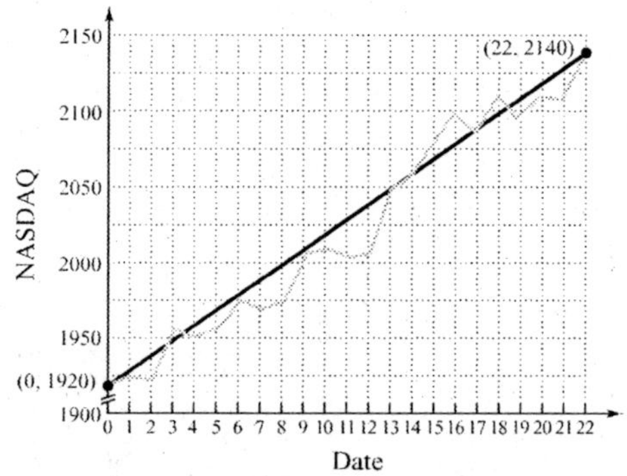

b) $m=\dfrac{y_2-y_1}{x_2-x_1}=\dfrac{2140-1920}{22-0}=\dfrac{220}{22}=10$

c) $2140+10=2150$

67. a)

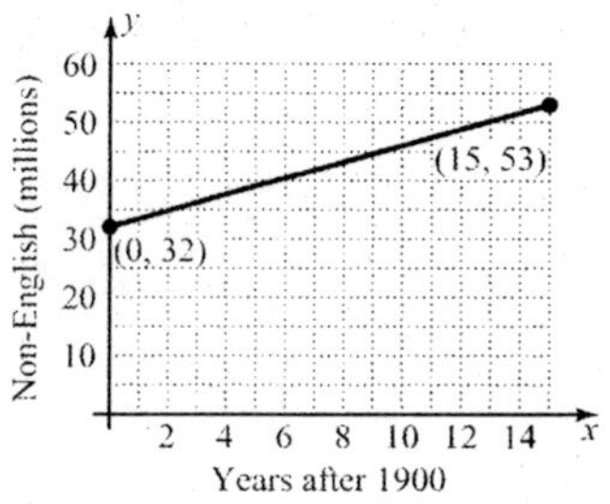

b) $m=\dfrac{y_2-y_1}{x_2-x_1}=\dfrac{53-32}{15-0}=\dfrac{21}{15}=1.4$

69. a)

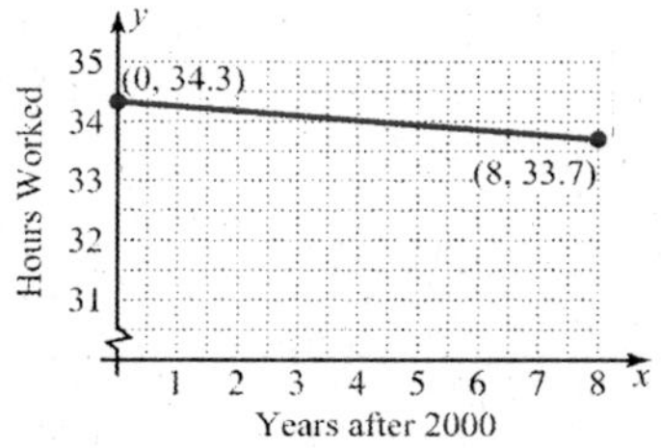

b) $m=\dfrac{y_2-y_1}{x_2-x_1}=\dfrac{33.7-34.3}{8-0}=-\dfrac{0.6}{8}=-0.075$

Review Exercises

1. $4n^2-m(n+2)=4(-2)^2-7(-2+2)$
$=4(4)-7(0)$
$=16-0$
$=16$

2. $4n^2-m(n+2)=4(3)^2-(-5)(3+2)$
$=4(9)+5(5)$
$=36+25$
$=61$

3. $16x-3y-15-x+3y-12$
$=16x-x-3y+3y-15-12$
$=15x-27$

4. $-5(3x-2)=-5(3x)-5(-2)=-15x+10$

5. $V=lwh$
$\dfrac{V}{lw}=\dfrac{lwh}{lw}$
$\dfrac{V}{lw}=h$

6. $Ax+By=C$
$By=C-Ax$
$\dfrac{By}{B}=\dfrac{C-Ax}{B}$
$y=\dfrac{C-Ax}{B}$

Exercise Set 3.3

1. The slope-intercept form.

3. Solve the equation for y.

5. The lines are parallel if their slopes are equal. Vertical lines are parallel.

7. $m=-4;\ (0,3)$
$y=mx+b$
$y=-4x+3$

9. $m=\dfrac{3}{5};\ (0,-2)$
$y=mx+b$
$y=\dfrac{3}{5}x-2$

11. $m=-0.2;\ (0,-1.5)$
$y=mx+b$
$y=-0.2x-1.5$

13. $m=\dfrac{1}{2};\ (0,2)$
$y=mx+b$
$y=\dfrac{1}{2}x+2$

15. $m=-2;(0,-4)$
$y=mx+b$
$y=-2x-4$

17. $m = \frac{4-2}{3-0} = \frac{2}{3}$; $(0,2)$

$y = mx + b$

$y = \frac{2}{3}x + 2$

19. $m = \frac{-3-(-6)}{-2-0} = -\frac{3}{2}$; $(0,-6)$

$y = mx + b$

$y = -\frac{3}{2}x - 6$

21. $y - y_1 = m(x - x_1)$

$y - 1 = 2(x-3)$

$y - 1 = 2x - 6$

$y = 2x - 5$

23. $y - y_1 = m(x - x_1)$

$y - 0 = -2(x-(-5))$

$y = -2(x+5)$

$y = -2x - 10$

25. $y - y_1 = m(x - x_1)$

$y - 0 = -3(x-0)$

$y = -3x$

27. $y - y_1 = m(x - x_1)$

$y - (-2) = \frac{2}{5}(x-0)$

$y + 2 = \frac{2}{5}x$

$y = \frac{2}{5}x - 2$

29. $y - y_1 = m(x - x_1)$

$y - (-2) = \frac{4}{5}(x-(-2))$

$y + 2 = \frac{4}{5}(x+2)$

$y + 2 = \frac{4}{5}x + \frac{8}{5}$

$y = \frac{4}{5}x - \frac{2}{5}$

31. $y - y_1 = m(x - x_1)$

$y - 3 = -\frac{3}{2}(x-(-1))$

$y - 3 = -\frac{3}{2}(x+1)$

$y - 3 = -\frac{3}{2}x - \frac{3}{2}$

$y = -\frac{3}{2}x + \frac{3}{2}$

33. Find the slope: $m = \frac{3-(-1)}{2-4} = \frac{3+1}{-2} = \frac{4}{-2} = -2$

a) $y - 3 = -2(x-2)$

$y - 3 = -2x + 4$

$y = -2x + 7$

b) $y = -2x + 7$

$2x + y = 7$

35. Find the slope: $m = \frac{-7-(-4)}{-2-1} = \frac{-7+4}{-3} = \frac{-3}{-3} = 1$

a) $y - y_1 = m(x - x_1)$

$y - (-4) = 1(x-1)$

$y + 4 = x - 1$

$y = x - 5$

b) $y = x - 5$

$-x + y = -5$

$x - y = 5$

37. Find the slope: $m = \frac{0-(-6)}{3-(-6)} = \frac{0+6}{3+6} = \frac{6}{9} = \frac{2}{3}$

a) $y - y_1 = m(x - x_1)$

$y - 0 = \frac{2}{3}(x-3)$

$y = \frac{2}{3}x - 2$

b) $y = \frac{2}{3}x - 2$

$3y = 2x - 6$

$-2x + 3y = -6$

$2x - 3y = 6$

39. Find the slope: $m=\dfrac{6-0}{0-(-5)}=\dfrac{6}{0+5}=\dfrac{6}{5}$

a) $y-y_1=m(x-x_1)$

$y-0=\dfrac{6}{5}(x-(-5))$

$y=\dfrac{6}{5}(x+5)$

$y=\dfrac{6}{5}x+6$

b) $y=\dfrac{6}{5}x+6$

$5y=6x+30$

$-6x+5y=30$

$6x-5y=-30$

41. Find the slope: $m=\dfrac{3-2}{-9-4}=\dfrac{1}{-13}=-\dfrac{1}{13}$

a) $y-y_1=m(x-x_1)$

$y-2=-\dfrac{1}{13}(x-4)$

$y-2=-\dfrac{1}{13}x+\dfrac{4}{13}$

$y=-\dfrac{1}{13}x+\dfrac{30}{13}$

b) $y=-\dfrac{1}{13}x+\dfrac{30}{13}$

$13y=-x+30$

$x+13y=30$

43. Find the slope: $m=\dfrac{9-2}{-5-(-2)}=\dfrac{7}{-5+2}=\dfrac{7}{-3}$

a) $y-y_1=m(x-x_1)$

$y-2=-\dfrac{7}{3}(x-(-2))$

$y-2=-\dfrac{7}{3}(x+2)$

$y-2=-\dfrac{7}{3}x-\dfrac{14}{3}$

$y=-\dfrac{7}{3}x-\dfrac{8}{3}$

b) $y=-\dfrac{7}{3}x-\dfrac{8}{3}$

$3y=-7x-8$

$7x+3y=-8$

45. Because these lines have equal slopes and different y-intercepts, they are parallel.

47. Because the slopes are $\dfrac{3}{4}$ and $-\dfrac{4}{3}$, these lines are perpendicular.

49. These slopes are neither equal nor of the form $\dfrac{a}{b}$ and $-\dfrac{b}{a}$. Therefore, these lines are neither parallel nor perpendicular.

51. $2x+7y=8$

$7y=-2x+8$

$y=-\dfrac{2}{7}x+\dfrac{8}{7}$

$4x+14y=-9$

$14y=-4x-9$

$y=-\dfrac{4}{14}x-\dfrac{9}{14}$

$y=-\dfrac{2}{7}x-\dfrac{9}{14}$

Because these lines have equal slopes and different y-intercepts, they are parallel.

53. $3x+5y=4$

$5y=-3x+4$

$y=-\dfrac{3}{5}x+\dfrac{4}{5}$

$5x-3y=2$

$5x-2=3y$

$\dfrac{5}{3}x-\dfrac{2}{3}=y$

Because the slopes are $-\dfrac{3}{5}$ and $\dfrac{5}{3}$, these lines are perpendicular.

55. The first line is vertical and the second line is horizontal. Horizontal and vertical lines are perpendicular.

57. a) $m=-5$

$y-y_1=m(x-x_1)$

$y-(-6)=-5(x-2)$

$y+6=-5x+10$

$y=-5x+4$

b) $y=-5x+4$

$5x+y=4$

59. a) $m=4$

$y-y_1=m(x-x_1)$

$y-(-2)=4(x-(-5))$

$y+2=4(x+5)$

$y+2=4x+20$

$y=4x+18$

b) $y=4x+18$

$4x-y=-18$

61. a) $m = \frac{2}{3}$

$$y - y_1 = m(x - x_1)$$
$$y - (-4) = \frac{2}{3}(x - 3)$$
$$y + 4 = \frac{2}{3}x - 2$$
$$y = \frac{2}{3}x - 6$$

b)
$$y = \frac{2}{3}x - 6$$
$$3y = 2x - 18$$
$$2x - 3y = 18$$

63. $4x + 6y = 3$
$$6y = -4x + 3$$
$$y = -\frac{2}{3}x + \frac{1}{2}$$

a) $m = -\frac{2}{3}$

$$y - y_1 = m(x - x_1)$$
$$y - (-3) = -\frac{2}{3}(x - 4)$$
$$y + 3 = -\frac{2}{3}x + \frac{8}{3}$$
$$y = -\frac{2}{3}x - \frac{1}{3}$$

b)
$$y = -\frac{2}{3}x - \frac{1}{3}$$
$$3y = -2x - 1$$
$$2x + 3y = -1$$

65. $2x + 5y - 30 = 0$
$$5y = -2x + 30$$
$$y = -\frac{2}{5}x + 6$$

a) $m = -\frac{2}{5}$

$$y - y_1 = m(x - x_1)$$
$$y - (-7) = -\frac{2}{5}(x - (-3))$$
$$y + 7 = -\frac{2}{5}(x + 3)$$
$$y + 7 = -\frac{2}{5}x - \frac{6}{5}$$
$$y = -\frac{2}{5}x - \frac{41}{5}$$

b)
$$y = -\frac{2}{5}x - \frac{41}{5}$$
$$5y = -2x - 41$$
$$2x + 5y = -41$$

67. a) The slope of the given equation is $m = \frac{1}{3}$, so the slope of a perpendicular line is $m = -3$.

$$y - y_1 = m(x - x_1)$$
$$y - (-1) = -3(x - 2)$$
$$y + 1 = -3x + 6$$
$$y = -3x + 5$$

b)
$$y = -3x + 5$$
$$3x + y = 5$$

69. a) The slope of the given equation is $m = -\frac{2}{5}$, so the slope of a perpendicular line is $m = \frac{5}{2}$.

$$y - y_1 = m(x - x_1)$$
$$y - (-3) = \frac{5}{2}(x - 0)$$
$$y + 3 = \frac{5}{2}x$$
$$y = \frac{5}{2}x - 3$$

b)
$$y = \frac{5}{2}x - 3$$
$$2y = 5x - 6$$
$$-5x + 2y = -6$$
$$5x - 2y = 6$$

71. a) The slope of the given equation is $m = -3$, so the slope of a perpendicular line is $m = \frac{1}{3}$.

$$y - y_1 = m(x - x_1)$$
$$y - (-9) = \frac{1}{3}(x - (-2))$$
$$y + 9 = \frac{1}{3}(x + 2)$$
$$y + 9 = \frac{1}{3}x + \frac{2}{3}$$
$$y = \frac{1}{3}x - \frac{25}{3}$$

b)
$$y = \frac{1}{3}x - \frac{25}{3}$$
$$3y = x - 25$$
$$-x + 3y = -25$$
$$x - 3y = 25$$

73. $x + 4y = -10$
$$4y = -x - 10$$
$$y = -\frac{1}{4}x - \frac{5}{2}$$

a) The slope of the given equation is $m = -\frac{1}{4}$, so the slope of a perpendicular line is $m = 4$.

$$y - y_1 = m(x - x_1)$$
$$y - (-3) = 4(x - (-2))$$
$$y + 3 = 4(x + 2)$$
$$y + 3 = 4x + 8$$
$$y = 4x + 5$$

b)
$$y = 4x + 5$$
$$-4x + y = 5$$
$$4x - y = -5$$

75. $2x - 3y = 15$
$$-3y = -2x + 15$$
$$y = \frac{2}{3}x - 5$$

a) The slope of the given equation is $m = \frac{2}{3}$, so the slope of a perpendicular line is $m = -\frac{3}{2}$.

$$y - y_1 = m(x - x_1)$$
$$y - 4 = -\frac{3}{2}(x - 1)$$
$$y - 4 = -\frac{3}{2}x + \frac{3}{2}$$
$$y = -\frac{3}{2}x + \frac{11}{2}$$

b)
$$y = -\frac{3}{2}x + \frac{11}{2}$$
$$2y = -3x + 11$$
$$3x + 2y = 11$$

77. $y = -4$

79. $y = 0$

81. a) $m = \frac{304.1 - 282.4}{8 - 0} = \frac{21.7}{8} = 2.7125$

b) $p = 2.7125t + 282.4$

c) 2015 would be year 15.
$$p = 2.7125(15) + 282.4$$
$$= 323.0875$$
The model predicts 323,087,500 people in 2015.

83. a) $m = \frac{5012 - 6465}{10 - 0} = \frac{-1453}{10} = -145.3$

b) $b = -145.3t + 6465$

c) 2012 would be the year 19
$$b = -145.3(19) + 6465$$
$$= 3704.3$$
3704.3 thousand or 3,704,300 barrels

85. a)

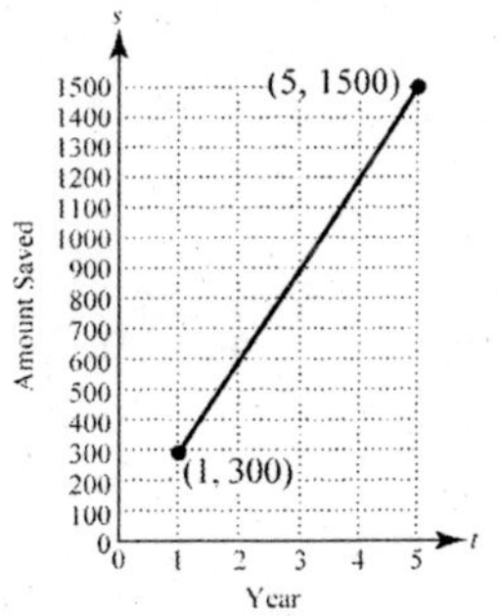

b) $m = \frac{1500 - 300}{5 - 1} = 300$

c) $s = 300t$

d) $s = 300(10)$
$$s = \$3000$$

87. a)

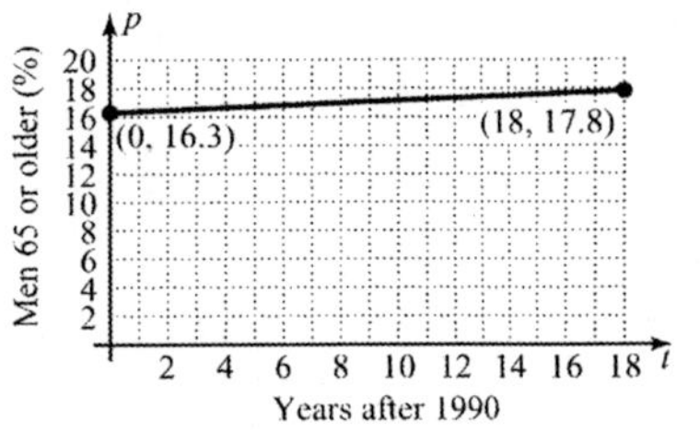

b) $m = \dfrac{17.8-16.3}{18-0} = \dfrac{1.5}{18} \approx 0.08$

c) $p = 0.08t + 16.3$

d) 2015 would be the year 25

$p = 0.08(25) + 16.3$

$= 18.3$

The percentage is 18.3%.

Review Exercises

1. $-0.8 < -\dfrac{1}{8}$

2. $4(2x+3) - 2(x-6) = 7 - (3x+1)$

$8x + 12 - 2x + 12 = 7 - 3x - 1$

$6x + 24 = 6 - 3x$

$9x = -18$

$x = -2$

3. $\dfrac{1}{4}x - \dfrac{5}{6} = \dfrac{3}{2}x - 1$

$\dfrac{1}{6} = \dfrac{5}{4}x$

$\dfrac{2}{15} = x$

4. $-4x - 7 \geq 13$ $\{x | x \leq -5\}$ $(-\infty, -5]$

$-4x \geq 20$

$x \leq -5$

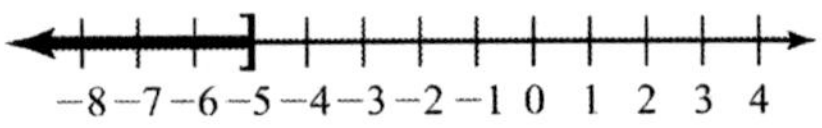

5. $6x - 2(x-1) \geq 9x - 8$

$6x - 2x + 2 \geq 9x - 8$

$4x + 2 \geq 9x - 8$

$-5x \geq -10$

$x \leq 2$

$\{x | x \leq 2\}$ $(-\infty, 2]$

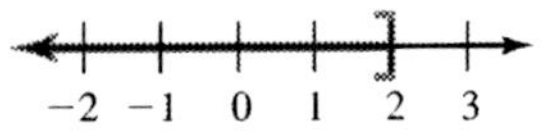

6. $50w \geq 2000$

$w \geq 40$ ft.

Exercise Set 3.4

1. Graph the related equation (the boundary line).

3. If the inequality symbol is $\leq$ or $\geq$, draw a solid boundary line. If the inequality symbol is $<$ or $>$, draw a dashed boundary line.

5. Test $(-1, 3)$.

$y \geq -x + 7$

$3 \overset{?}{\geq} -(-1) + 7$

$3 \overset{?}{\geq} 1 + 7$

$3 \geq 8$

This statement is false. No, $(-1, 3)$ is not a solution.

7. Test $(0, 0)$.

$2x + 3y \geq 0$

$2(0) + 3(0) \overset{?}{\geq} 0$

$0 + 0 \overset{?}{\geq} 0$

$0 \geq 0$

This statement is true. Yes, $(0, 0)$ is a solution.

9. Test $(3, 4)$.

$3y - x < 8$

$3(4) - 3 \overset{?}{<} 8$

$12 - 3 \overset{?}{<} 8$

$9 < 8$

This statement is false. No, $(3, 4)$ is not a solution.

11. Test $(5,3)$.

$$\begin{aligned} y &> \frac{3}{4}x-5 \\ 3 &\stackrel{?}{>} \frac{3}{4}(5)-5 \\ 3 &\stackrel{?}{>} \frac{15}{4}-5 \\ 3 &> -\frac{5}{4} \end{aligned}$$

This statement is true. Yes, $(5,3)$ is a solution.

13. $y \le x+3$

Begin by graphing the related equation $y = x+3$ with a solid line. Now choose $(0,0)$ as a test point.

$$\begin{aligned} y &\le x+3 \\ 0 &\stackrel{?}{\le} 0+3 \\ 0 &\le 3 \end{aligned}$$

Because $(0,0)$ satisfies the inequality, shade the region which includes $(0,0)$.

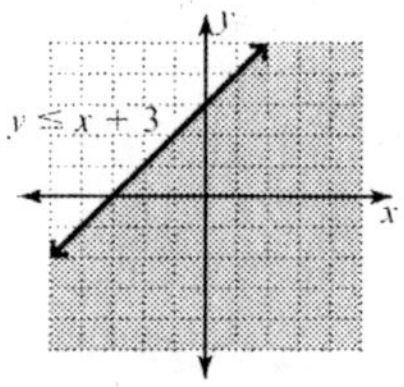

15. $y \ge -4x+6$

Begin by graphing the related equation $y = -4x+6$ with a solid line. Now choose $(0,0)$ as a test point.

$$\begin{aligned} y &\ge -4x+6 \\ 0 &\stackrel{?}{\ge} -4(0)+6 \\ 0 &\ge 6 \end{aligned}$$

Because $(0,0)$ does not satisfy the inequality, shade the side of the line on the opposite side from $(0,0)$.

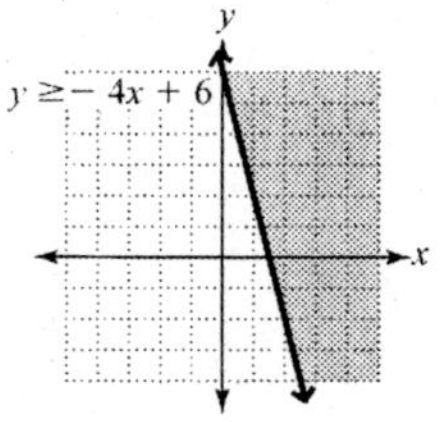

17. $y > x$

Begin by graphing the related equation $y = x$ with a dashed line. Now choose $(1,0)$ as a test point.

$$\begin{aligned} y &> x \\ 0 &\stackrel{?}{>} (1) \\ 0 &> 1 \end{aligned}$$

Because $(1,0)$ does not satisfy the inequality, shade the side of the line on the opposite side from $(1,0)$.

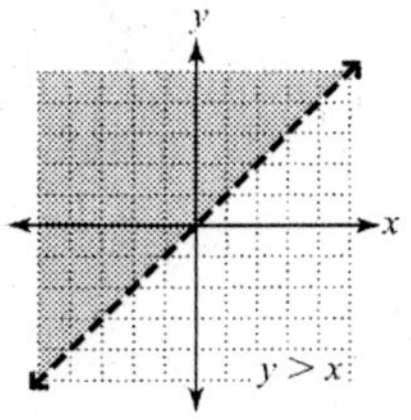

19. $y > -3x$

Begin by graphing the related equation $y = -3x$ with a dashed line. Now choose $(1,0)$ as a test point.

$$\begin{aligned} y &> -3x \\ 0 &\stackrel{?}{>} -3(1) \\ 0 &> -3 \end{aligned}$$

Because $(1,0)$ satisfies the inequality, shade the region which includes $(1,0)$.

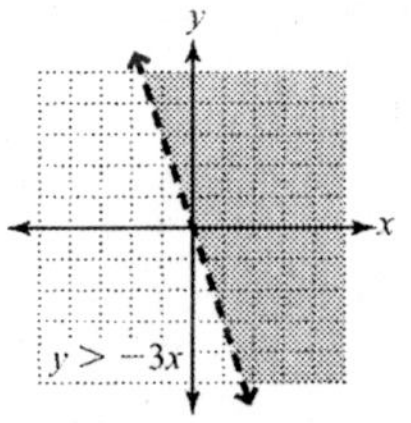

21. $y > \frac{2}{5}x$

Begin by graphing the related equation $y = \frac{2}{5}x$ with a dashed line. Now choose $(1,0)$ as a test point.

$$y > \frac{2}{5}x$$
$$0 \stackrel{?}{>} \frac{2}{5}(1)$$
$$0 > \frac{2}{5}$$

Because $(1,0)$ does not satisfy the inequality, shade the side of the line on the opposite side from $(1,0)$.

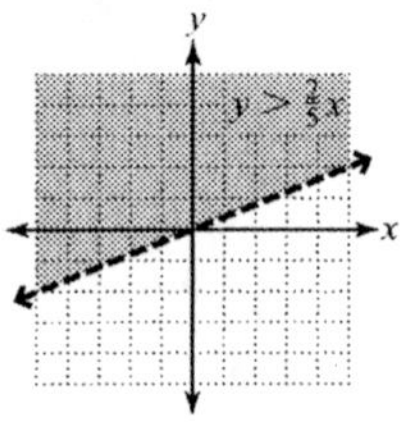

23. $x - y < 2$
$-y < -x + 2$
$y > x - 2$

Begin by graphing the related equation $y = x - 2$ with a dashed line. Now choose $(0,0)$ as a test point.

$$x - y < 2$$
$$0 - 0 \stackrel{?}{<} 2$$
$$0 < 2$$

Because $(0,0)$ satisfies the inequality, shade the region which includes $(0,0)$.

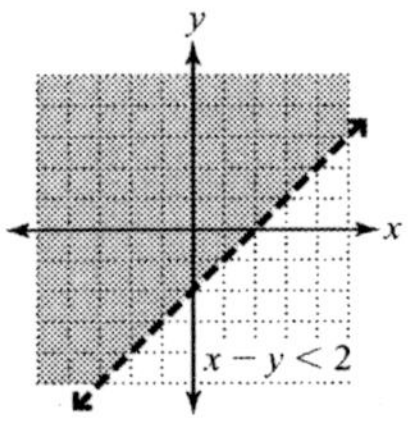

25. $x + 3y > -9$
$3y > -x - 9$
$y > -\frac{1}{3}x - 3$

Begin by graphing the related equation $y = -\frac{1}{3}x - 3$ with a dashed line. Now choose $(0,0)$ as a test point.

$$x + 3y > -9$$
$$0 - 3(0) \stackrel{?}{>} -9$$
$$0 - 0 \stackrel{?}{>} -9$$
$$0 > -9$$

Because $(0,0)$ satisfies the inequality, shade the region which includes $(0,0)$.

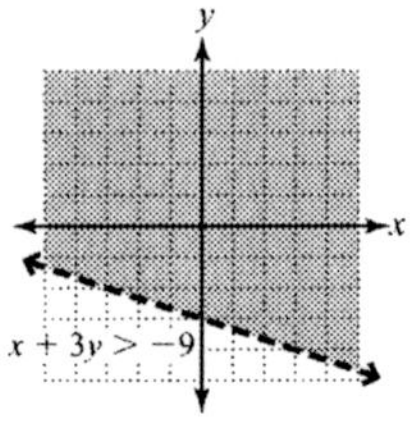

27. $x - 2y \geq -6$
$-2y \geq -x - 6$
$y \leq \frac{1}{2}x + 3$

Begin by graphing the related equation $y = \frac{1}{2}x + 3$ with a solid line. Now choose $(0,0)$ as a test point.

$$x - 2y \geq -6$$
$$0 - 2(0) \stackrel{?}{\geq} -6$$
$$0 - 0 \stackrel{?}{\geq} -6$$
$$0 \geq -6$$

Because $(0,0)$ satisfies the inequality, shade the region which includes $(0,0)$.

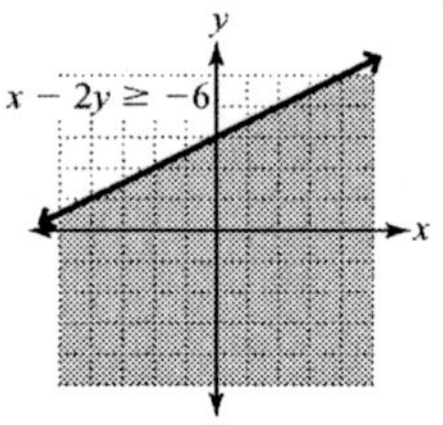

29. $3x-2y>6$

$$-2y>-3x+6$$

$$y<\frac{3}{2}x-3$$

Begin by graphing the related equation $y=\frac{3}{2}x-3$ with a dashed line. Now choose $(0,0)$ as a test point.

$$3x-2y > 6$$
$$3(0)-2(0) \overset{?}{>} 6$$
$$0-0 \overset{?}{>} 6$$
$$0 > 6$$

Because $(0,0)$ does not satisfy the inequality, shade the side of the line on the opposite side from $(0,0)$.

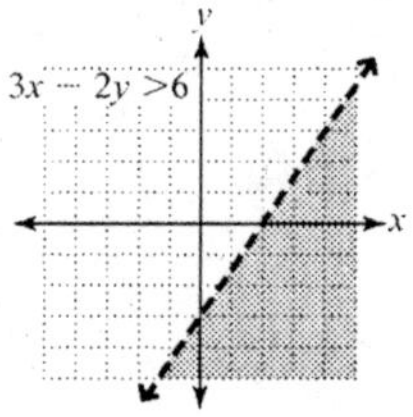

31. $5x-y\le 0$

$$-y\le -5x$$

$$y\ge 5x$$

Begin by graphing the related equation $y=5x$ with a solid line. Now choose $(1,0)$ as a test point.

$$5x-y \le 0$$
$$5(1)-0 \overset{?}{\le} 0$$
$$5-0 \overset{?}{\le} 0$$
$$5 \le 0$$

Because $(1,0)$ does not satisfy the inequality, shade the side of the line on the opposite side from $(1,0)$.

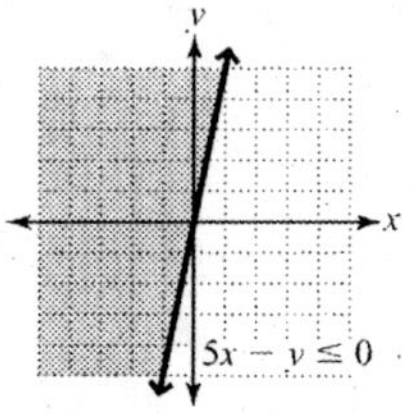

33. $4x+2y\le 3$

$$2y\le -4x+3$$

$$y\le -2x+\frac{3}{2}$$

Begin by graphing the related equation $y=-2x+\frac{3}{2}$ with a solid line. Now choose $(0,0)$ as a test point.

$$4x+2y \le 3$$
$$4(0)+2(0) \overset{?}{\le} 3$$
$$0+0 \overset{?}{\le} 3$$
$$0 \le 3$$

Because $(0,0)$ satisfies the inequality, shade the region which includes $(0,0)$.

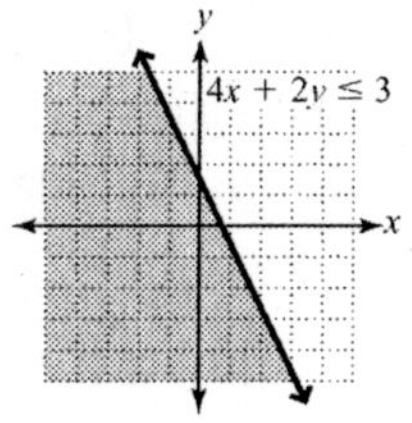

35. $x>6$

Begin by graphing the related equation $x=6$ with a dashed line. Now choose $(0,0)$ as a test point.

$$x > 6$$
$$0 \overset{?}{>} 6$$

Because $(0,0)$ does not satisfy the inequality, shade the side of the line on the opposite side from $(0,0)$.

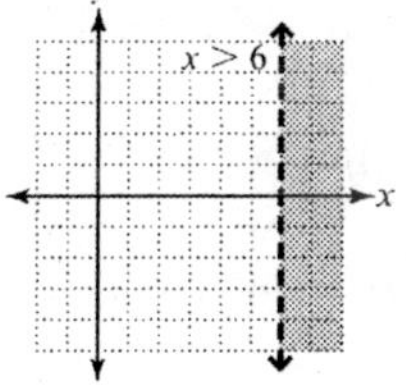

37. $y \le 7$

Begin by graphing the related equation $y = 7$ with a solid line. Now choose $(0,0)$ as a test point.

$$y < 7$$
$$0 \overset{?}{<} 7$$

Because $(0,0)$ satisfies the inequality, shade the region which includes $(0,0)$.

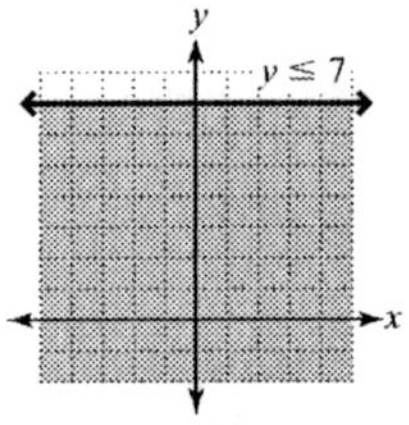

39. a) x represents the number of board games produced, and y represents the number of video games produced.

b) Because x and y represent nonnegative numbers, the graph should be in the first quadrant only $(x \ge 0 \text{ and } y \ge 0)$.

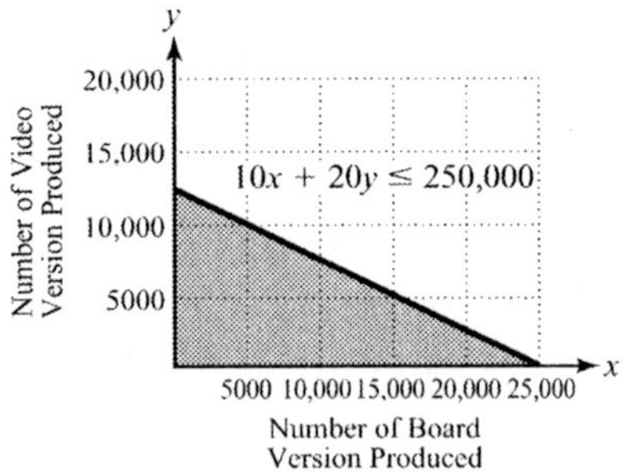

c) All combinations that cost exactly $250,000 to produce.
d) All combinations that cost less than $250,000 to produce.
e) Answers may vary. Two examples are (0, 12,500) and (25,000, 0).
f) Answers may vary. Two examples are (0, 500) and (10,000, 5000).
g) No, fractions of a game are not produced.

41. a) $2l + 2w \le 200$

b)

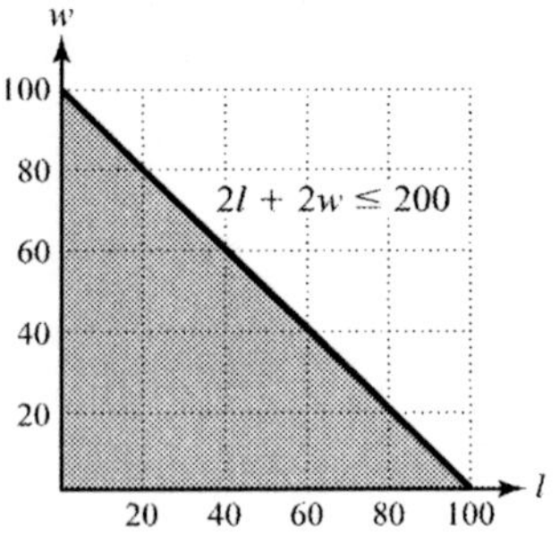

c) Combinations of length and width that make the perimeter exactly 200 ft.
d) All combinations of length and width that make the perimeter less than 200 ft.
e) Answers may vary. Two examples are (20, 80) and (60, 40).
f) Answers may vary. Two examples are (20, 40) and (80, 10).

43. a) $12x + 15y \ge 18,000$

b)

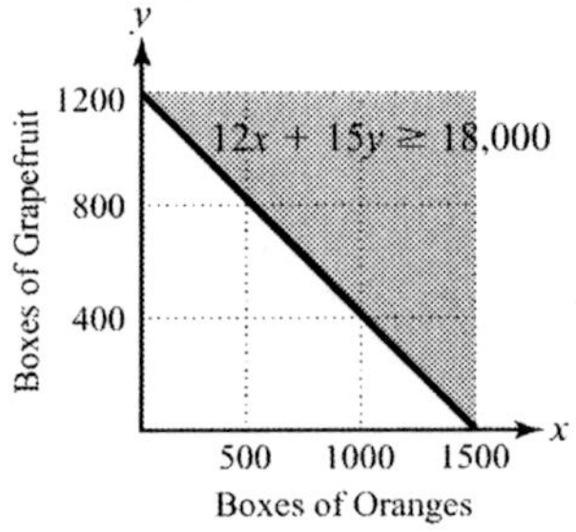

c) All sales combinations that raise exactly $18,000.
d) All sales combination that will raise more than $18,000.
e) Answers may vary. Two examples are (500, 800) and (1000, 400).
f) Answers may vary. Two examples are (500, 1000) and (800, 1200).

Review Exercises

1. $\{-2, 0, 3, 4\}$

2. $\{1, 2, 5, 7\}$

3. $$\begin{aligned} x^2 - 4x + 1 &= (-2)^2 - 4(-2) + 1 \\ &= 4 + 8 + 1 \\ &= 13 \end{aligned}$$

4. $$\begin{aligned} x - 3 &= 0 \\ x &= 3 \end{aligned}$$

The expression is undefined when the denominator equals 0, which occurs when $x = 3$.

5. $$\begin{aligned} 3x-4(x+8) &= 5(x-1)+21 \\ 3x-4x-32 &= 5x-5+21 \\ -x-32 &= 5x+16 \\ -48 &= 6x \\ -8 &= x \end{aligned}$$

6. x-intercept: $$\begin{aligned} 4x-5(0) &= 8 \\ 4x &= 8 \\ x &= 2 \\ &(2,0) \end{aligned}$$

y-intercept: $$\begin{aligned} 4(0)-5y &= 8 \\ -5y &= 8 \\ y &= -\frac{8}{5} \\ &\left(0,-\frac{8}{5}\right) \end{aligned}$$

Exercise Set 3.5

1. The domain is a set containing initial values of a relation; its input values; the first coordinates in ordered pairs.

3. A function is a relation in which every value in the domain is assigned to exactly one value in the range.

5. Given a function $f(x)$, to find $f(a)$, where a is a real number in the domain of f, replace x in the function with a and evaluate or simplify.

7. Domain: {George Washington University, Kenyon College, Bucknell University, Vassar College, Sarah Lawrence College}
Range: {$39,240, $38,140, $38,134, $38,115, $38,090}
It is a function because every element in the domain is paired with exactly one element in the range.

9. Domain: {1, 2, 3, 5}
Range: {San Francisco 49ers, Dallas Cowboys, Pittsburgh Steelers, Green Bay Packers, New England Patriots, Oakland/L.A. Raiders, Washington Redskins, New York Giants, Miami Dolphins, Denver Broncos, Indianapolis/Baltimore Colts, Baltimore Ravens, Chicago Bears, New York Jets, Tampa Bay Buccaneers, Kansas City Chiefs, St. Louis/L.A. Rams}
It is not a function because an element in the domain is paired with more than one element in the range.

11. Domain: {41%, 16%, 5%, 4%}
Range: {Savings, Sale of stock or bonds, Equity from other homes, Financial institution loan, Inheritance}
It is not a function because an element in the domain is paired with more than one element in the range.

13. Domain: {1, 2, 3, 4, 5}
Range: {*Sunday at Tiffany's, The Whole Truth, Twenty Wishes, Hold Tight, Unaccustomed Earth*}
It is a function because every element in the domain is paired with exactly one element in the range.

15. Domain: $\{-2, 6, 5, -1\}$
Range: $\{4, -2, 1, -6\}$
It is a function because every element in the domain is paired with exactly one element in the range.

17. Domain: $\{-6, 2, -5\}$
Range: $\{2, -3, 7, 4\}$
It is not a function because an element in the domain is paired with more than one element in the range.

19. Domain: $\{2, -4, 5, 3\}$
Range: $\{-4, 3, -1\}$
It is a function because every element in the domain is paired with exactly one element in the range.

21. Domain: {2000, 2001, 2002, 2003, 2004, 2005, 2006}
Range: {5920, 5915, 5534, 5574, 5764, 5734, 5840}
It is a function because every element in the domain is paired with exactly one element in the range.

23. Domain: $\mathbb{R}$ or $(-\infty,\infty)$
Range: $\{y \mid y \le 4\}$ or $(-\infty,4]$
It is a function because a vertical line at each x-value would intersect the graph at only one point.

25. Domain: $\{x \mid x \ge 0\}$ or $[0,\infty)$
Range: $\mathbb{R}$ or $(-\infty,\infty)$
It is not a function because there are values in the domain that correspond to two values in the range.

27. Domain: $\{x|-4 \le x \le 5\}$ or $[-4,5]$

Range: $\{-3,-2\}$

It is a function because a vertical line at each x-value would intersect the graph at only one point.

29. Domain: $\{x|-4 \le x \le 0\}$ or $[-4,0]$

Range: $\{y|-3 \le y \le 1\}$ or $[-3,1]$

It is not a function because there are values in the domain that correspond to two values in the range.

31. $f(x) = -2x - 9$

a) $f(0) = -2(0) - 9 = 0 - 9 = -9$

b) $f(1) = -2(1) - 9 = -2 - 9 = -11$

c) $f(-1) = -2(-1) - 9 = 2 - 9 = -7$

d) $f(a+1) = -2(a+1) - 9 = -2a - 2 - 9 = -2a - 11$

33. $f(x) = 2x^2 - x + 7$

a) $f(0) = 2(0)^2 - 0 + 7 = 0 - 0 + 7 = 7$

b) $f(1) = 2(1)^2 - 1 + 7 = 2(1) - 1 + 7 = 2 - 1 + 7 = 8$

c) $f(-1) = 2(-1)^2 - (-1) + 7 = 2(1) + 1 + 7 = 2 + 1 + 7 = 10$

d) $f(a) = 2(a)^2 - a + 7 = 2a^2 - a + 7$

35. $f(x) = \sqrt{3-x}$

a) $f(-1) = \sqrt{3-(-1)} = \sqrt{3+1} = \sqrt{4} = 2$

b) $f(12) = \sqrt{3-12} = \sqrt{-9}$ not real

c) $f(-2) = \sqrt{3-(-2)} = \sqrt{3+2} = \sqrt{5}$

d) $f(t) = \sqrt{3-t}$

37. $f(x) = \frac{2}{5}x + 1$

a) $f(0) = \frac{2}{5}(0) + 1 = 0 + 1 = 1$

b) $f(-5) = \frac{2}{5}(-5) + 1 = -2 + 1 = -1$

c) $f(-1) = \frac{2}{5}(-1) + 1 = -\frac{2}{5} + 1 = \frac{3}{5}$

d) $f(r) = \frac{2}{5}r + 1$

39. $f(x) = x^2 - 2.1x - 3$

a) $f(0) = (0)^2 - 2.1(0) - 3 = 0 - 0 - 3 = -3$

b) $f(-2.2) = (-2.2)^2 - 2.1(-2.2) - 3 = 4.84 + 4.62 - 3 = 6.46$

c) $f\left(\frac{2}{3}\right) = \left(\frac{2}{3}\right)^2 - 2.1\left(\frac{2}{3}\right) - 3 = \frac{4}{9} - \frac{4.2}{3} - 3 \approx -3.96$

d) $f(a) = (a)^2 - 2.1(a) - 3 = a^2 - 2.1a - 3$

41. $f(x)=\sqrt{x^2-4x}$

a) $f(4)=\sqrt{4^2-4(4)}$
$=\sqrt{16-16}$
$=\sqrt{0}$
$=0$

b) $f(7)=\sqrt{7^2-4(7)}$
$=\sqrt{49-28}$
$=\sqrt{21}$

c) $f(2)=\sqrt{2^2-4(2)}$
$=\sqrt{4-8}$
$=\sqrt{-4}$
not real

d) $f(n)=\sqrt{n^2-4(n)}$
$=\sqrt{n^2-4n}$

43. $f(x)=|3x^2+1|$

a) $f(0)=|3(0)^2+1|$
$=|3(0)+1|$
$=|0+1|$
$=|1|$
$=1$

b) $f\left(\frac{2}{3}\right)=\left|3\left(\frac{2}{3}\right)^2+1\right|$
$=\left|3\left(\frac{4}{9}\right)+1\right|$
$=\left|\frac{4}{3}+1\right|$
$=\left|\frac{7}{3}\right|=\frac{7}{3}$

c) $f(-1)=|3(-1)^2+1|$
$=|3(1)+1|$
$=|3+1|$
$=|4|$
$=4$

d) $f(-2)=|3(-2)^2+1|$
$=|3(4)+1|$
$=|12+1|$
$=|13|$
$=13$

45. $f(x)=\dfrac{3-x}{x-4}$

a) $f(5)=\dfrac{3-5}{5-4}$
$=-2$

b) $f(4)=\dfrac{3-4}{4-4}$
$=\dfrac{-1}{0}$
undefined

c) $f(-2)=\dfrac{3-(-2)}{-2-4}$
$=\dfrac{3+2}{-6}$
$=-\dfrac{5}{6}$

d) $f(3)=\dfrac{3-3}{3-4}$
$=\dfrac{0}{-1}$
$=0$

47. $f(x)=\dfrac{x}{x^2-1}$

a) $f(0)=\dfrac{0}{0^2-1}=\dfrac{0}{-1}=0$

b) $f(1)=\dfrac{1}{1^2-1}=\dfrac{1}{1-1}=\dfrac{1}{0}$ is undefined

c) $f(2)=\dfrac{2}{2^2-1}=\dfrac{2}{3}$

d) $f(m)=\dfrac{m}{m^2-1}$

49. $f(x)=\dfrac{2x}{\sqrt{2-x}}$

a) $f(-2)=\dfrac{2(-2)}{\sqrt{2-(-2)}}$
$=\dfrac{-4}{\sqrt{2+2}}$
$=\dfrac{-4}{\sqrt{4}}$
$=\dfrac{-4}{2}$
$=-2$

b) $f(1)=\dfrac{2(1)}{\sqrt{2-1}}$
$=\dfrac{2}{\sqrt{1}}$
$=\dfrac{2}{1}$
$=2$

c) $f(2)=\dfrac{2(2)}{\sqrt{2-2}}$
$=\dfrac{4}{\sqrt{0}}$
$=\dfrac{4}{0}$
undefined

d) $f(6)=\dfrac{2(6)}{\sqrt{2-6}}$
$=\dfrac{12}{\sqrt{-4}}$
not real

51. a) $f(-4)=2$
b) $f(0)=0$
c) $f(2)=-1$

53. a) $f(-2)=-4$
b) $f(0)=0$
c) $f(2)=-4$

55. a) $f(-4)=-3$

b) $f(0)=1$

c) $f(3)=2$

57. $f(x)=-3x+2$

Think of the function as the equation $y=-3x+2$ and use the fact that the slope is $m=-3$ and the y-intercept is 2.

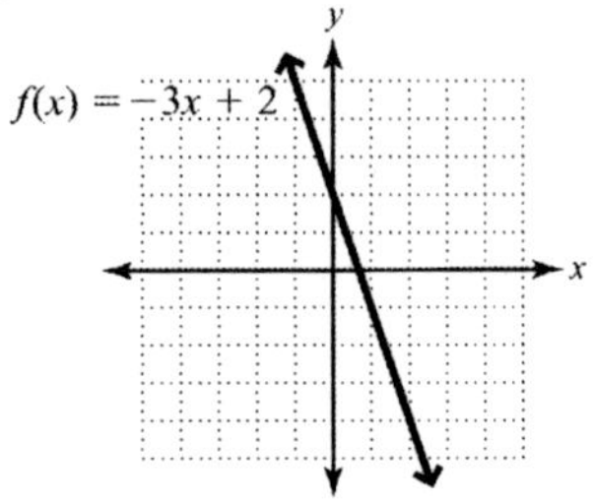

Domain: $\mathbb{R}$ or $(-\infty,\infty)$

Range: $\mathbb{R}$ or $(-\infty,\infty)$

59. $f(x)=\frac{2}{3}x-1$

Think of the function as the equation $y=\frac{2}{3}x-1$ and use the fact that the slope is $m=\frac{2}{3}$ and the y-intercept is -1.

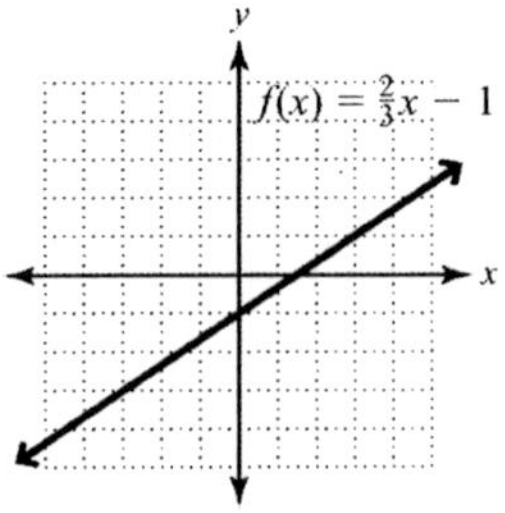

Domain: $\mathbb{R}$ or $(-\infty,\infty)$

Range: $\mathbb{R}$ or $(-\infty,\infty)$

61. $f(x)=-5x$

Think of the function as the equation $y=-5x$ and use the fact that the slope is $m=-5$ and the y-intercept is 0.

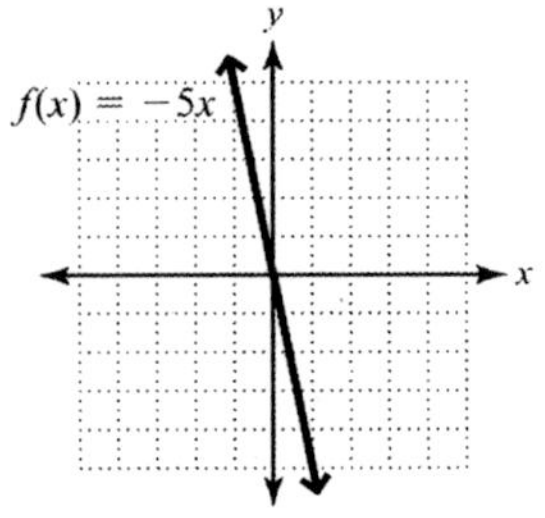

Domain: $\mathbb{R}$ or $(-\infty,\infty)$

Range: $\mathbb{R}$ or $(-\infty,\infty)$

63. $f(x)=x^2-1$

x	$f(x)$
−2	3
−1	0
0	−1
1	0
2	3

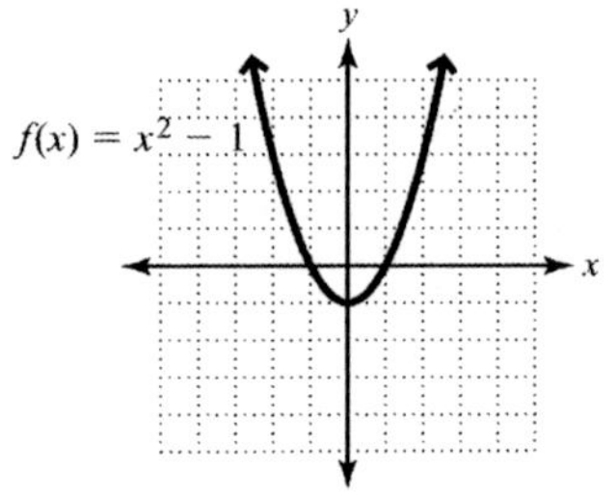

Domain: $\mathbb{R}$ or $(-\infty,\infty)$

Range: $\{y\,|\,-1\le y<\infty\}$ or $[-1,\infty)$

65. $f(x)=-x^2+3$

x	$f(x)$
−2	−1
−1	2
0	3
1	2
2	−1

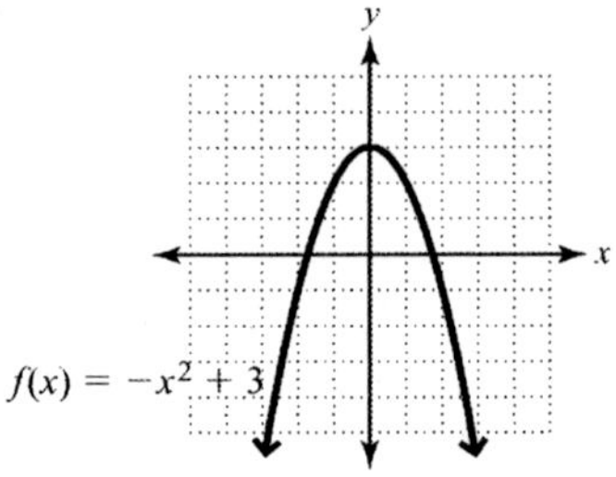

Domain: $\mathbb{R}$ or $(-\infty,\infty)$

Range: $\{y\,|\,-\infty<y\le 3\}$ or $(-\infty,3]$

67. a)

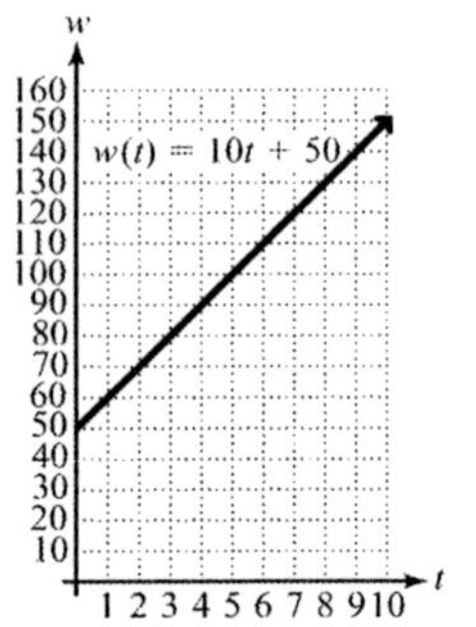

b) $w(t)=10t+50$

$w(7.5)=10(7.5)+50$

$=75+50$

$=\$125$

69. a) $C(V) = 1.225V + 6$

b)

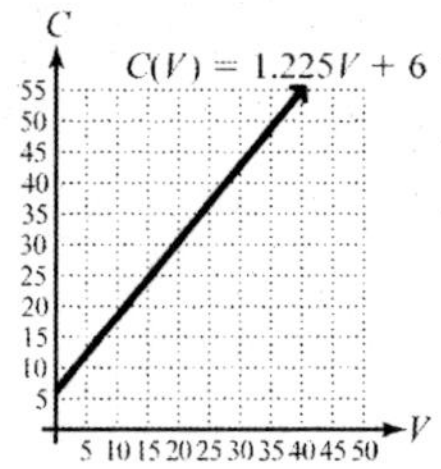

c) $C(V) = 1.225V + 6$

$$\begin{aligned} C(40) &= 1.225(40) + 6 \\ &= 49 + 6 \\ &= \$55 \end{aligned}$$

71. a)

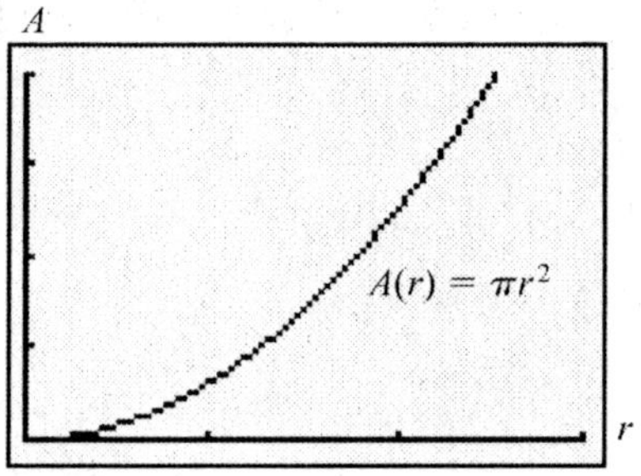

b) No, it is not a linear function.
c) The radius of the circle is 1.5 units.
d) $A(r) \approx \pi r^2$

$$\begin{aligned} A(1.5) &\approx 3.14(1.5)^2 \\ &\approx 7.07 \text{ sq. units} \end{aligned}$$

73. a)

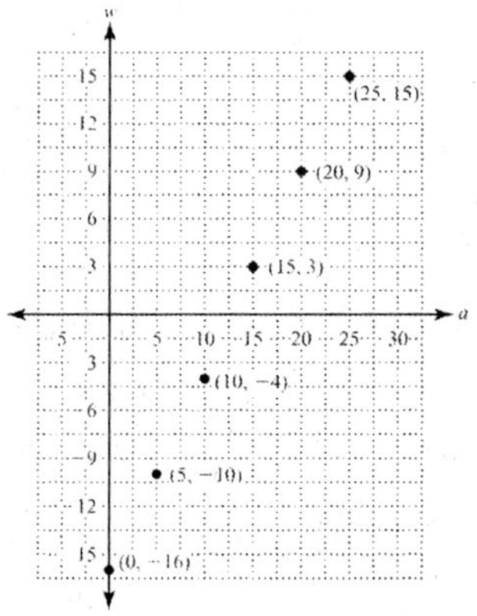

b) Yes.

$$m = \frac{15 - (-16)}{25 - 0} = \frac{15 + 16}{25} = \frac{31}{25} = 1.24$$

c) $w(a) = 1.24a - 16$

d) $w(a) = 1.24a - 16$

$$\begin{aligned} w(40) &= 1.24(40) - 16 \\ &= 49.6 - 16 \\ &= 33.6 \end{aligned}$$

When the actual temperature is 40° F, the windchill temperature is 33.6° F.

Review Exercises

1. $$\begin{aligned} & x + 9y - 3x + 7 - 3y - 9 \\ &= x - 3x + 9y - 3y + 7 - 9 \\ &= -2x + 6y - 2 \end{aligned}$$

2. $$\begin{aligned} 5(x+1) - x &= 3x - (x+1) \\ 5x + 5 - x &= 3x - x - 1 \\ 4x + 5 &= 2x - 1 \\ 2x + 5 &= -1 \\ 2x &= -6 \\ x &= -3 \end{aligned}$$

3. $$\begin{aligned} 4x - (6x+3) &\geq 8x - 23 \\ 4x - 6x - 3 &\geq 8x - 23 \\ -2x - 3 &\geq 8x - 23 \\ -10x - 3 &\geq -23 \\ -10x &\geq -20 \\ x &\leq 2 \end{aligned}$$

$\{x \mid x \leq 2\}$ $(-\infty, 2]$

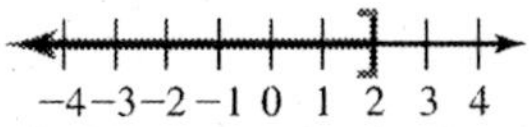

4. Let the first integer be x and the second integer be $x + 2$.

$$\begin{aligned} x + x + 2 &= 74 \\ 2x + 2 &= 74 \\ 2x &= 72 \\ x &= 36 \end{aligned}$$

The integers are 36 and 38.

5. If the width is x, then the length is $x + 6$.

$$\begin{aligned} 2(x+6) + 2x &= 108 \\ 2x + 12 + 2x &= 108 \\ 12 + 4x &= 108 \\ 4x &= 96 \\ x &= 24 \end{aligned}$$

The width is 24 in. and the length is 24 + 6 = 30 in.

6. Create a table.

	how many	value	total
large	$40-x$	2.25	$2.25(40-x)$
small	x	1.50	$1.50x$

Translate to an equation and solve.

$$\begin{aligned} 2.25(40-x)+1.50x &= 69 \\ 90-2.25x+1.50x &= 69 \\ 90-0.75x &= 69 \\ -0.75x &= -21 \\ 28 &= x \end{aligned}$$

28 small and 40 – 28 = 12 large

Chapter 3 Review Exercises

1. False; when writing coordinates, the horizontal coordinate is written first.
2. True
3. True
4. True
5. True
6. False; only two points are needed to determine a line.
7. $m=\dfrac{y_2-y_1}{x_2-x_1}$
8. vertical
9. upward, downward
10. A function is a relation in which each value in the domain is assigned to exactly one value in the range.
11. $A(3, 2)$, $B(-3, 4)$, $C(-4, -2)$, $D(0, -3)$
12.

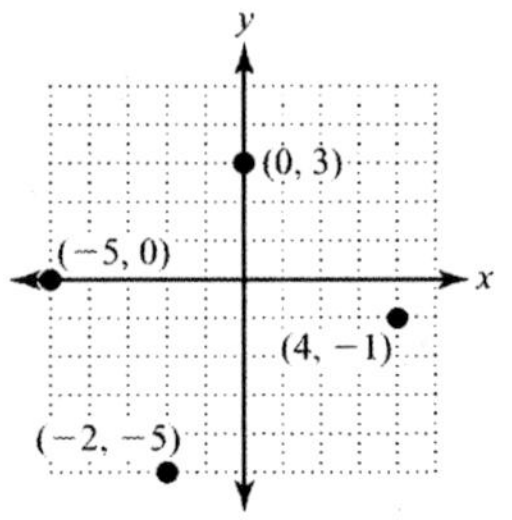

13. II

14. III

15. Replace x with 2 and y with 1 and see if the equation is true.

$$\begin{aligned} 2x-y &= -3 \\ 2(2)-1 &\stackrel{?}{=} -3 \\ 4-1 &\stackrel{?}{=} -3 \\ 3 &\neq -3 \end{aligned}$$

No, $(2,1)$ is not a solution.

16. Replace x with $\frac{4}{5}$ and y with 2 and see if the equation is true.

$$\begin{aligned} 5x+y &= 6 \\ 5\left(\frac{4}{5}\right)+2 &\stackrel{?}{=} 6 \\ 4+2 &\stackrel{?}{=} 6 \\ 6 &= 6 \end{aligned}$$

Yes, $\left(\frac{4}{5},2\right)$ is a solution.

17. Replace x with 0.4 and y with 1.2 and see if the equation is true.

$$\begin{aligned} y+2x &= 4.1 \\ 1.2+2(0.4) &\stackrel{?}{=} 4.1 \\ 1.2+0.8 &\stackrel{?}{=} 4.1 \\ 2 &\neq 4.1 \end{aligned}$$

No, $(0.4,1.2)$ is not a solution.

18. Replace x with 5 and y with 2 and see if the equation is true.

$$\begin{aligned} y &= \frac{2}{5}x \\ 2 &\stackrel{?}{=} \frac{2}{5}(5) \\ 2 &= 2 \end{aligned}$$

Yes, $(5,2)$ is a solution.

19. $y = x - 4$

x	y	Ordered Pair
0	−4	$(0,-4)$
4	0	$(4,0)$
2	−2	$(2,-2)$

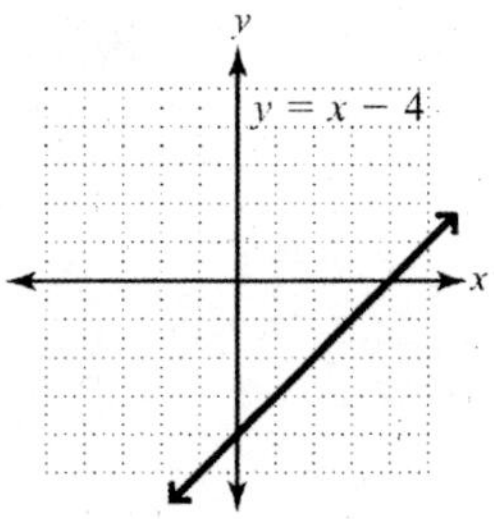

20. $y = 5x$

x	y	Ordered Pair
0	0	$(0,0)$
−1	−5	$(-1,-5)$
1	5	$(1,5)$

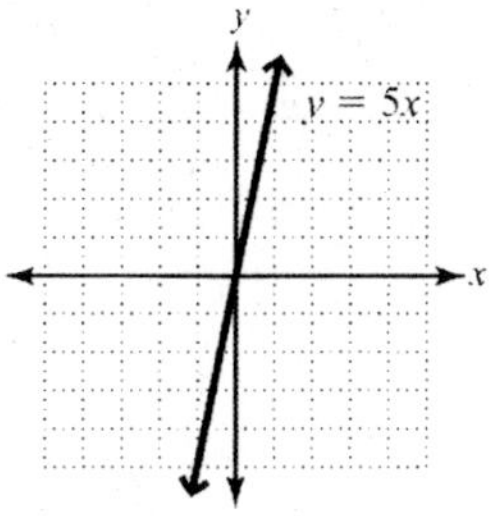

21. $y = \frac{2}{3}x - 3$

x	y	Ordered Pair
0	−3	$(0,-3)$
−3	−5	$(-3,-5)$
3	−1	$(3,-1)$

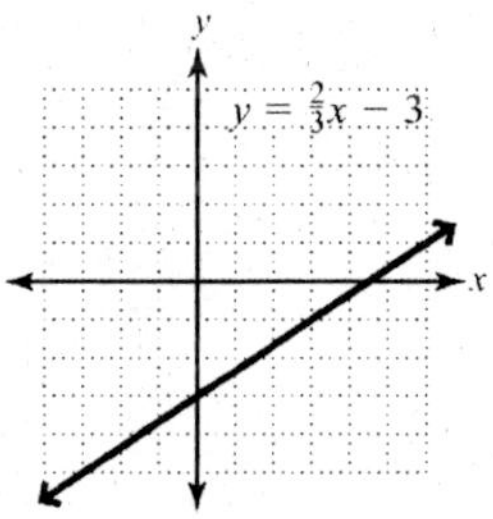

22. $y = -\frac{2}{7}x + 1$

x	y	Ordered Pair
0	1	$(0,1)$
7	−1	$(7,-1)$
−7	3	$(-7,3)$

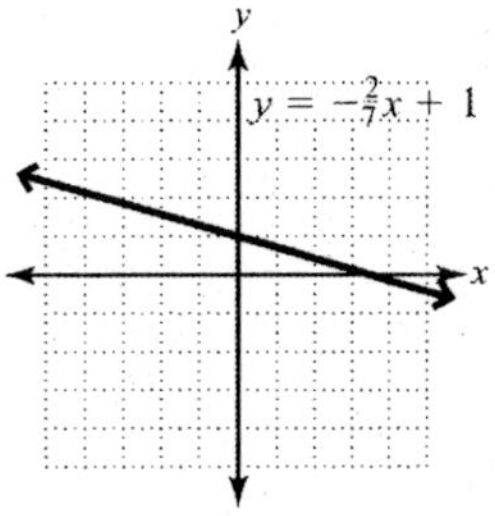

23. $2x - 3y = 6$

x	y	Ordered Pair
0	−2	$(0,-2)$
3	0	$(3,0)$
−3	−4	$(-3,-4)$

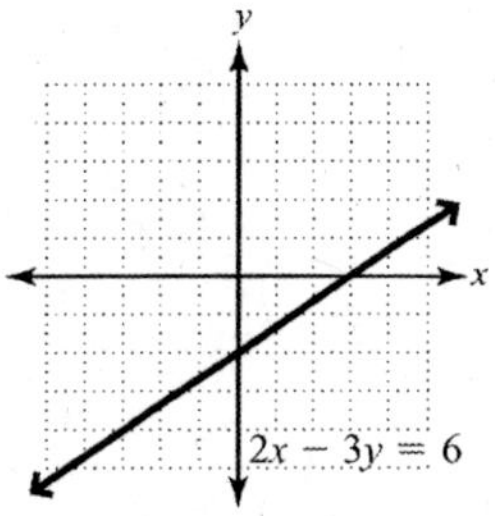

24. $3x + 4y = 28$

x	y	Ordered Pair
0	7	$(0,7)$
4	4	$(4,4)$
8	1	$(8,1)$

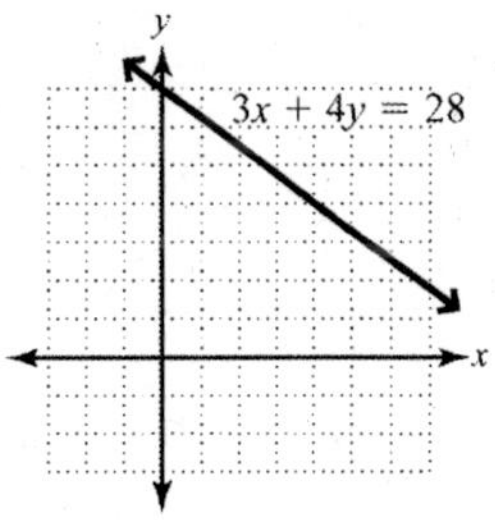

25. $y = -3x + 2$

$m = -3$; y-intercept: $(0, 2)$

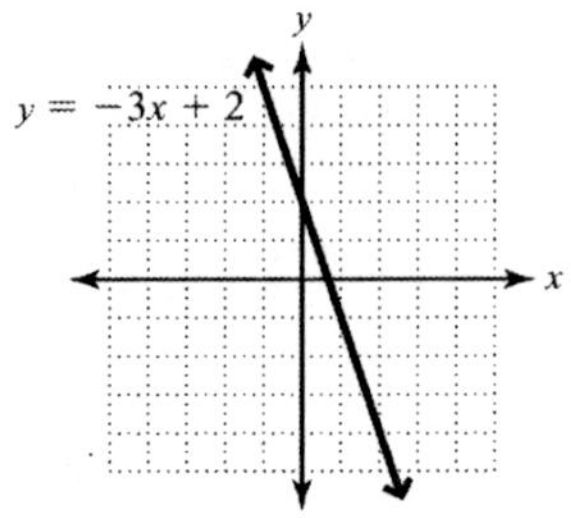

26. $y = \frac{5}{2}x + 3$

$m = \frac{5}{2}$; y-intercept: $(0, 3)$

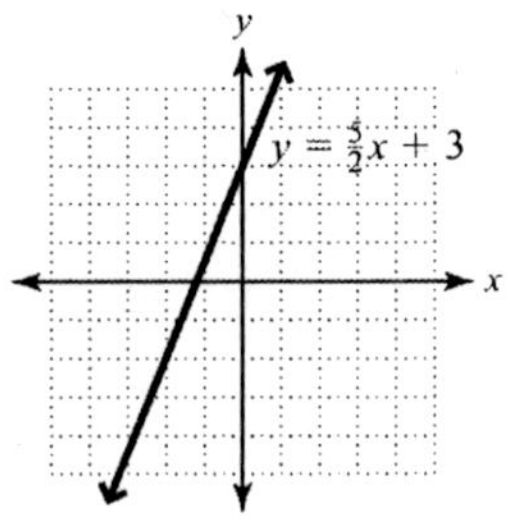

27. $2x + 5y = 15$

$$5y = -2x + 15$$

$$y = -\frac{2}{5}x + 3$$

$m = -\frac{2}{5}$; y-intercept: $(0, 3)$

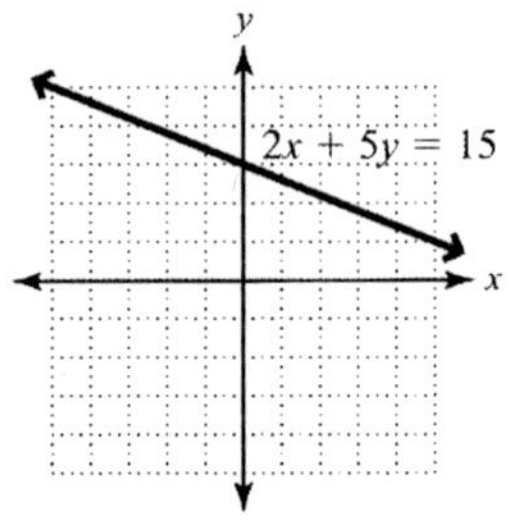

28. $y - 4x + 3 = 0$

$$y = 4x - 3$$

$m = 4$; y-intercept: $(0, -3)$

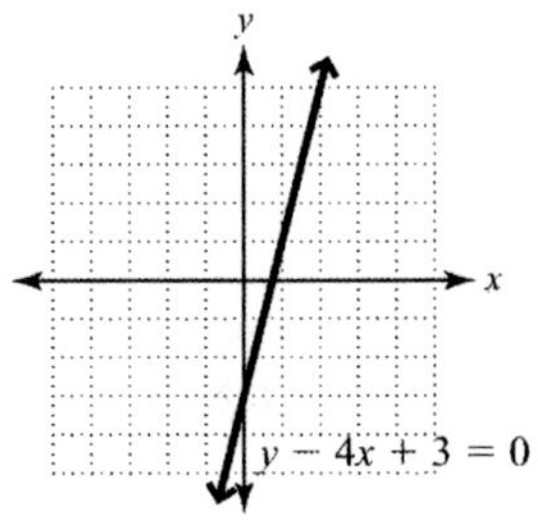

29. $(1, 8), (4, -1)$

$$m = \frac{y_2 - y_1}{x_2 - x_1} = \frac{-1 - 8}{4 - 1} = \frac{-9}{3} = -3$$

30. $(3, -1), (3, 2)$

$$m = \frac{y_2 - y_1}{x_2 - x_1} = \frac{2 - (-1)}{3 - 3} = \frac{2 + 1}{0} = \frac{3}{0} \text{ is undefined}$$

31. $(7, -4), (2, -9)$

$$m = \frac{y_2 - y_1}{x_2 - x_1} = \frac{-9 - (-4)}{2 - 7} = \frac{-9 + 4}{-5} = \frac{-5}{-5} = 1$$

32. $(-1, -1), (3, -1)$

$$m = \frac{y_2 - y_1}{x_2 - x_1} = \frac{-1 - (-1)}{3 - (-1)} = \frac{-1 + 1}{3 + 1} = \frac{0}{4} = 0$$

33. $m = 2$; $(0, -4)$

$$y = mx + b$$
$$y = 2x - 4$$

34. $m = -\frac{2}{5}$; $(0, 4)$

$$y = mx + b$$
$$y = -\frac{2}{5}x + 4$$

35. $m = -0.3$; $(0, -1)$

$$y = mx + b$$
$$y = -0.3x - 1$$

36. $m = -3$; $(0, 0)$

$$y = mx + b$$
$$y = -3x$$

37. $m = \frac{2}{5}$; $(0, 2)$; $y = \frac{2}{5}x + 2$

38. $m = -2$; $(0, -1)$; $y = -2x - 1$

39. Find the slope: $m = \frac{9 - 3}{5 - 3} = \frac{6}{2} = 3$

a) $y - y_1 = m(x - x_1)$

$$y - 3 = 3(x - 3)$$
$$y - 3 = 3x - 9$$
$$y = 3x - 6$$

b) $y = 3x - 6$

$$-3x + y = -6$$
$$3x - y = 6$$

40. Find the slope: $m = \dfrac{-5-2}{-4-(-2)} = \dfrac{-7}{-4+2} = \dfrac{-7}{-2} = \dfrac{7}{2}$

a) $y - y_1 = m(x - x_1)$

$$y - 2 = \frac{7}{2}(x - (-2))$$
$$y - 2 = \frac{7}{2}(x + 2)$$
$$y - 2 = \frac{7}{2}x + 7$$
$$y = \frac{7}{2}x + 9$$

b) $y = \dfrac{7}{2}x + 9$

$$2y = 7x + 18$$
$$-7x + 2y = 18$$
$$7x - 2y = -18$$

41. Find the slope: $m = \dfrac{-3-(-2)}{-2-4} = \dfrac{-3+2}{-6} = \dfrac{-1}{-6} = \dfrac{1}{6}$

a) $y - y_1 = m(x - x_1)$

$$y - (-2) = \frac{1}{6}(x - 4)$$
$$y + 2 = \frac{1}{6}x - \frac{4}{6}$$
$$y + 2 = \frac{1}{6}x - \frac{2}{3}$$
$$y = \frac{1}{6}x - \frac{8}{3}$$

b) $y = \dfrac{1}{6}x - \dfrac{8}{3}$

$$6y = x - 16$$
$$-x + 6y = -16$$
$$x - 6y = 16$$

42. Find the slope: $m = \dfrac{6-0}{0-6} = \dfrac{6}{-6} = -1$

a) $y - y_1 = m(x - x_1)$

$$y - 6 = -(x - 0)$$
$$y - 6 = -x$$
$$y = -x + 6$$

b) $y = -x + 6$

$$x + y = 6$$

43. Because these lines have equal slopes and different y-intercepts, they are parallel.

44. $7x + y = -9$ $\quad$ $7x - y = -6$

$$y = -7x - 9 \qquad -y = -7x - 6$$
$$y = 7x + 6$$

These slopes are neither equal nor of the form $\dfrac{a}{b}$ and $-\dfrac{b}{a}$. Therefore, these lines are neither parallel nor perpendicular.

45. Because the slopes are $\dfrac{4}{3}$ and $-\dfrac{3}{4}$, these lines are perpendicular.

46. $x + y = -1$ $\quad$ $x - y = 5$

$$y = -x - 1 \qquad -y = -x + 5$$
$$y = x - 5$$

Because the slopes are -1 and 1, these lines are perpendicular.

47. $m = -1$

$$y - y_1 = m(x - x_1)$$
$$y - 8 = -(x - 2)$$
$$y - 8 = -x + 2$$
$$y = -x + 10$$

48. $m = 6.2$

$$y - y_1 = m(x - x_1)$$
$$y - (-5) = 6.2(x - 3)$$
$$y + 5 = 6.2x - 18.6$$
$$y = 6.2x - 23.6$$

49. $m = \dfrac{2}{3}$

$$y - y_1 = m(x - x_1)$$
$$y - 0 = \frac{2}{3}(x - 4)$$
$$y = \frac{2}{3}x - \frac{8}{3}$$

50. $m = -3$

$$y - y_1 = m(x - x_1)$$
$$y - (-2) = -3(x - (-2))$$
$$y + 2 = -3(x + 2)$$
$$y + 2 = -3x - 6$$
$$y = -3x - 8$$

51. The slope of the given equation is $m = -\frac{3}{5}$, so the slope of a parallel line is $m = -\frac{3}{5}$. The *y*-intercept is $(0, -2)$.

$$y = mx + b$$
$$y = -\frac{3}{5}x - 2$$

52. Find *m*. $x - 5y = 10$

$$-5y = -x + 10$$
$$y = \frac{1}{5}x - 2$$

The slope of the given equation is $m = \frac{1}{5}$, so the slope of a parallel line is $m = \frac{1}{5}$.

$$y - y_1 = m(x - x_1)$$
$$y - 5 = \frac{1}{5}(x - 2)$$
$$y - 5 = \frac{1}{5}x - \frac{2}{5}$$
$$y = \frac{1}{5}x + \frac{23}{5}$$

53. The slope of the given equation is $m = -\frac{2}{3}$, so the slope of a perpendicular line is $m = \frac{3}{2}$.

$$y - y_1 = m(x - x_1)$$
$$y - (-2) = \frac{3}{2}(x - (-1))$$
$$y + 2 = \frac{3}{2}(x + 1)$$
$$y + 2 = \frac{3}{2}x + \frac{3}{2}$$
$$y = \frac{3}{2}x - \frac{1}{2}$$

54. The slope of the given equation is $m = 3$, so the slope of a perpendicular line is $m = -\frac{1}{3}$.

$$y - (-5) = -\frac{1}{3}(x - (-3))$$
$$y + 5 = -\frac{1}{3}(x + 3)$$
$$y + 5 = -\frac{1}{3}x - 1$$
$$y = -\frac{1}{3}x - 6$$

55. Test $(-3, -1)$.

$$3x + 5y > 1$$
$$3(-3) + 5(-1) \overset{?}{>} 1$$
$$-9 - 5 \overset{?}{>} 1$$
$$-14 > 1$$

This statement is false. No, $(-3, -1)$ is not a solution.

56. Test $(-4, 2)$.

$$y \geq 3x + 1$$
$$2 \overset{?}{\geq} 3(-4) + 1$$
$$2 \overset{?}{\geq} -12 + 1$$
$$2 \geq -11$$

This statement is true. Yes, $(-4, 2)$ is a solution.

57. $y < -2x + 5$

Begin by graphing the related equation $y = -2x + 5$ with a dashed line. Now choose $(0,0)$ as a test point.

$$\begin{aligned} y &< -2x+5 \\ 0 &\overset{?}{<} -2(0)+5 \\ 0 &\overset{?}{<} 0+5 \\ 0 &< 5 \end{aligned}$$

Because $(0,0)$ satisfies the inequality, shade the region which includes $(0,0)$.

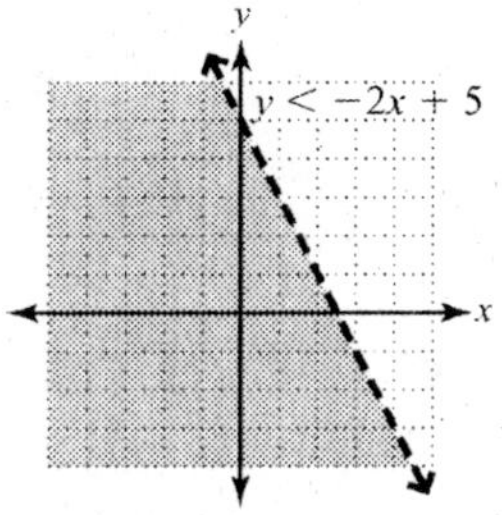

58. $3x - 4y > 12$

$$-4y > -3x + 12$$

$$y < \frac{3}{4}x - 3$$

Begin by graphing the related equation $y = \frac{3}{4}x - 3$ with a dashed line. Now choose $(0,0)$ as a test point.

$$\begin{aligned} 3x - 4y &> 12 \\ 3(0) - 4(0) &\overset{?}{>} 12 \\ 0 - 0 &\overset{?}{>} 12 \\ 0 &> 12 \end{aligned}$$

Because $(0,0)$ does not satisfy the inequality, shade the side of the line on the opposite side from $(0,0)$.

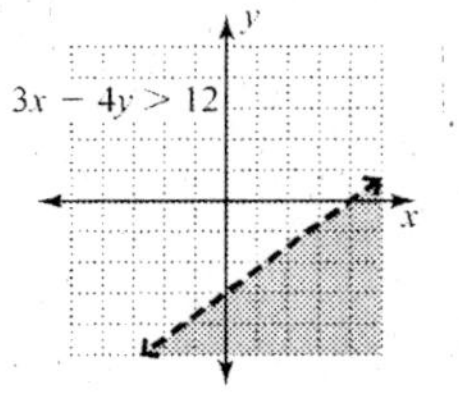

59. $-2x - 5y \le -10$

$$-5y \le 2x - 10$$

$$y \ge -\frac{2}{5}x + 2$$

Begin by graphing the related equation $y = -\frac{2}{5}x + 2$ with a solid line. Now choose $(0,0)$ as a test point.

$$\begin{aligned} -2x - 5y &\le -10 \\ -2(0) - 5(0) &\overset{?}{\le} -10 \\ 0 - 0 &\overset{?}{\le} -10 \\ 0 &\le -10 \end{aligned}$$

Because $(0,0)$ does not satisfy the inequality, shade the side of the line on the opposite side from $(0,0)$.

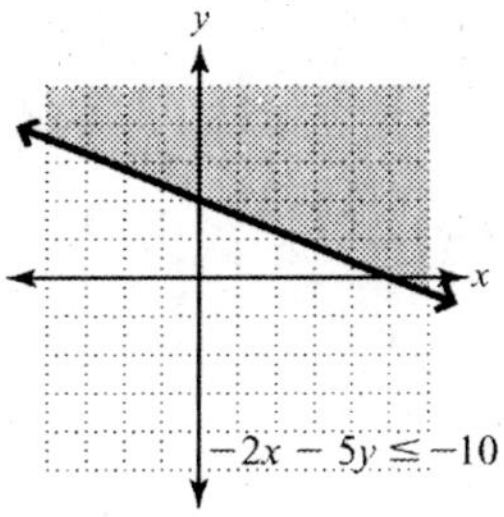

60. $y > \frac{4}{5}x$

Begin by graphing the related equation $y = \frac{4}{5}x$ with a dashed line. Now choose $(1,0)$ as a test point.

$$\begin{aligned} y &> \frac{4}{5}x \\ 0 &\overset{?}{>} \frac{4}{5}(1) \\ 0 &> \frac{4}{5} \end{aligned}$$

Because $(1,0)$ does not satisfy the inequality, shade the side of the line on the opposite side from $(1,0)$.

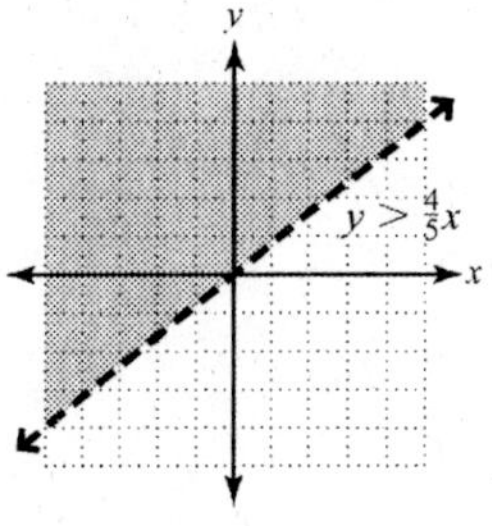

61. $x+y \geq 3$

$y \geq -x+3$

Begin by graphing the related equation $y=-x+3$ with a solid line. Now choose $(0,0)$ as a test point.

$$x+y \geq 3$$
$$0+0 \overset{?}{\geq} 3$$
$$0 \geq 3$$

Because $(0,0)$ does not satisfy the inequality, shade the side of the line on the opposite side from $(0,0)$.

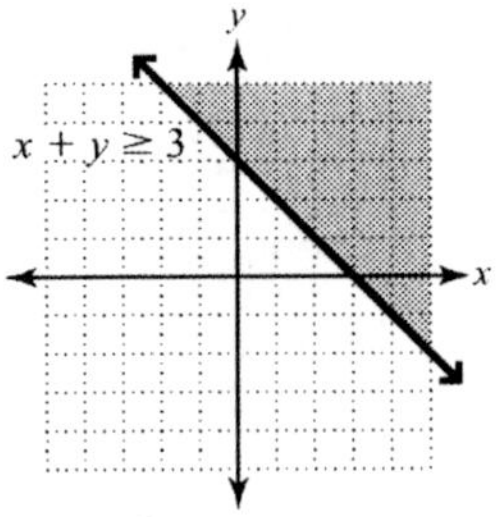

62. $y \geq -3$

Begin by graphing the related equation $y=-3$ with a solid line. Now choose $(0,0)$ as a test point.

$$y \geq -3$$
$$0 \overset{?}{\geq} -3$$

Because $(0,0)$ satisfies the inequality, shade the region which includes $(0,0)$.

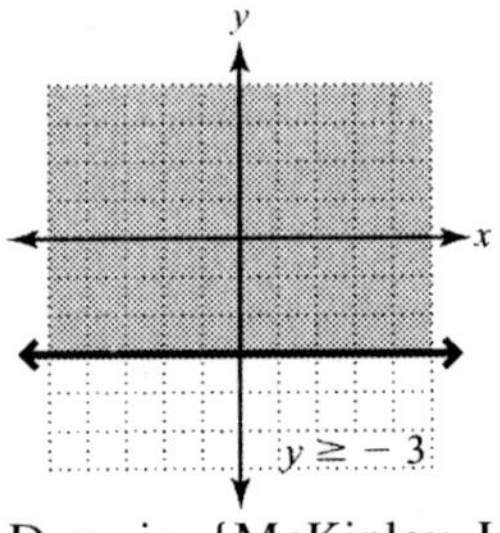

63. Domain: {McKinley, Logan, Pico de Orizaba, St. Elias, Popocatepetl}

Range: {20,320, 19,551, 18,555, 18,008, 17,930}

It is a function because every element in the domain is paired with exactly one element in the range.

64. Domain: {21, 23, 32, 35}

Range: {California, Indiana, New York, Ohio, Pennsylvania}

It is not a function because an element in the domain is paired with more than one element in the range.

65. Domain: $\{x|-4 \leq x \leq 5\}$ or $[-4,5]$

Range: $\{-2,2,3\}$

It is a function because a vertical line at each x-value would intersect the graph at only one point.

66. Domain: $\{x|-3 \leq x \leq 3\}$ or $[-3,3]$

Range: $\{y|0 \leq y \leq 3\}$ or $[0,3]$

It is a function because a vertical line at each x-value would intersect the graph at only one point.

67. Domain: $\mathbb{R}$ or $(-\infty,\infty)$

Range: $\mathbb{R}$ or $(-\infty,\infty)$

It is a function because a vertical line at each x-value would intersect the graph at only one point.

68. Domain: $\{x|x \leq 3\}$ or $(-\infty,3]$

Range: $\mathbb{R}$ or $(-\infty,\infty)$

It is not a function because there are values in the domain that correspond to two values in the range.

69. a) $f(-3)=-2$ b) $f(2)=3$

c) $f(0)=3$ d) $f(5)=2$

70. a) $f(-3)=0$ b) $f(0)=3$

c) $f(3)=0$ d) $f(4)$ undefined

71. $f(x)=x^2-4$

a) $f(2)=2^2-4$
$=4-4$
$=0$

b) $f(0)=0^2-4$
$=0-4$
$=-4$

c) $f(-3)=(-3)^2-4$
$=9-4$
$=5$

d) $f(n)=n^2-4$

72. $g(x)=\dfrac{x-3}{x+5}$

a) $g(2)=\dfrac{2-3}{2+5}$
$=-\dfrac{1}{7}$

b) $g(3)=\dfrac{3-3}{3+5}$
$=0$

c) $g(-5)=\dfrac{-5-3}{-5+5}$
$=\dfrac{-8}{0}$
undefined

73. $m=\dfrac{10}{12}=\dfrac{5}{6}$

74. a) $m=\dfrac{12.2-23.1}{7-0}=\dfrac{-10.9}{7}\approx -1.557$

b) $y=-1.557x+23.1$

c) Year 2012 would be $x = 13$
$y=-1.557x+23.1$
$y=-1.557(13)+23.1$
$=2.86$
The model predicts that only 2.86% of 12th-grade students will smoke daily in 2012.

75. a) $35x+50y\geq 70{,}000$

b)

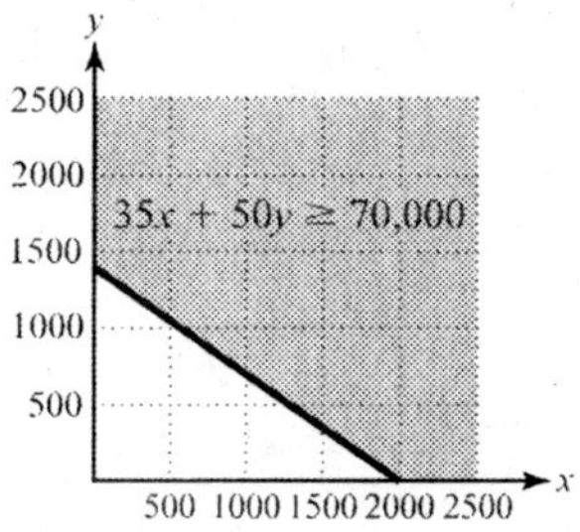

c) All combinations of chair sales that make the company break even ($70,000 revenue).
d) All combinations of chair sales that make the company a profit (more than $70,000 revenue).
e) Answers may vary. Some possible answers are (1000, 700) and (2000, 0).
f) Answers may vary. Two examples are (2000, 500) and (2000, 1000).

76. a) $c(t)=25t+75$

b) $c(1.5)=25(1.5)+75$
$=37.5+75$
$=\$112.50$

c) $150=25t+75$
$75=25t$
$3\text{ hr.}=t$

d)

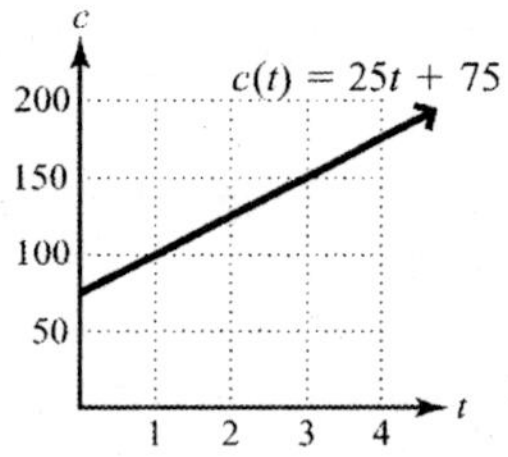

Chapter 3 Practice Test

1. $A(0, 5)$ because it is on the y-axis and is 5 units up from the origin.
$B(3, 3)$ because it is 3 units to the right and 3 units up from the origin.
$C(-2, -3)$ because it is 2 units to the left and 3 units down from the origin.
$D(4, -2)$ because it is 4 units to the right and 2 units down from the origin.

2. Quadrant IV because the first coordinate is positive and the second coordinate is negative.

3. No
$5 \stackrel{?}{=} -\dfrac{1}{4}(-3)+3$
$5 \stackrel{?}{=} \dfrac{3}{4}+3$
$5 \neq 3\dfrac{3}{4}$

4. $y=-\dfrac{4}{3}x+5$
$m=-\dfrac{4}{3}; b=(0,5)$

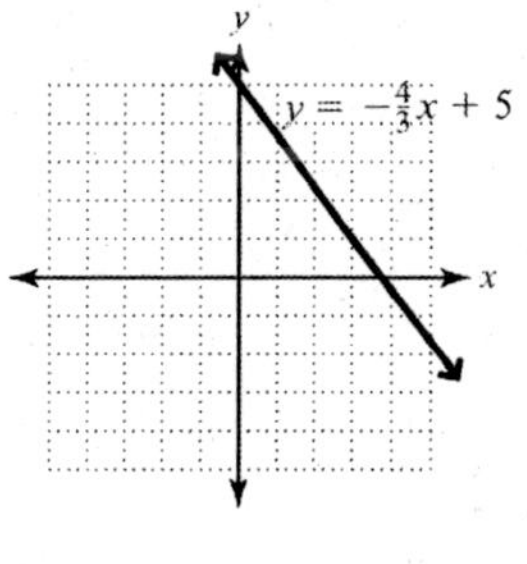

5. $x-2y=-8$

$-2y=-x-8$

$y=\frac{1}{2}x+4$

$m=\frac{1}{2};b=(0,4)$

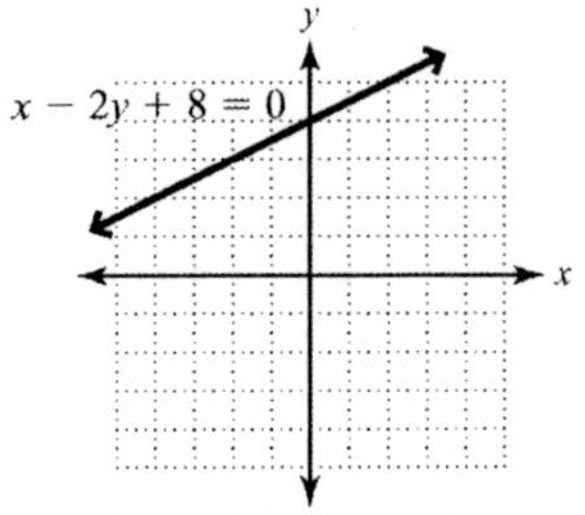

6. Let $(x_1,y_1)=(-1,-4)$ and $(x_2,y_2)=(-2,-4)$.

$m=\frac{y_2-y_1}{x_2-x_1}=\frac{-4-(-4)}{-2-(-1)}=\frac{-4+4}{-2+1}=\frac{0}{-1}=0$

7. Let $(x_1,y_1)=(-3,-9)$ and $(x_2,y_2)=(4,-1)$.

$m=\frac{y_2-y_1}{x_2-x_1}=\frac{-1-(-9)}{4-(-3)}=\frac{-1+9}{4+3}=\frac{8}{7}$

8. $m=\frac{2}{7}$; y-intercept: $(0,b)=(0,5)$

$y=mx+b$

$y=\frac{2}{7}x+5$

9. Let $(x_1,y_1)=(4,2)$ and $(x_2,y_2)=(-5,-1)$.

$m=\frac{y_2-y_1}{x_2-x_1}=\frac{-1-2}{-5-4}=\frac{-3}{-9}=\frac{1}{3}$

$y-y_1=m(x-x_1)$

$y-2=\frac{1}{3}(x-4)$

$y-2=\frac{1}{3}x-\frac{4}{3}$

$y=\frac{1}{3}x+\frac{2}{3}$

10. Let $(x_1,y_1)=(1,4)$ and $(x_2,y_2)=(-3,-1)$.

$m=\frac{y_2-y_1}{x_2-x_1}=\frac{-1-4}{-3-1}=\frac{-5}{-4}=\frac{5}{4}$

$y-y_1=m(x-x_1)$

$y-4=\frac{5}{4}(x-1)$

$4(y-4)=4\left(\frac{5}{4}(x-1)\right)$

$4(y-4)=5(x-1)$

$4y-16=5x-5$

$-5x+4y-16=-5$

$-5x+4y=11$

$-1(-5x+4y)=-1(11)$

$5x-4y=-11$

11. The slope of the first line is $m=\frac{3}{4}$ and the slope of the second line is $m=\frac{4}{3}$. Because these slopes are neither equal nor negative reciprocals of one another, the graphs are neither parallel nor perpendicular.

12.

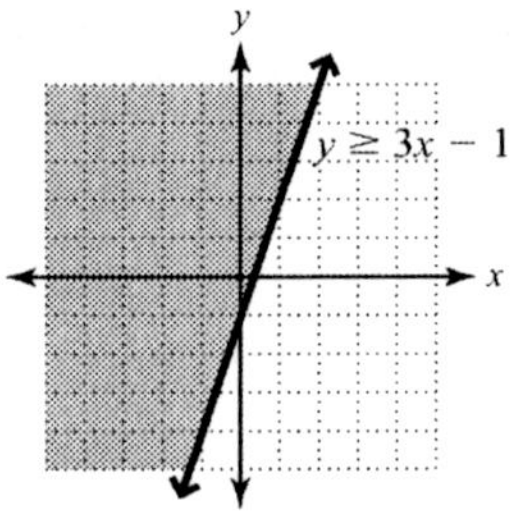

13.

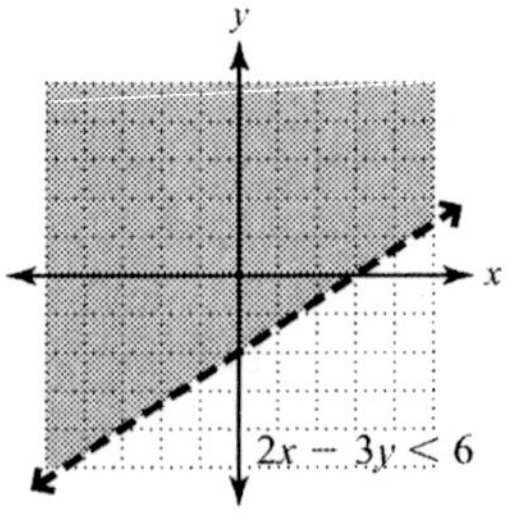

14. Domain: {Spanish, Chinese, French, German, Tagalog}
Range: {28.1, 2.0, 1.6, 1.4, 1.2}
Because every value in the domain is paired with exactly one value in the range, it is a function.

15. Domain: $\{x|0 \le x \le 3\}$, Range: $\{y|-3 \le y \le 3\}$
It is not a function because at least one element in the domain is paired with more than one element in the range.

16. $f(x) = 2x^2 - 7$

a) $f(-2) = 2(-2)^2 - 7$
$= 8 - 7$
$= 1$

b) $f(3) = 2(3)^2 - 7$
$= 18 - 7$
$= 11$

c) $f(t) = 2t^2 - 7$

17. $f(x) = \dfrac{4x}{x+5}$

a) $f(-1) = \dfrac{4(-1)}{-1+5}$
$= \dfrac{-4}{4}$
$= -1$

b) $f(5) = \dfrac{4(5)}{5+5}$
$= \dfrac{20}{10}$
$= 2$

c) $f(-5) = \dfrac{4(-5)}{-5+5}$
$= \dfrac{-20}{0}$
undefined

18. a) Domain: $\{x|-3 \le x \le 3\}$ or $[-3,3]$;
Range: $\{-2, 1, 3\}$
b) From the graph, we find that $f(1) = 3$.

19. a) $2l + 2w \le 1000$

b)

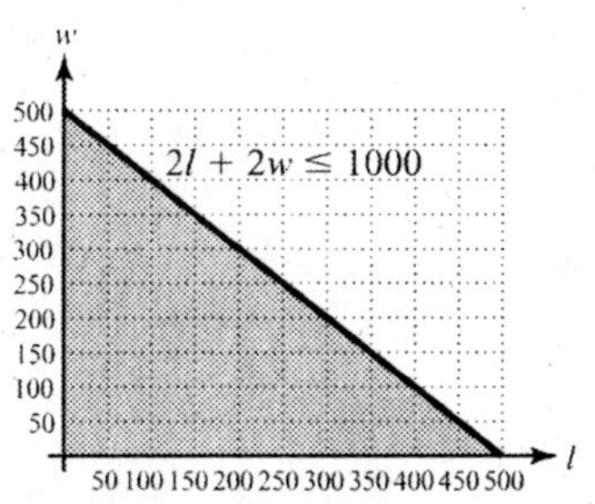

c) Answers may vary. Two possible answers are 300 ft. by 200 ft. or 200 ft. by 200 ft.

20. a) $c(w) = 0.45w + 3$

b)

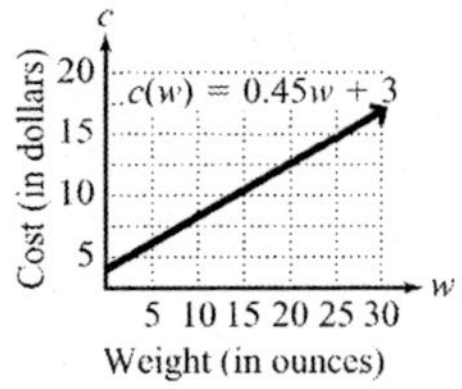

c) $c(32) = 0.45(32) + 3$
$= 14.4 + 3$
$= \$17.40$

Chapters 1 – 3 Cumulative Review

1. False; in interval notation, the inequality $x > 0$ is expressed as $(0, \infty)$.
2. True
3. True
4. 10
5. principal or non negative
6. We perform operations in the following order:
1) Within grouping symbols beginning with the innermost: parentheses ( ), brackets [ ], braces { }, absolute value $|\ |$, above and/or below fraction bars, and radicals $\sqrt{\ }$.
2) Exponents/roots from left to right, in order as they occur.
3) Multiplication/division from left to right, in order as they occur.
4) Addition/subtraction from left to right in order as they occur.
7. $A \cap B = \varnothing$
$A \cup B = \{w, e, l, o, v, m, a, t, h\}$
8. $(-3)^2 = (-3)(-3) = 9$
9. $\sqrt[3]{27} = 3$
10. $-|4+3| - 8(1-5)^2 = -|7| - 8(-4)^2$
$= -7 - 8(16)$
$= -7 - 128$
$= -135$
11. $\dfrac{3x}{x-5} = \dfrac{3(5)}{5-5} = \dfrac{15}{0}$ is undefined
12. a) $(-4n+8) - 5n$ b) $-9n + 8$

c) $-9(4)+8=-36+8=-28$

13. Multiplicative inverse

14. Distributive property

15. $6(5+y)-3y=2y+1$
$$30+6y-3y=2y+1$$
$$30+3y=2y+1$$
$$30+y=1$$
$$y=-29$$

16. $2x-1=x+8$ and $2x-1=-(x+8)$
$$x=9 \qquad 2x-1=-x-8$$
$$3x=-7$$
$$x=-\frac{7}{3}$$

17. $3+2x\le 8-4x$
$$3+6x\le 8$$
$$6x\le 5$$
$$x\le\frac{5}{6}$$

a) $\left\{x\middle|x\le\frac{5}{6}\right\}$

b) $\left(-\infty,\frac{5}{6}\right]$

c) 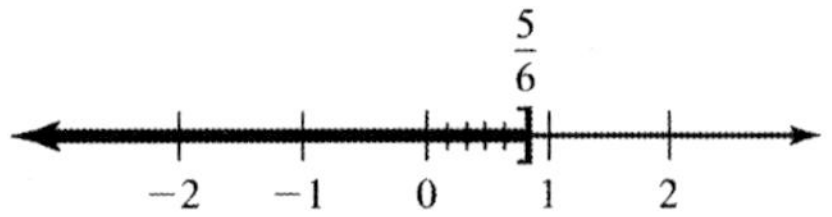

18. $3|x+1|-7>2$
$$3|x+1|>9$$
$$|x+1|>3$$
$$x+1<-3 \quad\text{or}\quad x+1>3$$
$$x<-4 \qquad\qquad x>2$$

a) $\{x|x<-4\text{ or }x>2\}$

b) $(-\infty,-4)\cup(2,\infty)$

c) 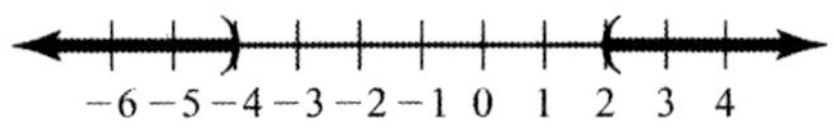

19. $d=rt$
$$\frac{d}{r}=t$$

20. $m=\dfrac{-1-(-1)}{-2-3}=\dfrac{-1+1}{-5}=\dfrac{0}{-5}=0$

21. $2x+y=6 \qquad y=-\frac{1}{2}x+1$
$$y=-2x+6$$

These slopes are neither equal nor of the form $\frac{a}{b}$ and $-\frac{b}{a}$. Therefore, these lines are neither parallel nor perpendicular.

22. $2x-y=3$
$$-y=-2x+3$$
$$y=2x-3$$

The slope of the given equation is $m=2$, so the slope of a parallel line is $m=2$.
$$y-y_1=m(x-x_1)$$
$$y-8=2(x-1)$$
$$y-8=2x-2$$
$$y=2x+6$$

23. $2x-y\ge 6$
$$-y\ge -2x+6$$
$$y\le 2x-6$$

Begin by graphing the related equation $y=2x-6$ with a solid line. Now choose $(0,0)$ as a test point.
$$2x-y \ge 6$$
$$2(0)-0 \overset{?}{\ge} 6$$
$$0-0 \overset{?}{\ge} 6$$
$$0 \ge 6$$

Because $(0,0)$ does not satisfy the inequality, shade the side of the line on the opposite side from $(0,0)$.

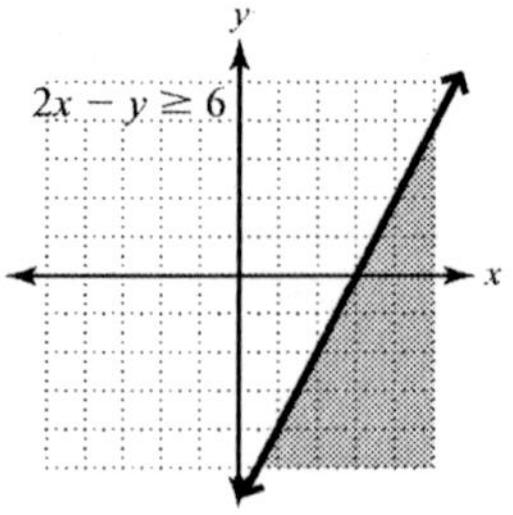

24. $f(x)=3x^2+2$
$$f(-1)=3(-1)^2+2$$
$$=3(1)+2$$
$$=3+2$$
$$=5$$

25. $f(x)=\frac{2}{5}x-1$

Think of the function as the equation $y=\frac{2}{5}x-1$ and use the fact that the slope is $m=\frac{2}{5}$ and the y-intercept is -1.

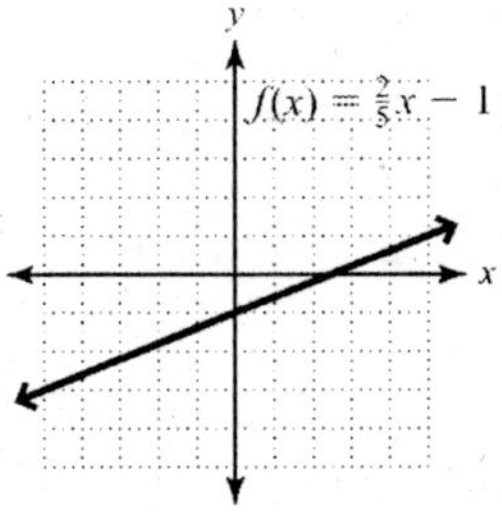

26. $\frac{3(4)+3(4)+4(3)}{10}=\frac{12+12+12}{10}=\frac{36}{10}=3.6$

27. Let the first angle be x and the second angle be $180-x$.

$x=2(180-x)$

$x=360-2x$

$3x=360$

$x=120$

The angles are 120 and 60 degrees.

28. Create a table.

	amount	value	total cost
standard	$2x$	35	$70x$
deluxe	x	75	$75x$

Translate into an equation and solve.

$70x+75x=580$

$145x=580$

$x=4$

He ordered 8 standard phones and 4 deluxe phones

29. Create a table.

Solutions	Concen- trate	Vol. of Solution	Vol. of HCl
10%	0.10	50	$0.10(50)$
25%	0.25	x	$0.25(x)$
15%	0.15	$x+50$	$0.15(x+50)$

Translate into an equation and solve.

$0.10(50)+0.25x=0.15(x+50)$

$5+0.25x=0.15x+7.5$

$5+0.10x=7.5$

$0.10x=2.5$

$x=25$

Janet will need 25 ml of 25% solution.

30. $60 – $40 = $20

$0.04t\le 20$

$t\le 500$

Maximum: 500 + 500 = 1000 minutes or less

Chapter 4

Systems of Linear Equations and Inequalities

Exercise Set 4.1

1. Replace each variable in each equation with its corresponding value. Verify that each equation is true.

3. The lines are parallel.

5. a) y in the second equation b) $x + 2$

7. Is $(-1, 2)$ a solution of $\begin{cases} x - y = -3 \\ x + y = 1 \end{cases}$?

Equation 1	Equation 2
$x - y = -3$	$x + y = 1$
$-1 - 2 \overset{?}{=} -3$	$-1 + 2 \overset{?}{=} 1$
$-3 = -3$	$1 = 1$

Yes, $(-1, 2)$ is a solution for the system because it satisfies both equations.

9. Is $(-2, 3)$ a solution of $\begin{cases} 3x + 4y = 6 \\ x - 4y = 8 \end{cases}$?

Equation 1	Equation 2
$3x + 4y = 6$	$x - 4y = 8$
$3(-2) + 4(3) \overset{?}{=} 6$	$(-2) - 4(3) \overset{?}{=} 8$
$-6 + 12 \overset{?}{=} 6$	$-2 - 12 \overset{?}{=} 8$
$6 = 6$	$-14 \neq 8$

No, $(-2, 3)$ is not a solution for the system because it does not satisfy both equations.

11. Is $\left(-\frac{3}{4}, -\frac{2}{3}\right)$ a solution of $\begin{cases} 4x + 3y = -5 \\ 12x - 6y = -5 \end{cases}$?

Equation 1	Equation 2
$4x + 3y = -5$	$12x - 6y = -5$
$4\left(-\frac{3}{4}\right) + 3\left(-\frac{2}{3}\right) \overset{?}{=} -5$	$12\left(-\frac{3}{4}\right) - 6\left(-\frac{2}{3}\right) \overset{?}{=} -5$
$-3 - 2 \overset{?}{=} -5$	$-9 + 4 \overset{?}{=} -5$
$-5 = -5$	$-5 = -5$

Yes, $\left(-\frac{3}{4}, -\frac{2}{3}\right)$ is a solution for the system because it satisfies both equations.

13. Is $(3, -4)$ a solution of $\begin{cases} 3x + 2y = 1 \\ \frac{1}{3}x + \frac{1}{2}y = 3 \end{cases}$?

Equation 1	Equation 2
$3x + 2y = 1$	$\frac{1}{3}x + \frac{1}{2}y = 3$
$3(3) + 2(-4) \overset{?}{=} 1$	$\frac{1}{3}(3) + \frac{1}{2}(-4) \overset{?}{=} 3$
$9 - 8 \overset{?}{=} 1$	$1 - 2 \overset{?}{=} 3$
$1 = 1$	$-1 \neq 3$

No, $(3, -4)$ is not a solution for the system because it does not satisfy both equations.

15. $\begin{cases} x + y = 5 \\ x - y = 3 \end{cases}$

17. $\begin{cases} 2l + 2w = 50 \\ w = l - 2 \end{cases}$

19. $\begin{cases} x + y = 5 \\ x - y = 3 \end{cases}$ Graph each equation.

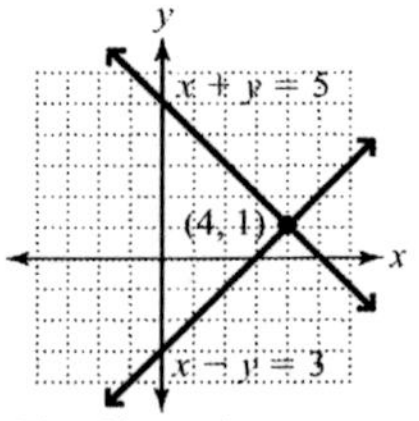

The lines intersect at a single point, which appears to be (4, 1).

21. $\begin{cases} y = 2x + 5 \\ y = -x - 4 \end{cases}$ Graph each equation.

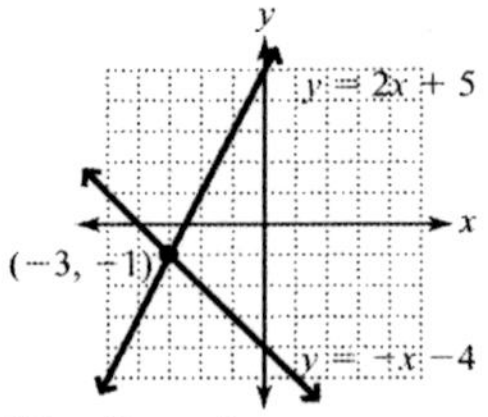

The lines intersect at a single point, which appears to be $(-3, -1)$.

23. $\begin{cases} 2x - y = 3 \\ 2x - y = 8 \end{cases}$ Graph each equation.

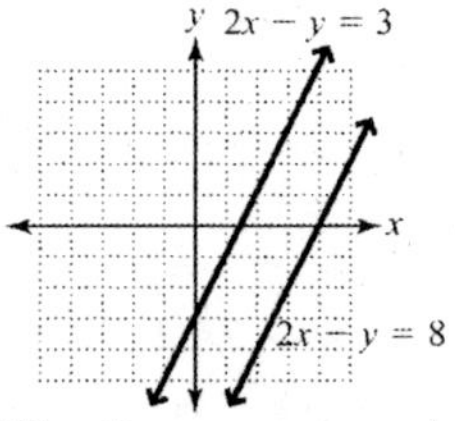

The lines appear to be parallel, so the system has no solution.

25. $\begin{cases} 3x + y = 4 \\ 6x + 2y = 8 \end{cases}$ Graph each equation.

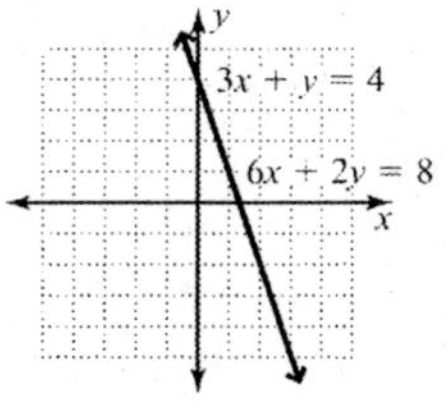

The lines appear to be identical, so the solution is the set of all ordered pairs that solve $3x + y = 4$.

27. $\begin{cases} x = 4 \\ y = -2 \end{cases}$ Graph each equation.

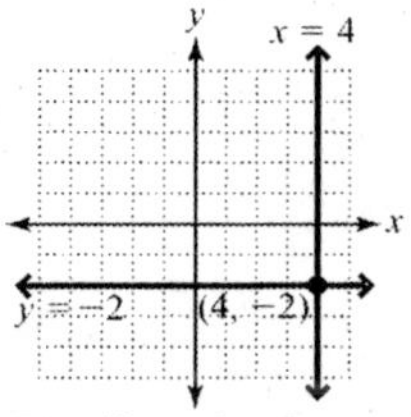

The lines intersect at a single point, which appears to be $(4,-2)$.

29. a) Because these lines intersect in a single point, this system is consistent with independent equations.
b) This system has one solution.

31. a) Because the lines are parallel, this system is inconsistent.
b) This system has no solution.

33. a) Because these lines coincide, this system is consistent with dependent equations.
b) This system has an infinite number of solutions.

35. $\begin{cases} y = -x \\ y - x = 6 \end{cases} \rightarrow \begin{cases} y = -x \\ y = x + 6 \end{cases}$

The lines have different slopes. The graphs are different. This system is consistent with independent equations.

37. $\begin{cases} x + 3y = 1 \\ 2x + 6y = 2 \end{cases} \rightarrow \begin{cases} y = -\frac{1}{3}x + \frac{1}{3} \\ y = -\frac{1}{3}x + \frac{1}{3} \end{cases}$

The lines have the same slope and the same y-intercept. The graphs are identical. This system is consistent with dependent equations.

39. $\begin{cases} 3x + 2y = 12 \\ 6x + 4y = -12 \end{cases} \rightarrow \begin{cases} y = -\frac{3}{2}x + 6 \\ y = -\frac{3}{2}x - 3 \end{cases}$

The lines have the same slope but different y-intercepts. The graphs are parallel lines. This system is inconsistent.

41. $\begin{cases} 3x - y = 2 \\ y = 2x \end{cases}$

$3x - (2x) = 2 \qquad y = 2(2)$

$x = 2 \qquad y = 4$

Solution: (2, 4)

43. $\begin{cases} x + y = 1 \\ y = -2x - 1 \end{cases}$

$x + y = 1 \qquad y = -2x - 1$

$x + (-2x - 1) = 1 \qquad y = -2(-2) - 1$

$-x - 1 = 1 \qquad y = 4 - 1$

$-x = 2 \qquad y = 3$

$x = -2$

Solution: $(-2, 3)$

45. $\begin{cases} y = -\frac{3}{4}x \\ x - 8y = -7 \end{cases}$

$x - 8(y) = -7 \qquad y = -\frac{3}{4}x$

$x - 8\left(-\frac{3}{4}x\right) = -7 \qquad y = -\frac{3}{4}(-1)$

$x + 6x = -7 \qquad y = \frac{3}{4}$

$7x = -7$

$x = -1$

Solution: $\left(-1, \frac{3}{4}\right)$

47. $\begin{cases} 3x+4y=11 \\ x+2y=5 \end{cases}$

Solve the second equation for x: $x=5-2y$

$$\begin{aligned} 3(x)+4y&=11 \\ 3(5-2y)+4y&=11 \\ 15-6y+4y&=11 \\ 15-2y&=11 \\ -2y&=-4 \\ y&=2 \end{aligned} \qquad \begin{aligned} x&=5-2(y) \\ x&=5-2(2) \\ x&=5-4 \\ x&=1 \end{aligned}$$

Solution: $(1,2)$

49. $\begin{cases} 4x-y=-13 \\ 3x-4y=-13 \end{cases}$

Solve the first equation for y: $y=4x+13$

$$\begin{aligned} 3x-4(4x+13)&=-13 \\ 3x-16x-52&=-13 \\ -13x-52&=-13 \\ -13x&=39 \\ x&=-3 \end{aligned} \qquad \begin{aligned} y&=4(-3)+13 \\ y&=-12+13 \\ y&=1 \end{aligned}$$

Solution: $(-3, 1)$

51. $\begin{cases} 4x+3y=2 \\ 3x+2y=2 \end{cases}$

Solve the second equation for y.

$$\begin{aligned} 3x+2y&=2 \\ 2y&=-3x+2 \\ y&=-\frac{3}{2}x+1 \end{aligned}$$

$$\begin{aligned} 4x+3y&=2 \\ 4x+3\left(-\frac{3}{2}x+1\right)&=2 \\ 4x-\frac{9}{2}x+3&=2 \\ -\frac{1}{2}x&=-1 \\ x&=2 \end{aligned} \qquad \begin{aligned} y&=-\frac{3}{2}x+1 \\ y&=-\frac{3}{2}(2)+1 \\ y&=-3+1 \\ y&=-2 \end{aligned}$$

Solution: $(2,-2)$

53. $\begin{cases} x-3y=-4 \\ 5x-15y=-6 \end{cases}$

Solve the first equation for x: $x=3y-4$

$$\begin{aligned} 5(3y-4)-15y&=-6 \\ 15y-20-15y&=-6 \\ -20&=-6 \quad \text{false} \end{aligned}$$

Because this last statement is false, the system is inconsistent and has no solution.

55. $\begin{cases} x-2y=6 \\ -2x+4y=-12 \end{cases}$

Solve the first equation for x: $x=2y+6$

$$\begin{aligned} -2(2y+6)+4y&=-12 \\ -4y-12+4y&=-12 \\ -12&=-12 \quad \text{true} \end{aligned}$$

Notice that $-12=-12$ no longer has a variable and is true. The equations in the system are dependent; there are an infinite number of solutions that are all of the ordered pairs along $x-2y=6$.

57. Mistake: Did not distribute properly.
Correct: (3, 2)

59. $\begin{cases} x+y=1 \\ 2x-y=2 \end{cases}$

$$\begin{array}{r} x+y=1 \\ \underline{2x-y=2} \\ 3x=3 \\ x=1 \end{array} \qquad \begin{aligned} 1+y&=1 \\ y&=0 \end{aligned}$$

Solution: $(1,0)$

61. $\begin{cases} 2x+3y=9 \\ 4x+3y=15 \end{cases}$

$$\begin{array}{r} -(2x+3y=9) \\ \underline{4x+3y=15} \\ -2x-3y=-9 \\ \underline{4x+3y=15} \\ 2x=6 \\ x=3 \end{array} \qquad \begin{aligned} 2(3)+3y&=9 \\ 6+3y&=9 \\ 3y&=3 \\ y&=1 \end{aligned}$$

Solution: $(3,1)$

63. $\begin{cases} 4x + y = 8 \\ 5x + 3y = 3 \end{cases}$

$$\begin{aligned} -3(4x + y &= 8) \\ 5x + 3y &= 3 \\ \hline -12x - 3y &= -24 \\ 5x + 3y &= 3 \\ \hline -7x &= -21 \\ x &= 3 \end{aligned} \qquad \begin{aligned} 4(3) + y &= 8 \\ 12 + y &= 8 \\ y &= -4 \end{aligned}$$

Solution: $(3, -4)$

65. $\begin{cases} x + 3y = -1 \\ 3x + 6y = -1 \end{cases}$

$$\begin{aligned} -3(x + 3y &= -1) \\ 3x + 6y &= -1 \\ \hline -3x - 9y &= 3 \\ 3x + 6y &= -1 \\ \hline -3y &= 2 \\ y &= -\frac{2}{3} \end{aligned} \qquad \begin{aligned} x + 3\left(-\frac{2}{3}\right) &= -1 \\ x - 2 &= -1 \\ x &= 1 \end{aligned}$$

Solution: $\left(1, -\frac{2}{3}\right)$

67. $\begin{cases} 3x = 2y + 7 \\ 4x - 3y = 10 \end{cases}$

$$\begin{aligned} -3(3x - 2y &= 7) \\ 2(4x - 3y &= 10) \\ \hline -9x + 6y &= -21 \\ 8x - 6y &= 20 \\ \hline -x &= -1 \\ x &= 1 \end{aligned} \qquad \begin{aligned} 3(1) &= 2y + 7 \\ -4 &= 2y \\ -2 &= y \end{aligned}$$

Solution: $(1, -2)$

69. $\begin{cases} 3x - 4y = 2 \\ 4x + 5y = 6 \end{cases}$

$$\begin{aligned} 5(3x - 4y &= 2) \\ 4(4x + 5y &= 6) \\ \hline 15x - 20y &= 10 \\ 16x + 20y &= 24 \\ \hline 31x &= 34 \\ x &= \frac{34}{31} \end{aligned} \qquad \begin{aligned} 3x - 4y &= 2 \\ 3\left(\frac{34}{31}\right) - 4y &= 2 \\ \frac{102}{31} - 4y &= 2 \\ -4y &= -\frac{40}{31} \\ y &= \frac{10}{31} \end{aligned}$$

Solution: $\left(\frac{34}{31}, \frac{10}{31}\right)$

71. $\begin{cases} \frac{1}{5}x + \frac{1}{2}y = \frac{1}{5} \\ \frac{1}{2}x + \frac{1}{3}y = -\frac{4}{3} \end{cases}$

$$\begin{aligned} 10\left(\frac{1}{5}x + \frac{1}{2}y = \frac{1}{5}\right) \\ 6\left(\frac{1}{2}x + \frac{1}{3}y = -\frac{4}{3}\right) \\ \hline 3(2x + 5y = 2) \\ -2(3x + 2y = -8) \\ \hline 6x + 15y = 6 \\ -6x - 4y = 16 \\ \hline 11y = 22 \\ y = 2 \end{aligned} \qquad \begin{aligned} \frac{1}{5}x + \frac{1}{2}(2) &= \frac{1}{5} \\ \frac{1}{5}x + 1 &= \frac{1}{5} \\ \frac{1}{5}x &= -\frac{4}{5} \\ x &= -4 \end{aligned}$$

Solution: $(-4, 2)$

73. $\begin{cases} 0.4x - 0.3y = 1.3 \\ 0.3x + 0.5y = 1.7 \end{cases}$

$$\begin{aligned} 10(0.4x - 0.3y = 1.3) \\ 10(0.3x + 0.5y = 1.7) \\ \hline 5(4x - 3y = 13) \\ 3(3x + 5y = 17) \\ \hline 20x - 15y = 65 \\ 9x + 15y = 51 \\ \hline 29x = 116 \\ x = 4 \end{aligned} \qquad \begin{aligned} 0.4(4) - 0.3y &= 1.3 \\ 1.6 - 0.3y &= 1.3 \\ -0.3y &= -0.3 \\ y &= 1 \end{aligned}$$

Solution: $(4, 1)$

75. $\begin{cases} 2x+3y=6 \\ 4x-18=-6y \end{cases}$

$$-2(2x+3y=6)$$
$$4x+6y=18$$
$$-4x-6y=-12$$
$$4x+6y=18$$
$$0=6 \quad \text{false}$$

Both variables have been eliminated, and the resulting equation, $0=6$, is false. This system of equations is inconsistent. Therefore, there is no solution.

77. $\begin{cases} x+2y=4 \\ x-4=-2y \end{cases}$

$$x+2y=4$$
$$-1(x+2y=4)$$
$$x+2y=4$$
$$-x-2y=-4$$
$$0=0$$

Both variables have been eliminated, and the resulting equation, 0 = 0, is true. This means that the equations are dependent. There are an infinite number of solutions, which are all of the ordered pairs along the line $x+2y=4$.

79. Mistake: Subtracted 7 from 8 instead of adding.

Correct: $\left(\frac{15}{2}, \frac{1}{2}\right)$

81. a) 5000 b) \$4000 c) $n>5000$

83. a) $\begin{cases} c=45 \\ c=0.10n+30 \end{cases}$

b)

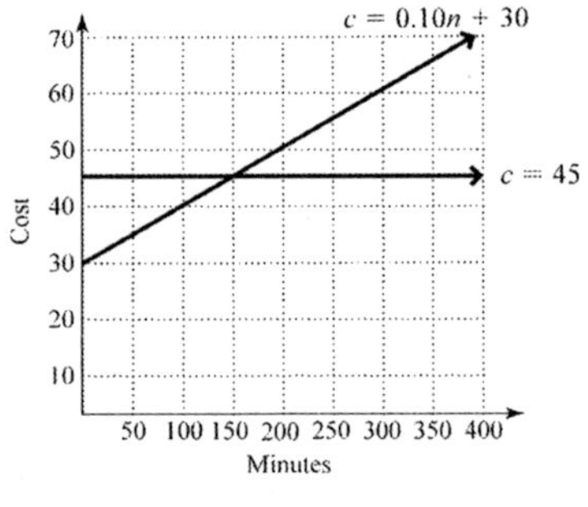

c) 150 min.
d) Plan 1
e) Plan 2

Review Exercises

1. $4(5x+7y-z)=20x+28y-4z$

2. $3x-4y+z-2x+y-z=x-3y$

3. $$x+3y=6$$
$$3y=-x+6$$
$$y=-\frac{1}{3}x+2$$

4. $$\frac{1}{3}x-2y=10$$
$$x-6y=30$$
$$x=6y+30$$

5. $$x+y+2z=7$$
$$x+(-1)+2(3)=7$$
$$x-1+6=7$$
$$x=2$$

6. $$x-3y+2z=6$$
$$-2-3y+2(4)=6$$
$$-2-3y+8=6$$
$$6-3y=6$$
$$-3y=0$$
$$y=0$$

Exercise Set 4.2

1. It doesn't matter which equations are chosen as long as the second pair is different from the first pair.

3. Answers may vary. A good choice is to eliminate z using equations 1 and 2 and 1 and 3. Multiplying equation 1 by –3 and then adding this result to both equations 2 and 3 takes very few steps.

5. Two parallel planes intersecting a third plane.

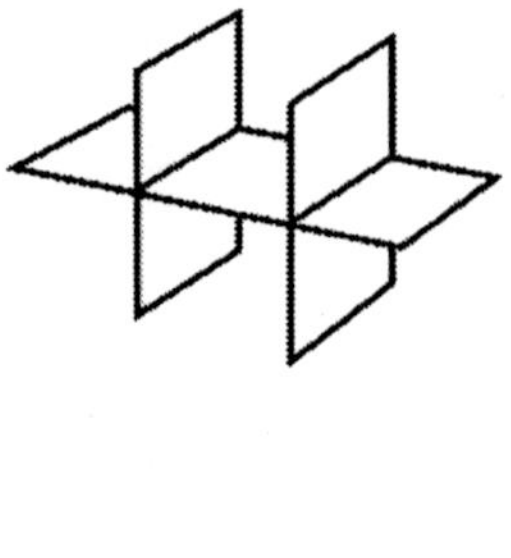

7. Is $(3,-1,1)$ a solution of $\begin{cases} x+y+z=3 \\ 2x-2y-z=7 \\ 2x+y-2z=3 \end{cases}$?

Equation 1: $x+y+z = 3$

$3+(-1)+1 \stackrel{?}{=} 3$

$3 = 3$

Equation 2: $2x-2y-z = 7$

$2(3)-2(-1)-1 \stackrel{?}{=} 7$

$6+2-1 \stackrel{?}{=} 7$

$7 = 7$

Equation 3: $2x+y-2z = 3$

$2(3)+(-1)-2(1) \stackrel{?}{=} 3$

$6-1-2 \stackrel{?}{=} 3$

$3 = 3$

Yes, $(3,-1,1)$ is a solution for the system because it satisfies all of the equations.

9. Is $(1,0,2)$ a solution of $\begin{cases} 2x+3y-3z=-4 \\ -2x+4y-z=-4 \\ 3x-4y+2z=5 \end{cases}$?

Equation 1: $2x+3y-3z = -4$

$2(1)+3(0)-3(2) \stackrel{?}{=} -4$

$2+0-6 \stackrel{?}{=} -4$

$-4 = -4$

Equation 2: $-2x+4y-z = -4$

$-2(1)+4(0)-(2) \stackrel{?}{=} -4$

$-2+0-2 \stackrel{?}{=} -4$

$-4 = -4$

Equation 3: $3x-4y+2z = 5$

$3(1)-4(0)+2(2) \stackrel{?}{=} 5$

$3-0+4 \stackrel{?}{=} 5$

$7 \neq 5$

No, $(1,0,2)$ is not a solution for the system because it does not satisfy all three equations.

11. Is $(2,-2,4)$ a solution of $\begin{cases} x+2y-z=-6 \\ 2x-3y+4z=26 \\ -x+2y-3z=-18 \end{cases}$?

Equation 1: $x+2y-z = -6$

$2+2(-2)-4 \stackrel{?}{=} -6$

$2-4-4 \stackrel{?}{=} -6$

$-6 = -6$

Equation 2: $2x-3y+4z = 26$

$2(2)-3(-2)+4(4) \stackrel{?}{=} 26$

$4+6+16 \stackrel{?}{=} 26$

$26 = 26$

Equation 3: $-x+2y-3z = -18$

$-(2)+2(-2)-3(4) \stackrel{?}{=} -18$

$-2-4-12 \stackrel{?}{=} -18$

$-18 = -18$

Yes, $(2,-2,4)$ is a solution for the system because it satisfies all of the equations.

13. $\begin{cases} x+y+z=5 & \text{Eqtn. 1} \\ 2x+y-2z=-5 & \text{Eqtn. 2} \\ x-2y+z=8 & \text{Eqtn. 3} \end{cases}$

Multiply equation 1 by -1 and add to equation 3. This makes equation 4.

$$\begin{array}{r} -x-y-z=-5 \\ \underline{x-2y+z=8} \\ -3y=3 \\ y=-1 \end{array} \quad \text{Eqtn. 4}$$

Multiply equation 1 by -2 and add to equation 2. This makes equation 5.

$$\begin{array}{r} -2x-2y-2z=-10 \\ \underline{2x+y-2z=-5} \\ -y-4z=-15 \end{array} \quad \text{Eqtn. 5}$$

Substitute equation 4 into equation 5 and solve for z.

$$-(-1)-4z=-15$$
$$1-4z=-15$$
$$-4z=-16$$
$$z=4$$

Substitute the values for y and z into equation 1 to solve for x.

$$x-1+4=5$$
$$x+3=5$$
$$x=2$$

Solution: $(2,-1,4)$

15. $\begin{cases} x+y+z=2 & \text{Eqtn. 1} \\ 4x-3y+2z=2 & \text{Eqtn. 2} \\ 2x+3y-2z=-8 & \text{Eqtn. 3} \end{cases}$

Multiply equation 1 by –2 and add to equation 3. This makes equation 4.

$$-2x-2y-2z=-4$$
$$\underline{2x+3y-2z=-8}$$
$$y-4z=-12 \quad \text{Eqtn. 4}$$

Multiply equation 1 by –4 and add to equation 2. This makes equation 5.

$$-4x-4y-4z=-8$$
$$\underline{4x-3y+2z=2}$$
$$-7y-2z=-6 \quad \text{Eqtn. 5}$$

Use equations 4 and 5 to make a system of equations in two variables. Solve for y.

$$y-4z=-12$$
$$\underline{-2(-7y-2z=-6)}$$
$$y-4z=-12$$
$$\underline{14y+4z=12}$$
$$15y=0$$
$$y=0$$

Substitute the value for y into equation 4 and solve for z.

$$0-4z=-12$$
$$-4z=-12$$
$$z=3$$

Substitute the values for y and z into equation 1 to solve for x.

$$x+0+3=2$$
$$x+3=2$$
$$x=-1$$

Solution: $(-1,0,3)$

17. $\begin{cases} 2x+y+2z=5 & \text{Eqtn. 1} \\ 3x-2y+3z=4 & \text{Eqtn. 2} \\ -2x+3y+z=8 & \text{Eqtn. 3} \end{cases}$

Multiply equation 1 by 2 and add to equation 2. This makes equation 4.

$$4x+2y+4z=10$$
$$\underline{3x-2y+3z=4}$$
$$7x+7z=14 \quad \text{Eqtn. 4}$$

Multiply equation 1 by –3 and add to equation 3. This makes equation 5.

$$-6x-3y-6z=-15$$
$$\underline{-2x+3y+z=8}$$
$$-8x-5z=-7 \quad \text{Eqtn. 5}$$

Use equations 4 and 5 to make a system of equations in two variables. Solve.

$$8(7x+7z=14)$$
$$7(\underline{-8x-5z=-7})$$
$$56x+56z=112$$
$$\underline{-56x-35z=-49}$$
$$21z=63$$
$$z=3$$

Substitute the value for z into equation 4 and solve for x.

$$7x+7\cdot 3=14$$
$$7x+21=14$$
$$7x=-7$$
$$x=-1$$

Substitute the values for x and z into equation 1 to solve for y.

$$2(-1)+y+2(3)=5$$
$$-2+y+6=5$$
$$y+4=5$$
$$y=1$$

Solution: $(-1,1,3)$

19. $\begin{cases} x+2y-z=1 & \text{Eqtn. 1} \\ 2x+4y-2z=-8 & \text{Eqtn. 2} \\ 3x+y-4z=6 & \text{Eqtn. 3} \end{cases}$

Multiply equation 1 by –2 and add to equation 2. This makes equation 4.

$$\begin{array}{r} -2x-4y+2z=-2 \\ \underline{2x+4y-2z=-8} \\ 0=-10 \end{array}$$

Because this is a false statement, the system is inconsistent and there is no solution.

21. $\begin{cases} 4x-2y+3z=6 & \text{Eqtn. 1} \\ 6x-3y+4.5z=9 & \text{Eqtn. 2} \\ 12x-6y+9z=18 & \text{Eqtn. 3} \end{cases}$

Multiply equation 2 by –2 and add to equation 3. This makes equation 4.

$$\begin{array}{rl} -12x+6y-9z=-18 & \\ \underline{12x-6y+9z=18} & \\ 0=0 & \text{Eqtn. 4} \end{array}$$

Multiply equation 1 by -3 and add to equation 3. This makes equation 5.

$$\begin{array}{rl} -12x+6y-9z=-18 & \\ \underline{12x-6y+9z=18} & \\ 0=0 & \text{Eqtn. 5} \end{array}$$

Because equation 4 and equation 5 are both true statements, the three equations are dependent. There are an infinite number of solutions.

23. $\begin{cases} x=2y+z+7 & \text{Eqtn. 1} \\ y=-3x+2z+1 & \text{Eqtn. 2} \\ 2x+y-z=0 & \text{Eqtn. 3} \end{cases}$

Multiply equation 2 by -1 and add to equation 3. This makes equation 4.

$$\begin{array}{rl} -3x-y+2z=-1 & \\ \underline{2x+y-z=0} & \\ -x+z=-1 & \text{Eqtn. 4} \end{array}$$

Multiply equation 2 by 2 and add to equation 1. This makes equation 5.

$$\begin{array}{rl} x-2y-z=7 & \\ \underline{6x+2y-4z=2} & \\ 7x-5z=9 & \text{Eqtn. 5} \end{array}$$

Use equations 4 and 5 to make a system of equations in two variables. Solve for x.

$$\begin{array}{rl} -x+z=-1 & \text{Multiply by 5} \\ \underline{7x-5z=9} & \\ -5x+5z=-5 & \\ \underline{7x-5z=9} & \\ 2x=4 & \\ x=2 & \end{array}$$

Substitute the value for x into equation 4 and solve for z.

$$\begin{aligned} -x+z&=-1 \\ -2+z&=-1 \\ z&=1 \end{aligned}$$

Substitute the values for x and z into equation 2 to solve for y.

$$\begin{aligned} y&=-3x+2z+1 \\ y&=-3(2)+2(1)+1 \\ y&=-6+2+1 \\ y&=-3 \end{aligned}$$

Solution: $(2,-3,1)$

25. $\begin{cases} x-4y+z=1 & \text{Eqtn. 1} \\ 3x+2y-z=-8 & \text{Eqtn. 2} \\ x+6y+2z=-3 & \text{Eqtn. 3} \end{cases}$

Add equations 1 and 2. This makes equation 4.

$$\begin{array}{rl} x-4y+z=1 & \\ \underline{3x+2y-z=-8} & \\ 4x-2y=-7 & \text{Eqtn. 4} \end{array}$$

Multiply equation 2 by 2 and add to equation 3. This makes equation 5.

$$\begin{array}{rl} x+6y+2z=-3 & \\ \underline{6x+4y-2z=-16} & \\ 7x+10y=-19 & \text{Eqtn. 5} \end{array}$$

Use equations 4 and 5 to make a system of equations in two variables. Solve for x.

$$\begin{array}{rl} 4x-2y=-7 & \text{Multiply by 5} \\ \underline{7x+10y=-19} & \\ 20x-10y=-35 & \\ \underline{7x+10y=-19} & \\ 27x=-54 & \\ x=-2 & \end{array}$$

Substitute the value for x into equation 4 and solve for y.

$$4(-2)-2y=-7$$
$$-8-2y=-7$$
$$-2y=1$$
$$y=-\frac{1}{2}$$

Substitute the values for x and y into equation 1 to solve for z.

$$-2-4\left(-\frac{1}{2}\right)+z=1$$
$$-2+2+z=1$$
$$z=1$$

Solution: $\left(-2,-\frac{1}{2},1\right)$

27. $\begin{cases} 4x+2y+3z=9 & \text{Eqtn. 1} \\ 2x-4y-z=7 & \text{Eqtn. 2} \\ 3x-2z=4 & \text{Eqtn. 3} \end{cases}$

Multiply equation 1 by 2 and add to equation 2. This makes equation 4.

$$8x+4y+6z=18$$
$$\underline{2x-4y-z=7}$$
$$10x+5z=25 \quad \text{Eqtn. 4}$$

Use equations 3 and 4 to make a system of equations in two variables. Solve for z.

$$-10(3x-2z=4)$$
$$\underline{3(10x+5z=25)}$$
$$-30x+20z=-40$$
$$\underline{30x+15z=75}$$
$$35z=35$$
$$z=1$$

Substitute the value of z into equation 3 and solve for x.

$$3x-2\cdot 1=4$$
$$3x-2=4$$
$$3x=6$$
$$x=2$$

Substitute the values for x and z into equation 1 to solve for y.

$$4\cdot 2+2y+3\cdot 1=9$$
$$8+2y+3=9$$
$$2y=-2$$
$$y=-1$$

Solution: $(2,-1,1)$

29. $\begin{cases} 3x \quad -2z=-1 & \text{Eqtn. 1} \\ 4x+5y \quad =23 & \text{Eqtn. 2} \\ y+2z=-1 & \text{Eqtn. 3} \end{cases}$

Add equation 1 and equation 3. This makes equation 4.

$$3x \quad -2z=-1$$
$$\underline{y+2z=-1}$$
$$3x+y \quad =-2 \quad \text{Eqtn. 4}$$

Use equation 2 and equation 4 to form a system. Multiply equation 4 by –5 and solve for x.

$$4x+5y=23$$
$$\underline{3x+y=-2}$$
$$4x+5y=23$$
$$\underline{-15x-5y=10}$$
$$-11x=33$$
$$x=-3$$

Substitute the value for x in equation 1 and solve for z.

$$3x-2z=-1$$
$$3(-3)-2z=-1$$
$$-9-2z=-1$$
$$-2z=8$$
$$z=-4$$

Substitute the value for z in equation 3 and solve for y.

$$y+2(-4)=-1$$
$$y-8=-1$$
$$y=7$$

Solution: (–3, 7, –4)

31. $\begin{cases} 3x+2y=-2 & \text{Eqtn. 1} \\ 2x-3z=1 & \text{Eqtn. 2} \\ 0.4y-0.5z=-2.1 & \text{Eqtn. 3} \end{cases}$

Multiply equation 1 by 2 and multiply equation 2 by -3. Add the equations to make equation 4.

$$\begin{aligned} 6x+4y \quad &= -4 \\ -6x \quad +9z &= -3 \\ \hline 4y+9z &= -7 \quad \text{Eqtn. 4} \end{aligned}$$

Use equations 3 and 4 to make a system of equations in two variables. Solve for z.

$$\begin{aligned} 0.4y-0.5z &= -2.1 \quad \text{Multiply by -10} \\ 4y+9z &= -7 \\ \hline -4y+5z &= 21 \\ 4y+9z &= -7 \\ \hline 14z &= 14 \\ z &= 1 \end{aligned}$$

Substitute the value of z into equation 4 and solve for y.

$$\begin{aligned} 4y+9\cdot 1 &= -7 \\ 4y+9 &= -7 \\ 4y &= -16 \\ y &= -4 \end{aligned}$$

Substitute the value for y into equation 1 to solve for x.

$$\begin{aligned} 3x+2(-4) &= -2 \\ 3x-8 &= -2 \\ 3x &= 6 \\ x &= 2 \end{aligned}$$

Solution: $(2,-4,1)$

33. $\begin{cases} \frac{3}{2}x+y-z=0 & \text{Eqtn. 1} \\ 4y-3z=-22 & \text{Eqtn. 2} \\ -0.2x+0.3y=-2 & \text{Eqtn. 3} \end{cases}$

Multiply equation 1 by -3 and add to equation 2 to make equation 4.

$$\begin{aligned} -\frac{9}{2}x-3y+3z &= 0 \\ 4y-3z &= -22 \\ \hline -\frac{9}{2}x+y &= -22 \quad \text{Eqtn. 4} \end{aligned}$$

Use equations 3 and 4 to make a system of equations in two variables. Solve for x.

$$\begin{aligned} -\frac{9}{2}x+y \quad &= -22 \quad \text{Multiply by -0.3} \\ -0.2x+0.3y &= -2 \\ \hline 1.35x-0.3y &= 6.6 \\ -0.2x+0.3y &= -2 \\ \hline 1.15x \quad &= 4.6 \\ x &= 4 \end{aligned}$$

Substitute the value of x into equation 4 and solve for y.

$$\begin{aligned} -\frac{9}{2}\cdot 4+y &= -22 \\ -18+y &= -22 \\ y &= -4 \end{aligned}$$

Substitute the value for y into equation 2 to solve for z.

$$\begin{aligned} 4(-4)-3z &= -22 \\ -16-3z &= -22 \\ -3z &= -6 \\ z &= 2 \end{aligned}$$

Solution: $(4,-4,2)$

Review Exercises

1. Let one number be x and the other be $x + 62$.

$$\begin{aligned} 5x-2(x+62) &= 155 \\ 5x-2x-124 &= 155 \\ 3x &= 279 \\ x &= 93 \end{aligned}$$

The numbers are 93 and 155.

2. If the width is w, then the length is $w + 50$.

$$\begin{aligned} 2(w+50)+2w &= 400 \\ 2w+100+2w &= 400 \\ 4w+100 &= 400 \\ 4w &= 300 \\ w &= 75 \end{aligned}$$

The dimensions of the rectangle are 75 ft. by 125 ft.

3. Let one of the odd integers be x and the other be $x + 2$.

$$x+(x+2)=124$$
$$2x+2=124$$
$$2x=122$$
$$x=61$$

The consecutive odd integers are 61 and 63.

4. Since the two angles are supplementary, their sum is 180 degrees. So, let one of the angles be x and the other be $180 - x$.

$$180-x=15+2x$$
$$165=3x$$
$$55=x$$

The angles are 55 and 125 degrees.

5. Complete a table.

	rate	time	distance
Dora	5	t	$5t$
Jose	13	t	$13t$

$$5t+13t=3$$
$$18t=3$$
$$t=\frac{1}{6}\text{ hr. or 10 min.}$$

6. Complete a table.

Solutions	Concen-trate	Vol. of Solution	Vol. of HCl
10%	0.1	x	$0.1x$
30%	0.3	$200-x$	$0.3(200-x)$
20%	0.2	200	$0.2(200)$

$$0.1x+0.3(200-x)=0.2(200)$$
$$0.1x+60-0.3x=40$$
$$-0.2x=-20$$
$$x=100$$

100 ml of 10% and 100 ml of 30%

Exercise Set 4.3

1. $y = x + 4$

3. three

5. Let x = amount of money in million dollars that *Shrek II* grossed and y = the amount of money in million dollars that *Shrek the Third* grossed.

$$\begin{cases} x+y=1236 \\ x+362=y \end{cases}$$

$$x+y=1236 \qquad x+362=y$$
$$x+(x+362)=1236 \qquad 437+362=y$$
$$2x+362=1236 \qquad 799=y$$
$$2x=874$$
$$x=437$$

Shrek II grossed \$437 million and *Shrek the Third* grossed \$799 million.

7. Let x = number of women in millions who are enrolled in a community college and y = number of men in millions who are enrolled in a community college.

$$\begin{cases} x+y=11.6 \\ y=x-1.8 \end{cases}$$

$$x+y=11.6$$
$$x+(x-1.8)=11.6$$
$$2x-1.8=11.6$$
$$2x=13.4$$
$$x=6.7$$

6.7 million women are enrolled in a community college.

9. Let x = salary in dollars of a math teacher and y = salary in dollars of a mathematician.

$$\begin{cases} x+y=134,720 \\ x=y-21,760 \end{cases}$$

$$x+y=134,720 \qquad x=y-21,760$$
$$(y-21,760)+y=134,720 \qquad x=78,240-21,760$$
$$2y-21,760=134,720 \qquad x=56,480$$
$$2y=156,480$$
$$y=78,240$$

The average annual salary of a mathematician is \$78,240, and the average annual salary of a math teacher is \$56,480.

11. Let x = percent of shark attacks occurring in deep water and y = percent of shark attacks occurring in shallower waters.

$$\begin{cases} x+y=100 \\ x=6y \end{cases}$$

$$\begin{aligned} x+y&=100 \\ 6y+y&=100 \\ 7y&=100 \\ y&\approx 14.3 \end{aligned} \qquad \begin{aligned} x+y&=100 \\ x+14.3&=100 \\ x&\approx 85.7 \end{aligned}$$

About 85.7% of shark attacks occur in deep water.

13. Let x = number of 18-in. wreaths sold and y = number of 22-in. wreaths sold.

Categories	Selling Price	Number Sold	Revenue
18-inch	20	x	$20x$
22-inch	25	y	$25y$

$$\begin{cases} 20x+25y=1100 \\ 6+3x=y \end{cases}$$

$$\begin{aligned} 20x+25y&=1100 \\ 20x+25(6+3x)&=1100 \\ 20x+150+75x&=1100 \\ 95x&=950 \\ x&=10 \end{aligned} \qquad \begin{aligned} 6+3x&=y \\ 6+3(10)&=y \\ 36&=y \end{aligned}$$

A total of 10 18-in. wreaths and 36 22-in. wreaths were sold.

15. Let u = number of Union soldiers and c = number of Confederate soldiers.

$$\begin{cases} u+c=498{,}000 \\ u=3c-38{,}000 \end{cases}$$

$$\begin{aligned} u+c&=498{,}000 \\ (3c-38{,}000)+c&=498{,}000 \\ 4c-38{,}000&=498{,}000 \\ 4c&=536{,}000 \\ c&=134{,}000 \end{aligned}$$

$$u=3c-38{,}000$$
$$u=3(134{,}000)-38{,}000$$
$$u=364{,}000$$

A total of 364,000 Union soldiers and 134,000 Confederate soldiers were lost.

17. Let w = the width of the base and l = the length of the base.

$$\begin{cases} 2l+2w=220.5 \\ w=l \end{cases}$$

$$\begin{aligned} 2l+2w&=220.5 \\ 2w+2w&=220.5 \\ 4w&=220.5 \\ w&=55.125 \end{aligned}$$

The width and length are 55.125 ft.

19. Let x = the measure of the larger angle and y = the measure of the smaller angle. The angles are supplementary.

$$\begin{cases} x+y=180 \\ x=2y \end{cases}$$

$$\begin{aligned} x+y&=180 \\ 2y+y&=180 \\ 3y&=180 \\ y&=60 \end{aligned} \qquad \begin{aligned} x&=2y \\ x&=2(60) \\ x&=120 \end{aligned}$$

The angles are 60 degrees and 120 degrees.

21. Let j = Josh's time in hours and d = Dolph's time in hours.

$$\begin{cases} j=d+0.05 \\ 12j=15d \end{cases}$$

$$\begin{aligned} 12j&=15d \\ 12(d+0.05)&=15d \\ 12d+0.6&=15d \\ 0.6&=3d \\ 0.2&=d \end{aligned}$$

It will take Dolph 0.2 hours to catch up to Josh.

23. Let p = Poloma's time in hours and f = Francis' time in hours.

$$\begin{cases} p=f+0.5 \\ 65p=70f \end{cases}$$

$$\begin{aligned} 65p&=70f \\ 65(f+0.5)&=70f \\ 65f+32.5&=70f \\ 32.5&=5f \\ 6.5&=f \end{aligned}$$

Francis will catch up with Poloma in 6.5 hours at 10:00 P.M.

25. Let x = the boat's speed in mph and y = the speed of the current in mph.

$$\begin{cases} 3(x-y)=36 \\ 3(x+y)=48 \end{cases} \Rightarrow \begin{cases} x-y=12 \\ x+y=16 \end{cases}$$

$$\begin{aligned} x-y&=12 \\ \underline{x+y}&\underline{=16} \\ 2x&=28 \\ x&=14 \end{aligned} \qquad \begin{aligned} x+y&=16 \\ 14+y&=16 \\ y&=2 \end{aligned}$$

The boat's speed is 14 mph and the current's speed is 2 mph.

27. Let x = the speed of the plane in mph and y = the wind speed in mph.

$$\begin{cases} x+y=650 \\ x-y=550 \end{cases}$$

$$\begin{aligned} x+y&=650 \\ \underline{x-y}&\underline{=550} \\ 2x&=1200 \\ x&=600 \end{aligned} \qquad \begin{aligned} x+y&=650 \\ 600+y&=650 \\ y&=50 \end{aligned}$$

The plane's speed is 600 mph and the wind's speed is 50 mph.

29. Let x = the volume of 5% HCl mixture and y = the volume of 20% HCl.

Solution	Concen-tration	Volume of Solution	Volume of HCl
5%	0.05	x	$0.05x$
20%	0.20	y	$0.20y$
12.5%	0.125	10	$0.125(10)$

$$\begin{cases} x+y=10 \\ 0.05x+0.20y=0.125(10) \end{cases}$$

Solve the first equation for x: $x=10-y$

$$\begin{aligned} 0.05x+0.20y&=0.125(10) \\ 0.05(10-y)+0.20y&=1.25 \\ 0.5-0.05y+0.20y&=1.25 \\ 0.15y&=0.75 \\ y&=5 \end{aligned}$$

$x=10-y$

$x=10-5$

$x=5$

Combine 5 ml of 20% and 5 ml of 5%.

31. Let x = the amount in the account at 9% and y = the amount in the account at 5%

Categories	Interest Rate	Amount in Account	Interest
9% account	0.09	x	$0.09x$
5% account	0.05	y	$0.05y$

$$\begin{cases} x=4y \\ 0.09x+0.05y=1435 \end{cases}$$

$$\begin{aligned} 0.09x+0.05y&=1435 \\ 0.09(4y)+0.05y&=1435 \\ 0.36y+0.05y&=1435 \\ 0.41y&=1435 \\ y&=3500 \end{aligned} \qquad \begin{aligned} x&=4y \\ x&=4(3500) \\ x&=14{,}000 \end{aligned}$$

He should invest \$3500 at 5% and \$14,000 at 9%.

33. Let x, y, and z represent the three numbers.

$$\begin{cases} x+y+z=16 & \text{Eqtn. 1} \\ x-2y=2 & \text{Eqtn. 2} \\ x-z=-2 & \text{Eqtn. 3} \end{cases}$$

Multiply equation 2 by -1 and add to equation 3 to make equation 4.

$$\begin{aligned} -x+2y&=-2 \\ \underline{x-z}&\underline{=-2} \\ 2y-z&=-4 \quad \text{Eqtn. 4} \end{aligned}$$

Multiply equation 1 by -1 and add to equation 2. This makes equation 5.

$$\begin{aligned} -x-y-z&=-16 \\ \underline{x-2y}&\underline{=2} \\ -3y-z&=-14 \quad \text{Eqtn. 5} \end{aligned}$$

Use equations 4 and 5 to make a system of equations in two variables. Solve.

$$\begin{aligned} -1(2y-z&=-4) \\ \underline{-3y-z}&\underline{=-14} \\ -2y+z&=4 \\ \underline{-3y-z}&\underline{=-14} \\ -5y&=-10 \\ y&=2 \end{aligned}$$

Substitute the value of y into equation 2 and solve for x.

$x=2+2\cdot 2$

$x=6$

Substitute the value for x into equation 3 to solve for z.

$6 = z - 2$

$8 = z$

The numbers are 6, 2, and 8.

35. Let x, y, and z represent the measures of the three angles of the triangle.

$$\begin{cases} x + y + z = 180 & \text{Eqtn. 1} \\ x = 3y & \text{Eqtn. 2} \\ y = z - 5 & \text{Eqtn. 3} \end{cases}$$

Substitute equation 2 into equation 1. This becomes equation 4.

$3y + y + z = 180$

$4y + z = 180$ Eqtn. 4

Use equations 3 and 4 to make a system of equations in two variables. Solve.

$y - z = -5$ Eqtn. 3

$\underline{4y + z = 180}$ Eqtn. 4

$5y = 175$

$y = 35$

Substitute the value of y into equation 2 and solve for x.

$x = 3 \cdot 35$

$x = 105$

Substitute the value for y into equation 3 and solve for z.

$35 = z - 5$

$40 = z$

The angles are 105°, 35°, and 40°.

37. Let b = the cost of a burger, f = the cost of an order of fries, and d = the cost of a drink.

$$\begin{cases} b + f + d = 5 & \text{Eqtn. 1} \\ 3b + 2f + 2d = 12.50 & \text{Eqtn. 2} \\ 2b + 4f + 3d = 14 & \text{Eqtn. 3} \end{cases}$$

Multiply equation 1 by −2 and add to equation 3 to make equation 4.

$-2b - 2f - 2d = -10$

$\underline{2b + 4f + 3d = 14}$

$2f + d = 4$ Eqtn. 4

Multiply equation 1 by −3 and add to equation 2 to make equation 5.

$-3b - 3f - 3d = -15$

$\underline{3b + 2f + 2d = 12.50}$

$-f - d = -2.5$ Eqtn. 5

Use equations 4 and 5 to make a system of equations in two variables. Solve.

$2f + d = 4$

$\underline{-f - d = -2.5}$

$f = 1.5$

Substitute into equation 5 to solve for d.

$-1.5 - d = -2.5$

$-d = -1$

$d = 1$

Substitute into equation 1 to solve for b.

$b + 1.5 + 1 = 5$

$b + 2.5 = 5$

$b = 2.5$

A burger costs \$2.50, fries cost \$1.50, and a drink costs \$1.00.

39. Let c = the number of children's tickets sold, s = the number of student tickets sold, and a = the number of adult tickets sold.

$$\begin{cases} c + s + a = 500 & \text{Eqtn. 1} \\ 3c + 5s + 8a = 2500 & \text{Eqtn. 2} \\ a = s - 150 & \text{Eqtn. 3} \end{cases}$$

Multiply equation 1 by −3 and add to equation 2. This makes equation 4.

$-3c - 3s - 3a = -1500$

$\underline{3c + 5s + 8a = 2500}$

$2s + 5a = 1000$ Eqtn. 4

Use equations 3 and 4 to make a systems of equations and solve.

$2s + 5a = 1000$

$\underline{-s + a = -150}$ Multiply by 2

$2s + 5a = 1000$

$\underline{-2s + 2a = -300}$

$7a = 700$

$a = 100$

Substitute into equation 3 to solve for s.

$-s + 100 = -150$

$-s = -250$

$s = 250$

Substitute into equation 1 to solve for c.

$$c+250+100=500$$
$$c=150$$

150 children, 250 students, and 100 adults

41. Let a = then number of 2-point shots, b = the number of 3-point shots, and let c = the number of 1-point shots.

$$\begin{cases} 2a+3b+c=14 & \text{Eqtn. 1} \\ a+b+c=7 & \text{Eqtn. 2} \\ c=a+2 & \text{Eqtn. 3} \end{cases}$$

Substitute equation 3 into equation 1. This makes equation 4.

$$2a+3b+c=14$$
$$2a+3b+(a+2)=14$$
$$3a+3b=12 \quad \text{Eqtn. 4}$$

Substitute equation 3 into equation 2. This makes equation 5.

$$a+b+c=7$$
$$a+b+a+2=7$$
$$2a+b=5 \quad \text{Eqtn. 5}$$

Set up a system using equation 4 and equation 5.

$$3a+3b=12$$
$$\underline{-3(2a+b=5)}$$
$$3a+3b=12$$
$$\underline{-6a-3b=-15}$$
$$-3a=-3$$
$$a=1$$

Substitute 1 for a in equation 5 and solve for b.

$$2a+b=5$$
$$2(1)+b=5$$
$$2+b=5$$
$$b=3$$

Substitute 1 for a in equation 3 and solve for c.

$$c=a+2$$
$$c=1+2$$
$$c=3$$

John made a = 1 two-point shots, b = 3 three-point shots, and c = 3 one-point free-throw shots.

43. Let z = the number of pounds of zinc, t = the number of pounds of tin, and c = the number of pounds of copper.

$$\begin{cases} z+t+c=1000 & \text{Eqtn. 1} \\ t=3z & \text{Eqtn. 2} \\ c=20+15t & \text{Eqtn. 3} \end{cases}$$

Substitute equation 2 into equations 1 and 3 to make into a system of equations in two variables.

$$z+3z+c=1000 \qquad t=3\cdot 20$$
$$\underline{c=20+15(3z)} \qquad t=60$$
$$4z+c=1000$$
$$\underline{45z-c=-20}$$
$$49z=980$$
$$z=20$$

$$c=20+15\cdot 60$$
$$c=920$$

20 lbs. of zinc, 60 lbs. of tin, and 920 lbs. of copper were used.

45. Let h = the number of pounds of Black Forest ham, t = the number of pounds of turkey breast, and r = the number of pounds of roast beef.

$$\begin{cases} h+t+r=10 & \text{Eqtn. 1} \\ 11.96h+8.76t+9.16r=10(9.80) & \text{Eqtn. 2} \\ t=h+r & \text{Eqtn. 3} \end{cases}$$

Substitute equation 3 into equations 1 and 2 to make a system of linear equations in two variables.

$$h+h+r+r=10$$
$$\underline{11.96h+8.76(h+r)+9.16r=98}$$
$$2h+2r=10 \qquad \text{Multiply by -8.96}$$
$$\underline{20.72h+17.92r=98}$$
$$-17.92h-17.92r=-89.6$$
$$\underline{20.72h+17.92r=98}$$
$$2.8h=8.4$$
$$h=3$$

$$2\cdot 3+2r=10 \qquad 3+t+2=10$$
$$2r=4 \qquad t=5$$
$$r=2$$

3 lbs. of ham, 5 lbs. of turkey, and 2 lbs. of roast beef were used for the party tray.

47. Let x = the amount invested at 4%, y = the amount invested at 6%, and z = the amount invested at 7%.

$$\begin{cases} x+y+z=8000 & \text{Eqtn. 1} \\ 0.04x+0.06y+0.07z=475 & \text{Eqtn. 2} \\ z-x=1500 & \text{Eqtn. 3} \end{cases}$$

Multiply equation 1 by –0.06 and add to equation 2. This makes equation 4.

$$\begin{aligned} -0.06x-0.06y-0.06z &= -480 \\ 0.04x+0.06y+0.07z &= 475 \\ \hline -0.02x+0.01z &= -5 \quad \text{Eqtn. 4} \end{aligned}$$

Multiply equation 3 by –0.01 and add to equation 4.

$$\begin{aligned} 0.01x-0.01z &= -15 \\ -0.02x+0.01z &= -5 \\ \hline -0.01x &= -20 \\ x &= 2000 \end{aligned}$$

$$\begin{aligned} z-x &= 1500 & x+y+z &= 8000 \\ z-2000 &= 1500 & 2000+y+3500 &= 8000 \\ z &= 3500 & y &= 2500 \end{aligned}$$

A total of \$2000 is invested at 4%, \$2500 is invested at 6%, and \$3500 is invested at 7%.

49. Let b = the number of calories burned while bicycling, w = the number of calories burned while walking, and c = the number of calories burned while climbing stairs.

$$\begin{cases} b=120+w & \text{Eqtn. 1} \\ c=180+b & \text{Eqtn. 2} \\ c=2w & \text{Eqtn. 3} \end{cases}$$

Substitute equation 3 into equation 2. Use this new equation with equation 1 to form a system of equations in two variables.

$$\begin{aligned} b-w &= 120 & c &= 2\cdot 300 \\ -b+2w &= 180 & c &= 600 \\ \hline w &= 300 \end{aligned}$$

$$\begin{aligned} b &= 120+300 \\ b &= 420 \end{aligned}$$

The number of calories burned while bicycling is 420, while walking is 300, and while stair climbing is 600.

51. $$\begin{cases} a+v_0+h_0=234 & \text{Eqtn. 1} \\ 9a+3v_0+h_0=306 & \text{Eqtn. 2} \\ 36a+6v_0+h_0=174 & \text{Eqtn. 3} \end{cases}$$

Multiply equation 1 by –1 and add it to equation 2. This will make equation 4.

$$\begin{aligned} -a-v_0-h_0 &= -234 \\ 9a+3v_0+h_0 &= 306 \\ \hline 8a+2v_0 &= 72 \quad \text{Eqtn. 4} \end{aligned}$$

Multiply equation 1 by –1 and add it to equation 3. This will make equation 5.

$$\begin{aligned} -a-v_0-h_0 &= -234 \\ 36a+6v_0+h_0 &= 174 \\ \hline 35a+5v_0 &= -60 \quad \text{Eqtn. 5} \end{aligned}$$

Use equations 4 and 5 to make a system of linear equations in two variables. Solve.

$$\begin{aligned} 8a+2v_0 &= 72 & &\text{Multiply by -5} \\ 35a+5v_0 &= -60 & &\text{Multiply by 2} \\ \hline -40a-10v_0 &= -360 \\ 70a+10v_0 &= -120 \\ \hline 30a &= -480 \\ a &= -16 \end{aligned}$$

$$\begin{aligned} 8(-16)+2v_0 &= 72 & -16+100+h_0 &= 234 \\ -128+2v_0 &= 72 & 84+h_0 &= 234 \\ 2v_0 &= 200 & h_0 &= 150 \\ v_0 &= 100 \end{aligned}$$

$$h=-16t^2+100t+150$$

Review Exercises

1. $-3(2x-4y-z+9)=-6x+12y+3z-27$

2. $$\begin{aligned} x-0.5y &= 8 \\ 6-0.5y &= 8 \\ -0.5y &= 2 \\ y &= -4 \end{aligned}$$

3. $$\begin{aligned} f(x) &= -2x+1 \\ f(1) &= -2(1)+1 \\ &= -2+1 \\ &= -1 \end{aligned}$$

4. The y-intercept is (0, 1).

5. The slope is $m=-2$.

6.

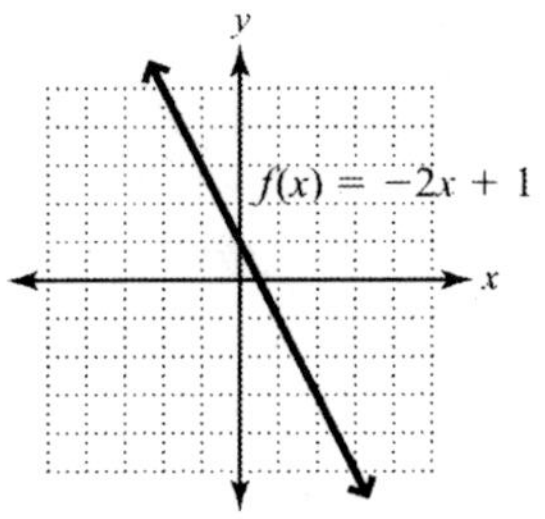

Exercise Set 4.4

1. 4 rows, 2 columns

3. The dashed line corresponds to the equal signs in the equations.

5. A matrix is in row echelon form when the coefficient portion of the augmented matrix has 1s on the diagonal from the upper left to lower right and 0s below the 1s.

7. $\left[\begin{array}{cc|c} 14 & 7 & 6 \\ 7 & 6 & 8 \end{array}\right]$

9. $\left[\begin{array}{cc|c} 7 & -6 & 1 \\ 0 & -2 & 5 \end{array}\right]$

11. $\left[\begin{array}{ccc|c} 1 & -3 & 1 & 4 \\ 2 & -4 & 2 & -4 \\ 6 & -2 & 5 & -4 \end{array}\right]$

13. $\left[\begin{array}{ccc|c} 4 & 6 & -2 & -1 \\ 8 & 3 & 0 & -12 \\ 0 & -1 & 2 & 4 \end{array}\right]$

15. $\left[\begin{array}{cc|c} 1 & -3 & -2 \\ 0 & 1 & 2 \end{array}\right]$ represents the system $\begin{cases} x-3y=-2 \\ y=2 \end{cases}$

$$x-3y=-2$$
$$x-3(2)=-2$$
$$x-6=-2$$
$$x=4$$

Solution: $(4,2)$

17. $\left[\begin{array}{ccc|c} 1 & -4 & -8 & 6 \\ 0 & 1 & -2 & -7 \\ 0 & 0 & 1 & 1 \end{array}\right]$ represents $\begin{cases} x-4y-8z=6 \\ y-2z=-7 \\ z=1 \end{cases}$

$$y-2z=-7$$
$$y-2(1)=-7$$
$$y-2=-7$$
$$y=-5$$

$$x-4y-8z=6$$
$$x-4(-5)-8(1)=6$$
$$x+20-8=6$$
$$x=-6$$

Solution: $(-6,-5,1)$

19. $\left[\begin{array}{cc|c} 1 & 3 & -1 \\ -2 & 5 & 6 \end{array}\right]$

$2R_1+R_2 \to \left[\begin{array}{cc|c} 1 & 3 & -1 \\ 0 & 11 & 4 \end{array}\right]$

21. $\left[\begin{array}{ccc|c} 1 & -2 & 4 & 6 \\ 0 & 2 & -1 & -5 \\ 0 & 8 & -6 & -3 \end{array}\right]$

$-4R_2+R_3 \to \left[\begin{array}{ccc|c} 1 & -2 & 4 & 6 \\ 0 & 2 & -1 & -5 \\ 0 & 0 & -2 & 17 \end{array}\right]$

23. $\left[\begin{array}{cc|c} 4 & 8 & -10 \\ -1 & 3 & 2 \end{array}\right]$

$\frac{1}{4}R_1 \to \left[\begin{array}{cc|c} 1 & 2 & -2.5 \\ -1 & 3 & 2 \end{array}\right]$

25. Replace R_2 with $3R_1+R_2$.

27. Replace R_3 with $-2R_2+R_3$.

29. $\begin{cases} x-y=5 \\ x+y=-1 \end{cases}$

$\left[\begin{array}{cc|c} 1 & -1 & 5 \\ 1 & 1 & -1 \end{array}\right]$

$-1R_1+R_2 \to \left[\begin{array}{cc|c} 1 & -1 & 5 \\ 0 & 2 & -6 \end{array}\right]$

$\frac{1}{2}R_2 \to \left[\begin{array}{cc|c} 1 & -1 & 5 \\ 0 & 1 & -3 \end{array}\right]$

$\begin{cases} x-y=5 \\ y=-3 \end{cases}$

$$x-y=5$$
$$x-(-3)=5$$
$$x+3=5$$
$$x=2$$

Solution: $(2,-3)$

31. $\begin{cases} x+y=3 \\ 3x-y=1 \end{cases}$

$$\left[\begin{array}{cc|c} 1 & 1 & 3 \\ 3 & -1 & 1 \end{array}\right]$$

$$-3R_1+R_2 \to \left[\begin{array}{cc|c} 1 & 1 & 3 \\ 0 & -4 & -8 \end{array}\right]$$

$$-\frac{1}{4}R_2 \to \left[\begin{array}{cc|c} 1 & 1 & 3 \\ 0 & 1 & 2 \end{array}\right]$$

$$\begin{cases} x+y=3 \\ \quad y=2 \end{cases}$$

$x+y=3$ Solution: $(1,2)$

$x+2=3$

$x=1$

33. $\begin{cases} x+2y=-7 \\ 2x-4y=2 \end{cases}$

$$\left[\begin{array}{cc|c} 1 & 2 & -7 \\ 2 & -4 & 2 \end{array}\right]$$

$$-2R_1+R_2 \to \left[\begin{array}{cc|c} 1 & 2 & -7 \\ 0 & -8 & 16 \end{array}\right]$$

$$-\frac{1}{8}R_2 \to \left[\begin{array}{cc|c} 1 & 2 & -7 \\ 0 & 1 & -2 \end{array}\right]$$

$$\begin{cases} x+2y=-7 \\ \quad y=-2 \end{cases}$$

$x+2y=-7$ Solution: $(-3,-2)$

$x+2(-2)=-7$

$x-4=-7$

$x=-3$

35. $\begin{cases} -2x+5y=4 \\ x-2y=-2 \end{cases}$

$$\left[\begin{array}{cc|c} -2 & 5 & 4 \\ 1 & -2 & -2 \end{array}\right]$$

$$R_1+2R_2 \to \left[\begin{array}{cc|c} -2 & 5 & 4 \\ 0 & 1 & 0 \end{array}\right]$$

$$-\frac{1}{2}R_1 \to \left[\begin{array}{cc|c} 1 & -5/2 & -2 \\ 0 & 1 & 0 \end{array}\right]$$

$$\begin{cases} x-\frac{5}{2}y=-2 \\ \quad y=0 \end{cases}$$

$x-\frac{5}{2}y=-2$ Solution: $(-2,0)$

$x-\frac{5}{2}(0)=-2$

$x-0=-2$

$x=-2$

37. $\begin{cases} 4x-3y=-2 \\ 2x-3y=-10 \end{cases}$

$$\left[\begin{array}{cc|c} 4 & -3 & -2 \\ 2 & -3 & -10 \end{array}\right]$$

$$R_1-2R_2 \to \left[\begin{array}{cc|c} 4 & -3 & -2 \\ 0 & 3 & 18 \end{array}\right]$$

$$\frac{1}{4}R_1 \to \left[\begin{array}{cc|c} 1 & -3/4 & -1/2 \\ 0 & 3 & 18 \end{array}\right]$$

$$\frac{1}{3}R_2 \to \left[\begin{array}{cc|c} 1 & -3/4 & -1/2 \\ 0 & 1 & 6 \end{array}\right]$$

$$\begin{cases} x-\frac{3}{4}y=-\frac{1}{2} \\ \quad y=6 \end{cases}$$

$x-\frac{3}{4}y=-\frac{1}{2}$ Solution: $(4,6)$

$x-\frac{3}{4}(6)=-\frac{1}{2}$

$x-\frac{9}{2}=-\frac{1}{2}$

$x=4$

39. $\begin{cases} 5x+2y=12 \\ 2x+3y=-4 \end{cases}$

$$\left[\begin{array}{cc|c} 5 & 2 & 12 \\ 2 & 3 & -4 \end{array}\right]$$

$$-2R_1+5R_2 \rightarrow \left[\begin{array}{cc|c} 5 & 2 & 12 \\ 0 & 11 & -44 \end{array}\right]$$

$$\frac{1}{5}R_1 \rightarrow \left[\begin{array}{cc|c} 1 & 2/5 & 12/5 \\ 0 & 11 & -44 \end{array}\right]$$

$$\frac{1}{11}R_2 \rightarrow \left[\begin{array}{cc|c} 1 & 2/5 & 12/5 \\ 0 & 1 & -4 \end{array}\right]$$

$$\begin{cases} x+\frac{2}{5}y=\frac{12}{5} \\ y=-4 \end{cases}$$

$$x+\frac{2}{5}y=\frac{12}{5}$$
$$x+\frac{2}{5}(-4)=\frac{12}{5}$$
$$x-\frac{8}{5}=\frac{12}{5}$$
$$x=4$$

Solution: $(4,-4)$

41. $\begin{cases} x+y+z=3 \\ 2x+y-3z=-10 \\ 2x+2y+z=3 \end{cases}$

$$\left[\begin{array}{ccc|c} 1 & 1 & 1 & 3 \\ 2 & 1 & -3 & -10 \\ 2 & 2 & 1 & 3 \end{array}\right]$$

$$-2R_1+R_2 \rightarrow \left[\begin{array}{ccc|c} 1 & 1 & 1 & 3 \\ 0 & -1 & -5 & -16 \\ 2 & 2 & 1 & 3 \end{array}\right]$$

$$-2R_1+R_3 \rightarrow \left[\begin{array}{ccc|c} 1 & 1 & 1 & 3 \\ 0 & -1 & -5 & -16 \\ 0 & 0 & -1 & -3 \end{array}\right]$$

$$\begin{array}{l} \\ -1R_2 \rightarrow \\ -1R_3 \rightarrow \end{array} \left[\begin{array}{ccc|c} 1 & 1 & 1 & 3 \\ 0 & 1 & 5 & 16 \\ 0 & 0 & 1 & 3 \end{array}\right]$$

$$\begin{cases} x+y+z=3 \\ y+5z=16 \\ z=3 \end{cases}$$

$$y+5z=16$$
$$y+5(3)=16$$
$$y+15=16$$
$$y=1$$

$$x+y+z=3$$
$$x+1+3=3$$
$$x+4=3$$
$$x=-1$$

Solution: $(-1,1,3)$

43. $\begin{cases} x-y+z=-1 \\ 2x-2y+z=0 \\ x+3y+2z=1 \end{cases}$

$$\left[\begin{array}{ccc|c} 1 & -1 & 1 & -1 \\ 2 & -2 & 1 & 0 \\ 1 & 3 & 2 & 1 \end{array}\right]$$

$$-2R_1+R_2 \rightarrow \left[\begin{array}{ccc|c} 1 & -1 & 1 & -1 \\ 0 & 0 & -1 & 2 \\ 1 & 3 & 2 & 1 \end{array}\right]$$

$$\begin{array}{l} \\ R_3 \rightarrow \\ R_2 \rightarrow \end{array} \left[\begin{array}{ccc|c} 1 & -1 & 1 & -1 \\ 1 & 3 & 2 & 1 \\ 0 & 0 & -1 & 2 \end{array}\right]$$

$$-R_1+R_2 \left[\begin{array}{ccc|c} 1 & -1 & 1 & -1 \\ 0 & 4 & 1 & 2 \\ 0 & 0 & -1 & 2 \end{array}\right]$$

$$\begin{array}{l} \\ \frac{1}{4}R_2 \rightarrow \\ -1R_3 \rightarrow \end{array} \left[\begin{array}{ccc|c} 1 & -1 & 1 & -1 \\ 0 & 1 & \frac{1}{4} & \frac{1}{2} \\ 0 & 0 & 1 & -2 \end{array}\right]$$

$$\begin{cases} x-y+z=-1 \\ y+\frac{1}{4}z=\frac{1}{2} \\ z=-2 \end{cases}$$

$$y+\frac{1}{4}z=\frac{1}{2}$$
$$y+\frac{1}{4}(-2)=\frac{1}{2}$$
$$y-\frac{1}{2}=\frac{1}{2}$$
$$y=1$$

$$x-y+z=-1$$
$$x-1-2=-1$$
$$x-3=-1$$
$$x=2$$

Solution: $(2,1,-2)$

45. $\begin{cases} 2x - y + z = 8 \\ x - 2y + 3z = 11 \\ 2x + 3y - z = -6 \end{cases}$

$$\left[\begin{array}{ccc|c} 2 & -1 & 1 & 8 \\ 1 & -2 & 3 & 11 \\ 2 & 3 & -1 & -6 \end{array}\right]$$

$$\begin{array}{r} R_2 \to \\ R_1 \to \\ \end{array}\left[\begin{array}{ccc|c} 1 & -2 & 3 & 11 \\ 2 & -1 & 1 & 8 \\ 2 & 3 & -1 & -6 \end{array}\right]$$

$$\begin{array}{r} \\ -2R_1 + R_2 \to \\ \end{array}\left[\begin{array}{ccc|c} 1 & -2 & 3 & 11 \\ 0 & 3 & -5 & -14 \\ 2 & 3 & -1 & -6 \end{array}\right]$$

$$\begin{array}{r} \\ \\ -2R_1 + R_3 \to \end{array}\left[\begin{array}{ccc|c} 1 & -2 & 3 & 11 \\ 0 & 3 & -5 & -14 \\ 0 & 7 & -7 & -28 \end{array}\right]$$

$$\begin{array}{r} \\ \\ \frac{1}{7}R_3 \to \end{array}\left[\begin{array}{ccc|c} 1 & -2 & 3 & 11 \\ 0 & 3 & -5 & -14 \\ 0 & 1 & -1 & -4 \end{array}\right]$$

$$\begin{array}{r} \\ R_3 \to \\ R_2 \to \end{array}\left[\begin{array}{ccc|c} 1 & -2 & 3 & 11 \\ 0 & 1 & -1 & -4 \\ 0 & 3 & -5 & -14 \end{array}\right]$$

$$\begin{array}{r} \\ \\ -3R_2 + R_3 \to \end{array}\left[\begin{array}{ccc|c} 1 & -2 & 3 & 11 \\ 0 & 1 & -1 & -4 \\ 0 & 0 & -2 & -2 \end{array}\right]$$

$$\begin{array}{r} \\ \\ -\frac{1}{2}R_3 \to \end{array}\left[\begin{array}{ccc|c} 1 & -2 & 3 & 11 \\ 0 & 1 & -1 & -4 \\ 0 & 0 & 1 & 1 \end{array}\right]$$

$$\begin{cases} x - 2y + 3z = 11 \\ y - z = -4 \\ z = 1 \end{cases}$$

$y - z = -4$

$y - 1 = -4$

$y = -3$

$x - 2y + 3z = 11$

$x - 2(-3) + 3(1) = 11$

$x + 6 + 3 = 11$

$x = 2$

Solution: $(2, -3, 1)$

47. $\begin{cases} 3x + 2y - 3z = 1 \\ -2x + 3y - 4z = 7 \\ 5x - 2y + z = -5 \end{cases}$

$$\left[\begin{array}{ccc|c} 3 & 2 & -3 & 1 \\ -2 & 3 & -4 & 7 \\ 5 & -2 & 1 & -5 \end{array}\right]$$

$$\begin{array}{r} R_2 + R_1 \to \\ \\ \end{array}\left[\begin{array}{ccc|c} 1 & 5 & -7 & 8 \\ -2 & 3 & -4 & 7 \\ 5 & -2 & 1 & -5 \end{array}\right]$$

$$\begin{array}{r} \\ 2R_1 + R_2 \to \\ \end{array}\left[\begin{array}{ccc|c} 1 & 5 & -7 & 8 \\ 0 & 13 & -18 & 23 \\ 5 & -2 & 1 & -5 \end{array}\right]$$

$$\begin{array}{r} \\ \\ -5R_1 + R_3 \to \end{array}\left[\begin{array}{ccc|c} 1 & 5 & -7 & 8 \\ 0 & 13 & -18 & 23 \\ 0 & -27 & 36 & -45 \end{array}\right]$$

$$\begin{array}{r} \\ -R_3 - 2R_2 \to \\ \frac{1}{9}R_3 \to \end{array}\left[\begin{array}{ccc|c} 1 & 5 & -7 & 8 \\ 0 & 1 & 0 & -1 \\ 0 & -3 & 4 & -5 \end{array}\right]$$

$$\begin{array}{r} \\ \\ 3R_2 + R_3 \to \end{array}\left[\begin{array}{ccc|c} 1 & 5 & -7 & 8 \\ 0 & 1 & 0 & -1 \\ 0 & 0 & 4 & -8 \end{array}\right]$$

$$\begin{array}{r} \\ \\ \frac{1}{4}R_3 \to \end{array}\left[\begin{array}{ccc|c} 1 & 5 & -7 & 8 \\ 0 & 1 & 0 & -1 \\ 0 & 0 & 1 & -2 \end{array}\right]$$

$$\begin{cases} x + 5y - 7z = 8 \\ y = -1 \\ z = -2 \end{cases}$$

$x + 5y - 7z = 8$

$x + 5(-1) - 7(-2) = 8$

$x - 5 + 14 = 8$

$x = -1$

Solution: $(-1, -1, -2)$

49. $\begin{cases} 3x-6y+z=-10 \\ 7y-z=2 \\ 2x+4z=14 \end{cases}$

$$\left[\begin{array}{ccc|c} 3 & -6 & 1 & -10 \\ 0 & 7 & -1 & 2 \\ 2 & 0 & 4 & 14 \end{array}\right]$$

$$\begin{array}{l} R_3 \to \\ \\ R_1 \to \end{array}\left[\begin{array}{ccc|c} 2 & 0 & 4 & 14 \\ 0 & 7 & -1 & 2 \\ 3 & -6 & 1 & -10 \end{array}\right]$$

$$\tfrac{1}{2}R_1 \to \left[\begin{array}{ccc|c} 1 & 0 & 2 & 7 \\ 0 & 7 & -1 & 2 \\ 3 & -6 & 1 & -10 \end{array}\right]$$

$$\begin{array}{r} \\ \\ -3R_1+R_3 \to \end{array}\left[\begin{array}{ccc|c} 1 & 0 & 2 & 7 \\ 0 & 7 & -1 & 2 \\ 0 & -6 & -5 & -31 \end{array}\right]$$

$$\begin{array}{r} \\ \\ 6R_2+7R_3 \to \end{array}\left[\begin{array}{ccc|c} 1 & 0 & 2 & 7 \\ 0 & 7 & -1 & 2 \\ 0 & 0 & -41 & -205 \end{array}\right]$$

$$\begin{array}{r} \\ \tfrac{1}{7}R_2 \to \\ -\tfrac{1}{41}R_3 \to \end{array}\left[\begin{array}{ccc|c} 1 & 0 & 2 & 7 \\ 0 & 1 & -\frac{1}{7} & \frac{2}{7} \\ 0 & 0 & 1 & 5 \end{array}\right]$$

$$\begin{cases} x+2z=7 \\ y-\frac{1}{7}z=\frac{2}{7} \\ z=5 \end{cases}$$

$$y-\frac{1}{7}z=\frac{2}{7}$$
$$y-\frac{1}{7}(5)=\frac{2}{7}$$
$$y-\frac{5}{7}=\frac{2}{7}$$
$$y=1$$

$$x+2z=7$$
$$x+2(5)=7$$
$$x+10=7$$
$$x=-3$$

Solution: $(-3,1,5)$

51. $\begin{cases} 2x+5y-z=10 \\ x-y-z=14 \\ x-6y=20 \end{cases}$

$$\left[\begin{array}{ccc|c} 2 & 5 & -1 & 10 \\ 1 & -1 & -1 & 14 \\ 1 & -6 & 0 & 20 \end{array}\right]$$

$$\begin{array}{r} R_2 \to \\ R_1 \to \\ \\ \end{array}\left[\begin{array}{ccc|c} 1 & -1 & -1 & 14 \\ 2 & 5 & -1 & 10 \\ 1 & -6 & 0 & 20 \end{array}\right]$$

$$\begin{array}{r} \\ -2R_1+R_2 \to \\ -R_1+R_3 \to \end{array}\left[\begin{array}{ccc|c} 1 & -1 & -1 & 14 \\ 0 & 7 & 1 & -18 \\ 0 & -5 & 1 & 6 \end{array}\right]$$

$$\begin{array}{r} \\ \tfrac{1}{7}R_2 \to \\ \\ \end{array}\left[\begin{array}{ccc|c} 1 & -1 & -1 & 14 \\ 0 & 1 & 1/7 & -18/7 \\ 0 & -5 & 1 & 6 \end{array}\right]$$

$$\begin{array}{r} \\ \\ 5R_2+R_3 \to \end{array}\left[\begin{array}{ccc|c} 1 & -1 & -1 & 14 \\ 0 & 1 & 1/7 & -18/7 \\ 0 & 0 & 12/7 & -48/7 \end{array}\right]$$

$$\begin{array}{r} \\ \\ \tfrac{7}{12}R_3 \to \end{array}\left[\begin{array}{ccc|c} 1 & -1 & -1 & 14 \\ 0 & 1 & 1/7 & -18/7 \\ 0 & 0 & 1 & -4 \end{array}\right]$$

$$\begin{cases} x-y-z=14 \\ y+\frac{1}{7}z=-\frac{18}{7} \\ z=-4 \end{cases}$$

$$y+\frac{1}{7}z=-\frac{18}{7}$$
$$y+\frac{1}{7}(-4)=-\frac{18}{7}$$
$$y-\frac{4}{7}=-\frac{18}{7}$$
$$y=-2$$

$$x-y-z=14$$
$$x-(-2)-(-4)=14$$
$$x+2+4=14$$
$$x=8$$

Solution: $(8,-2,-4)$

53. $(3,-4)$

55. $(-6,8)$

57. $(4,-3,5)$

59. $(7,-6,5)$

61. Mistake: In the second step, $4R_1+R_2$ is calculated incorrectly.

Correct: The correct calculation is $\left[\begin{array}{cc|c} 1 & 3 & 13 \\ 0 & 11 & 26 \end{array}\right]$.

The solution is $\left(\frac{65}{11},\frac{26}{11}\right)$.

63. Let x = the cost of a grilled chicken sandwich and y = the cost of a drink.

$$\begin{cases} 3x+2y=12.90 \\ 7x+4y=29.30 \end{cases}$$

$$\left[\begin{array}{cc|c} 3 & 2 & 12.9 \\ 7 & 4 & 29.3 \end{array}\right]$$

$$-7R_1+3R_2 \rightarrow \left[\begin{array}{cc|c} 3 & 2 & 12.9 \\ 0 & -2 & -2.4 \end{array}\right]$$

$$\begin{array}{r} \frac{1}{3}R_1 \rightarrow \\ -\frac{1}{2}R_2 \rightarrow \end{array} \left[\begin{array}{cc|c} 1 & 2/3 & 4.3 \\ 0 & 1 & 1.2 \end{array}\right]$$

$$\begin{cases} x+\frac{2}{3}y=4.3 \\ y=1.2 \end{cases}$$

$$x+\frac{2}{3}y=4.3$$
$$x+\frac{2}{3}(1.2)=4.3$$
$$x+0.8=4.3$$
$$x=3.5$$

The chicken sandwich costs \$3.50 and the drink costs \$1.20.

65. Let n = the length of the Nile and a = the length of the Amazon.

$$\begin{cases} n+a=8050 \\ n=250+a \end{cases}$$

$$\left[\begin{array}{cc|c} 1 & 1 & 8050 \\ 1 & -1 & 250 \end{array}\right]$$

$$-R_1+R_2 \rightarrow \left[\begin{array}{cc|c} 1 & 1 & 8050 \\ 0 & -2 & -7800 \end{array}\right]$$

$$-\frac{1}{2}R_2 \rightarrow \left[\begin{array}{cc|c} 1 & 1 & 8050 \\ 0 & 1 & 3900 \end{array}\right]$$

$$\begin{cases} n+a=8050 \\ a=3900 \end{cases}$$

$$n+a=8050$$
$$n+3900=8050$$
$$n=4150$$

The Amazon River is 3900 mi. and the Nile River is 4150 mi.

67. Let c = the amount invested in the CD at 5% interest and m = the amount invested in the money market account at 6% interest.

$$\begin{cases} c+m=10,000 \\ 0.05c+0.06m=536 \end{cases}$$

$$\left[\begin{array}{cc|c} 1 & 1 & 10,000 \\ 0.05 & 0.06 & 536 \end{array}\right]$$

$$-0.05R_1+R_2 \rightarrow \left[\begin{array}{cc|c} 1 & 1 & 10,000 \\ 0 & 0.01 & 36 \end{array}\right]$$

$$100R_2 \rightarrow \left[\begin{array}{cc|c} 1 & 1 & 10,000 \\ 0 & 1 & 3600 \end{array}\right]$$

$$\begin{cases} c+m=10,000 \\ m=3600 \end{cases}$$

$$c+m=10,000$$
$$c+3600=10,000$$
$$c=6400$$

\$6400 was invested in the CD and \$3600 was invested in the money market account.

69. Let c = the cost of a CD, b = the cost of a book, and d = the cost of a DVD.

$$\begin{cases} 2c+4b+3d=164 \\ 5c+2b+2d=160 \\ c+b=d \end{cases}$$

$$\left[\begin{array}{ccc|c} 2 & 4 & 3 & 164 \\ 5 & 2 & 2 & 160 \\ 1 & 1 & -1 & 0 \end{array}\right]$$

$$\begin{array}{r} R_3 \rightarrow \\ \\ R_1 \rightarrow \end{array} \left[\begin{array}{ccc|c} 1 & 1 & -1 & 0 \\ 5 & 2 & 2 & 160 \\ 2 & 4 & 3 & 164 \end{array}\right]$$

$$\begin{array}{r} \\ -5R_1+R_2 \rightarrow \\ -2R_1+R_3 \rightarrow \end{array} \left[\begin{array}{ccc|c} 1 & 1 & -1 & 0 \\ 0 & -3 & 7 & 160 \\ 0 & 2 & 5 & 164 \end{array}\right]$$

$$\begin{array}{r} \\ -\frac{1}{3}R_2 \rightarrow \\ 2R_2+3R_3 \rightarrow \end{array} \left[\begin{array}{ccc|c} 1 & 1 & -1 & 0 \\ 0 & 1 & -7/3 & -160/3 \\ 0 & 0 & 29 & 812 \end{array}\right]$$

$$\begin{array}{r} \\ \\ \frac{1}{29}R_3 \rightarrow \end{array} \left[\begin{array}{ccc|c} 1 & 1 & -1 & 0 \\ 0 & 1 & -7/3 & -160/3 \\ 0 & 0 & 1 & 28 \end{array}\right]$$

$$\begin{cases} c+b-d=0 \\ b-\frac{7}{3}d=-\frac{160}{3} \\ d=28 \end{cases}$$

$$b - \frac{7}{3}b = -\frac{160}{3}$$
$$b - \frac{7}{3}(28) = -\frac{160}{3}$$
$$b - \frac{196}{3} = -\frac{160}{3}$$
$$b = 12$$

$$c + b - d = 0$$
$$c + 12 - 28 = 0$$
$$c - 16 = 0$$
$$c = 16$$

The CDs cost \$16 each, the books cost \$12 each, and the DVDs cost \$28 each.

71. Let t = the number of touchdowns, e = the number of extra points, and f = the number of field goals.

$$\begin{cases} 6t + e + 3f = 17 \\ t = e \\ t = f + 1 \end{cases} = \begin{cases} 6t + e + 3f = 17 \\ t - e = 0 \\ t - f = 1 \end{cases}$$

$$\left[\begin{array}{ccc|c} 6 & 1 & 3 & 17 \\ 1 & -1 & 0 & 0 \\ 1 & 0 & -1 & 1 \end{array}\right]$$

$$\begin{array}{r} R_3 \rightarrow \\ \\ R_1 \rightarrow \end{array} \left[\begin{array}{ccc|c} 1 & 0 & -1 & 1 \\ 1 & -1 & 0 & 0 \\ 6 & 1 & 3 & 17 \end{array}\right]$$

$$\begin{array}{r} \\ -R_1 + R_2 \rightarrow \\ -6R_1 + R_3 \rightarrow \end{array} \left[\begin{array}{ccc|c} 1 & 0 & -1 & 1 \\ 0 & -1 & 1 & -1 \\ 0 & 1 & 9 & 11 \end{array}\right]$$

$$\begin{array}{r} \\ -1R_2 \rightarrow \\ R_2 + R_3 \rightarrow \end{array} \left[\begin{array}{ccc|c} 1 & 0 & -1 & 1 \\ 0 & 1 & -1 & 1 \\ 0 & 0 & 10 & 10 \end{array}\right]$$

$$\begin{array}{r} \\ \\ \frac{1}{10}R_3 \rightarrow \end{array} \left[\begin{array}{ccc|c} 1 & 0 & -1 & 1 \\ 0 & 1 & -1 & 1 \\ 0 & 0 & 1 & 1 \end{array}\right]$$

$$\begin{cases} t - f = 1 \\ e - f = 1 \\ f = 1 \end{cases}$$

$$e - f = 1 \qquad t - f = 1$$
$$e - 1 = 1 \qquad t - 1 = 1$$
$$e = 2 \qquad t = 2$$

New York scored 2 touchdowns, 2 extra points, and 1 field goal.

73. $$\begin{cases} 4v_1 - v_2 = 30 \\ -2v_1 + 5v_2 - v_3 = 10 \\ -v_2 + 5v_3 = 4 \end{cases}$$

$$\left[\begin{array}{ccc|c} 4 & -1 & 0 & 30 \\ -2 & 5 & -1 & 10 \\ 0 & -1 & 5 & 4 \end{array}\right]$$

$$\begin{array}{r} \\ R_3 \rightarrow \\ R_2 \rightarrow \end{array} \left[\begin{array}{ccc|c} 4 & -1 & 0 & 30 \\ 0 & -1 & 5 & 4 \\ -2 & 5 & -1 & 10 \end{array}\right]$$

$$\begin{array}{r} \\ -1R_2 \rightarrow \\ R_1 + 2R_3 \rightarrow \end{array} \left[\begin{array}{ccc|c} 4 & -1 & 0 & 30 \\ 0 & 1 & -5 & -4 \\ 0 & 9 & -2 & 50 \end{array}\right]$$

$$\begin{array}{r} \\ \\ -9R_2 + R_3 \rightarrow \end{array} \left[\begin{array}{ccc|c} 4 & -1 & 0 & 30 \\ 0 & 1 & -5 & -4 \\ 0 & 0 & 43 & 86 \end{array}\right]$$

$$\begin{array}{r} \frac{1}{4}R_1 \rightarrow \\ \\ \frac{1}{43}R_3 \rightarrow \end{array} \left[\begin{array}{ccc|c} 1 & -\frac{1}{4} & 0 & \frac{15}{2} \\ 0 & 1 & -5 & -4 \\ 0 & 0 & 1 & 2 \end{array}\right]$$

$$\begin{cases} v_1 - \frac{1}{4}v_2 = \frac{15}{2} \\ v_2 - 5v_3 = -4 \\ v_3 = 2 \end{cases}$$

$$v_2 - 5v_3 = -4 \qquad v_1 - \frac{1}{4}v_2 = \frac{15}{2}$$
$$v_2 - 5(2) = -4 \qquad v_1 - \frac{1}{4}(6) = \frac{15}{2}$$
$$v_2 - 10 = -4 \qquad v_1 - \frac{3}{2} = \frac{15}{2}$$
$$v_2 = 6 \qquad v_1 = 9$$

Review Exercises

1. $(-3)(4) - (-4)(5) = -12 + 20 = 8$

2. $2(2-8) + 3[-1-(-6)] - 4(4-6)$
$= 2(-6) + 3(5) - 4(-2)$
$= -12 + 15 + 8$
$= 11$

3. $\dfrac{(-5)(-1)-(9)(3)}{(2)(-1)-(3)(3)} = \dfrac{5-27}{-2-9} = \dfrac{-22}{-11} = 2$

4. $$\frac{2\left(\frac{5}{4}\right)-8\left(\frac{1}{4}\right)}{\left(\frac{1}{2}\right)\left(\frac{5}{4}\right)-\left(\frac{3}{2}\right)\left(\frac{1}{4}\right)}=\frac{\frac{5}{2}-2}{\frac{5}{8}-\frac{3}{8}}=\frac{\frac{1}{2}}{\frac{2}{8}}=2$$

5. $$\frac{(1.1)(-0.2)-(1.7)(-0.5)}{(0.2)(-0.2)-(0.5)(-0.5)}=\frac{-0.22+0.85}{-0.04+0.25}$$
$$=\frac{0.63}{0.21}$$
$$=3$$

6. $$\frac{1[-45-(-9)]+7(-9-2)-2(-27-30)}{1(6-3)-2(-9-2)-2[9-(-4)]}$$
$$=\frac{1[-45+9]+7(-11)-2(-57)}{1(3)-2(-11)-2[9+4]}$$
$$=\frac{1[-36]-77+114}{3+22-2\cdot 13}$$
$$=\frac{-36-77+114}{3+22-26}$$
$$=\frac{1}{-1}$$
$$=-1$$

Exercise Set 4.5

1. No, because it is not a square matrix.

3. Eliminate the elements in the same row and column as the element and evaluate the determinant of the remaining matrix.

5. $(3)(5)-(1)(2)=15-2=13$

7. $(-3)(4)-(2)(5)=-12-10=-22$

9. $(3)(4)-(2)(-6)=12+12=24$

11. $(-3)(5)-(-2)(4)=-15+8=-7$

13. $(-2)(5)-(-4)(-3)=-10-12=-22$

15. $(0)(7)-(-5)(3)=0+15=15$

17. $$1\begin{vmatrix}1 & 4\\ 3 & 2\end{vmatrix}-(3)\begin{vmatrix}2 & 1\\ 3 & 2\end{vmatrix}+(2)\begin{vmatrix}2 & 1\\ 1 & 4\end{vmatrix}$$
$$=1(2-12)-3(4-3)+2(8-1)$$
$$=-10-3+14$$
$$=1$$

19. $$-1\begin{vmatrix}2 & 4\\ 2 & 3\end{vmatrix}-(-3)\begin{vmatrix}2 & 0\\ 2 & 3\end{vmatrix}+(-4)\begin{vmatrix}2 & 0\\ 2 & 4\end{vmatrix}$$
$$=-1(6-8)+3(6-0)-4(8-0)$$
$$=2+18-32$$
$$=-12$$

21. $$2\begin{vmatrix}-3 & 2\\ 1 & -3\end{vmatrix}-(0)\begin{vmatrix}1 & -3\\ 1 & -3\end{vmatrix}+(4)\begin{vmatrix}1 & -3\\ -3 & 2\end{vmatrix}$$
$$=2(9-2)-0(-3+3)+4(2-9)$$
$$=14-0-28$$
$$=-14$$

23. $$0\begin{vmatrix}2 & 0\\ 4 & 3\end{vmatrix}-(3)\begin{vmatrix}4 & -2\\ 4 & 3\end{vmatrix}+(-1)\begin{vmatrix}4 & -2\\ 2 & 0\end{vmatrix}$$
$$=0(6-0)-3(12+8)-1(0+4)$$
$$=0-60-4$$
$$=-64$$

25. $(0.3)(-0.6)-(1.3)(-0.5)=-0.18+0.65=0.47$

27. $$-0.4\begin{vmatrix}1.5 & -3.2\\ -2.2 & -1.5\end{vmatrix}-(3.1)\begin{vmatrix}0.7 & -1.2\\ -2.2 & -1.5\end{vmatrix}$$
$$+(1.6)\begin{vmatrix}0.7 & -1.2\\ 1.5 & -3.2\end{vmatrix}$$
$$=-0.4(-2.25-7.04)-3.1(-1.05-2.64)$$
$$+1.6(-2.24+1.8)$$
$$=3.716+11.439-0.704$$
$$=14.451$$

29. $$\left(\frac{1}{2}\right)\left(\frac{3}{5}\right)-\left(\frac{2}{5}\right)\left(-\frac{1}{3}\right)=\frac{3}{10}+\frac{2}{15}=\frac{13}{30}$$

31. $$\frac{1}{2}\begin{vmatrix}\frac{1}{5} & -\frac{3}{2}\\ \frac{1}{2} & \frac{3}{5}\end{vmatrix}-\left(\frac{1}{3}\right)\begin{vmatrix}-\frac{3}{4} & \frac{2}{5}\\ \frac{1}{2} & \frac{3}{5}\end{vmatrix}+\left(-\frac{3}{4}\right)\begin{vmatrix}-\frac{3}{4} & \frac{2}{5}\\ \frac{1}{5} & -\frac{3}{2}\end{vmatrix}$$
$$=\frac{1}{2}\left(\frac{3}{25}+\frac{3}{4}\right)-\frac{1}{3}\left(-\frac{9}{20}-\frac{1}{5}\right)-\frac{3}{4}\left(\frac{9}{8}-\frac{2}{25}\right)$$
$$=\frac{87}{200}+\frac{13}{60}-\frac{627}{800}$$
$$=-\frac{317}{2400}$$

33. $x\begin{vmatrix} -1 & 3 \\ 0 & 1 \end{vmatrix} - (2)\begin{vmatrix} y & 1 \\ 0 & 1 \end{vmatrix} + (-2)\begin{vmatrix} y & 1 \\ -1 & 3 \end{vmatrix}$
$= x(-1-0) - 2(y-0) - 2(3y+1)$
$= -x - 2y - 6y - 2$
$= -x - 8y - 2$

35. $D = \begin{vmatrix} 1 & 1 \\ 1 & -2 \end{vmatrix} = (1)(-2) - (1)(1) = -3$

$D_x = \begin{vmatrix} -5 & 1 \\ -2 & -2 \end{vmatrix} = (-5)(-2) - (-2)(1) = 12$

$D_y = \begin{vmatrix} 1 & -5 \\ 1 & -2 \end{vmatrix} = (1)(-2) - (1)(-5) = 3$

$x = \frac{D_x}{D} = \frac{12}{-3} = -4 \qquad y = \frac{D_y}{D} = \frac{3}{-3} = -1$

$(-4, -1)$

37. $D = \begin{vmatrix} 2 & -3 \\ 1 & -1 \end{vmatrix} = (2)(-1) - (1)(-3) = 1$

$D_x = \begin{vmatrix} -6 & -3 \\ -1 & -1 \end{vmatrix} = (-6)(-1) - (-1)(-3) = 3$

$D_y = \begin{vmatrix} 2 & -6 \\ 1 & -1 \end{vmatrix} = (2)(-1) - (1)(-6) = 4$

$x = \frac{D_x}{D} = \frac{3}{1} = 3 \qquad y = \frac{D_y}{D} = \frac{4}{1} = 4$

$(3, 4)$

39. $D = \begin{vmatrix} -1 & 2 \\ 2 & -3 \end{vmatrix} = (-1)(-3) - (2)(2) = -1$

$D_x = \begin{vmatrix} -12 & 2 \\ 20 & -3 \end{vmatrix} = (-12)(-3) - (20)(2) = -4$

$D_y = \begin{vmatrix} -1 & -12 \\ 2 & 20 \end{vmatrix} = (-1)(20) - (2)(-12) = 4$

$x = \frac{D_x}{D} = \frac{-4}{-1} = 4 \qquad y = \frac{D_y}{D} = \frac{4}{-1} = -4$

$(4, -4)$

41. $D = \begin{vmatrix} 4 & -6 \\ 6 & -9 \end{vmatrix} = (4)(-9) - (6)(-6) = 0$

$D_x = \begin{vmatrix} 7 & -6 \\ 8 & -9 \end{vmatrix} = (7)(-9) - (8)(-6) = -15$

$D_y = \begin{vmatrix} 4 & 7 \\ 6 & 8 \end{vmatrix} = (4)(8) - (6)(7) = -10$

Because $D = 0$, $D_x \neq 0$, and $D_y \neq 0$, the system is inconsistent and has no solution.

43. $D = \begin{vmatrix} 8 & -3 \\ 4 & 3 \end{vmatrix} = (8)(3) - (4)(-3) = 36$

$D_x = \begin{vmatrix} 10 & -3 \\ 14 & 3 \end{vmatrix} = (10)(3) - (14)(-3) = 72$

$D_y = \begin{vmatrix} 8 & 10 \\ 4 & 14 \end{vmatrix} = (8)(14) - (4)(10) = 72$

$x = \frac{D_x}{D} = \frac{72}{36} = 2 \qquad y = \frac{D_y}{D} = \frac{72}{36} = 2$

$(2, 2)$

45. $D = \begin{vmatrix} 1/2 & -1/4 \\ 3/4 & 5/2 \end{vmatrix} = \left(\frac{1}{2}\right)\left(\frac{5}{2}\right) - \left(\frac{3}{4}\right)\left(-\frac{1}{4}\right) = \frac{23}{16}$

$D_x = \begin{vmatrix} 0 & -1/4 \\ 23/2 & 5/2 \end{vmatrix} = (0)\left(\frac{5}{2}\right) - \left(\frac{23}{2}\right)\left(-\frac{1}{4}\right) = \frac{23}{8}$

$D_y = \begin{vmatrix} 1/2 & 0 \\ 3/4 & 23/2 \end{vmatrix} = \left(\frac{1}{2}\right)\left(\frac{23}{2}\right) - \left(\frac{3}{4}\right)(0) = \frac{23}{4}$

$x = \frac{D_x}{D} = \frac{23/8}{23/16} = 2 \qquad y = \frac{D_y}{D} = \frac{23/4}{23/16} = 4$

$(2, 4)$

47. $D=\begin{vmatrix}0.2 & 0.5\\0.7 & -0.3\end{vmatrix}=(0.2)(-0.3)-(0.7)(0.5)$
$=-0.41$

$D_x=\begin{vmatrix}3.4 & 0.5\\-0.4 & -0.3\end{vmatrix}=(3.4)(-0.3)-(-0.4)(0.5)$
$=-0.82$

$D_y=\begin{vmatrix}0.2 & 3.4\\0.7 & -0.4\end{vmatrix}=(0.2)(-0.4)-(0.7)(3.4)$
$=-2.46$

$x=\dfrac{D_x}{D}=\dfrac{-0.82}{-0.41}=2$

$y=\dfrac{D_y}{D}=\dfrac{-2.46}{-0.41}=6$

$(2,6)$

49. $D=\begin{vmatrix}1 & 1 & 1\\2 & -4 & 2\\3 & 2 & 1\end{vmatrix}=1\begin{vmatrix}-4 & 2\\2 & 1\end{vmatrix}-2\begin{vmatrix}1 & 1\\2 & 1\end{vmatrix}+3\begin{vmatrix}1 & 1\\-4 & 2\end{vmatrix}$
$=1(-4-4)-2(1-2)+3(2+4)$
$=-8+2+18$
$=12$

$D_x=\begin{vmatrix}6 & 1 & 1\\6 & -4 & 2\\11 & 2 & 1\end{vmatrix}=6\begin{vmatrix}-4 & 2\\2 & 1\end{vmatrix}-6\begin{vmatrix}1 & 1\\2 & 1\end{vmatrix}+11\begin{vmatrix}1 & 1\\-4 & 2\end{vmatrix}$
$=6(-4-4)-6(1-2)+11(2+4)$
$=-48+6+66$
$=24$

$D_y=\begin{vmatrix}1 & 6 & 1\\2 & 6 & 2\\3 & 11 & 1\end{vmatrix}=1\begin{vmatrix}6 & 2\\11 & 1\end{vmatrix}-2\begin{vmatrix}6 & 1\\11 & 1\end{vmatrix}+3\begin{vmatrix}6 & 1\\6 & 2\end{vmatrix}$
$=1(6-22)-2(6-11)+3(12-6)$
$=-16+10+18$
$=12$

$D_z=\begin{vmatrix}1 & 1 & 6\\2 & -4 & 6\\3 & 2 & 11\end{vmatrix}=1\begin{vmatrix}-4 & 6\\2 & 11\end{vmatrix}-2\begin{vmatrix}1 & 6\\2 & 11\end{vmatrix}$
$+3\begin{vmatrix}1 & 6\\-4 & 6\end{vmatrix}$
$=1(-44-12)-2(11-12)+3(6+24)$
$=-56+2+90$
$=36$

$x=\dfrac{D_x}{D}=\dfrac{24}{12}=2 \qquad y=\dfrac{D_y}{D}=\dfrac{12}{12}=1$

$z=\dfrac{D_z}{D}=\dfrac{36}{12}=3 \qquad (2,1,3)$

51. $D=\begin{vmatrix}3 & 1 & -1\\2 & -1 & 2\\1 & -3 & 1\end{vmatrix}=3\begin{vmatrix}-1 & 2\\-3 & 1\end{vmatrix}-2\begin{vmatrix}1 & -1\\-3 & 1\end{vmatrix}$
$+1\begin{vmatrix}1 & -1\\-1 & 2\end{vmatrix}$
$=3(-1+6)-2(1-3)+1(2-1)$
$=15+4+1$
$=20$

$D_x=\begin{vmatrix}-4 & 1 & -1\\-7 & -1 & 2\\-6 & -3 & 1\end{vmatrix}=-4\begin{vmatrix}-1 & 2\\-3 & 1\end{vmatrix}-(-7)\begin{vmatrix}1 & -1\\-3 & 1\end{vmatrix}$
$-6\begin{vmatrix}1 & -1\\-1 & 2\end{vmatrix}$
$=-4(-1+6)+7(1-3)-6(2-1)$
$=-20-14-6$
$=-40$

$D_y=\begin{vmatrix}3 & -4 & -1\\2 & -7 & 2\\1 & -6 & 1\end{vmatrix}=3\begin{vmatrix}-7 & 2\\-6 & 1\end{vmatrix}-2\begin{vmatrix}-4 & -1\\-6 & 1\end{vmatrix}$
$+1\begin{vmatrix}-4 & -1\\-7 & 2\end{vmatrix}$
$=3(-7+12)-2(-4-6)+1(-8-7)$
$=15+20-15$
$=20$

$D_z=\begin{vmatrix}3 & 1 & -4\\2 & -1 & -7\\1 & -3 & -6\end{vmatrix}=3\begin{vmatrix}-1 & -7\\-3 & -6\end{vmatrix}-2\begin{vmatrix}1 & -4\\-3 & -6\end{vmatrix}$
$+1\begin{vmatrix}1 & -4\\-1 & -7\end{vmatrix}$
$=3(6-21)-2(-6-12)+1(-7-4)$
$=-45+36-11$
$=-20$

$x=\dfrac{D_x}{D}=\dfrac{-40}{20}=-2 \qquad y=\dfrac{D_y}{D}=\dfrac{20}{20}=1$

$z=\dfrac{D_z}{D}=\dfrac{-20}{20}=-1 \qquad (-2,1,-1)$

53. $D=\begin{vmatrix}4&2&3\\2&-4&-1\\3&0&-2\end{vmatrix}=4\begin{vmatrix}-4&-1\\0&-2\end{vmatrix}-2\begin{vmatrix}2&3\\0&-2\end{vmatrix}$

$+3\begin{vmatrix}2&3\\-4&-1\end{vmatrix}$

$=4(8-0)-2(-4-0)+3(-2+12)$

$=32+8+30$

$=70$

$D_x=\begin{vmatrix}9&2&3\\7&-4&-1\\4&0&-2\end{vmatrix}=9\begin{vmatrix}-4&-1\\0&-2\end{vmatrix}-7\begin{vmatrix}2&3\\0&-2\end{vmatrix}$

$+4\begin{vmatrix}2&3\\-4&-1\end{vmatrix}$

$=9(8-0)-7(-4-0)+4(-2+12)$

$=72+28+40$

$=140$

$D_y=\begin{vmatrix}4&9&3\\2&7&-1\\3&4&-2\end{vmatrix}=4\begin{vmatrix}7&-1\\4&-2\end{vmatrix}-2\begin{vmatrix}9&3\\4&-2\end{vmatrix}$

$+3\begin{vmatrix}9&3\\7&-1\end{vmatrix}$

$=4(-14+4)-2(-18-12)+3(-9-21)$

$=-40+60-90$

$=-70$

$D_z=\begin{vmatrix}4&2&9\\2&-4&7\\3&0&4\end{vmatrix}=4\begin{vmatrix}-4&7\\0&4\end{vmatrix}-2\begin{vmatrix}2&9\\0&4\end{vmatrix}+3\begin{vmatrix}2&9\\-4&7\end{vmatrix}$

$=4(-16-0)-2(8-0)+3(14+36)$

$=-64-16+150$

$=70$

$x=\dfrac{D_x}{D}=\dfrac{140}{70}=2 \qquad y=\dfrac{D_y}{D}=\dfrac{-70}{70}=-1$

$z=\dfrac{D_z}{D}=\dfrac{70}{70}=1 \qquad (2,-1,1)$

55. $D=\begin{vmatrix}3&2&0\\0&3&10\\6&0&-2\end{vmatrix}=3\begin{vmatrix}3&10\\0&-2\end{vmatrix}-0+6\begin{vmatrix}2&0\\3&10\end{vmatrix}$

$=3(-6-0)-0+6(20-0)$

$=-18-0+120$

$=102$

$D_x=\begin{vmatrix}-12&2&0\\-16&3&10\\3&0&-2\end{vmatrix}=-12\begin{vmatrix}3&10\\0&-2\end{vmatrix}+16\begin{vmatrix}2&0\\0&-2\end{vmatrix}$

$+3\begin{vmatrix}2&0\\3&10\end{vmatrix}$

$=-12(-6-0)+16(-4-0)+3(20-0)$

$=72-64+60$

$=68$

$D_y=\begin{vmatrix}3&-12&0\\0&-16&10\\6&3&-2\end{vmatrix}=3\begin{vmatrix}-16&10\\3&-2\end{vmatrix}-0+6\begin{vmatrix}-12&0\\-16&10\end{vmatrix}$

$=3(32-30)-0+6(-120-0)$

$=6-720$

$=-714$

$D_z=\begin{vmatrix}3&2&-12\\0&3&-16\\6&0&3\end{vmatrix}=3\begin{vmatrix}3&-16\\0&3\end{vmatrix}-0+6\begin{vmatrix}2&-12\\3&-16\end{vmatrix}$

$=3(9+0)-0+6(-32+36)$

$=27-0+24$

$=51$

$x=\dfrac{D_x}{D}=\dfrac{68}{102}=\dfrac{2}{3} \qquad y=\dfrac{D_y}{D}=\dfrac{-714}{102}=-7$

$z=\dfrac{D_z}{D}=\dfrac{51}{102}=\dfrac{1}{2} \qquad \left(\dfrac{2}{3},-7,\dfrac{1}{2}\right)$

57. $$D = \begin{vmatrix} \frac{1}{2} & \frac{1}{3} & \frac{3}{4} \\ \frac{1}{3} & \frac{2}{9} & \frac{1}{2} \\ \frac{3}{4} & \frac{1}{4} & -\frac{2}{3} \end{vmatrix} = \frac{1}{2}\begin{vmatrix} \frac{2}{9} & \frac{1}{2} \\ \frac{1}{4} & -\frac{2}{3} \end{vmatrix} - \frac{1}{3}\begin{vmatrix} \frac{1}{3} & \frac{3}{4} \\ \frac{1}{4} & -\frac{2}{3} \end{vmatrix} + \frac{3}{4}\begin{vmatrix} \frac{1}{3} & \frac{3}{4} \\ \frac{2}{9} & \frac{1}{2} \end{vmatrix}$$
$$= \frac{1}{2}\left(-\frac{4}{27} - \frac{1}{8}\right) - \frac{1}{3}\left(-\frac{2}{9} - \frac{3}{16}\right) + \frac{3}{4}\left(\frac{1}{6} - \frac{1}{6}\right)$$
$$= -\frac{59}{432} + \frac{59}{432} + 0$$
$$= 0$$

$$D_x = \begin{vmatrix} \frac{25}{12} & \frac{1}{3} & \frac{3}{4} \\ \frac{25}{18} & \frac{2}{9} & \frac{1}{2} \\ -\frac{7}{4} & \frac{1}{4} & -\frac{2}{3} \end{vmatrix} = \frac{25}{12}\begin{vmatrix} \frac{2}{9} & \frac{1}{2} \\ \frac{1}{4} & -\frac{2}{3} \end{vmatrix} - \frac{25}{18}\begin{vmatrix} \frac{1}{3} & \frac{3}{4} \\ \frac{1}{4} & -\frac{2}{3} \end{vmatrix} - \frac{7}{4}\begin{vmatrix} \frac{1}{3} & \frac{3}{4} \\ \frac{2}{9} & \frac{1}{2} \end{vmatrix}$$
$$= \frac{25}{12}\left(-\frac{4}{27} - \frac{1}{8}\right) - \frac{25}{18}\left(-\frac{2}{9} - \frac{3}{16}\right) - \frac{7}{4}\left(\frac{1}{6} - \frac{1}{6}\right)$$
$$= -\frac{1475}{2592} + \frac{1475}{2592} + 0$$
$$= 0$$

$$D_y = \begin{vmatrix} \frac{1}{2} & \frac{25}{12} & \frac{3}{4} \\ \frac{1}{3} & \frac{25}{18} & \frac{1}{2} \\ \frac{3}{4} & -\frac{7}{4} & -\frac{2}{3} \end{vmatrix} = \frac{1}{2}\begin{vmatrix} \frac{25}{18} & \frac{1}{2} \\ -\frac{7}{4} & -\frac{2}{3} \end{vmatrix} - \frac{1}{3}\begin{vmatrix} \frac{25}{12} & \frac{3}{4} \\ -\frac{7}{4} & -\frac{2}{3} \end{vmatrix} + \frac{3}{4}\begin{vmatrix} \frac{25}{12} & \frac{3}{4} \\ \frac{25}{18} & \frac{1}{2} \end{vmatrix}$$
$$= \frac{1}{2}\left(-\frac{50}{54} + \frac{7}{8}\right) - \frac{1}{3}\left(-\frac{50}{36} + \frac{21}{16}\right) + \frac{3}{4}\left(\frac{25}{24} - \frac{75}{72}\right)$$
$$= -\frac{11}{432} + \frac{11}{432} + \frac{3}{4}(0)$$
$$= 0$$

$$D_z = \begin{vmatrix} \frac{1}{2} & \frac{1}{3} & \frac{25}{12} \\ \frac{1}{3} & \frac{2}{9} & \frac{25}{18} \\ \frac{3}{4} & \frac{1}{4} & -\frac{7}{4} \end{vmatrix} = \frac{1}{2}\begin{vmatrix} \frac{2}{9} & \frac{25}{18} \\ \frac{1}{4} & -\frac{7}{4} \end{vmatrix} - \frac{1}{3}\begin{vmatrix} \frac{1}{3} & \frac{25}{12} \\ \frac{1}{4} & -\frac{7}{4} \end{vmatrix} + \frac{3}{4}\begin{vmatrix} \frac{1}{3} & \frac{25}{12} \\ \frac{2}{9} & \frac{25}{18} \end{vmatrix}$$
$$= \frac{1}{2}\left(-\frac{14}{36} - \frac{25}{72}\right) - \frac{1}{3}\left(-\frac{7}{12} - \frac{25}{48}\right) + \frac{3}{4}\left(\frac{25}{54} - \frac{50}{108}\right)$$
$$= -\frac{53}{144} + \frac{53}{144} + \frac{3}{4}(0)$$
$$= 0$$

Because $D = 0$, $D_x = 0$, $D_y = 0$, and $D_z = 0$, the system is dependent and has infinitely many solutions.

59. $$D = \begin{vmatrix} 0.3 & 0.4 & -0.6 \\ 0.5 & -0.2 & 0.7 \\ 1.4 & 1.3 & -2.2 \end{vmatrix} = 0.3\begin{vmatrix} -0.2 & 0.7 \\ 1.3 & -2.2 \end{vmatrix} - 0.5\begin{vmatrix} 0.4 & -0.6 \\ 1.3 & -2.2 \end{vmatrix} + 1.4\begin{vmatrix} 0.4 & -0.6 \\ -0.2 & 0.7 \end{vmatrix}$$
$$= 0.3(0.44 - 0.91) - 0.5(-0.88 + 0.78) + 1.4(0.28 - 0.12)$$
$$= -0.141 + 0.05 + 0.224$$
$$= 0.133$$

$$D_x = \begin{vmatrix} 2.6 & 0.4 & -0.6 \\ -0.8 & -0.2 & 0.7 \\ 9.8 & 1.3 & -2.2 \end{vmatrix} = 2.6\begin{vmatrix} -0.2 & 0.7 \\ 1.3 & -2.2 \end{vmatrix} + 0.8\begin{vmatrix} 0.4 & -0.6 \\ 1.3 & -2.2 \end{vmatrix} + 9.8\begin{vmatrix} 0.4 & -0.6 \\ -0.2 & 0.7 \end{vmatrix}$$
$$= 2.6(0.44 - 0.91) + 0.8(-0.88 + 0.78) + 9.8(0.28 - 0.12)$$
$$= -1.222 - 0.08 + 1.568$$
$$= 0.266$$

$$D_y = \begin{vmatrix} 0.3 & 2.6 & -0.6 \\ 0.5 & -0.8 & 0.7 \\ 1.4 & 9.8 & -2.2 \end{vmatrix} = 0.3\begin{vmatrix} -0.8 & 0.7 \\ 9.8 & -2.2 \end{vmatrix}$$

$$-0.5\begin{vmatrix} 2.6 & -0.6 \\ 9.8 & -2.2 \end{vmatrix} + 1.4\begin{vmatrix} 2.6 & -0.6 \\ -0.8 & 0.7 \end{vmatrix}$$

$$= 0.3(1.76 - 6.86) - 0.5(-5.72 + 5.88)$$
$$+1.4(1.82 - 0.48)$$
$$= -1.53 - 0.08 + 1.876$$
$$= 0.266$$

$$D_z = \begin{vmatrix} 0.3 & 0.4 & 2.6 \\ 0.5 & -0.2 & -0.8 \\ 1.4 & 1.3 & 9.8 \end{vmatrix} = 0.3\begin{vmatrix} -0.2 & -0.8 \\ 1.3 & 9.8 \end{vmatrix}$$

$$-0.5\begin{vmatrix} 0.4 & 2.6 \\ 1.3 & 9.8 \end{vmatrix} + 1.4\begin{vmatrix} 0.4 & 2.6 \\ -0.2 & -0.8 \end{vmatrix}$$

$$= 0.3(-1.96 + 1.04) - 0.5(3.92 - 3.38)$$
$$+1.4(-0.32 + 0.52)$$
$$= -0.276 - 0.27 + 0.28$$
$$= -0.266$$

$$x = \frac{D_x}{D} = \frac{0.266}{0.133} = 2 \qquad y = \frac{D_y}{D} = \frac{0.266}{0.133} = 2$$

$$z = \frac{D_z}{D} = \frac{-0.266}{0.133} = -2 \quad (2, 2, -2)$$

61. $(9 \cdot 5) - (-6 \cdot x) = 21$

$$45 + 6x = 21$$
$$6x = -24$$
$$x = -4$$

63. $2\begin{vmatrix} x & 1 \\ 0 & 4 \end{vmatrix} - 0 - 2\begin{vmatrix} -1 & 0 \\ x & 1 \end{vmatrix} = -38$

$$2(4x - 0) - 0 - 2(-1 - 0) = -38$$
$$8x + 2 = -38$$
$$8x = -40$$
$$x = -5$$

65. $A = \frac{1}{2}\left|\det\begin{bmatrix} 2 & 4 & 1 \\ 4 & 0 & 1 \\ 6 & 5 & 1 \end{bmatrix}\right|$

$$= \frac{1}{2}\left|2\det\begin{bmatrix} 0 & 1 \\ 5 & 1 \end{bmatrix} - 4\det\begin{bmatrix} 4 & 1 \\ 5 & 1 \end{bmatrix} + 6\det\begin{bmatrix} 4 & 1 \\ 0 & 1 \end{bmatrix}\right|$$

$$= \frac{1}{2}\left|2(0-5) - 4(4-5) + 6(4-0)\right|$$
$$= \frac{1}{2}\left|-10 + 4 + 24\right|$$
$$= \frac{1}{2}\left|18\right|$$
$$= 9$$

The area is 9 square units.

67. $A = \frac{1}{2}\left|\det\begin{bmatrix} -3 & 1 & 1 \\ 2 & -3 & 1 \\ 4 & 4 & 1 \end{bmatrix}\right|$

$$= \frac{1}{2}\left|\begin{matrix} -3\det\begin{bmatrix} -3 & 1 \\ 4 & 1 \end{bmatrix} - 2\det\begin{bmatrix} 1 & 1 \\ 4 & 1 \end{bmatrix} \\ +4\det\begin{bmatrix} 1 & 1 \\ -3 & 1 \end{bmatrix} \end{matrix}\right|$$

$$= \frac{1}{2}\left|-3(-3-4) - 2(1-4) + 4(1+3)\right|$$
$$= \frac{1}{2}\left|21 + 6 + 16\right|$$
$$= \frac{1}{2}\left|43\right|$$
$$= 21.5$$

The area is 21.5 square units.

69. $A = \frac{1}{2}\left|\det\begin{bmatrix} -4 & -1 & 1 \\ 1 & 3 & 1 \\ 3 & -3 & 1 \end{bmatrix}\right|$

$$= \frac{1}{2}\left|\begin{matrix} -4\det\begin{bmatrix} 3 & 1 \\ -3 & 1 \end{bmatrix} - 1\det\begin{bmatrix} -1 & 1 \\ -3 & 1 \end{bmatrix} \\ +3\det\begin{bmatrix} -1 & 1 \\ 3 & 1 \end{bmatrix} \end{matrix}\right|$$

$$= \frac{1}{2}\left|-4(3+3) - 1(-1+3) + 3(-1-3)\right|$$
$$= \frac{1}{2}\left|-24 - 2 - 12\right|$$
$$= \frac{1}{2}\left|-38\right|$$
$$= 19$$

The area is 19 square units.

71. Let h be the weight of Hoba West and a be the weight of Ahnighito. Translate to a system of equations and solve.

$$\begin{cases} h+a=90 \\ h=2a \end{cases} = \begin{cases} a+h=90 \\ 2a-h=0 \end{cases}$$

$$D=\begin{vmatrix} 1 & 1 \\ 2 & -1 \end{vmatrix}=1(-1)-2(1)=-3$$

$$D_a=\begin{vmatrix} 90 & 1 \\ 0 & -1 \end{vmatrix}=(90)(-1)-0\cdot 1=-90$$

$$D_h=\begin{vmatrix} 1 & 90 \\ 2 & 0 \end{vmatrix}=(1)(0)-(2)(90)=-180$$

$$a=\frac{D_a}{D}=\frac{-90}{-3}=30 \qquad h=\frac{D_h}{D}=\frac{-180}{-3}=60$$

Hoba is 60 tons and Ahnighito is 30 tons.

73. Let g be the weight of the garbanzo beans and b be the weight of the black turtle beans. Translate to a system of equations and solve.

$$\begin{cases} g+b=10 \\ g+0.7b=8.8 \end{cases}$$

$$D=\begin{vmatrix} 1 & 1 \\ 1 & 0.7 \end{vmatrix}=1(0.7)-1(1)=-0.3$$

$$D_g=\begin{vmatrix} 10 & 1 \\ 8.8 & 0.7 \end{vmatrix}=(10)(0.7)-8.8(1)=-1.8$$

$$D_b=\begin{vmatrix} 1 & 10 \\ 1 & 8.8 \end{vmatrix}=(1)(8.8)-(1)(10)=-1.2$$

$$g=\frac{D_g}{D}=\frac{-1.8}{-0.3}=6 \qquad b=\frac{D_b}{D}=\frac{-1.2}{-0.3}=4$$

He purchased 6 lbs. of garbanzo beans and 4 lbs. of black turtle beans.

75. Let z be the percent of zinc, t be the percent of tin, and c be the percent of copper.

$$\begin{cases} t=4z \\ c=19(z+t) \\ z+t+c=100 \end{cases} \text{ so } \begin{cases} 4z-t=0 \\ 19z+19t-c=0 \\ z+t+c=100 \end{cases}$$

$$D=\begin{vmatrix} 4 & -1 & 0 \\ 19 & 19 & -1 \\ 1 & 1 & 1 \end{vmatrix}=4\begin{vmatrix} 19 & -1 \\ 1 & 1 \end{vmatrix}-19\begin{vmatrix} -1 & 0 \\ 1 & 1 \end{vmatrix}+1\begin{vmatrix} -1 & 0 \\ 19 & -1 \end{vmatrix}$$
$$=4(19+1)-19(-1-0)+1(1-0)$$
$$=80+19+1$$
$$=100$$

$$D_z=\begin{vmatrix} 0 & -1 & 0 \\ 0 & 19 & -1 \\ 100 & 1 & 1 \end{vmatrix}=0-0+100\begin{vmatrix} -1 & 0 \\ 19 & -1 \end{vmatrix}$$
$$=100(1-0)$$
$$=100$$

$$D_t=\begin{vmatrix} 4 & 0 & 0 \\ 19 & 0 & -1 \\ 1 & 100 & 1 \end{vmatrix}=4\begin{vmatrix} 0 & -1 \\ 100 & 1 \end{vmatrix}-19\begin{vmatrix} 0 & 0 \\ 100 & 1 \end{vmatrix}+1\begin{vmatrix} 0 & 0 \\ 0 & -1 \end{vmatrix}$$
$$=4(0+100)-19(0-0)+1(0-0)$$
$$=400$$

$$D_c=\begin{vmatrix} 4 & -1 & 0 \\ 19 & 19 & 0 \\ 1 & 1 & 100 \end{vmatrix}=4\begin{vmatrix} 19 & 0 \\ 1 & 100 \end{vmatrix}-19\begin{vmatrix} -1 & 0 \\ 1 & 100 \end{vmatrix}+1\begin{vmatrix} -1 & 0 \\ 19 & 0 \end{vmatrix}$$
$$=4(1900-0)-19(-100-0)+1(0-0)$$
$$=7600+1900$$
$$=9500$$

$$z=\frac{D_z}{D}=\frac{100}{100}=1 \qquad t=\frac{D_t}{D}=\frac{400}{100}=4$$
$$c=\frac{D_c}{D}=\frac{9500}{100}=95$$

1% zinc, 4% tin, 95% copper

Review Exercises

1. $x > 2x - 3$

$-x > -3$

$x < 3$

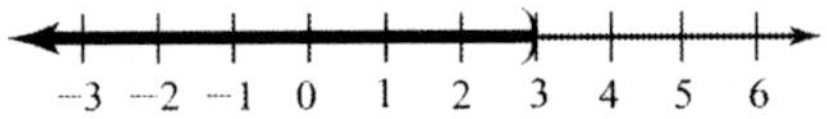

2. $2x + 3 \geq 10 - 5x$

$7x + 3 \geq 10$

$7x \geq 7$

$x \geq 1$

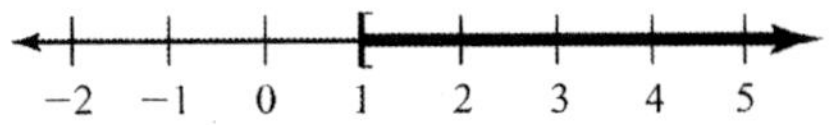

3. Let x be the first integer and $x + 1$ be the second consecutive integer.

$x + (x + 1) = 39$

$2x + 1 = 39$

$2x = 38$

$x = 19$

The integers are 19 and 19 + 1 = 20.

4. $y > 3x + 2$

Begin by graphing the related equation $y = 3x + 2$ with a dashed line. Now choose $(0,0)$ as a test point.

$$y > 3x + 2$$

$$0 \overset{?}{>} 3(0) + 2$$

$$0 \overset{?}{>} 0 + 2$$

$$0 > 2$$

Because $(0,0)$ does not satisfy the inequality, shade the side of the line on the opposite side from $(0,0)$.

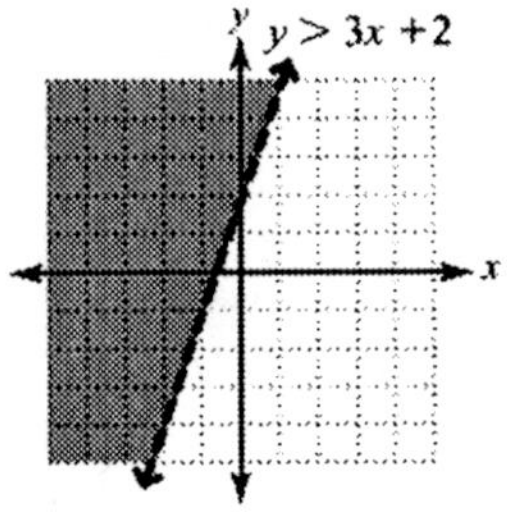

5. $x + y \leq -2$

$y \leq -x - 2$

Begin by graphing the related equation $y = -x - 2$ with a solid line. Now choose $(0,0)$ as a test point.

$$x + y \leq -2$$

$$0 + 0 \overset{?}{\leq} -2$$

$$0 \leq -2$$

Because $(0,0)$ does not satisfy the inequality, shade the side of the line on the opposite side from $(0,0)$.

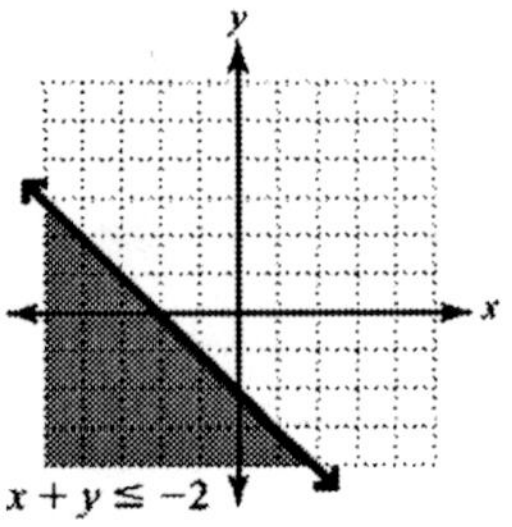

6. $\begin{cases} x = 3 - y \\ 5x + 3y = 5 \end{cases}$

$5x + 3y = 5$

$5(3 - y) + 3y = 5$

$15 - 5y + 3y = 5$

$-2y = -10$

$y = 5$

$x = 3 - y$

$x = 3 - 5$

$x = -2$

Solution: $(-2, 5)$

Exercise Set 4.6

1. Substitute the coordinates of the ordered pair into each inequality of the system. If it makes every inequality in the system true, then it is a solution for the system.

3. The boundary lines must be parallel.

5. $\begin{cases} x < 0 \\ y < 0 \end{cases}$

7. $\begin{cases} y > 2x \\ y < -2x + 4 \end{cases}$

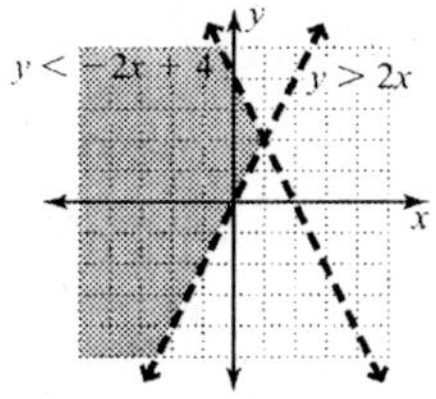

9. $\begin{cases} x - y < -5 \\ x + y < 3 \end{cases}$

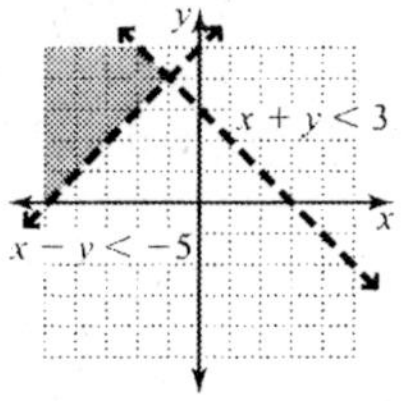

11. $\begin{cases} 2x + 3y \le 12 \\ 2x + y \le 8 \end{cases}$

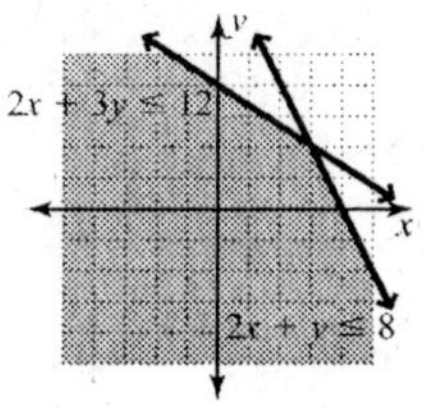

13. $\begin{cases} x + 3y \ge 6 \\ 2x - y \le 1 \end{cases}$

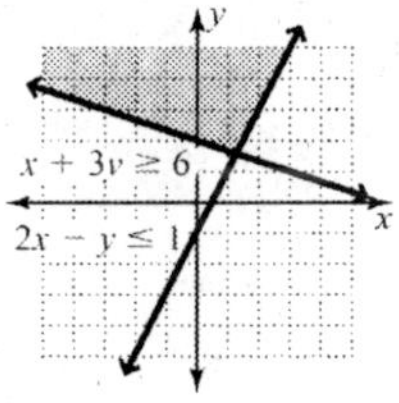

15. $\begin{cases} 2x + 3y < 1 \\ x - 4y \ge 3 \end{cases}$

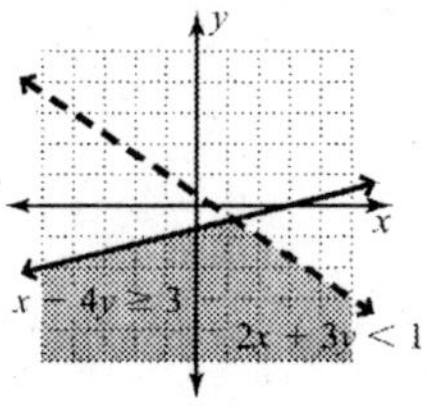

17. $\begin{cases} 2x + 5y \le 7 \\ 3x + y > -9 \end{cases}$

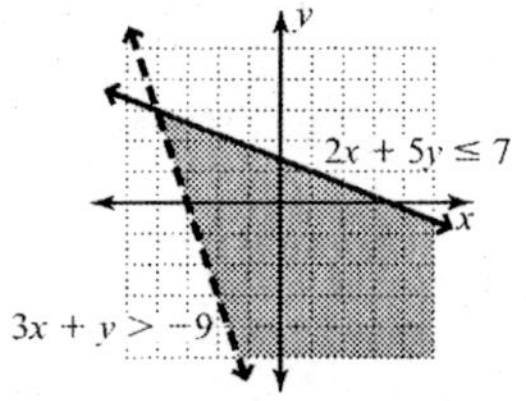

19. $\begin{cases} x + 2y < 6 \\ x + 2y \ge -2 \end{cases}$

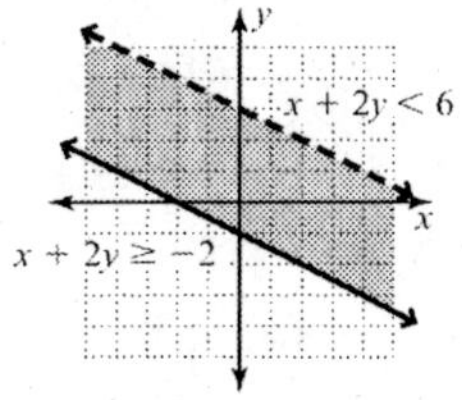

21. $\begin{cases} 4x - 2y < 8 \\ 2x - y < -4 \end{cases}$

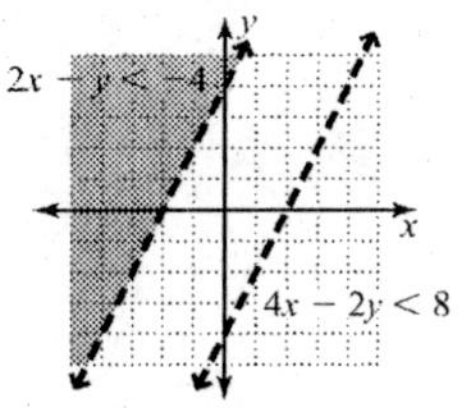

23. $\begin{cases} x + 2y > 4 \\ 3x + 6y \le -6 \end{cases}$

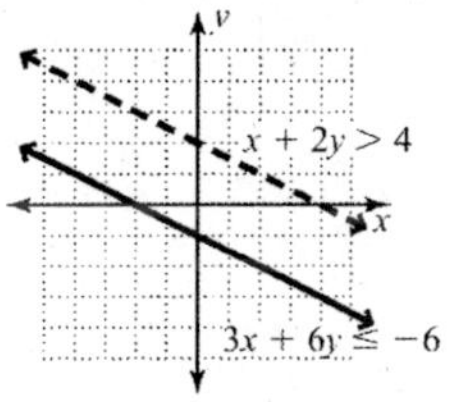

This system has no solution.

25. $\begin{cases} x + y \ge 4 \\ y \ge 2 \end{cases}$

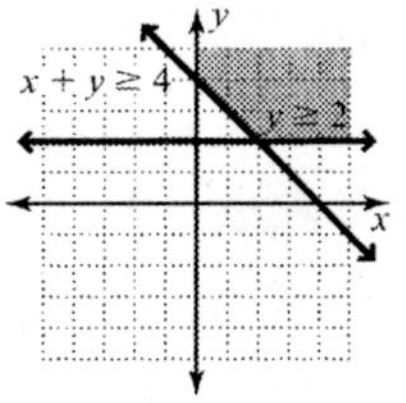

27. $\begin{cases} 2x + y \geq 0 \\ x < 3 \end{cases}$

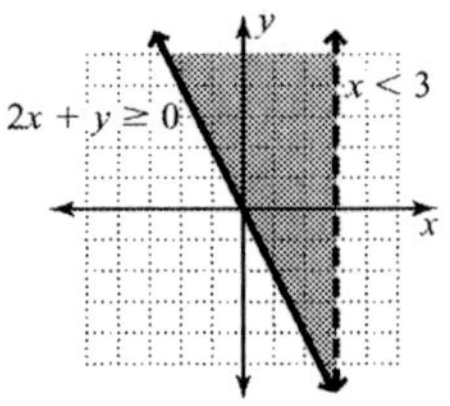

29. $\begin{cases} 5x + 3y < -4 \\ 2y \geq 4 \end{cases}$

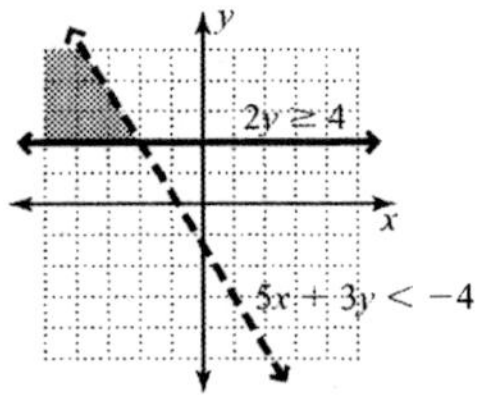

31. $\begin{cases} x \geq -2 \\ y < 4 \end{cases}$

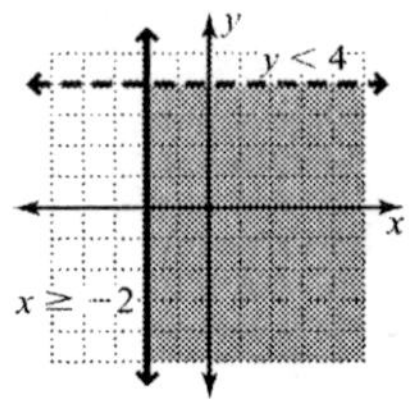

33. $\begin{cases} x < 1 \\ y \geq 0 \end{cases}$

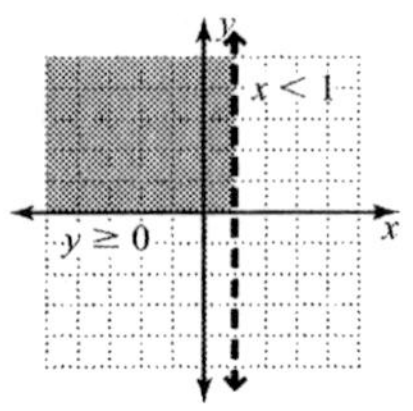

35. $\begin{cases} x + 2y \leq 4 \\ 3x + 2y \geq 6 \\ 2x - 4y < 8 \end{cases}$

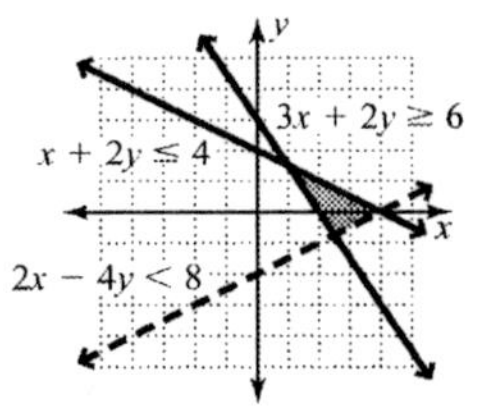

37. $\begin{cases} x + y \leq 1 \\ y - x > -5 \\ y > -3 \end{cases}$

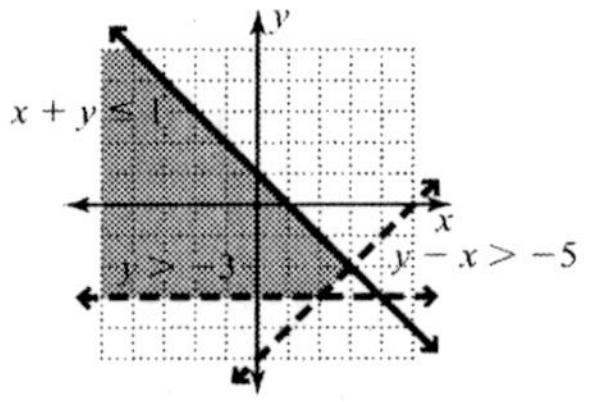

39. The boundary line for $x - y > -1$ should be a dashed line.

41. The wrong region is shaded (instead, the region containing the point (5, 0) should be shaded).

43. a) $\begin{cases} S + G \geq 75 \\ 150S + 200G \geq 12{,}000 \end{cases}$

b)

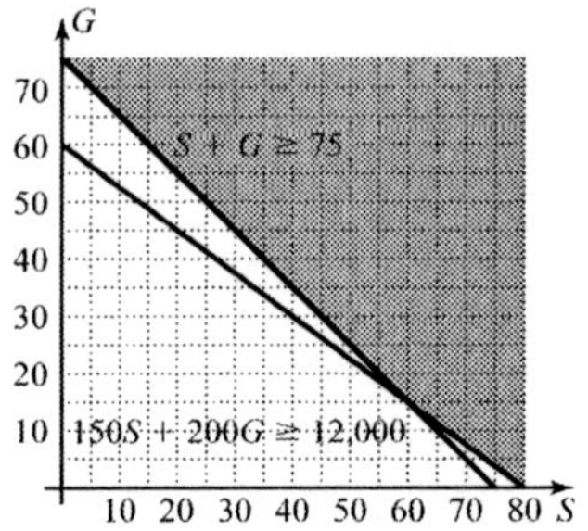

c) Answers may vary. Two examples are (20, 70) and (40, 50).

45. a) $\begin{cases} C + B \leq 5000 \\ 0.04C + 0.05B \geq 100 \end{cases}$

b)

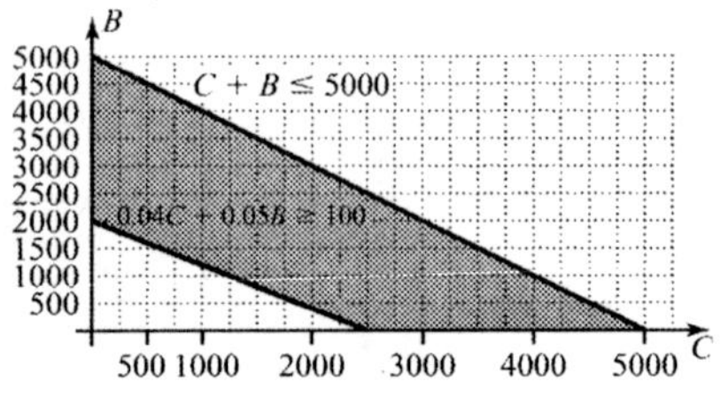

c) Answers may vary. Two examples are (2000, 3000) and (3500, 1000).

Review Exercises

1. $3^4 = 3 \cdot 3 \cdot 3 \cdot 3 = 81$

2. $-10^2 = -1(10 \cdot 10) = -100$

3. $-6(5x - 4) = -6(5x) + (-6)(-4)$
$= -30x + 24$

4. $3x - 2(2x + 7) = 3x - 4x - 14 = -x - 14$

5. $4x-9=x+3$

$3x-9=3$

$3x=12$

$x=4$

6. $3(x-5)=10-(x+1)$

$3x-15=10-x-1$

$3x-15=-x+9$

$4x-15=9$

$4x=24$

$x=6$

Chapter 4 Review Exercises

1. False; it is possible that a system of two linear equations could be satisfied by infinitely many ordered pairs.
2. True
3. True
4. True
5. False; the determinant of a matrix is a single value.
6. Replace
7. slope; y-intercepts
8. the intersection (overlap) of the two solution sets
9. the coefficient portion of the augmented matrix has 1's on the diagonal from upper left to lower right and 0's below the 1's
10. determinant
11. Is $(4,3)$ a solution of $\begin{cases} x-y=1 \\ -x+y=-1 \end{cases}$?

Equation 1	Equation 2
$x-y=1$	$-x+y=-1$
$4-3\overset{?}{=}1$	$-4+3\overset{?}{=}-1$
$1=1$	$-1=-1$

Yes, $(4,3)$ is a solution for the system because it satisfies both equations.

12. Is $(1,1)$ a solution of $\begin{cases} 2x-y=7 \\ x+y=8 \end{cases}$?

Equation 1	Equation 2
$2x-y=7$	$x+y=8$
$2(1)-1\overset{?}{=}7$	$1+1\overset{?}{=}8$
$2-1\overset{?}{=}7$	$2\neq 8$
$1\neq 7$	

No, $(1,1)$ is not a solution for the system because it satisfies neither equation.

13. a) Because these lines intersect in a single point, this system is consistent with independent equations.
b) This system has one solution.

14. a) Because the lines are parallel, this system is inconsistent.
b) This system has no solution.

15. $\begin{cases} x-3y=10 \\ 2x+3y=5 \end{cases} \rightarrow \begin{cases} y=\frac{1}{3}x-\frac{10}{3} \\ y=-\frac{2}{3}x+\frac{5}{3} \end{cases}$

a) Because these lines have different slopes, the graphs are different and the lines intersect in a single point. This system is consistent with independent equations.
b) This system has one solution.

16. $\begin{cases} x-5y=10 \\ 2x-10y=20 \end{cases} \rightarrow \begin{cases} y=\frac{1}{5}x-2 \\ y=\frac{1}{5}x-2 \end{cases}$

a) The lines have the same slope and the same y-intercept, so the graphs are identical. Because these lines coincide, this system is consistent with dependent equations.
b) This system has an infinite number of solutions.

17. $\begin{cases} 4x=y \\ 3x+y=-7 \end{cases}$ Graph each equation.

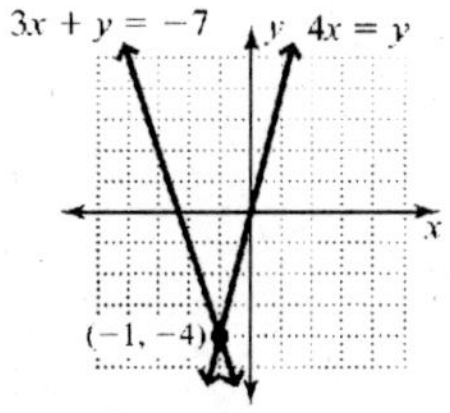

The lines intersect at a single point, which appears to be $(-1,-4)$.

18. $\begin{cases} x - y = 4 \\ 2x + 3y = 3 \end{cases}$ Graph each equation.

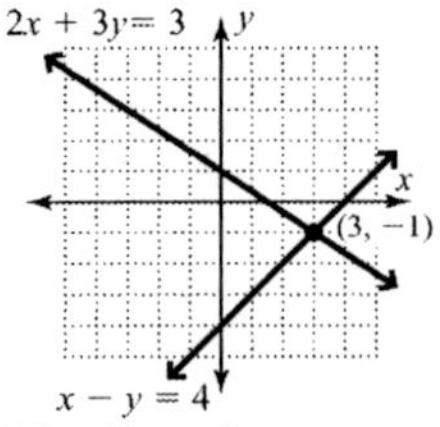

The lines intersect at a single point, which appears to be $(3, -1)$.

19. $\begin{cases} y = -2x - 3 \\ 4x + 2y = 5 \end{cases}$ Graph each equation.

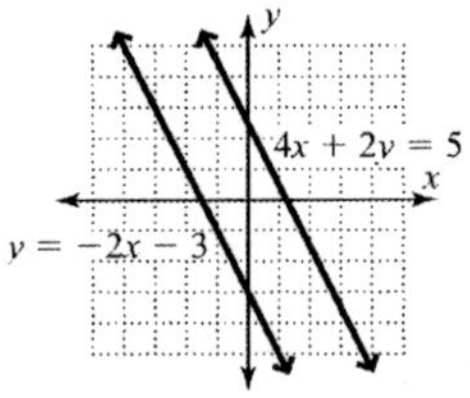

The lines appear to be parallel, so the system has no solution.

20. $\begin{cases} 3x - 2y = 6 \\ y = \frac{3}{2}x - 3 \end{cases}$ Graph each equation.

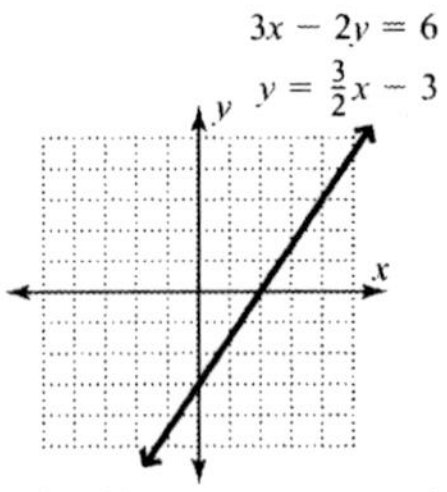

The lines appear to be identical, so the solution is the set of all ordered pairs that solve $3x - 2y = 6$.

21. $\begin{cases} 3x + 10y = 2 \\ x - 2y = 6 \end{cases}$

Solve the second equation for x: $x = 6 + 2y$

$$3x + 10y = 2$$
$$3(6 + 2y) + 10y = 2$$
$$18 + 6y + 10y = 2$$
$$16y = -16$$
$$y = -1$$

$$x = 6 + 2(-1)$$
$$x = 6 - 2$$
$$x = 4$$

Solution: $(4, -1)$

22. $\begin{cases} 2x + 5y = 8 \\ x - 10y = 9 \end{cases}$

Solve the second equation for x: $x = 9 + 10y$

$$2x + 5y = 8$$
$$2(9 + 10y) + 5y = 8$$
$$18 + 20y + 5y = 8$$
$$25y = -10$$
$$y = -\frac{2}{5}$$

$$x = 9 + 10\left(-\frac{2}{5}\right)$$
$$x = 9 - 4$$
$$x = 5$$

Solution: $\left(5, -\frac{2}{5}\right)$

23. $\begin{cases} 3y - 4x = 6 \\ y = \frac{4}{3}x + 2 \end{cases}$

The second equation is solved for y.

$$3y - 4x = 6$$
$$3\left(\frac{4}{3}x + 2\right) - 4x = 6$$
$$4x + 6 - 4x = 6$$
$$6 = 6$$

Notice that $6 = 6$ no longer has a variable and is true. The equations in the system are dependent; there are an infinite number of solutions that are all of the ordered pairs that solve $3y - 4x = 6$.

24. $\begin{cases} 8x - 2y = 12 \\ y - 4x = 3 \end{cases}$

Solve the second equation for y: $y = 4x + 3$

$$8x - 2y = 12$$
$$8x - 2(4x + 3) = 12$$
$$8x - 8x - 6 = 12$$
$$-6 \neq 12$$

Because this last statement is false, the system is inconsistent and has no solution.

25. $\begin{cases} x + y = 4 \\ x - y = -2 \end{cases}$

$$\begin{array}{l} x + y = 4 \\ \underline{x - y = -2} \\ 2x \quad = 2 \\ x \quad = 1 \end{array}$$

$$x + y = 4$$
$$1 + y = 4$$
$$y = 3$$

Solution: $(1, 3)$

26. $\begin{cases} 3x+2y=4 \\ 2x-3y=7 \end{cases}$

$$\begin{aligned} 3(3x+2y=4) \\ \underline{2(2x-3y=7)} \\ 9x+6y=12 \\ \underline{4x-6y=14} \\ 13x \quad =26 \\ x \quad =2 \end{aligned} \qquad \begin{aligned} 3x+2y&=4 \\ 3\cdot 2+2y&=4 \\ 6+2y&=4 \\ 2y&=-2 \\ y&=-1 \end{aligned}$$

Solution: $(2,-1)$

27. $\begin{cases} 0.25x+0.75y=4 \\ x-y=-4 \end{cases}$

$$\begin{aligned} -4(0.25x+0.75y=4) \\ \underline{x-y=-4} \\ -x-3y=-16 \\ \underline{x-y=-4} \\ -4y=-20 \\ y=5 \end{aligned} \qquad \begin{aligned} x-y&=-4 \\ x-5&=-4 \\ x&=1 \end{aligned}$$

Solution: $(1,5)$

28. $\begin{cases} \frac{1}{5}x-\frac{1}{3}y=2 \\ x+y=2 \end{cases}$

$$\begin{aligned} 3\left(\frac{1}{5}x-\frac{1}{3}y=2\right) \\ \underline{x+y=2} \\ \frac{3}{5}x-y=6 \\ \underline{x+y=2} \\ \frac{8}{5}x=8 \\ x=5 \end{aligned} \qquad \begin{aligned} x+y&=2 \\ 5+y&=2 \\ y&=-3 \end{aligned}$$

Solution: $(5,-3)$

29. $\begin{cases} x+y+z=-2 & \text{Eqtn. 1} \\ 2x+3y+4z=-10 & \text{Eqtn. 2} \\ 3x-2y-3z=12 & \text{Eqtn. 3} \end{cases}$

Multiply equation 1 by –2 and add to equation 2. This makes equation 4.

$$\begin{aligned} -2x-2y-2z&=4 \\ \underline{2x+3y+4z}&\underline{=-10} \\ y+2z&=-6 \quad \text{Eqtn. 4} \end{aligned}$$

Multiply equation 1 by –3 and add to equation 3. This makes equation 5.

$$\begin{aligned} -3x-3y-3z&=6 \\ \underline{3x-2y-3z}&\underline{=12} \\ -5y-6z&=18 \quad \text{Eqtn. 5} \end{aligned}$$

Use equations 4 and 5 to make a system of equations in two variables. Solve.

$$\begin{aligned} y+2z&=-6 \qquad \text{Multiply by 5} \\ \underline{-5y-6z}&\underline{=18} \\ 5y+10z&=-30 \\ \underline{-5y-6z}&\underline{=18} \\ 4z&=-12 \\ z&=-3 \end{aligned}$$

$$\begin{aligned} y+2(-3)&=-6 \\ y-6&=-6 \\ y&=0 \end{aligned} \qquad \begin{aligned} x+0-3&=-2 \\ x-3&=-2 \\ x&=1 \end{aligned}$$

Solution: $(1,0,-3)$

30. $\begin{cases} x+y+z=0 & \text{Eqtn. 1} \\ 2x-4y+3z=-12 & \text{Eqtn. 2} \\ 3x-3y+4z=2 & \text{Eqtn. 3} \end{cases}$

Multiply equation 1 by –2 and add to equation 2. This makes equation 4.

$$\begin{aligned} -2x-2y-2z&=0 \\ \underline{2x-4y+3z}&\underline{=-12} \\ -6y+z&=-12 \quad \text{Eqtn 4} \end{aligned}$$

Multiply equation 1 by –3 and add to equation 3. This makes equation 5

$$\begin{aligned} -3x-3y-3z&=0 \\ \underline{3x-3y+4z}&\underline{=2} \\ -6y+z&=2 \quad \text{Eqtn. 5} \end{aligned}$$

Use equations 4 and 5 to form a system of equations in two variables. Solve.

$$\begin{aligned} -6y+z&=-12 \qquad \text{Multiply by } -1 \\ \underline{-6y+z}&\underline{=2} \\ 6y-z&=12 \\ \underline{-6y+z}&\underline{=2} \\ 0&\neq 14 \end{aligned}$$

Because this last statement is false, the system is inconsistent and has no solution.

31. $\begin{cases} 3x+4y=-3 & \text{Eqtn. 1} \\ -2y+3z=12 & \text{Eqtn. 2} \\ 4x-3z=6 & \text{Eqtn. 3} \end{cases}$

Multiply equation 2 by 2 and add to equation 1. This makes equation 4.

$$\begin{aligned} 3x+4y&=-3 \\ \underline{-4y+6z}&\underline{=24} \\ 3x+6z&=21 \quad \text{Eqtn. 4} \end{aligned}$$

Use equations 3 and 4 to make a system of equations in two variables. Solve.

$$\begin{aligned} 4x-3z&=6 \qquad \text{Multiply by 2} \\ \underline{3x+6z}&\underline{=21} \\ 8x-6z&=12 \\ \underline{3x+6z}&\underline{=21} \\ 11x&=33 \\ x&=3 \end{aligned}$$

$$\begin{aligned} 4\cdot3-3z&=6 & 3\cdot3+4y&=-3 \\ 12-3z&=6 & 9+4y&=-3 \\ -3z&=-6 & 4y&=-12 \\ z&=2 & y&=-3 \end{aligned}$$

Solution: $(3,-3,2)$

32. $\begin{cases} 2x+2y-3z=5 & \text{Eqtn. 1} \\ x+3y-4z=0 & \text{Eqtn. 2} \\ x+2y+z=5 & \text{Eqtn. 3} \end{cases}$

Multiply equation 2 by –2 and add to equation 1. This makes equation 4.

$$\begin{aligned} 2x+2y-3z&=5 \\ \underline{-2x-6y+8z}&\underline{=0} \\ -4y+5z&=5 \quad \text{Eqtn. 4} \end{aligned}$$

Multiply equation 2 by –1 and add to equation 3. This makes equation 5.

$$\begin{aligned} -x-3y+4z&=0 \\ \underline{x+2y+z}&\underline{=5} \\ -y+5z&=5 \quad \text{Eqtn. 5} \end{aligned}$$

Use equations 4 and 5 to make a system of equations in two variables. Solve.

$$\begin{aligned} -4y+5z&=5 \qquad \text{Multiply by } -1 \\ \underline{-y+5z}&\underline{=5} \\ 4y-5z&=-5 \\ \underline{-y+5z}&\underline{=5} \\ 3y&=0 \\ y&=0 \end{aligned}$$

$$\begin{aligned} -y+5z&=5 & x+2y+z&=5 \\ 0+5z&=5 & x+2(0)+1&=5 \\ 5z&=5 & x+1&=5 \\ z&=1 & x&=4 \end{aligned}$$

Solution: $(4,0,1)$

33. $\begin{cases} x-4y=8 \\ x+2y=2 \end{cases}$

$$\left[\begin{array}{cc|c} 1 & -4 & 8 \\ 1 & 2 & 2 \end{array}\right]$$

$$-1R_1+R_2 \to \left[\begin{array}{cc|c} 1 & -4 & 8 \\ 0 & 6 & -6 \end{array}\right]$$

$$\tfrac{1}{6}R_2 \to \left[\begin{array}{cc|c} 1 & -4 & 8 \\ 0 & 1 & -1 \end{array}\right]$$

$$\begin{cases} x-4y=8 \\ \qquad y=-1 \end{cases}$$

$$\begin{aligned} x-4y&=8 \\ x-4(-1)&=8 \\ x+4&=8 \\ x&=4 \end{aligned}$$

Solution: $(4,-1)$

34. $\begin{cases} 2x - y = -4 \\ 2x - 3y = 0 \end{cases}$

$$\left[\begin{array}{cc|c} 2 & -1 & -4 \\ 2 & -3 & 0 \end{array}\right]$$

$$-R_1 + R_2 \rightarrow \left[\begin{array}{cc|c} 2 & -1 & -4 \\ 0 & -2 & 4 \end{array}\right]$$

$$\begin{array}{r} \frac{1}{2}R_1 \rightarrow \\ -\frac{1}{2}R_2 \rightarrow \end{array} \left[\begin{array}{cc|c} 1 & -\frac{1}{2} & -2 \\ 0 & 1 & -2 \end{array}\right]$$

$$\begin{cases} x - \frac{1}{2}y = -2 \\ y = -2 \end{cases}$$

$$\begin{aligned} x - \frac{1}{2}y &= -2 \\ x - \frac{1}{2}(-2) &= -2 \\ x + 1 &= -2 \\ x &= -3 \end{aligned}$$

Solution: $(-3, -2)$

35. $\begin{cases} x + y + z = 2 \\ 2x + y + 2z = 1 \\ 3x + 2y + z = 1 \end{cases}$

$$\left[\begin{array}{ccc|c} 1 & 1 & 1 & 2 \\ 2 & 1 & 2 & 1 \\ 3 & 2 & 1 & 1 \end{array}\right]$$

$$\begin{array}{r} \\ -2R_1 + R_2 \rightarrow \\ -3R_1 + R_3 \rightarrow \end{array} \left[\begin{array}{ccc|c} 1 & 1 & 1 & 2 \\ 0 & -1 & 0 & -3 \\ 0 & -1 & -2 & -5 \end{array}\right]$$

$$-1R_2 \rightarrow \left[\begin{array}{ccc|c} 1 & 1 & 1 & 2 \\ 0 & 1 & 0 & 3 \\ 0 & -1 & -2 & -5 \end{array}\right]$$

$$R_2 + R_3 \rightarrow \left[\begin{array}{ccc|c} 1 & 1 & 1 & 2 \\ 0 & 1 & 0 & 3 \\ 0 & 0 & -2 & -2 \end{array}\right]$$

$$-\frac{1}{2}R_3 \rightarrow \left[\begin{array}{ccc|c} 1 & 1 & 1 & 2 \\ 0 & 1 & 0 & 3 \\ 0 & 0 & 1 & 1 \end{array}\right]$$

$$\begin{cases} x + y + z = 2 \\ y = 3 \\ z = 1 \end{cases}$$

$$\begin{aligned} x + y + z &= 2 \\ x + 3 + 1 &= 2 \\ x + 4 &= 2 \\ x &= -2 \end{aligned}$$

Solution: $(-2, 3, 1)$

36. $\begin{cases} x + 2y + z = -1 \\ 2x - 2y + z = -10 \\ -x + 3y - 2z = 15 \end{cases}$

$$\left[\begin{array}{ccc|c} 1 & 2 & 1 & -1 \\ 2 & -2 & 1 & -10 \\ -1 & 3 & -2 & 15 \end{array}\right]$$

$$\begin{array}{r} \\ -2R_1 + R_2 \rightarrow \\ R_1 + R_3 \rightarrow \end{array} \left[\begin{array}{ccc|c} 1 & 2 & 1 & -1 \\ 0 & -6 & -1 & -8 \\ 0 & 5 & -1 & 14 \end{array}\right]$$

$$5R_2 + 6R_3 \rightarrow \left[\begin{array}{ccc|c} 1 & 2 & 1 & -1 \\ 0 & -6 & -1 & -8 \\ 0 & 0 & -11 & 44 \end{array}\right]$$

$$\begin{array}{r} \\ -\frac{1}{6}R_2 \rightarrow \\ -\frac{1}{11}R_3 \rightarrow \end{array} \left[\begin{array}{ccc|c} 1 & 2 & 1 & -1 \\ 0 & 1 & \frac{1}{6} & \frac{4}{3} \\ 0 & 0 & 1 & -4 \end{array}\right]$$

$$\begin{cases} x + 2y + z = -1 \\ y + \frac{1}{6}z = \frac{4}{3} \\ z = -4 \end{cases}$$

$$\begin{aligned} y + \frac{1}{6}z &= \frac{4}{3} \\ y + \frac{1}{6}(-4) &= \frac{4}{3} \\ y - \frac{2}{3} &= \frac{4}{3} \\ y &= 2 \end{aligned}$$

$$\begin{aligned} x + 2y + z &= -1 \\ x + 2(2) + (-4) &= -1 \\ x + 4 - 4 &= -1 \\ x &= -1 \end{aligned}$$

Solution: $(-1, 2, -4)$

37. $D = \begin{vmatrix} 9 & 4 \\ 5 & -2 \end{vmatrix} = 9(-2) - 5 \cdot 4 = -38$

$D_x = \begin{vmatrix} 18 & 4 \\ -28 & -2 \end{vmatrix} = (18)(-2) - (-28)(4) = 76$

$D_y = \begin{vmatrix} 9 & 18 \\ 5 & -28 \end{vmatrix} = 9(-28) - 5(18) = -342$

$x = \frac{D_x}{D} = \frac{76}{-38} = -2 \qquad y = \frac{D_y}{D} = \frac{-342}{-38} = 9$

Solution: $(-2, 9)$

38. $D = \begin{vmatrix} 5 & -1 \\ 10 & -2 \end{vmatrix} = 5(-2) - 10(-1) = 0$

$D_x = \begin{vmatrix} 3 & -1 \\ 5 & -2 \end{vmatrix} = 3(-2) - (5)(-1) = -1$

$D_y = \begin{vmatrix} 5 & 3 \\ 10 & 5 \end{vmatrix} = 5(5) - 10(3) = -5$

Because $D = 0$, $D_x \neq 0$, and $D_y \neq 0$, the system is inconsistent and has no solution.

39. $D = \begin{vmatrix} 1 & 1 & 1 \\ 3 & -2 & 1 \\ 1 & 3 & 4 \end{vmatrix} = 1\begin{vmatrix} -2 & 1 \\ 3 & 4 \end{vmatrix} - 3\begin{vmatrix} 1 & 1 \\ 3 & 4 \end{vmatrix} + 1\begin{vmatrix} 1 & 1 \\ -2 & 1 \end{vmatrix}$

$= 1(-8-3) - 3(4-3) + 1(1+2)$

$= -11 - 3 + 3$

$= -11$

$D_x = \begin{vmatrix} 5 & 1 & 1 \\ -3 & -2 & 1 \\ 10 & 3 & 4 \end{vmatrix} = 5\begin{vmatrix} -2 & 1 \\ 3 & 4 \end{vmatrix} + 3\begin{vmatrix} 1 & 1 \\ 3 & 4 \end{vmatrix} + 10\begin{vmatrix} 1 & 1 \\ -2 & 1 \end{vmatrix}$

$= 5(-8-3) + 3(4-3) + 10(1+2)$

$= -55 + 3 + 30$

$= -22$

$D_y = \begin{vmatrix} 1 & 5 & 1 \\ 3 & -3 & 1 \\ 1 & 10 & 4 \end{vmatrix} = 1\begin{vmatrix} -3 & 1 \\ 10 & 4 \end{vmatrix} - 3\begin{vmatrix} 5 & 1 \\ 10 & 4 \end{vmatrix} + 1\begin{vmatrix} 5 & 1 \\ -3 & 1 \end{vmatrix}$

$= 1(-12-10) - 3(20-10) + 1(5+3)$

$= -22 - 30 + 8$

$= -44$

$D_z = \begin{vmatrix} 1 & 1 & 5 \\ 3 & -2 & -3 \\ 1 & 3 & 10 \end{vmatrix} = 1\begin{vmatrix} -2 & -3 \\ 3 & 10 \end{vmatrix} - 3\begin{vmatrix} 1 & 5 \\ 3 & 10 \end{vmatrix}$

$+1\begin{vmatrix} 1 & 5 \\ -2 & -3 \end{vmatrix}$

$= 1(-20+9) - 3(10-15) + 1(-3+10)$

$= -11 + 15 + 7$

$= 11$

$x = \frac{D_x}{D} = \frac{-22}{-11} = 2 \qquad y = \frac{D_y}{D} = \frac{-44}{-11} = 4$

$z = \frac{D_z}{D} = \frac{11}{-11} = -1$

Solution: $(2, 4, -1)$

40. $D = \begin{vmatrix} 1 & 3 & 2 \\ 2 & -3 & 1 \\ -3 & 2 & 1 \end{vmatrix} = 1\begin{vmatrix} -3 & 1 \\ 2 & 1 \end{vmatrix} - 2\begin{vmatrix} 3 & 2 \\ 2 & 1 \end{vmatrix} - 3\begin{vmatrix} 3 & 2 \\ -3 & 1 \end{vmatrix}$

$= 1(-3-2) - 2(3-4) - 3(3+6)$

$= -5 + 2 - 27$

$= -30$

$D_x = \begin{vmatrix} 6 & 3 & 2 \\ -18 & -3 & 1 \\ 12 & 2 & 1 \end{vmatrix} = 6\begin{vmatrix} -3 & 1 \\ 2 & 1 \end{vmatrix} + 18\begin{vmatrix} 3 & 2 \\ 2 & 1 \end{vmatrix}$

$+12\begin{vmatrix} 3 & 2 \\ -3 & 1 \end{vmatrix}$

$= 6(-3-2) + 18(3-4) + 12(3+6)$

$= -30 - 18 + 108$

$= 60$

$D_y = \begin{vmatrix} 1 & 6 & 2 \\ 2 & -18 & 1 \\ -3 & 12 & 1 \end{vmatrix} = 1\begin{vmatrix} -18 & 1 \\ 12 & 1 \end{vmatrix} - 2\begin{vmatrix} 6 & 2 \\ 12 & 1 \end{vmatrix}$

$-3\begin{vmatrix} 6 & 2 \\ -18 & 1 \end{vmatrix}$

$= 1(-18-12) - 2(6-24) - 3(6+36)$

$= -30 + 36 - 126$

$= -120$

$$D_z = \begin{vmatrix} 1 & 3 & 6 \\ 2 & -3 & -18 \\ -3 & 2 & 12 \end{vmatrix} = 1\begin{vmatrix} -3 & -18 \\ 2 & 12 \end{vmatrix} - 2\begin{vmatrix} 3 & 6 \\ 2 & 12 \end{vmatrix}$$

$$-3\begin{vmatrix} 3 & 6 \\ -3 & -18 \end{vmatrix}$$

$$= 1(-36+36) - 2(36-12) - 3(-54+18)$$

$$= 0 - 48 + 108$$

$$= 60$$

$$x = \frac{D_x}{D} = \frac{60}{-30} = -2 \qquad y = \frac{D_y}{D} = \frac{-120}{-30} = 4$$

$$z = \frac{D_z}{D} = \frac{60}{-30} = -2$$

Solution: $(-2, 4, -2)$

41. $\begin{cases} x + y > -5 \\ x - y \geq -1 \end{cases}$

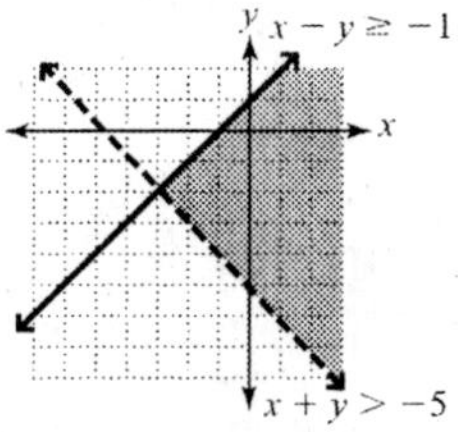

42. $\begin{cases} 2x + y \geq -4 \\ -3x + y \geq -1 \end{cases}$

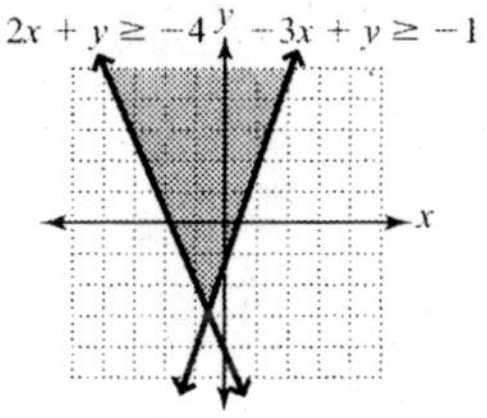

43. $\begin{cases} 2x + y > 1 \\ 3x - y \leq -1 \end{cases}$

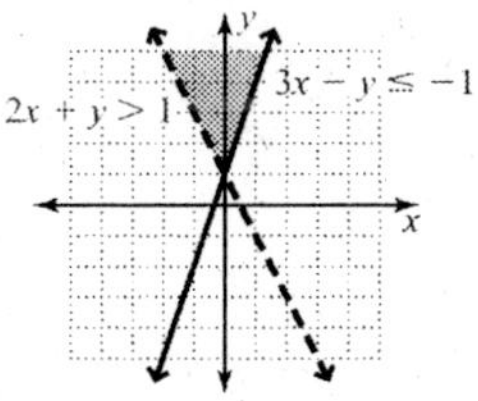

44. $\begin{cases} -3x + 4y < 12 \\ 2x - y \leq -3 \end{cases}$

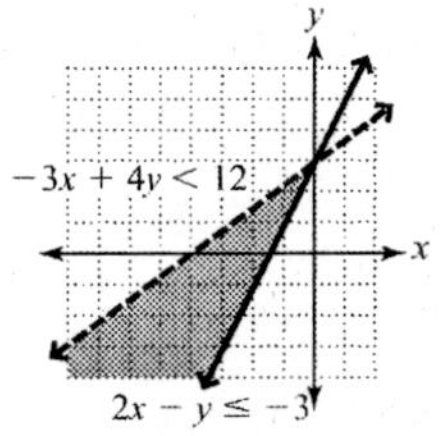

45. Let x = one number and y = the other number. Translate to a system of equations and solve.

$$\begin{cases} x + y = 16 \\ x - y = 4 \end{cases}$$

$$\begin{aligned} x + y &= 16 \\ x - y &= 4 \\ \hline 2x &= 20 \\ x &= 10 \end{aligned} \qquad \begin{aligned} x + y &= 16 \\ 10 + y &= 16 \\ y &= 6 \end{aligned}$$

The numbers are 10 and 6.

46. Complete a table.

categories	price	number	revenue
adults	7	x	$7x$
children	5.50	y	$5.50y$

Translate to a system of equations.

$$\begin{cases} x + y = 139 \\ 7x + 5.50y = 835 \end{cases}$$

Solve the first equation for y. $y = 139 - x$

$$\begin{aligned} 7x + 5.50y &= 835 \\ 7x + 5.50(139 - x) &= 835 \\ 7x + 764.5 - 5.50x &= 835 \\ 1.5x &= 70.5 \\ x &= 47 \end{aligned} \qquad \begin{aligned} y &= 139 - x \\ y &= 139 - 47 \\ y &= 92 \end{aligned}$$

47 adult tickets and 92 children's ticket were sold.

47. Let c = the number of visitors from Canada (in thousands) and m = the number of visitors from Mexico (in thousands).

$$\begin{cases} c + m = 3330 \\ c - m = 1670 \end{cases}$$

$$\begin{aligned} c + m &= 3330 \\ c - m &= 1670 \\ \hline 2c &= 5000 \\ c &= 2500 \end{aligned} \qquad \begin{aligned} c + m &= 3330 \\ 2500 + m &= 3330 \\ m &= 830 \end{aligned}$$

The United States was visited by 2500 thousand visitors from Canada and 830 thousand visitors from Mexico.

48. Let w be the percent of carbon dioxide emissions contributed by Western Europe and n be the percent of carbon dioxide emissions contributed by North America. Translate to a system of equations and solve.

$$\begin{cases} w+n=43 \\ n=2w-4 \end{cases}$$

$$w+n=43 \qquad w+n=43$$
$$w+(2w-4)=43 \qquad 15\frac{2}{3}+n=43$$
$$3w=47$$
$$w=\frac{47}{3}=15\frac{2}{3} \qquad n=27\frac{1}{3}$$

Western Europe: $15\frac{2}{3}\%$

North America: $27\frac{1}{3}\%$

49. Let x = the measure of one angle and y = the measure of the other angle. If two angles are complementary, the sum of their measures is 90°. Translate to a system of equations and solve.

$$\begin{cases} x+y=90 \\ x=\frac{1}{3}y-10 \end{cases}$$

$$x+y=90 \qquad x+y=90$$
$$\left(\frac{1}{3}y-10\right)+y=90 \qquad x+75=90$$
$$x=15$$
$$\frac{4}{3}y=100$$
$$y=75$$

The angles are 75° and 15°.

50. Create a table.

categories	APR	principal	interest
fund 1	0.07	x	$0.07x$
fund 2	0.09	y	$0.09y$

Translate to a system of equations.

$$\begin{cases} x+y=5000 \\ 0.07x+0.09y=380 \end{cases}$$

Solve the first equation for x. $x=5000-y$

$$0.07x+0.09y=380 \qquad x=5000-y$$
$$0.07(5000-y)+0.09y=380 \qquad x=5000-1500$$
$$350-0.07y+0.09y=380 \qquad x=3500$$
$$0.02y=30$$
$$y=1500$$

He invested \$1500 at 9% and \$3500 at 7%.

51. Complete a table.

Categories	Rate	Time	Distance
Yolanda	60	x	$60x$
Dee	70	y	$70y$

Translate to a system of equations.

$$\begin{cases} 60x=70y \\ x=y+\frac{1}{4} \end{cases}$$

$$60x=70y$$
$$60\left(y+\frac{1}{4}\right)=70y$$
$$60y+15=70y$$
$$15=10y$$
$$1.5=y$$

Yolanda will catch up with Dee 1.5 hours after 4:15, at 5:45 pm.

52. Let x be the speed of the plane in still air and y be the speed of the jet stream.
Complete a table.

Categories	Rate	Time	Distance
against	$x-y$	5	$5(x-y)$
with	$x+y$	3	$3(x+y)$

Translate to a system of equations.

$$\begin{cases} 3(x+y)=450 \\ 5(x-y)=450 \end{cases} \rightarrow \begin{cases} x+y=150 \\ x-y=90 \end{cases}$$

$$x+y=150$$
$$\underline{x-y=90}$$
$$2x=240$$
$$x=120$$

The speed in still air is 120 mph.

53. Create a table.

Solutions	Concentration	Volume	Amt of Alcohol
10% sol'n	0.10	x	$0.10x$
60% sol'n	0.60	y	$0.60y$
30% sol'n	0.30	50	$0.30(50)$

Translate to a system of equations.

$$\begin{cases} x+y=50 \\ 0.10x+0.60y=0.30(50) \end{cases}$$

Solve the first equation for y: $y=50-x$

$$\begin{aligned} 0.10x+0.60y&=0.30(50) \\ 0.10x+0.60(50-x)&=15 \\ 0.10x+30-0.60x&=15 \\ -0.50x&=-15 \\ x&=30 \end{aligned}$$

30 ml of the 10% solution will be needed.

54. Let v = the number of vitamin supplements, f = the number of rolls of film, and c = the number of bags of candy.

$$\begin{cases} 8v+4f+3c=32 & \text{Eqtn. 1} \\ v+f+c=8 & \text{Eqtn. 2} \\ f=c-1 & \text{Eqtn. 3} \end{cases}$$

Multiply equation 2 by -8 and add it to equation 1. This will make equation 4.

$$\begin{aligned} 8v+4f+3c&=32 \\ -8v-8f-8c&=-64 \\ \hline -4f-5c&=-32 \quad \text{Eqtn. 4} \end{aligned}$$

Use equations 3 and 4 to make a system of linear equations in two variables. Solve.

$$\begin{aligned} 4(f-c&=-1) \\ -4f-5c&=-32 \\ \hline 4f-4c&=-4 \\ -4f-5c&=-32 \\ \hline -9c&=-36 \\ c&=4 \end{aligned}$$

$$f-4=-1 \qquad v+3+4=8$$

$$f=3 \qquad v+7=8$$

$$v=1$$

He purchased 1 vitamin supplement, 3 rolls of film, and 4 bags of candy.

55. Let f = the number of first-place finishes, s = the number of second-place finishes, and t = the number of third-place finishes.

$$\begin{cases} 5f+3s+t=38 & \text{Eqtn. 1} \\ f=1+s & \text{Eqtn. 2} \\ t=3s & \text{Eqtn. 3} \end{cases}$$

Substitute equations 2 and 3 into equation 1 and solve for s.

$$\begin{aligned} 5f+3s+t&=38 \\ 5(1+s)+3s+3s&=38 \\ 5+5s+3s+3s&=38 \\ 11s&=33 \\ s&=3 \end{aligned} \qquad \begin{aligned} t&=3s \\ t&=3(3) \\ t&=9 \end{aligned} \qquad \begin{aligned} f&=1+s \\ f&=1+3 \\ f&=4 \end{aligned}$$

There were 4 first-place finishes, 3 second-place finishes, and 9 third-place finishes.

56. a. $\begin{cases} x+y\le 16 \\ 250x+400y\ge 2750 \\ x\ge 0 \\ y\ge 0 \end{cases}$

b. 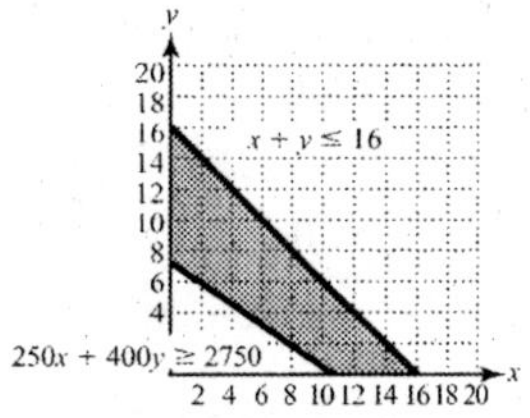

c. Answers may vary. One example is (8, 4).

Chapter 4 Practice Test

1. Is $(-1,3)$ a solution of $\begin{cases} 3x+2y=3 \\ 4x-y=-7 \end{cases}$?

Equation 1	Equation 2
$3x+2y=3$	$4x-y=-7$
$3(-1)+2(3)\overset{?}{=}3$	$4(-1)-3\overset{?}{=}-7$
$-3+6\overset{?}{=}3$	$-4-3\overset{?}{=}-7$
$3=3$	$-7=-7$

Yes, $(-1,3)$ is a solution for the system because it satisfies both equations.

2. Is $(2,2,3)$ a solution of $\begin{cases} 2x-3y+z=1 \\ -2x+y-3z=11 \\ 3x+y+3z=14 \end{cases}$?

Equation 1: $2x-3y+z = 1$

$$2(2)-3(2)+(3) \stackrel{?}{=} 1$$
$$4-6+3 \stackrel{?}{=} 1$$
$$1 = 1$$

Equation 2: $-2x+y-3z = 11$

$$-2(2)+(2)-3(3) \stackrel{?}{=} 11$$
$$-4+2-9 \stackrel{?}{=} 11$$
$$-11 \neq 11$$

Equation 3: $3x+y+3z = 14$

$$3(2)+(2)+3(3) \stackrel{?}{=} 14$$
$$6+2+9 \stackrel{?}{=} 14$$
$$17 \neq 14$$

No, $(2,2,3)$ is not a solution for the system because it does not satisfy all three equations.

3. $\begin{cases} x+3y=1 \\ -2x+y=5 \end{cases}$ Graph each equation.

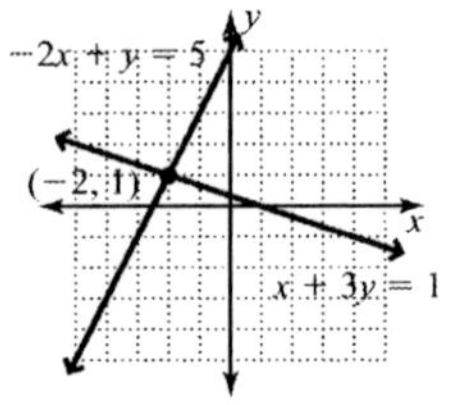

The lines intersect at a single point, which appears to be (–2, 1).

4. $\begin{cases} 2x+y=15 \\ y=7-x \end{cases}$

$$2x+y=15 \qquad y=7-x$$
$$2x+(7-x)=15 \qquad y=7-8$$
$$x+7=15 \qquad y=-1$$
$$x=8$$

Solution: $(8,-1)$

5. $\begin{cases} x-2y=1 \\ 3x-5y=4 \end{cases}$

Solve the first equation for x. $x=2y+1$

$$3x-5y=4 \qquad x=2y+1$$
$$3(2y+1)-5y=4 \qquad x=2\cdot1+1$$
$$6y+3-5y=4 \qquad x=3$$
$$y=1$$

Solution: $(3,1)$

6. $\begin{cases} 3x-2y=-8 \\ 2x+3y=-14 \end{cases}$

$3x-2y=-8$ Muliply by 3
$2x+3y=-14$ Multiply by 2

$$9x-6y=-24$$
$$4x+6y=-28$$
$$13x = -52$$
$$x = -4$$
$$3x-2y=-8$$
$$3(-4)-2y=-8$$
$$-12-2y=-8$$
$$-2y=4$$
$$y=-2$$

Solution: $(-4,-2)$

7. $\begin{cases} 4x+6y=2 \\ 6x+9y=3 \end{cases}$

$4x+6y=2$ Muliply by -3
$6x+9y=3$ Multiply by 2

$$-12x-18y=-6$$
$$12x+18y=6$$
$$0=0$$

Notice that $0=0$ no longer has a variable and is true. The equations in the system are dependent; there are an infinite number of solutions that are all ordered pairs that solve $4x+6y=2$.

8. $\begin{cases} x+2y+z=2 & \text{Eqtn. 1} \\ x+4y-z=12 & \text{Eqtn. 2} \\ 3x-3y-2z=-11 & \text{Eqtn. 3} \end{cases}$

Multiply equation 1 by –3 and add to equation 3. This makes equation 4.

$$\begin{array}{r} -3x-6y-3z=-6 \\ \underline{3x-3y-2z=-11} \\ -9y-5z=-17 \end{array} \quad \text{Eqtn. 4}$$

Multiply equation 1 by –1 and add to equation 2. This makes equation 5.

$$\begin{array}{r} -x-2y-z=-2 \\ \underline{x+4y-z=12} \\ 2y-2z=10 \end{array} \quad \text{Eqtn. 5}$$

Use equations 4 and 5 to make a system of equations in two variables. Solve.

$$\begin{array}{ll} -9y-5z=-17 & \text{Multiply by 2} \\ \underline{2y-2z=10} & \text{Multipy by 9} \end{array}$$

$$\begin{array}{r} -18y-10z=-34 \\ \underline{18y-18z=90} \\ -28z=56 \\ z=-2 \end{array}$$

$$\begin{array}{r} 2y-2(-2)=10 \quad \text{Eqtn. 5} \\ 2y+4=10 \\ 2y=6 \\ y=3 \end{array}$$

$$\begin{array}{r} x+2\cdot 3-2=2 \quad \text{Eqtn. 1} \\ x+6-2=2 \\ x+4=2 \\ x=-2 \end{array}$$

Solution: $(-2,3,-2)$

9. $\begin{cases} x+2y-3z=-9 & \text{Eqtn. 1} \\ 3x-y+2z=-8 & \text{Eqtn. 2} \\ 4x-3y+3z=-13 & \text{Eqtn. 3} \end{cases}$

Multiply equation 1 by –3 and add to equation 2. This makes equation 4.

$$\begin{array}{r} -3x-6y+9z=27 \\ \underline{3x-y+2z=-8} \\ -7y+11z=19 \end{array} \quad \text{Eqtn. 4}$$

Multiply equation 1 by –4 and add to equation 3. This makes equation 5.

$$\begin{array}{r} -4x-8y+12z=36 \\ \underline{4x-3y+3z=-13} \\ -11y+15z=23 \end{array} \quad \text{Eqtn. 5}$$

Use equations 4 and 5 to make a system of equations in two variables. Solve.

$$\begin{array}{ll} -7y+11z=19 & \text{Multiply by -11} \\ \underline{-11y+15z=23} & \text{Multipy by 7} \end{array}$$

$$\begin{array}{r} 77y-121z=-209 \\ \underline{-77y+105z=161} \\ -16z=-48 \\ z=3 \end{array}$$

$$\begin{array}{r} -7y+11\cdot 3=19 \quad \text{Eqtn. 4} \\ -7y+33=19 \\ -7y=-14 \\ y=2 \end{array}$$

$$\begin{array}{l} x+2\cdot 2-3\cdot 3=-9 \quad \text{Eqtn. 1} \\ x+4-9=-9 \\ x-5=-9 \\ x=-4 \end{array}$$

Solution: $(-4,2,3)$

10. $\begin{cases} x+2y=-6 \\ 3x+4y=-10 \end{cases}$

$$\left[\begin{array}{cc|c} 1 & 2 & -6 \\ 3 & 4 & -10 \end{array}\right]$$

$$-3R_1+R_2 \quad \left[\begin{array}{cc|c} 1 & 2 & -6 \\ 0 & -2 & 8 \end{array}\right]$$

$$-\frac{1}{2}R_2 \quad \left[\begin{array}{cc|c} 1 & 2 & -6 \\ 0 & 1 & -4 \end{array}\right]$$

$$\begin{cases} x+2y=-6 \\ \quad\;\; y=-4 \end{cases}$$

$$\begin{array}{r} x+2(-4)=-6 \\ x-8=-6 \\ x=2 \end{array}$$

Solution: $(2,-4)$

11. $\begin{cases} x+y+z=6 & \text{Eqtn. 1} \\ 3x-2y+3z=-7 & \text{Eqtn. 2} \\ 4x-2y+z=-12 & \text{Eqtn. 3} \end{cases}$

$$\left[\begin{array}{ccc|c} 1 & 1 & 1 & 6 \\ 3 & -2 & 3 & -7 \\ 4 & -2 & 1 & -12 \end{array}\right]$$

$-3R_1+R_2, -4R_1+R_2$ $\left[\begin{array}{ccc|c} 1 & 1 & 1 & 6 \\ 0 & -5 & 0 & -25 \\ 0 & -6 & -3 & -36 \end{array}\right]$

$-\frac{1}{5}R_2$ $\left[\begin{array}{ccc|c} 1 & 1 & 1 & 6 \\ 0 & 1 & 0 & 5 \\ 0 & -6 & -3 & -36 \end{array}\right]$

$6R_2+R_3$ $\left[\begin{array}{ccc|c} 1 & 1 & 1 & 6 \\ 0 & 1 & 0 & 5 \\ 0 & 0 & -3 & -6 \end{array}\right]$

$-\frac{1}{3}R_3$ $\left[\begin{array}{ccc|c} 1 & 1 & 1 & 6 \\ 0 & 1 & 0 & 5 \\ 0 & 0 & 1 & 2 \end{array}\right]$

$$\begin{cases} x+y+z=-6 \\ y=5 \\ z=2 \end{cases}$$

$x+5+2=6$

$x=-1$

$(-1,5,2)$

12. $D=\begin{vmatrix} 3 & 4 \\ 2 & -3 \end{vmatrix}=3(-3)-2\cdot 4=-17$

$D_x=\begin{vmatrix} 14 & 4 \\ -19 & -3 \end{vmatrix}=14(-3)-(-19)(4)=34$

$D_y=\begin{vmatrix} 3 & 14 \\ 2 & -19 \end{vmatrix}=3(-19)-2(14)=-85$

$x=\frac{D_x}{D}=\frac{34}{-17}=-2$ $\quad y=\frac{D_y}{D}=\frac{-85}{-17}=5$

Solution: $(-2,5)$

13. $D=\begin{vmatrix} 1 & 1 & 1 \\ 3 & -2 & 4 \\ 2 & 5 & -1 \end{vmatrix}=1\begin{vmatrix} -2 & 4 \\ 5 & -1 \end{vmatrix}-3\begin{vmatrix} 1 & 1 \\ 5 & -1 \end{vmatrix}+2\begin{vmatrix} 1 & 1 \\ -2 & 4 \end{vmatrix}$

$=1(2-20)-3(-1-5)+2(4+2)$

$=-18+18+12$

$=12$

$D_x=\begin{vmatrix} -1 & 1 & 1 \\ 0 & -2 & 4 \\ -11 & 5 & -1 \end{vmatrix}=-1\begin{vmatrix} -2 & 4 \\ 5 & -1 \end{vmatrix}+0-11\begin{vmatrix} 1 & 1 \\ -2 & 4 \end{vmatrix}$

$=-1(2-20)+0-11(4+2)$

$=18+0-66$

$=-48$

$D_y=\begin{vmatrix} 1 & -1 & 1 \\ 3 & 0 & 4 \\ 2 & -11 & -1 \end{vmatrix}=1\begin{vmatrix} 0 & 4 \\ -11 & -1 \end{vmatrix}-3\begin{vmatrix} -1 & 1 \\ -11 & -1 \end{vmatrix}+2\begin{vmatrix} -1 & 1 \\ 0 & 4 \end{vmatrix}$

$=1(0+44)-3(1+11)+2(-4-0)$

$=44-36-8$

$=0$

$D_z=\begin{vmatrix} 1 & 1 & -1 \\ 3 & -2 & 0 \\ 2 & 5 & -11 \end{vmatrix}=1\begin{vmatrix} -2 & 0 \\ 5 & -11 \end{vmatrix}-3\begin{vmatrix} 1 & -1 \\ 5 & -11 \end{vmatrix}+2\begin{vmatrix} 1 & -1 \\ -2 & 0 \end{vmatrix}$

$=1(22-0)-3(-11+5)+2(0-2)$

$=22+18-4$

$=36$

$x=\frac{D_x}{D}=\frac{-48}{12}=-4$ $\quad y=\frac{D_y}{D}=\frac{0}{12}=0$

$z=\frac{D_z}{D}=\frac{36}{12}=3$

Solution: $(-4,0,3)$

14. $\begin{cases} 2x-3y<1 \\ x+2y\le -2 \end{cases}$

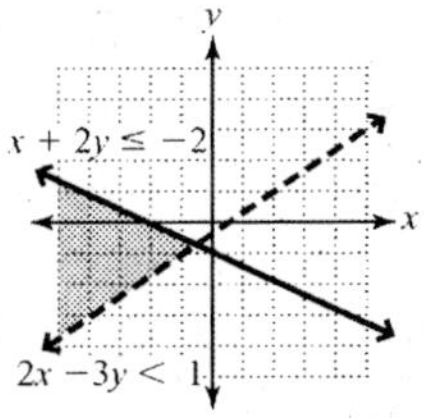

15. Let a be the number of accountants and let w be the number of waiters/waitresses. We are given that the combined number of accountants and waiters/waitresses who got headaches is 163. We are also given that 9 more accountants than waiters/waitresses got headaches. Translate to a system of equations.

$\begin{cases} a+w=163 \\ 9+w=a \end{cases}$

Substitute $9+w$ for a in the first equation.

$a+w=163$	$9+w=a$
$9+w+w=163$	$9+77=a$
$2w=154$	$86=a$
$w=77$	

Therefore, 77 waiters and 86 accountants got headaches.

16. Let l be the number of people considering the internet to be a library and h be the number of people considering the internet to be a highway. We are given that a total of 240 people were polled, and that 3 times as many people responded "library" as opposed to "highway," Translate to a system of equations.

$\begin{cases} l+h=240 \\ 3h=l \end{cases}$

Substitute $3h$ for l in the first equation.

$l+h=240$	$3h=l$
$3h+h=240$	$3(60)=l$
$4h=240$	$180=l$
$h=60$	

So, 180 people considered the internet to be a library.

17. Let x represent the rate of the boat in still water and y represent the amount that the current increases or decreases the boat's rate. We are told that with the current, the boat took 3 hours to go 30 miles. We are also given that against the current, the boat took 3 hours to go 12 miles. Create a table.

Categories	Rate	Time	Distance
With	$x+y$	3	30
Against	$x-y$	3	12

Translate to a system of equations. $\begin{cases} 3(x+y)=30 \\ 3(x-y)=12 \end{cases}$

Divide each equation by 3 to get the following system.

$\begin{cases} x+y=10 \\ x-y=4 \end{cases}$

Using elimination, solve the system.

$x+y=10$	$x+y=10$
$x-y=4$	$7+y=10$
$2x=14$	$y=3$
$x=7$	

The speed of the boat in still water is 7 mph.

18. Let x represent the amount invested at 6% interest (fund 1) and let y represent the amount invested at 8% interest (fund 2). We are given that a total of \$12,000 was invested in the two funds. We are also given that the interest earned after one year was \$880. Create a table.

Categories	APR	Principal	Interest
fund 1	0.06	x	$0.06x$
fund 2	0.08	y	$0.08y$

Translate to a system of equations.

$\begin{cases} x+y=12{,}000 \\ 0.06x+0.08y=880 \end{cases}$

Solve the first equation for x. $x=12{,}000-y$, and then substitute into the second equation.

$$0.06x+0.08y=880$$
$$0.06(12{,}000-y)+0.08y=880$$
$$720-0.06y+0.08y=880$$
$$0.02y=160$$
$$y=8000$$

$x=12{,}000-y$

$x=12{,}000-8{,}000$

$x=4000$

So, \$4000 was invested in fund 1 at 6% and \$8000 was invested in fund 2 at 8%.

19. Let a represent the number of adult tickets, c the number of child tickets, and s the number of student tickets. Translate to a system of equations.

$$\begin{cases} c+s+a=800 & \text{Eqtn. 1} \\ 3c+5s+8a=4750 & \text{Eqtn. 2} \\ a=50+2c & \text{Eqtn. 3} \end{cases}$$

Multiply equation 1 by –5 and add to equation 2. This makes equation 4.

$$\begin{aligned} -5c-5s-5a &= -4000 \\ \underline{3c+5s+8a} &\underline{= 4750} \\ -2c+3a &= 750 \quad \text{Eqtn. 4} \end{aligned}$$

Use equations 3 and 4 to form a system of equations in two variables. Solve.

$$\begin{aligned} -1(-2c+a &= 50) \\ \underline{-2c+3a} &\underline{= 750} \\ 2c-a &= -50 \\ \underline{-2c+3a} &\underline{= 750} \\ 2a &= 700 \\ a &= 350 \end{aligned}$$

$$\begin{aligned} 350 &= 50+2c \text{ Eqtn. 3} \\ 300 &= 2c \\ 150 &= c \end{aligned}$$

$$\begin{aligned} 150+s+350 &= 800 \quad \text{Eqtn. 1} \\ s+500 &= 800 \\ s &= 300 \end{aligned}$$

Therefore, there were 150 child tickets, 300 student tickets, and 350 adult tickets sold.

20. a. Let l represent the length of the garden and let w represent the width. Translate to a system of equations.

$$\begin{cases} l>0 \\ w>0 \\ 2l+2w \le 200 \\ l \ge w+10 \end{cases}$$

b. 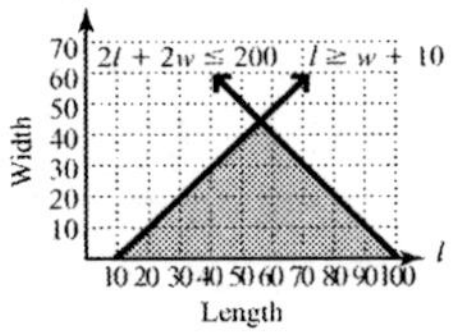

c. Answers may vary. One example is (60, 20).

Chapters 1-4 Cumulative Review

1. True

2. False; the given absolute value inequality is rewritten as $2x+4 \ge -6$ or $2x+4 \le 6$.

3. False; the slope of the first line is $m=-2$ and the slope of the second line is $m=-\dfrac{1}{2}$. Because these slopes are not of the form $m=a$ and $m=-\dfrac{1}{a}$, the lines are not perpendicular.

4. True

5. $\dfrac{4}{5}$

6. $[-3,5)$

7. 8

8. $-3^4 = -(3\cdot 3\cdot 3\cdot 3) = -81$

9. $\sqrt[3]{-64} = -4$

10. $$\begin{aligned} &8+2(10-2^2\cdot 3)+\sqrt{100-64} \\ &= 8+2(10-4\cdot 3)+\sqrt{36} \\ &= 8+2(10-12)+6 \\ &= 8+2(-2)+6 \\ &= 8-4+6 \\ &= 10 \end{aligned}$$

11. $$\begin{aligned} &3x^2-6+2x+12-8x^2+5x-3 \\ &= 3x^2-8x^2+2x+5x-6+12-3 \\ &= -5x^2+7x+3 \end{aligned}$$

12. $$\begin{aligned} 5(a-1)-9(a-2) &= -3(2a+1)-2 \\ 5a-5-9a+18 &= -6a-3-2 \\ -4a+13 &= -6a-5 \\ 2a+13 &= -5 \\ 2a &= -18 \\ a &= -9 \end{aligned}$$

13. $8-(u+2)<4(u-1)$
$$8-u-2<4u-4$$
$$6-u<4u-4$$
$$6-5u<-4$$
$$-5u<-10$$
$$u>2$$
The solution set is $(2,\infty)$.

14. $2|3x-4|+6=10$
$$2|3x-4|=4$$
$$|3x-4|=2$$
$$3x-4=-2 \quad \text{or} \quad 3x-4=2$$
$$3x=2 \qquad 3x=6$$
$$x=\frac{2}{3} \qquad x=2$$

15. $|2x-3|>7$
$$2x-3<-7 \quad \text{or} \quad 2x-3>7$$
$$2x<-4 \qquad 2x>10$$
$$x<-2 \qquad x>5$$
Solution: $\{x \mid x<-2 \text{ or } x>5\}$ or $(-\infty,-2)\cup(5,\infty)$

16. Is $(4,5)$ a solution of $5x-2y=10$?
$$5x-2y=10$$
$$5(4)-2(5)\stackrel{?}{=}10$$
$$20-10\stackrel{?}{=}10$$
$$10=10$$
Yes, $(4,5)$ is a solution.

17. $-2x+5y=-10$

x	y	Ordered Pair
0	−2	$(0,-2)$
5	0	$(5,0)$
10	2	$(10,2)$

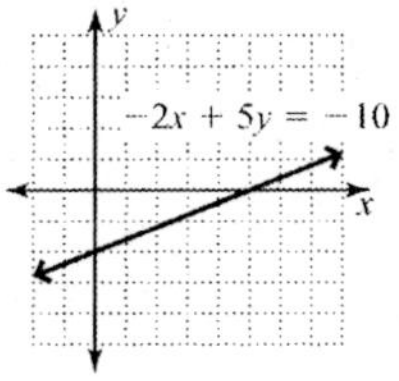

18. $y=-\frac{2}{3}x-6$

The slope is $m=-\frac{2}{3}$ and the y-intercept is $(0,-6)$.

19. $(-5,2),(-3,-1)$
$$m=\frac{y_2-y_1}{x_2-x_1}=\frac{-1-2}{-3-(-5)}=\frac{-3}{-3+5}=-\frac{3}{2}$$

20.
$$y-y_1=m(x-x_1)$$
$$y-(-3)=\frac{3}{5}(x-5)$$
$$y+3=\frac{3}{5}x-3$$
$$y=\frac{3}{5}x-6$$

21. $3x-4y\ge-12$
$$-4y\ge-3x-12$$
$$y\le\frac{3}{4}x+3$$
Begin by graphing the related equation $y=\frac{3}{4}x+3$ with a solid line. Now choose $(0,0)$ as a test point.
$$3x-4y \ge -12$$
$$3(0)-4(0) \stackrel{?}{\ge} -12$$
$$0-0 \stackrel{?}{\ge} -12$$
$$0 \ge -12$$
Because $(0,0)$ satisfies the inequality, shade the region which includes $(0,0)$.

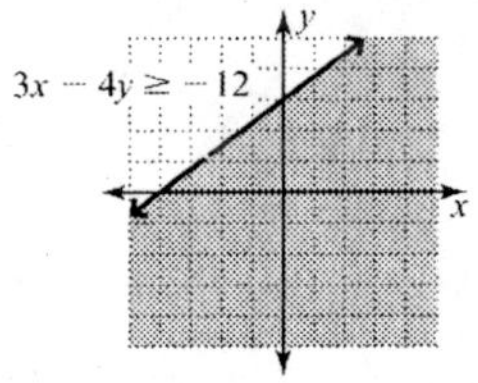

22. $f(x)=\frac{x+3}{x-4}$

a. $f(5)=\frac{5+3}{5-4}=\frac{8}{1}=8$

b. $f(4)=\frac{4+3}{4-4}=\frac{7}{0}$ undefined

23. $\begin{cases} 2x + y = 2 \\ x + 2y = -5 \end{cases}$ Graph each equation.

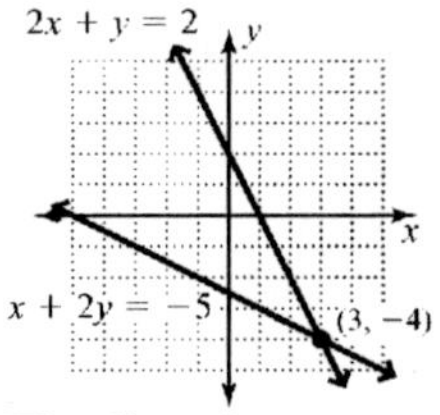

The lines intersect at a single point, which appears to be $(3, -4)$.

24. $\begin{cases} 4x + 5y = 2 \\ 3x - 2y = -10 \end{cases}$

$$\begin{aligned} 2(4x + 5y &= 2) \\ 5(3x - 2y &= -10) \\ \hline 8x + 10y &= 4 \\ 15x - 10y &= -50 \\ \hline 23x &= -46 \\ x &= -2 \end{aligned}$$

$$\begin{aligned} 4x + 5y &= 2 \\ 4(-2) + 5y &= 2 \\ -8 + 5y &= 2 \\ 5y &= 10 \\ y &= 2 \end{aligned}$$

Solution: $(-2, 2)$

25. $\begin{cases} 3x + 4y = 6 \\ x - 3y = -11 \end{cases}$

$$\left[\begin{array}{cc|c} 3 & 4 & 6 \\ 1 & -3 & -11 \end{array}\right]$$

$$\begin{array}{r} R_2 \rightarrow \\ R_1 \rightarrow \end{array} \left[\begin{array}{cc|c} 1 & -3 & -11 \\ 3 & 4 & 6 \end{array}\right]$$

$$-3R_1 + R_2 \rightarrow \left[\begin{array}{cc|c} 1 & -3 & -11 \\ 0 & 13 & 39 \end{array}\right]$$

$$\tfrac{1}{13}R_2 \rightarrow \left[\begin{array}{cc|c} 1 & -3 & -11 \\ 0 & 1 & 3 \end{array}\right]$$

$$\begin{cases} x - 3y = -11 \\ y = 3 \end{cases}$$

$$\begin{aligned} x - 3y &= -11 \\ x - 3(3) &= -11 \\ x - 9 &= -11 \\ x &= -2 \end{aligned}$$

Solution: $(-2, 3)$

26. $\begin{cases} 2x + y - 3z = 3 \\ x + 3y + 2z = 3 \\ 3x + 2y - z = 2 \end{cases}$

$$D = \begin{vmatrix} 2 & 1 & -3 \\ 1 & 3 & 2 \\ 3 & 2 & -1 \end{vmatrix} = 2\begin{vmatrix} 3 & 2 \\ 2 & -1 \end{vmatrix} - 1\begin{vmatrix} 1 & -3 \\ 2 & -1 \end{vmatrix} + 3\begin{vmatrix} 1 & -3 \\ 3 & 2 \end{vmatrix}$$

$$\begin{aligned} &= 2(-3-4) - 1(-1+6) + 3(2+9) \\ &= -14 - 5 + 33 \\ &= 14 \end{aligned}$$

$$D_x = \begin{vmatrix} 3 & 1 & -3 \\ 3 & 3 & 2 \\ 2 & 2 & -1 \end{vmatrix} = 3\begin{vmatrix} 3 & 2 \\ 2 & -1 \end{vmatrix} - 3\begin{vmatrix} 1 & -3 \\ 2 & -1 \end{vmatrix} + 2\begin{vmatrix} 1 & -3 \\ 3 & 2 \end{vmatrix}$$

$$\begin{aligned} &= 3(-3-4) - 3(-1+6) + 2(2+9) \\ &= -21 - 15 + 22 \\ &= -14 \end{aligned}$$

$$D_y = \begin{vmatrix} 2 & 3 & -3 \\ 1 & 3 & 2 \\ 3 & 2 & -1 \end{vmatrix} = 2\begin{vmatrix} 3 & 2 \\ 2 & -1 \end{vmatrix} - 1\begin{vmatrix} 3 & -3 \\ 2 & -1 \end{vmatrix} + 3\begin{vmatrix} 3 & -3 \\ 3 & 2 \end{vmatrix}$$

$$\begin{aligned} &= 2(-3-4) - 1(-3+6) + 3(6+9) \\ &= -14 - 3 + 45 \\ &= 28 \end{aligned}$$

$$D_z = \begin{vmatrix} 2 & 1 & 3 \\ 1 & 3 & 3 \\ 3 & 2 & 2 \end{vmatrix} = 2\begin{vmatrix} 3 & 3 \\ 2 & 2 \end{vmatrix} - 1\begin{vmatrix} 1 & 3 \\ 2 & 2 \end{vmatrix} + 3\begin{vmatrix} 1 & 3 \\ 3 & 3 \end{vmatrix}$$

$$\begin{aligned} &= 2(6-6) - 1(2-6) + 3(3-9) \\ &= 0 + 4 - 18 \\ &= -14 \end{aligned}$$

$$x = \frac{D_x}{D} = \frac{-14}{14} = -1 \qquad y = \frac{D_y}{D} = \frac{28}{14} = 2$$

$$z = \frac{D_z}{D} = \frac{-14}{14} = -1 \qquad (-1, 2, -1)$$

27. Let p represent the wholesale price. Because \$32,450 is the retail price, it is the result of adding 18% of the wholesale price to the wholesale price. Translate to an equation and solve for p.

$$0.18p + p = 32,450$$
$$1.18p = 32,450$$
$$\frac{1.18p}{1.18} = \frac{32,450}{1.18}$$
$$p = 27,500$$

The wholesale price is \$27,500.

28. Let x = the amount in the account at 4% and y = the amount in the account at 5%

Categories	Interest Rate	Amount in Account	Interest
4% account	0.04	x	$0.04x$
5% account	0.05	y	$0.05y$

$$\begin{cases} x - y = 6000 \\ 0.04x + 0.05y = 1860 \end{cases}$$

Solve the first equation for x: $x = y + 6000$

$$0.04x + 0.05y = 1860$$
$$0.04(y + 6000) + 0.05y = 1860$$
$$0.04y + 240 + 0.05y = 1860$$
$$0.09y + 240 = 1860$$
$$0.09y = 1620$$
$$y = 18,000$$

$$x = y + 6000$$
$$x = 18,000 + 6000$$
$$x = 24,000$$

He invested \$18,000 at 5% and \$24,000 at 4%.

29. Let x = the boat's speed in mph and y = the speed of the current in mph.

$$\begin{cases} 2(x - y) = 40 \\ 2(x + y) = 56 \end{cases} \Rightarrow \begin{cases} x - y = 20 \\ x + y = 28 \end{cases}$$

$$x - y = 20$$
$$\underline{x + y = 28}$$
$$2x = 48$$
$$x = 24$$

$$x + y = 28$$
$$24 + y = 28$$
$$y = 4$$

The boat's speed is 24 mph and the current's speed is 4 mph.

30. Create a table.

	concentration	vol. of solution	vol. of saline
15%	0.15	x	$0.15x$
20%	0.20	40	$0.20(40)$
17%	0.17	$x + 40$	$0.17(x + 40)$

Translate the information in the table to an equation and solve.

$$0.15x + 0.20(40) = 0.17(x + 40)$$
$$0.15x + 8 = 0.17x + 6.8$$
$$-0.02x + 8 = 6.8$$
$$-0.02x = -1.2$$
$$x = 60$$

60 ml of the 15% solution should be added.

Chapter 5

Exponents, Polynomials, and Polynomial Functions

Exercise Set 5.1

1. Keep the same base and add the exponents.

3. The result is positive because $5^{-2} = \frac{1}{5^2}$, which is positive.

5. a. Locate the new decimal point position, which is to the right of the first nonzero digit in the number.
b. Determine the power of 10. The power is the number of digits between the old decimal point position and the new position expressed as a positive power.

7. $mn^2 \cdot m^3n^3 = m^{1+3}n^{2+3} = m^4n^5$

9. $3^{10} \cdot 3^2 = 3^{10+2} = 3^{12}$

11. $(-3)^5(-3)^3 = (-3)^{5+3} = (-3)^8$ or 3^8

13. $(-4p^4q^3)(3p^2q^2) = -12p^{4+2}q^{3+2} = -12p^6q^5$

15. $(6r^3st^4)(-3r^4s^5t^8) = 6(-3)r^{3+4}s^{1+5}t^{4+8}$
$= -18r^7s^6t^{12}$

17. $(1.2u^3t^9)(3.1u^4t^2) = (1.2)(3.1)u^{3+4}t^{9+2} = 3.72u^7t^{11}$

19. $15^0 = 1$

21. $4x^0 = 4(1) = 4$

23. $(3xy^2)^0 = 1$

25. $-5^2 = -(25) = -25$

27. $2^{-3} = \frac{1}{2^3} = \frac{1}{8}$

29. $-3^{-4} = -\frac{1}{3^4} = -\frac{1}{81}$

31. $\frac{1}{5^{-3}} = 5^3 = 125$

33. $\frac{4}{x^{-3}} = 4x^3$

35. $\frac{1}{4a^{-6}} = \frac{a^6}{4}$

37. Mistake: Assigned the minus sign to the base.
Correct: It should be $-(2\cdot2\cdot2\cdot2) = -16$

39. Mistake: Multiplied 4 by -1.
Correct: It should be $\frac{1}{4}$.

41. $\frac{h^5}{h^2} = h^{5-2} = h^3$

43. $\frac{a^2}{a^7} = a^{2-7} = a^{-5} = \frac{1}{a^5}$

45. $\frac{6x^2y^6}{3xy^3} = 2x^{2-1}y^{6-3} = 2xy^3$

47. $\frac{18r^3s^7t}{-12r^6s^4t} = -\frac{3}{2}r^{3-6}s^{7-4}t^{1-1} = -\frac{3}{2}r^{-3}s^3t^0 = -\frac{3s^3}{2r^3}$

49. $\frac{15u^{-8}v^3w^4}{-21u^3v^3w^{-2}} = -\frac{5}{7}u^{-8-3}v^{3-3}w^{4-(-2)}$
$= -\frac{5}{7}u^{-11}v^0w^6$
$= -\frac{5w^6}{7u^{11}}$

51. $\frac{8a^{-2}b^3c^8}{2a^3b^{-4}c^5} = 4a^{-2-3}b^{3-(-4)}c^{8-5} = 4a^{-5}b^7c^3 = \frac{4b^7c^3}{a^5}$

53. $(x^3)^4 = x^{3\cdot4} = x^{12}$

55. $(2x^5)^4 = 2^4x^{5\cdot4} = 16x^{20}$

57. $(-5x^3y)^2 = (-5)^2x^{3\cdot2}y^{1\cdot2} = 25x^6y^2$

59. $\left(\frac{3}{4}a^2b^4\right)^3 = \left(\frac{3}{4}\right)^3 a^{2\cdot3}b^{4\cdot3} = \frac{27}{64}a^6b^{12}$

61. $(-0.3r^2t^4u)^3 = (-0.3)^3 r^{2\cdot3}t^{4\cdot3}u^{1\cdot3} = -0.027r^6t^{12}u^3$

63. $\left(\frac{c}{d}\right)^4 = \frac{c^4}{d^4}$

65. $\left(\frac{3}{x}\right)^3 = \frac{3^3}{x^3} = \frac{27}{x^3}$

67. $\left(-\frac{4}{x^3}\right)^2 = \frac{(-4)^2}{(x^3)^2} = \frac{16}{x^6}$

69. $\left(\frac{a}{b}\right)^{-6} = \left(\frac{b}{a}\right)^6 = \frac{b^6}{a^6}$

71. $\left(\frac{x^2}{y^3}\right)^{-4} = \left(\frac{y^3}{x^2}\right)^4 = \frac{y^{12}}{x^8}$

73. $\left(\frac{4}{x^{-2}}\right)^{-3} = \left(\frac{x^{-2}}{4}\right)^3 = \frac{x^{-6}}{4^3} = \frac{1}{64x^6}$

75. $\left(\frac{3x^2}{y^3}\right)^{-4} = \left(\frac{y^3}{3x^2}\right)^4 = \frac{(y^3)^4}{(3)^4(x^2)^4} = \frac{y^{12}}{81x^8}$

77. $(4x^2y^3)^2(-2x^3y^4)^3$
$= 4^2(x^2)^2(y^3)^2 \cdot (-2)^3(x^3)^3(y^4)^3$
$= 16x^4y^6 \cdot (-8)x^9y^{12}$
$= -128x^{13}y^{18}$

79. $\frac{(3q^4p^2)^2}{(5q^2p^2)^3} = \frac{3^2(q^4)^2(p^2)^2}{5^3(q^2)^3(p^2)^3} = \frac{9q^8p^4}{125q^6p^6} = \frac{9q^2}{125p^2}$

81. $(3h^3t^5)^{-2}(9h^2t^4)^2$
$= \frac{1}{(3)^2(h^3)^2(t^5)^2} \cdot (9)^2(h^2)^2(t^4)^2$
$= \frac{1}{9h^6t^{10}} \cdot 81h^4t^8$
$= \frac{81h^4t^8}{9h^6t^{10}}$
$= \frac{9}{h^2t^2}$

83. $\frac{(9u^3v^2)^{-1}}{(2u^4v^2)^4} = \frac{1}{9u^3v^2} \cdot \frac{1}{2^4(u^4)^4(v^2)^4}$
$= \frac{1}{9u^3v^2} \cdot \frac{1}{16u^{16}v^8}$
$= \frac{1}{144u^{19}v^{10}}$

85. $\frac{(2u^2v^3)^{-3}(4u^{-2}v^3)}{(3u^2v^{-3})^{-2}} = \frac{(4u^{-2}v^3)(3u^2v^{-3})^2}{(2u^2v^3)^3}$
$= \frac{(4u^{-2}v^3)(3)^2(u^2)^2(v^{-3})^2}{(2)^3(u^2)^3(v^3)^3}$
$= \frac{(4u^{-2}v^3) \cdot 9u^4v^{-6}}{8u^6v^9}$
$= \frac{36u^2v^{-3}}{8u^6v^9}$
$= \frac{9}{2u^4v^{12}}$

87. $\frac{(-4a^{-2}b^3c^2)^{-1}(2abc)^{-2}}{(3a^5b^2)^3(2a^{-1}b^{-3}c)^{-3}}$
$= \frac{(2a^{-1}b^{-3}c)^3}{(3a^5b^2)^3(-4a^{-2}b^3c^2)(2abc)^2}$
$= \frac{8a^{-3}b^{-9}c^3}{(27a^{15}b^6)(-4a^{-2}b^3c^2)(4a^2b^2c^2)}$
$= \frac{8a^{-3}b^{-9}c^3}{-432a^{15}b^{11}c^4}$
$= -\frac{1}{54a^{18}b^{20}c}$

89. $2.9 \times 10^6 = 2{,}900{,}000$

91. $2 \times 10^{11} = 200{,}000{,}000{,}000$

93. $5.6 \times 10^{10} = 56{,}000{,}000{,}000$

95. $1.7 \times 10^{-7} = 0.00000017$

97. $1.675 \times 10^{-24} =$
$0.000000000000000000000001675$

99. $6.645 \times 10^{-27} =$
$0.000000000000000000000000006645$

101. $50{,}150{,}000 = 5.015 \times 10^7$

103. $9{,}567{,}000{,}000{,}000 = 9.567 \times 10^{12}$

105. $8{,}909{,}000{,}000 = 8.909 \times 10^9$

107. $0.00000055 = 5.5 \times 10^{-7}$

109. $0.0000001 = 1 \times 10^{-7}$

111. $0.00000000000000209 = 2.09\times10^{-15}$

113. $6\times10^5, 7.4\times10^6, 8.3\times10^6, 1.2\times10^7, 2.4\times10^8$

115. $(2100)(30,000) = (2.1\times10^3)(3.0\times10^4)$
$= 2.1\times3.0\times10^3\times10^4$
$= 6.3\times10^{3+4}$
$= 6.3\times10^7$

117. $(-32,000)(410,000) = (-3.2\times10^4)(4.1\times10^5)$
$= -3.2\times4.1\times10^4\times10^5$
$= -13.12\times10^{4+5}$
$= -13.12\times10^9$
$= -1.312\times10^{10}$

119. $(0.00081)(220,000) = (8.1\times10^{-4})(2.2\times10^5)$
$= 8.1\times2.2\times10^{-4}\times10^5$
$= 17.82\times10^{-4+5}$
$= 17.82\times10^1$
$= 1.782\times10^2$

121. $\dfrac{8,400,000}{210} = \dfrac{8.4\times10^6}{2.1\times10^2}$
$= \dfrac{8.4}{2.1}\times\dfrac{10^6}{10^2}$
$= 4\times10^{6-2}$
$= 4\times10^4$

123. $\dfrac{930,000,000,000,000}{-3,000,000} = \dfrac{9.3\times10^{14}}{-3.0\times10^6}$
$= \dfrac{9.3}{-3.0}\times\dfrac{10^{14}}{10^6}$
$= -3.1\times10^{14-6}$
$= -3.1\times10^8$

125. $\dfrac{0.0057}{0.0000095} = \dfrac{5.7\times10^{-3}}{9.5\times10^{-6}}$
$= \dfrac{5.7}{9.5}\times\dfrac{10^{-3}}{10^{-6}}$
$= 0.6\times10^{-3-(-6)}$
$= 0.6\times10^3$
$= 6\times10^2$

127. $30,000^3 = (3\times10^4)^3$
$= 3^3\times(10^4)^3$
$= 27\times10^{12}$
$= 2.7\times10^{13}$

129. $20,000,000^3 = (2\times10^7)^3 = 2^3\times(10^7)^3 = 8\times10^{21}$

131. $0.0005^2 = (5\times10^{-4})^2$
$= 5^2\times(10^{-4})^2$
$= 25\times10^{-8}$
$= 2.5\times10^{-7}$

133. $(-0.000004)^{-2} = (-4\times10^{-6})^{-2}$
$= (-4)^{-2}\times(10^{-6})^{-2}$
$= \dfrac{1}{16}\times10^{12}$
$= 0.0625\times10^{12}$
$= 6.25\times10^{10}$

135. $E = hf$
$E = (6.626\times10^{-34})(4.2\times10^{14})$
$\approx 27.8\times10^{-20}$
$= 2.78\times10^{-19}$ joules

137. $E = hf$
$1.4\times10^{-19} = (6.626\times10^{-34})f$
$\dfrac{1.4\times10^{-19}}{6.626\times10^{-34}} = f$
$0.211\times10^{15} \approx f$
2.11×10^{14} Hz $= f$

139. $E = mc^2$
$E = (4.2\times10^{-12})(3\times10^8)^2$
$= (4.2\times10^{-12})(9\times10^{16})$
$= 37.8\times10^4$
$= 3.78\times10^5$ joules

Review Exercises

1. Change the operation from subtraction to addition and the subtrahend, –6, to its additive inverse, 6.

2. Associative property of addition

3. $-8(n-4) = -8 \cdot n - 8 \cdot (-4) = -8n + 32$

4. $7x + 8y - 4x - 12 + 13y - 5$
$= 7x - 4x + 8y + 13y - 12 - 5$
$= 3x + 21y - 17$

5. $10x^2 - x + 9 - 12x + 3x^2$
$= 10x^2 + 3x^2 - x - 12x + 9$
$= 13x^2 - 13x + 9$

6. $2(u+8) - (5u-3) - 7$
$= 2u + 16 - 5u + 3 - 7$
$= 2u - 5u + 16 + 3 - 7$
$= -3u + 12$

Exercise Set 5.2

1. Add the exponents of all the variables in the monomial.

3. The degree of a polynomial is the greatest degree of any of the terms in the polynomial.

5. To add polynomials, combine like terms.

7. The degree of a quadratic function is 2. The graph of a quadratic function is a parabola.

9. $-7uv^3$
$d = 1 + 3 = 4$; monomial

11. $25 - x^2$
$d = 2$; binomial

13. $5p^3 + 2p^2 - 1$
$d = 3$; trinomial

15. $4g^2 - 5g - 11g^3 - 8g^4 + 7$
$d = 4$; this polynomial has more than three terms, so it has no special polynomial name.

17. $7x + 5x^3 - 19$
$d = 3$; trinomial

19. -7
$d = 0$; monomial

21. $\frac{1}{3}pq - 3p^2q$
$d = 2 + 1 = 3$, binomial

23. $(5x^2 - 3x + 1) + (2x^2 + 7x - 3) = 7x^2 + 4x - 2$

25. $(p^4 - 3p^3 + 4p - 1) + (p^4 - 2p^3 - 4p + 7)$
$= 2p^4 - 5p^3 + 6$

27. $(4u^3 - 6u^2 + u + 11) + (-5u^3 - 3u^2 + u - 5)$
$= -u^3 - 9u^2 + 2u + 6$

29. $\left(\frac{2}{3}u^4 + \frac{3}{4}u^3 - u^2 - u + 3\right)$
$+\left(\frac{2}{3}u^4 - \frac{1}{4}u^3 + 3u^2 + 4u - 1\right)$
$= \frac{4}{3}u^4 + \frac{1}{2}u^3 + 2u^2 + 3u + 2$

31. $(3.1t^4 - 2.1t^3 + 7t^2 + 5.8t + 4)$
$+(4.2t^4 + 3.6t^3 - 8t^2 - 3.1t + 3)$
$= 7.3t^4 + 1.5t^3 - t^2 + 2.7t + 7$

33. $(7a^3 - a^2 - a) - (8a^3 - 3a + 1)$
$= (7a^3 - a^2 - a) + (-8a^3 + 3a - 1)$
$= -a^3 - a^2 + 2a - 1$

35. $(3m^4 + 2m^3 - m^2 + 1) - (6m^3 - 2m^2 - m + 3)$
$= (3m^4 + 2m^3 - m^2 + 1) + (-6m^3 + 2m^2 + m - 3)$
$= 3m^4 - 4m^3 + m^2 + m - 2$

37. $(-7r^3 + 3r^2 - 7r - 4) - (-5r^3 + 2r^2 - 6r - 1)$
$= (-7r^3 + 3r^2 - 7r - 4) + (5r^3 - 2r^2 + 6r + 1)$
$= -2r^3 + r^2 - r - 3$

39. $\left(\frac{4}{5}y^3 - \frac{1}{3}y^2 + 7y - \frac{2}{7}\right) - \left(\frac{1}{5}y^3 + \frac{2}{3}y^2 + \frac{3}{4}y - \frac{1}{7}\right)$
$= \left(\frac{4}{5}y^3 - \frac{1}{3}y^2 + 7y - \frac{2}{7}\right)$
$+\left(-\frac{1}{5}y^3 - \frac{2}{3}y^2 - \frac{3}{4}y + \frac{1}{7}\right)$
$= \frac{3}{5}y^3 - y^2 + \frac{25}{4}y - \frac{1}{7}$

41. $\left(-4.5w^3-5.1w^2+2.7w+4.1\right)$
$-\left(3.8w^3-1.4w^2+3.4w-2.6\right)$
$=\left(-4.5w^3-5.1w^2+2.7w+4.1\right)$
$+\left(-3.8w^3+1.4w^2-3.4w+2.6\right)$
$=-8.3w^3-3.7w^2-0.7w+6.7$

43. $\left(-5w^4-3w^3-8w^2+w-14\right)$
$+\left(-3w^4+6w^3+w^2+12w+5\right)$
$=-8w^4+3w^3-7w^2+13w-9$

45. $\left(6x^4+4x^3-3x+5\right)-\left(-2x^3-5x^2+9x-8\right)$
$=\left(6x^4+4x^3-3x+5\right)+\left(2x^3+5x^2-9x+8\right)$
$=6x^4+6x^3+5x^2-12x+13$

47. $\left(-\frac{7}{3}g^4+\frac{4}{5}g^3+\frac{7}{5}\right)+\left(\frac{4}{3}g^4-\frac{2}{3}g^2+\frac{2}{5}\right)$
$=-g^4+\frac{4}{5}g^3-\frac{2}{3}g^2+\frac{9}{5}$

49. $\left(4y^2-8y+1\right)+\left(5y^2+3y+2\right)-\left(6y^2-9y+2\right)$
$=\left(4y^2-8y+1\right)+\left(5y^2+3y+2\right)+\left(-6y^2+9y-2\right)$
$=3y^2+4y+1$

51. $\left(-6a^3-5a^2+10\right)-\left(9a^3+5a^2-3a-1\right)$
$+\left(-4a^3-2a^2+6a-2\right)$
$=\left(-6a^3-5a^2+10\right)+\left(-9a^3-5a^2+3a+1\right)$
$+\left(-4a^3-2a^2+6a-2\right)$
$=-19a^3-12a^2+9a+9$

53. $\left(8a^2+5ab-2b^2\right)-\left(10ab-6a^2-8b^2\right)$
$=\left(8a^2+5ab-2b^2\right)+\left(-10ab+6a^2+8b^2\right)$
$=14a^2-5ab+6b^2$

55. $\left(7x^3y^4-2x^2y^3-xy+4\right)$
$+\left(-5x^3y^4+7x^2y^3+8xy-12\right)$
$=2x^3y^4+5x^2y^3+7xy-8$

57. $\left(x^2y^2+8xy^2-12x^2y-4xy+7y^2-9\right)$
$-\left(-3x^2y^2+2xy^2+4x^2y-xy+6y^2+4\right)$
$=\left(x^2y^2+8xy^2-12x^2y-4xy+7y^2-9\right)$
$+\left(3x^2y^2-2xy^2-4x^2y+xy-6y^2-4\right)$
$=4x^2y^2+6xy^2-16x^2y-3xy+y^2-13$

59. $(x+10)+(x-4)+(2x-1)=4x+5$

61. $2(a-10)+2(3a+12)=2a-20+6a+24$
$=8a+4$

63. Because the degree of this polynomial is 0, the function is a constant function.

65. Because the degree of this polynomial is 2, the function is a quadratic function.

67. Because the degree of this polynomial is 1, the function is a linear function.

69. Because the degree of this polynomial is 3, the function is a cubic function.

71. $h(x)=-2$

x	$h(x)$
-2	-2
0	-2
2	-2

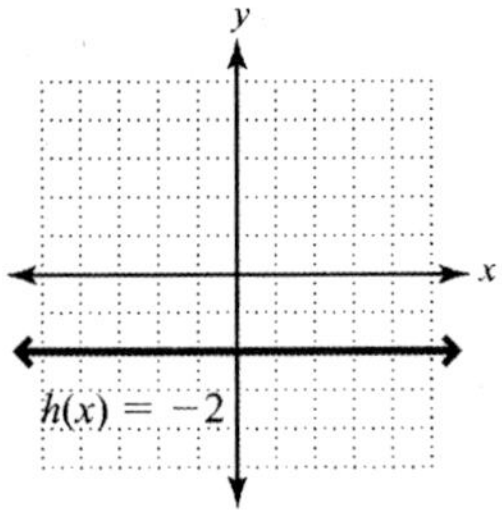

Domain: $\{x\mid -\infty<x<\infty\}$ or $(-\infty,\infty)$
Range: $\{y\mid y=-2\}$ or $[-2,-2]$

73. $c(x) = -5x + 1$

x	$c(x)$
-1	6
0	1
1	-4

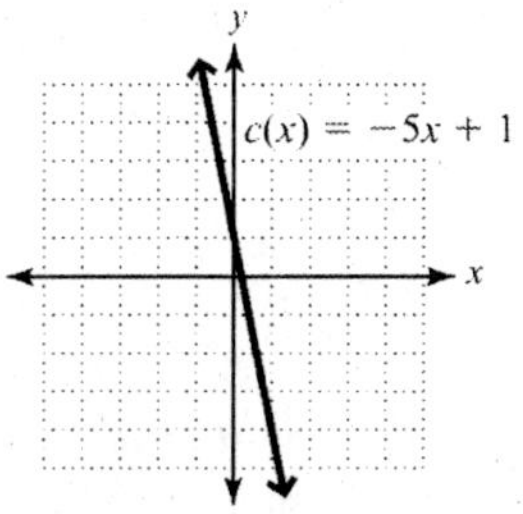

Domain: $\{x \mid -\infty < x < \infty\}$ or $(-\infty, \infty)$

Range: $\{y \mid -\infty < y < \infty\}$ or $(-\infty, \infty)$

75. $g(x) = x^2 + 2x - 3$

x	$g(x)$
-3	0
-2	-3
-1	-4
0	-3
1	0

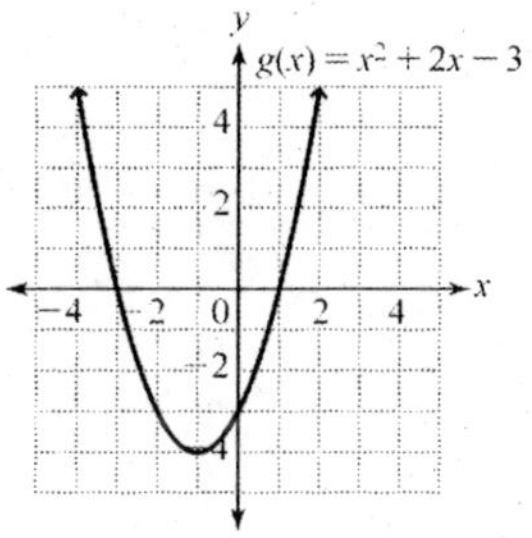

Domain: $\{x \mid -\infty < x < \infty\}$ or $(-\infty, \infty)$

Range: $\{y \mid y \geq -4\}$ or $[-4, \infty)$

77. $f(x) = x^3 - 2$

x	$f(x)$
-2	-10
-1	-3
0	-2
1	-1
2	6

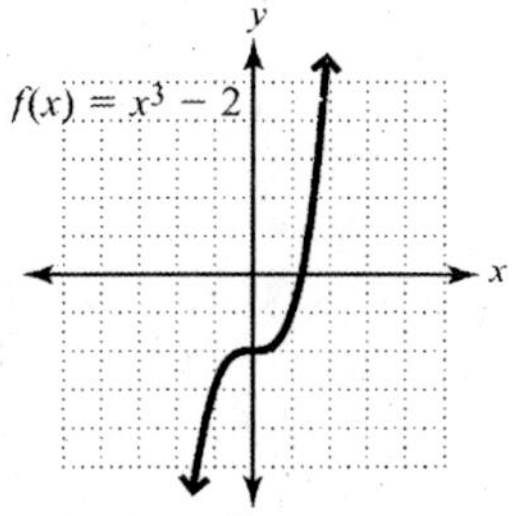

Domain: $\{x \mid -\infty < x < \infty\}$ or $(-\infty, \infty)$

Range: $\{y \mid -\infty < y < \infty\}$ or $(-\infty, \infty)$

79. $f(x) = -4$; $g(x) = 3$

a) $(f + g)(x) = -4 + 3 = -1$

b) $(f - g)(x) = -4 - 3 = -7$

81. $f(x) = 4x + 3$; $g(x) = -x - 1$

a) $(f + g)(x) = (4x + 3) + (-x - 1)$
$= 3x + 2$

b) $(f - g)(x) = (4x + 3) - (-x - 1)$
$= (4x + 3) + (x + 1)$
$= 5x + 4$

83. $f(x) = x^2 - 1$; $g(x) = 3x^2 + 2$

a) $(f + g)(x) = (x^2 - 1) + (3x^2 + 2)$
$= 4x^2 + 1$

b) $(f - g)(x) = (x^2 - 1) - (3x^2 + 2)$
$= (x^2 - 1) + (-3x^2 - 2)$
$= -2x^2 - 3$

85. $f(x) = x^3 + 2x + 5$; $g(x) = -2x^3 + 2$

a) $(f + g)(x) = (x^3 + 2x + 5) + (-2x^3 + 2)$
$= -x^3 + 2x + 7$

b) $(f - g)(x) = (x^3 + 2x + 5) - (-2x^3 + 2)$
$= (x^3 + 2x + 5) + (2x^3 - 2)$
$= 3x^3 + 2x + 3$

87. a) $P(x) = R(x) - C(x)$

$= (0.2x^2 + x + 3000) - (18x + 2000)$

$= (0.2x^2 + x + 3000) + (-18x - 2000)$

$= 0.2x^2 - 17x + 1000$

b) $P(1000) = 0.2x^2 - 17x + 1000$

$= 0.2(1000)^2 - 17(1000) + 1000$

$= 184,000$

The profit is \$184,000.

89. a) $R(x) = P(x) + C(x)$

$= (190x + 260) + (230x + 580)$

$= 420x + 840$

b) $R(x) = 420x + 840$

$R(300) = 420(300) + 840$

$= 126,000 + 840$

$= 126,840$

The revenue is \$126,840.

91. $h(t) = -16t^2 + 200$

$h(0.5) = -16(0.5)^2 + 200$

$= -4 + 200$

$= 196$

After 0.5 second, the height is 196 feet.

93. $V(r) = 1.5r^2 - 0.8r + 9.5$

$V(6) = 1.5(6)^2 - 0.8(6) + 9.5$

$= 58.7$

The voltage is 58.7 V.

95. a) The completed table is shown below.

Year	Years since 1990	Fatalities
1990	0	31.4
1992	2	26.92
1994	4	24.2
1996	6	23.24
1998	8	24.04
2000	10	26.6
2002	12	30.92
2004	14	37
2006	16	44.84

b)

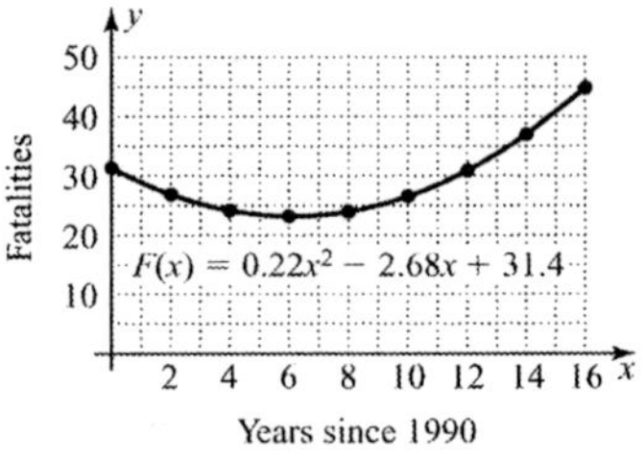

c) $F(x) = 0.22x^2 - 2.68x + 31.4$

$F(24) = 0.22(24)^2 - 2.68(24) + 31.4$

$= 126.72 - 64.32 + 31.4$

$= 93.8$

The model predicts that the fatality rate will be 93.8 per 100 million vehicle miles in 2014.

Review Exercises

1. $8(3x - 9) = 24x - 72$

2. $-6(2y + 7) = -12y - 42$

3. $5x - 3(x + 2) = 7x - 8$

$5x - 3x - 6 = 7x - 8$

$2x - 6 = 7x - 8$

$2 = 5x$

$\frac{2}{5} = x$

4. $12x + 9 < 4(2x - 5)$

$12x + 9 < 8x - 20$

$4x < -29$

$x < -\frac{29}{4}$

5. $-11 < 2x - 3 \le 9$

$-8 < 2x \le 12$

$-4 < x \le 6$

a) $\{x \mid -4 < x \le 6\}$ b) $(-4, 6]$

c) 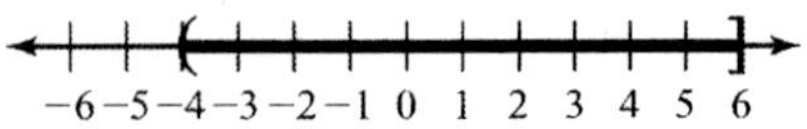

6. $2x - 7 \le -5$ or $2x - 7 \ge 5$

$2x \le 2$ $\quad$ $2x \ge 12$

$x \le 1$ $\quad$ $x \ge 6$

a) $\{x \mid x \le 1 \text{ or } x \ge 6\}$

b) $(-\infty, 1] \cup [6, \infty)$

c) 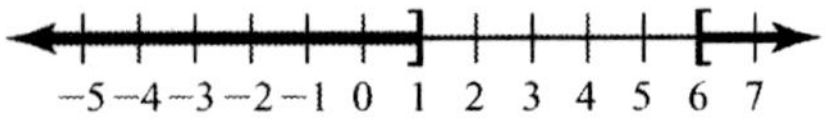

Exercise Set 5.3

1. We apply the distributive property.

3. Multiply both terms in the second binomial by both terms in the first binomial (FOIL); then combine like terms.

5. The product is a difference of squares.

7. $5x^3(x^2+3x-2) = 5x^3 \cdot x^2 + 5x^3 \cdot 3x + 5x^3 \cdot (-2)$
$= 5x^5 + 15x^4 - 10x^3$

9. $-6x^3(2x^2+4x-3)$
$= -6x^3 \cdot 2x^2 - 6x^3 \cdot 4x - 6x^3(-3)$
$= -12x^5 - 24x^4 + 18x^3$

11. $4n^4(n^3+7n^2-2n-3)$
$= 4n^4 \cdot n^3 + 4n^4 \cdot 7n^2 + 4n^4 \cdot (-2n) + 4n^4 \cdot (-3)$
$= 4n^7 + 28n^6 - 8n^5 - 12n^4$

13. $9a^2b^4(3a^4b^2 - 2ab^8)$
$= 9a^2b^4 \cdot 3a^4b^2 + 9a^2b^4 \cdot (-2ab^8)$
$= 27a^6b^6 - 18a^3b^{12}$

15. $\frac{1}{4}m^2np^3(2mn^5 - 5m^2n^2 + 8m^2n^3p^2)$
$= \frac{1}{4}m^2np^3 \cdot 2mn^5 + \frac{1}{4}m^2np^3 \cdot (-5m^2n^2)$
$+ \frac{1}{4}m^2np^3 \cdot 8m^2n^3p^2$
$= \frac{1}{2}m^3n^6p^3 - \frac{5}{4}m^4n^3p^3 + 2m^4n^4p^5$

17. $-0.3p^2q^2(8p^3q^7 - 2q + 7p^6)$
$= -0.3p^2q^2 \cdot 8p^3q^7 - 0.3p^2q^2 \cdot (-2q)$
$- 0.3p^2q^2 \cdot 7p^6$
$= -2.4p^5q^9 + 0.6p^2q^3 - 2.1p^8q^2$

19. $-5x^5y^2(4x^2y + 2xy^4 - 5x^8y^5)$
$= -5x^5y^2 \cdot 4x^2y - 5x^5y^2 \cdot 2xy^4$
$- 5x^5y^2 \cdot (-5x^8y^5)$
$= -20x^7y^3 - 10x^6y^6 + 25x^{13}y^7$

21. $-3r^2s(2r^4s^3 - 6r^2s^2 - 3rs + 3)$
$= -3r^2s \cdot 2r^4s^3 - 3r^2s \cdot (-6r^2s^2) - 3r^2s \cdot (-3rs)$
$- 3r^2s \cdot 3$
$= -6r^6s^4 + 18r^4s^3 + 9r^3s^2 - 9r^2s$

23. $-0.2a^2b^2(2.1a - 6ab + 3a^2b - a^2b^2)$
$= -0.2a^2b^2 \cdot 2.1a - 0.2a^2b^2 \cdot (-6ab)$
$- 0.2a^2b^2 \cdot 3a^2b - 0.2a^2b^2 \cdot (-a^2b^2)$
$= -0.42a^3b^2 + 1.2a^3b^3 - 0.6a^4b^3 + 0.2a^4b^4$

25. $\frac{1}{3}abc^6(9a^2b^2c^2 - 4a^2bc + 12a)$
$= \frac{1}{3}abc^6 \cdot 9a^2b^2c^2 + \frac{1}{3}abc^6 \cdot (-4a^2bc)$
$+ \frac{1}{3}abc^6 \cdot 12a$
$= 3a^3b^3c^8 - \frac{4}{3}a^3b^2c^7 + 4a^2bc^6$

27. a) $x+5$
b) $2x$
c) $(x+5)(2x)$
d) $x^2 + 5x + x^2 + 5x = 2x^2 + 10x$
e) They are equivalent because they both describe the area of the figure.

29. a) $x+3$
b) $x+2$
c) $(x+3)(x+2)$
d) $x^2 + 3x + 2x + 6 = x^2 + 5x + 6$
e) They are equivalent because they both describe the area of the figure.

31. $(2x+3)(3x+4) = 2x \cdot 3x + 2x \cdot 4 + 3 \cdot 3x + 3 \cdot 4$
$= 6x^2 + 8x + 9x + 12$
$= 6x^2 + 17x + 12$

33. $(3x-1)(5x+2) = 3x \cdot 5x + 3x \cdot 2 + (-1) \cdot 5x + (-1) \cdot 2$
$= 15x^2 + 6x - 5x - 2$
$= 15x^2 + x - 2$

35. $(2m-5n)(3m-2n)$
$= 2m \cdot 3m + 2m \cdot (-2n) + (-5n) \cdot 3m + (-5n) \cdot (-2n)$
$= 6m^2 - 4mn - 15mn + 10n^2$
$= 6m^2 - 19mn + 10n^2$

37. $(5m-3n)(3m+4n)$
$= 5m\cdot 3m+5m\cdot 4n+(-3n)\cdot 3m+(-3n)\cdot 4n$
$= 15m^2+20mn-9mn-12n^2$
$= 15m^2+11mn-12n^2$

39. $(t^2-5)(t^2-2)$
$= t^2\cdot t^2+t^2\cdot(-2)+(-5)\cdot t^2+(-5)(-2)$
$= t^4-2t^2-5t^2+10$
$= t^4-7t^2+10$

41. $(a^2+6b^2)(a^2-b^2)$
$= a^2\cdot a^2+a^2\cdot(-b^2)+6b^2\cdot a^2+6b^2\cdot(-b^2)$
$= a^4-a^2b^2+6a^2b^2-6b^4$
$= a^4+5a^2b^2-6b^4$

43. $(3x-1)(4x^2-2x+1)$
$= 3x\cdot 4x^2+3x\cdot(-2x)+3x\cdot 1$
$\quad +(-1)\cdot 4x^2+(-1)\cdot(-2x)+(-1)\cdot 1$
$= 12x^3-6x^2+3x-4x^2+2x-1$
$= 12x^3-10x^2+5x-1$

45. $(7a+1)(3a^2-2a-2)$
$= 7a\cdot 3a^2+7a\cdot(-2a)+7a\cdot(-2)$
$\quad +1\cdot 3a^2+1\cdot(-2a)+1\cdot(-2)$
$= 21a^3-14a^2-14a+3a^2-2a-2$
$= 21a^3-11a^2-16a-2$

47. $(7c+2)(2c^2-4c-3)$
$= 7c\cdot 2c^2+7c\cdot(-4c)+7c\cdot(-3)$
$\quad +2\cdot 2c^2+2\cdot(-4c)+2\cdot(-3)$
$= 14c^3-28c^2-21c+4c^2-8c-6$
$= 14c^3-24c^2-29c-6$

49. $(4p+10q)(3p^2+2pq+q^2)$
$= 4p\cdot 3p^2+4p\cdot 2pq+4p\cdot q^2$
$\quad +10q\cdot 3p^2+10q\cdot 2pq+10q\cdot q^2$
$= 12p^3+8p^2q+4pq^2+30p^2q+20pq^2+10q^3$
$= 12p^3+38p^2q+24pq^2+10q^3$

51. $(2y-3z)(6y^2-2yz+4z^2)$
$= 2y\cdot 6y^2+2y\cdot(-2yz)+2y\cdot 4z^2$
$\quad +(-3z)\cdot 6y^2+(-3z)\cdot(-2yz)+(-3z)\cdot 4z^2$
$= 12y^3-4y^2z+8yz^2-18y^2z+6yz^2-12z^3$
$= 12y^3-22y^2z+14yz^2-12z^3$

53. $(3u^2-2u-1)(3u^2+2u+1)$
$= 3u^2\cdot 3u^2+3u^2\cdot 2u+3u^2\cdot 1$
$\quad +(-2u)\cdot 3u^2+(-2u)\cdot 2u+(-2u)\cdot 1$
$\quad +(-1)\cdot 3u^2+(-1)\cdot 2u+(-1)\cdot 1$
$= 9u^4+6u^3+3u^2-6u^3-4u^2-2u-3u^2-2u-1$
$= 9u^4-4u^2-4u-1$

55. $(x+y)^2 = x^2+2xy+y^2$

57. $(4w+3)^2 = (4w)^2+2(4w)(3)+3^2$
$= 16w^2+24w+9$

59. $(4t-3w)^2 = (4t)^2-2(4t)(3w)+(3w)^2$
$= 16t^2-24tw+9w^2$

61. $(9-5y^2)^2 = 9^2-2(9)(5y^2)+(5y^2)^2$
$= 81-90y^2+25y^4$
$= 25y^4-90y^2+81$

63. $[(x+1)-y]^2$
$= (x+1)^2-2y(x+1)+y^2$
$= x^2+2x+1-2xy-2y+y^2$
$= x^2+y^2-2xy+2x-2y+1$

65. $[p-(q+5)]^2$
$= p^2-2p(q+5)+(q+5)^2$
$= p^2-2pq-10p+q^2+10q+25$
$= p^2+q^2-2pq-10p+10q+25$

67. $x-8$

69. $3m-2n$

71. m^2-n^2

73. $-2j+5k$

75. $(2x-7)(2x+7) = (2x)^2-7^2 = 4x^2-49$

77. $(2q-5)(2q+5)=(2q)^2-5^2=4q^2-25$

79. $(x+2y)(x-2y)=x^2-(2y)^2=x^4-4y^2$

81. $[(s+1)+t][(s+1)-t]=(s+1)^2-t^2$
$=s^2+2s+1-t^2$
$=s^2-t^2+2s+1$

83. $[3b+(c+2)][3b-(c+2)]=(3b)^2-(c+2)^2$
$=9b^2-(c^2+4c+4)$
$=9b^2-c^2-4c-4$

85. $A=yz(3x+y)$
$=3xyz+y^2z$

87. $A=\frac{1}{2}h[(h+1)+(h+6)]$
$=\frac{1}{2}h(2h+7)$
$=h^2+\frac{7}{2}h$

89. Mistake: Combined like terms incorrectly.
Correct: $6x^2-37x+45$

91. Mistake: Multiplied $2x$ by $2x$ incorrectly.
Correct: $4x^2-49$

93. $-4wv(-w^4v^3-4x^2wv^7+5xw^2v)$
$=4v^4w^5+16v^8w^2x^2-20v^2w^3x$

95. $(8r^2-3s)(8r^2+3s)=(8r^2)^2-(3s)^2$
$=64r^4-9s^2$

97. $(2u^2+3v^2)^2=(2u^2)^2+2(2u^2)(3v^2)+(3v^2)^2$
$=4u^4+12u^2v^2+9v^4$

99. $(2a^2-5b^2)(a^2-4b^2)$
$=2a^4-8a^2b^2-5a^2b^2+20b^4$
$=2a^4-13a^2b^2+20b^4$

101. $(5q^2-3t)(3q^2+4t)$
$=15q^4+20tq^2-9tq^2-12t^2$
$=15q^4+11tq^2-12t^2$

103. $3r^2s^3\left(r^4-\frac{1}{9}r^3s-\frac{1}{6}r^2s^2+\frac{1}{3}rs-1\right)$
$=3r^6s^3-\frac{1}{3}r^5s^4-\frac{1}{2}r^4s^5+r^3s^4-3r^2s^3$

105. $-0.1t^2r^3(3.5t^2r^3-8t^3r^2+2.2t^3r-2tr)$
$=-0.35t^4r^6+0.8t^5r^5-0.22t^5r^4+0.2t^3r^4$

107. $[(3m-4)+n][(3m-4)-n]$
$=(3m-4)^2-n^2$
$=9m^2-24m+16-n^2$
$=9m^2-n^2-24m+16$

109. $(x^2+3xy+y^2)(4x^2+2xy-y^2)$
$=4x^4+2x^3y-x^2y^2+12x^3y+6x^2y^2$
$-3xy^3+4x^2y^2+2xy^3-y^4$
$=4x^4+14x^3y+9x^2y^2-xy^3-y^4$

111. $[(x^2-4)+3]^2$
$=(x^2-4)^2+6(x^2-4)+9$
$=x^4-8x^2+16+6x^2-24+9$
$=x^4-2x^2+1$

113. $f(x)=3x-2;\quad g(x)=2x-1$

a) $(f\cdot g)(x)=(3x-2)(2x-1)$
$=6x^2-3x-4x+2$
$=6x^2-7x+2$

b) $(f\cdot g)(-4)=6(-4)^2-7(-4)+2$
$=96+28+2$
$=126$

c) $(f\cdot g)(2)=6(2)^2-7(2)+2$
$=24-14+2$
$=12$

115. $f(x)=3x+4;\quad g(x)=4x^2-3x+2$

a) $(f\cdot g)(x)=(3x+4)(4x^2-3x+2)$
$=12x^3-9x^2+6x+16x^2-12x+8$
$=12x^3+7x^2-6x+8$

b) $(f \cdot g)(5) = 12(5)^3 + 7(5)^2 - 6(5) + 8$
$= 12(125) + 7(25) - 30 + 8$
$= 1500 + 175 - 30 + 8$
$= 1653$

c) $(f \cdot g)(-3) = 12(-3)^3 + 7(-3)^2 - 6(-3) + 8$
$= 12(-27) + 7(9) + 18 + 8$
$= -324 + 63 + 18 + 8$
$= -235$

117. $f(x) = x^2 + 3x - 1$
$f(n+1) = (n+1)^2 + 3(n+1) - 1$
$= n^2 + 2n + 1 + 3n + 3 - 1$
$= n^2 + 5n + 3$

119. $f(x) = 2x^2 - 1$
$f(t-6) = 2(t-6)^2 - 1$
$= 2(t^2 - 12t + 36) - 1$
$= 2t^2 - 24t + 72 - 1$
$= 2t^2 - 24t + 71$

121. $f(x) = 2x - 3$
$f(x+h) - f(x) = 2(x+h) - 3 - (2x - 3)$
$= 2x + 2h - 3 - 2x + 3$
$= 2h$

123. $f(x) = 2x^2 + 3x - 4$
$f(x+h) - f(x)$
$= 2(x+h)^2 + 3(x+h) - 4 - (2x^2 + 3x - 4)$
$= 2(x^2 + 2xh + h^2) + 3x + 3h - 4 - 2x^2 - 3x + 4$
$= 2x^2 + 4xh + 2h^2 + 3x + 3h - 4 - 2x^2 - 3x + 4$
$= 4xh + 2h^2 + 3h$

125. Let l = the length of the room and $l - 2$ = the width of the room.
Area = $l(l-2) = l^2 - 2l$
The area of the room is $(l^2 - 2l)$ square feet.

127. Volume $= 4x(x+1)(x+5)$
$= 4x(x^2 + 6x + 5)$
$= 4x^3 + 24x^2 + 20x$
The volume is $(4x^3 + 24x^2 + 20x)$ cubic centimeters.

129. The width is w. The length is $3w$. The height is $(3w + 5)$.
$V = w(3w)(3w+5) = 3w^2(3w+5)$
$= 9w^3 + 15w^2$
The volume is $(9w^3 + 15w^2)$ cubic feet.

131. The radius is r. The height is $(r + 3)$.
$V = \pi r^2(r+3) = \pi r^3 + 3\pi r^2$
The volume is $(\pi r^3 + 3\pi r^2)$ cubic inches.

Review Exercises

1. $(-4.26 \times 10^6)(2.1 \times 10^5) = (-4.26 \cdot 2.1) \times (10^{6+5})$
$= -8.946 \times 10^{11}$

2. $\dfrac{1.5 \times 10^5}{2.4 \times 10^{-4}} = \dfrac{1.5}{2.4} \times \dfrac{10^5}{10^{-4}}$
$= 0.625 \times 10^{5-(-4)}$
$= 0.625 \times 10^9$
$= 6.25 \times 10^8$

3. $\frac{5}{6}x - \frac{1}{8} = \frac{3}{4}x + \frac{1}{2}$
$24\left(\frac{5}{6}x - \frac{1}{8}\right) = 24\left(\frac{3}{4}x + \frac{1}{2}\right)$
$20x - 3 = 18x + 12$
$2x - 3 = 12$
$2x = 15$
$x = \frac{15}{2}$

4. $d = 30t - n$
$d + n = 30t$
$\frac{d+n}{30} = t$

5. Use $m = -\frac{1}{5}$.
$y - 5 = -\frac{1}{5}(x - (-3))$
$y - 5 = -\frac{1}{5}(x + 3)$
$y - 5 = -\frac{1}{5}x - \frac{3}{5}$
$y = -\frac{1}{5}x + \frac{22}{5}$

6. Let s = the number of smaller boxes sold and l = the number of larger boxes sold.

$$\begin{cases} s+l=84 \\ 3.5s+4.8l=349.9 \end{cases}$$

Solve the first equation for s: $s=84-l$

$$\begin{aligned} 3.5(84-l)+4.8l &= 349.9 \\ 294-3.5l+4.8l &= 349.9 \\ 1.3l &= 55.9 \\ l &= 43 \end{aligned} \qquad \begin{aligned} s &= 84-43 \\ &= 41 \end{aligned}$$

43 large boxes and 41 small boxes were sold.

Exercise Set 5.4

1. To divide a polynomial by a monomial, divide each term in the polynomial by the monomial.

3. Divide $6x^2$ by $2x$.

5. Write a sum of the quotient and the remainder over the divisor, as in $3x^2-5x+2+\dfrac{7}{6x-1}$.

7. $$\frac{28a^4-7a^3+21a^2}{7a^2}=\frac{28a^4}{7a^2}-\frac{7a^3}{7a^2}+\frac{21a^2}{7a^2}=4a^2-a+3$$

9. $$\frac{12u^5-6u^4-15u^3+3u^2}{3u^2}=\frac{12u^5}{3u^2}-\frac{6u^4}{3u^2}-\frac{15u^3}{3u^2}+\frac{3u^2}{3u^2}=4u^3-2u^2-5u+1$$

11. $$\frac{24a^5b^4-8a^4b^2+16a^2b}{8ab}=\frac{24a^5b^4}{8ab}-\frac{8a^4b^2}{8ab}+\frac{16a^2b}{8ab}=3a^4b^3-a^3b+2a$$

13. $$\frac{36u^3v^4+12uv^5-15u^2v^2}{3uv^9}=\frac{36u^3v^4}{3uv^9}+\frac{12uv^5}{3uv^9}-\frac{15u^2v^2}{3uv^9}=\frac{12u^2}{v^5}+\frac{4}{v^4}-\frac{5u}{v^7}$$

15. $$\frac{30x^3y^2-45xy^3-xy}{-5xy}=\frac{30x^3y^2}{-5xy}-\frac{45xy^3}{-5xy}-\frac{xy}{-5xy}=-6x^2y+9y^2+\frac{1}{5}$$

17. $$\frac{12t^6u^5v-2t^4u^4-16t^3u^2+3tu^2}{-4t^3u^2}=\frac{12t^6u^5v}{-4t^3u^2}-\frac{2t^4u^4}{-4t^3u^2}-\frac{16t^3u^2}{-4t^3u^2}+\frac{3tu^2}{-4t^3u^2}=-3t^3u^3v+\frac{1}{2}tu^2+4-\frac{3}{4t^2}$$

19. $$\frac{18a^3b^2c^2+12a^2b^3c-3a^2b}{9a^2bc}=\frac{18a^3b^2c^2}{9a^2bc}+\frac{12a^2b^3c}{9a^2bc}-\frac{3a^2b}{9a^2bc}=2abc+\frac{4}{3}b^2-\frac{1}{3c}$$

21. $$\frac{25xy^2}{10x}=\frac{5y^2}{2}$$

23. a) $$\frac{6x^2}{3x}-\frac{12x}{3x}+\frac{18}{3x}=2x-4+\frac{6}{x}$$

b) $$\begin{aligned}\text{Area}&=6(5)^2-12(5)+18\\&=150-60+18\\&=108\end{aligned}\qquad \text{length}=3(5)=15$$

$$\text{width}=2(5)-4+\frac{6}{5}=7.2$$

25. $$\begin{array}{r} x+4 \\ x+6\overline{\big)\,x^2+10x+24} \\ \underline{x^2+6x} \\ 4x+24 \\ \underline{4x+24} \\ 0 \end{array}$$

27. $$\begin{array}{r} 3p+1 \\ p+1\overline{\big)\,3p^2+4p+1} \\ \underline{3p^2+3p} \\ p+1 \\ \underline{p+1} \\ 0 \end{array}$$

29. $$\begin{array}{r} 5n-8 \\ n+1\overline{)5n^2-3n-8} \\ \underline{5n^2+5n} \\ -8n-8 \\ \underline{-8n-8} \\ 0 \end{array}$$

31. $$\begin{array}{r} 3x-2 \\ 5x+4\overline{)15x^2+2x-8} \\ \underline{15x^2+12x} \\ -10x-8 \\ \underline{-10x-8} \\ 0 \end{array}$$

33. $$\begin{array}{r} 2y-1 \\ 3y+4\overline{)6y^2+5y-4} \\ \underline{6y^2+8y} \\ -3y-4 \\ \underline{-3y-4} \\ 0 \end{array}$$

35. $$\begin{array}{r} x^2-5x+6 \\ 2x-3\overline{)2x^3-13x^2+27x-18} \\ \underline{2x^3-3x^2} \\ -10x^2+27x \\ \underline{-10x^2+15x} \\ 12x-18 \\ \underline{12x-18} \\ 0 \end{array}$$

37. $$\begin{array}{r} x^2-2x+4 \\ x+2\overline{)x^3+0x^2+0x+8} \\ \underline{x^3+2x^2} \\ -2x^2+0x \\ \underline{-2x^2-4x} \\ 4x+8 \\ \underline{4x+8} \\ 0 \end{array}$$

39. $$\begin{array}{r} y^3+2y^2+4y+8 \\ y-2\overline{)y^4+0y^3+0y^2+0y-16} \\ \underline{y^4-2y^3} \\ 2y^3+0y^2 \\ \underline{2y^3-4y^2} \\ 4y^2+0y \\ \underline{4y^2-8y} \\ 8y-16 \\ \underline{8y-16} \\ 0 \end{array}$$

41. $$\begin{array}{r} 4z^2-10z+25 \\ 2z+5\overline{)8z^3+0z^2+0z+125} \\ \underline{8z^3+20z^2} \\ -20z^2+0z \\ \underline{-20z^2-50z} \\ 50z+125 \\ \underline{50z+125} \\ 0 \end{array}$$

43. $$\begin{array}{r} v^2+3v+2 \\ 2v+1\overline{)2v^3+7v^2+7v+2} \\ \underline{2v^3+v^2} \\ 6v^2+7v \\ \underline{6v^2+3v} \\ 4v+2 \\ \underline{4v+2} \\ 0 \end{array}$$

45. $$\begin{array}{r} 7a^2-2a+3 \\ 3a-2\overline{)21a^3-20a^2+13a-6} \\ \underline{21a^3-14a^2} \\ -6a^2+13a \\ \underline{-6a^2+4a} \\ 9a-6 \\ \underline{9a-6} \\ 0 \end{array}$$

47.
$$\begin{array}{r} 6q^2+5q+9 \\ 2q-1\overline{\big)12q^3+4q^2+13q+9} \\ \underline{12q^3-6q^2} \\ 10q^2+13q \\ \underline{10q^2-5q} \\ 18q+9 \\ \underline{18q-9} \\ 18 \end{array}$$

Answer: $6q^2+5q+9+\frac{18}{2q-1}$

49.
$$\begin{array}{r} 3x^2-6x+5 \\ x^2+0x+4\overline{\big)3x^4-6x^3+17x^2-24x+20} \\ \underline{3x^4+0x^3+12x^2} \\ -6x^3+5x^2-24x \\ \underline{-6x^3-0x^2-24x} \\ 5x^2+0x+20 \\ \underline{5x^2+0x+20} \\ 0 \end{array}$$

51.
$$\begin{array}{r} c^3-2c^2+c-4 \\ 3c+2\overline{\big)3c^4-4c^3-c^2-10c-8} \\ \underline{3c^4+2c^3} \\ -6c^3-c^2 \\ \underline{-6c^3-4c^2} \\ 3c^2-10c \\ \underline{3c^2+2c} \\ -12c-8 \\ \underline{-12c-8} \\ 0 \end{array}$$

53.
$$\begin{array}{r} 5u^2-11u+32 \\ u+2\overline{\big)5u^3-u^2+10u+2} \\ \underline{5u^3+10u^2} \\ -11u^2+10u \\ \underline{-11u^2-22u} \\ 32u+2 \\ \underline{32u+64} \\ -62 \end{array}$$

Answer: $5u^2-11u+32-\frac{62}{u+2}$

55. The width is 6, the height is $t+2$, and the volume is $6t^3+30t^2+12t-48$. We are looking for the length.

$$whl=V$$
$$6(t+2)l=6t^3+30t^2+12t-48$$
$$l=\left(6t^3+30t^2+12t-48\right)\div(6t+12)$$

a)
$$\begin{array}{r} t^2+3t-4 \\ 6t+12\overline{\big)6t^3+30t^2+12t-48} \\ \underline{6t^3+12t^2} \\ 18t^2+12t \\ \underline{18t^2+36t} \\ -24t-48 \\ \underline{-24t-48} \end{array}$$

b) height: $2+2=4$ ft.;
length: $2^2+3(2)-4=4+6-4=6$ ft.;
volume: $6(2)^3+30(2)^2+12(2)-48=144$ ft.^{3}

57. a) $(f/g)(x)=\frac{24x^4-42x^3-30x^2}{6x^2}$
$$=\frac{24x^4}{6x^2}-\frac{42x^3}{6x^2}-\frac{30x^2}{6x^2}$$
$$=4x^2-7x-5$$

b) $(f/g)(4)=4(4)^2-7(4)-5=64-28-5=31$

c) $(f/g)(-2)=4(-2)^2-7(-2)-5$
$$=16+14-5$$
$$=25$$

59. a) $(h/k)(x)=\frac{6x^3+7x^2-9x+2}{2x-1}$
$$\begin{array}{r} 3x^2+5x-2 \\ 2x-1\overline{\big)6x^3+7x^2-9x+2} \\ \underline{6x^3-3x^2} \\ 10x^2-9x \\ \underline{10x^2-5x} \\ -4x+2 \\ \underline{-4x+2} \\ 0 \end{array}$$
$$(h/k)(x)=3x^2+5x-2$$

b) $(h/k)(3)=3(3)^2+5(3)-2$
$$=27+15-2$$
$$=40$$

c) $(h/k)(-1) = 3(-1)^2 + 5(-1) - 2$
$= 3 - 5 - 2$
$= -4$

61. a) $(n/p)(x) = \dfrac{x^3 - x^2 - 7x + 4}{x - 3}$

$$\begin{array}{r} x^2 + 2x - 1 \\ x - 3\overline{)x^3 - x^2 - 7x + 4} \\ \underline{x^3 - 3x^2} \quad\quad\quad \\ 2x^2 - 7x \quad\;\; \\ \underline{2x^2 - 6x} \quad\;\; \\ -x + 4 \\ \underline{-x + 3} \\ 1 \end{array}$$

$(n/p)(x) = x^2 + 2x - 1 + \dfrac{1}{x-3}$

b) $(n/p)(4) = 4^2 + 2(4) - 1 + \dfrac{1}{4-3}$
$= 16 + 8 - 1 + 1$
$= 24$

c) $(n/p)(x) = (-2)^2 + 2(-2) - 1 + \dfrac{1}{-2-3}$
$= 4 - 4 - 1 - \dfrac{1}{5}$
$= -\dfrac{6}{5}$

Review Exercises

1. $9x - (6x + 2) = 5x - 2[4x - (6 - x)]$
$9x - 6x - 2 = 5x - 2(4x - 6 + x)$
$3x - 2 = 5x - 8x + 12 - 2x$
$3x - 2 = -5x + 12$
$8x = 14$
$x = \dfrac{7}{4}$

2. $7x - 5 \geq 8 - (2x + 1)$
$7x - 5 \geq 8 - 2x - 1$
$7x - 5 \geq 7 - 2x$
$9x \geq 12$
$x \geq \dfrac{4}{3}$

3. $5x - 2y = 10$
$-2y = -5x + 10$
$y = \dfrac{5}{2}x - 5$

Use $m = -\dfrac{2}{5}$

$y - (-1) = -\dfrac{2}{5}(x - 3)$
$y + 1 = -\dfrac{2}{5}x + \dfrac{6}{5}$
$y = -\dfrac{2}{5}x + \dfrac{1}{5}$ or $2x + 5y = 1$

4. Yes, because every value in the domain corresponds to only one value in the range. (It passes the vertical line test.)

5. $\begin{cases} 2x + 3y - z = 7 & \text{Eqtn 1} \\ x + y - 5z = -14 & \text{Eqtn 2} \\ -3x - 5y + z = -12 & \text{Eqtn 3} \end{cases}$

Multiply equation 2 by – 2 and add to equation 1, forming equation 4.

$$\begin{array}{r} 2x + 3y - z = 7 \\ \underline{-2x - 2y + 10z = 28} \\ y + 9z = 35 \end{array} \quad \text{Eqtn 4}$$

Multiply equation 2 by 3 and add to equation 3, forming equation 5.

$$\begin{array}{r} 3x + 3y - 15z = -42 \\ \underline{-3x - 5y + z = -12} \\ -2y - 14z = -54 \end{array} \quad \text{Eqtn 5}$$

Multiply equation 4 by 2 and add to equation 5.

$$\begin{array}{r} 2y + 18z = 70 \\ \underline{-2y - 14z = -54} \\ 4z = 16 \\ z = 4 \end{array}$$

$y + 9z = 35$
$y + 9(4) = 35$
$y + 36 = 35$
$y = -1$

$x + y - 5z = -14$
$x + (-1) - 5(4) = -14$
$x - 1 - 20 = -14$
$x = 7$

Solution: $(7, -1, 4)$

6. $\begin{cases} x - y > 3 \\ 2x + y \le 4 \end{cases}$

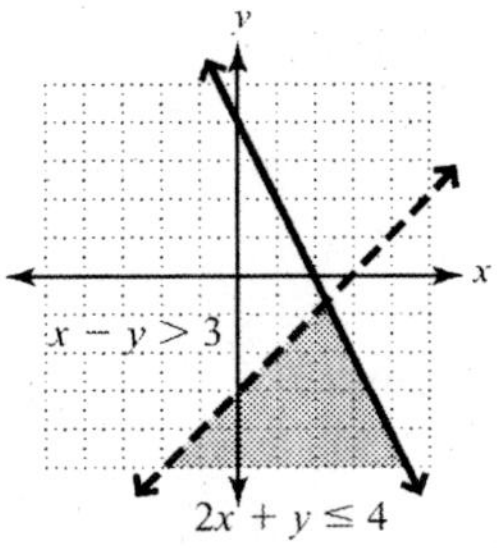

Exercise Set 5.5

1. $x - c$ 3. 1, 0, –5, –3, 11

5. It is the last number in the last row.

7. Given a polynomial $P(x)$, the remainder of $\dfrac{P(x)}{x-c}$ is equal to $P(c)$.

9. a) $x+3$ b) $x^2+7x+12$ c) $x+4$

11. a) $x-3$ b) $x^3+4x^2-25x+7$
 c) $x^2+7x-4-\dfrac{5}{x-3}$

13. a) $x-4$ b) $2x^4-8x^3-5x^2+17x+10$
 c) $2x^3-5x-3-\dfrac{2}{x-4}$

15.
$$\begin{array}{r|rrr} 2 & 1 & -5 & 6 \\ & & 2 & -6 \\ \hline & 1 & -3 & 0 \end{array}$$

Answer: $x-3$

17.
$$\begin{array}{r|rrr} 4 & 2 & -1 & -5 \\ & & 8 & 28 \\ \hline & 2 & 7 & 23 \end{array}$$

Answer: $2x+7+\dfrac{23}{x-4}$

19.
$$\begin{array}{r|rrrr} 2 & 1 & -1 & -5 & 6 \\ & & 2 & 2 & -6 \\ \hline & 1 & 1 & -3 & 0 \end{array}$$

Answer: x^2+x-3

21.
$$\begin{array}{r|rrrr} -2 & 3 & 8 & 3 & -2 \\ & & -6 & -4 & 2 \\ \hline & 3 & 2 & -1 & 0 \end{array}$$

Answer: $3x^2+2x-1$

23.
$$\begin{array}{r|rrrr} -3 & 3 & -1 & -22 & 24 \\ & & -9 & 30 & -24 \\ \hline & 3 & -10 & 8 & 0 \end{array}$$

Answer: $3x^2-10x+8$

25.
$$\begin{array}{r|rrrr} -2 & 2 & -5 & -1 & 6 \\ & & -4 & 18 & -34 \\ \hline & 2 & -9 & 17 & -28 \end{array}$$

Answer: $2x^2-9x+17-\dfrac{28}{x+2}$

27.
$$\begin{array}{r|rrrr} 2 & 1 & -2 & 1 & -3 \\ & & 2 & 0 & 2 \\ \hline & 1 & 0 & 1 & -1 \end{array}$$

Answer: $x^2+1-\dfrac{1}{x-2}$

29.
$$\begin{array}{r|rrrr} -3 & 2 & 1 & 0 & -4 \\ & & -6 & 15 & -45 \\ \hline & 2 & -5 & 15 & -49 \end{array}$$

Answer: $2x^2-5x+15-\dfrac{49}{x+3}$

31.
$$\begin{array}{r|rrrr} 3 & 3 & -4 & 0 & 2 \\ & & 9 & 15 & 45 \\ \hline & 3 & 5 & 15 & 47 \end{array}$$

Answer: $3x^2+5x+15+\dfrac{47}{x-3}$

33.
$$\begin{array}{r|rrrr} 2 & 1 & 0 & -7 & 6 \\ & & 2 & 4 & -6 \\ \hline & 1 & 2 & -3 & 0 \end{array}$$

Answer: x^2+2x-3

35.
$$\begin{array}{r|rrrr} 3 & 1 & 0 & 0 & 27 \\ & & 3 & 9 & 27 \\ \hline & 1 & 3 & 9 & 54 \end{array}$$

Answer: $x^2+3x+9+\dfrac{54}{x-3}$

37.

-2	1	0	0	0	16
		-2	4	-8	16
	1	-2	4	-8	32

Answer: $x^3-2x^2+4x-8+\dfrac{32}{x+2}$

39.

$-\frac{2}{3}$	6	1	-11	-6
		-4	2	6
	6	-3	-9	0

Answer: $6x^2-3x-9$

41.

3	3	-8	8	-3
		9	3	33
	3	1	11	30

$P(3)=30$

43.

2	6	-13	1	2
		12	-2	-2
	6	-1	-1	0

$P(2)=0$

45.

-3	3	0	-11	-7
		-9	27	-48
	3	-9	16	-55

$P(-3)=-55$

47.

1	1	-1	-2	1	-2
		1	0	-2	-1
	1	0	-2	-1	-3

$P(1)=-3$

49.

-1	2	0	-6	-5	4
		-2	2	4	1
	2	-2	-4	-1	5

$P(-1)=5$

51.

-2	1	-1	0	3	-3	-3
		-2	6	-12	18	-30
	1	-3	6	-9	15	-33

$P(-2)=-33$

53. a)

2	9	-10	-16
		18	16
	9	8	0

$9x+8$

b) Base: $9(12)+8=108+8=116$ in.

Height: $12-2=10$ in.

Area: $116(10)=1160$ in.^{2}

55. a)

-4	2	-7	-38	88
		-8	60	-88
	2	-15	22	0

2	2	-15	22
		4	-22
	2	-11	0

The height of the room is $2y-11$.

b) height $=2(10)-11=9$ ft.

length $=10+4=14$ ft.

width $=10-2=8$ ft.

Volume $=9(14)(8)=1008$ ft.^{3}

Review Exercises

1. {2, 3, 5, 7, 11}

2. $m=\dfrac{5-(-5)}{-2-3}=\dfrac{10}{-5}=-2$

3. $2x-3<1$ or $2x-3>-5$

$2x<4$ $\quad$ $2x>-2$

$x<2$ $\quad$ $x>-1$

a) $\{x \mid x \text{ is a real number}\}$ $\quad$ b) $(-\infty,\infty)$

c) 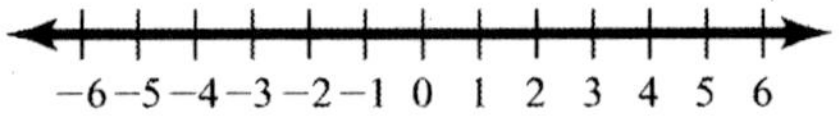

4. $-8\le 3x-1\le 8$

$-7\le 3x\le 9$

$-\dfrac{7}{3}\le x\le 3$

a) $\left\{x \,\middle|\, -\dfrac{7}{3}\le x\le 3\right\}$ $\quad$ b) $\left[-\dfrac{7}{3},3\right]$

c) 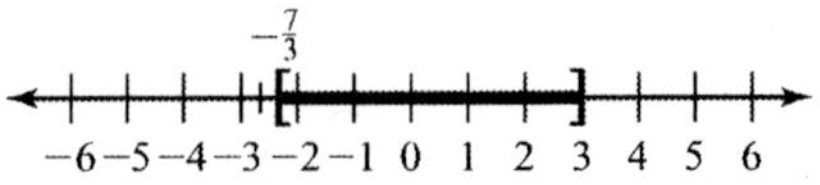

5. $y>2$

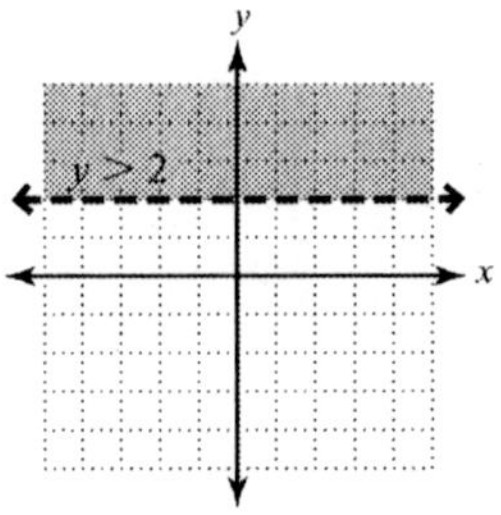

6. $\begin{cases} l = 2w \\ 100 = 2l + 2w \end{cases}$

$100 = 2(2w) + 2w$

$100 = 4w + 2w$

$100 = 6w$

$16\frac{2}{3} \text{ ft.} = w$

$l = 2\left(16\frac{2}{3}\right)$

$= 33\frac{1}{3} \text{ ft.}$

Chapter 5 Review Exercises

1. True
2. True
3. False; the degree of the monomial is 3 + 1 = 4.
4. False; FOIL can only be used to multiply two binomials.
5. False; conjugates have the form $a+b$ and $a-b$.
6. False; $a^0 = 1$
7. monomial
8. $m-n$
9. $a \cdot b$
10. distributive
11. $\left(\frac{2}{3}\right)^{-3} = \left(\frac{3}{2}\right)^3 = \frac{27}{8}$
12. $-4^2 = -(4\cdot 4) = -16$
13. $2^{-5} = \frac{1}{2^5} = \frac{1}{32}$
14. $14x^0 = 14\cdot 1 = 14$
15. $1.6736\times 10^{-24} =$ 0.0000000000000000000000016737
16. $1.65\times 10^{10} = 16{,}500{,}000{,}000$
17. $0.000000000753 = 7.53\times 10^{-10}$
18. $300{,}000{,}000 = 3\times 10^8$
19. $(4x^2)(5xy^3) = 20x^{2+1}y^3 = 20x^3y^3$
20. $\left(-\frac{2}{3}m^2n^4\right)(3mn^5) = -2m^{2+1}n^{4+5} = -2m^3n^9$
21. $\frac{6x^4}{2x^9} = 3x^{4-9} = 3x^{-5} = \frac{3}{x^5}$
22. $-\frac{7a^3bc^9}{3abc^{15}} = -\frac{7}{3}a^{3-1}b^{1-1}c^{9-15}$

$= -\frac{7}{3}a^2b^0c^{-6}$

$= -\frac{7a^2}{3c^6}$

23. $(3m^5n)^2 = 3^2m^{5\cdot 2}n^2 = 9m^{10}n^2$
24. $\frac{(6m^2n^3p)^{-1}}{(2m^6n^2)^3} = \frac{1}{(6m^2n^3p)(2m^6n^2)^3}$

$= \frac{1}{6m^2n^3p\cdot(2)^3(m^6)^3(n^2)^3}$

$= \frac{1}{6m^2n^3p\cdot 8m^{18}n^6}$

$= \frac{1}{48m^{20}n^9p}$

25. $(3j^2k^5)^2(0.1jk^3)^3$

$= 3^2(j^2)^2(k^5)^2\cdot(0.1)^3(j)^3(k^3)^3$

$= 9j^4k^{10}\cdot 0.001j^3k^9$

$= 0.009j^7k^{19}$

26. $\frac{(2s^3t)^{-1}}{(3st)^2(s^4t)^8} = \frac{1}{(2s^3t)(3st)^2(s^4t)^8}$

$= \frac{1}{(2s^3t)(9s^2t^2)(s^{32}t^8)}$

$= \frac{1}{18s^{37}t^{11}}$

27. $\dfrac{(5m^7n^3)^{-2}}{(3m)^{-1}(3m)^2} = \dfrac{(5m^7n^3)^{-2}}{(3m)^{-1+2}}$

$= \dfrac{1}{(3m)^1(5m^7n^3)^2}$

$= \dfrac{1}{3m(25m^{14}n^6)}$

$= \dfrac{1}{75m^{15}n^6}$

28. $\left(\dfrac{2a}{b^4}\right)^{-3} = \left(\dfrac{b^4}{2a}\right)^3 = \dfrac{b^{12}}{8a^3}$

29. $(5.1\times10^4)(-2\times10^6) = 5.1\times(-2)\times(10^{4+6})$

$= -10.2\times10^{10}$

$= -1.02\times10^{11}$

30. $\dfrac{8.12\times10^{-8}}{2\times10^{-5}} = \dfrac{8.12}{2}\times\dfrac{10^{-8}}{10^{-5}}$

$= 4.06\times10^{-8-(-5)}$

$= 4.06\times10^{-3}$

31. $-\dfrac{1}{4}+2c^2+8.7c^3+\dfrac{2}{5}c$

$d = 3$; this polynomial has more than three terms, so it has no special polynomial name.

32. $7m^2+1$

$d = 2$; binomial

33. $-8xy^4$

$d = 1 + 4 = 5$, monomial

34. $9h^5+4h^3-h^2+1$

$d = 5$; this polynomial has more than three terms, so it has no special polynomial name.

35. $(3c^3+2c^2-c-1)+(8c^2+c-10)$

$= 3c^3+10c^2-11$

36. $(y^2+3y+6)-(-5y^2+3y-8)$

$= (y^2+3y+6)+(5y^2-3y+8)$

$= 6y^2+14$

37. $(x^2y^3-xy^2+2x^2y-4xy+7y^2-9)$

$+(-3x^2y^2+2xy^2+4x^2y-xy+6y^2+4)$

$= x^2y^3-3x^2y^2+xy^2+6x^2y-5xy+13y^2-5$

38. $(4hk-8k^3)-(5kh+3k-2k^3)$

$= (4hk-8k^3)+(-5hk-3k+2k^3)$

$= -hk-3k-6k^3$

39. linear

40. quadratic

41. cubic

42. constant

43. $f(x) = x^2-3$

x	$f(x)$
−2	1
−1	−2
0	−3
1	−2
2	1

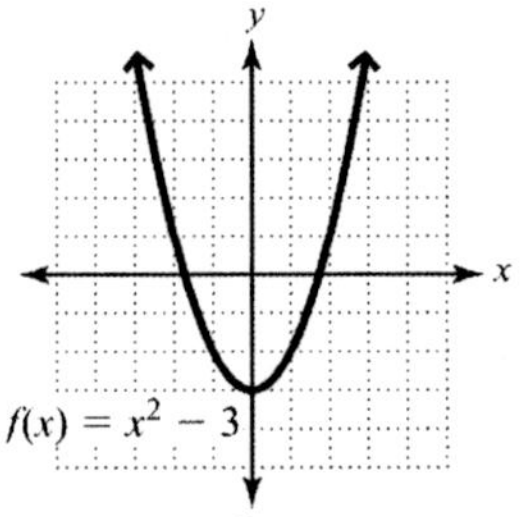

Domain: $\{x\mid-\infty<x<\infty\}$ or $(-\infty,\infty)$

Range: $\{y\mid y\geq-3\}$ or $[-3,\infty)$

44. $g(x) = -\dfrac{1}{3}x+2$

x	$g(x)$
−3	3
0	2
3	1

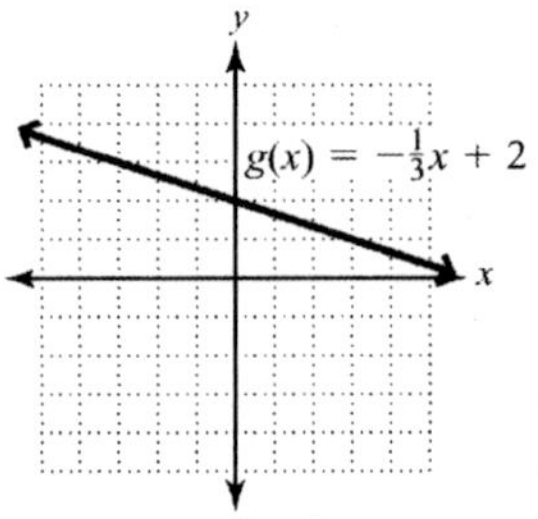

Domain: $\{x\mid-\infty<x<\infty\}$ or $(-\infty,\infty)$

Range: $\{y\mid-\infty<y<\infty\}$ or $(-\infty,\infty)$

45. $h(x) = x^3 - 1$

x	$h(x)$
-2	-9
-1	-2
0	-1
1	0
2	7

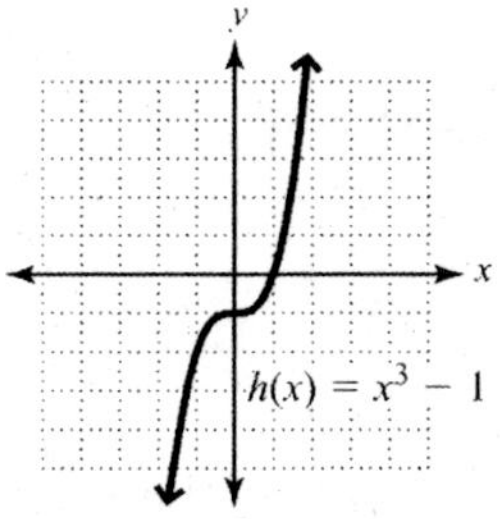

Domain: $\{x \mid -\infty < x < \infty\}$ or $(-\infty, \infty)$

Range: $\{y \mid -\infty < y < \infty\}$ or $(-\infty, \infty)$

46. $w(x) = 7$

x	$w(x)$
-2	7
0	7
2	7

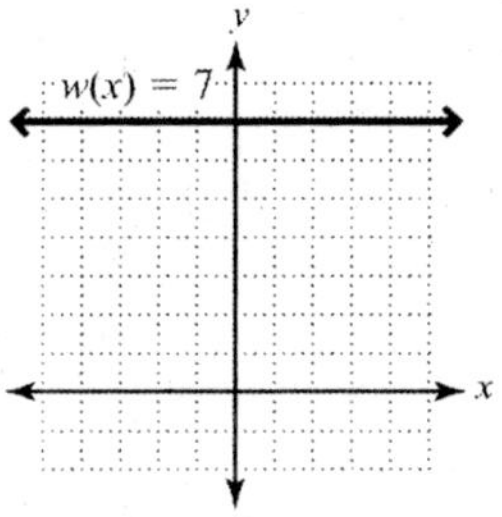

Domain: $\{x \mid -\infty < x < \infty\}$ or $(-\infty, \infty)$

Range: $\{y \mid y = 7\}$ or $[7,7]$

47. $f(x) = x^3 + 2x + 5;\quad g(x) = -2x^3 + 2$

a) $(f+g)(x) = (x^3 + 2x + 5) + (-2x^3 + 2)$
$= -x^3 + 2x + 7$

b) $(f+g)(2) = -2^3 + 2(2) + 7 = -8 + 4 + 7 = 3$

c) $(f-g)(x) = (x^3 + 2x + 5) - (-2x^3 + 2)$
$= (x^3 + 2x + 5) + (2x^3 - 2)$
$= 3x^3 + 2x + 3$

d) $(f-g)(-4) = 3(-4)^3 + 2(-4) + 3$
$= 3(-64) - 8 + 3$
$= -192 - 8 + 3$
$= -197$

48. $3x^2y^3(2x^2 - 3x + 1)$
$= 3x^2y^3 \cdot 2x^2 + 3x^2y^3 \cdot (-3x) + 3x^2y^3 \cdot 1$
$= 6x^4y^3 - 9x^3y^3 + 3x^2y^3$

49. $-pq(2p^2 - 3pq + 3q^2)$
$= -pq \cdot 2p^2 - pq \cdot (-3pq) - pq \cdot 3q^2$
$= -2p^3q + 3p^2q^2 - 3pq^3$

50. $(x-3)(2x+8) = 2x^2 + 8x - 6x - 24$
$= 2x^2 + 2x - 24$

51. $(9w-1)(3w+2) = 27w^2 + 18w - 3w - 2$
$= 27w^2 + 15w - 2$

52. $(3r-s)(6r+7s) = 18r^2 + 21rs - 6rs - 7s^2$
$= 18r^2 + 15rs - 7s^2$

53. $(x^2-1)(x^2+2) = x^4 + 2x^2 - x^2 - 2$
$= x^4 + x^2 - 2$

54. $(x-1)(2x^3 - 3x^2 + 4x + 7)$
$= 2x^4 - 3x^3 + 4x^2 + 7x - 2x^3 + 3x^2 - 4x - 7$
$= 2x^4 - 5x^3 + 7x^2 + 3x - 7$

55. $(3t^2 - t + 1)(4t^2 + t - 1)$
$= 12t^4 + 3t^3 - 3t^2 - 4t^3 - t^2 + t + 4t^2 + t - 1$
$= 12t^4 - t^3 + 2t - 1$

56. $(3a-5)(3a+5) = (3a)^2 - 5^2 = 9a^2 - 25$

57. $(2p-1)^2 = (2p)^2 - 2(2p)(1) + 1^2 = 4p^2 - 4p + 1$

58. $(8k+3)^2 = (8k)^2 + 2(8k)(3) + 3^2 = 64k^2 + 48k + 9$

59. $(4h^2+7)(4h^2-7) = (4h^2)^2 - 7^2 = 16h^4 - 49$

60. $5m + 2$

61. $f(x)=2x+7;\quad g(x)=x^2+4x-1$

a) $(f\cdot g)(x)=(2x+7)(x^2+4x-1)$

$$=2x^3+8x^2-2x+7x^2+28x-7$$
$$=2x^3+15x^2+26x-7$$

b) $(f\cdot g)(-1)=2(-1)^3+15(-1)^2+26(-1)-7$

$$=2(-1)+15(1)-26-7$$
$$=-2+15-26-7$$
$$=-20$$

62. $f(x)=4x^2-x$

$$f(x-4)=4(x-4)^2-(x-4)$$
$$=4(x^2-8x+16)-x+4$$
$$=4x^2-32x+64-x+4$$
$$=4x^2-33x+68$$

63. $\dfrac{20m^5-5m^4+15m^3-5m^2}{5m^2}$

$$=\frac{20m^5}{5m^2}-\frac{5m^4}{5m^2}+\frac{15m^3}{5m^2}-\frac{5m^2}{5m^2}$$
$$=4m^3-m^2+3m-1$$

64. $\dfrac{8x^2-2x+3}{2x}=\dfrac{8x^2}{2x}-\dfrac{2x}{2x}+\dfrac{3}{2x}=4x-1+\dfrac{3}{2x}$

65.
$$\begin{array}{r} x+3 \\ x+2\overline{)x^2+5x+7} \\ \underline{x^2+2x} \\ 3x+7 \\ \underline{3x+6} \\ 1 \end{array}$$

Answer: $x+3+\dfrac{1}{x+2}$

66.
$$\begin{array}{r} x-2 \\ x-2\overline{)x^2-4x+4} \\ \underline{x^2-2x} \\ -2x+4 \\ \underline{-2x+4} \\ 0 \end{array}$$

67.
$$\begin{array}{r} 2x+3 \\ 2x-1\overline{)4x^2+4x-3} \\ \underline{4x^2-2x} \\ 6x-3 \\ \underline{6x-3} \\ 0 \end{array}$$

68. $(f/g)(x)=\dfrac{9x^4-7x^2-10x-5}{3x-1}$

$$\begin{array}{r} 3x^3+x^2-2x-4 \\ 3x-1\overline{)9x^4+0x^3-7x^2-10x-5} \\ \underline{9x^4-3x^3} \\ 3x^3-7x^2 \\ \underline{3x^3-x^2} \\ -6x^2-10x \\ \underline{-6x^2+2x} \\ -12x-5 \\ \underline{-12x+4} \\ -9 \end{array}$$

$$(f/g)(x)=3x^3+x^2-2x-4-\frac{9}{3x-1}$$

69.
$$\begin{array}{r|rrr} 1 & 5 & -3 & 2 \\ & & 5 & 2 \\ \hline & 5 & 2 & 4 \end{array}$$

$5x+2+\dfrac{4}{x-1}$

70.
$$\begin{array}{r|rrrr} 3 & 2 & -6 & -5 & 15 \\ & & 6 & 0 & -15 \\ \hline & 2 & 0 & -5 & 0 \end{array}$$

$2x^2-5$

71.
$$\begin{array}{r|rrrr} 2 & 4 & 0 & -2 & 5 \\ & & 8 & 16 & 28 \\ \hline & 4 & 8 & 14 & 33 \end{array}$$

$4x^2+8x+14+\dfrac{33}{x-2}$

72.
$$\begin{array}{r|rrrr} 2 & 1 & 0 & 0 & -8 \\ & & 2 & 4 & 8 \\ \hline & 1 & 2 & 4 & 0 \end{array}$$

x^2+2x+4

73. $\underline{1|}\ \ 5\ \ -1\ \ 8\ \ 3$ $\quad P(1)=15$

$$\begin{array}{r|rrrr} 1 & 5 & -1 & 8 & 3 \\ & & 5 & 4 & 12 \\ \hline & 5 & 4 & 12 & 15 \end{array}$$

74. $P(-2)=-51$

$$\begin{array}{r|rrrr} -2 & 7 & 0 & -2 & 1 \\ & & -14 & 28 & -52 \\ \hline & 7 & -14 & 26 & -51 \end{array}$$

75. $E = hf$

$= (6.626\times10^{-34})(6.1\times10^{14})$

$= 40.4186\times10^{-20}$

$= 4.04186\times10^{-19}$

The energy is 4.04186×10^{-19} joules.

76. $h(t) = -16(t)^2 + 200$

$h(3) = -16(3)^2 + 200$

$= 56$

The height after 3 seconds is 56 feet.

77. a) $a + (5a - b) + b + (6a + b) = 12a + b$

b) $12a + b = 12(12) + 9$

$= 144 + 9$

$= 153$

The perimeter is 153 cm.

78. a) $P(x) = R(x) - C(x)$

$P(x) = (0.2x^2 + x + 1000) - (10x + 2000)$

$= (0.2x^2 + x + 1000) + (-10x - 2000)$

$= 0.2x^2 - 9x - 1000$

b) $P(200) = 0.2(200)^2 - 9(200) - 1000$

$= 5200$

The profit is \$5200.

79. The width is w. The length is $w + 10$. The height is $w + 5$.

$w(w+10)(w+5) = w(w^2 + 15w + 50)$

$= w^3 + 15w^2 + 50w$

The volume of the tank is $(w^3 + 15w^2 + 50w)$ cubic inches.

80.

a)
$$\begin{array}{r} y^3 + 2y^2 + 4y + 1 \\ 3y+5\overline{\big)\,3y^4 + 11y^3 + 22y^2 + 23y + 5} \\ \underline{3y^4 + 5y^3} \\ 6y^3 + 22y^2 \\ \underline{6y^3 + 10y^2} \\ 12y^2 + 23y \\ \underline{12y^2 + 20y} \\ 3y + 5 \\ \underline{3y + 5} \\ 0 \end{array}$$

The current is $(y^3 + 2y^2 + 4y + 1)$ amps.

b) $y^3 + 2y^2 + 4y + 1 = 3^3 + 2(3)^2 + 4(3) + 1 = 58$

The current is 58 amps if $y = 3$.

Chapter 5 Practice Test

1. $7.2\times10^{-3} = 0.0072$

2. $0.00357 = 3.57\times10^{-3}$

3. $5x^3(3x^2y) = 15x^{3+2}y = 15x^5y$

4. $(3x^4y)^{-2} = \dfrac{1}{(3x^4y)^2} = \dfrac{1}{9x^8y^2}$

5. $\dfrac{8u^7v}{2u^4v^5} = 4u^{7-4}v^{1-5} = 4u^3v^{-4} = \dfrac{4u^3}{v^4}$

6. $\left(-\dfrac{2}{3}t^5u^2v\right)^4 = \left(-\dfrac{2}{3}\right)^4(t^5)^4(u^2)^4(v)^4$

$= \dfrac{16}{81}t^{20}u^8v^4$

7. $(6\times10^5)(2.1\times10^4) = (6\cdot2.1)\times10^{5+4}$

$= 12.6\times10^9$

$= 1.26\times10^{10}$

8. $\dfrac{8.4\times10^{10}}{4.2\times10^3} = \dfrac{8.4}{4.2}\times\dfrac{10^{10}}{10^3}$

$= 2\times10^{10-3}$

$= 2\times10^7$

9. The graph of a constant function is a horizontal line.

10. The graph of a cubic function resembles an S-shaped curve.

11. Domain: $\{x \mid -\infty < x < \infty\}$ or $(-\infty, \infty)$;
Range: $\{y \mid y \geq -1\}$ or $[-1, \infty)$

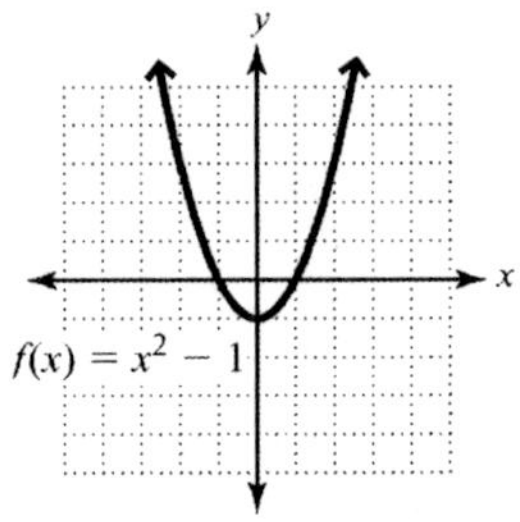

12. $\left(a^2 - 4ab + b^2\right) - \left(a^2 + 4ab + b^2\right)$
$= \left(a^2 - 4ab + b^2\right) + \left(-a^2 - 4ab - b^2\right)$
$= -8ab$

13. $\left(6r^5 - 9r^4 - 2r^2 + 8\right)$
$+ \left(5r^5 + 2r^4 + 7r^3 - 8r^2 - r - 9\right)$
$= 11r^5 - 7r^4 + 7r^3 - 10r^2 - r - 1$

14. $(3x - 4y)(5x + 2y)$
$= 3x \cdot 5 + 3x \cdot 2y + (-4y) \cdot 5x + (-4y) \cdot 2y$
$= 15x^2 + 6xy - 20xy - 8y^2$
$= 15x^2 - 14xy - 8y^2$

15. $4m^3n^9\left(3m^2 - 2mn + 5n^2\right)$
$= 4m^3n^9 \cdot 3m^2 + 4m^3n^9 \cdot (-2mn) + 4m^3n^9 \cdot 5n^2$
$= 12m^5n^9 - 8m^4n^{10} + 20m^3n^{11}$

16. $(7k - 2j)^2 = (7k)^2 - 2(7k)(2j) + (2j)^2$
$= 49k^2 - 28jk + 4j^2$

17. $(2x - 3)\left(7x^2 + 3x - 5\right)$
$= 2x \cdot 7x^2 + 2x \cdot 3x + 2x \cdot (-5) + (-3) \cdot 7x^2$
$+ (-3) \cdot 3x + (-3) \cdot (-5)$
$= 14x^3 + 6x^2 - 10x - 21x^2 - 9x + 15$
$= 14x^3 - 15x^2 - 19x + 15$

18. $(4h - 3)(4h + 3) = (4h)^2 - (3)^2$
$= 16h^2 - 9$

19. $(f \cdot g)(x) = (5x - 8)(3x + 2)$
$= 15x^2 + 10x - 24x - 16$
$= 15x^2 - 14x - 16$

20. $(f \cdot g)(x) = (5x - 8)(3x + 2)$
$= 15x^2 + 10x - 24x - 16$
$= 15x^2 - 14x - 16$
$(f \cdot g)(2) = 15(2)^2 - 14(2) - 16$
$= 60 - 28 - 16$
$= 16$

21. $\dfrac{8k^3 - 4k^2 + 2k}{2k} = \dfrac{8k^3}{2k} - \dfrac{4k^2}{2k} + \dfrac{2k}{2k}$
$= 4k^2 - 2k + 1$

22.
$$\begin{array}{r}
m^2 + 3m + 2 \\
3m + 1\overline{)3m^3 + 10m^2 + 9m + 2} \\
\underline{3m^3 + m^2} \qquad\qquad \\
9m^2 + 9m \qquad \\
\underline{9m^2 + 3m} \qquad \\
6m + 2 \\
\underline{6m + 2} \\
0
\end{array}$$

23.

−3	2	5	0	−8
		−6	3	−9
	2	−1	3	−17

$2x^2 - x + 3 - \dfrac{17}{x + 3}$

24.

−3	1	0	−5	4
		−3	9	−12
	1	−3	4	−8

$P(-3) = -8$

25. a) $P(x) = R(x) - C(x)$
$= \left(0.1x^2 + x + 5000\right) - (15x + 3000)$
$= \left(0.1x^2 + x + 5000\right) + (-15x - 3000)$
$= 0.1x^2 - 14x + 2000$

b) $P(1000) = 0.1(1000)^2 - 14(1000) + 2000$
$= 100{,}000 - 14{,}000 + 2000$
$= \$88{,}000$

Chapters 1—5 Cumulative Review

1. False; the identity for multiplication is 1.

2. True

3. False; the boundary line of the graph of $3x+4y<15$ is a dashed line.

4. True

5. identity

6. $3x+7=5$; $3x+7=-5$

7. inconsistent

8. $x-c$

9. Let $x_2=6,\ x_1=-2,\ y_2=-3,\ y_1=3$.

$$\begin{aligned}\sqrt{(x_2-x_1)^2+(y_2-y_1)^2}&=\sqrt{(6-(-2))^2+(-3-3)^2}\\&=\sqrt{(6+2)^2+(-6)^2}\\&=\sqrt{(8)^2+36}\\&=\sqrt{64+36}\\&=\sqrt{100}\\&=10\end{aligned}$$

10. The expression is undefined if the denominator equals 0.

$$\begin{aligned}x+5&=0\\x&=-5\end{aligned}$$

The expression is undefined if $x=-5$.

11. $$\begin{aligned}\begin{vmatrix}-4 & 5\\ -3 & 2\end{vmatrix}&=(-4)(2)-(-3)(5)\\&=-8-(-15)\\&=-8+15\\&=7\end{aligned}$$

12. $$\begin{aligned}\left(-2a^4b^{-3}\right)\left(4a^{-2}b^{-1}\right)&=-8a^{4+(-2)}b^{-3+(-1)}\\&=-8a^2b^{-4}\\&=-\frac{8a^2}{b^4}\end{aligned}$$

13. $$\begin{aligned}\frac{8x^{-2}y^3z^8}{2x^3y^{-4}z^4}&=4x^{-2-3}y^{3-(-4)}z^{8-4}\\&=4x^{-5}y^7z^4\\&=\frac{4y^7z^4}{x^5}\end{aligned}$$

14. $$\begin{aligned}\left(-3x^4y^{-2}\right)^3&=(-3)^3\left(x^4\right)^3\left(y^{-2}\right)^3\\&=-27x^{4\cdot 3}y^{-2\cdot 3}\\&=-27x^{12}y^{-6}\\&=-\frac{27x^{12}}{y^6}\end{aligned}$$

15. $$\begin{aligned}&\left(5x^3-7x^2+2x-3\right)+\left(-8x^3+7x^2+3x-5\right)\\&=-3x^3+5x-8\end{aligned}$$

16. $$\begin{aligned}&\left(3x^2+6x-2\right)-\left(-4x^2-7x+1\right)\\&=\left(3x^2+6x-2\right)+\left(4x^2+7x-1\right)\\&=7x^2+13x-3\end{aligned}$$

17. $$\begin{aligned}(5x+2y)(3x-2y)&=15x^2-10xy+6xy-4y^2\\&=15x^2-4xy-4y^2\end{aligned}$$

18. $$\begin{array}{r} 2x^2-\ \ 2x+1\\ 3x+10\overline{)6x^3+14x^2-17x+10}\\ \underline{6x^3+20x^2}\qquad\qquad\\ -6x^2-17x\qquad\\ \underline{-6x^2-20x}\qquad\\ 3x+10\\ \underline{3x+10}\\ 0\end{array}$$

19. $$\begin{aligned}P&=2\pi r+2d\\P-2\pi r&=2d\\\frac{P-2\pi r}{2}&=\frac{2d}{2}\\\frac{P-2\pi r}{2}&=d\end{aligned}$$

20. $$\begin{aligned}-6\le -4x&-2\le 6\\-4\le -4&x\le 8\\1\ge x&\ge -2\end{aligned}$$

The solution set is $\{x\mid -2\le x\le 1\}$ or $[-2,1]$.

21. $$\begin{aligned}3|2x-5|&>9\\|2x-5|&>3\end{aligned}$$

$2x-5<-3$ or $2x-5>3$

$2x<2$ $\quad$ $2x>8$

$x<1$ $\quad$ $x>4$

The solution set is $\{x\mid x<1 \text{ or } x>4\}$ or $(-\infty,1)\cup(4,\infty)$.

22. $\begin{cases} 5x - 3y = 2 \\ 3x - y = -2 \end{cases}$

Solve the second equation for y:

$3x - y = -2$

$3x + 2 = y$

Substitute for y in the first equation.

$$\begin{aligned} 5x - 3y &= 2 \\ 5x - 3(3x+2) &= 2 \\ 5x - 9x - 6 &= 2 \\ -4x - 6 &= 2 \\ -4x &= 8 \\ x &= -2 \end{aligned} \qquad \begin{aligned} 3x + 2 &= y \\ 3(-2) + 2 &= y \\ -6 + 2 &= y \\ -4 &= y \end{aligned}$$

The solution is $(-2, -4)$.

23. $y = \frac{3}{4}x - 2$

x	y	Ordered Pair
0	−2	$(0, -2)$
$\frac{8}{3}$	0	$\left(\frac{8}{3}, 0\right)$
4	1	$(4, 1)$

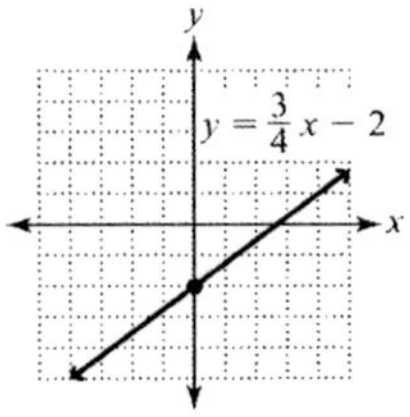

24. $2x - 3y > -6$

$-3y > -2x - 6$

$y < \frac{2}{3}x + 2$

Begin by graphing the related equation $y = \frac{2}{3}x + 2$ with a dashed line. Now choose $(0,0)$ as a test point.

$$\begin{aligned} 2x - 3y &> -6 \\ 2(0) - 3(0) &\overset{?}{>} -6 \\ 0 - 0 &\overset{?}{>} -6 \\ 0 &> -6 \end{aligned}$$

Because $(0,0)$ satisfies the inequality, shade the region which includes $(0,0)$.

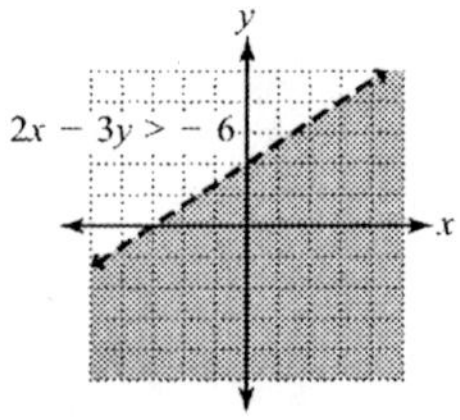

25. $5x - 3y = 8$

$-3y = -5x + 8$

$y = \frac{5}{3}x - \frac{8}{3}$

The slope of the given equation is $m = \frac{5}{3}$, so the slope of a parallel line is $m = \frac{5}{3}$.

$$\begin{aligned} y - y_1 &= m(x - x_1) \\ y - 2 &= \frac{5}{3}(x - (-6)) \\ y - 2 &= \frac{5}{3}(x + 6) \\ y - 2 &= \frac{5}{3}x + 10 \\ y &= \frac{5}{3}x + 12 \\ 3y &= 5x + 36 \\ -5x + 3y &= 36 \\ 5x - 3y &= -36 \end{aligned}$$

26. No, it is not a function because there are values in the domain that correspond to two values in the range. (It fails the vertical line test.)

27. The consecutive integers are x, $x + 1$, and $x + 2$.

$$\begin{aligned} x + (x+1) + (x+2) &= 162 \\ 3x + 3 &= 162 \\ 3x &= 159 \\ x &= 53 \end{aligned}$$

The seat numbers are 53, 53 + 1 = 54, and 53 + 2 = 55.

28. Create a table.

	rate	time	distance
Franz	50	x	$50x$
Hanz	60	$x-1$	$60(x-1)$

Translate the information in the table to an equation and solve.

$$50x = 60(x-1)$$
$$50x = 60x - 60$$
$$-10x = -60$$
$$x = 6$$

Hanz will catch up in 6 – 1 = 5 hours.

29. Create a table.

Categories	how many	value	total amount
azaleas	x	6	$6x$
rosebushes	$2x+7$	10	$10(2x+7)$

Translate the information in the table to an equation and solve.

$$6x+10(2x+7) = 798$$
$$6x + 20x + 70 = 798$$
$$26x + 70 = 798$$
$$26x = 728$$
$$x = 28$$

A total of 28 azaleas and 2(28) + 7 = 63 rosebushes were sold.

30. Let w represent the width of the rectangle and $3w+2$ represent the length.

$$P = 2w + 2l$$
$$84 = 2w + 2(3w+2)$$
$$84 = 2w + 6w + 4$$
$$80 = 8w$$
$$10 = w$$

The width is 10 feet and the length is 3(10) + 2 = 32 feet.

Chapter 6
Factoring

Exercise Set 6.1

1. No, factored form is a number or an expression written as a product of factors.

3. Four-term polynomials

5. $4x^3y^2 = 2^2 \cdot x^3 \cdot y^2$
$24x^2y = 2^3 \cdot 3 \cdot x^2 \cdot y$
$\text{GCF} = 2^2 \cdot x^2 \cdot y = 4x^2y$

7. $12u^3v^6 = 2^2 \cdot 3 \cdot u^3 \cdot v^6$
$28u^8v^8 = 2^2 \cdot 7 \cdot u^8 \cdot v^8$
$\text{GCF} = 2^2 \cdot u^3 \cdot v^6 = 4u^3v^6$

9. $10a^3b^2c^8 = 2 \cdot 5 \cdot a^3 \cdot b^2 \cdot c^8$
$14abc^8 = 2 \cdot 7 \cdot a \cdot b \cdot c^8$
$20a^7b^3c^4 = 2^2 \cdot 5 \cdot a^7 \cdot b^3 \cdot c^4$
$\text{GCF} = 2 \cdot a \cdot b \cdot c^4 = 2abc^4$

11. $5(a+b) = 5 \cdot (a+b)$
$7(a+b) = 7 \cdot (a+b)$
$\text{GCF} = a+b$

13. $$15c^4d - 20c^2 = 5c^2\left(\frac{15c^4d - 20c^2}{5c^2}\right) = 5c^2\left(\frac{15c^4d}{5c^2} - \frac{20c^2}{5c^2}\right) = 5c^2\left(3c^2d - 4\right)$$

15. $$x^5 - x^3 + x^2 = x^2\left(\frac{x^5 - x^3 + x^2}{x^2}\right) = x^2\left(\frac{x^5}{x^2} - \frac{x^3}{x^2} + \frac{x^2}{x^2}\right) = x^2\left(x^3 - x + 1\right)$$

17. $25xy - 50xz + 100x^2$
$$= 25x\left(\frac{25xy - 50xz + 100x^2}{25x}\right)$$
$$= 25x\left(\frac{25xy}{25x} - \frac{50xz}{25x} + \frac{100x^2}{25x}\right)$$
$$= 25x(y - 2z + 4x)$$

19. $-14u^2v^2 - 7uv^2 + 7uv$
$$= -7uv\left(\frac{-14u^2v^2 - 7uv^2 + 7uv}{-7uv}\right)$$
$$= -7uv\left(\frac{-14u^2v^2}{-7uv} - \frac{7uv^2}{-7uv} + \frac{7uv}{-7uv}\right)$$
$$= -7uv(2uv + v - 1)$$

21. $9a^7b^3 + 3a^4b^2 - 6a^2b$
$$= 3a^2b\left(\frac{9a^7b^3 + 3a^4b^2 - 6a^2b}{3a^2b}\right)$$
$$= 3a^2b\left(\frac{9a^7b^3}{3a^2b} + \frac{3a^4b^2}{3a^2b} - \frac{6a^2b}{3a^2b}\right)$$
$$= 3a^2b\left(3a^5b^2 + a^2b - 2\right)$$

23. $3w^3v^4 + 39w^2v + 18wv^2$
$$= 3wv\left(\frac{3w^3v^4 + 39w^2v + 18wv^2}{3wv}\right)$$
$$= 3wv\left(\frac{3w^3v^4}{3wv} + \frac{39w^2v}{3wv} + \frac{18wv^2}{3wv}\right)$$
$$= 3wv\left(w^2v^3 + 13w + 6v\right)$$

25. $18ab^3c - 36a^2b^2c + 24a^5b^2c^8$
$$= 6ab^2c\left(\frac{18ab^3c - 36a^2b^2c + 24a^5b^2c^8}{6ab^2c}\right)$$
$$= 6ab^2c\left(\frac{18ab^3c}{6ab^2c} - \frac{36a^2b^2c}{6ab^2c} + \frac{24a^5b^2c^8}{6ab^2c}\right)$$
$$= 6ab^2c\left(3b - 6a + 4a^4c^7\right)$$

27. $-8x^2y + 16xy^2 - 12xy$
$$= -4xy\left(\frac{-8x^2y + 16xy^2 - 12xy}{-4xy}\right)$$
$$= -4xy\left(\frac{-8x^2y}{-4xy} + \frac{16xy^2}{-4xy} - \frac{12xy}{-4xy}\right)$$
$$= -4xy(2x - 4y + 3)$$

29. $m(n-3) + 4(n-3)$
$$= (n-3)\left(\frac{m(n-3) + 4(n-3)}{(n-3)}\right)$$
$$= (n-3)(m+4)$$

31. $6(b+2c)-a(b+2c)$
$=(b+2c)\left(\dfrac{6(b+2c)-a(b+2c)}{(b+2c)}\right)$
$=(b+2c)(6-a)$

33. $ax+ay+bx+by=(ax+ay)+(bx+by)$
$=a(x+y)+b(x+y)$
$=(x+y)(a+b)$

35. $u^3+3u^2+3u+9=(u^3+3u^2)+(3u+9)$
$=u^2(u+3)+3(u+3)$
$=(u+3)(u^2+3)$

37. $mn+np-3m-3p=(mn+np)+(-3m-3p)$
$=n(m+p)-3(m+p)$
$=(m+p)(n-3)$

39. $cd+d+c+1=(cd+d)+(c+1)$
$=d(c+1)+1(c+1)$
$=(c+1)(d+1)$

41. $2a^2-a+2a-1=(2a^2-a)+(2a-1)$
$=a(2a-1)+1(2a-1)$
$=(2a-1)(a+1)$

43. $3ax+6ay+8by+4bx=(3ax+6ay)+(8by+4bx)$
$=3a(x+2y)+4b(2y+x)$
$=(x+2y)(3a+4b)$

45. $h^2+8h-hk-8k=(h^2+8h)+(-hk-8k)$
$=h(h+8)-k(h+8)$
$=(h+8)(h-k)$

47. $3x^2+3y^2-ax^2-ay^2$
$=(3x^2+3y^2)+(-ax^2-ay^2)$
$=3(x^2+y^2)-a(x^2+y^2)$
$=(x^2+y^2)(3-a)$

49. $3p^3-6p^2q+2pq^2-4q^3$
$=(3p^3-6p^2q)+(2pq^2-4q^3)$
$=3p^2(p-2q)+2q^2(p-2q)$
$=(p-2q)(3p^2+2q^2)$

51. $2x^3-8x^2y-3xy^2+12y^3$
$=(2x^3-8x^2y)-(3xy^2-12y^3)$
$=2x^2(x-4y)-3y^2(x-4y)$
$=(x-4y)(2x^2-3y^2)$

53. $2ab+2bx-2ac-2cx$
$=2(ab+bx-ac-cx)$
$=2[(ab+bx)+(-ac-cx)]$
$=2[b(a+x)-c(a+x)]$
$=2(b-c)(a+x)$

55. $12xy+4y+30x+10$
$=2(6xy+2y+15x+5)$
$=2[(6xy+2y)+(15x+5)]$
$=2[2y(3x+1)+5(3x+1)]$
$=2(3x+1)(2y+5)$

57. $3a^2y-12a^2+9ay-36a$
$=3a(ay-4a+3y-12)$
$=3a[(ay-4a)+(3y-12)]$
$=3a[a(y-4)+3(y-4)]$
$=3a(y-4)(a+3)$

59. $3m^3+6m^2n-10m^2-20mn$
$=m(3m^2+6mn-10m-20n)$
$=m[(3m^2+6mn)-(10m+20n)]$
$=m[3m(m+2n)-10(m+2n)]$
$=m(m+2n)(3m-10)$

61. $15st^2-5t^2-30st+10t$
$=5t(3st-t-6s+2)$
$=5t[(3st-t)-(6s-2)]$
$=5t[t(3s-1)-2(3s-1)]$
$=5t(3s-1)(t-2)$

63. $5x^3y - 20x^2y + 5xy - 20y$
$= 5y\left(x^3 - 4x^2 + x - 4\right)$
$= 5y\left[\left(x^3 - 4x^2\right) + (x-4)\right]$
$= 5y\left[x^2(x-4) + 1(x-4)\right]$
$= 5y(x-4)\left(x^2+1\right)$

65. Mistake: Did not factor out the GCF, which is $12x^2y$.
Correct: $12x^2y\left(2y^2+3x\right)$

67. Mistake: Incorrect power of b in the parentheses.
Correct: $9a^2b\left(b^2c-2a^2\right)$

69. It is correct.

71. It is not correct.
Correct form: $2x\left(2x^2+7x+4\right)$

73. $10x(x+4) - 4x(x) = 10x^2 + 40x - 4x^2$
$= 6x^2 + 40x$
$= 2x(3x+20)$

75. Divide the figure into 2 rectangles by drawing a horizontal line at the bottom of the side measuring x. Add the area of each rectangle and then subtract the area of the hearth.

Area of top rectangle : $A = 3x \cdot x = 3x^2$

Area of bottom rectangle:
$A = (2x+4)(6x+1) = 12x^2 + 26x + 4$

Area of hearth: $1 \cdot 4 = 4$

a) $3x^2 + \left(12x^2 + 26x + 4\right) - 4 = 15x^2 + 26x$

b) $15x^2 + 26x = x(15x+26)$

c) $A = 3(15\cdot 3 + 26) = 3\cdot 71 = 213 \text{ ft.}^2$

77. Add the volume of the cylinder and the sphere. R is the radius of the sphere and $R = 3r$.

a) $\frac{4}{3}\pi R^3 + \pi r^2 h = \frac{4}{3}\pi(3r)^3 + \pi r^2(18)$
$= \frac{4}{3}\pi\left(27r^3\right) + 18\pi r^2$
$= 36\pi r^3 + 18\pi r^2$

b) $18\pi r^2(2r+1)$

c) $18\pi(5)^2(2\cdot 5+1) = 18\pi(25)(11)$
$\approx 15{,}550.9 \text{ ft.}^3$

79. a) $p - pr$ b) $p(1-r)$
c) $54.95(1-0.4) = \$32.97$

81. a) $p + prt$ b) $p(1+rt)$
c) $850(1+(0.03)(0.5)) = \$862.75$

83. a) $2\pi r^2 + 2\pi rh$ b) $2\pi r(r+h)$
c) $2\pi(15)(15+5) \approx 1884.96 \text{ in.}^2$

Review Exercises

1. $-2.45\times 10^7 = -24{,}500{,}000$

2. $0.000092 = 9.2\times 10^{-5}$

3. $(x+3)(x+5) = x^2 + 5x + 3x + 15$
$= x^2 + 8x + 15$

4. $(x-6)(x-4) = x^2 - 4x - 6x + 24$
$= x^2 - 10x + 24$

5. $(2x+7)(3x-1) = 6x^2 - 2x + 21x - 7$
$= 6x^2 + 19x - 7$

6. $2x(4x-5)(x+3) = 2x\left(4x^2 + 12x - 5x - 15\right)$
$= 2x\left(4x^2 + 7x - 15\right)$
$= 8x^3 + 14x^2 - 30x$

Exercise Set 6.2

1. Both will be positive.

3. Both will be negative.

5. Find two factors of the product ac whose sum is b.

7. $r^2 + 8r + 7 = (r+7)(r+1)$

9. $w^2 - 2w - 3 = (w+1)(w-3)$

11. $a^2 + 9a + 18 = (a+6)(a+3)$

13. $y^2 - 13y + 36 = (y-9)(y-4)$

15. $m^2 + 2n - 8 = (m+4)(m-2)$

17. $b^2 - 6b - 40 = (b-10)(b+4)$

19. $x^2 - 5x - 18$ is prime

21. $3st^2 + 24st + 21s = 3s(t^2 + 8t + 7)$
$= 3s(t + 7)(t + 1)$

23. $5y^3 - 65y^2 + 60y = 5y(y^2 - 13y + 12)$
$= 5y(y - 12)(y - 1)$

25. $6au^3 + 6au^2 - 36au = 6au(u^2 + u - 6)$
$= 6au(u + 3)(u - 2)$

27. $3x^2y^3 - 12x^2y^2 + 30x^2y = 3x^2y(y^2 - 4y + 10)$

29. $p^2 + 11pq + 18q^2 = (p + 9q)(p + 2q)$

31. $u^2 - 13uv + 42v^2 = (u - 7v)(u - 6v)$

33. $x^2 - 5xy - 14y^2 = (x - 7y)(x + 2y)$

35. $a^2 - ab - 42b^2 = (a + 6b)(a - 7b)$

37. $3a^2 + 10a + 7 = (3a + 7)(a + 1)$

39. $2w^2 - 3w - 2 = (2w + 1)(w - 2)$

41. $4r^2 - r + 2$ is prime

43. $4q^2 - 9q + 2 = (4q - 1)(q - 2)$

45. $6b^2 + 7b - 3 = (3b - 1)(2b + 3)$

47. $16m^2 + 24m + 9 = (4m + 3)(4m + 3) = (4m + 3)^2$

49. $4x^2 + 5x - 6 = (4x - 3)(x + 2)$

51. $2w^2 + 15wv + 7v^2 = (2w + v)(w + 7v)$

53. $5x^2 - 16xy + 3y^2 = (5x - y)(x - 3y)$

55. $16x^2 - 10xy + y^2 = (2x - y)(8x - y)$

57. $6a^2 - 13ab - 10b^2$ is prime

59. $3t^2 + 19tu - 14u^2 = (3t - 2u)(t + 7u)$

61. $3m^2 - 10mn - 8n^2 = (3m + 2n)(m - 4n)$

63. $12a^2 - 17ab + 6b^2 = (3a - 2b)(4a - 3b)$

65. $22m^3 + 200m^2 + 18m = 2m(11m^2 + 100m + 9)$
$= 2m(11m + 1)(m + 9)$

67. $4u^2v + 2uv^2 - 30v^3 = 2v(2u^2 + uv - 15v^2)$
$= 2v(2u - 5v)(u + 3v)$

69. $3y^2 + 16y + 5 = 3y^2 + 15y + y + 5$
$= (3y^2 + 15y) + (y + 5)$
$= 3y(y + 5) + 1(y + 5)$
$= (y + 5)(3y + 1)$

71. $6c^2 + 11c + 6$ is prime

73. $3t^2 - 17t + 10 = 3t^2 - 15t - 2t + 10$
$= (3t^2 - 15t) - (2t - 10)$
$= 3t(t - 5) - 2(t - 5)$
$= (t - 5)(3t - 2)$

75. $6x^2 - x - 15 = 6x^2 - 10x + 9x - 15$
$= (6x^2 - 10x) + (9x - 15)$
$= 2x(3x - 5) + 3(3x - 5)$
$= (3x - 5)(2x + 3)$

77. $32x^2y + 24xy - 36y$
$= 4y(8x^2 + 6x - 9)$
$= 4y[(8x^2 + 12x) + (-6x - 9)]$
$= 4y[4x(2x + 3) - 3(2x + 3)]$
$= 4y(2x + 3)(4x - 3)$

79. $15a^2b - 25ab^2 - 10b^3$
$= 5b(3a^2 - 5ab - 2b^2)$
$= 5b[(3a^2 - 6ab) + (ab - 2b^2)]$
$= 5b[3a(a - 2b) + b(a - 2b)]$
$= 5b(a - 2b)(3a + b)$

81. $x^4 - x^2 - 2$
Let $u = x^2$.
$u^2 - u - 2 = (u + 1)(u - 2)$
$= (x^2 + 1)(x^2 - 2)$

83. $8r^4 + 2r^2 - 3$
Let $u = r^2$.
$$8u^2 + 2u - 3 = (4u+3)(2u-1)$$
$$= (4r^2+3)(2r^2-1)$$

85. $15x^4 - 11x^2 + 2$
Let $u = x^2$.
$$15u^2 - 11u + 2 = (5u-2)(3u-1)$$
$$= (5x^2-2)(3x^2-1)$$

87. $y^6 - 16y^3 + 48$
Let $u = y^3$.
$$u^2 - 16u + 48 = (u-12)(u-4)$$
$$= (y^3-12)(y^3-4)$$

89. $7(x+1)^2 + 8(x+1) + 1$
Let $u = x+1$.
$$7u^2 + 8u + 1 = (7u+1)(u+1)$$
$$= [7(x+1)+1][(x+1)+1]$$
$$= (7x+7+1)(x+2)$$
$$= (7x+8)(x+2)$$

91. $3(a+2)^2 - 10(a+2) - 8$
Let $u = a+2$.
$$3u^2 - 10u - 8 = (3u+2)(u-4)$$
$$= [3(a+2)+2][(a+2)-4]$$
$$= (3a+6+2)(a-2)$$
$$= (3a+8)(a-2)$$

93. Mistake: The signs are incorrect.
Correct: $(x+2)(x-3)$

95. Mistake: The GCF monomial was not factored out.
Correct: $4(x+2)^2$

97. $x^2 + bx + 16$
To make the trinomial factorable, the value of b must be the sum of factor pairs of 16. Because b must be a natural number, both factors of 16 must be positive. Factor pairs of 16, and their resulting sums, are:
$1+16=17$
$2+8=10$
$4+4=8$
The natural number values of b that make the trinomial factorable are 8, 10, and 17.

99. $x^2 + bx - 63$
To make the trinomial factorable, the value of b must be the sum of factor pairs of -63. Because we wish b to be a natural number, we can disregard all negative values of b. Factor pairs of -63, and their resulting sums, are:
$-1+63=62$
$1+(-63)=-62$
$-3+21=18$
$3+(-21)=-18$
$-7+9=2$
$7+(-9)=-2$
The natural number values of b that make the trinomial factorable are 2, 18, and 62.

101. $x^2 + 9x + c$
To make the trinomial factorable, the value of c must be the product of pairs of numbers whose sum is 9. Because c must be a natural number, the value of c must be positive, which means both numbers must be positive. Possibilities are:
$1(8)=8$
$2(7)=14$
$3(6)=18$
$4(5)=20$
Possible natural number values of c that make the trinomial factorable are 8, 14, 18, or 20. Answers may vary.

103. $x^2 + x - c$
To make the trinomial factorable, the value of c must be the product of pairs of numbers whose sum is 1. The value of $-c$ is negative, which means that in the number pairs, one number must be positive and one must be negative. Possibilities are:
$-1(2)=-2$
$-2(3)=-6$
$-3(4)=-12$
$-4(5)=-20$
A few possible natural number values of c that make the trinomial factorable are 2, 6, 12, and 20. Answers may vary.

Review Exercises

1. $(2x+3)(2x+3) = 4x^2 + 12x + 9$

2. $(4y-1)(4y-1)=16y^2-1$

3. $(n-2)(n^2+2n+4)$

$=n^3+2n^2+4n-2n^2-4n-8$

$=n^3-8$

4. $$\begin{array}{r} 6x^2+x+4 \\ x-1\overline{)6x^3-5x^2+3x-2} \\ \underline{6x^3-6x^2} \\ x^2+3x \\ \underline{x^2-x} \\ 4x-2 \\ \underline{4x-4} \\ 2 \end{array}$$

Answer: $6x^2+x+4+\dfrac{2}{x-1}$

5. $5(x-6)+2x\ge 3x-4$

$5x-30+2x\ge 3x-4$

$7x-30\ge 3x-4$

$4x\ge 26$

$x\ge\dfrac{13}{2}$

a) $\left\{x \middle| x\ge\dfrac{13}{2}\right\}$ b) $\left[\dfrac{13}{2},\infty\right)$

c) 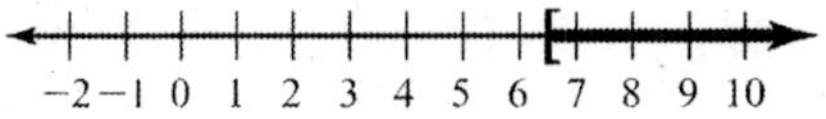

6. $-5<2x+7<5$

$-12<2x<-2$

$-6<x<-1$

a) $\{x|-6<x<-1\}$ b) $(-6,-1)$

c) 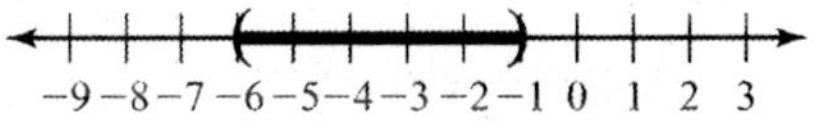

Exercise Set 6.3

1. The pattern in a perfect square trinomial is that the first and last terms are perfect squares and the middle term is twice the product of the square roots of the first and last terms.

3. The minus sign is placed in the binomial factor.

5. They are the cube roots of a^3 and b^3.

7. A sum of squares cannot be factored.

9. $x^2+10x+25=(x+5)^2$

11. $b^2-4b+4=(b-2)^2$

13. $25u^2-30u+9=(5u-3)^2$

15. $n^2+24mn+144m^2=(n+12m)^2$

17. $9q^2-30pq+25p^2=(3q-5p)^2$

19. $4p^2-28pq+49q^2=(2p-7q)^2$

21. $a^2-y^2=(a+y)(a-y)$

23. $25x^2-4=(5x+2)(5x-2)$

25. $100u^2-49v^2=(10u+7v)(10u-7v)$

27. $9x^2-36=9(x^2-4)=9(x+2)(x-2)$

29. $x^4-16=(x^2+4)(x^2-4)=(x^2+4)(x+2)(x-2)$

31. $9(x-3)^2-16$

$=[3(x-3)+4][3(x-3)-4]$

$=[3x-9+4][3x-9-4]$

$=(3x-5)(3x-13)$

33. $16z^2-9(x-y)^2$

$=[4z+3(x-y)][4z-3(x-y)]$

$=(4z+3x-3y)(4z-3x+3y)$

35. $m^3-27=(m-3)(m^2+3m+9)$

37. $125x^3+27=(5x+3)(25x^2-15x+9)$

39. $27x^3-8=(3x-2)(9x^2+6x+4)$

41. $u^3+125v^3=(u+5v)(u^2-5uv+25v^2)$

43. $27m^3-125m^6n^3$

$=m^3(27-125m^3n^3)$

$=m^3(3-5mn)(9+15mn+25m^2n^2)$

45. $(u+3)^3+8$
$=\left[(u+3)+2\right]\left[(u+3)^2-2(u+3)+4\right]$
$=(u+5)\left[(u^2+6u+9)-2u-6+4\right]$
$=(u+5)(u^2+4u+7)$

47. $27-(a+b)^3$
$=\left[3-(a+b)\right]\left[9+3(a+b)+(a+b)^2\right]$
$=(3-a-b)(9+3a+3b+a^2+2ab+b^2)$

49. $64x^3+27(y+z)^3$
$=\left[4x+3(y+z)\right]$
$\cdot\left[16x^2-12x(y+z)+9(y+z)^2\right]$
$=(4x+3y+3z)$
$\cdot\left(16x^2-12xy-12xz+9(y^2+2yz+z^2)\right)$
$=(4x+3y+3z)$
$\cdot\left(16x^2-12xy-12xz+9y^2+18yz+9z^2\right)$

51. $64d^3-27(x+y)^3$
$=\left[4d-3(x+y)\right]$
$\cdot\left[16d^2+12d(x+y)+9(x+y)^2\right]$
$=(4d-3x-3y)$
$\cdot\left[16d^2+12dx+12dy+9(x^2+2xy+y^2)\right]$
$=(4d-3x-3y)$
$\cdot\left(16d^2+12dx+12dy+9x^2+18xy+9y^2\right)$

53. $16x^2+bx+25$ will be a perfect square trinomial if $b=2\sqrt{a}\sqrt{c}=2\sqrt{16}\sqrt{25}=2(4)(5)=40$.

55. $4x^2-bx+81$ will be a perfect square trinomial if $b=2\sqrt{a}\sqrt{c}=2\sqrt{4}\sqrt{81}=2(2)(9)=36$.

57. $b=2\sqrt{a}\sqrt{c}$
$\frac{b}{2\sqrt{a}}=\sqrt{c}$
$\left(\frac{b}{2\sqrt{a}}\right)^2=c$
x^2+8x+c will be a perfect square trinomial if
$c=\left(\frac{b}{2\sqrt{a}}\right)^2=\left(\frac{8}{2\sqrt{1}}\right)^2=\left(\frac{8}{2}\right)^2=4^2=16$

59. $b=2\sqrt{a}\sqrt{c}$
$\frac{b}{2\sqrt{a}}=\sqrt{c}$
$\left(\frac{b}{2\sqrt{a}}\right)^2=c$
$9x^2-24x+c$ will be a perfect square trinomial if
$c=\left(\frac{b}{2\sqrt{a}}\right)^2=\left(\frac{24}{2\sqrt{9}}\right)^2=\left(\frac{24}{2\cdot 3}\right)^2$
$=\left(\frac{24}{6}\right)^2=4^2=16$

61. $12m^3n^2+20m^2n^4=4m^2n^2(3m+5n^2)$

63. $x^2+8x+15=(x+5)(x+3)$

65. $(x^2-16)=(x+4)(x-4)$

67. $12c^2-8c-15=(2c-3)(6c+5)$

69. $ax-xy-ay+y^2=(ax-xy)-(ay-y^2)$
$=x(a-y)-y(a-y)$
$=(a-y)(x-y)$

71. $4x^2-28x+49=(2x-7)^2$

73. $25x^2+36y^2$ is prime

75. $12a^3b^2c+3a^2b^2c^2+9abc^3$
$=3abc(4a^2b+abc+3c^2)$

77. $b^3+125=(b+5)(b^2-5b+25)$

79. $15x^2-12x-2$ is prime

81. $6b^2+b-2=(3b+2)(2b-1)$

83. $ab-36ab^3=ab\left(1-36b^2\right)=ab(1+6b)(1-6b)$

85. $16c^2-24c+9=(4c-3)^2$

87. $18x^3-3x^2-36x=3x\left(6x^2-x-12\right)$
$=3x(2x-3)(3x+4)$

89. $x^4-16=\left(x^2+4\right)\left(x^2-4\right)$
$=\left(x^2+4\right)(x+2)(x-2)$

91. $20a^2-9a+20$ is prime

93. $12u^2-84uv+147v^2=3\left(4u^2-28uv+49v^2\right)$
$=3(2u-7v)^2$

95. $36a^4b^3-39a^3b^4-12a^2b^5$
$=3a^2b^3\left(12a^2-13ab-4b^2\right)$
$=3a^2b^3(4a+b)(3a-4b)$

97. $6x^5y+5x^4y^2-12x^3y^3=x^3y\left(6x^2+5xy-12y^2\right)$

99. $x^2y+3x-xy^2-3y=\left(x^2y+3x\right)-\left(xy^2+3y\right)$
$=x(xy+3)-y(xy+3)$
$=(xy+3)(x-y)$

101. $27+(x-1)^3$
$=\left[3+(x-1)\right]\left[9-3(x-1)+(x-1)^2\right]$
$=(3+x-1)\left(9-3x+3+x^2-2x+1\right)$
$=(x+2)\left(x^2-5x+13\right)$

103. $5m^5+10m^3n^2+5mn^4=5m\left(m^4+2m^2n^2+n^4\right)$
$=5m\left(m^2+n^2\right)^2$

105. $(n-3)^2+6(n-3)+9$
Let $u=n-3$.
$u^2+6u+9=(u+3)^2$
$=(n-3+3)^2$
$=n^2$

107. $64t^6-t^3u^3=t^3\left(64t^3-u^3\right)$
$=t^3(4t-u)\left(16t^2+4tu+u^2\right)$

109. $2x^3+3x^2-2xy^2-3y^2$
$=\left(2x^3+3x^2\right)+\left(-2xy^2-3y^2\right)$
$=x^2(2x+3)-y^2(2x+3)$
$=(2x+3)\left(x^2-y^2\right)$
$=(2x+3)(x+y)(x-y)$

111. $36y^2-\left(x^2-8x+16\right)$
$=36y^2-(x-4)^2$
$=(6y+(x-4))(6y-(x-4))$
$=(6y+x-4)(6y-x+4)$

113. $(x\cdot 12x)-(8\cdot 5)=12x^2-40=4\left(3x^2-10\right)$

115. $(6x\cdot 3x\cdot x)-(9\cdot 3\cdot 1)=18x^3-27=9\left(2x^3-3\right)$

117. $6x^2-11x+3=(3x-1)(2x-3)$
Length: $3x-1$ Width: $2x-3$

119. $15x^3+55x^2+30x=5x\left(3x^2+11x+6\right)$
$=5x(3x+2)(x+3)$
Length: $3x+2$; Width: $x+3$; Height: $5x$

Review Exercises

1. $9x-5(3x+2)=12x-(4x-3)$
$9x-15x-10=12x-4x+3$
$-6x-10=8x+3$
$-13=14x$
$-\frac{13}{14}=x$

2. $2x-5=-7$ or $2x-5=7$
$2x=-2$ $2x=12$
$x=-1$ $x=6$

3. $\begin{cases} x+y+z=6000 & \text{Eqtn, 1} \\ x+y-2z=0 & \text{Eqtn. 2} \\ 0.04x+0.06y+0.08z=370 & \text{Eqtn. 3} \end{cases}$

Multiply equation 2 by –1 and add to equation 1.
Multiply equation 3 by –25 and add to equation 1.

$$\begin{array}{r} -x-y+2z=0 \\ \underline{x+y+z=6000} \\ 3z=6000 \end{array} \qquad \begin{array}{r} -x-1.5y-2z=-9250 \\ \underline{x+y+z=6000} \\ -0.5y-z=-3250 \end{array}$$

$$\begin{cases} x+y+z=6000 \\ 3z=6000 \\ -0.5y-z=-3250 \end{cases}$$

$3z=6000 \qquad -0.5y-2000=-3250$

$z=2000 \qquad -0.5y=-1250$

$y=2500$

$x+2500+2000=6000$

$x=1500$

\$1500 at 4%, \$2500 at 6%, \$2000 at 8%

4. $f(x)=-\frac{3}{4}x-2$

Think of the function as the equation

$y=-\frac{3}{4}x-2$ and use the fact that the slope is

$m=-\frac{3}{4}$ and the y-intercept is -2.

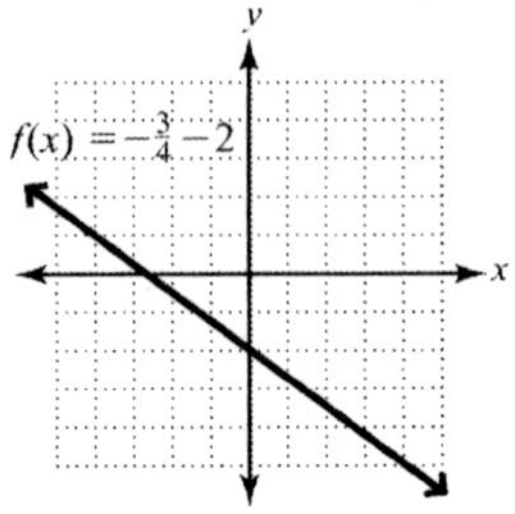

5. $y>2x-3$

Begin by graphing the related equation $y=2x-3$ with a dashed line. Now choose $(0,0)$ as a test point.

$y > 2x-3$

$0 \overset{?}{>} 2(0)-3$

$0 \overset{?}{>} 0-3$

$0 > -3$

Because $(0,0)$ satisfies the inequality, shade the region which includes $(0,0)$.

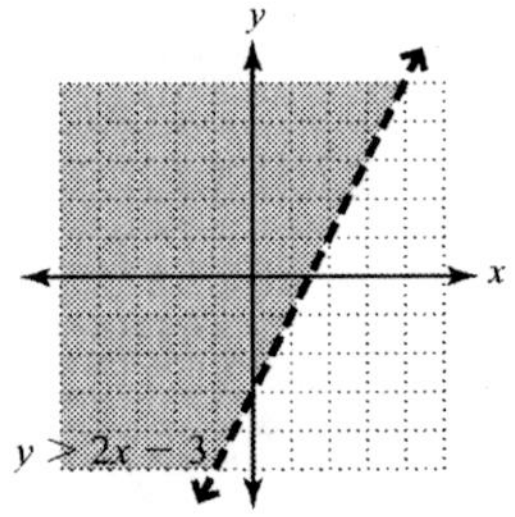

6. Find the slope: $m=\frac{1-(-2)}{-4-3}=-\frac{3}{7}$

$y-1=-\frac{3}{7}(x-(-4))$

$y-1=-\frac{3}{7}(x+4)$

$y-1=-\frac{3}{7}x-\frac{12}{7}$

$y=-\frac{3}{7}x-\frac{5}{7}$

$3x+7y=-5$

Exercise Set 6.4

1. If a and b are real numbers and $ab=0$, then $a=0$ or $b=0$.

3. 2

5. Answers may vary.

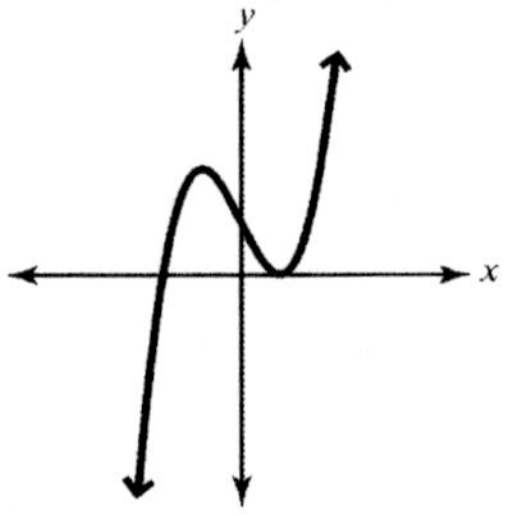

7. $x(x+4)=0$

$x=0 \quad \text{or} \quad x+4=0$

$x=-4$

9. $(x-3)(x+2)=0$

$x-3=0 \quad \text{or} \quad x+2=0$

$x=3 \qquad x=-2$

11. $(3b-2)(2b+5)=0$

$3b-2=0 \quad \text{or} \quad 2b+5=0$

$3b=2 \qquad 2b=-5$

$b=\frac{2}{3} \qquad b=-\frac{5}{2}$

13. $x(x-3)(2x+5)=0$

$x=0 \quad \text{or} \quad x-3=0 \quad \text{or} \quad 2x+5=0$

$x=3 \qquad x=-\frac{5}{2}$

15. $(x+2)(x-5)(x+7)=0$

$x+2=0$ or $x-5=0$ or $x+7=0$

$x=-2$ $x=5$ $x=-7$

17. $(2x+3)(3x-1)(4x+7)=0$

$2x+3=0$ or $3x-1=0$ or $4x+7=0$

$x=-\frac{3}{2}$ $x=\frac{1}{3}$ $x=-\frac{7}{4}$

19. $d^2-4d=0$

$d(d-4)=0$

$d=0$ or $d-4=0$

$d=4$

21. $x^2-9=0$

$(x+3)(x-3)=0$

$x+3=0$ or $x-3=0$

$x=-3$ $x=3$

23. $m^2+14m+45=0$

$(m+5)(m+9)=0$

$m+5=0$ or $m+9=0$

$m=-5$ $m=-9$

25. $2x^2+3x-2=0$

$(x+2)(2x-1)=0$

$x+2=0$ or $2x-1=0$

$x=-2$ $x=\frac{1}{2}$

27. $6a^2+13a-5=0$

$(3a-1)(2a+5)=0$

$3a-1=0$ or $2a+5=0$

$a=\frac{1}{3}$ $a=-\frac{5}{2}$

29. $x^2+6x+9=0$

$(x+3)^2=0$

$x+3=0$

$x=-3$

31. $6y^2=3y$

$6y^2-3y=0$

$3y(2y-1)=0$

$3y=0$ or $2y-1=0$

$y=0$ $y=\frac{1}{2}$

33. $x^2=25$

$x^2-25=0$

$(x+5)(x-5)=0$

$x+5=0$ or $x-5=0$

$x=-5$ $x=5$

35. $n^2+6n=27$

$n^2+6n-27=0$

$(n+9)(n-3)=0$

$n+9=0$ or $n-3=0$

$n=-9$ $n=3$

37. $p^2=3p-2$

$p^2-3p+2=0$

$(p-2)(p-1)=0$

$p-2=0$ or $p-1=0$

$p=2$ $p=1$

39. $2v^2+5=-7v$

$2v^2+7v+5=0$

$(2v+5)(v+1)=0$

$2v+5=0$ or $v+1=0$

$v=-\frac{5}{2}$ $v=-1$

41. $a(a-5)=14$

$a^2-5a=14$

$a^2-5a-14=0$

$(a-7)(a+2)=0$

$a-7=0$ or $a+2=0$

$a=7$ $a=-2$

43. $4x(x+7)=-49$

$4x^2+28x=-49$

$4x^2+28x+49=0$

$(2x+7)^2=0$

$2x+7=0$

$2x=-7$

$x=-\frac{7}{2}$

45. $(x+1)(x+2)=20$

$x^2+3x+2=20$

$x^2+3x-18=0$

$(x+6)(x-3)=0$

$x+6=0$ or $x-3=0$

$x=-6$ $\quad x=3$

47. $x^3+x^2-6x=0$

$x(x^2+x-6)=0$

$x(x+3)(x-2)=0$

$x=0$ or $x+3=0$ or $x-2=0$

$x=-3$ $\quad x=2$

49. $12x(x+1)+3=5(2x+1)+2$

$12x^2+12x+3=10x+5+2$

$12x^2+2x-4=0$

$2(6x^2+x-2)=0$

$2(3x+2)(2x-1)=0$

$3x+2=0$ or $2x-1=0$

$x=-\frac{2}{3}$ $\quad x=\frac{1}{2}$

51. $(2v+1)(v-2)=-v(v+2)$

$2v^2-3v-2=-v^2-2v$

$3v^2-v-2=0$

$(3v+2)(v-1)=0$

$3v+2=0$ or $v-1=0$

$v=-\frac{2}{3}$ $\quad v=1$

53. $9x(x^2+x)=3x(x-2)+5x$

$9x^3+9x^2=3x^2-6x+5x$

$9x^3+6x^2+x=0$

$x(9x^2+6x+1)=0$

$x(3x+1)^2=0$

$x=0$ or $3x+1=0$

$3x=-1$

$x=-\frac{1}{3}$

55. $x^3+2x^2-15=13x-(4x-3)$

$x^3+2x^2-15=13x-4x+3$

$x^3+2x^2-9x-18=0$

$(x^3+2x^2)-(9x+18)=0$

$x^2(x+2)-9(x+2)=0$

$(x+2)(x^2-9)=0$

$(x+2)(x+3)(x-3)=0$

$x+2=0$ or $x+3=0$ or $x-3=0$

$x=-2$ $\quad x=-3$ $\quad x=3$

57. $x=-3$ $\quad x=2$

$x+3=0$ $\quad x-2=0$

$(x+3)(x-2)=0$

$x^2+x-6=0$

59. $x=-\frac{2}{3}$ $\quad x=4$

$3x=-2$ $\quad x-4=0$

$3x+2=0$

$(3x+2)(x-4)=0$

$3x^2-10x-8=0$

61. $x=-1$ $\quad x=0$ $\quad x=3$

$x+1=0$ $\quad x-0=0$ $\quad x-3=0$

$(x+1)(x-0)(x-3)=0$

$(x^2+x)(x-3)=0$

$x^3-3x^2+x^2-3x=0$

$x^3-2x^2-3x=0$

63. b 65. a 67. c

69. $f(x) = x^2 - 25$

$x^2 - 25 = 0$

$(x+5)(x-5) = 0$

$x+5=0$ or $x-5=0$

$x=-5$ $\quad x=5$

$(-5,0)(5,0)$

x	$f(x)$
0	−25
−3	−16
3	−16

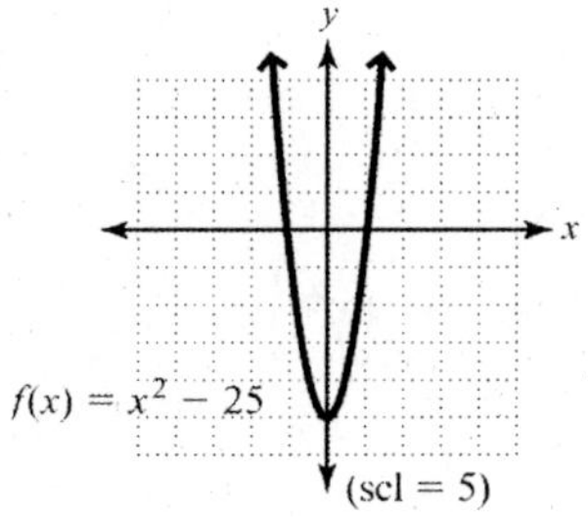

71. $f(x) = x^2 - 6x + 5$

$x^2 - 6x + 5 = 0$

$(x-5)(x-1) = 0$

$x-5=0$ or $x-1=0$

$x=5$ $\quad x=1$

$(5,0),(1,0)$

x	$f(x)$
2	−3
3	−4
4	−3

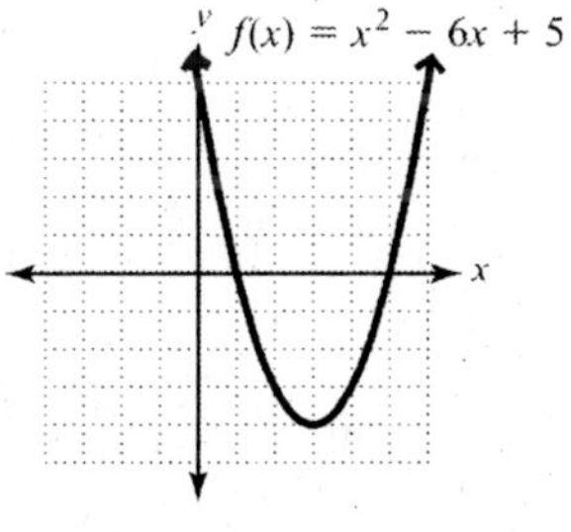

73. $f(x) = x(x+4)(x-1)$

$x(x+4)(x-1) = 0$

$x=0$ or $x+4=0$ or $x-1=0$

$x=-4$ $\quad x=1$

$(0,0),(-4,0),(1,0)$

x	$f(x)$
−3	12
−2	12
−1	6
2	12

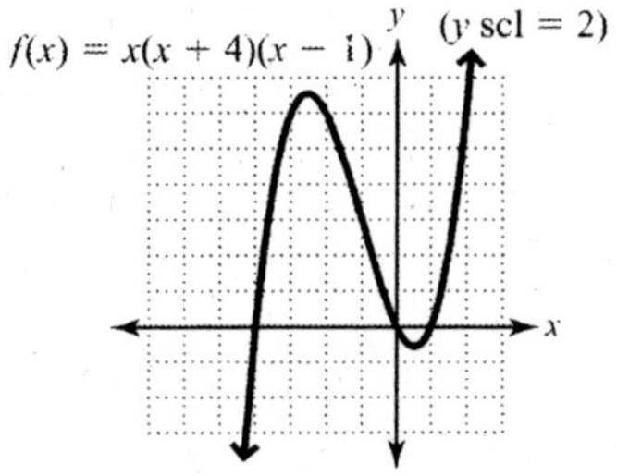

75. $f(x) = x^3 - x^2 - 6x$

$x^3 - x^2 - 6x = 0$

$x(x^2 - x - 6) = 0$

$x(x-3)(x+2) = 0$

$x=0$ or $x-3=0$ or $x+2=0$

$x=3$ $\quad x=-2$

$(0,0),(3,0),(-2,0)$

x	$f(x)$
−1	4
1	−6
2	−8

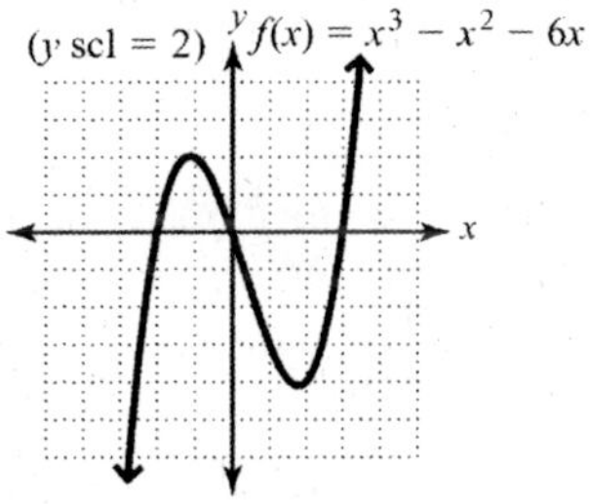

77. (3, 0), (6, 0)

79. $\left(-\frac{5}{3},0\right),\left(\frac{1}{2},0\right)$

81. (−1, 0),(0, 0), (2, 0)

83. −5, 8

85. −0.25, 2.5

87. −4, 0, 3

89. $x^2 + (x+2)^2 = (x+4)^2$

$x^2 + x^2 + 4x + 4 = x^2 + 8x + 16$

$x^2 - 4x - 12 = 0$

$(x-6)(x+2) = 0$

$x - 6 = 0$ or $x + 2 = 0$

$x = 6$ $\quad$ $x = -2$

x cannot be negative

The lengths of the sides are 6, 6 + 2 = 8, and 6 + 4 = 10.

91. $n^2 + (3n+3)^2 = (4n-3)^2$

$n^2 + 9n^2 + 18n + 9 = 16n^2 - 24n + 9$

$-6n^2 + 42n = 0$

$-6n(n-7) = 0$

$-6n = 0$ or $n - 7 = 0$

$n = 0$ $\quad$ $n = 7$

n cannot be zero

The lengths of the sides are 7, 3(7) + 3 = 24, and $4(7) - 3 = 25$.

93. Let x represent the number.

$x^2 + 55 = 16x$

$x^2 - 16x + 55 = 0$

$(x-5)(x-11) = 0$

$x - 5 = 0$ or $x - 11 = 0$

$x = 5$ $\quad$ $x = 11$

There are two numbers; 5 and 11.

95. Let x represent the first consecutive positive odd integer. Then $x + 2$ and $x + 4$ represent the other two consecutive positive odd integers.

$x^2 + (x+2)^2 + (x+4)^2 = 155$

$x^2 + x^2 + 4x + 4 + x^2 + 8x + 16 = 155$

$3x^2 + 12x + 20 = 155$

$3x^2 + 12x - 135 = 0$

$3(x^2 + 4x - 45) = 0$

$3(x+9)(x-5) = 0$

$x + 9 = 0$ or $x - 5 = 0$

$x = -9$ $\quad$ $x = 5$

x cannot be negative

The numbers are 5, 5 + 2 = 7, and 5 + 4 = 9.

97. Let x be the width and $x + 4$ be the length.

$x(x+4) = 320$

$x^2 + 4x - 320 = 0$

$(x+20)(x-16) = 0$

$x + 20 = 0$ or $x - 16 = 0$

$x = -20$ $\quad$ $x = 16$

x cannot be negative

The dimensions are 16m by 20m.

99. The area is $(22\text{ ft}) \cdot (28\text{ ft}) = 616\text{ ft}^2$.

$\frac{22}{7}r^2 = 616$

$r^2 = 196$

$r^2 - 196 = 0$

$(r+14)(r-14) = 0$

$r + 14 = 0$ or $r - 14 = 0$

$r = -14$ $\quad$ $r = 14$

r cannot be negative

The radius is 14 ft.

101. $\frac{1}{2}h(10+h) = 85.5$

$5h + 0.5h^2 = 85.5$

$0.5h^2 + 5h - 85.5 = 0$

$0.5(h^2 + 10h - 171) = 0$

$0.5(h+19)(h-9) = 0$

$h + 19 = 0$ or $h - 9 = 0$

$h = -19$ $\quad$ $h = 9$

h cannot be negative

The height is 9 in.

103. $h^2 + 10^2 = (h+2)^2$

$h^2 + 100 = h^2 + 4h + 4$

$96 = 4h$

$24 = h$

Height: 24 ft. $\quad$ Wire length: 26 ft.

105. $9 = -16t^2 + 4t + 29$

$0 = -16t^2 + 4t + 20$

$0 = -4(4t^2 - t - 5)$

$0 = -4(4t-5)(t+1)$

$4t - 5 = 0$ or $t + 1 = 0$

$t = 1.25$ $\quad$ $t = \times$

Time would be 1.25 sec.

107. $4840 = 4000\left(1+\frac{r}{1}\right)^{2\cdot 1}$

$$1.21 = (1+r)^2$$
$$1.21 = 1+2r+r^2$$
$$0 = r^2+2r-0.21$$
$$0 = 100r^2+200r-21$$
$$0 = (10r+21)(10r-1)$$
$$10r+21=0 \quad \text{or} \quad 10r-1=0$$
$$r=-2.1 \qquad\qquad r=0.1$$

r is not negative

$r = 0.1 = 10\%$

Review Exercises

1. $0.0004203 = 4.203\times 10^{-4}$

2. $x+4\overline{)x^3+x^2-10x+11}$ with quotient x^2-3x+2

$$\underline{x^3+4x^2}$$
$$-3x^2-10x$$
$$\underline{-3x^2-12x}$$
$$2x+11$$
$$\underline{2x+8}$$
$$3$$

Answer: $x^2-3x+2+\dfrac{3}{x+4}$

3. $2x-5\le -1 \quad \text{or} \quad 2x-5\ge 1$

$$2x\le 4 \qquad\qquad 2x\ge 6$$
$$x\le 2 \qquad\qquad x\ge 3$$

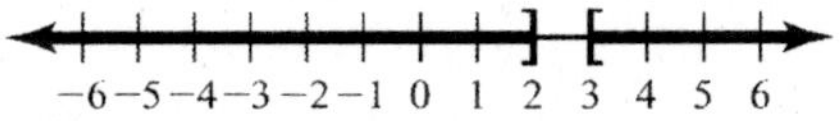

4. 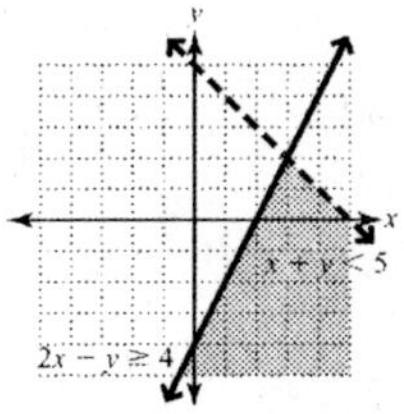

5. No

6. $(f+g)(x) = (5x^3+6x-19)+(3x^2-6x-8)$

$$= 5x^3+3x^2-27$$

Chapter 6 Review Exercises

1. False; factoring by grouping is used when the polynomial contains four terms.
2. True
3. True
4. True
5. False; x^2+25 is prime and cannot be factored.
6. False; the described technique can only be used if the factors on the left side of the equation are equal to 0 on the right side of the equation.
7. greatest common factor (GCF)
8. b
9. $(a+b)^2$
10. 0, 0
11. $16x^4y^2 = 2^4\cdot x^4\cdot y^2$

    $24x^3y^9 = 2^3\cdot 3\cdot x^3\cdot y^9$

    $\text{GCF} = 2^3\cdot x^3\cdot y^2 = 8x^3y^2$
12. $20mn^3 = 2^2\cdot 5\cdot m\cdot n^3$

    $15m^4n^2 = 3\cdot 5\cdot m^4\cdot n^2$

    $\text{GCF} = 5\cdot m\cdot n^2 = 5mn^2$
13. $2u^4v^3 = 2\cdot u^4\cdot v^3$

    $3xy^9 = 3\cdot x\cdot y^9$

    $\text{GCF} = 1$
14. $4(x+1) = 2^2\cdot(x+1)$

    $6(x+1) = 2\cdot 3\cdot(x+1)$

    $\text{GCF} = 2\cdot(x+1) = 2(x+1)$
15. $u^6-u^4-u^2 = u^2\left(\dfrac{u^6-u^4-u^2}{u^2}\right)$

    $= u^2\left(u^4-u^2-1\right)$
16. $13d^2-26de = 13d\left(\dfrac{13d^2-26de}{13d}\right)$

    $= 13d(d-2e)$
17. $4h^2k^6-2h^5k = 2h^2k\left(\dfrac{4h^2k^6-2h^5k}{2h^2k}\right)$

    $= 2h^2k\left(2k^5-h^3\right)$

18. $12cd - 4c^2d^2 + 10c^4d^4$

$= 2cd\left(\dfrac{12cd - 4c^2d^2 + 10c^4d^4}{2cd}\right)$

$= 2cd\left(6 - 2cd + 5c^3d^3\right)$

19. $16p^6q^3 - 12p^8q^5 + 13p^7q^2$

$= p^6q^2\left(\dfrac{16p^6q^3 - 12p^8q^5 + 13p^7q^2}{p^6q^2}\right)$

$= p^6q^2\left(16q - 12p^2q^3 + 13p\right)$

20. $9w^6v^8 + 6wv^6 - 12w^3v^3$

$= 3wv^3\left(\dfrac{9w^6v^8 + 6wv^6 - 12w^3v^3}{3wv^3}\right)$

$= 3wv^3\left(3w^5v^5 + 2v^3 - 4w^2\right)$

21. $17(w+3) - m(w+3)$

$= (w+3)\left(\dfrac{17(w+3) - m(w+3)}{(w+3)}\right)$

$= (w+3)(17-m)$

22. $2y(x+5) + (x+5)$

$= (x+5)\left(\dfrac{2y(x+5) + (x+5)}{(x+5)}\right)$

$= (x+5)(2y+1)$

23. $3m + mn + 6 + 2n = (3m + mn) + (6 + 2n)$

$= m(3+n) + 2(3+n)$

$= (3+n)(m+2)$

$= (n+3)(m+2)$

24. $a^3 + 3a^2 + 3a + 9 = \left(a^3 + 3a^2\right) + (3a + 9)$

$= a^2(a+3) + 3(a+3)$

$= (a+3)\left(a^2 + 3\right)$

25. $2x + 2y - ax - ay = (2x + 2y) + (-ax - ay)$

$= 2(x+y) - a(x+y)$

$= (x+y)(2-a)$

26. $bc^2 + bd^2 - 5c^2 - 5d^2$

$= \left(bc^2 + bd^2\right) + \left(-5c^2 - 5d^2\right)$

$= b\left(c^2 + d^2\right) - 5\left(c^2 + d^2\right)$

$= \left(c^2 + d^2\right)(b-5)$

27. $x^2y - x^2s - ry + rs = \left(x^2y - x^2s\right) + (-ry + rs)$

$= x^2(y-s) - r(y-s)$

$= (y-s)\left(x^2 - r\right)$

28. $4k^2 - k + 4k - 1 = \left(4k^2 - k\right) + (4k - 1)$

$= k(4k-1) + 1(4k-1)$

$= (4k-1)(k+1)$

29. $8uv^2 - 4v^2 + 20uv - 10v$

$= 2v(4uv - 2v + 10u - 5)$

$= 2v\left[(4uv - 2v) + (10u - 5)\right]$

$= 2v\left[2v(2u-1) + 5(2u-1)\right]$

$= 2v(2u-1)(2v+5)$

30. $5c^2d^2 - 5d^3 + 20c^2d - 20d^2$

$= 5d\left(c^2d - d^2 + 4c^2 - 4d\right)$

$= 5d\left[\left(c^2d - d^2\right) + \left(4c^2 - 4d\right)\right]$

$= 5d\left[d\left(c^2 - d\right) + 4\left(c^2 - d\right)\right]$

$= 5d\left(c^2 - d\right)(d+4)$

31. $a^2 - 10a + 9 = (a-9)(a-1)$

32. $m^2 + 20m + 51 = (m+3)(m+17)$

33. $y^2 + 2y - 48 = (y+8)(y-6)$

34. $x^2 - 7x - 30 = (x-10)(x+3)$

35. $3x^2 - x - 14 = (3x-7)(x+2)$

36. $16h^2 + 10h + 1 = (8h+1)(2h+1)$

37. $4u^2 - 2u + 3$ is prime

38. $6t^2 + t - 15 = (3t+5)(2t-3)$

39. $s^2 - 11st + 10t^2 = (s-10t)(s-t)$

40. $6u^2+13uv+6v^2=(3u+2v)(2u+3v)$

41. $5m^2-16mn+3n^2=(5m-n)(m-3n)$

42. $4x^2+5xy-8y^2$ is prime

43. $b^4-7b^3-18b^2=b^2(b^2-7b-18)$
$=b^2(b-9)(b+2)$

44. $2x^3-16x^2-40x=2x(x^2-8x-20)$
$=2x(x-10)(x+2)$

45. $2u^2+9u+10=(2u^2+5u)+(4u+10)$
$=u(2u+5)+2(2u+5)$
$=(2u+5)(u+2)$

46. $3m^2-8m+4=(3m^2-2m)+(-6m+4)$
$=m(3m-2)-2(3m-2)$
$=(3m-2)(m-2)$

47. $10u^2+7u-3=(10u^2+10u)+(-3u-3)$
$=10u(u+1)-3(u+1)$
$=(u+1)(10u-3)$

48. $6y^2-13y-5=6y^2+2y-15y-5$
$=(6y^2+2y)+(-15y-5)$
$=2y(3y+1)-5(3y+1)$
$=(3y+1)(2y-5)$

49. x^4+5x^2+4
Let $u=x^2$.
$u^2+5u+4=(u+4)(u+1)$
$=(x^2+4)(x^2+1)$

50. $3c^4+13c^2+4$
Let $u=c^2$.
$3u^2+13u+4=(3u+1)(u+4)$
$=(3c^2+1)(c^2+4)$

51. $2h^6+9h^3+4$
Let $u=h^3$.
$2u^2+9u+4=(2u+1)(u+4)$
$=(2h^3+1)(h^3+4)$

52. $3(k+1)^2-2(k+1)-5$
Let $u=k+1$.
$3u^2-2u-5=(3u-5)(u+1)$
$=[3(k+1)-5][(k+1)+1]$
$=(3k+3-5)(k+2)$
$=(3k-2)(k+2)$

53. $x^2+6x+9=(x+3)^2$

54. $y^2+12y+36=(y+6)^2$

55. $m^2-4m+4=(m-2)^2$

56. $w^2-14w+49=(w-7)^2$

57. $9d^2+30d+25=(3d+5)^2$

58. $4c^2-28c+49=(2c-7)^2$

59. $h^2-9=(h+3)(h-3)$

60. $p^2-64=(p+8)(p-8)$

61. $9d^2-4=(3d+2)(3d-2)$

62. $81k^2-100=(9k+10)(9k-10)$

63. $2w^2-50=2(w^2-25)$
$=2(w+5)(w-5)$

64. $4q^2-36=4(q^2-9)$
$=4(q+3)(q-3)$

65. $25y^2-9z^2=(5y+3z)(5y-3z)$

66. $x^4-81=(x^2+9)(x^2-9)$
$=(x^2+9)(x+3)(x-3)$

67. $c^3-27=(c-3)(c^2+3c+9)$

68. $m^3+64=(m+4)(m^2-4m+16)$

69. $27b^3+8a^3=(3b+2a)(9b^2-6ab+4a^2)$

70. $64d^3-8c^3=8(8d^3-c^3)$
$=8(2d-c)(4d^2+2cd+c^2)$

71. $9d^2 - 6d + 1 = (3d - 1)^2$

72. $3m^2 - 3n^2 = 3(m^2 - n^2) = 3(m+n)(m-n)$

73. $2a^2 - 5a - 12 = (2a+3)(a-4)$

74. $x^2 + 9$ is prime

75. $3p^4 + 3p^3 - 90p^2 = 3p^2(p^2 + p - 30)$
$= 3p^2(p+6)(p-5)$

76. $x^4 - 16 = (x^2+4)(x^2-4)$
$= (x^2+4)(x+2)(x-2)$

77. $15a^3b^2c^7 + 3a^2b^4c^2 + 5a^9bc^3$
$= a^2bc^2(15abc^5 + 3b^3 + 5a^7c)$

78. $w^3 - (2+y)^3$
$= [w-(2+y)][w^2 + w(2+y) + (2+y)^2]$
$= (w-2-y)[w^2 + 2w + wy + 4 + 4y + y^2]$

79. $(x+4)(x-1) = 0$
$x+4=0$ or $x-1=0$
$x=-4$ $x=1$

80. $x^2 - 64 = 0$
$(x+8)(x-8) = 0$
$x+8=0$ or $x-8=0$
$x=-8$ $x=8$

81. $w^2 + 2w - 3 = 0$
$(w+3)(w-1) = 0$
$w+3=0$ or $w-1=0$
$w=-3$ $w=1$

82. $6y^2 = 7y - 1$
$6y^2 - 7y + 1 = 0$
$(6y-1)(y-1) = 0$
$6y-1=0$ or $y-1=0$
$y = \frac{1}{6}$ $y=1$

83. $4x(x+7) + 9 = -40$
$4x^2 + 28x + 49 = 0$
$(2x+7)^2 = 0$
$2x+7=0$
$x = -\frac{7}{2}$

84. $b^3 - 2b^2 + 20 = 2b^2 - (16 - 9b)$
$b^3 - 2b^2 + 20 = 2b^2 - 16 + 9b$
$b^3 - 4b^2 - 9b + 36 = 0$
$(b^3 - 4b^2) + (-9b + 36) = 0$
$b^2(b-4) - 9(b-4) = 0$
$(b-4)(b^2-9) = 0$
$(b-4)(b+3)(b-3) = 0$
$b-4=0$ or $b+3=0$ or $b-3=0$
$b=4$ $b=-3$ $b=3$

85. $f(x) = x^2 - 4$
$x^2 - 4 = 0$
$(x+2)(x-2) = 0$
$x+2=0$ or $x-2=0$
$x=-2$ $x=2$
(−2, 0), (2, 0)

x	$f(x)$
0	−4
−1	−3
1	−3

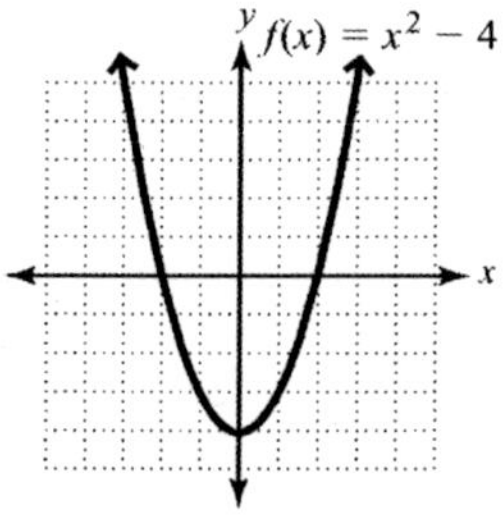

86. $f(x) = x(x+3)(x-2)$

$x(x+3)(x-2) = 0$

$x = 0$ or $x+3=0$ or $x-2=0$

$x = -3$ $\quad x = 2$

(0, 0), (–3, 0), (2, 0)

x	$f(x)$
–2	8
–1	6
1	–4

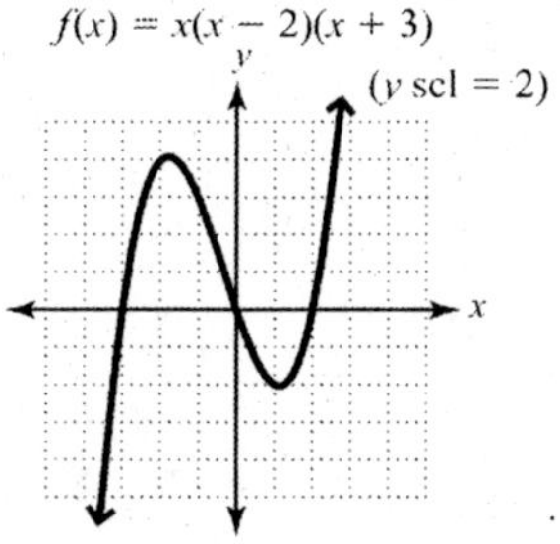

87. Let x be the first natural number and let $x + 10$ be the second natural number.

$x(x+10) = 56$

$x^2 + 10x - 56 = 0$

$(x+14)(x-4) = 0$

$x + 14 = 0$ or $x - 4 = 0$

$x = -14$ $\quad x = 4$

x cannot be negative

The two numbers are 4 and 4 + 10 = 14.

88. Let x be the first consecutive positive integer and let $x + 1$ be the second consecutive positive integer.

$x(x+1) = 110$

$x^2 + x - 110 = 0$

$(x+11)(x-10) = 0$

$x + 11 = 0$ or $x - 10 = 0$

$x = -11$ $\quad x = 10$

x cannot be negative

The two numbers are 10 and 10 + 1 = 11.

89. Let x be the first consecutive positive even integer and let $x + 2$ be the second consecutive positive even integer.

$x(x+2) = 288$

$x^2 + 2x - 288 = 0$

$(x+18)(x-16) = 0$

$x + 18 = 0$ or $x - 16 = 0$

$x = -18$ $\quad x = 16$

x cannot be negative

The two numbers are 16 and 16 + 2 = 18.

90. Let x be the width of the room and let $x + 3$ be the length.

$x(x+3) = 88$

$x^2 + 3x - 88 = 0$

$(x+11)(x-8) = 0$

$x + 11 = 0$ or $x - 8 = 0$

$x = -11$ $\quad x = 8$

x cannot be negative

The dimensions of the room are 8 ft. by 8 + 3 = 11 feet.

91. $\frac{1}{2}h(3h+36) = 672$

$1.5h^2 + 18h - 672 = 0$

$15h^2 + 180h - 6720 = 0$

$15(h^2 + 12h - 448) = 0$

$15(h+28)(h-16) = 0$

$h + 28 = 0$ or $h - 16 = 0$

$h = -28$ $\quad h = 16$

h cannot be negative

The height is 16 ft. and the base is 48 ft.

92. $x^2 + (x+7)^2 = (2x+3)^2$

$x^2 + x^2 + 14x + 49 = 4x^2 + 12x + 9$

$2x^2 - 2x - 40 = 0$

$2(x^2 - x - 20) = 0$

$2(x-5)(x+4) = 0$

$x - 5 = 0$ or $x + 4 = 0$

$x = 5$ $\quad x = -4$

x cannot be negative

The sides of the triangle are 5, 12, and 13.

93. The figure shows a rectangle with a base of 35 ft. and a height of 5 ft. with a triangle on top of it. The triangle has a height of $h - 5$.

$$35^2 + (h-5)^2 = (2h+3)^2$$
$$1225 + h^2 - 10h + 25 = 4h^2 + 12h + 9$$
$$3h^2 + 22h - 1241 = 0$$
$$(h-17)(3h+73) = 0$$
$$h - 17 = 0 \quad \text{or} \quad 3h + 73 = 0$$
$$h = 17 \qquad h = -\frac{73}{3}$$

h cannot be negative

The height is 17 ft. The distance from the ear is $2h + 3 = 2(17) + 3 = 37$ ft.

94. $$6 = -16t^2 + 12t + 60$$
$$16t^2 - 12t - 54 = 0$$
$$2(8t^2 - 6t - 27) = 0$$
$$2(4t-9)(2t+3) = 0$$
$$4t - 9 = 0 \quad \text{or} \quad 2t + 3 = 0$$
$$t = \frac{9}{4} \qquad t = -\frac{3}{2}$$
$$t = 2\frac{1}{4} \text{ sec.}$$

t cannot be negative

Chapter 6 Practice Test

1. $14m^7n^2 = 2 \cdot 7 \cdot m^7 \cdot n^2$
$$21m^3n^9 = 3 \cdot 7 \cdot m^3 \cdot n^9$$
$$\text{GCF} = 7 \cdot m^3 \cdot n^2 = 7m^3n^2$$

2. $$3m + 6m^3 - 9m^6 = 3m\left(\frac{3m + 6m^3 - 9m^6}{3m}\right)$$
$$= 3m\left(\frac{3m}{3m} + \frac{6m^3}{3m} - \frac{9m^6}{3m}\right)$$
$$= 3m(1 + 2m^2 - 3m^5)$$

3. $$3m + 3n - m - n = 2m + 2n$$
$$= 2(m+n)$$

4. $$9n^2 - 16 = (3n)^2 - (4)^2$$
$$= (3n+4)(3n-4)$$

5. $$8x^3 - 27 = (2x)^3 - (3)^3$$
$$= (2x-3)\left[(2x)^2 + (2x)(3) + (3)^2\right]$$
$$= (2x-3)(4x^2 + 6x + 9)$$

6. $y^2 + 14y + 49 = (y+7)^2$

7. $q^2 - 2q - 48 = (q-8)(q+6)$

8. $3ab^2 - 30ab + 24a = 3a(b^2 - 10b + 8)$

9. $$5 - 125t^2 = 5(1 - 25t^2)$$
$$= 5(1+5t)(1-5t)$$

10. $6d^2 + d - 2 = (3d+2)(2d-1)$

11. $$8c^3 + 8d^3 = 8(c^3 + d^3)$$
$$= 8(c+d)(c^2 - cd + d^2)$$

12. $$w^3 + 2w^2 + 3w + 6 = (w^3 + 2w^2) + (3w + 6)$$
$$= w^2(w+2) + 3(w+2)$$
$$= (w+2)(w^2 + 3)$$

13. $$s^4 - 81 = (s^2+9)(s^2-9)$$
$$= (s^2+9)(s+3)(s-3)$$

14. $5p^2 - 7pq - 12q^2 = (5p - 12q)(p+q)$

15. $$8x^2 - 14x + 5 = 0$$
$$(2x-1)(4x-5) = 0$$
$$2x - 1 = 0 \quad \text{or} \quad 4x - 5 = 0$$
$$2x = 1 \qquad 4x = 5$$
$$x = \frac{1}{2} \qquad x = \frac{5}{4}$$

16. $$x(x+3) - 6 = 12$$
$$x^2 + 3x - 18 = 0$$
$$(x+6)(x-3) = 0$$
$$x + 6 = 0 \quad \text{or} \quad x - 3 = 0$$
$$x = -6 \qquad x = 3$$

17. $$2x^3 + x^2 = 15x$$
$$2x^3 + x^2 - 15x = 0$$
$$x\left(2x^2 + x - 15\right) = 0$$
$$x(2x-5)(x+3) = 0$$
$$x = 0 \quad \text{or} \quad 2x - 5 = 0 \quad \text{or} \quad x + 3 = 0$$
$$2x = 5 \qquad x = -3$$
$$x = \frac{5}{2}$$

18. $$x^2 + (x+7)^2 = (2x+1)^2$$
$$x^2 + x^2 + 14x + 49 = 4x^2 + 4x + 1$$
$$2x^2 + 14x + 49 = 4x^2 + 4x + 1$$
$$-2x^2 + 10x + 48 = 0$$
$$-2\left(x^2 - 5x - 24\right) = 0$$
$$-2(x-8)(x+3) = 0$$
$$x - 8 = 0 \quad \text{or} \quad x + 3 = 0$$
$$x = 8 \qquad x = -3$$
Because the length of the side cannot be negative, choose $x = 8$ as the length of the shortest side. The other sides are $x + 7 = 8 + 7 = 15$ and $2x + 1 = 2(8) + 1 = 17$.

19. Let $v_0 = 4$ and $h_0 = 56$. Set the formula equal to 0 and solve for t to find the time it takes for the rock to hit the ground.
$$0 = -16t^2 + 4t + 56$$
$$0 = -4\left(4t^2 - t - 14\right)$$
$$0 = -4(4t+7)(t-2)$$
$$4t + 7 = 0 \quad \text{or} \quad t - 2 = 0$$
$$4t = -7 \qquad t = 2$$
$$t = -\frac{7}{4}$$
Disregard the negative solution because time cannot be negative in this example. It takes the rock 2 seconds to hit the ground.

20. To find the x-intercepts, set the function equal to 0 and solve for x.
$$x^2 - 9 = 0$$
$$(x+3)(x-3) = 0$$
$$x + 3 = 0 \quad \text{or} \quad x - 3 = 0$$
$$x = -3 \qquad x = 3$$
The x-intercepts are (–3, 0) and (3, 0).

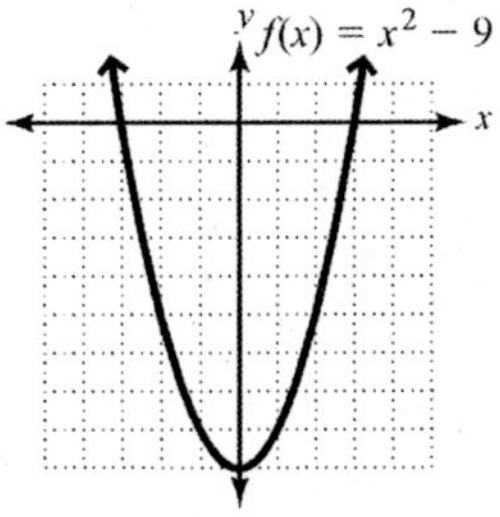

Chapters 1 – 6 Cumulative Review

1. True
2. False; this is an example of a quadratic function.
3. True
4. False; $a^2 + b^2$ is prime and cannot be factored further.
5. the same
6. coefficients, add
7. $(a+b)\left(a^2 - ab + b^2\right)$
8. $$\frac{4\left(3^2 - 10\right) - 4}{3 - \sqrt{25 - 16}} = \frac{4(9-10) - 4}{3 - \sqrt{9}}$$
$$= \frac{4(-1) - 4}{3 - 3}$$
$$= \frac{-4 - 4}{0} \text{ is undefined}$$
9. $$\left(2x^3 + 5x^2 - 19x + 14\right) - \left(17x^2 + 2x - 6\right)$$
$$= \left(2x^3 + 5x^2 - 19x + 14\right) + \left(-17x^2 - 2x + 6\right)$$
$$= 2x^3 - 12x^2 - 21x + 20$$
10. $$(2x+1)\left(4x^2 - 3x + 2\right)$$
$$= 2x\left(4x^2\right) + 2x(-3x) + 2x(2) + 1\left(4x^2\right) + 1(-3x) + 1(2)$$
$$= 8x^3 - 6x^2 + 4x + 4x^2 - 3x + 2$$
$$= 8x^3 - 2x^2 + x + 2$$
11. $$\left(\frac{2}{x}\right)^{-3} = \left(\frac{x}{2}\right)^3 = \frac{x^3}{2^3} = \frac{x^3}{8}$$
12. $4m - 16m^2 = 4m(1 - 4m)$
13. $8x^2 + 29x - 12 = (8x - 3)(x + 4)$

14. $49a^2+84a+36=(7a+6)^2$

15. $64p^3+8=8(8p^3+1)=8(2p+1)(4p^2-2p+1)$

16. $$\begin{aligned}3x-6y+ax-2ay&=(3x-6y)+(ax-2ay)\\&=3(x-2y)+a(x-2y)\\&=(x-2y)(3+a)\end{aligned}$$

17. $$\begin{aligned}9x-[14+2(x-3)]&=8x-2(3x-1)\\9x-[14+2x-6]&=8x-6x+2\\9x-[8+2x]&=2x+2\\9x-8-2x&=2x+2\\5x&=10\\x&=2\end{aligned}$$

18. $$\begin{aligned}&|2x-3|-9=16\\&|2x-3|=25\end{aligned}$$
$2x-3=-25$ or $2x-3=25$
$2x=-22 \qquad 2x=28$
$x=-11 \qquad x=14$

19. $$\begin{aligned}5x^2+26x&=24\\5x^2+26x-24&=0\\(5x-4)(x+6)&=0\end{aligned}$$
$5x-4=0$ or $x+6=0$
$x=\frac{4}{5} \qquad x=-6$

20. $$\begin{aligned}\frac{3}{4}x+\frac{2}{3}&\le\frac{5}{6}x+\frac{1}{2}\\12\left(\frac{3}{4}x+\frac{2}{3}\right)&\le12\left(\frac{5}{6}x+\frac{1}{2}\right)\\9x+8&\le10x+6\\-x&\le-2\\x&\ge2\end{aligned}$$

a) $\{x|x\ge2\}$ b) $[2,\infty)$

c) 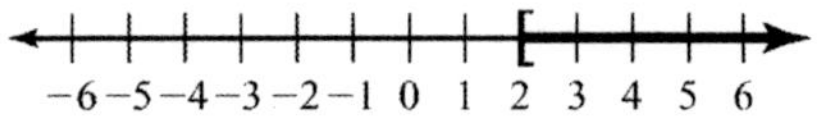

21. $$\begin{aligned}2|x-3|-2&<10\\2|x-3|&<12\\|x-3|&<6\end{aligned}$$
$$\begin{aligned}-6<x-3&<6\\-3<x&<9\end{aligned}$$

a) $\{x|-3<x<9\}$ b) $(-3,9)$

c) 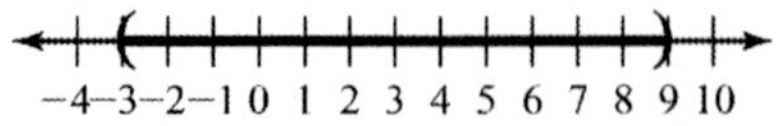

22. $$\begin{cases}5x-2y=-1\\3x+2y=-7\end{cases}$$
$$\begin{aligned}5x-2y&=-1\\\underline{3x+2y}&\underline{=-7}\\8x&=-8\\x&=-1\end{aligned}\qquad\begin{aligned}5x-2y&=-1\\5(-1)-2y&=-1\\-5-2y&=-1\\-2y&=4\\y&=-2\end{aligned}$$
Solution: $(-1,-2)$

23. $$\begin{cases}x-2y+z=8\\x+y+z=5\\2x+y-2z=-5\end{cases}$$
Add equation 1 and equation 2 multiplied by –1.
$$\begin{aligned}x-2y+z&=8\\\underline{-x-y-z}&\underline{=-5}\\-3y&=3\\y&=-1\end{aligned}$$
Substitute $y=-1$ into equations 1 and 3.
$$\begin{aligned}x-2(-1)+z&=8\\x+2+z&=8\\x+z&=6\end{aligned}\qquad\begin{aligned}2x-1-2z&=-5\\2x-2z&=-4\end{aligned}$$
Solve the new system.
$$\begin{cases}x+z=6\\2x-2z=-4\end{cases}$$
$$\begin{aligned}-2x-2z&=-12\\\underline{2x-2z}&\underline{=-4}\\-4z&=-16\\z&=4\end{aligned}\qquad\begin{aligned}x+z&=6\\x+4&=6\\x&=2\end{aligned}$$
Solution: $(2,-1,4)$

24. $$\begin{vmatrix}3&-2\\x&4\end{vmatrix}=16$$
$$\begin{aligned}3(4)-x(-2)&=16\\12+2x&=16\\2x&=4\\x&=2\end{aligned}$$

25. $\begin{cases} x+y<6 \\ y \geq 2x \end{cases}$

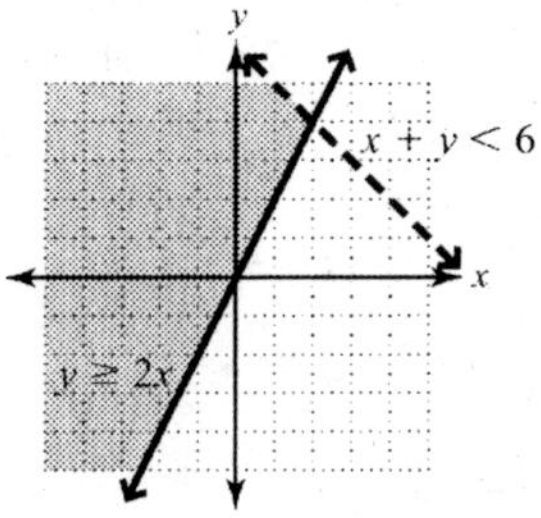

26. $f(x)=x^2+x-6$

$x^2+x-6=0$

$(x+3)(x-2)=0$

$x+3=0$ or $x-2=0$

$x=-3$ $x=2$

The x-intercepts are (–3, 0) and (2, 0).

$f(0)=(0)^2+(0)-6=-6$

The y-intercept is $(0,-6)$.

x	$f(x)$
–2	–4
–1	–6
1	–4

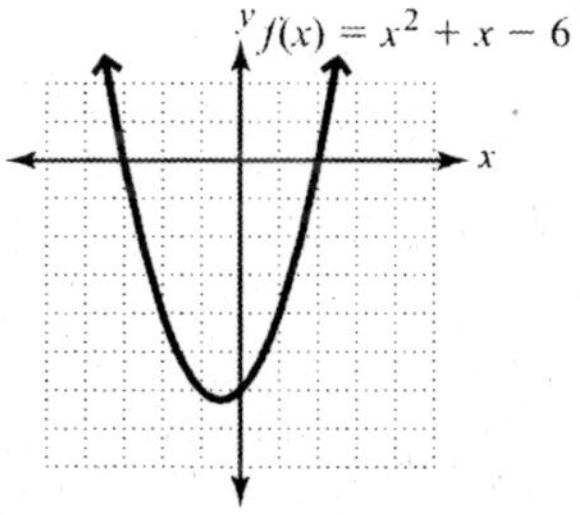

27. a) $50 \leq 10n < 80$
$5 \leq n < 8$
b) $n \geq 8$

28. $\begin{cases} a=2+2t \\ a+t=77 \end{cases}$

$a+t=77$

$(2+2t)+t=77$

$2+3t=77$

$3t=75$

$t=25$

Teens drive on average 25 minutes per day. Adults drive on average 2(25) + 2 = 52 minutes per day.

29. Since the angles are complementary, we know that their sum is 90 degrees.

$\begin{cases} x+y=90 \\ x=2y-6 \end{cases}$

$(2y-6)+y=90$

$3y=96$

$y=32$

The angles are 32° and $2(32)-6=$ 58°.

30. $60=\frac{1}{2}(a+3)(a+a+5)$

$120=(a+3)(2a+5)$

$120=2a^2+11a+15$

$0=2a^2+11a-105$

$0=(2a+21)(a-5)$

$2a+21=0$ or $a-5=0$

$a=-\frac{21}{2}$ $a=5$

a cannot be negative

a = 5 in., a + 3 = 8 in., a + 5 = 10 in.

Chapter 7

Rational Expressions and Equations

Exercise Set 7.1

1. No, since $f(5)$ yields 0 in the denominator, $x = 5$ is not in the domain of the function.

3. They are not factors of the numerator and denominator.

5. Factor each numerator and denominator completely, divide out common factors, multiply the numerators and multiply the denominators, and simplify as needed.

7. $\dfrac{-28m^3n^5}{16m^7n^6}$

$$= \frac{-1\cdot\cancel{2}\cdot\cancel{2}\cdot 7\cdot\cancel{m}\cdot\cancel{m}\cdot\cancel{m}\cdot\cancel{n}\cdot\cancel{n}\cdot\cancel{n}\cdot\cancel{n}\cdot\cancel{n}}{\cancel{2}\cdot\cancel{2}\cdot 2\cdot 2\cdot\cancel{m}\cdot\cancel{m}\cdot\cancel{m}\cdot m\cdot m\cdot m\cdot m\cdot\cancel{n}\cdot\cancel{n}\cdot\cancel{n}\cdot\cancel{n}\cdot\cancel{n}\cdot n}$$

$$= -\frac{7}{4m^4n}$$

9. $\dfrac{-32x^4y^3z^2}{24x^2y^2z^2}$

$$= \frac{-\cancel{2}\cdot\cancel{2}\cdot\cancel{2}\cdot 2\cdot 2\cdot\cancel{x}\cdot\cancel{x}\cdot x\cdot x\cdot\cancel{y}\cdot\cancel{y}\cdot y\cdot\cancel{z}\cdot\cancel{z}}{\cancel{2}\cdot\cancel{2}\cdot\cancel{2}\cdot 3\cdot\cancel{x}\cdot\cancel{x}\cdot\cancel{y}\cdot\cancel{y}\cdot\cancel{z}\cdot\cancel{z}}$$

$$= -\frac{4x^2y}{3}$$

11. $\dfrac{2x+12}{3x+18} = \dfrac{2\cancel{(x+6)}}{3\cancel{(x+6)}} = \dfrac{2}{3}$

13. $\dfrac{8x^2+12x}{20x^2+30x} = \dfrac{4\cancel{x}\cancel{(2x+3)}}{10\cancel{x}\cancel{(2x+3)}} = \dfrac{4}{10} = \dfrac{2}{5}$

15. $\dfrac{-3x-15}{4x+20} = \dfrac{-3\cancel{(x+5)}}{4\cancel{(x+5)}} = -\dfrac{3}{4}$

17. $\dfrac{a^2+3a-10}{a^2+2a-15} = \dfrac{\cancel{(a+5)}(a-2)}{\cancel{(a+5)}(a-3)} = \dfrac{a-2}{a-3}$

19. $\dfrac{x^2-6x}{x^2-8x+12} = \dfrac{x\cancel{(x-6)}}{\cancel{(x-6)}(x-2)} = \dfrac{x}{x-2}$

21. $\dfrac{4x^2-9y^2}{6x^2-xy-15y^2} = \dfrac{\cancel{(2x+3y)}(2x-3y)}{(3x-5y)\cancel{(2x+3y)}}$

$$= \frac{2x-3y}{3x-5y}$$

23. $\dfrac{4a^2-4ab-3b^2}{6a^2-ab-2b^2} = \dfrac{\cancel{(2a+b)}(2a-3b)}{\cancel{(2a+b)}(3a-2b)}$

$$= \frac{2a-3b}{3a-2b}$$

25. $\dfrac{x^3-8}{3x^2-2x-8} = \dfrac{\cancel{(x-2)}(x^2+2x+4)}{(3x+4)\cancel{(x-2)}}$

$$= \frac{x^2+2x+4}{3x+4}$$

27. $\dfrac{x^3+27}{3x^2-27} = \dfrac{\cancel{(x+3)}(x^2-3x+9)}{3\cancel{(x+3)}(x-3)}$

$$= \frac{x^2-3x+9}{3(x-3)}$$

29. $\dfrac{xy-4x+3y-12}{xy-6x+3y-18} = \dfrac{x(y-4)+3(y-4)}{x(y-6)+3(y-6)}$

$$= \frac{(y-4)\cancel{(x+3)}}{(y-6)\cancel{(x+3)}}$$

$$= \frac{y-4}{y-6}$$

31. $\dfrac{6a^3+4a^2+9a+6}{6a^2+13a+6} = \dfrac{2a^2(3a+2)+3(3a+2)}{(2a+3)(3a+2)}$

$$= \frac{\cancel{(3a+2)}(2a^2+3)}{(2a+3)\cancel{(3a+2)}}$$

$$= \frac{2a^2+3}{2a+3}$$

33. $\dfrac{2b-a}{a-2b} = \dfrac{-1\cancel{(a-2b)}}{\cancel{(a-2b)}} = -1$

35. $\dfrac{x^2+4x-21}{6-2x} = \dfrac{(x+7)\cancel{(x-3)}}{-2\cancel{(x-3)}} = -\dfrac{x+7}{2}$

37. $\dfrac{8x^2y^4}{12x^3y^2}\cdot\dfrac{4x^3y^3}{6x^4y^3}$

$=\dfrac{\cancel{2}\cdot\cancel{2}\cdot 2\cdot\cancel{x}\cdot\cancel{x}\cdot\cancel{y}\cdot\cancel{y}\cdot y\cdot y}{\cancel{2}\cdot\cancel{2}\cdot 3\cdot\cancel{x}\cdot\cancel{x}\cdot x\cdot\cancel{y}\cdot\cancel{y}}$

$\cdot\dfrac{\cancel{2}\cdot 2\cdot\cancel{x}\cdot\cancel{x}\cdot\cancel{x}\cdot\cancel{y}\cdot\cancel{y}\cdot\cancel{y}}{\cancel{2}\cdot 3\cdot\cancel{x}\cdot\cancel{x}\cdot\cancel{x}\cdot x\cdot\cancel{y}\cdot\cancel{y}\cdot\cancel{y}}$

$=\dfrac{4y^2}{9x^2}$

39. $-\dfrac{24a^2bc^3}{16ab^3c^2}\cdot\dfrac{32a^3b^2c^3}{18a^4b^3c^3}$

$=-\dfrac{\cancel{2}\cdot\cancel{2}\cdot\cancel{2}\cdot\cancel{3}\cdot\cancel{a}\cdot\cancel{a}\cdot\cancel{b}\cdot\cancel{c}\cdot\cancel{c}\cdot c}{\cancel{2}\cdot\cancel{2}\cdot\cancel{2}\cdot\cancel{2}\cdot\cancel{a}\cdot\cancel{b}\cdot b\cdot b\cdot\cancel{c}\cdot\cancel{c}}$

$\cdot\dfrac{\cancel{2}\cdot\cancel{2}\cdot 2\cdot 2\cdot 2\cdot\cancel{a}\cdot\cancel{a}\cdot\cancel{a}\cdot\cancel{b}\cdot\cancel{b}\cdot\cancel{c}\cdot\cancel{c}\cdot\cancel{c}}{\cancel{2}\cdot\cancel{3}\cdot 3\cdot\cancel{a}\cdot\cancel{a}\cdot\cancel{a}\cdot\cancel{a}\cdot\cancel{b}\cdot\cancel{b}\cdot b\cdot\cancel{c}\cdot\cancel{c}\cdot\cancel{c}}$

$=-\dfrac{8c}{3b^3}$

41. $\dfrac{3x^2y^4}{15x+10y}\cdot\dfrac{24x+16y}{12xy^2}$

$=\dfrac{\cancel{3}\cdot\cancel{x}\cdot x\cdot\cancel{y}\cdot\cancel{y}\cdot y\cdot y}{5\left(\cancel{3x+2y}\right)}\cdot\dfrac{\cancel{2}\cdot\cancel{2}\cdot 2\cdot\left(\cancel{3x+2y}\right)}{\cancel{2}\cdot\cancel{2}\cdot\cancel{3}\cdot\cancel{x}\cdot\cancel{y}\cdot\cancel{y}}$

$=\dfrac{2xy^2}{5}$

43. $\dfrac{8x+12}{18x-24}\cdot\dfrac{12-9x}{10x+15}=\dfrac{4\left(\cancel{2x+3}\right)}{6\left(\cancel{3x-4}\right)}\cdot\dfrac{-3\left(\cancel{3x-4}\right)}{5\left(\cancel{2x+3}\right)}$

$=\dfrac{-12}{30}=-\dfrac{2}{5}$

45. $\dfrac{4a-16}{3a-3}\cdot\dfrac{a^2-1}{a^2-16}=\dfrac{4\left(\cancel{a-4}\right)}{3\left(\cancel{a-1}\right)}\cdot\dfrac{(a+1)\left(\cancel{a-1}\right)}{(a+4)\left(\cancel{a-4}\right)}$

$=\dfrac{4(a+1)}{3(a+4)}$

47. $\dfrac{4x^2y^3}{2x-6}\cdot\dfrac{12-4x}{6x^3y}$

$=\dfrac{\cancel{2}\cdot\cancel{2}\cdot\cancel{x}\cdot\cancel{x}\cdot\cancel{y}\cdot y\cdot y}{\cancel{2}\left(\cancel{x-3}\right)}\cdot\dfrac{-2\cdot 2\cdot\left(\cancel{x-3}\right)}{\cancel{2}\cdot 3\cdot\cancel{x}\cdot\cancel{x}\cdot x\cdot\cancel{y}}$

$=-\dfrac{4y^2}{3x}$

49. $\dfrac{x^2-4}{x^2-3x-10}\cdot\dfrac{x^2-8x+15}{x^2-9}$

$=\dfrac{\left(\cancel{x+2}\right)(x-2)}{\left(\cancel{x-5}\right)\left(\cancel{x+2}\right)}\cdot\dfrac{\left(\cancel{x-5}\right)\left(\cancel{x-3}\right)}{(x+3)\left(\cancel{x-3}\right)}$

$=\dfrac{x-2}{x+3}$

51. $\dfrac{6x^2+x-12}{2x^2-5x-12}\cdot\dfrac{3x^2-14x+8}{9x^2-18x+8}$

$=\dfrac{\left(\cancel{2x+3}\right)\left(\cancel{3x-4}\right)}{\left(\cancel{2x+3}\right)\left(\cancel{x-4}\right)}\cdot\dfrac{\left(\cancel{3x-2}\right)\left(\cancel{x-4}\right)}{\left(\cancel{3x-2}\right)\left(\cancel{3x-4}\right)}$

$=1$

53. $\dfrac{2x^2-7x+6}{8x^2-6x-9}\cdot\dfrac{3x^2-13x-10}{x^2-7x+10}$

$=\dfrac{\left(\cancel{2x-3}\right)\left(\cancel{x-2}\right)}{(4x+3)\left(\cancel{2x-3}\right)}\cdot\dfrac{(3x+2)\left(\cancel{x-5}\right)}{\left(\cancel{x-5}\right)\left(\cancel{x-2}\right)}$

$=\dfrac{3x+2}{4x+3}$

55. $\dfrac{2x^2-11x+12}{2x^2+11x-21}\cdot\dfrac{3x^2+20x-7}{4-x}$

$=\dfrac{\left(\cancel{2x-3}\right)\left(\cancel{x-4}\right)}{\left(\cancel{2x-3}\right)\left(\cancel{x+7}\right)}\cdot\dfrac{(3x-1)\left(\cancel{x+7}\right)}{-1\left(\cancel{x-4}\right)}$

$=-(3x-1)$

57. $\dfrac{ac+3a+2c+6}{ad+a+2d+2}\cdot\dfrac{ad-5a+2d-10}{bc+3b-4c-12}$

$=\dfrac{a(c+3)+2(c+3)}{a(d+1)+2(d+1)}\cdot\dfrac{a(d-5)+2(d-5)}{b(c+3)-4(c+3)}$

$=\dfrac{\left(\cancel{c+3}\right)\left(\cancel{a+2}\right)}{(d+1)\left(\cancel{a+2}\right)}\cdot\dfrac{(d-5)(a+2)}{\left(\cancel{c+3}\right)(b-4)}$

$=\dfrac{(d-5)(a+2)}{(d+1)(b-4)}$

59. $\dfrac{4x^3-3x^2+4x-3}{4x^3-3x^2-8x+6}\cdot\dfrac{2x^3+x^2-4x-2}{3x^3+2x^2+3x+2}$

$=\dfrac{x^2(4x-3)+1(4x-3)}{x^2(4x-3)-2(4x-3)}\cdot\dfrac{x^2(2x+1)-2(2x+1)}{x^2(3x+2)+1(3x+2)}$

$=\dfrac{(4x-3)(x^2+1)}{(4x-3)(x^2-2)}\cdot\dfrac{(2x+1)(x^2-2)}{(3x+2)(x^2+1)}$

$=\dfrac{2x+1}{3x+2}$

61. $\dfrac{a^3-b^3}{a^2+ab+b^2}\cdot\dfrac{2a^2+ab-b^2}{a^2-b^2}$

$=\dfrac{(a-b)(a^2+ab+b^2)}{(a^2+ab+b^2)}\cdot\dfrac{(2a-b)(a+b)}{(a+b)(a-b)}$

$=2a-b$

63. $\dfrac{8x^7g^3}{15x^2g^4}\div\dfrac{4xg^2}{3x^2g^5}$

$=\dfrac{8x^7g^3}{15x^2g^4}\cdot\dfrac{3x^2g^5}{4xg^2}$

$=\dfrac{2\cdot 2\cdot 2\cdot x\cdot x\cdot x\cdot x\cdot x\cdot x\cdot x\cdot g\cdot g\cdot g}{3\cdot 5\cdot x\cdot x\cdot g\cdot g\cdot g\cdot g}$

$\cdot\dfrac{3\cdot x\cdot x\cdot g\cdot g\cdot g\cdot g\cdot g}{2\cdot 2\cdot x\cdot g\cdot g}=\dfrac{2x^6g^2}{5}$

65. $\dfrac{4a^3-8a^2}{15a}\div\dfrac{3a^2-6a}{5a^4}$

$=\dfrac{4a^3-8a^2}{15a}\cdot\dfrac{5a^4}{3a^2-6a}$

$=\dfrac{2\cdot 2\cdot a\cdot a(a-2)}{3\cdot 5\cdot a}\cdot\dfrac{5\cdot a\cdot a\cdot a\cdot a}{3a(a-2)}$

$=\dfrac{4a^4}{9}$

67. $\dfrac{10x^3+15x^2}{18x-6}\div\dfrac{20x+30}{9x^2-3x}$

$=\dfrac{10x^3+15x^2}{18x-6}\cdot\dfrac{9x^2-3x}{20x+30}$

$=\dfrac{5\cdot x\cdot x(2x+3)}{2\cdot 3\cdot(3x-1)}\cdot\dfrac{3x(3x-1)}{2\cdot 5\cdot(2x+3)}$

$=\dfrac{x^3}{4}$

69. $\dfrac{3a-4b}{6a-9b}\div\dfrac{12b-9a}{4a-6b}$

$=\dfrac{3a-4b}{6a-9b}\cdot\dfrac{4a-6b}{12b-9a}$

$=\dfrac{(3a-4b)}{3(2a-3b)}\cdot\dfrac{2(2a-3b)}{-3(3a-4b)}$

$=-\dfrac{2}{9}$

71. $\dfrac{d^2-d-12}{2d^2}\div\dfrac{3d^2+13d+12}{d}$

$=\dfrac{d^2-d-12}{2d^2}\cdot\dfrac{d}{3d^2+13d+12}$

$=\dfrac{(d-4)(d+3)}{2\cdot d\cdot d}\cdot\dfrac{d}{(3d+4)(d+3)}$

$=\dfrac{d-4}{2d(3d+4)}$

73. $\dfrac{m^2-25n^2}{2m-12n}\div\dfrac{3m+15n}{m^2-36n^2}$

$=\dfrac{m^2-25n^2}{2m-12n}\cdot\dfrac{m^2-36n^2}{3m+15n}$

$=\dfrac{(m+5n)(m-5n)}{2(m-6n)}\cdot\dfrac{(m+6n)(m-6n)}{3(m+5n)}$

$=\dfrac{(m-5n)(m+6n)}{6}$

75. $\dfrac{3x+24}{4x-32}\div\dfrac{x^2+16x+64}{x^2-16x+64}$

$=\dfrac{3x+24}{4x-32}\cdot\dfrac{x^2-16x+64}{x^2+16x+64}$

$=\dfrac{3(x+8)}{4(x-8)}\cdot\dfrac{(x-8)(x-8)}{(x+8)(x+8)}$

$=\dfrac{3(x-8)}{4(x+8)}$

77. $\dfrac{2x^2-7xy+6y^2}{2x^2+7xy+6y^2} \div \dfrac{4x^2-9y^2}{4x^2+12xy+9y^2}$

$= \dfrac{2x^2-7xy+6y^2}{2x^2+7xy+6y^2} \cdot \dfrac{4x^2+12xy+9y^2}{4x^2-9y^2}$

$= \dfrac{(x-2y)\cancel{(2x-3y)}}{(x+2y)\cancel{(2x+3y)}} \cdot \dfrac{\cancel{(2x+3y)}\cancel{(2x+3y)}}{\cancel{(2x+3y)}\cancel{(2x-3y)}}$

$= \dfrac{x-2y}{x+2y}$

79. $\dfrac{2x^2-7x-4}{3x^2-14x+8} \div \dfrac{2x^2+7x+3}{3x^2-8x+4}$

$= \dfrac{2x^2-7x-4}{3x^2-14x+8} \cdot \dfrac{3x^2-8x+4}{2x^2+7x+3}$

$= \dfrac{\cancel{(2x+1)}\cancel{(x-4)}}{\cancel{(3x-2)}\cancel{(x-4)}} \cdot \dfrac{\cancel{(3x-2)}(x-2)}{\cancel{(2x+1)}(x+3)}$

$= \dfrac{x-2}{x+3}$

81. $\dfrac{27a^3-8b^3}{9a^2-4b^2} \div \dfrac{2a^2-5ab-12b^2}{6a^2+13ab+6b^2}$

$= \dfrac{27a^3-8b^3}{9a^2-4b^2} \cdot \dfrac{6a^2+13ab+6b^2}{2a^2-5ab-12b^2}$

$= \dfrac{\cancel{(3a-2b)}(9a^2+6ab+4b^2)}{\cancel{(3a+2b)}\cancel{(3a-2b)}}$

$\cdot \dfrac{\cancel{(2a+3b)}\cancel{(3a+2b)}}{\cancel{(2a+3b)}(a-4b)}$

$= \dfrac{9a^2+6ab+4b^2}{a-4b}$

83. $\dfrac{ab+3a+2b+6}{bc+4b+3c+12} \div \dfrac{ac-3a+2c-6}{bc+4b-4c-16}$

$= \dfrac{ab+3a+2b+6}{bc+4b+3c+12} \cdot \dfrac{bc+4b-4c-16}{ac-3a+2c-6}$

$= \dfrac{a(b+3)+2(b+3)}{b(c+4)+3(c+4)} \cdot \dfrac{b(c+4)-4(c+4)}{a(c-3)+2(c-3)}$

$= \dfrac{\cancel{(b+3)}\cancel{(a+2)}}{\cancel{(c+4)}\cancel{(b+3)}} \cdot \dfrac{\cancel{(c+4)}(b-4)}{(c-3)\cancel{(a+2)}}$

$= \dfrac{b-4}{c-3}$

85. $\dfrac{x^2-3x-18}{2x^2+13x+20} \cdot \dfrac{3x^2+10x-8}{4x^2-24x} \div \dfrac{3x^2+7x-6}{6x^2+11x-10}$

$= \dfrac{\cancel{(x-6)}\cancel{(x+3)}}{\cancel{(2x+5)}\cancel{(x+4)}} \cdot \dfrac{(3x-2)\cancel{(x+4)}}{4x\cancel{(x-6)}}$

$\cdot \dfrac{\cancel{(3x-2)}\cancel{(2x+5)}}{\cancel{(3x-2)}\cancel{(x+3)}} = \dfrac{3x-2}{4x}$

87. $\dfrac{5x^2-17x-12}{12x^2-29x+15}$

$\cdot \left(\dfrac{6x^2+x-12}{3x^2+14x-24} \div \dfrac{2x^2-5x-12}{3x^2+13x-30}\right)$

$= \dfrac{(5x+3)\cancel{(x-4)}}{(4x-3)\cancel{(3x-5)}} \cdot \dfrac{\cancel{(2x+3)}\cancel{(3x-4)}}{\cancel{(3x-4)}\cancel{(x+6)}}$

$\cdot \dfrac{\cancel{(3x-5)}\cancel{(x+6)}}{\cancel{(2x+3)}\cancel{(x-4)}}$

$= \dfrac{5x+3}{4x-3}$

89. a^2+b^2 does not factor.

Correct: $\dfrac{a^2+b^2}{(a+b)\cancel{(a-b)}} \cdot \dfrac{\cancel{(a-b)}\cancel{(a+3b)}}{\cancel{(a+3b)}(a-2b)}$

$= \dfrac{a^2+b^2}{(a+b)(a-2b)}$

91. $f(x) = \dfrac{x+3}{x-4}$

a) $f(3) = \dfrac{3+3}{3-4} = \dfrac{6}{-1} = -6$

b) $f(-3) = \dfrac{-3+3}{-3-4} = \dfrac{0}{-7} = 0$

c) $f(4) = \dfrac{4+3}{4-4} = \dfrac{7}{0}$ is undefined

93. $f(x) = \dfrac{2x-3}{3x+2}$

a) $f\left(\dfrac{3}{2}\right) = \dfrac{2\left(\dfrac{3}{2}\right)-3}{3\left(\dfrac{3}{2}\right)+2} = \dfrac{0}{\frac{13}{2}} = 0$

b) $f(-1)=\dfrac{2(-1)-3}{3(-1)+2}=\dfrac{-5}{-1}=5$

c) $f(2)=\dfrac{2(2)-3}{3(2)+2}=\dfrac{1}{8}$

95. $f(x)=\dfrac{3x-4}{2x^2-x-6}$

a) $f(3)=\dfrac{3(3)-4}{2(3)^2-(3)-6}=\dfrac{5}{9}$

b) $f\left(\dfrac{4}{3}\right)=\dfrac{3\left(\dfrac{4}{3}\right)-4}{2\left(\dfrac{4}{3}\right)^2-\left(\dfrac{4}{3}\right)-6}=\dfrac{0}{\left(-\dfrac{34}{9}\right)}=0$

c) $f(2)=\dfrac{3(2)-4}{2(2)^2-(2)-6}=\dfrac{2}{0}$ is undefined

97. $f(x)=\dfrac{x^2+x-6}{x^2+x-12}$

a) $f(-3)=\dfrac{(-3)^2+(-3)-6}{(-3)^2+(-3)-12}=\dfrac{0}{-6}=0$

b) $f(5)=\dfrac{(5)^2+(5)-6}{(5)^2+(5)-12}=\dfrac{24}{18}=\dfrac{4}{3}$

c) $f(3)=\dfrac{(3)^2+(3)-6}{(3)^2+(3)-12}=\dfrac{6}{0}$ is undefined

99. $f(x)=\dfrac{x+6}{3x+12}$

$3x+12=0$

$3x=-12$

$x=-4$

The domain is $\{x\,|\,x\neq -4\}$.

101. $f(x)=\dfrac{3x-6}{x^2-25}$

$x^2-25=0$

$x^2=25$

$x=\pm 5$

The domain is $\{x|x\neq -5,5\}$.

103. $g(c)=\dfrac{3c-9}{4c^2-81}$

$4c^2-81=0$

$4c^2=81$

$c^2=\dfrac{81}{4}$

$c=\pm\dfrac{9}{2}$

The domain is $\left\{c\middle|c\neq -\dfrac{9}{2},\dfrac{9}{2}\right\}$.

105. $h(t)=\dfrac{2t^2+4t-8}{6t^2+11t-10}$

$6t^2+11t-10=0$

$(3t-2)(2t+5)=0$

$3t-2=0$ or $2t+5=0$

$t=\dfrac{2}{3}$ $\qquad t=-\dfrac{5}{2}$

The domain is $\left\{t\middle|t\neq -\dfrac{5}{2},\dfrac{2}{3}\right\}$.

107. $g(x)=\dfrac{4x-8}{x^3+4x^2-21x}$

$x^3+4x^2-21x=0$

$x(x+7)(x-3)=0$

$x=0$ or $x+7=0$ or $x-3=0$

$x=-7$ $\qquad x=3$

The domain is $\{x|x\neq 0,-7,3\}$.

109. $f(x)=\dfrac{x-7}{x^3-8}$

$x^3-8=0$

$(x-2)\left(x^2+2x+4\right)=0$

$x-2=0$ or $x^2+2x+4=0$

$x=2$

When solved for x, $x^2+2x+4=0$ yields complex solutions.

The domain is $\{x|x\neq 2\}$.

111. c $\qquad$ 113. d

115. $f(x)=\frac{1}{x-2}$
Note that $x \neq 2$, so the graph has a vertical asymptote at $x=2$.

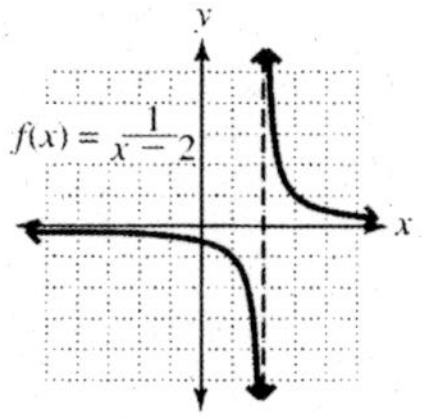

117. $f(x)=\frac{-2}{x+1}$
Note that $x \neq -1$, so the graph has a vertical asymptote at $x=-1$.

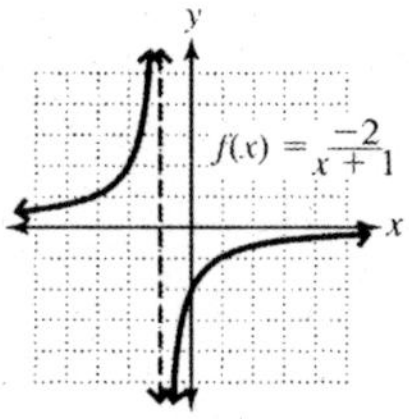

119. Substitute 2 for d and solve for R.
$$R=\frac{5}{d^2}=\frac{5}{2^2}=\frac{5}{4} \text{ ohms}$$

121. Substitute 100 for x and solve for C.
$$C=\frac{15x+1800}{x}$$
$$=\frac{15(200)+1800}{200}$$
$$=\frac{4800}{200}=\$24$$

123. Substitute 50 for x and solve for C.
$$C=200\left(\frac{300}{x^2}+\frac{x}{x+50}\right)$$
$$=200\left(\frac{300}{50^2}+\frac{50}{50+50}\right)$$
$$=200\left(\frac{300}{2500}+\frac{50}{100}\right)$$
$$=200(0.62)$$
$$=124$$
The cost of the order is \$124,000.

125. Substitute 70 for v and solve for F.
$$F=\frac{130,000}{330-v}=\frac{130,000}{330-70}=500$$
The frequency is 500 cycles/sec.

Review Exercises

1. $\frac{3}{11}+\frac{4}{11}=\frac{7}{11}$

2. $\frac{3}{5}-\frac{5}{6}=\frac{18}{30}-\frac{25}{30}=-\frac{7}{30}$

3. $\frac{5}{12}y+\frac{7}{30}y=\frac{25}{60}y+\frac{14}{60}y=\frac{39}{60}y=\frac{13}{20}y$

4. $\frac{11}{15}x+4-\frac{2}{15}x-3=\frac{9}{15}x+1$
$$=\frac{3}{5}x+1$$

5. $6x^2-15x=3x(2x-5)$

6. $x^2+x-20=(x+5)(x-4)$

Exercise Set 7.2

1. The rational expression must be multiplied by 1 so as not to change its value.

3. Factoring $6x^2+5x-6=(3x-2)(2x+3)$ shows that it is necessary to multiply by $(x-4)$ to obtain the LCD.

5. Both are correct because
$$\frac{-6}{3-a}=\frac{6}{-(3-a)}=\frac{6}{a-3}.$$

7. $\frac{3c}{2d}+\frac{5c}{2d}=\frac{3c+5c}{2d}=\frac{8c}{2d}=\frac{4c}{d}$

9. $\frac{4a+3b}{2a-5b}+\frac{3a-4b}{2a-5b}=\frac{4a+3b+3a-4b}{2a-5b}=\frac{7a-b}{2a-5b}$

11. $\frac{2x-3y}{x+2y}-\frac{4x+2y}{x+2y}=\frac{2x-3y-(4x+2y)}{x+2y}$
$$=\frac{2x-3y-4x-2y}{x+2y}$$
$$=\frac{-2x-5y}{x+2y}$$

13. $\frac{c^2}{c^2-5c+4}-\frac{-2c+24}{c^2-5c+4}=\frac{c^2-(-2c+24)}{c^2-5c+4}$
$=\frac{c^2+2c-24}{c^2-5c+4}$
$=\frac{(c+6)(c-4)}{(c-4)(c-1)}$
$=\frac{c+6}{c-1}$

15. $\frac{y^2-2y}{y^2-12y+36}+\frac{-5y+6}{y^2-12y+36}$
$=\frac{y^2-2y-5y+6}{y^2-12y+36}$
$=\frac{y^2-7y+6}{y^2-12y+36}$
$=\frac{(y-6)(y-1)}{(y-6)(y-6)}$
$=\frac{y-1}{y-6}$

17. $\frac{h^2-4h}{h^2+10h+21}-\frac{-6h+3}{h^2+10h+21}$
$=\frac{h^2-4h-(-6h+3)}{h^2+10h+21}$
$=\frac{h^2-4h+6h-3}{h^2+10h+21}$
$=\frac{h^2+2h-3}{h^2+10h+21}$
$=\frac{(h+3)(h-1)}{(h+7)(h+3)}$
$=\frac{h-1}{h+7}$

19. $\frac{2x^2+x-3}{x^2+6x+5}+\frac{x^2-2x+4}{x^2+6x+5}-\frac{2x^2+x+4}{x^2+6x+5}$
$=\frac{2x^2+x-3+x^2-2x+4-(2x^2+x+4)}{x^2+6x+5}$
$=\frac{2x^2+x-3+x^2-2x+4-2x^2-x-4}{x^2+6x+5}$
$=\frac{x^2-2x-3}{x^2+6x+5}$
$=\frac{(x-3)(x+1)}{(x+5)(x+1)}$
$=\frac{x-3}{x+5}$

21. $\frac{x^3}{x^2-xy+y^2}+\frac{y^3}{x^2-xy+y^2}$
$=\frac{x^3+y^3}{x^2-xy+y^2}$
$=\frac{(x+y)(x^2-xy+y^2)}{(x^2-xy+y^2)}$
$=x+y$

23. $15a^6b^8=3\cdot 5\cdot a^6\cdot b^8$
$20a^4b^5=2^2\cdot 5\cdot a^4\cdot b^5$
$\text{LCD}=2^2\cdot 3\cdot 5\cdot a^6\cdot b^8=60a^6b^8$
$\frac{4}{15a^6b^8}=\frac{4\cdot 4}{15a^6b^8\cdot 4}=\frac{16}{60a^6b^8}$
$\frac{9}{20a^4b^5}=\frac{9\cdot 3a^2b^3}{20a^4b^5\cdot 3a^2b^3}=\frac{27a^2b^3}{60a^6b^8}$

25. $\text{LCD}=(x+7)(x-5)$
$\frac{x}{x+7}=\frac{x\cdot(x-5)}{(x+7)\cdot(x-5)}=\frac{x^2-5x}{(x+7)(x-5)}$
$\frac{x}{x-5}=\frac{x\cdot(x+7)}{(x-5)\cdot(x+7)}=\frac{x^2+7x}{(x+7)(x-5)}$

27. $9r+18=3^2(r+2)$
$15r+30=3\cdot 5(r+2)$
$\text{LCD}=3^2\cdot 5(r+2)=45(r+2)$
$\frac{7r}{9r+18}=\frac{7r\cdot 5}{9(r+2)\cdot 5}=\frac{35r}{45(r+2)}$
$\frac{8r}{15r+30}=\frac{8r\cdot 3}{15(r+2)\cdot 3}=\frac{24r}{45(r+2)}$

29. $b^2-1=(b+1)(b-1)$

$b^2+3b-4=(b+4)(b-1)$

$\text{LCD}=(b+1)(b-1)(b+4)$

$$\frac{3b}{b^2-1}=\frac{3b\cdot(b+4)}{(b+1)(b-1)\cdot(b+4)}=\frac{3b^2+12b}{(b+1)(b-1)(b+4)}$$

$$\frac{b}{b^2+3b-4}=\frac{b\cdot(b+1)}{(b+4)(b-1)\cdot(b+1)}=\frac{b^2+b}{(b+1)(b-1)(b+4)}$$

31. $c^2-2c-3=(c-3)(c+1)$

$c^2-5c+6=(c-2)(c-3)$

$\text{LCD}=(c-3)(c+1)(c-2)$

$$\frac{c-5}{c^2-2c-3}=\frac{(c-5)\cdot(c-2)}{(c-3)(c+1)\cdot(c-2)}=\frac{c^2-7c+10}{(c-3)(c+1)(c-2)}$$

$$\frac{3+c}{c^2-5c+6}=\frac{(c+3)\cdot(c+1)}{(c-3)(c-2)\cdot(c+1)}=\frac{c^2+4c+3}{(c-3)(c+1)(c-2)}$$

33. $n^2+8n+16=(n+4)^2$

$n^2+5n+4=(n+4)(n+1)$

$\text{LCD}=(n+4)^2(n+1)$

$$\frac{n+1}{n^2+8n+16}=\frac{(n+1)\cdot(n+1)}{(n+4)^2\cdot(n+1)}=\frac{n^2+2n+1}{(n+4)^2(n+1)}$$

$$\frac{n-4}{n^2+5n+4}=\frac{(n-4)\cdot(n+4)}{(n+4)(n-1)\cdot(n+4)}=\frac{n^2-16}{(n+4)^2(n-1)}$$

35. $x^3+x^2-6x=x(x^2+x-6)=x(x+3)(x-2)$

$x^4+7x^3+12x^2=x^2(x^2+7x+12)=x^2(x+3)(x+4)$

$\text{LCD}=x^2(x+3)(x-2)(x+4)$

$$\frac{x+5}{x^3+x^2-6x}=\frac{(x+5)\cdot x(x+4)}{x(x+3)(x-2)\cdot x(x+4)}=\frac{x^3+9x^2+20x}{x^2(x+3)(x-2)(x+4)}$$

$$\frac{x-3}{x^4+7x^3+12x^2}=\frac{(x-3)\cdot(x-2)}{x^2(x+3)(x+4)\cdot(x-2)}=\frac{x^2-5x+6}{x^2(x+3)(x-2)(x+4)}$$

37. $x=x$

$x^2+4x=x(x+4)$

$x+4=(x+4)$

$\text{LCD}=x(x+4)$

$$\frac{3}{x}=\frac{3\cdot(x+4)}{x\cdot(x+4)}=\frac{3x+12}{x(x+4)}$$

$$\frac{4}{x^2+4x}=\frac{4}{x(x+4)}$$

$$\frac{6}{x+4}=\frac{6\cdot x}{(x+4)\cdot x}=\frac{6x}{x(x+4)}$$

39. $\text{LCD}=24u$

$$\frac{5}{8u}-\frac{7}{12u}=\frac{5(3)}{8u(3)}-\frac{7(2)}{12u(2)}=\frac{15}{24u}-\frac{14}{24u}=\frac{1}{24u}$$

41. $\text{LCD}=20x$

$$\frac{3z}{10x}+\frac{5z}{4x}=\frac{3z(2)}{10x(2)}+\frac{5z(5)}{4x(5)}=\frac{6z}{20x}+\frac{25z}{20x}=\frac{31z}{20x}$$

43. $\text{LCD}=12y$

$$\frac{y+6}{4y}+\frac{2y-3}{3y}=\frac{(y+6)(3)}{4y(3)}+\frac{(2y-3)(4)}{3y(4)}=\frac{3y+18}{12y}+\frac{8y-12}{12y}=\frac{3y+18+8y-12}{12y}=\frac{11y+6}{12y}$$

45. LCD = $6m$

$$\frac{m+8}{2m}-\frac{m-7}{3m}=\frac{(m+8)(3)}{2m(3)}-\frac{(m-7)(2)}{3m(2)}$$
$$=\frac{3m+24}{6m}-\frac{2m-14}{6m}$$
$$=\frac{3m+24-2m+14}{6m}$$
$$=\frac{m+38}{6m}$$

47. LCD = $30p^3q^2$

$$\frac{3p+1}{10p^3q^2}+\frac{9p-2}{6p^2q}$$
$$=\frac{(3p+1)(3)}{10p^3q^2(3)}+\frac{(9p-2)(5pq)}{6p^2q(5pq)}$$
$$=\frac{9p+3}{30p^3q^2}+\frac{45p^2q-10pq}{30p^3q^2}$$
$$=\frac{9p+3+45p^2q-10pq}{30p^3q^2}$$

49. LCD = $24a^3b^4$

$$\frac{2a+b}{8a^2b^4}-\frac{a-3b}{12a^3b^3}$$
$$=\frac{(2a+b)(3a)}{8a^2b^4(3a)}-\frac{(a-3b)(2b)}{12a^3b^3(2b)}$$
$$=\frac{6a^2+3ab}{24a^3b^4}-\frac{2ab-6b^2}{24a^3b^4}$$
$$=\frac{6a^2+3ab-2ab+6b^2}{24a^3b^4}$$
$$=\frac{6a^2+ab+6b^2}{24a^3b^4}$$

51. LCD = $k(k+2)$

$$\frac{4}{k}-\frac{6}{k+2}=\frac{4(k+2)}{k(k+2)}-\frac{6(k)}{(k+2)(k)}$$
$$=\frac{4k+8}{k(k+2)}-\frac{6k}{k(k+2)}$$
$$=\frac{4k+8-6k}{k(k+2)}$$
$$=\frac{8-2k}{k(k+2)}$$

53. LCD = $(w-3)(w+7)$

$$\frac{4}{w-3}+\frac{5}{w+7}$$
$$=\frac{4(w+7)}{(w-3)(w+7)}+\frac{5(w-3)}{(w+7)(w-3)}$$
$$=\frac{4w+28}{(w-3)(w+7)}+\frac{5w-15}{(w-3)(w+7)}$$
$$=\frac{9w+13}{(w-3)(w+7)}$$

55. LCD = $6(x+3)$

$$\frac{x-4}{3x+9}-\frac{x+5}{6x+18}$$
$$=\frac{(x-4)(2)}{3(x+3)(2)}-\frac{x+5}{6(x+3)}$$
$$=\frac{2x-8}{6(x+3)}-\frac{x+5}{6(x+3)}$$
$$=\frac{2x-8-x-5}{6(x+3)}$$
$$=\frac{x-13}{6(x+3)}$$

57. LCD = $(t-4)^2$

$$\frac{2t}{t^2-8t+16}+\frac{6}{t-4}$$
$$=\frac{2t}{(t-4)^2}+\frac{6(t-4)}{(t-4)(t-4)}$$
$$=\frac{2t}{(t-4)^2}+\frac{6t-24}{(t-4)^2}$$
$$=\frac{8t-24}{(t-4)^2}$$

59. LCD = $(u-3)^2$

$$\frac{u^2-2u}{u^2-6u+9}+\frac{4}{4u-12}$$
$$=\frac{u^2-2u}{(u-3)^2}+\frac{\cancel{4}}{\cancel{4}(u-3)}$$
$$=\frac{u^2-2u}{(u-3)^2}+\frac{1(u-3)}{(u-3)(u-3)}$$
$$=\frac{u^2-2u}{(u-3)^2}+\frac{u-3}{(u-3)^2}$$
$$=\frac{u^2-2u+u-3}{(u-3)^2}$$
$$=\frac{u^2-u-3}{(u-3)^2}$$

61. LCD = $(x-6)(x+2)$

$$\frac{x}{x+2}-\frac{16}{x^2-4x-12}$$
$$=\frac{x(x-6)}{(x+2)(x-6)}-\frac{16}{(x+2)(x-6)}$$
$$=\frac{x^2-6x}{(x+2)(x-6)}-\frac{16}{(x+2)(x-6)}$$
$$=\frac{x^2-6x-16}{(x+2)(x-6)}$$
$$=\frac{\cancel{(x+2)}(x-8)}{\cancel{(x+2)}(x-6)}$$
$$=\frac{x-8}{x-6}$$

63. LCD = $(x-3)(x+2)(x+4)$

$$\frac{x+1}{x^2-x-6}+\frac{2x+8}{x^2+6x+8}$$
$$=\frac{(x+1)(x+4)}{(x-3)(x+2)(x+4)}+\frac{(2x+8)(x-3)}{(x+2)(x+4)(x-3)}$$
$$=\frac{x^2+5x+4}{(x-3)(x+2)(x+4)}+\frac{2x^2+2x-24}{(x-3)(x+2)(x+4)}$$
$$=\frac{x^2+5x+4+2x^2+2x-24}{(x-3)(x+2)(x+4)}$$
$$=\frac{3x^2+7x-20}{(x-3)(x+2)(x+4)}$$
$$=\frac{(3x-5)\cancel{(x+4)}}{(x-3)(x+2)\cancel{(x+4)}}$$
$$=\frac{3x-5}{(x-3)(x+2)}$$

65. LCD = $(v+4)(v-4)(v+3)$

$$\frac{2v+5}{v^2-16}-\frac{2v+6}{v^2-v-12}$$
$$=\frac{(2v+5)(v+3)}{(v+4)(v-4)(v+3)}-\frac{(2v+6)(v+4)}{(v-4)(v+3)(v+4)}$$
$$=\frac{2v^2+11v+15}{(v+4)(v-4)(v+3)}-\frac{2v^2+14v+24}{(v+4)(v-4)(v+3)}$$
$$=\frac{2v^2+11v+15-2v^2-14v-24}{(v+4)(v-4)(v+3)}$$
$$=\frac{-3v-9}{(v+4)(v-4)(v+3)}$$
$$=\frac{-3\cancel{(v+3)}}{(v+4)(v-4)\cancel{(v+3)}}$$
$$=\frac{-3}{(v+4)(v-4)}$$

67. LCD = $(z+3)^2(z-2)$

$$\frac{z-3}{z^2+6z+9}+\frac{z+1}{z^2+z-6}$$
$$=\frac{(z-3)(z-2)}{(z+3)^2(z-2)}+\frac{(z+1)(z+3)}{(z+3)(z-2)(z+3)}$$
$$=\frac{z^2-5z+6}{(z+3)^2(z-2)}+\frac{z^2+4z+3}{(z+3)^2(z-2)}$$
$$=\frac{z^2-5z+6+z^2+4z+3}{(z+3)^2(z-2)}$$
$$=\frac{2z^2-z+9}{(z+3)^2(z-2)}$$

69. LCD = $x(x+6)(x-6)(x-3)$

$$\frac{x+4}{x^3-36x}-\frac{1}{x^2+3x-18}$$
$$=\frac{(x+4)(x-3)}{x(x+6)(x-6)(x-3)}-\frac{1x(x-6)}{(x+6)(x-3)x(x-6)}$$
$$=\frac{x^2+x-12}{x(x+6)(x-6)(x-3)}-\frac{x^2-6x}{x(x+6)(x-6)(x-3)}$$
$$=\frac{x^2+x-12-x^2+6x}{x(x+6)(x-6)(x-3)}$$
$$=\frac{7x-12}{x(x+6)(x-6)(x-3)}$$

71. LCD = $(x+2)(x-2)$

$$\frac{3x}{x-2}-\frac{4}{x+2}-\frac{3}{x^2-4}$$
$$=\frac{3x(x+2)}{(x-2)(x+2)}-\frac{4(x-2)}{(x+2)(x-2)}-\frac{3}{(x+2)(x-2)}$$
$$=\frac{3x^2+6x}{(x-2)(x+2)}-\frac{4x-8}{(x+2)(x-2)}-\frac{3}{(x+2)(x-2)}$$
$$=\frac{3x^2+6x-4x+8-3}{(x-2)(x+2)}$$
$$=\frac{3x^2+2x+5}{x^2-4}$$

73. LCD = $a(a-4)$

$$\frac{5}{a^2-4a}+\frac{6}{a}+\frac{4}{a-4}$$
$$=\frac{5}{a(a-4)}+\frac{6(a-4)}{a(a-4)}+\frac{4a}{(a-4)a}$$
$$=\frac{5}{a(a-4)}+\frac{6a-24}{a(a-4)}+\frac{4a}{a(a-4)}$$
$$=\frac{5+6a-24+4a}{a(a-4)}$$
$$=\frac{10a-19}{a(a-4)}$$

75. LCD = $x(x-4)$

$$\frac{3x+1}{x-4}-\frac{7}{x}+\frac{2}{x^2-4x}$$
$$=\frac{(3x+1)x}{(x-4)x}-\frac{7(x-4)}{x(x-4)}+\frac{2}{x(x-4)}$$
$$=\frac{3x^2+x}{x(x-4)}-\frac{7x-28}{x(x-4)}+\frac{2}{x(x-4)}$$
$$=\frac{3x^2+x-7x+28+2}{x(x-4)}$$
$$=\frac{3x^2-6x+30}{x(x-4)}$$

77. LCD = $(r+3)(r-3)$

$$\frac{3r}{r+3}-\frac{5r}{r-3}+\frac{2}{r^2-9}$$
$$=\frac{3r(r-3)}{(r+3)(r-3)}-\frac{5r(r+3)}{(r-3)(r+3)}+\frac{2}{(r+3)(r-3)}$$
$$=\frac{3r^2-9r}{(r+3)(r-3)}-\frac{5r^2+15r}{(r+3)(r-3)}+\frac{2}{(r+3)(r-3)}$$
$$=\frac{3r^2-9r-5r^2-15r+2}{(r+3)(r-3)}$$
$$=\frac{-2r^2-24r+2}{(r+3)(r-3)}$$
$$=\frac{-2r^2-24r+2}{r^2-9}$$

79. LCD = $u - v$

$$\frac{v}{u-v}-\frac{3v}{v-u}=\frac{v}{u-v}-\frac{3v}{-1(u-v)}$$
$$=\frac{v}{u-v}+\frac{3v}{(u-v)}$$
$$=\frac{4v}{u-v}$$

81. LCD = $m - n$

$$\frac{2m-n}{m-n}-\frac{m-3n}{n-m}=\frac{2m-n}{m-n}-\frac{m-3n}{-1(m-n)}$$
$$=\frac{2m-n}{m-n}+\frac{m-3n}{(m-n)}$$
$$=\frac{3m-4n}{m-n}$$

83. LCD = $3a - b$

$$\frac{2a-3b}{3a-b}+\frac{3a+2b}{b-3a}=\frac{2a-3b}{3a-b}+\frac{3a+2b}{-1(3a-b)}$$
$$=\frac{2a-3b}{3a-b}-\frac{3a+2b}{(3a-b)}$$
$$=\frac{2a-3b-3a-2b}{3a-b}$$
$$=\frac{-a-5b}{3a-b}$$

85. Mistake: Added denominators instead of finding the LCD.

Correct: $\frac{4(5)}{a(5)}+\frac{a(a)}{5(a)}=\frac{20}{5a}+\frac{a^2}{5a}=\frac{a^2+20}{5a}$

87. Mistake: Added denominators.

Correct: $\frac{9v}{2x}$

89. Mistake: Did not distribute the subtraction sign to both terms in $2c + 3$.

Correct: $\frac{5c+2-2c-3}{3c-5}=\frac{3c-1}{3c-5}$

91. $f(x)=\frac{9x+7}{6x}$ and $g(x)=\frac{x-3}{6x}$

a) $$f(x)+g(x)=\frac{9x+7}{6x}+\frac{x-3}{6x}$$
$$=\frac{10x+4}{6x}$$
$$=\frac{\cancel{2}(5x+2)}{\cancel{2}\cdot 3x}$$
$$=\frac{5x+2}{3x}$$

b) $$f(x)-g(x)=\frac{9x+7}{6x}-\frac{x-3}{6x}$$
$$=\frac{9x+7-x+3}{6x}$$
$$=\frac{8x+10}{6x}$$
$$=\frac{\cancel{2}(4x+5)}{\cancel{2}\cdot 3x}$$
$$=\frac{4x+5}{3x}$$

c) Using part a.

$$(f+g)(4)=\frac{5(4)+2}{3(4)}=\frac{22}{12}=\frac{11}{6}$$

d) Using part b.

$$(f-g)(-2)=\frac{4(-2)+5}{3(-2)}=\frac{-3}{-6}=\frac{1}{2}$$

93. $f(x)=\frac{x-14}{x^2-4}$ and $g(x)=\frac{x+1}{x-2}$

$$=\frac{x-14}{(x+2)(x-2)}$$

a) $$f(x)+g(x)=\frac{x-14}{(x+2)(x-2)}+\frac{(x+1)(x+2)}{(x-2)(x+2)}$$
$$=\frac{x-14}{(x+2)(x-2)}+\frac{x^2+3x+2}{(x-2)(x+2)}$$
$$=\frac{x^2+4x-12}{(x+2)(x-2)}$$
$$=\frac{(x+6)(x-2)}{(x+2)(x-2)}$$
$$=\frac{x+6}{x+2}$$

b) $$f(x)-g(x)=\frac{x-14}{(x+2)(x-2)}-\frac{(x+1)(x+2)}{(x-2)(x+2)}$$
$$=\frac{x-14}{(x+2)(x-2)}-\frac{x^2+3x+2}{(x-2)(x+2)}$$
$$=\frac{x-14-x^2-3x-2}{(x+2)(x-2)}$$
$$=\frac{-x^2-2x-16}{(x+2)(x-2)}$$
$$=-\frac{x^2+2x+16}{x^2-4} \text{ or } \frac{x^2+2x+16}{4-x^2}$$

c) Using part a.

$$(f+g)(1)=\frac{1+6}{1+2}=\frac{7}{3}$$

d) Using part b.

$$(f-g)(-1)=-\frac{(-1)^2+2(-1)+16}{(-1)^2-4}$$
$$=-\frac{1-2+16}{1-4}$$
$$=-\frac{15}{-3}$$
$$=5$$

95. $\frac{t}{5}+\frac{t}{3}=\frac{t(3)}{5(3)}+\frac{t(5)}{3(5)}=\frac{3t}{15}+\frac{5t}{15}=\frac{8t}{15}$

97. perimeter = 2(length) + 2(width)

$$2\left(\frac{3}{x+4}\right)+2\left(\frac{1}{x-2}\right)$$
$$=\frac{6}{x+4}+\frac{2}{x-2}$$
$$=\frac{6(x-2)}{(x+4)(x-2)}+\frac{2(x+4)}{(x-2)(x+4)}$$
$$=\frac{6x-12}{(x+4)(x-2)}+\frac{2x+8}{(x-2)(x+4)}$$
$$=\frac{8x-4}{(x+4)(x-2)}$$

99. Let x represent the number, then $\left(\frac{1}{x}\right)$ represents the reciprocal.

$$3x+2\left(\frac{1}{x}\right)=3x+\frac{2}{x}=\frac{3x(x)}{1(x)}+\frac{2}{x}=\frac{3x^2+2}{x}$$

101. $\left(\frac{x}{5}+\frac{x}{3}\right)\div 2=\left(\frac{x(3)}{5(3)}+\frac{x(5)}{3(5)}\right)\div 2$

$$=\left(\frac{3x}{15}+\frac{5x}{15}\right)\cdot\frac{1}{2}$$
$$=\frac{8x}{15}\cdot\frac{1}{2}$$
$$=\frac{8x}{30}$$
$$=\frac{4x}{15}$$

Review Exercises

1. $\frac{1}{3}+\frac{3}{4}=\frac{1(4)}{3(4)}+\frac{3(3)}{4(3)}=\frac{4}{12}+\frac{9}{12}=\frac{13}{12}$

2. $\frac{5}{12}\div\frac{3}{8}=\frac{5}{12}\cdot\frac{8}{3}=\frac{40}{36}=\frac{10}{9}$

3. $\left(\frac{2}{3}+\frac{1}{2}\right)\div\left(\frac{3}{4}-\frac{1}{3}\right)$

$$=\left(\frac{2(2)}{3(2)}+\frac{1(3)}{2(3)}\right)\div\left(\frac{3(3)}{4(3)}-\frac{1(4)}{3(4)}\right)$$
$$=\left(\frac{4}{6}+\frac{3}{6}\right)\div\left(\frac{9}{12}-\frac{4}{12}\right)$$
$$=\frac{7}{6}\div\frac{5}{12}$$
$$=\frac{7}{\cancel{6}}\cdot\frac{\cancel{12}^{2}}{5}$$
$$=\frac{14}{5}$$

4. $12\left(\frac{2}{3}+\frac{5}{6}\right)=\frac{12\cdot 2}{3}+\frac{12\cdot 5}{6}$

$$=\frac{24}{3}+\frac{60}{6}$$
$$=8+10$$
$$=18$$

5. $\frac{x^2-9x+8}{x^2-6x-16}=\frac{\cancel{(x-8)}(x-1)}{\cancel{(x-8)}(x+2)}=\frac{x-1}{x+2}$

6. $\frac{3y+2}{27y^3+8}=\frac{3y+2}{(3y+2)(9y^2-6y+4)}$

$$=\frac{1}{9y^2-6y+4}$$

Exercise Set 7.3

1. $(x-2)(x-3)$

3. $\frac{x+2}{x-4}\div\frac{x-3}{x+5}$

5. Method 2 because the numerator and denominator are not monomials.

7. $\dfrac{\frac{5}{21}}{\frac{3}{14}} = \dfrac{5}{{}_3\cancel{21}}\cdot\dfrac{\cancel{14}^2}{3} = \dfrac{10}{9}$

9. $\dfrac{\frac{u}{v^3}}{\frac{w}{v^4}} = \dfrac{u}{{}_1\cancel{v^3}}\cdot\dfrac{\cancel{v^4}^{v}}{w} = \dfrac{uv}{w}$

11. $\dfrac{\frac{u^7v^2}{w^5}}{\frac{u^3v^4}{w}} = \dfrac{\cancel{u^7}^{u^4}\cancel{v^2}}{{}_{w^4}\cancel{w^5}}\cdot\dfrac{\cancel{w}^1}{\cancel{u^3}\cancel{v^4}_{v^2}} = \dfrac{u^4}{w^4v^2}$

13. $\dfrac{\frac{15}{8}}{20} = \dfrac{\cancel{15}^3}{8}\cdot\dfrac{1}{\cancel{20}_4} = \dfrac{3}{32}$

15. $\dfrac{a}{\frac{b}{c}} = \dfrac{a}{1}\cdot\dfrac{c}{b} = \dfrac{ac}{b}$

17. $\dfrac{\frac{16}{2x-2}}{\frac{8}{x-3}} = \dfrac{\cancel{16}}{\cancel{2}(x-1)}\cdot\dfrac{x-3}{\cancel{8}} = \dfrac{x-3}{x-1}$

19. $\dfrac{\frac{x+4}{6}}{\frac{2x+8}{18}} = \dfrac{\cancel{(x+4)}}{\cancel{6}}\cdot\dfrac{\cancel{18}^3}{2\cancel{(x+4)}} = \dfrac{3}{2}$

21. $\dfrac{\frac{3x^2}{x+2}}{\frac{2x}{x+2}} = \dfrac{3x^2}{\cancel{(x+2)}}\cdot\dfrac{\cancel{(x+2)}}{2x} = \dfrac{3x^2}{2x} = \dfrac{3x}{2}$

23. $\dfrac{\frac{4}{9}-\frac{1}{3}}{\frac{7}{12}-\frac{5}{18}} = \dfrac{\frac{4}{9}-\frac{3}{9}}{\frac{21}{36}-\frac{10}{36}} = \dfrac{\frac{1}{9}}{\frac{11}{36}} = \dfrac{1}{{}_1\cancel{9}}\cdot\dfrac{\cancel{36}^4}{11} = \dfrac{4}{11}$

25. $\dfrac{1+\frac{x}{2}}{1-\frac{x}{2}} = \dfrac{\frac{2}{2}+\frac{x}{2}}{\frac{2}{2}-\frac{x}{2}} = \dfrac{\frac{2+x}{2}}{\frac{2-x}{2}} = \dfrac{2+x}{\cancel{2}}\cdot\dfrac{\cancel{2}}{2-x} = \dfrac{2+x}{2-x}$

27. $\dfrac{\left(\frac{1}{t}-1\right)\cdot t^3}{\left(\frac{1}{t^3}-1\right)\cdot t^3} = \dfrac{t^2-t^3}{1-t^3}$

$= \dfrac{t^2\cancel{(1-t)}}{\cancel{(1-t)}\left(1+t+t^2\right)}$

$= \dfrac{t^2}{1+t+t^2}$

29. $\dfrac{\left(1-\frac{2}{v}-\frac{3}{v^2}\right)\cdot v^2}{\left(1-\frac{4}{v}+\frac{3}{v^2}\right)\cdot v^2} = \dfrac{v^2-2v-3}{v^2-4v+3}$

$= \dfrac{(v+1)\cancel{(v-3)}}{(v-1)\cancel{(v-3)}}$

$= \dfrac{v+1}{v-1}$

31. $\dfrac{\left(\frac{2}{x}-\frac{7}{x^2}-\frac{30}{x^3}\right)\cdot x^3}{\left(2+\frac{11}{x}+\frac{15}{x^2}\right)\cdot x^3} = \dfrac{2x^2-7x-30}{2x^3+11x^2+15x}$

$= \dfrac{(2x+5)(x-6)}{x\left(2x^2+11x+15\right)}$

$= \dfrac{\cancel{(2x+5)}(x-6)}{x\cancel{(2x+5)}(x+3)}$

$= \dfrac{x-6}{x(x+3)}$

33. $\dfrac{\left(5+\frac{4}{r+8}\right)\cdot(r+8)}{(r+8)\cdot(r+8)} = \dfrac{5(r+8)+4}{(r+8)^2}$

$= \dfrac{5r+40+4}{(r+8)^2}$

$= \dfrac{5r+44}{(r+8)^2}$

35. $$\frac{\left(t+\frac{9}{t-7}\right)\cdot(t-7)}{\left(11-\frac{3}{t-7}\right)\cdot(t-7)}=\frac{t(t-7)+9}{11(t-7)-3}$$
$$=\frac{t^2-7t+9}{11t-77-3}$$
$$=\frac{t^2-7t+9}{11t-80}$$

37. $$\frac{\left(\frac{x+4}{(x+3)(x-3)}\right)\cdot(x+3)(x-3)}{\left(2+\frac{1}{x+3}\right)\cdot(x+3)(x-3)}$$
$$=\frac{x+4}{2(x+3)(x-3)+1(x-3)}$$
$$=\frac{x+4}{2x^2-18+x-3}$$
$$=\frac{x+4}{2x^2+x-21}$$

39. $$\frac{\left(\frac{5}{a+3}-\frac{4}{a-3}\right)\cdot(a+3)(a-3)}{\left(\frac{3}{a+3}-\frac{2}{a-3}\right)\cdot(a+3)(a-3)}$$
$$=\frac{5(a-3)-4(a+3)}{3(a-3)-2(a+3)}$$
$$=\frac{5a-15-4a-12}{3a-9-2a-6}$$
$$=\frac{a-27}{a-15}$$

41. $$\frac{\left(\frac{r+3}{r-3}-\frac{r-3}{r+3}\right)\cdot(r-3)(r+3)}{\left(\frac{r+3}{r-3}+\frac{r-3}{r+3}\right)\cdot(r-3)(r+3)}$$
$$=\frac{(r+3)(r+3)-(r-3)(r-3)}{(r+3)(r+3)+(r-3)(r-3)}$$
$$=\frac{\left(r^2+6r+9\right)-\left(r^2-6r+9\right)}{\left(r^2+6r+9\right)+\left(r^2-6r+9\right)}$$
$$=\frac{r^2+6r+9-r^2+6r-9}{2r^2+18}$$
$$=\frac{12r}{2r^2+18}$$
$$=\frac{\cancel{2}(6r)}{\cancel{2}\left(r^2+9\right)}$$
$$=\frac{6r}{r^2+9}$$

43. $$\frac{\left(\frac{3x-9}{x-3}-\frac{x-3}{x-5}\right)\cdot(x-3)(x-5)}{\left(\frac{x+5}{x-5}+\frac{2x-6}{x-3}\right)\cdot(x-3)(x-5)}$$
$$=\frac{(3x-9)(x-5)-(x-3)(x-3)}{(x+5)(x-3)+(2x-6)(x-5)}$$
$$=\frac{\left(3x^2-24x+45\right)-\left(x^2-6x+9\right)}{\left(x^2+2x-15\right)+\left(2x^2-16x+30\right)}$$
$$=\frac{3x^2-24x+45-x^2+6x-9}{3x^2-14x+15}$$
$$=\frac{2x^2-18x+36}{3x^2-14x+15}$$
$$=\frac{2\left(x^2-9x+18\right)}{(3x-5)(x-3)}$$
$$=\frac{2(x-6)\cancel{(x-3)}}{(3x-5)\cancel{(x-3)}}$$
$$=\frac{2(x-6)}{3x-5}$$

45. $$\frac{\left(\frac{2}{y^2}-\frac{5}{xy}-\frac{3}{x^2}\right)\cdot\left(x^2y^2\right)}{\left(\frac{2}{y^2}+\frac{5}{xy}+\frac{2}{x^2}\right)\cdot\left(x^2y^2\right)}=\frac{2x^2-5xy-3y^2}{2x^2+5xy+2y^2}$$
$$=\frac{\cancel{(2x+y)}(x-3y)}{\cancel{(2x+y)}(x+2y)}$$
$$=\frac{x-3y}{x+2y}$$

47. $$\frac{\left(\frac{2a}{a+6}+\frac{1}{a}\right)\cdot a(a+6)}{\left(\frac{9a}{1}-\frac{3}{a+6}\right)\cdot a(a+6)}=\frac{2a\cdot a+1(a+6)}{9a\cdot a(a+6)-3a}$$
$$=\frac{2a^2+a+6}{9a^3+54a^2-3a}$$

49. $$\frac{3x^{-1}}{3x^{-1}+1}=\frac{\frac{3}{x}}{\frac{3}{x}+1}=\frac{\left(\frac{3}{x}\right)\cdot x}{\left(\frac{3}{x}+1\right)\cdot x}=\frac{3}{3+x}$$

51. $$\frac{1-9x^{-2}}{1+3x^{-1}}=\frac{1-\frac{9}{x^2}}{1+\frac{3}{x}}$$
$$=\frac{\left(1-\frac{9}{x^2}\right)\cdot x^2}{\left(1+\frac{3}{x}\right)\cdot x^2}$$
$$=\frac{x^2-9}{x^2+3x}$$
$$=\frac{\cancel{(x+3)}(x-3)}{x\cancel{(x+3)}}$$
$$=\frac{x-3}{x}$$

53. $$\frac{3x^{-2}+5y^{-1}}{x^{-1}+y^{-1}}=\frac{\frac{3}{x^2}+\frac{5}{y}}{\frac{1}{x}+\frac{1}{y}}$$
$$=\frac{\left(\frac{3}{x^2}+\frac{5}{y}\right)\cdot x^2y}{\left(\frac{1}{x}+\frac{1}{y}\right)\cdot x^2y}$$
$$=\frac{3y+5x^2}{xy+x^2}$$

55. $$\frac{x^{-2}y^{-2}}{4x^{-1}+3y^{-1}}=\frac{\frac{1}{x^2y^2}}{\frac{4}{x}+\frac{3}{y}}$$
$$=\frac{\left(\frac{1}{x^2y^2}\right)\cdot x^2y^2}{\left(\frac{4}{x}+\frac{3}{y}\right)\cdot x^2y^2}$$
$$=\frac{1}{4xy^2+3x^2y}$$

57. $$\frac{36a^{-2}-25b^{-2}}{6a^{-1}+5b^{-1}}=\frac{\frac{36}{a^2}-\frac{25}{b^2}}{\frac{6}{a}+\frac{5}{b}}$$
$$=\frac{\left(\frac{36}{a^2}-\frac{25}{b^2}\right)\cdot a^2b^2}{\left(\frac{6}{a}+\frac{5}{b}\right)\cdot a^2b^2}$$
$$=\frac{36b^2-25a^2}{6ab^2+5a^2b}$$
$$=\frac{\cancel{(6b+5a)}(6b-5a)}{ab\cancel{(6b+5a)}}$$
$$=\frac{6b-5a}{ab}$$

59. $\dfrac{(4a)^{-1}+2b^{-2}}{2a^{-1}+b^{-2}}=\dfrac{\dfrac{1}{4a}+\dfrac{2}{b^2}}{\dfrac{2}{a}+\dfrac{1}{b^2}}$

$=\dfrac{\left(\dfrac{1}{4a}+\dfrac{2}{b^2}\right)\cdot 4ab^2}{\left(\dfrac{2}{a}+\dfrac{1}{b^2}\right)\cdot 4ab^2}$

$=\dfrac{b^2+8a}{8b^2+4a}$

61. Mistake: The +1 was omitted in the multiplication of $\dfrac{1}{3}\cdot 3$.

Correct: $\dfrac{a+\dfrac{1}{3}}{b+\dfrac{1}{3}}=\dfrac{\left(a+\dfrac{1}{3}\right)\cdot 3}{\left(b+\dfrac{1}{3}\right)\cdot 3}=\dfrac{3a+1}{3b+1}$

63. Mistake: Did not multiply n by n.

Correct: $\dfrac{n+\dfrac{3}{n}}{\dfrac{3}{n}}=\dfrac{\left(n+\dfrac{3}{n}\right)\cdot n}{\left(\dfrac{3}{n}\right)\cdot n}=\dfrac{n^2+3}{3}$

65. Average rate $=\dfrac{20+20}{\dfrac{1}{3}+\dfrac{1}{4}}$

$=\dfrac{(40)\cdot 12}{\left(\dfrac{1}{3}+\dfrac{1}{4}\right)\cdot 12}$

$=\dfrac{480}{4+3}$

$=\dfrac{480}{7}$

≈ 68.6 mph

67. $w=\dfrac{A}{l}=\dfrac{\dfrac{3x^2-11x-4}{18}}{\dfrac{x-4}{2}}$

$=\dfrac{\left(\dfrac{3x^2-11x-4}{18}\right)\cdot 18}{\left(\dfrac{x-4}{2}\right)\cdot 18}$

$=\dfrac{3x^2-11x-4}{9x-36}$

$=\dfrac{(3x+1)\cancel{(x-4)}}{9\cancel{(x-4)}}$

$=\dfrac{3x+1}{9}$ in.

69. a) $\dfrac{1}{\dfrac{1}{R_1}+\dfrac{1}{R_2}}=\dfrac{1\cdot R_1R_2}{\left(\dfrac{1}{R_1}+\dfrac{1}{R_2}\right)\cdot R_1R_2}=\dfrac{R_1R_2}{R_2+R_1}$

b) $\dfrac{R_1R_2}{R_2+R_1}=\dfrac{60\cdot 40}{60+40}=\dfrac{2400}{100}=24$ ohms

c) $\dfrac{R_1R_2}{R_2+R_1}=\dfrac{40\cdot 40}{40+40}=\dfrac{1600}{80}=20$ ohms

Review Exercises

1. $3x+5=2$
$3x=-3$
$x=-1$

2. $4x-6=2x-14$
$2x-6=-14$
$2x=-8$
$x=-4$

3. $3x^2+7x=0$
$x(3x+7)=0$
$x=0$ or $3x+7=0$
$x=0$ $\quad x=-\dfrac{7}{3}$

4. $2x^2-7x-15=0$
$(2x+3)(x-5)=0$
$2x+3=0$ or $x-5=0$
$x=-\dfrac{3}{2}$ $\quad x=5$

5. $d = rt$

$2.6 = r \cdot \frac{3}{4}$

$\frac{4}{3} \cdot \frac{13}{5} = r \cdot \frac{3}{4} \cdot \frac{4}{3}$

$\frac{52}{15} = r$

$3.4\overline{6} = r$

6.

	Rate in mph	Time	Distance
Northbound	40	t	$40t$
Southbound	60	t	$60t$

$40t + 60t = 20$

$100t = 20$

$t = 0.2$ hr. or 12 min.

Exercise Set 7.4

1. Adding rational expressions yields another rational expression. Solving an equation involving a rational expression yields a numerical value for the variable.

3. Multiply by the LCD, $(x+1)(x-1)(x+2)$.

5. Possible extraneous solutions are 1, –2, –1. These are the values of x that cause expressions in the equation to be undefined.

7. The value of x that causes expressions in the equation to be undefined is 4.

9. The values of x that cause expressions in the equation to be undefined are 2 and –6.

11. $x^2 - 4 = (x+2)(x-2)$

$x^2 + 5x + 6 = (x+3)(x+2)$

$x^2 + x - 6 = (x+3)(x-2)$

The values of x that cause expressions in the equation to be undefined are 2, –2, and –3.

13. $\frac{4}{3u} - \frac{1}{2u} = \frac{5}{6}$

$6u \cdot \left(\frac{4}{3u} - \frac{1}{2u}\right) = 6u \cdot \frac{5}{6}$

$8 - 3 = 5u$

$5 = 5u$

$1 = u$

Check: $\frac{4}{3} - \frac{1}{2} \stackrel{?}{=} \frac{5}{6}$

$\frac{8}{6} - \frac{3}{6} \stackrel{?}{=} \frac{5}{6}$

$\frac{5}{6} = \frac{5}{6}$

15. $\frac{5}{4w} - \frac{3}{5w} = -\frac{13}{40}$

$40w \cdot \left(\frac{5}{4w} - \frac{3}{5w}\right) = 40w \cdot \left(-\frac{13}{40}\right)$

$50 - 24 = -13w$

$26 = -13w$

$-2 = w$

Check: $\frac{5}{-8} - \frac{3}{-10} \stackrel{?}{=} -\frac{13}{40}$

$-\frac{25}{40} + \frac{12}{40} \stackrel{?}{=} -\frac{13}{40}$

$-\frac{13}{40} = -\frac{13}{40}$

17. $\frac{7}{x-3} = 8 - \frac{1}{x-3}$

$(x-3)\left(\frac{7}{x-3}\right) = (x-3)\left(8 - \frac{1}{x-3}\right)$

$7 = 8(x-3) - 1$

$7 = 8x - 24 - 1$

$7 = 8x - 25$

$32 = 8x$

$4 = x$

Check: $\frac{7}{1} \stackrel{?}{=} 8 - \frac{1}{1}$

$7 \stackrel{?}{=} 8 - 1$

$7 = 7$

19. $$\frac{t}{t+4} = 3 - \frac{12}{t+4}$$
$$(t+4)\left(\frac{t}{t+4}\right) = (t+4)\left(3 - \frac{12}{t+4}\right)$$
$$t = 3(t+4) - 12$$
$$t = 3t + 12 - 12$$
$$t - 3t = 0$$
$$-2t = 0$$
$$t = 0$$

Check: $\frac{0}{4} \stackrel{?}{=} 3 - \frac{12}{4}$

$0 \stackrel{?}{=} 3 - 3$

$0 = 0$

21. $$\frac{6a}{a+5} = \frac{3}{a+5} + 3$$
$$(a+5)\left(\frac{6a}{a+5}\right) = (a+5)\left(\frac{3}{a+5} + 3\right)$$
$$6a = 3 + 3(a+5)$$
$$6a = 3 + 3a + 15$$
$$6a = 3a + 18$$
$$3a = 18$$
$$a = 6$$

Check: $\frac{36}{11} \stackrel{?}{=} \frac{3}{11} + 3$

$\frac{36}{11} \stackrel{?}{=} \frac{3}{11} + \frac{33}{11}$

$\frac{36}{11} = \frac{36}{11}$

23. $$\frac{4x}{2x-3} = 3 + \frac{6}{2x-3}$$
$$(2x-3)\left(\frac{4x}{2x-3}\right) = (2x-3)\left(3 + \frac{6}{2x-3}\right)$$
$$4x = 3(2x-3) + 6$$
$$4x = 6x - 9 + 6$$
$$4x = 6x - 3$$
$$-2x = -3$$
$$x = \frac{3}{2}$$

Notice that if $x = \frac{3}{2}$, then $\frac{4x}{2x-3}$ and $\frac{6}{2x-3}$ are undefined. Therefore, $x = \frac{3}{2}$ is extraneous and the equation has no solution.

25. $$\frac{3x}{x+3} = \frac{6}{x+7}$$
$$3x(x+7) = 6(x+3)$$
$$3x^2 + 21x = 6x + 18$$
$$3x^2 + 15x - 18 = 0$$
$$3(x^2 + 5x - 6) = 0$$
$$3(x-1)(x+6) = 0$$
$x - 1 = 0$ or $x + 6 = 0$

$x = 1$ $\qquad$ $x = -6$

Check $x = 1$:

$\frac{3(1)}{1+3} \stackrel{?}{=} \frac{6}{1+7}$

$\frac{3}{4} \stackrel{?}{=} \frac{6}{8}$

$\frac{3}{4} = \frac{3}{4}$

Check $x = -6$:

$\frac{3(-6)}{-6+3} \stackrel{?}{=} \frac{6}{-6+7}$

$\frac{-18}{-3} \stackrel{?}{=} \frac{6}{1}$

$6 = 6$

27. $$\frac{m}{m-1} = \frac{3m}{4m-3}$$
$$(4m-3)\cdot m = (m-1)\cdot 3m$$
$$4m^2 - 3m = 3m^2 - 3m$$
$$m^2 = 0$$
$$m = 0$$

Check: $\frac{0}{-1} \stackrel{?}{=} \frac{0}{-3}$

$0 = 0$

29.
$$\frac{2y}{3y-6}+\frac{3}{4y-8}=\frac{1}{4}$$
$$12(y-2)\left(\frac{2y}{3(y-2)}+\frac{3}{4(y-2)}\right)=12(y-2)\left(\frac{1}{4}\right)$$
$$8y+9=3y-6$$
$$5y+9=-6$$
$$5y=-15$$
$$y=-3$$

Check: $\frac{-6}{-15}+\frac{3}{-20}\overset{?}{=}\frac{1}{4}$
$$\frac{24}{60}-\frac{9}{60}\overset{?}{=}\frac{1}{4}$$
$$\frac{15}{60}\overset{?}{=}\frac{1}{4}$$
$$\frac{1}{4}=\frac{1}{4}$$

31.
$$\frac{a^2}{a+2}-3=\frac{a+6}{a+2}-4$$
$$(a+2)\left(\frac{a^2}{a+2}-3\right)=(a+2)\left(\frac{a+6}{a+2}-4\right)$$
$$a^2-3(a+2)=(a+6)-4(a+2)$$
$$a^2-3a-6=a+6-4a-8$$
$$a^2-3a-6=-3a-2$$
$$a^2-4=0$$
$$(a+2)(a-2)=0$$
$$a+2=0 \quad\text{or}\quad a-2=0$$
$$a=-2 \qquad\qquad a=2$$

Notice that if $a=-2$, then $\frac{a^2}{a+2}$ and $\frac{a+6}{a+2}$ are undefined. Therefore, $a=-2$ is extraneous, and 2 is the only solution.

Check: $\frac{4}{4}-3\overset{?}{=}\frac{8}{4}-4$
$$1-3\overset{?}{=}2-4$$
$$-2=-2$$

33.
$$2x-\frac{12}{x}=5$$
$$x\left(2x-\frac{12}{x}\right)=x\cdot 5$$
$$2x^2-12=5x$$
$$2x^2-5x-12=0$$
$$(2x+3)(x-4)=0$$
$$2x+3=0 \quad\text{or}\quad x-4=0$$
$$2x=-3 \qquad\qquad x=4$$
$$x=-\frac{3}{2}$$

Check $x=-\frac{3}{2}$:
$$2\left(-\frac{3}{2}\right)-\frac{12}{\left(-\frac{3}{2}\right)}\overset{?}{=}5$$
$$-3-(-8)\overset{?}{=}5$$
$$5=5$$

Check $x=4$:
$$2(4)-\frac{12}{(4)}\overset{?}{=}5$$
$$8-3\overset{?}{=}5$$
$$5=5$$

35.
$$\frac{x+1}{x+2}+\frac{5}{2x}=\frac{3x+2}{x+2}$$
$$2x(x+2)\left(\frac{x+1}{x+2}+\frac{5}{2x}\right)=2x(x+2)\left(\frac{3x+2}{x+2}\right)$$
$$2x(x+1)+5(x+2)=2x(3x+2)$$
$$2x^2+2x+5x+10=6x^2+4x$$
$$-4x^2+3x+10=0$$
$$4x^2-3x-10=0$$
$$(4x+5)(x-2)=0$$
$$4x+5=0 \quad\text{or}\quad x-2=0$$
$$4x=-5 \qquad\qquad x=2$$
$$x=-\frac{5}{4}$$

Check $x = -\frac{5}{4}$:

$$\frac{-\frac{5}{4}+1}{-\frac{5}{4}+2} + \frac{5}{2\left(-\frac{5}{4}\right)} \stackrel{?}{=} \frac{-\frac{15}{4}+2}{-\frac{5}{4}+2}$$

$$\frac{-\frac{1}{4}}{\frac{3}{4}} + \frac{5}{-\frac{5}{2}} \stackrel{?}{=} \frac{-\frac{7}{4}}{\frac{3}{4}}$$

$$-\frac{1}{3} - 2 \stackrel{?}{=} -\frac{7}{3}$$

$$-\frac{1}{3} - \frac{6}{3} \stackrel{?}{=} -\frac{7}{3}$$

$$-\frac{7}{3} = -\frac{7}{3}$$

Check $x = 2$:

$$\frac{3}{4} + \frac{5}{4} \stackrel{?}{=} \frac{8}{4}$$

$$2 = 2$$

37. $\dfrac{2x+1}{x+2} - \dfrac{x+1}{3x-2} = \dfrac{x}{x+2}$

$$(x+2)(3x-2)\left(\frac{2x+1}{x+2} - \frac{x+1}{3x-2}\right) = (x+2)(3x-2)\left(\frac{x}{x+2}\right)$$

$$(3x-2)(2x+1) - (x+2)(x+1) = (3x-2)(x)$$

$$6x^2 - x - 2 - x^2 - 3x - 2 = 3x^2 - 2x$$

$$2x^2 - 2x - 4 = 0$$

$$2\left(x^2 - x - 2\right) = 0$$

$$2(x-2)(x+1) = 0$$

$$x - 2 = 0 \quad \text{or} \quad x + 1 = 0$$

$$x = 2 \qquad\qquad x = -1$$

Check $x = 2$: $\dfrac{5}{4} - \dfrac{3}{4} \stackrel{?}{=} \dfrac{2}{4}$

$$\frac{1}{2} = \frac{1}{2}$$

Check $x = -1$: $\dfrac{-1}{1} - \dfrac{0}{-5} \stackrel{?}{=} \dfrac{-1}{1}$

$$-1 - 0 \stackrel{?}{=} -1$$

$$-1 = -1$$

39. $\dfrac{4}{x+5} - \dfrac{2}{x-5} = \dfrac{4x-20}{x^2-25}$

$$(x+5)(x-5)\left(\frac{4}{x+5} - \frac{2}{x-5}\right) = (x+5)(x-5)\left(\frac{4x-20}{(x+5)(x-5)}\right)$$

$$4(x-5) - 2(x+5) = 4x - 20$$

$$4x - 20 - 2x - 10 = 4x - 20$$

$$2x - 30 = 4x - 20$$

$$-2x = 10$$

$$x = -5$$

Notice that if $x = -5$, then $\dfrac{4}{x+5}$ and $\dfrac{4x-20}{x^2-25}$ are undefined. Therefore, $x = -5$ is extraneous and the equation has no solution.

41. $\dfrac{x}{x-2} - \dfrac{4}{x-1} = \dfrac{2}{x^2-3x+2}$

$$(x-2)(x-1)\left(\frac{x}{x-2} - \frac{4}{x-1}\right) = (x-2)(x-1)\left(\frac{2}{(x-2)(x-1)}\right)$$

$$(x-1)(x) - (x-2)(4) = 2$$

$$x^2 - x - 4x + 8 = 2$$

$$x^2 - 5x + 6 = 0$$

$$(x-2)(x-3) = 0$$

$$x - 2 = 0 \quad \text{or} \quad x - 3 = 0$$

$$x = 2 \qquad\qquad x = 3$$

Notice that if $x = 2$, then $\dfrac{x}{x-2}$ and $\dfrac{2}{x^2-3x+2}$ are undefined. Therefore, $x = 2$ is extraneous, and the only solution is $x = 3$.

Check: $\dfrac{3}{1} - \dfrac{4}{2} \stackrel{?}{=} \dfrac{2}{2}$

$$3 - 2 \stackrel{?}{=} 1$$

$$1 = 1$$

43. $\dfrac{6}{t^2+t-12}=\dfrac{4}{t^2-t-6}+\dfrac{1}{t^2+6t+8}$

$$(t+4)(t-3)(t+2)\left(\frac{6}{(t+4)(t-3)}\right)$$
$$=(t+4)(t-3)(t+2)$$
$$\cdot\left(\frac{4}{(t-3)(t+2)}+\frac{1}{(t+4)(t+2)}\right)$$
$$6(t+2)=4(t+4)+1(t-3)$$
$$6t+12=4t+16+t-3$$
$$6t+12=5t+13$$
$$t=1$$

Check: $\dfrac{6}{-10}\stackrel{?}{=}\dfrac{4}{-6}+\dfrac{1}{15}$

$$-\frac{3}{5}\stackrel{?}{=}-\frac{10}{15}+\frac{1}{15}$$
$$-\frac{3}{5}\stackrel{?}{=}-\frac{9}{15}$$
$$-\frac{3}{5}=-\frac{3}{5}$$

45. $\dfrac{5}{x^2-x-6}-\dfrac{8}{x^2+2x-15}=\dfrac{3}{x^2+7x+10}$

$$(x-3)(x+2)(x+5)$$
$$\cdot\left(\frac{5}{(x-3)(x+2)}-\frac{8}{(x+5)(x-3)}\right)$$
$$=(x-3)(x+2)(x+5)\left(\frac{3}{(x+5)(x+2)}\right)$$
$$5(x+5)-8(x+2)=3(x-3)$$
$$5x+25-8x-16=3x-9$$
$$-3x+9=3x-9$$
$$-6x=-18$$
$$x=3$$

Notice that if $x=3$, then $\dfrac{5}{x^2-x-6}$ and $\dfrac{8}{x^2+2x-15}$ are undefined. Therefore, $x=3$ is extraneous and the equation has no solution.

47. $\dfrac{p+1}{p^2+8p+15}-\dfrac{2}{p^2+7p+10}=\dfrac{1}{p^2+5p+6}$

$$(p+3)(p+5)(p+2)\left(\frac{p+1}{(p+3)(p+5)}-\frac{2}{(p+5)(p+2)}\right)$$
$$=(p+3)(p+5)(p+2)\left(\frac{1}{(p+3)(p+2)}\right)$$
$$(p+2)(p+1)-2(p+3)=(p+5)(1)$$
$$p^2+3p+2-2p-6=p+5$$
$$p^2+p-4=p+5$$
$$p^2=9$$
$$p^2-9=0$$
$$(p+3)(p-3)=0$$
$$p+3=0 \quad \text{or} \quad p-3=0$$
$$p=-3 \qquad\qquad p=3$$

Notice that if $p=-3$, then $\dfrac{p+1}{p^2+8p+15}$ and $\dfrac{1}{p^2+5p+6}$ are undefined. Therefore, $p=-3$ is extraneous, and the only solution is $p=3$.

Check: $\dfrac{4}{48}-\dfrac{2}{40}\stackrel{?}{=}\dfrac{1}{30}$

$$\frac{1}{12}-\frac{1}{20}\stackrel{?}{=}\frac{1}{30}$$
$$\frac{5}{60}-\frac{3}{60}\stackrel{?}{=}\frac{1}{30}$$
$$\frac{2}{60}\stackrel{?}{=}\frac{1}{30}$$
$$\frac{1}{30}=\frac{1}{30}$$

49. $\dfrac{a-5}{a+5}+\dfrac{a+15}{a-5}=2-\dfrac{10}{a^2-25}$

$$(a+5)(a-5)\left(\frac{a-5}{a+5}+\frac{a+15}{a-5}\right)$$
$$=(a+5)(a-5)\left(2-\frac{10}{(a+5)(a-5)}\right)$$
$$(a-5)^2+(a+5)(a+15)=2(a+5)(a-5)-10$$
$$a^2-10a+25+a^2+20a+75=2a^2-50-10$$
$$2a^2+10a+100=2a^2-60$$
$$10a=-160$$
$$a=-16$$

Check: $\dfrac{-21}{-11}+\dfrac{-1}{-21}\stackrel{?}{=}2-\dfrac{10}{231}$

$$\frac{441}{231}+\frac{11}{231}\stackrel{?}{=}\frac{462}{231}-\frac{10}{231}$$
$$\frac{452}{231}=\frac{452}{231}$$

51. $\dfrac{3n}{n^2-2n-15}=\dfrac{2n}{n-5}+\dfrac{n}{n+3}$

$$(n-5)(n+3)\left(\frac{3n}{(n-5)(n+3)}\right)$$
$$=(n-5)(n+3)\left(\frac{2n}{n-5}+\frac{n}{n+3}\right)$$
$$3n=2n(n+3)+n(n-5)$$
$$3n=2n^2+6n+n^2-5n$$
$$0=3n^2-2n$$
$$0=n(3n-2)$$
$$n=0 \quad \text{or} \quad 3n-2=0$$
$$3n=2$$
$$n=\frac{2}{3}$$

Check $n=0$: $\dfrac{0}{-15}\stackrel{?}{=}\dfrac{0}{-5}+\dfrac{0}{3}$

$$0=0$$

Check $n=\dfrac{2}{3}$:

$$\frac{\frac{2}{3}\cdot 2}{-\frac{143}{9}}\ \ \text{i.e.}\ \ \frac{2}{-\frac{143}{9}}\stackrel{?}{=}\frac{\frac{4}{3}}{-\frac{13}{3}}+\frac{\frac{2}{3}}{\frac{11}{3}}$$
$$-\frac{18}{143}\stackrel{?}{=}-\frac{4}{13}+\frac{2}{11}$$
$$-\frac{18}{143}\stackrel{?}{=}-\frac{44}{143}+\frac{26}{143}$$
$$-\frac{18}{143}=-\frac{18}{143}$$

53. $\dfrac{x+5}{x-5}-\dfrac{x-5}{x+5}=\dfrac{40}{x^2-25}$

$$(x+5)(x-5)\left(\frac{x+5}{x-5}-\frac{x-5}{x+5}\right)$$
$$=(x+5)(x-5)\left(\frac{40}{(x+5)(x-5)}\right)$$
$$(x+5)^2-(x-5)^2=40$$
$$x^2+10x+25-x^2+10x-25=40$$
$$20x=40$$
$$x=2$$

Check: $\dfrac{2+5}{2-5}-\dfrac{2-5}{2+5}\stackrel{?}{=}\dfrac{40}{4-25}$

$$\frac{7}{-3}-\frac{-3}{7}\stackrel{?}{=}\frac{40}{-21}$$
$$-\frac{49}{21}+\frac{9}{21}\stackrel{?}{=}-\frac{40}{21}$$
$$-\frac{40}{21}=-\frac{40}{21}$$

55. $$\frac{2}{2x^2-4x-10}=\frac{3}{3x^2+6x-3}$$
$$\frac{2}{2(x^2-2x-5)}=\frac{3}{3(x^2+2x-1)}$$
$$\frac{1}{x^2-2x-5}=\frac{1}{x^2+2x-1}$$
$$x^2+2x-1=x^2-2x-5$$
$$4x-1=-5$$
$$4x=-4$$
$$x=-1$$

Check:
$$\frac{2}{2(-1)^2-4(-1)-10}\stackrel{?}{=}\frac{3}{3(-1)^2+6(-1)-3}$$
$$\frac{2}{2+4-10}\stackrel{?}{=}\frac{3}{3-6-3}$$
$$\frac{2}{-4}\stackrel{?}{=}\frac{3}{-6}$$
$$-\frac{1}{2}=-\frac{1}{2}$$

57. $$1=\frac{3}{a-2}-\frac{12}{a^2-4}$$
$$(a+2)(a-2)(1)$$
$$=(a+2)(a-2)\left(\frac{3}{a-2}-\frac{12}{(a+2)(a-2)}\right)$$
$$a^2-4=3(a+2)-12$$
$$a^2-4=3a+6-12$$
$$a^2-3a+2=0$$
$$(a-2)(a-1)=0$$
$$a-2=0 \quad \text{or} \quad a-1=0$$
$$a=2 \qquad\qquad a=1$$

Notice that if $a=2$, then $\frac{3}{a-2}$ and $\frac{12}{a^2-4}$ are undefined. Therefore, $a=2$ is extraneous, and the only solution is $a=1$.

Check: $$1\stackrel{?}{=}\frac{3}{-1}-\frac{12}{-3}$$
$$1\stackrel{?}{=}-3-(-4)$$
$$1=1$$

59. $$\frac{w+2}{w^2-5w+6}+\frac{2w}{w^2-w-2}=-\frac{2}{w^2-2w-3}$$
$$(w-3)(w-2)(w+1)$$
$$\cdot\left(\frac{w+2}{(w-3)(w-2)}+\frac{2w}{(w-2)(w+1)}\right)$$
$$=(w-3)(w-2)(w+1)\left(-\frac{2}{(w-3)(w+1)}\right)$$
$$(w+2)(w+1)+(2w)(w-3)=-2(w-2)$$
$$w^2+3w+2+2w^2-6w=-2w+4$$
$$3w^2-3w+2=-2w+4$$
$$3w^2-w-2=0$$
$$(3w+2)(w-1)=0$$
$$3w+2=0 \quad \text{or} \quad w-1=0$$
$$3w=-2 \qquad\qquad w=1$$
$$w=-\frac{2}{3}$$

Check $w=-\frac{2}{3}$:
$$\frac{-\frac{2}{3}+2}{\frac{4}{9}-5\left(-\frac{2}{3}\right)+6}+\frac{2\left(-\frac{2}{3}\right)}{\frac{4}{9}+\frac{2}{3}-2}\stackrel{?}{=}-\frac{2}{\frac{4}{9}-2\left(-\frac{2}{3}\right)-3}$$
$$\frac{\frac{4}{3}}{\frac{88}{9}}+\frac{-\frac{4}{3}}{-\frac{8}{9}}\stackrel{?}{=}-\frac{2}{\frac{-11}{9}}$$
$$\frac{3}{22}+\frac{3}{2}\stackrel{?}{=}\frac{18}{11}$$
$$\frac{18}{11}=\frac{18}{11}$$

Check $w=1$:
$$\frac{3}{2}+\frac{2}{-2}\stackrel{?}{=}-\frac{2}{-4}$$
$$\frac{1}{2}=\frac{1}{2}$$

61. $\dfrac{2p}{2p-3}=\dfrac{15-32p^2}{4p^2-9}+\dfrac{3p}{2p+3}$

$$(2p+3)(2p-3)\left(\frac{2p}{2p-3}\right)$$
$$=(2p+3)(2p-3)\cdot\left(\frac{15-32p^2}{(2p+3)(2p-3)}+\frac{3p}{2p+3}\right)$$
$$2p(2p+3)=15-32p^2+3p(2p-3)$$
$$4p^2+6p=15-32p^2+6p^2-9p$$
$$30p^2+15p-15=0$$
$$15\left(2p^2+p-1\right)=0$$
$$15(2p-1)(p+1)=0$$
$2p-1=0$ or $p+1=0$

$2p=1$ $\quad p=-1$

$p=\dfrac{1}{2}$

Check $p=\dfrac{1}{2}$: $\dfrac{1}{1-3}\overset{?}{=}\dfrac{15-32\left(\frac{1}{4}\right)}{4\left(\frac{1}{4}\right)-9}+\dfrac{\frac{3}{2}}{1+3}$

$$-\frac{1}{2}\overset{?}{=}\frac{7}{-8}+\frac{3}{8}$$
$$-\frac{1}{2}\overset{?}{=}-\frac{4}{8}$$
$$-\frac{1}{2}=-\frac{1}{2}$$

Check $p=-1$: $\dfrac{-2}{-2-3}\overset{?}{=}\dfrac{15-32}{4-9}+\dfrac{-3}{-2+3}$

$$\frac{2}{5}\overset{?}{=}\frac{-17}{-5}-3$$
$$\frac{2}{5}\overset{?}{=}\frac{17}{5}-\frac{15}{5}$$
$$\frac{2}{5}=\frac{2}{5}$$

63. $\dfrac{4x}{x^2+4x}-\dfrac{2x}{x^2+2x}=\dfrac{3x-2}{x^2+6x+8}$

$$x(x+2)(x+4)\left(\frac{4x}{x(x+4)}-\frac{2x}{x(x+2)}\right)$$
$$=x(x+2)(x+4)\left(\frac{3x-2}{(x+4)(x+2)}\right)$$
$$4x(x+2)-2x(x+4)=x(3x-2)$$
$$4x^2+8x-2x^2-8x=3x^2-2x$$
$$-x^2+2x=0$$
$$-x(x-2)=0$$
$-x=0$ or $x-2=0$

$x=0$ $\quad x=2$

Notice that if $x=0$, then $\dfrac{4x}{x^2+4x}$ and $\dfrac{2x}{x^2+2x}$ are undefined. Therefore, $x=0$ is extraneous, and the only solution is $x=2$.

Check $x=2$: $\dfrac{8}{12}-\dfrac{4}{8}\overset{?}{=}\dfrac{4}{24}$

$$\frac{16}{24}-\frac{12}{24}\overset{?}{=}\frac{4}{24}$$
$$\frac{4}{24}=\frac{4}{24}$$

65. $$\frac{6}{x^3+2x^2-x-2}=\frac{6}{x+2}-\frac{3}{x^2-1}$$

$$(x+1)(x-1)(x+2)\left(\frac{6}{(x+1)(x-1)(x+2)}\right)$$

$$=(x+1)(x-1)(x+2)\left(\frac{6}{x+2}-\frac{3}{(x+1)(x-1)}\right)$$

$$6=6(x+1)(x-1)-3(x+2)$$

$$6=6x^2-6-3x-6$$

$$0=6x^2-3x-18$$

$$0=3(2x^2-x-6)$$

$$0=3(2x+3)(x-2)$$

$$2x+3=0 \quad \text{or} \quad x-2=0$$

$$2x=-3 \qquad x=2$$

$$x=-\frac{3}{2}$$

Check $x=-\frac{3}{2}$:

$$\frac{6}{\left(-\frac{3}{2}\right)^3+2\left(-\frac{3}{2}\right)^2-\left(-\frac{3}{2}\right)-2}$$

$$\stackrel{?}{=}\frac{6}{\left(-\frac{3}{2}\right)+2}-\frac{3}{\left(-\frac{3}{2}\right)^2-1}$$

$$\frac{48}{5}\stackrel{?}{=}12-\frac{12}{5}$$

$$\frac{48}{5}\stackrel{?}{=}\frac{60}{5}-\frac{12}{5}$$

$$\frac{48}{5}=\frac{48}{5}$$

Check $x=2$:

$$\frac{6}{(2)^3+2(2)^2-(2)-2}\stackrel{?}{=}\frac{6}{(2)+2}-\frac{3}{(2)^2-1}$$

$$\frac{1}{2}\stackrel{?}{=}\frac{3}{2}-1$$

$$\frac{1}{2}=\frac{1}{2}$$

67. $$C=\frac{90,000p}{100-p}$$

$$(100-p)C=(100-p)\left(\frac{90,000p}{100-p}\right)$$

$$100C-pC=90,000p$$

$$100C=90,000p+pC$$

$$100C=p(90,000+C)$$

$$\frac{100C}{90,000+C}=p$$

69. $$I=\frac{2E}{R+2r}$$

$$(R+2r)I=(R+2r)\left(\frac{2E}{R+2r}\right)$$

$$RI+2rI=2E$$

$$2rI=2E-RI$$

$$r=\frac{2E-RI}{2I}$$

71. $$\frac{1}{s}+\frac{1}{S}=\frac{1}{f}$$

$$sSf\left(\frac{1}{s}+\frac{1}{S}\right)=sSf\left(\frac{1}{f}\right)$$

$$Sf+sf=sS$$

$$f(S+s)=sS$$

$$f=\frac{sS}{S+s}$$

73. $$R=\frac{1}{\frac{1}{R_1}+\frac{1}{R_2}}$$

$$R=\frac{1(R_1R_2)}{\left(\frac{1}{R_1}+\frac{1}{R_2}\right)(R_1R_2)}$$

$$R=\frac{R_1R_2}{R_2+R_1}$$

$$R(R_2+R_1)=(R_2+R_1)\frac{R_1R_2}{R_2+R_1}$$

$$RR_2+RR_1=R_1R_2$$

$$RR_2=R_1R_2-RR_1$$

$$RR_2=R_1(R_2-R)$$

$$\frac{RR_2}{R_2-R}=R_1$$

75. 4 is extraneous. There is no solution.

77. Use the formula found in exercise #67.

$$p = \frac{100C}{90,000 + C}$$
$$= \frac{100(22,500)}{90,000 + 22,500}$$
$$= \frac{2,250,000}{112,500}$$
$$= 20\%$$

79.
$$F = \frac{132,000}{330 - s}$$
$$440 = \frac{132,000}{330 - s}$$
$$440(330 - s) = 132,000$$
$$330 - s = 300$$
$$-s = -30$$
$$s = 30$$

The speed is 30 m/sec.

81.
$$T = \frac{80}{x} + \frac{80}{x - 20}$$
$$6 = \frac{80}{x} + \frac{80}{x - 20}$$
$$x(x - 20)(6) = x(x - 20)\left(\frac{80}{x} + \frac{80}{x - 20}\right)$$
$$6x^2 - 120x = 80(x - 20) + 80x$$
$$6x^2 - 120x = 80x - 1600 + 80x$$
$$6x^2 - 280x + 1600 = 0$$
$$2\left(3x^2 - 140x + 800\right) = 0$$
$$2(3x - 20)(x - 40) = 0$$
$$3x - 20 = 0 \quad \text{or} \quad x - 40 = 0$$
$$3x = 20 \qquad x = 40$$
$$x = \frac{20}{3}$$

The solution is 40 mph. Although $\frac{20}{3}$ is also a solution, it is not reasonable because it leads to a time of –6 hours for the return trip.

83.
$$I = \frac{E}{R + r}$$
$$2 = \frac{110}{50 + r}$$
$$2(50 + r) = 110$$
$$100 + 2r = 110$$
$$2r = 10$$
$$r = 5$$

The internal resistance is 5 ohms.

Review Exercises

1. Let n be the number of nickels and d be the number of dimes. Translate to a system of equations.

$$\begin{cases} d + 2 = n \\ 0.05n + 0.10d = 1.30 \end{cases}$$

$$0.05(d + 2) + 0.10d = 1.30$$
$$0.05d + 0.10 + 0.10d = 1.30$$
$$0.15d = 1.20$$
$$d = 8$$

$$8 + 2 = n$$
$$10 = n$$

There are 8 dimes and 10 nickels.

2. Let l be the length of the painting and w be the width of the painting. Translate to a system of equations.

$$\begin{cases} l = 2 + w \\ lw = 80 \end{cases}$$

$$(2 + w)w = 80$$
$$2w + w^2 = 80$$
$$w^2 + 2w - 80 = 0$$
$$(w + 10)(w - 8) = 0$$
$$w = -10 \quad \text{or} \quad w = 8$$

$$l = 2 + 8$$
$$l = 10$$

Since length cannot be negative, –10 is not a reasonable solution. The length is 10 inches and the width is 8 inches.

3. Let one number be x and the other be $(x + 3)$.

$$(x+3)^2 - x^2 = 39$$
$$x^2 + 6x + 9 - x^2 = 39$$
$$6x = 30$$
$$x = 5$$

One number is 5 and the other is 8.

4. Let the larger number be x and the smaller be y. Translate to a system of equations.

$$\begin{cases} x + y = 28 \\ x - y = 4 \end{cases}$$

$$x + y = 28$$
$$x - y = 4$$
$$2x = 32$$
$$x = 16$$

$$16 + y = 28$$
$$y = 12$$

The numbers are 16 and 12.

5. Let s be the price of the saw and d be the price of the drill. Translate to a system of equations.

$$\begin{cases} d = s - 54 \\ d + s = 238 \end{cases}$$

$$(s-54) + s = 238$$
$$2s - 54 = 238$$
$$2s = 292$$
$$s = 146$$

$$d = 146 - 54$$
$$d = 92$$

The drill cost \$92 and the saw \$146.

6.

	rate	time	distance
rv	55	x	d
truck	75	$x-4$	d

$$\begin{cases} 55x = d \\ 75(x-4) = d \end{cases}$$

$$55x = 75(x-4)$$
$$55x = 75x - 300$$
$$-20x = -300$$
$$x = 15$$

It will take the truck $15 - 4 =$ 11 hours to catch up to the recreational vehicle.

Exercise Set 7.5

1. $\dfrac{1}{x}$

3. $\dfrac{100}{r}$, inverse

5. p decreases

7. Complete a table.

Categories	Rate of Work	Time at Work	Amt of Task Completed
Jason	$\frac{1}{4}$	t	$\frac{t}{4}$
sister	$\frac{1}{6}$	t	$\frac{t}{6}$

$$\frac{t}{4} + \frac{t}{6} = 1$$
$$24\left(\frac{t}{4} + \frac{t}{6}\right) = 24 \cdot 1$$
$$6t + 4t = 24$$
$$10t = 24$$
$$t = 2\frac{2}{5}$$

Working together, it takes Jason and his sister $2\frac{2}{5}$ hours to wash and wax the car.

9. Complete a table.

Categories	Rate of Work	Time at Work	Amt of Task Completed
Alicia	$\frac{1}{25}$	t	$\frac{t}{25}$
Geraldine	$\frac{1}{35}$	t	$\frac{t}{35}$

$$\frac{t}{25} + \frac{t}{35} = 1$$
$$175\left(\frac{t}{25} + \frac{t}{35}\right) = 175 \cdot 1$$
$$7t + 5t = 175$$
$$12t = 175$$
$$t = 14\frac{7}{12}$$

Working together it will take Alicia and Geraldine $14\frac{7}{12}$ days to make the quilt.

11. Complete a table.

Categories	Rate of Work	Time at Work	Amt of Task Completed
1st roofer	$\frac{1}{6}$	4	$\frac{2}{3}$
2nd roofer	$\frac{1}{t}$	4	$\frac{4}{t}$

$$\frac{2}{3}+\frac{4}{t}=1$$
$$3t\left(\frac{2}{3}+\frac{4}{t}\right)=3t\cdot 1$$
$$2t+12=3t$$
$$12=t$$

It would take the second roofer 12 hr. to do the job.

13. Complete a table.

Categories	Rate of Work	Time at Work	Amt of Task Completed
cold	$\frac{1}{15}$	9	$\frac{3}{5}$
hot	$\frac{1}{t}$	9	$\frac{9}{t}$

$$\frac{3}{5}+\frac{9}{t}=1$$
$$5t\left(\frac{3}{5}+\frac{9}{t}\right)=5t\cdot 1$$
$$3t+45=5t$$
$$45=2t$$
$$22.5=t$$

It would take the hot water faucet 22.5 minutes to fill the tub alone.

15. Complete a table.

Categories	Rate	Time	Distance
northbound bus	63	$\frac{x}{63}$	x
southbound car	72	$\frac{675-x}{72}$	$675-x$

$$\frac{x}{63}=\frac{675-x}{72}$$
$$72x=42{,}525-63x$$
$$135x=42{,}525$$
$$x=315$$

$$\text{time}=\frac{x}{63}=\frac{315}{63}=5\text{ hrs.}$$

They will be 675 miles apart after 5 hours, at 3 p.m.

17. Complete a table.

Categories	Rate	Time	Distance
freight train	$\frac{3}{5}r$	3	$\frac{9}{5}r$
passenger train	r	3	$3r$

$$\frac{9}{5}r+3r=360$$
$$\frac{24}{5}r=360$$
$$r=75$$

The passenger train travels at 75 mph.

19. Complete a table.

Categories	Rate	Time	Distance
Jack	45	$\frac{x}{45}$	x
Frances	60	$\frac{x}{60}$	x

$$\frac{x}{45}-\frac{x}{60}=2$$
$$180\left(\frac{x}{45}-\frac{x}{60}\right)=180\cdot 2$$
$$4x-3x=360$$
$$x=360$$

Frances' time is $\frac{x}{60}=\frac{360}{60}=6$.

It takes Frances 6 hours to catch Jack.

21. Complete a table.

Categories	Rate	Time	Distance
against	$r-21$	5.5	$5.5(r-21)$
with	$r+21$	5	$5(r+21)$

$$5.5(r-21)=5(r+21)$$
$$5.5r-115.5=5r+105$$
$$0.5r=220.5$$
$$r=441$$

The speed of the airliner in still air is 441 mph.

23. Translate "a varies directly as b."

$a = k \cdot b$ Now, $a = \frac{4}{9}b$

$4 = k \cdot 9$ $a = \frac{4}{9}(27)$

$\frac{4}{9} = k$ $a = 12$

25. Translate "y varies directly as the square of x."

$y = k \cdot x^2$ Now, $y = 4x^2$

$100 = k \cdot 5^2$ $y = 4(3)^2$

$100 = 25k$ $y = 36$

$4 = k$

27. Translate "m varies directly as n."

$m = k \cdot n$ Now, $m = \frac{3}{4}n$

$6 = k \cdot 8$ $9 = \frac{3}{4}n$

$\frac{3}{4} = k$ $12 = n$

29. Translating "The price increases with the quantity purchased," we write $p = kq$, where p represents purchase price and q represents the quantity. $p = kq$

$16.25 = k \cdot 2.5$

$6.5 = k$

Now, $p = 6.5q$

$p = 6.5(6)$

$p = 39$

The salmon will cost $39 for 6 pounds.

31. Translating "the volume of a gas varies directly with temperature," we write $v = kt$ where v represents the volume and t represents the temperature. $v = kt$

$288 = k \cdot 80$

$3.6 = k$

Now, $v = 3.6t$

$v = 3.6(50)$

$v = 180$

The volume is 180 cubic cm.

33. $d = k \cdot t^2$ Now, $d = 16t^2$

$144 = k \cdot 3^2$ $400 = 16t^2$

$\frac{144}{9} = k$ $25 = t^2$

$16 = k$ $5 = t$

In 5 seconds it will fall 400 feet.

35. $d = k \cdot t^2$ Now, $d = 1.62t^2$

$6.48 = k \cdot 2^2$ $d = 1.62(5)^2$

$\frac{6.48}{4} = k$ $d = 40.5$

$1.62 = k$

In 5 seconds it will fall 40.5 meters.

37. Translating "The circumference varies directly with its diameter," we write $c = kd$, where c represents circumference and d represents the diameter.

$c = k \cdot d$ Now, $c = 3.14d$

$12.56 = k \cdot 4$ $21.98 = 3.14d$

$3.14 = k$ $7 = d$

The diameter will be 7 ft. when the circumference is 21.98 feet.

39. Translate "a varies inversely as b."

$a = \frac{k}{b}$ Now, $a = \frac{16}{b}$

$3.2 = \frac{k}{5}$ $8 = \frac{16}{b}$

$16 = k$ $8b = 16$

$b = 2$

41. Translate "m varies inversely as n."

$m = \frac{k}{n}$ Now, $m = \frac{66}{n}$

$11 = \frac{k}{6}$ $8 = \frac{66}{n}$

$66 = k$ $8n = 66$

$n = 8.25$

43. Translating "The pressure that a gas exerts against the walls of a container is inversely proportional to the volume of the container," we write $p = \frac{k}{v}$ where p represents the pressure the gas exerts and v is the volume of the container.

$p = \frac{k}{v}$ Now, $p = \frac{800}{v}$

$40 = \frac{k}{20}$ $p = \frac{800}{16}$

$800 = k$ $p = 50$

The pressure is 50 psi.

45. Translating "the current I varies inversely as the resistance R," we write $I = \frac{k}{R}$ where I represents the current in amperes and R represents the resistance in ohms.

$I = \frac{k}{R}$ Now, $I = \frac{150}{R}$

$10 = \frac{k}{15}$ $25 = \frac{150}{R}$

$150 = k$ $25R = 150$

$R = 6$

The resistance is 6 ohms.

47. $l = \frac{k}{f}$ Now, $l = \frac{360,000}{f}$

$400 = \frac{k}{900}$ $l = \frac{360,000}{600}$

$360,000 = k$ $l = 600$

The wavelength is 600 m.

49. $f = \frac{k}{a}$ Now, $f = \frac{280}{a}$

$5.6 = \frac{k}{50}$ $2.8 = \frac{280}{a}$

$280 = k$ $2.8a = 280$

$a = 100$

The aperture is 100 mm.

51. Translate "a varies jointly with b and c."

$a = kbc$ Now, $a = 4bc$

$96 = k \cdot 6 \cdot 4$ $a = 4(2)(8)$

$96 = 24k$ $a = 64$

$4 = k$

53. Translate "a varies jointly as the square of b and c."

$a = k \cdot b^2 \cdot c$ Now, $a = 4b^2c$

$96 = k \cdot 2^2 \cdot 6$ $a = 4(3)^2(2)$

$96 = 24k$ $a = 72$

$4 = k$

55. Translating "the volume of a rectangular solid varies jointly with the length and the height," we write $v = klh$, where v represents volume, l represents length, and h represents height.

$v = k \cdot l \cdot h$ Now, $v = 4lh$

$192 = k \cdot 8 \cdot 6$ $v = 4(7)(12)$

$192 = 48k$ $v = 336$

$4 = k$

The volume is 336 cubic inches.

57. Translating "the volume of a right circular cylinder varies jointly as the square of the radius and the height," we write $v = kr^2h$, where v represents volume, r represents radius, and h represents height.

$v = k \cdot r^2 \cdot h$ Now, $v = 3.14167 \cdot r^2 \cdot h$

$301.6 = k \cdot 4^2 \cdot 6$ $v = 3.14167 \cdot 3^2 \cdot 6$

$301.6 = 96k$ $v \approx 169.65$

$3.14167 \approx k$

The volume is approximately 169.65 cubic cm.

59. Translate "y varies directly with x and inversely as z."

$y = \frac{kx}{z}$ Now, $y = \frac{12x}{z}$

$8 = \frac{k \cdot 4}{6}$ $y = \frac{12 \cdot 5}{10}$

$48 = 4k$ $y = 6$

$12 = k$

61. Translate "y varies jointly with x and z and inversely with n."

$y = \frac{kxz}{n}$ Now, $y = \frac{18xz}{n}$

$81 = \frac{k(4)(9)}{8}$ $8 = \frac{18(6)(12)}{n}$

$648 = 36k$ $8n = 1296$

$18 = k$ $n = 162$

63. Translating "the resistance of a wire varies directly with the length and inversely with the square of the diameter," we write $R = \frac{kl}{d^2}$, where R represents resistance, l represents length, and d represents diameter.

$R = \frac{kl}{d^2}$ Now, $R = \frac{0.0005l}{d^2}$

$7.5 = \frac{k \cdot 6}{0.02^2}$ $R = \frac{0.0005 \cdot 10}{0.04^2}$

$0.003 = 6k$ $0.0016R = 0.005$

$0.0005 = k$ $R = 3.125$

The resistance is 3.125 ohms.

65. Translating "the force of attraction between two bodies is jointly proportional to their masses and inversely proportional to the square of the distance between them," we write

$f = \dfrac{k \cdot m_1 \cdot m_2}{d^2}$, where f is the force of attraction, m_1 is the mass of one body, m_2 is the mass of the other body, and d is the distance between them.

$$f = \frac{k \cdot m_1 \cdot m_2}{d^2} \qquad \text{Now, } f = \frac{18m_1m_2}{d^2}$$
$$48 = \frac{k \cdot 4 \cdot 6}{3^2} \qquad f = \frac{18(2)(12)}{6^2}$$
$$432k = 24k \qquad f = 12$$
$$18 = k$$

The force of attraction is 12 dynes.

67. a)

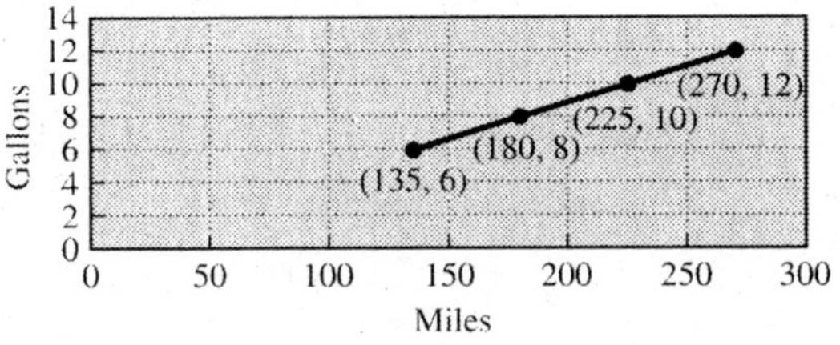

b) They are directly proportional. As the number of miles increases, so does the number of gallons.

c) $d = k \cdot g$

$135 = k \cdot 6$

$22.5 = k$, which represents the miles per gallon the car gets.

d) Yes, the data represent a function because for any given number of miles, there will be one quantity of gasoline (assuming that the miles per gallon stays constant).

Review Exercises

1. $2^3 = 8$

2. $-12^2 = -144$

3. $(6a^3b^2)(-5ab^4) = -30a^4b^6$

4. $\dfrac{-48x^3y^6}{-12x^7y^4} = \dfrac{4y^2}{x^4}$

5. $-3(x+4) - 2(x-5) = -3x - 12 - 2x + 10$
$= -5x - 2$

6. $3x^2 - 13x - 10 = 0$

$(3x+2)(x-5) = 0$

$3x + 2 = 0$ or $x - 5 = 0$

$x = -\dfrac{2}{3}$ $\quad x = 5$

Chapter 7 Review Exercises

1. True

2. False; to add two rational expressions, it is necessary to write each expression with the LCD as its denominator.

3. False; multiply the numerator and the denominator of the complex rational expression by the LCD of the fractions in the numerator and denominator.

4. True

5. True

6. True

7. $\dfrac{32x^2y^5}{8xy^3} = 4xy^2$

8. $\dfrac{-42p^6q^2}{27p^3q^4} = \dfrac{-2\cdot\cancel{3}\cdot 7\cdot p^6q^2}{\cancel{3}\cdot 3\cdot 3\cdot p^3q^4} = -\dfrac{14p^3}{9q^2}$

9. $-\dfrac{18m^3n^2}{54m^5n^5} = -\dfrac{1}{3m^2n^3}$

10. $\dfrac{14x+63}{10x+45} = \dfrac{7\cancel{(2x+9)}}{5\cancel{(2x+9)}} = \dfrac{7}{5}$

11. $\dfrac{4a-20}{7a-35} = \dfrac{4\cancel{(a-5)}}{7\cancel{(a-5)}} = \dfrac{4}{7}$

12. $\dfrac{a^2+4a-21}{a^2+9a+14} = \dfrac{\cancel{(a+7)}(a-3)}{\cancel{(a+7)}(a+2)} = \dfrac{a-3}{a+2}$

13. $\dfrac{2x-3}{6x^2-x-12} = \dfrac{\cancel{(2x-3)}}{\cancel{(2x-3)}(3x+4)} = \dfrac{1}{3x+4}$

14. $\dfrac{25x^2-9}{10x^2+x-3} = \dfrac{\cancel{(5x+3)}(5x-3)}{\cancel{(5x+3)}(2x-1)} = \dfrac{5x-3}{2x-1}$

15. $\dfrac{2c+3d}{8c^2+26cd+21d^2}=\dfrac{\cancel{(2c+3d)}}{\cancel{(2c+3d)}(4c+7d)}$

$=\dfrac{1}{4c+7d}$

16. $\dfrac{5-2x}{4x^2-25}=\dfrac{-1\cancel{(2x-5)}}{(2x+5)\cancel{(2x-5)}}=-\dfrac{1}{2x+5}$

17. $\dfrac{8ac-2a+12c-3}{6ac-8a+9c-12}=\dfrac{2a(4c-1)+3(4c-1)}{2a(3c-4)+3(3c-4)}$

$=\dfrac{(4c-1)\cancel{(2a+3)}}{(3c-4)\cancel{(2a+3)}}$

$=\dfrac{4c-1}{3c-4}$

18. $\dfrac{8x^3+27}{2x^2-7x-15}=\dfrac{\cancel{(2x+3)}(4x^2-6x+9)}{\cancel{(2x+3)}(x-5)}$

$=\dfrac{4x^2-6x+9}{x-5}$

19. $\dfrac{{}^{2}\cancel{28}m^2n^3}{{}_{3}\cancel{15}p^2q^6}\cdot\dfrac{\cancel{25}^{5}p^4q^3}{\cancel{14}mn}=\dfrac{10m^2n^3p^4q^3}{3mnp^2q^6}$

$=\dfrac{10mn^2p^2}{3q^3}$

20. $\dfrac{8m-12n}{32m^3n}\cdot\dfrac{36mn^3}{6m-9n}=\dfrac{\cancel{4}\cancel{(2m-3n)}}{{}_{2}\cancel{8}\cancel{32}m^3n}\cdot\dfrac{\cancel{36}^{\cancel{12}^{3}}mn^3}{\cancel{3}\cancel{(2m-3n)}}$

$=\dfrac{3mn^3}{2m^3n}$

$=\dfrac{3n^2}{2m^2}$

21. $\dfrac{14x-21}{30x-40}\cdot\dfrac{15x-20}{8x-12}=\dfrac{7\cancel{(2x-3)}}{{}_{2}\cancel{10}\cancel{(3x-4)}}\cdot\dfrac{\cancel{5}\cancel{(3x-4)}}{4\cancel{(2x-3)}}$

$=\dfrac{7}{8}$

22. $\dfrac{25x^2-16}{16x+24}\cdot\dfrac{12x+18}{5x-4}$

$=\dfrac{(5x+4)\cancel{(5x-4)}}{{}_{4}\cancel{8}\cancel{(2x+3)}}\cdot\dfrac{\cancel{6}^{3}\cancel{(2x+3)}}{\cancel{(5x-4)}}$

$=\dfrac{3(5x+4)}{4}$

23. $\dfrac{\cancel{(2x-7)}}{{}_{6}\cancel{12}}\cdot\dfrac{\cancel{10}^{5}}{-1\cancel{(2x-7)}}=-\dfrac{5}{6}$

24. $\dfrac{x^2-16}{x^2+6x+8}\cdot\dfrac{x^2-3x-10}{x^2-25}$

$=\dfrac{\cancel{(x+4)}(x-4)}{\cancel{(x+4)}\cancel{(x+2)}}\cdot\dfrac{\cancel{(x-5)}\cancel{(x+2)}}{(x+5)\cancel{(x-5)}}$

$=\dfrac{x-4}{x+5}$

25. $\dfrac{y^2-9}{y^2-3y-18}\cdot\dfrac{y^2-4y-12}{y^2-6y+9}$

$=\dfrac{\cancel{(y+3)}\cancel{(y-3)}}{\cancel{(y-6)}\cancel{(y+3)}}\cdot\dfrac{\cancel{(y-6)}(y+2)}{\cancel{(y-3)}(y-3)}$

$=\dfrac{y+2}{y-3}$

26. $\dfrac{8x^2+2x-15}{6x^2+x-12}\cdot\dfrac{3x^2-13x+12}{3x^2-7x-6}$

$=\dfrac{(4x-5)\cancel{(2x+3)}}{\cancel{(3x-4)}\cancel{(2x+3)}}\cdot\dfrac{\cancel{(3x-4)}\cancel{(x-3)}}{(3x+2)\cancel{(x-3)}}$

$=\dfrac{4x-5}{3x+2}$

27. $\dfrac{ab+2ad-3bc-6cd}{ab-4ad+2bc-8cd}\cdot\dfrac{ab+5ad+2bc+10cd}{ab+5ad-3bc-15cd}$

$=\dfrac{a(b+2d)-3c(b+2d)}{a(b-4d)+2c(b-4d)}\cdot\dfrac{a(b+5d)+2c(b+5d)}{a(b+5d)-3c(b+5d)}$

$=\dfrac{(b+2d)\cancel{(a-3c)}}{(b-4d)\cancel{(a+2c)}}\cdot\dfrac{\cancel{(b+5d)}\cancel{(a+2c)}}{\cancel{(b+5d)}\cancel{(a-3c)}}$

$=\dfrac{b+2d}{b-4d}$

28. $\dfrac{8x^3+y^3}{4x^2-y^2}\cdot\dfrac{6x^2+5xy-4y^2}{4x^2-2xy+y^2}$

$=\dfrac{\cancel{(2x+y)}\cancel{(4x^2-2xy+y^2)}}{\cancel{(2x+y)}\cancel{(2x-y)}}\cdot\dfrac{(3x+4y)\cancel{(2x-y)}}{\cancel{(4x^2-2xy+y^2)}}$

$=3x+4y$

29. $\dfrac{39a^2b^4}{27x^3y^2}\div\dfrac{26a^3b}{28xy^4}=\dfrac{{}^{13}\cancel{39}a^2b^4}{{}_{9}\cancel{27}x^3y^2}\cdot\dfrac{\cancel{28}^{14}xy^4}{\cancel{26}^{13}a^3b}$

$=\dfrac{14a^2b^4xy^4}{9a^3bx^3y^2}$

$=\dfrac{14b^3y^2}{9ax^2}$

30. $\dfrac{8y^3-20y^2}{9y}\div\dfrac{6y^4-15y^3}{10y^2}$

$=\dfrac{4y^2\cancel{(2y-5)}}{9y}\cdot\dfrac{10y^2}{3y^3\cancel{(2y-5)}}$

$=\dfrac{40y^4}{27y^4}=\dfrac{40}{27}$

31. $\dfrac{21a+7b}{16a-24b}\div\dfrac{42a+14b}{24a-36b}$

$=\dfrac{\cancel{7}\cancel{(3a+b)}}{{}_2\cancel{8}\cancel{(2a-3b)}}\cdot\dfrac{\cancel{12}^3\cancel{(2a-3b)}}{\cancel{14}^2\cancel{(3a+b)}}$

$=\dfrac{3}{4}$

32. $\dfrac{z^2}{z^2-2z-8}\div\dfrac{9z^3+3z^4}{z^2+5z+6}$

$=\dfrac{z^2}{(z-4)\cancel{(z+2)}}\cdot\dfrac{\cancel{(z+3)}\cancel{(z+2)}}{3z^3\cancel{(z+3)}}$

$=\dfrac{z^2}{3z^3(z-4)}$

$=\dfrac{1}{3z(z-4)}$

33. $\dfrac{4p^2-9}{24p+28}\div\dfrac{10p-15}{36p^2-49}$

$=\dfrac{(2p+3)\cancel{(2p-3)}}{4\cancel{(6p+7)}}\cdot\dfrac{\cancel{(6p+7)}(6p-7)}{5\cancel{(2p-3)}}$

$=\dfrac{(2p+3)(6p-7)}{20}$

34. $\dfrac{16p^2-8pq-3q^2}{8p^2+22pq+5q^2}\div\dfrac{8p^2-10pq+3q^2}{10p^2+pq-3q^2}$

$=\dfrac{\cancel{(4p+q)}\cancel{(4p-3q)}}{\cancel{(4p+q)}(2p+5q)}\cdot\dfrac{(5p+3q)\cancel{(2p-q)}}{\cancel{(4p-3q)}\cancel{(2p-q)}}$

$=\dfrac{5p+3q}{2p+5q}$

35. $\dfrac{x^3-8y^3}{x^2-36y^2}\div\dfrac{x^2+2xy-8y^2}{x^2-2xy-24y^2}$

$=\dfrac{\cancel{(x-2y)}\left(x^2+2xy+4y^2\right)}{(x+6y)\cancel{(x-6y)}}\cdot\dfrac{\cancel{(x-6y)}\cancel{(x+4y)}}{\cancel{(x+4y)}\cancel{(x-2y)}}$

$=\dfrac{x^2+2xy+4y^2}{x+6y}$

36. $\dfrac{xz+2xw-3yz-6yw}{4xz-2xw+6yz-3yw}\div\dfrac{xz-2xw-3yz+6yw}{2xz-4xw+3yz-6yw}$

$=\dfrac{x(z+2w)-3y(z+2w)}{2x(2z-w)+3y(2z-w)}\cdot\dfrac{2x(z-2w)+3y(z-2w)}{x(z-2w)-3y(z-2w)}$

$=\dfrac{(z+2w)\cancel{(x-3y)}}{(2z-w)\cancel{(2x+3y)}}\cdot\dfrac{\cancel{(z-2w)}\cancel{(2x+3y)}}{\cancel{(z-2w)}\cancel{(x-3y)}}$

$=\dfrac{z+2w}{2z-w}$

37. $f(x)=\dfrac{2x}{5-x}$

a) $f(3)=\dfrac{2(3)}{5-3}=\dfrac{6}{2}=3$

b) $f(0)=\dfrac{2(0)}{5-0}=\dfrac{0}{5}=0$

c) $f(-1)=\dfrac{2(-1)}{5-(-1)}=\dfrac{-2}{6}=-\dfrac{1}{3}$

38. $g(x)=\dfrac{x+2}{x^2-8x+15}$

a) $g(4)=\dfrac{4+2}{4^2-8(4)+15}=\dfrac{6}{-1}=-6$

b) $g(-2)=\dfrac{-2+2}{(-2)^2-8(-2)+15}=\dfrac{0}{35}=0$

c) $g(5)=\dfrac{5+2}{5^2-8(5)+15}=\dfrac{7}{0}$ is undefined

39. $f(x)=\dfrac{2x+4}{3x-5}$

$3x-5=0$

$3x=5$

$x=\dfrac{5}{3}$

The domain is $\left\{x \middle| x \neq \dfrac{5}{3}\right\}$.

40. $f(x)=\dfrac{3x-4}{x^2+4x-12}$

$x^2+4x-12=0$

$(x+6)(x-2)=0$

$x+6=0$ or $x-2=0$

$x=-6$ $\quad$ $x=2$

The domain is $\{x | x \neq -6, 2\}$.

41. $\dfrac{5r}{14x}+\dfrac{3r}{14x}=\dfrac{8r}{14x}=\dfrac{4r}{7x}$

42. $\dfrac{3a+4b}{2a-3b}-\dfrac{a-2b}{2a-3b}=\dfrac{3a+4b-(a-2b)}{2a-3b}$

$=\dfrac{3a+4b-a+2b}{2a-3b}$

$=\dfrac{2a+6b}{2a-3b}$

43. $\dfrac{2p^2-p+2}{p^2-16}+\dfrac{p^2+4p-8}{p^2-16}=\dfrac{3p^2+3p-6}{p^2-16}$

44. $\dfrac{x^2+2x-5}{x^2-3x-18}-\dfrac{3x+7}{x^2-3x-18}$

$=\dfrac{x^2+2x-5-(3x+7)}{x^2-3x-18}$

$=\dfrac{x^2+2x-5-3x-7}{x^2-3x-18}$

$=\dfrac{x^2-x-12}{x^2-3x-18}$

$=\dfrac{(x-4)\cancel{(x+3)}}{(x-6)\cancel{(x+3)}}$

$=\dfrac{x-4}{x-6}$

45. $9p^3q^4=3^2\cdot p^3\cdot q^4$

$12p^2q^5=2^2\cdot 3\cdot p^2\cdot q^5$

$\text{LCD}=2^2\cdot 3^2\cdot p^3\cdot q^5$

$=36p^3q^5$

$\dfrac{8b}{9p^3q^4}=\dfrac{8b(4q)}{9p^3q^4(4q)}=\dfrac{32bq}{36p^3q^5}$

$\dfrac{11c}{12p^2q^5}=\dfrac{11c(3p)}{12p^2q^5(3p)}=\dfrac{33cp}{36p^3q^5}$

46. $\text{LCD}=(t-4)(t+2)$

$\dfrac{7}{t-4}=\dfrac{7(t+2)}{(t-4)(t+2)}=\dfrac{7t+14}{(t-4)(t+2)}$

$\dfrac{9}{t+2}=\dfrac{9(t-4)}{(t+2)(t-4)}=\dfrac{9t-36}{(t-4)(t+2)}$

47. $8u+12=4(2u+3)$

$14u+21=7(2u+3)$

$\text{LCD}=4\cdot 7\cdot(2u+3)$

$=28(2u+3)$

$\dfrac{5u}{8u+12}=\dfrac{5u(7)}{4(2u+3)(7)}=\dfrac{35u}{28(2u+3)}$

$\dfrac{9u}{14u+21}=\dfrac{9u(4)}{7(2u+3)(4)}=\dfrac{36u}{28(2u+3)}$

48. $w^2-16=(w+4)(w-4)$

$w^2+5w+4=(w+4)(w+1)$

$\text{LCD}=(w+4)(w-4)(w+1)$

$\dfrac{2w}{w^2-16}=\dfrac{2w(w+1)}{(w+4)(w-4)(w+1)}$

$=\dfrac{2w^2+2w}{(w+4)(w-4)(w+1)}$

$\dfrac{6w}{w^2+5w+4}=\dfrac{6w(w-4)}{(w+4)(w+1)(w-4)}$

$=\dfrac{6w^2-24w}{(w+4)(w-4)(w+1)}$

49. $a^2+6a+9=(a+3)^2$

$a^2-2a-15=(a-5)(a+3)$

$\text{LCD}=(a-5)(a+3)^2$

$\dfrac{3a}{a^2+6a+9}=\dfrac{3a(a-5)}{(a+3)^2(a-5)}=\dfrac{3a^2-15a}{(a+3)^2(a-5)}$

$\dfrac{10a}{a^2-2a-15}=\dfrac{10a(a+3)}{(a+3)(a-5)(a+3)}$

$=\dfrac{10a^2+30a}{(a+3)^2(a-5)}$

50. $m^2-2m-8=(m-4)(m+2)$

$m^2-3m-4=(m-4)(m+1)$

$\text{LCD}=(m-4)(m+2)(m+1)$

$\dfrac{m+1}{m^2-2m-8}=\dfrac{(m+1)(m+1)}{(m-4)(m+2)(m+1)}$

$=\dfrac{m^2+2m+1}{(m-4)(m+2)(m+1)}$

$\dfrac{m-1}{m^2-3m-4}=\dfrac{(m-1)(m+2)}{(m-4)(m+1)(m+2)}$

$=\dfrac{m^2+m-2}{(m-4)(m+2)(m+1)}$

51. $x^3+6x^2+8x=x(x+4)(x+2)$

$x^5+2x^4-8x^3=x^3(x+4)(x-2)$

$\text{LCD}=x^3(x+4)(x-2)(x+2)$

$\dfrac{2x-3}{x^3+6x^2+8x}=\dfrac{(2x-3)x^2(x-2)}{x(x+4)(x+2)x^2(x-2)}$

$=\dfrac{2x^4-7x^3+6x^2}{x^3(x+4)(x+2)(x-2)}$

$\dfrac{6x+3}{x^5+2x^4-8x^3}=\dfrac{(6x+3)(x+2)}{x^3(x+4)(x-2)(x+2)}$

$=\dfrac{6x^2+15x+6}{x^3(x+4)(x+2)(x-2)}$

52. $x^3=x^3$

$x^2+10+24=(x+6)(x+4)$

$4x^3+16x^2=4x^2(x+4)$

$\text{LCD}=4x^3(x+4)(x+6)$

$\dfrac{6}{x^3}=\dfrac{6(4)(x+6)(x+4)}{x^3(4)(x+6)(x+4)}=\dfrac{24x^2+240x+576}{4x^3(x+4)(x+6)}$

$\dfrac{4x-3}{x^2+10x+24}=\dfrac{(4x-3)4x^3}{(x+6)(x+4)4x^3}$

$=\dfrac{16x^4-12x^3}{4x^3(x+4)(x+6)}$

$\dfrac{3x}{4x^3+16x^2}=\dfrac{3x\cdot x(x+6)}{4x^2(x+4)\cdot x(x+6)}$

$=\dfrac{3x^3+18x^2}{4x^3(x+6)(x+4)}$

53. $\text{LCD}=45x$

$\dfrac{8a}{15x}-\dfrac{4a}{9x}=\dfrac{(8a)\cdot 3}{(15x)\cdot 3}-\dfrac{(4a)\cdot 5}{(9x)\cdot 5}$

$=\dfrac{24a}{45x}-\dfrac{20a}{45x}$

$=\dfrac{4a}{45x}$

54. $\text{LCD}=20y$

$\dfrac{y-3}{4y}-\dfrac{y+2}{5y}=\dfrac{(y-3)\cdot 5}{(4y)\cdot 5}-\dfrac{(y+2)\cdot 4}{(5y)\cdot 4}$

$=\dfrac{5y-15}{20y}-\dfrac{4y+8}{20y}$

$=\dfrac{5y-15-4y-8}{20y}$

$=\dfrac{y-23}{20y}$

55. $\text{LCD}=36t^4u^5$

$\dfrac{2t-3}{18t^4u^2}+\dfrac{5t+1}{12t^3u^5}$

$=\dfrac{(2t-3)\cdot 2u^3}{18t^4u^2\cdot 2u^3}+\dfrac{(5t+1)\cdot 3t}{12t^3u^5\cdot 3t}$

$=\dfrac{4tu^3-6u^3}{36t^4u^5}+\dfrac{15t^2+3t}{36t^4u^5}$

$=\dfrac{4tu^3-6u^3+15t^2+3t}{36t^4u^5}$

56. LCD = $w(w-3)$

$$\frac{8}{w-3}-\frac{-3}{w}=\frac{8\cdot w}{(w-3)\cdot w}-\frac{-3\cdot(w-3)}{w\cdot(w-3)}$$
$$=\frac{8w}{w(w-3)}-\frac{-3w+9}{w(w-3)}$$
$$=\frac{8w+3w-9}{w(w-3)}$$
$$=\frac{11w-9}{w(w-3)}$$

57. LCD = $4(v+2)(v-3)$

$$\frac{v+4}{4v+8}-\frac{v-2}{2v-6}$$
$$=\frac{(v+4)\cdot(v-3)}{4(v+2)\cdot(v-3)}-\frac{(v-2)\cdot 2(v+2)}{2(v-3)\cdot 2(v+2)}$$
$$=\frac{v^2+v-12}{4(v+2)(v-3)}-\frac{2v^2-8}{4(v+2)(v-3)}$$
$$=\frac{v^2+v-12-2v^2+8}{4(v+2)(v-3)}$$
$$=\frac{-v^2+v-4}{4(v+2)(v-3)}$$

58. LCD = $(t+4)^2$

$$\frac{t^2-5t}{t^2+8t+16}+\frac{6}{t+4}=\frac{t^2-5t}{(t+4)^2}+\frac{6(t+4)}{(t+4)\cdot(t+4)}$$
$$=\frac{t^2-5t}{(t+4)^2}+\frac{6t+24}{(t+4)^2}$$
$$=\frac{t^2+t+24}{(t+4)^2}$$

59. LCD = $(w+5)(w-5)(w+2)$

$$\frac{2w+5}{w^2-25}+\frac{6w}{w^2-3w-10}$$
$$=\frac{(2w+5)\cdot(w+2)}{(w+5)(w-5)\cdot(w+2)}+\frac{6w\cdot(w+5)}{(w-5)(w+2)\cdot(w+5)}$$
$$=\frac{2w^2+9w+10}{(w+5)(w-5)(w+2)}+\frac{6w^2+30w}{(w+5)(w-5)(w+2)}$$
$$=\frac{8w^2+39w+10}{(w+5)(w-5)(w+2)}$$

60. LCD = $(4z+5)(z+7)^2$

$$\frac{z+9}{4z^2+33z+35}-\frac{z-12}{z^2+14z+49}$$
$$=\frac{(z+9)\cdot(z+7)}{(4z+5)(z+7)\cdot(z+7)}-\frac{(z-12)\cdot(4z+5)}{(z+7)^2\cdot(4z+5)}$$
$$=\frac{z^2+16z+63}{(4z+5)(z+7)^2}-\frac{4z^2-43z-60}{(4z+5)(z+7)^2}$$
$$=\frac{z^2+16z+63-4z^2+43z+60}{(4z+5)(z+7)^2}$$
$$=\frac{-3z^2+59z+123}{(4z+5)(z+7)^2}$$

61. LCD = $a(a+3)$

$$\frac{3}{a}-\frac{3a+5}{a+3}+\frac{5a-3}{a^2+3a}$$
$$=\frac{3\cdot(a+3)}{a\cdot(a+3)}-\frac{(3a+5)\cdot a}{(a+3)\cdot a}+\frac{5a-3}{a(a+3)}$$
$$=\frac{3a+9}{a(a+3)}-\frac{3a^2+5a}{a(a+3)}+\frac{5a-3}{a(a+3)}$$
$$=\frac{3a+9-3a^2-5a+5a-3}{a(a+3)}$$
$$=\frac{-3a^2+3a+6}{a(a+3)}$$

62. LCD = $3x-y$

$$\frac{4x-3y}{3x-y}-\frac{3x-4y}{-1(3x-y)}=\frac{4x-3y}{3x-y}-\frac{-1(3x-4y)}{3x-y}$$
$$=\frac{4x-3y}{3x-y}-\frac{-3x+4y}{3x-y}$$
$$=\frac{4x-3y+3x-4y}{3x-y}$$
$$=\frac{7x-7y}{3x-y}$$

63. $$\frac{\frac{4}{5}}{\frac{3}{10}}=\frac{4}{5}\cdot\frac{10}{3}=\frac{40}{15}=\frac{8}{3}$$

64. $$\frac{\frac{4}{3}-\frac{8}{9}}{\frac{5}{6}-\frac{4}{12}}=\frac{\left(\frac{4}{3}-\frac{8}{9}\right)\cdot 36}{\left(\frac{5}{6}-\frac{4}{12}\right)\cdot 36}=\frac{48-32}{30-12}=\frac{16}{18}=\frac{8}{9}$$

65. $\dfrac{\frac{u^4v^2}{w}}{\frac{uv^3}{w^2}} = \dfrac{u^4v^2}{w}\cdot\dfrac{w^2}{uv^3} = \dfrac{u^4v^2w^2}{uv^3w} = \dfrac{u^3w}{v}$

66. $\dfrac{\frac{x^6y^3}{t^4}}{\frac{x^2y^2}{t}} = \dfrac{x^6y^3}{t^4}\cdot\dfrac{t}{x^2y^2} = \dfrac{x^6y^3t}{x^2y^2t^4} = \dfrac{x^4y}{t^3}$

67. $\dfrac{2w-\frac{w}{4}}{6-\frac{w}{4}} = \dfrac{\left(2w-\frac{w}{4}\right)\cdot 4}{\left(6-\frac{w}{4}\right)\cdot 4} = \dfrac{8w-w}{24-w} = \dfrac{7w}{24-w}$

68. $\dfrac{\frac{7}{b-15}-6}{b-15} = \dfrac{\left(\frac{7}{b-15}-6\right)\cdot(b-15)}{(b-15)\cdot(b-15)}$

$= \dfrac{7-6(b-15)}{(b-15)^2}$

$= \dfrac{7-6b+90}{(b-15)^2}$

$= \dfrac{-6b+97}{(b-15)^2}$

69. $\dfrac{1-\frac{1}{y}-\frac{12}{y}}{1-\frac{6}{y}+\frac{8}{y}} = \dfrac{\left(1-\frac{1}{y}-\frac{12}{y}\right)\cdot y}{\left(1-\frac{6}{y}+\frac{8}{y}\right)\cdot y}$

$= \dfrac{y-1-12}{y-6+8}$

$= \dfrac{y-13}{y+2}$

70. $\dfrac{x+\frac{25}{x+10}}{1-\frac{5}{x+10}} = \dfrac{\left(x+\frac{25}{x+10}\right)\cdot(x+10)}{\left(1-\frac{5}{x+10}\right)\cdot(x+10)}$

$= \dfrac{x(x+10)+25}{x+10-5}$

$= \dfrac{x^2+10x+25}{x+5}$

$= \dfrac{\cancel{(x+5)}(x+5)}{\cancel{(x+5)}}$

$= x+5$

71. $\dfrac{\frac{6}{y^2}-\frac{1}{xy}-\frac{12}{x^2}}{\frac{4}{y^2}-\frac{4}{xy}-\frac{3}{x^2}} = \dfrac{\left(\frac{6}{y^2}-\frac{1}{xy}-\frac{12}{x^2}\right)\cdot x^2y^2}{\left(\frac{4}{y^2}-\frac{4}{xy}-\frac{3}{x^2}\right)\cdot x^2y^2}$

$= \dfrac{6x^2-xy-12y^2}{4x^2-4xy-3y^2}$

$= \dfrac{(3x+4y)\cancel{(2x-3y)}}{\cancel{(2x-3y)}(2x+y)}$

$= \dfrac{3x+4y}{2x+y}$

72. $\dfrac{\frac{x+4}{x-4}-\frac{x-4}{x+4}}{\frac{x+4}{x-4}+\frac{x-4}{x+4}} = \dfrac{\left(\frac{x+4}{x-4}-\frac{x-4}{x+4}\right)\cdot(x-4)(x+4)}{\left(\frac{x+4}{x-4}+\frac{x-4}{x+4}\right)\cdot(x-4)(x+4)}$

$= \dfrac{(x+4)^2-(x-4)^2}{(x+4)^2+(x-4)^2}$

$= \dfrac{x^2+8x+16-x^2+8x-16}{x^2+8x+16+x^2-8x+16}$

$= \dfrac{16x}{2x^2+32}$

$= \dfrac{2\cdot 8x}{2\cdot\left(x^2+16\right)}$

$= \dfrac{8x}{x^2+16}$

73. $\dfrac{5}{4t}-\dfrac{3}{8t}=\dfrac{1}{2}$

$8t\cdot\left(\dfrac{5}{4t}-\dfrac{3}{8t}\right)=8t\cdot\dfrac{1}{2}$

$10-3=4t$

$7=4t$

$\dfrac{7}{4}=t$

Check: $\dfrac{5}{4\left(\frac{7}{4}\right)}-\dfrac{3}{8\left(\frac{7}{4}\right)}\stackrel{?}{=}\dfrac{1}{2}$

$\dfrac{5}{7}-\dfrac{3}{14}\stackrel{?}{=}\dfrac{1}{2}$

$\dfrac{10}{14}-\dfrac{3}{14}\stackrel{?}{=}\dfrac{1}{2}$

$\dfrac{1}{2}=\dfrac{1}{2}$

74. $$\frac{12}{m-2}=9+\frac{m}{m-2}$$
$$(m-2)\left(\frac{12}{m-2}\right)=(m-2)\left(9+\frac{m}{m-2}\right)$$
$$12=9(m-2)+m$$
$$12=9m-18+m$$
$$30=10m$$
$$3=m$$

Check: $$\frac{12}{3-2}\stackrel{?}{=}9+\frac{3}{3-2}$$
$$12\stackrel{?}{=}9+3$$
$$12=12$$

75. $$\frac{x^2}{x-4}=\frac{3x+4}{x-4}-3$$
$$(x-4)\left(\frac{x^2}{x-4}\right)=(x-4)\left(\frac{3x+4}{x-4}-3\right)$$
$$x^2=3x+4-3(x-4)$$
$$x^2=3x+4-3x+12$$
$$x^2=16$$
$$x^2-16=0$$
$$(x+4)(x-4)=0$$
$$x+4=0 \quad \text{or} \quad x-4=0$$
$$x=-4 \qquad\qquad x=4$$

Notice that if $x=4$, then $\frac{x^2}{x-4}$ and $\frac{3x+4}{x-4}$ are undefined. Therefore, $x=4$ is extraneous, and the only solution is $x=-4$.

Check: $$\frac{(-4)^2}{-4-4}\stackrel{?}{=}\frac{3(-4)+4}{-4-4}-3$$
$$\frac{16}{-8}\stackrel{?}{=}\frac{-8}{-8}-3$$
$$-2\stackrel{?}{=}1-3$$
$$-2=-2$$

76. $$\frac{2a-2}{a+1}-\frac{a}{2a-3}=\frac{a-2}{2a-3}$$
$$(a+1)(2a-3)\left(\frac{2a-2}{a+1}-\frac{a}{2a-3}\right)$$
$$=(a+1)(2a-3)\left(\frac{a-2}{2a-3}\right)$$
$$(2a-2)(2a-3)-a(a+1)=(a-2)(a+1)$$
$$4a^2-10a+6-a^2-a=a^2-a-2$$
$$3a^2-11a+6=a^2-a-2$$
$$2a^2-10a+8=0$$
$$2(a-1)(a-4)=0$$
$$a=1 \quad \text{or} \quad a=4$$

Check: $a=1$
$$\frac{2(1)-2}{1+1}-\frac{(1)}{2(1)-3}\stackrel{?}{=}\frac{1-2}{2(1)-3}$$
$$0-\frac{1}{-1}\stackrel{?}{=}\frac{-1}{-1}$$
$$0+1\stackrel{?}{=}1$$
$$1=1$$

$a=4$
$$\frac{2(4)-2}{4+1}-\frac{(4)}{2(4)-3}\stackrel{?}{=}\frac{4-2}{2(4)-3}$$
$$\frac{6}{5}-\frac{4}{5}\stackrel{?}{=}\frac{2}{5}$$
$$\frac{2}{5}=\frac{2}{5}$$

77. $$\frac{1}{q+4}=\frac{q}{3q+2}$$
$$3q+2=q^2+4q$$
$$0=q^2+q-2$$
$$0=(q+2)(q-1)$$
$$q=-2 \quad \text{or} \quad q=1$$

Check: $q=-2$
$$\frac{1}{-2+4}\stackrel{?}{=}\frac{-2}{3(-2)+2}$$
$$\frac{1}{2}\stackrel{?}{=}\frac{-2}{-4}$$
$$\frac{1}{2}=\frac{1}{2}$$

Check: $q=1$
$$\frac{1}{1+4}\stackrel{?}{=}\frac{1}{3(1)+2}$$
$$\frac{1}{5}=\frac{1}{5}$$

78. $$\frac{5}{v^2+5v+6}-\frac{2}{v^2-2v-8}=\frac{3}{v^2-16}$$

$$(v+2)(v+3)(v-4)(v+4)\cdot\left(\frac{5}{(v+2)(v+3)}-\frac{2}{(v-4)(v+2)}\right)$$
$$=(v+2)(v+3)(v-4)(v+4)\left(\frac{3}{(v-4)(v+4)}\right)$$
$$5(v-4)(v+4)-2(v+3)(v+4)=3(v+2)(v+3)$$
$$5\left(v^2-16\right)-2\left(v^2+7v+12\right)=3\left(v^2+5v+6\right)$$
$$5v^2-80-2v^2-14v-24=3v^2+15v+18$$
$$-29v=122$$
$$v=-\frac{122}{29}$$

Check:

$$\frac{5}{\left(-\frac{122}{29}+2\right)\left(-\frac{122}{29}+3\right)}-\frac{2}{\left(-\frac{122}{29}-4\right)\left(-\frac{122}{29}+2\right)}\stackrel{?}{=}\frac{3}{\left(-\frac{122}{29}-4\right)\left(-\frac{122}{29}+4\right)}$$
$$\frac{5}{\left(-\frac{64}{29}\right)\left(\frac{-35}{29}\right)}-\frac{2}{\left(\frac{-238}{29}\right)\left(\frac{-64}{29}\right)}\stackrel{?}{=}\frac{3}{\left(\frac{-238}{29}\right)\left(\frac{-6}{29}\right)}$$
$$\frac{14,297}{7616}-\frac{841}{7616}\stackrel{?}{=}\frac{841}{476}$$
$$\frac{841}{476}=\frac{841}{476}$$

79. $$\frac{3}{x^2+5x+6}=\frac{2}{x+3}+\frac{x-3}{x^2+x-2}$$

$$(x+2)(x+3)(x-1)\cdot\left(\frac{3}{(x+2)(x+3)}\right)$$
$$=(x+2)(x+3)(x-1)\cdot\left(\frac{2}{x+3}+\frac{x-3}{(x+2)(x-1)}\right)$$
$$3(x-1)=2(x+2)(x-1)+(x-3)(x+3)$$
$$3x-3=2x^2+2x-4+x^2-9$$
$$0=3x^2-x-10$$
$$0=(3x+5)(x-2)$$
$$x=-\frac{5}{3}\quad\text{or}\quad x=2$$

Check: $x=2$

$$\frac{3}{(2+2)(2+3)}\stackrel{?}{=}\frac{2}{2+3}+\frac{2-3}{(2+2)(2-1)}$$
$$\frac{3}{20}\stackrel{?}{=}\frac{2}{5}+\frac{-1}{4}$$
$$\frac{3}{20}\stackrel{?}{=}\frac{8}{20}-\frac{5}{20}$$
$$\frac{3}{20}=\frac{3}{20}$$

Check: $x=-\frac{5}{3}$

$$\frac{3}{\left(-\frac{5}{3}+2\right)\left(-\frac{5}{3}+3\right)}\stackrel{?}{=}\frac{2}{-\frac{5}{3}+3}+\frac{-\frac{5}{3}-3}{\left(-\frac{5}{3}+2\right)\left(-\frac{5}{3}-1\right)}$$
$$\frac{3}{\left(\frac{1}{3}\right)\left(\frac{4}{3}\right)}\stackrel{?}{=}\frac{2}{\left(\frac{4}{3}\right)}+\frac{-\frac{14}{3}}{\left(\frac{1}{3}\right)\left(\frac{-8}{3}\right)}$$
$$\frac{27}{4}\stackrel{?}{=}\frac{6}{4}+\frac{21}{4}$$
$$\frac{27}{4}=\frac{27}{4}$$

80. $\dfrac{7x}{x^2-2x-8}-\dfrac{4x}{x^2-5x+4}=\dfrac{3}{x-4}$

$$(x-4)(x+2)(x-1)\cdot\left(\frac{7x}{(x-4)(x+2)}-\frac{4x}{(x-4)(x-1)}\right)$$
$$=(x-4)(x+2)(x-1)\left(\frac{3}{x-4}\right)$$
$$7x(x-1)-4x(x+2)=3(x+2)(x-1)$$
$$7x^2-7x-4x^2-8x=3x^2+3x-6$$
$$3x^2-15x=3x^2+3x-6$$
$$-18x=-6$$
$$x=\frac{1}{3}$$

Check:

$$\frac{7\left(\frac{1}{3}\right)}{\left(\frac{1}{3}-4\right)\left(\frac{1}{3}+2\right)}-\frac{4\left(\frac{1}{3}\right)}{\left(\frac{1}{3}-4\right)\left(\frac{1}{3}-1\right)}\stackrel{?}{=}\frac{3}{\frac{1}{3}-4}$$
$$\frac{\frac{7}{3}}{\left(-\frac{11}{3}\right)\left(\frac{7}{3}\right)}-\frac{\frac{4}{3}}{\left(-\frac{11}{3}\right)\left(-\frac{2}{3}\right)}\stackrel{?}{=}\frac{3}{\frac{-11}{3}}$$
$$-\frac{3}{11}-\frac{6}{11}\stackrel{?}{=}-\frac{9}{11}$$
$$-\frac{9}{11}=-\frac{9}{11}$$

81. Complete a table.

Categories	Rate of Work	Time at Work	Amt of Task Completed
George	$\frac{1}{30}$	t	$\frac{t}{30}$
Lucille	$\frac{1}{45}$	t	$\frac{t}{45}$

$$\frac{t}{30}+\frac{t}{45}=1$$
$$90\left(\frac{t}{30}+\frac{t}{45}\right)=90\cdot 1$$
$$3t+2t=90$$
$$5t=90$$
$$t=18$$

George and Lucille can write a chapter in 18 days working together.

82. Complete a table.

Categories	Rate	Time	Distance
against current	$r-3$	$\frac{200}{r-3}$	200
with current	$r+3$	$\frac{260}{r+3}$	260

$$\frac{200}{r-3}=\frac{260}{r+3}$$
$$260(r-3)=200(r+3)$$
$$260r-780=200r+600$$
$$60r=1380$$
$$r=23$$

The cutter can sail in still water at 23 mph.

83. Complete a table.

Categories	Rate	Time	Distance
truck	45	t	$45t$
car	55	t	$55t$

$$45t+55t=510$$
$$100t=510$$
$$t=5.1$$
$$0.1(60)=6 \text{ min.}$$

It will be 5 hours and 6 minutes before they meet at 9:06 am.

84. Translate "p varies directly as q."

$p=kq$ Now, $p=6q$

$24=k\cdot 4$ $p=6(7)$

$6=k$ $p=42$

85. Translate "s varies directly as the square of t."

$s=k\cdot t^2$ Now, $s=8t^2$

$72=k\cdot 9$ $s=8(5)^2$

$8=k$ $s=200$

86. Translate "y varies inversely as x."

$y=\dfrac{k}{x}$ Now, $y=\dfrac{24}{x}$

$8=\dfrac{k}{3}$ $4=\dfrac{24}{x}$

$24=k$ $4x=24$

$x=6$

87. Translate "suppose m varies inversely as the square of n."

$$m=\frac{k}{n^2} \qquad \text{Now, } m=\frac{144}{n^2}$$
$$16=\frac{k}{3^2} \qquad m=\frac{144}{4^2}$$
$$144=k \qquad m=9$$

88. Translate "y varies jointly with x and z."

$$y=kxz \qquad \text{Now, } y=4xz$$
$$40=k\cdot 2\cdot 5 \qquad y=4\cdot 4\cdot 2$$
$$4=k \qquad y=32$$

89. Translate "m varies directly with n and inversely with p."

$$m=\frac{kn}{p} \qquad \text{Now, } m=\frac{6n}{p}$$
$$2=\frac{k\cdot 3}{9} \qquad m=\frac{6\cdot 6}{4}$$
$$18=3k \qquad m=9$$
$$6=k$$

90. Translating "the distance a car can travel varies directly with the amount of gas it carries," we write $d=kg$ where d is the distance traveled and g is the amount of gas.

$$d=kg \qquad \text{Now, } d=26g$$
$$156=k\cdot 6 \qquad 234=26g$$
$$26=k \qquad 9=g$$

9 gallons are required to travel 234 miles.

91. $$v=\frac{k}{T} \qquad \text{Now, } v=\frac{28}{T}$$
$$4=\frac{k}{7} \qquad v=\frac{28}{12}$$
$$28=k \qquad v=\frac{7}{3}$$

The velocity is $\frac{7}{3}$ cm/sec.

92. $$v=kr^2h \qquad \text{Now, } v=3.14r^2h$$
$$62.8=k\cdot 2^2\cdot 5 \qquad v=3.14(4)^2(2)$$
$$62.8=20k \qquad v=100.48$$
$$3.14=k$$

The volume is 100.48 in.^{3}.

Chapter 7 Practice Test

1. $$\frac{42a^3b^4}{16ab^6}=\frac{\cancel{2}\cdot 3\cdot 7\cdot \cancel{a}\cdot a\cdot a\cdot \cancel{b}\cdot \cancel{b}\cdot \cancel{b}\cdot \cancel{b}}{\cancel{2}\cdot 2\cdot 2\cdot 2\cdot \cancel{a}\cdot \cancel{b}\cdot \cancel{b}\cdot \cancel{b}\cdot \cancel{b}\cdot b\cdot b}=\frac{21a^2}{8b^2}$$

2. $$\frac{6x^2+11x-10}{4x^2+4x-15}=\frac{(3x-2)\cancel{(2x+5)}}{(2x-3)\cancel{(2x+5)}}=\frac{3x-2}{2x-3}$$

3. $$\frac{m^3-64n^3}{3m^2-14mn+8n^2}=\frac{\cancel{(m-4n)}(m^2+4mn+16n^2)}{(3m-2n)\cancel{(m-4n)}}=\frac{m^2+4mn+16n^2}{3m-2n}$$

4. $$\frac{2bn-4bm-3cn+6cm}{2b^2+7bc-15c^2}=\frac{2b(n-2m)-3c(n-2m)}{(2b-3c)(b+5c)}=\frac{\cancel{(2b-3c)}(n-2m)}{\cancel{(2b-3c)}(b+5c)}=\frac{n-2m}{b+5c}$$

5. $$\frac{27a^2b^4}{14x^4y}\cdot\frac{35xy^3}{18a^5b^2}=\frac{\cancel{3}\cdot\cancel{3}\cdot 3\cdot\cancel{a}\cdot\cancel{a}\cdot\cancel{b}\cdot\cancel{b}\cdot b\cdot b}{2\cdot\cancel{7}\cdot\cancel{x}\cdot x\cdot x\cdot x\cdot\cancel{y}}\cdot\frac{5\cdot\cancel{7}\cdot\cancel{x}\cdot\cancel{y}\cdot y\cdot y}{2\cdot\cancel{3}\cdot\cancel{3}\cdot\cancel{a}\cdot\cancel{a}\cdot a\cdot a\cdot a\cdot\cancel{b}\cdot\cancel{b}}=\frac{15b^2y^2}{4a^3x^3}$$

6. $$\frac{16y^2-25}{6y^2-17y-14}\cdot\frac{3y^2+2y}{8y^2-2y-15}=\frac{\cancel{(4y+5)}(4y-5)}{\cancel{(3y+2)}(2y-7)}\cdot\frac{y\cancel{(3y+2)}}{\cancel{(4y+5)}(2y-3)}=\frac{y(4y-5)}{(2y-7)(2y-3)}$$

7. $\dfrac{6q-8p}{8p-2q} \div \dfrac{4p-3q}{12p-3q} = \dfrac{6q-8p}{8p-2q} \cdot \dfrac{12p-3q}{4p-3q}$

$= \dfrac{-2\cancel{(4p-3q)}}{2\cancel{(4p-q)}} \cdot \dfrac{3\cancel{(4p-q)}}{\cancel{(4p-3q)}}$

$= \dfrac{-6}{2} = -3$

8. $\dfrac{2n^3-5n^2-6n+15}{3n^3+2n^2-9n-6} \div \dfrac{2n^3-5n^2-8n+20}{3n^3+2n^2+12n+8}$

$= \dfrac{2n^3-5n^2-6n+15}{3n^3+2n^2-9n-6} \cdot \dfrac{3n^3+2n^2+12n+8}{2n^3-5n^2-8n+20}$

$= \dfrac{n^2(2n-5)-3(2n-5)}{n^2(3n+2)-3(3n+2)} \cdot \dfrac{n^2(3n+2)+4(3n+2)}{n^2(2n-5)-4(2n-5)}$

$= \dfrac{\cancel{(2n-5)}\cancel{(n^2-3)}}{\cancel{(3n+2)}\cancel{(n^2-3)}} \cdot \dfrac{\cancel{(3n+2)}(n^2+4)}{\cancel{(2n-5)}(n^2-4)}$

$= \dfrac{n^2+4}{n^2-4}$

9. $f(x) = \dfrac{2x-4}{3x^2-7x-6}$

a) $f(-1) = \dfrac{2(-1)-4}{3(-1)^2-7(-1)-6}$

$= \dfrac{-2-4}{3+7-6}$

$= \dfrac{-6}{4}$

$= -\dfrac{3}{2}$

b) $f(2) = \dfrac{2(2)-4}{3(2)^2-7(2)-6}$

$= \dfrac{4-4}{12-14-6}$

$= \dfrac{0}{-8}$

$= 0$

c) $f(3) = \dfrac{2(3)-4}{3(3)^2-7(3)-6}$

$= \dfrac{6-4}{27-21-6}$

$= \dfrac{2}{0}$ is undefined

10. To find the domain of the function $f(x) = \dfrac{3x-4}{2x^2-x-10}$, set the denominator equal to zero and solve for x to find any values that should be excluded from the domain.

$2x^2-x-10=0$

$(2x-5)(x+2)=0$

$2x-5=0 \quad$ or $\quad x+2=0$

$2x=5 \qquad x=-2$

$x=\dfrac{5}{2}$

The domain of $f(x) = \dfrac{3x-4}{2x^2-x-10}$ is $\left\{x \mid x \neq -2, \dfrac{5}{2}\right\}$.

11. $a^2-9=(a+3)(a-3)$

$2a^2+13a+21=(2a+7)(a+3)$

$\text{LCD} = (a+3)(a-3)(2a+7)$

$\dfrac{3a}{a^2-9} = \dfrac{3a}{(a+3)(a-3)}$

$= \dfrac{3a(2a+7)}{(a+3)(a-3)(2a+7)}$

$= \dfrac{6a^2+21a}{(a+3)(a-3)(2a+7)}$

$\dfrac{6a}{2a^2+13a+21} = \dfrac{6a}{(a+3)(2a+7)}$

$= \dfrac{6a(a-3)}{(a+3)(a-3)(2a+7)}$

$= \dfrac{6a^2-18a}{(a+3)(a-3)(2a+7)}$

12. LCD $= 36y^2$

$$\frac{2y^2+3}{9y^2}-\frac{y-6}{12y}=\frac{(2y^2+3)(4)}{9y^2(4)}-\frac{(y-6)(3y)}{12y(3y)}$$
$$=\frac{8y^2+12}{36y^2}-\frac{3y^2-18y}{36y^2}$$
$$=\frac{(8y^2+12)-(3y^2-18y)}{36y^2}$$
$$=\frac{(8y^2+12)+(-3y^2+18y)}{36y^2}$$
$$=\frac{5y^2+18y+12}{36y^2}$$

13. $\frac{2r^2-2r-5}{r^2-16}-\frac{-7r+7}{r^2-16}$

$$=\frac{(2r^2-2r-5)-(-7r+7)}{r^2-16}$$
$$=\frac{(2r^2-2r-5)+(7r-7)}{r^2-16}$$
$$=\frac{2r^2+5r-12}{r^2-16}$$
$$=\frac{(2r-3)\cancel{(r+4)}}{(r-4)\cancel{(r+4)}}$$
$$=\frac{2r-3}{r-4}$$

14. LCD $= 2(t+5)(t-5)^2$

$$\frac{t+3}{t^2-10t+25}-\frac{t-4}{2t^2-50}$$
$$=\frac{t+3}{(t-5)^2}-\frac{t-4}{2(t+5)(t-5)}$$
$$=\frac{(t+3)\cdot 2(t+5)}{(t-5)^2\cdot 2(t+5)}-\frac{(t-4)\cdot(t-5)}{2(t+5)(t-5)\cdot(t-5)}$$
$$=\frac{2t^2+16t+30}{2(t+5)(t-5)^2}-\frac{t^2-9t+20}{2(t+5)(t-5)^2}$$
$$=\frac{(2t^2+16t+30)-(t^2-9t+20)}{2(t+5)(t-5)^2}$$
$$=\frac{(2t^2+16t+30)+(-t^2+9t-20)}{2(t+5)(t-5)^2}$$
$$=\frac{t^2+25t+10}{2(t+5)(t-5)^2}$$

15. LCD $= 3a-5b$

$$\frac{3a+4b}{3a-5b}+\frac{2a-b}{5b-3a}=\frac{3a+4b}{3a-5b}+\frac{2a-b}{-1(3a-5b)}$$
$$=\frac{3a+4b}{3a-5b}+\frac{-2a+b}{3a-5b}$$
$$=\frac{3a+4b-2a+b}{3a-5b}$$
$$=\frac{a+5b}{3a-5b}$$

16. LCD $= x(x+5)$

$$\frac{5}{x}-\frac{3x-1}{x^2+5x}-\frac{2x-5}{x+5}$$
$$=\frac{5(x+5)}{x(x+5)}-\frac{3x-1}{x(x+5)}-\frac{(2x-5)x}{(x+5)x}$$
$$=\frac{5x+25}{x(x+5)}-\frac{3x-1}{x(x+5)}-\frac{2x^2-5x}{x(x+5)}$$
$$=\frac{(5x+25)-(3x-1)-(2x^2-5x)}{x(x+5)}$$
$$=\frac{(5x+25)+(-3x+1)+(-2x^2+5x)}{x(x+5)}$$
$$=\frac{5x+25-3x+1-2x^2+5x}{x(x+5)}$$
$$=\frac{-2x^2+7x+26}{x(x+5)}$$

17. $$\frac{\frac{6a^3b^2}{c^3}}{\frac{9a^2b^4}{c^6}} = \frac{6a^3b^2}{c^3} \div \frac{9a^2b^4}{c^6}$$
$$= \frac{6a^3b^2}{c^3} \cdot \frac{c^6}{9a^2b^4}$$
$$= \frac{6a^3b^2c^6}{9a^2b^4c^3}$$
$$= \frac{2ac^3}{3b^2}$$

18. $$\frac{6 - \frac{1}{x} - \frac{15}{x^2}}{4 + \frac{4}{x} - \frac{3}{x^2}} = \frac{\left(6 - \frac{1}{x} - \frac{15}{x^2}\right) \cdot x^2}{\left(4 + \frac{4}{x} - \frac{3}{x^2}\right) \cdot x^2}$$
$$= \frac{6 \cdot x^2 - \frac{1}{x} \cdot x^2 - \frac{15}{x^2} \cdot x^2}{4 \cdot x^2 + \frac{4}{x} \cdot x^2 - \frac{3}{x^2} \cdot x^2}$$
$$= \frac{6x^2 - x - 15}{4x^2 + 4x - 3}$$
$$= \frac{(3x-5)\cancel{(2x+3)}}{(2x-1)\cancel{(2x+3)}}$$
$$= \frac{3x-5}{2x-1}$$

19. $$\frac{t - \frac{14}{t-5}}{2 - \frac{4}{t-5}} = \frac{\left(t - \frac{14}{t-5}\right) \cdot (t-5)}{\left(2 - \frac{4}{t-5}\right) \cdot (t-5)}$$
$$= \frac{t \cdot (t-5) - \frac{14}{t-5} \cdot (t-5)}{2 \cdot (t-5) - \frac{4}{t-5} \cdot (t-5)}$$
$$= \frac{t^2 - 5t - 14}{2t - 10 - 4}$$
$$= \frac{t^2 - 5t - 14}{2t - 14}$$
$$= \frac{\cancel{(t-7)}(t+2)}{2\cancel{(t-7)}}$$
$$= \frac{t+2}{2}$$

20. $$\frac{\frac{9}{a^2} - \frac{4}{b^2}}{\frac{3}{a} + \frac{2}{b}} = \frac{\left(\frac{9}{a^2} - \frac{4}{b^2}\right) \cdot a^2b^2}{\left(\frac{3}{a} + \frac{2}{b}\right) \cdot a^2b^2}$$
$$= \frac{\frac{9}{a^2} \cdot a^2b^2 - \frac{4}{b^2} \cdot a^2b^2}{\frac{3}{a} \cdot a^2b^2 + \frac{2}{b} \cdot a^2b^2}$$
$$= \frac{9b^2 - 4a^2}{3ab^2 + 2a^2b}$$
$$= \frac{\cancel{(3b+2a)}(3b-2a)}{ab\cancel{(3b+2a)}}$$
$$= \frac{3b-2a}{ab}$$

21. Multiply both sides by the LCD, $18x$.
$$\frac{8}{6x} + \frac{5}{2x} = \frac{13}{9}$$
$$18x\left(\frac{8}{6x} + \frac{5}{2x}\right) = 18x\left(\frac{13}{9}\right)$$
$$18x\left(\frac{8}{6x}\right) + 18x\left(\frac{5}{2x}\right) = 26x$$
$$24 + 45 = 26x$$
$$69 = 26x$$
$$\frac{69}{26} = x$$

22. Multiply both sides by the LCD, $(x-4)$. Notice that if $x = 4$, then parts of the equation are undefined, so 4 cannot be a solution.
$$\frac{3x}{x-4} = 6 + \frac{12}{x-4}$$
$$(x-4)\left(\frac{3x}{x-4}\right) = (x-4)\left(6 + \frac{12}{x-4}\right)$$
$$3x = (x-4)(6) + (x-4)\frac{12}{x-4}$$
$$3x = 6x - 24 + 12$$
$$-3x = -12$$
$$x = 4$$
We already noted that 4 causes expressions in the equation to be undefined, so 4 is extraneous. Since 4 was the only possible solution, this equation has no solution.

23. Multiply both sides by the LCD, $(x+1)(x+2)$.

$$\frac{x}{x+1}+\frac{9x}{x^2+3x+2}=\frac{12}{x+2}$$
$$\frac{x}{x+1}+\frac{9x}{(x+2)(x+1)}=\frac{12}{x+2}$$
$$(x+1)(x+2)\left(\frac{x}{x+1}+\frac{9x}{(x+2)(x+1)}\right)=(x+1)(x+2)\left(\frac{12}{x+2}\right)$$
$$x(x+2)+9x=12(x+1)$$
$$x^2+2x+9x=12x+12$$
$$x^2+11x=12x+12$$
$$x^2-x-12=0$$
$$(x-4)(x+3)=0$$
$$x-4=0 \quad \text{or} \quad x+3=0$$
$$x=4 \qquad\qquad x=-3$$

24. Multiply both sides by the LCD, $(n+1)(n+2)$. Notice that if $n=-1$ or if $n=-2$, then parts of the equation are undefined, so neither of these values can be a solution.

$$\frac{n}{n+1}+\frac{n+1}{n+2}=\frac{6n+5}{n^2+3n+2}$$
$$\frac{n}{n+1}+\frac{n+1}{n+2}=\frac{6n+5}{(n+1)(n+2)}$$
$$(n+1)(n+2)\left(\frac{n}{n+1}+\frac{n+1}{n+2}\right)=(n+1)(n+2)\left(\frac{6n+5}{(n+1)(n+2)}\right)$$
$$n(n+2)+(n+1)^2=6n+5$$
$$n^2+2n+n^2+2n+1=6n+5$$
$$2n^2+4n+1=6n+5$$
$$2n^2-2n-4=0$$
$$2(n^2-n-2)=0$$
$$2(n-2)(n+1)=0$$
$$n-2=0 \quad \text{or} \quad n+1=0$$
$$n=2 \qquad\qquad n=-1$$

We already noted that -1 causes expressions in the equation to be undefined, so -1 is extraneous. The only solution is 2.

25. Let t represent the number of days it would take Eduardo to do the project if he worked by himself. Complete a table.

Categories	Rate of Work	Time at Work	Part of Task Completed
Elena	$\frac{1}{15}$	10	$\frac{10}{15}$
Eduardo	$\frac{1}{t}$	10	$\frac{10}{t}$

The total job in this case is 1 project completed, so we can write an equation that combines Elena's part of the task and Eduardo's part of the task and set this sum equal to 1.

$$\frac{10}{15}+\frac{10}{t}=1$$
$$15t\left(\frac{10}{15}+\frac{10}{t}\right)=15t\cdot 1$$
$$10t+150=15t$$
$$150=5t$$
$$30=t$$

It would take Eduardo 30 days to complete the project by himself.

26. Let d represent the distance flown by the cargo plane. Because the sum of the distances flown by each plane is 1900, the distance flown by the airliner is $1900-d$. The times that the planes are flying are equal. Using the relationship $rt=d$, complete a table.

Categories	Rate	Time	Distance
cargo plane	420	$\frac{d}{420}$	d
airliner	530	$\frac{1900-d}{530}$	$1900-d$

$$\frac{d}{420}=\frac{1900-d}{530}$$
$$530d=420(1900-d)$$
$$530d=798{,}000-420d$$
$$950d=798{,}000$$
$$d=840$$

Time is $\frac{d}{420}=\frac{840}{420}=2$ hr.

It takes 2 hours for the airplanes to be 1900 miles apart.

27. $r=ks$ so, $42=7s$

$14=k\cdot 2 \qquad 6=s$

$7=k$

28. $p = kqr$ so, $p = 2\cdot 3\cdot 5$

$48 = k\cdot 4\cdot 6$ $p = 30$

$2 = k$

29. Translate "intensity of light is inversely proportional to the square of the distance from the source" as $I = \frac{k}{d^2}$ where I is the intensity of light and d is the distance from the source.

$I = \frac{k}{d^2}$ so, $I = \frac{576}{20^2}$

$16 = \frac{k}{6^2}$ $I = 1.44$ foot-candles

$576 = k$

The intensity 20 feet from the bulb is 1.44 foot-candles.

30. $d = kg$ so, $d = kg$

$255 = k\cdot 5$ $d = 51\cdot 8$

$51 = k$ $d = 408$ mi.

Chapters 1-7 Cumulative Review

1. False; the given statement is an example of the commutative property.

2. False; the statement is false for $x = -2$.

3. True

4. True

5. $\frac{1}{12}$

6. no

7. -2

8. $\frac{2(-4)^2 - 7^2}{5^2 - 3(-3)^2} = \frac{2(16)-49}{25-3(9)} = \frac{32-49}{25-27} = \frac{-17}{-2} = \frac{17}{2}$

9. $-5 - 2\cdot 6^2 \div (-36)(-4) + 8$

$= -5 - 2\cdot 36 \div (-36)(-4) + 8$

$= -5 - 72 \div (-36)(-4) + 8$

$= -5 - (-2)(-4) + 8$

$= -5 - 8 + 8$

$= -13 + 8$

$= -5$

10. $x - y\left(x^4 - 2z\right) = (-2) - (-3)\left((-2)^4 - 2(6)\right)$

$= (-2) - (-3)\left(16 - 2(6)\right)$

$= (-2) - (-3)(16 - 12)$

$= (-2) - (-3)(4)$

$= (-2) - (-12)$

$= -2 + 12$

$= 10$

11. $\left(3x^3 - 2x^2 + 4\right) - \left(-4x^3 + 3x^2 - 5x + 6\right)$

$= \left(3x^3 - 2x^2 + 4\right) + \left(4x^3 - 3x^2 + 5x - 6\right)$

$= 7x^3 - 5x^2 + 5x - 2$

12. $(2a - 3b)\left(4a^2 + 6ab + 9b^2\right)$

$= 2a\cdot 4a^2 + 2a\cdot 6ab + 2a\cdot 9b^2 - 3b\cdot 4a^2$

$- 3b\cdot 6ab - 3b\cdot 9b^2$

$= 8a^3 + 12a^2b + 18ab^2 - 12a^2b - 18ab^2 - 27b^3$

$= 8a^3 - 27b^3$

13. $\frac{12x^3y^4 - 16x^4y^3 + 3x^2y^5}{4x^3y^4}$

$= \frac{12x^3y^4}{4x^3y^4} - \frac{16x^4y^3}{4x^3y^4} + \frac{3x^2y^5}{4x^3y^4}$

$= 3 - \frac{4x}{y} + \frac{3y}{4x}$

14. $\frac{6x^2 + x - 12}{2x^2 - 5x - 12} \cdot \frac{3x^2 - 14x + 8}{9x^2 - 18x + 8}$

$= \frac{\cancel{(3x-4)}\cancel{(2x+3)}}{\cancel{(x-4)}\cancel{(2x+3)}} \cdot \frac{\cancel{(3x-2)}\cancel{(x-4)}}{\cancel{(3x-4)}\cancel{(3x-2)}}$

$= 1$

15. $\frac{2x+5}{x^2-16} - \frac{x-9}{x^2-x-12}$

$= \frac{(2x+5)(x+3)}{(x+4)(x-4)(x+3)} - \frac{(x-9)(x+4)}{(x+3)(x-4)(x+4)}$

$= \frac{2x^2 + 11x + 15}{(x+4)(x-4)(x+3)} - \frac{x^2 - 5x - 36}{(x+4)(x-4)(x+3)}$

$= \frac{2x^2 + 11x + 15 - x^2 + 5x + 36}{(x+4)(x-4)(x+3)}$

$= \frac{x^2 + 16x + 51}{(x+4)(x-4)(x+3)}$

16. $$\frac{1-\frac{3}{x}-\frac{4}{x^2}}{\frac{3}{x}-\frac{13}{x^2}+\frac{4}{x^3}}=\frac{\left(1-\frac{3}{x}-\frac{4}{x^2}\right)\cdot x^3}{\left(\frac{3}{x}-\frac{13}{x^2}+\frac{4}{x^3}\right)\cdot x^3}$$
$$=\frac{x^3-3x^2-4x}{3x^2-13x+4}$$
$$=\frac{x(x+1)\cancel{(x-4)}}{(3x-1)\cancel{(x-4)}}$$
$$=\frac{x(x+1)}{3x-1}$$

17. $36a^4b^3-39a^3b^4-12a^2b^5$
$=3a^2b^3\left(12a^2-13ab-4b^2\right)$
$=3a^2b^3(4a+b)(3a-4b)$

18. $108c^3-32d^3=4\left(27c^3-8d^3\right)$
$=4(3c-2d)\left(9c^2+6cd+4d^2\right)$

19. $$\frac{3}{4}x+3=\frac{2}{3}(x-6)$$
$$12\cdot\left(\frac{3}{4}x+3\right)=12\cdot\left(\frac{2}{3}(x-6)\right)$$
$$9x+36=8(x-6)$$
$$9x+36=8x-48$$
$$x+36=-48$$
$$x=-84$$

20. $8+|3+2x|\geq 4$
$|3+2x|\geq -4$
This inequality indicates that the absolute value is greater than a negative number. Because the absolute value of every real number is either positive or 0, the solution set is $\mathbb{R}$.
$\{x|x \text{ is a real number}\}$ or $(-\infty,\infty)$

21. $$\begin{cases}2x+y-z=2 & \text{Eqtn. 1}\\ x+3y+2z=1 & \text{Eqtn. 2}\\ x+y+z=2 & \text{Eqtn. 3}\end{cases}$$

Multiply equation 1 by 2 and add to equation 2. This makes equation 4.

$$4x+2y-2z=4$$
$$\underline{x+3y+2z=1}$$
$$5x+5y=5 \quad \rightarrow \quad x+y=1 \quad \text{Eqtn. 4}$$

Add equation 1 to equation 3. This makes equation 5.

$$2x+y-z=2$$
$$\underline{x+y+z=2}$$
$$3x+2y=4 \quad \text{Eqtn. 5}$$

Use equations 4 and 5 to make a system of equations in two variables. Solve for y.

$$x+y=1 \quad \text{Multiply by } -3$$
$$\underline{3x+2y=4}$$
$$-3x-3y=-3$$
$$\underline{3x+2y=4}$$
$$-y=1$$
$$y=-1$$

Substitute the value for y into equation 4 and solve for x.

$$x+y=1$$
$$x+(-1)=1$$
$$x=2$$

Substitute the values for x and y into equation 3 to solve for z.

$$x+y+z=2$$
$$2+(-1)+z=2$$
$$1+z=2$$
$$z=1$$

Solution: $(2,-1,1)$

22. $$6x^2-10=11x$$
$$6x^2-11x-10=0$$
$$(3x+2)(2x-5)=0$$
$$3x+2=0 \quad \text{or} \quad 2x-5=0$$
$$3x=-2 \qquad 2x=5$$
$$x=-\frac{2}{3} \qquad x=\frac{5}{2}$$

23. $\dfrac{3}{x^2+2x-24}+\dfrac{x-5}{x^2-16}=\dfrac{x}{x^2+10x+24}$

$$(x+4)(x-4)(x+6)$$
$$\cdot\left(\frac{3}{(x+6)(x-4)}+\frac{x-5}{(x+4)(x-4)}\right)$$
$$=(x+4)(x-4)(x+6)\cdot\left(\frac{x}{(x+6)(x+4)}\right)$$
$$3(x+4)+(x-5)(x+6)=x(x-4)$$
$$3x+12+x^2+x-30=x^2-4x$$
$$x^2+4x-18=x^2-4x$$
$$4x-18=-4x$$
$$8x=18$$
$$x=\frac{18}{8}=\frac{9}{4}$$

24. $2x-5y=-10$

x	y	Ordered Pair
0	2	$(0,2)$
–5	0	$(-5,0)$
5	4	$(5,4)$

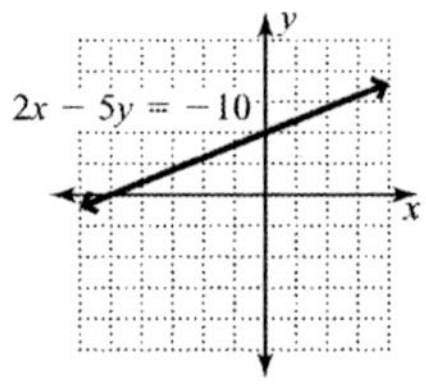

25. $3x-y\le -6$
$$-y\le -6-3x$$
$$y\ge 3x+6$$
Begin by graphing the related equation $y=3x+6$ with a solid line. Now choose $(0,0)$ as a test point.

$$3x-y \le -6$$
$$3(0)-0 \overset{?}{\le} -6$$
$$0-0 \overset{?}{\le} -6$$
$$0 \le -6$$

Because $(0,0)$ does not satisfy the inequality, shade the side of the line on the opposite side from $(0,0)$.

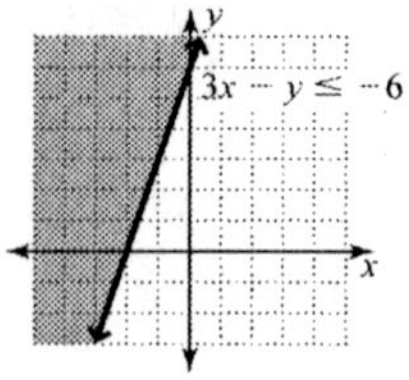

26. Find the slope: $m=\dfrac{1-(-3)}{-1-(-3)}=\dfrac{1+3}{-1+3}=\dfrac{4}{2}=2$
$$y-y_1=m(x-x_1)$$
$$y-1=2(x-(-1))$$
$$y-1=2(x+1)$$
$$y-1=2x+2$$
$$y=2x+3$$

27. If two angles are supplementary their sum is 180 degrees. Let one of the angles be x and the other angle be $3x+20$.
$$x+3x+20=180$$
$$4x+20=180$$
$$4x=160$$
$$x=40$$
The angles are 40° and $3(40)+20=140°$.

28. Complete a table.

Categories	Selling Price	Number of pounds	Revenue
peppermint	1.80	x	$1.80x$
butterscotch	2.30	15	$2.30(15)$
mixture	2.10	$x+15$	$2.10(x+15)$

$$1.80x+2.30(15)=2.10(x+15)$$
$$1.8x+34.5=2.1x+31.5$$
$$34.5=0.3x+31.5$$
$$3=0.3x$$
$$10=x$$
A total of 10 pounds of butterscotch candy will be needed for the mixture.

29. Let x be the width and $2x-3$ be the length.

$$x(2x-3)=54$$
$$2x^2-3x=54$$
$$2x^2-3x-54=0$$
$$(2x+9)(x-6)=0$$
$$2x+9=0 \quad \text{or} \quad x-6=0$$
$$2x=-9 \qquad x=6$$
$$x=-\frac{9}{2}$$

x cannot be negative

The width is 6 feet and the length is $2(6)-3=9$ feet.

30. Complete a table.

Categories	Rate of Work	Time at Work	Amt of Task Completed
Hal	$\frac{1}{7}$	t	$\frac{t}{7}$
Frank	$\frac{1}{10}$	t	$\frac{t}{10}$

$$\frac{t}{7}+\frac{t}{10}=1$$
$$70\left(\frac{t}{7}+\frac{t}{10}\right)=70\cdot 1$$
$$10t+7t=70$$
$$17t=70$$
$$t=\frac{70}{17}$$

Working together it will take Hal and Frank $\frac{70}{17}$ hours to paint the room.

Chapter 8

Rational Exponents, Radicals, and Complex Numbers

Exercise Set 8.1

1. Answers will vary. For example, 9 has rational square roots whereas 17 has irrational square roots. The square roots of 17 are irrational because they cannot be expressed in the form $\frac{a}{b}$ where a and b are integers and $b \neq 0$.

3. Squaring a number or its additive inverse results in the same positive number.

5. You cannot raise a number to an even power and get a negative value.

7. The square roots of 36 are ± 6.

9. The square roots of 121 are ± 11.

11. The square roots of 196 are ± 14.

13. The square roots of 225 are ± 15.

15. $\sqrt{25} = 5$

17. $\sqrt{-64}$ is not a real number

19. $-\sqrt{25} = -5$

21. $\pm\sqrt{25} = \pm 5$

23. $\sqrt{1.44} = 1.2$

25. $\sqrt{-10.64}$ is not a real number

27. $-\sqrt{0.0121} = -0.11$

29. $\sqrt{\frac{49}{81}} = \frac{7}{9}$

31. $-\sqrt{\frac{144}{169}} = -\frac{12}{13}$

33. $\sqrt[3]{27} = 3$

35. $\sqrt[3]{-64} = -4$

37. $-\sqrt[3]{-216} = -(-6) = 6$

39. $\sqrt[4]{256} = 4$

41. $\sqrt[4]{-625}$ is not a real number

43. $-\sqrt[4]{16} = -2$

45. $\sqrt[5]{32} = 2$

47. $\sqrt[5]{-243} = -3$

49. $-\sqrt[5]{-32} = -(-2) = 2$

51. $\sqrt[6]{64} = 2$

53. $\sqrt[3]{-\frac{8}{27}} = -\frac{2}{3}$

55. $\sqrt[4]{\frac{16}{81}} = \frac{2}{3}$

57. $\sqrt{7} \approx 2.646$

59. $-\sqrt{11} \approx -3.317$

61. $\sqrt[3]{50} \approx 3.684$

63. $\sqrt[3]{-53} \approx -3.756$

65. $\sqrt[4]{189} \approx 3.708$

67. $-\sqrt[4]{85} \approx -3.036$

69. $\sqrt[5]{89} \approx 2.454$

71. $\sqrt[6]{146} \approx 2.295$

73. $\sqrt{b^4} = b^2$

75. $\sqrt{16x^2} = 4x$

77. $\sqrt{100r^8s^6} = 10r^4s^3$

79. $\sqrt{0.25a^6b^{12}} = 0.5a^3b^6$

81. $\sqrt[3]{m^3} = m$

83. $\sqrt[3]{27a^9b^6} = 3a^3b^2$

85. $\sqrt[3]{-64a^3b^{12}} = -4ab^4$

87. $\sqrt[3]{0.008x^{18}} = 0.2x^6$

89. $\sqrt[4]{a^4} = a$

91. $\sqrt[4]{16x^{16}} = 2x^4$

93. $\sqrt[5]{32x^{10}} = 2x^2$

95. $\sqrt[6]{x^{12}y^6} = x^2y$

97. $\sqrt{36m^2} = 6|m|$

99. $\sqrt{(r-1)^2} = |r-1|$

101. $\sqrt[4]{256y^{12}} = 4|y^3|$

103. $\sqrt[3]{27y^3} = 3y$

105. $\sqrt{(y-3)^4} = (y-3)^2$

107. $\sqrt[3]{(y-4)^6} = (y-4)^2$

109. $f(x) = \sqrt{2x+4}$

$$\begin{aligned} f(0) &= \sqrt{2\cdot 0+4} \\ &= \sqrt{0+4} \\ &= \sqrt{4} \\ &= 2 \end{aligned}$$

111. $f(x) = \sqrt{4x+3}$

$$\begin{aligned} f(3) &= \sqrt{4\cdot 3+3} \\ &= \sqrt{12+3} \\ &= \sqrt{15} \end{aligned}$$

113. Because the index is even, the radicand must be nonnegative.

$2x-8 \geq 0$

$2x \geq 8$

$x \geq 4$

Domain: $\{x|x \geq 4\}$, or $[4, \infty)$

115. Because the index is even, the radicand must be nonnegative.

$-4x+16 \geq 0$

$-4x \geq -16$

$x \leq 4$

Domain: $\{x|x \leq 4\}$, or $(-\infty, 4]$

117. a)

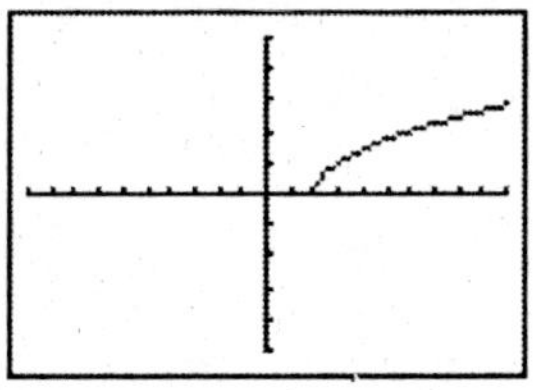

b) $\{x|x \geq 2\}$, or $[2, \infty)$

119. a)

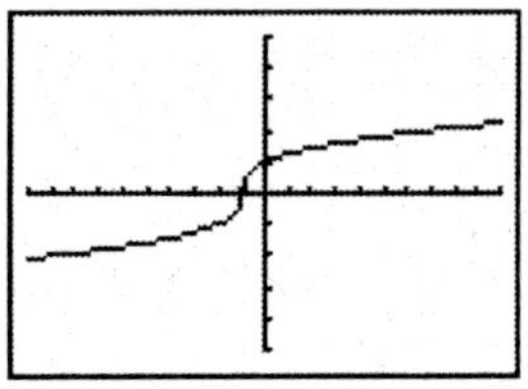

b) $\mathbb{R}$, or $(-\infty, \infty)$

121. $v = -\sqrt{19.6h}$

$$\begin{aligned} v &= -\sqrt{19.6(16)} \\ &= -\sqrt{313.6} \\ &\approx -17.709 \text{ m/sec.} \end{aligned}$$

123. $T = 2\pi\sqrt{\dfrac{L}{9.8}}$

$$\begin{aligned} T &= 2\pi\sqrt{\frac{3}{9.8}} \\ &\approx 2\pi\sqrt{0.306} \\ &\approx 3.476 \text{ sec.} \end{aligned}$$

125. $S = \dfrac{7}{2}\sqrt{2D}$

$$\begin{aligned} S &= \frac{7}{2}\sqrt{2\cdot 15} \\ &= \frac{7}{2}\sqrt{30} \\ &\approx \frac{7}{2}\cdot 5.477 \\ &\approx 19.170 \text{ mph} \end{aligned}$$

127. $c = \sqrt{a^2+b^2}$

$$\begin{aligned} c &= \sqrt{5^2+12^2} \\ &= \sqrt{25+144} \\ &= \sqrt{169} \\ &= 13 \text{ ft.} \end{aligned}$$

129. $R = \sqrt{F_1^2+F_2^2}$

$$\begin{aligned} R &= \sqrt{9^2+12^2} \\ &= \sqrt{81+144} \\ &= \sqrt{225} \\ &= 15 \text{ N} \end{aligned}$$

131. a) $f(x)=1570\sqrt{x}+4784$

$f(4)=1570\sqrt{4}+4784$
$=1570\cdot 2+4784$
$=7924$

Approximately 7924 earthquakes occurred in 2008.

b) $f(x)=1570\sqrt{x}+4784$

$f(2)=1570\sqrt{2}+4784$
$\approx 2220.3+4784$
≈ 7004

Approximately 7004 earthquakes occurred in 2006.

Review Exercises

1. a) $\sqrt{16\cdot 9}=\sqrt{144}$
$=12$

b) $\sqrt{16}\cdot\sqrt{9}=4\cdot 3$
$=12$

2. $x^5\cdot x^3=x^{5+3}=x^8$

3. $(-9m^3n)(5mn^2)=-9\cdot 5m^{3+1}n^{1+2}=-45m^4n^3$

4. $4x^2(3x^2-5x+1)$
$=4x^2\cdot 3x^2+4x^2\cdot(-5x)+4x^2\cdot 1$
$=12x^4-20x^3+4x^2$

5. $(7y-4)(3y+5)$
$=7y\cdot 3y+7y\cdot 5-4\cdot 3y-4\cdot 5$
$=21y^2+35y-12y-20$
$=21y^2+23y-20$

6. $A=(x-9)(2x+1)$
$=x\cdot 2x+x\cdot 1-9\cdot 2x-9\cdot 1$
$=2x^2+x-18x-9$
$=2x^2-17x-9$

Exercise Set 8.2

1. 4

3. a must be nonnegative because with an even index, the radicand must be nonnegative.

5. Yes, because $100^{1/4}=(10^2)^{1/4}=10^{1/2}$.

7. $25^{1/2}=\sqrt{25}=5$

9. $-100^{1/2}=-\sqrt{100}=-10$

11. $27^{1/3}=\sqrt[3]{27}=3$

13. $(-64)^{1/3}=\sqrt[3]{-64}=-4$

15. $y^{1/4}=\sqrt[4]{y}$

17. $(144x^8)^{1/2}=\sqrt{144x^8}=12x^4$

19. $18r^{1/2}=18\sqrt{r}$

21. $\left(\frac{x^4}{81}\right)^{1/2}=\sqrt{\frac{x^4}{81}}=\frac{x^2}{9}$

23. $64^{2/3}=\left(\sqrt[3]{64}\right)^2=(4)^2=16$

25. $-81^{3/4}=-\left(\sqrt[4]{81}\right)^3=-(3)^3=-27$

27. $(-8)^{4/3}=\left(\sqrt[3]{-8}\right)^4=(-2)^4=16$

29. $16^{-3/4}=\frac{1}{16^{3/4}}=\frac{1}{\left(\sqrt[4]{16}\right)^3}=\frac{1}{2^3}=\frac{1}{8}$

31. $x^{4/5}=\sqrt[5]{x^4}$

33. $8n^{2/3}=8\sqrt[3]{n^2}$

35. $(-32)^{-2/5}=\frac{1}{(-32)^{2/5}}$
$=\frac{1}{\left(\sqrt[5]{-32}\right)^2}$
$=\frac{1}{(-2)^2}$
$=\frac{1}{4}$

37. $\left(\frac{1}{25}\right)^{3/2}=\left(\sqrt{\frac{1}{25}}\right)^3=\left(\frac{1}{5}\right)^3=\frac{1}{125}$

39. $(2a+4)^{5/6}=\sqrt[6]{(2a+4)^5}$

41. $\sqrt[4]{25}=25^{1/4}$

43. $\sqrt[6]{z^5}=z^{5/6}$

45. $\frac{1}{\sqrt[6]{5^5}}=\frac{1}{5^{5/6}}=5^{-5/6}$

47. $\dfrac{5}{\sqrt[5]{x^4}} = \dfrac{5}{x^{4/5}} = 5x^{-4/5}$

49. $\left(\sqrt[3]{5}\right)^7 = 5^{7/3}$

51. $\left(\sqrt[7]{x}\right)^2 = x^{2/7}$

53. $\sqrt[4]{(4a-5)^7} = (4a-7)^{7/4}$

55. $\left(\sqrt[5]{2r-5}\right)^8 = (2r-5)^{8/5}$

57. $x^{1/5} \cdot x^{3/5} = x^{1/5+3/5} = x^{4/5}$

59. $x^{3/2} \cdot x^{-1/3} = x^{3/2+(-1/3)} = x^{9/6-2/6} = x^{7/6}$

61. $a^{2/3} \cdot a^{3/4} = a^{2/3+3/4} = a^{8/12+9/12} = a^{17/12}$

63. $\left(3w^{1/7}\right)\left(7w^{3/7}\right) = 21w^{1/7+3/7} = 21w^{4/7}$

65. $\left(-3a^{2/3}\right)\left(4a^{3/4}\right) = -12a^{2/3+3/4}$
$= -12a^{8/12+9/12}$
$= -12a^{17/12}$

67. $\dfrac{7^{7/3}}{7^{2/3}} = 7^{7/3-2/3} = 7^{5/3}$

69. $\dfrac{x^{1/6}}{x^{5/6}} = x^{1/6-5/6} = x^{-4/6} = x^{-2/3} = \dfrac{1}{x^{2/3}}$

71. $\dfrac{x^{3/4}}{x^{1/2}} = x^{3/4-1/2} = x^{3/4-2/4} = x^{1/4}$

73. $\dfrac{r^{3/4}}{r^{2/3}} = r^{3/4-2/3} = r^{9/12-8/12} = r^{1/12}$

75. $\dfrac{x^{-3/7}}{x^{2/7}} = x^{-3/7-2/7} = x^{-5/7} = \dfrac{1}{x^{5/7}}$

77. $\dfrac{a^{3/4}}{a^{-3/2}} = a^{3/4-(-3/2)} = a^{3/4+6/4} = a^{9/4}$

79. $\left(5s^{-2/7}\right)\left(4s^{5/7}\right) = 20s^{-2/7+5/7} = 20s^{3/7}$

81. $\left(-6b^{-5/4}\right)\left(4b^{3/2}\right) = -24b^{-5/4+3/2}$
$= -24b^{-5/4+6/4}$
$= -24b^{1/4}$

83. $\left(x^{2/3}\right)^3 = x^{(2/3)\cdot 3} = x^2$

85. $\left(a^{5/6}\right)^2 = a^{(5/6)\cdot 2} = a^{10/6} = a^{5/3}$

87. $\left(b^{2/3}\right)^{3/5} = b^{(2/3)\cdot(3/5)} = b^{6/15} = b^{2/5}$

89. $\left(2x^{2/3}y^{1/2}\right)^6 = 2^6 x^{(2/3)\cdot 6} y^{(1/2)\cdot 6}$
$= 64x^{12/3}y^{6/2}$
$= 64x^4y^3$

91. $\left(8q^{3/2}t^{3/4}\right)^{1/3} = 8^{1/3}q^{(3/2)\cdot(1/3)}t^{(3/4)\cdot(1/3)}$
$= 2q^{3/6}t^{3/12}$
$= 2q^{1/2}t^{1/4}$

93. $\dfrac{\left(3a^{5/4}\right)^4}{a^2} = \dfrac{3^4 a^{(5/4)\cdot 4}}{a^2}$
$= \dfrac{81a^{20/4}}{a^2}$
$= \dfrac{81a^5}{a^2}$
$= 81a^{5-2}$
$= 81a^3$

95. $\dfrac{\left(9z^{7/3}\right)^{1/2}}{z^{5/6}} = \dfrac{9^{1/2}z^{(7/3)\cdot(1/2)}}{z^{5/6}}$
$= \dfrac{3z^{7/6}}{z^{5/6}}$
$= 3z^{7/6-5/6}$
$= 3z^{2/6}$
$= 3z^{1/3}$

97. $\sqrt[4]{4} = 4^{1/4} = \left(2^2\right)^{1/4} = 2^{2\cdot(1/4)} = 2^{1/2} = \sqrt{2}$

99. $\sqrt[6]{49} = \left(7^2\right)^{1/6} = 7^{2\cdot(1/6)} = 7^{1/3} = \sqrt[3]{7}$

101. $\sqrt[4]{x^2} = \left(x^2\right)^{1/4} = x^{2\cdot(1/4)} = x^{1/2} = \sqrt{x}$

103. $\sqrt[8]{r^6} = \left(r^6\right)^{1/8} = r^{3/4} = \sqrt[4]{r^3}$

105. $\sqrt[8]{x^6y^2} = \left(x^6y^2\right)^{1/8}$
$= x^{6\cdot(1/8)}y^{2\cdot(1/8)}$
$= x^{3/4}y^{1/4}$
$= \left(x^3y\right)^{1/4}$
$= \sqrt[4]{x^3y}$

107. $\sqrt[10]{m^4n^6} = \left(m^4n^6\right)^{1/10}$
$= m^{4\cdot(1/10)}n^{6\cdot(1/10)}$
$= m^{2/5}n^{3/5}$
$= \left(m^2m^3\right)^{1/5}$
$= \sqrt[5]{m^2n^3}$

109. $\sqrt[3]{x}\cdot\sqrt{x} = x^{1/3}\cdot x^{1/2}$
$= x^{1/3+1/2}$
$= x^{2/6+3/6}$
$= x^{5/6}$
$= \sqrt[6]{x^5}$

111. $\sqrt[4]{y^2}\cdot\sqrt[3]{y^2} = y^{2/4}\cdot y^{2/3}$
$= y^{1/2+2/3}$
$= y^{3/6+4/6}$
$= y^{7/6}$
$= \sqrt[6]{y^7}$

113. $\dfrac{\sqrt[3]{x^4}}{\sqrt[4]{x^2}} = \dfrac{x^{4/3}}{x^{1/2}} = x^{4/3-1/2} = x^{8/6-3/6} = x^{5/6} = \sqrt[6]{x^5}$

115. $\dfrac{\sqrt[5]{n^4}}{\sqrt[3]{n^2}} = \dfrac{n^{4/5}}{n^{2/3}} = n^{4/5-2/3} = n^{12/15-10/15} = n^{2/15} = \sqrt[15]{n^2}$

117. $\sqrt{5}\cdot\sqrt[3]{3} = 5^{1/2}\cdot 3^{1/3}$
$= 5^{3/6}\cdot 3^{2/6}$
$= \left(5^3\cdot 3^2\right)^{1/6}$
$= (125\cdot 9)^{1/6}$
$= (1125)^{1/6}$
$= \sqrt[6]{1125}$

119. $\sqrt[4]{6}\cdot\sqrt[3]{2} = 6^{1/4}\cdot 2^{1/3}$
$= 6^{3/12}\cdot 2^{4/12}$
$= \left(6^3\cdot 2^4\right)^{1/12}$
$= (216\cdot 16)^{1/12}$
$= (3456)^{1/12}$
$= \sqrt[12]{3456}$

121. $\sqrt[3]{\sqrt[3]{x}} = \left(x^{1/3}\right)^{1/3} = x^{(1/3)\cdot(1/3)} = x^{1/9} = \sqrt[9]{x}$

123. $\sqrt{\sqrt[3]{n}} = \left(n^{1/3}\right)^{1/2} = n^{(1/3)\cdot(1/2)} = n^{1/6} = \sqrt[6]{n}$

Review Exercises

1. $2\cdot 2\cdot 2\cdot 2\cdot x\cdot x\cdot x\cdot y\cdot y = 2^4x^3y^2$

2. $\sqrt{16}\cdot\sqrt{9} = 4\cdot 3 = 12$

3. $\sqrt[3]{27}\cdot\sqrt[3]{125} = 3\cdot 5 = 15$

4. $\left(2.5\times 10^6\right)\left(3.2\times 10^5\right) = 8\times 10^{6+5} = 8\times 10^{11}$

5. $\left(\dfrac{3}{4}x^3y\right)\left(-\dfrac{5}{6}xyz^2\right) = -\dfrac{5}{8}x^{3+1}y^{1+1}z^2$
$= -\dfrac{5}{8}x^4y^2z^2$

6. $$\begin{array}{r} 2x^2-8x+5 \\ x+3\overline{)2x^3-2x^2-19x+18} \\ \underline{2x^3+6x^2} \\ -8x^2-19x \\ \underline{-8x^2-24x} \\ 5x+18 \\ \underline{5x+15} \\ 3 \end{array}$$

Answer: $2x^2-8x+5+\dfrac{3}{x+3}$

Exercise Set 8.3

1. Multiplying the approximate roots gives $\sqrt{8}\cdot\sqrt{18}\approx 2.828\cdot 4.243 = 11.999204$, which is tedious and inexact. Using the product rule for radicals gives $\sqrt{8}\cdot\sqrt{18} = \sqrt{8\cdot 18} = \sqrt{144} = 12$, which is fast and exact.

3. Rewrite the expression as a product of two radicals, the first containing the perfect cube and the second containing no perfect cubes. Then simplify the first radical. For example, $\sqrt[3]{54x^8}$ can be rewritten as $\sqrt[3]{27x^6}\cdot\sqrt[3]{2x^2}$, then simplified to $3x^2\sqrt[3]{2x^2}$.

5. $\sqrt{2}\cdot\sqrt{32} = \sqrt{64} = 8$

7. $\sqrt{3x}\cdot\sqrt{27x^5} = \sqrt{3x\cdot 27x^5} = \sqrt{81x^6} = 9x^3$

9. $\sqrt{6xy^3}\cdot\sqrt{24xy} = \sqrt{6xy^3\cdot 24xy}$
$= \sqrt{144x^2y^4}$
$= 12xy^2$

11. $\sqrt{5}\cdot\sqrt{13} = \sqrt{5\cdot 13} = \sqrt{65}$

13. $\sqrt{15}\cdot\sqrt{x}=\sqrt{15\cdot x}=\sqrt{15x}$

15. $\sqrt[3]{3}\cdot\sqrt[3]{9}=\sqrt[3]{3\cdot 9}=\sqrt[3]{27}=3$

17. $\sqrt[3]{5y}\cdot\sqrt[3]{2y}=\sqrt[3]{5y\cdot 2y}=\sqrt[3]{10y^2}$

19. $\sqrt[4]{3}\cdot\sqrt[4]{7}=\sqrt[4]{3\cdot 7}=\sqrt[4]{21}$

21. $\sqrt[4]{12w^3}\cdot\sqrt[4]{6w}=\sqrt[4]{12w^3\cdot 6w}$
$=\sqrt[4]{72w^4}$
$=\sqrt[4]{w^4\cdot 72}$
$=\sqrt[4]{w^4}\cdot\sqrt[4]{72}$
$=w\sqrt[4]{72}$

23. $\sqrt[4]{3x^2y}\cdot\sqrt[4]{5xy^2}=\sqrt[4]{3x^2y\cdot 5xy^2}=\sqrt[4]{15x^3y^3}$

25. $\sqrt[5]{6x^3}\cdot\sqrt[5]{5x^4}=\sqrt[5]{6x^3\cdot 5x^4}$
$=\sqrt[5]{30x^7}$
$=\sqrt[5]{x^5\cdot 30x^2}$
$=\sqrt[5]{x^5}\cdot\sqrt[5]{30x^2}$
$=x\sqrt[5]{30x^2}$

27. $\sqrt[6]{4x^2y^3}\cdot\sqrt[6]{2x^3y}=\sqrt[6]{4x^2y^3\cdot 2x^3y}=\sqrt[6]{8x^5y^4}$

29. $\sqrt{\frac{7}{2}}\cdot\sqrt{\frac{3}{5}}=\sqrt{\frac{7}{2}\cdot\frac{3}{5}}=\sqrt{\frac{21}{10}}$

31. $\sqrt{\frac{6}{x}}\cdot\sqrt{\frac{y}{5}}=\sqrt{\frac{6}{x}\cdot\frac{y}{5}}=\sqrt{\frac{6y}{5x}}$

33. $\sqrt{\frac{25}{36}}=\frac{\sqrt{25}}{\sqrt{36}}=\frac{5}{6}$

35. $\sqrt{\frac{10}{9}}=\frac{\sqrt{10}}{\sqrt{9}}=\frac{\sqrt{10}}{3}$

37. $\frac{\sqrt{180}}{\sqrt{5}}=\sqrt{\frac{180}{5}}=\sqrt{36}=6$

39. $\frac{\sqrt{15}}{\sqrt{5}}=\sqrt{\frac{15}{5}}=\sqrt{3}$

41. $\sqrt[3]{\frac{4}{w^6}}=\frac{\sqrt[3]{4}}{\sqrt[3]{w^6}}=\frac{\sqrt[3]{4}}{w^2}$

43. $\sqrt[3]{\frac{5y^2}{27x^9}}=\frac{\sqrt[3]{5y^2}}{\sqrt[3]{27x^9}}=\frac{\sqrt[3]{5y^2}}{3x^3}$

45. $\frac{\sqrt[3]{320}}{\sqrt[3]{5}}=\sqrt[3]{\frac{320}{5}}=\sqrt[3]{64}=4$

47. $\sqrt[4]{\frac{3u^3}{16x^8}}=\frac{\sqrt[4]{3u^3}}{\sqrt[4]{16x^8}}=\frac{\sqrt[4]{3u^3}}{2x^2}$

49. $\sqrt{98}=\sqrt{49\cdot 2}=\sqrt{49}\cdot\sqrt{2}=7\sqrt{2}$

51. $\sqrt{128}=\sqrt{64\cdot 2}=\sqrt{64}\cdot\sqrt{2}=8\sqrt{2}$

53. $6\sqrt{80}=6\sqrt{16\cdot 5}=6\sqrt{16}\cdot\sqrt{5}=6\cdot 4\sqrt{5}=24\sqrt{5}$

55. $5\sqrt{112}=5\sqrt{16\cdot 7}=5\sqrt{16}\cdot\sqrt{7}=5\cdot 4\sqrt{7}=20\sqrt{7}$

57. $\sqrt{a^7}=\sqrt{a^6\cdot a}=\sqrt{a^6}\cdot\sqrt{a}=a^3\sqrt{a}$

59. $\sqrt{x^2y^4}=xy^2$

61. $\sqrt{x^6y^8z^{10}}=x^3y^4z^5$

63. $rs^2\sqrt{r^9s^5}=rs^2\sqrt{r^8s^4\cdot rs}$
$=rs^2\sqrt{r^8s^4}\cdot\sqrt{rs}$
$=rs^2\cdot r^4s^2\cdot\sqrt{rs}$
$=r^5s^4\sqrt{rs}$

65. $3\sqrt{72x^5}=3\sqrt{36x^4\cdot 2x}$
$=3\sqrt{36x^4}\cdot\sqrt{2x}$
$=3\cdot 6x^2\cdot\sqrt{2x}$
$=18x^2\sqrt{2x}$

67. $\sqrt[3]{32}=\sqrt[3]{8\cdot 4}=\sqrt[3]{8}\cdot\sqrt[3]{4}=2\sqrt[3]{4}$

69. $\sqrt[3]{x^7}=\sqrt[3]{x^6\cdot x}=\sqrt[3]{x^6}\cdot\sqrt[3]{x}=x^2\sqrt[3]{x}$

71. $\sqrt[3]{x^6y^5}=\sqrt[3]{x^6y^3\cdot y^2}=\sqrt[3]{x^6y^3}\cdot\sqrt[3]{y^2}=x^2y\sqrt[3]{y^2}$

73. $\sqrt[3]{128z^8}=\sqrt[3]{64z^6\cdot 2z^2}$
$=\sqrt[3]{64z^6}\cdot\sqrt[3]{2z^2}$
$=4z^2\sqrt[3]{2z^2}$

75. $2\sqrt[3]{40}=2\sqrt[3]{8\cdot 5}=2\sqrt[3]{8}\cdot\sqrt[3]{5}=2\cdot 2\cdot\sqrt[3]{5}=4\sqrt[3]{5}$

77. $\sqrt[4]{80}=\sqrt[4]{16\cdot 5}=\sqrt[4]{16}\cdot\sqrt[4]{5}=2\sqrt[4]{5}$

79. $3x^2\sqrt[4]{243x^9}=3x^2\sqrt[4]{81x^8\cdot 3x}$
$=3x^2\sqrt[4]{81x^8}\cdot\sqrt[4]{3x}$
$=3x^2\cdot 3x^2\sqrt[4]{3x}$
$=9x^4\sqrt[4]{3x}$

81. $\sqrt[5]{486x^{16}} = \sqrt[5]{243x^{15} \cdot 2x}$
$= \sqrt[5]{243x^{15}} \cdot \sqrt[5]{2x}$
$= 3x^3\sqrt[5]{2x}$

83. $\sqrt[6]{x^8y^{14}z^{11}} = \sqrt[6]{x^6y^{12}z^6 \cdot x^2y^2z^5}$
$= \sqrt[6]{x^6y^{12}z^6} \cdot \sqrt[6]{x^2y^2z^5}$
$= xy^2z\sqrt[6]{x^2y^2z^5}$

85. $\sqrt{3} \cdot \sqrt{21} = \sqrt{63} = \sqrt{9 \cdot 7} = 3\sqrt{7}$

87. $5\sqrt{10} \cdot 3\sqrt{6} = 15\sqrt{60}$
$= 15\sqrt{4 \cdot 15}$
$= 15 \cdot 2\sqrt{15}$
$= 30\sqrt{15}$

89. $\sqrt{y^3} \cdot \sqrt{y^2} = \sqrt{y^5} = \sqrt{y^4 \cdot y} = y^2\sqrt{y}$

91. $x\sqrt{x^2y^3} \cdot y^2\sqrt{x^4y^4} = xy^2\sqrt{x^6y^7}$
$= xy^2\sqrt{x^6y^6 \cdot y}$
$= xy^2 \cdot x^3y^3\sqrt{y}$
$= x^4y^5\sqrt{y}$

93. $4\sqrt{6c^3} \cdot 3\sqrt{10c^5} = 12\sqrt{60c^8}$
$= 12\sqrt{4c^8 \cdot 15}$
$= 12 \cdot 2c^4\sqrt{15}$
$= 24c^4\sqrt{15}$

95. $4\sqrt{3} \cdot 5\sqrt{6} = 20\sqrt{18}$
$= 20\sqrt{9 \cdot 2}$
$= 20 \cdot 3\sqrt{2}$
$= 60\sqrt{2}$

97. $\dfrac{\sqrt{48}}{\sqrt{6}} = \sqrt{\dfrac{48}{6}} = \sqrt{8} = \sqrt{4 \cdot 2} = 2\sqrt{2}$

99. $\dfrac{9\sqrt{160}}{3\sqrt{8}} = 3\sqrt{\dfrac{160}{8}}$
$= 3\sqrt{20}$
$= 3\sqrt{4 \cdot 5}$
$= 3 \cdot 2\sqrt{5}$
$= 6\sqrt{5}$

101. $\dfrac{\sqrt{c^5d^6}}{\sqrt{cd^3}} = \sqrt{\dfrac{c^5d^6}{cd^3}} = \sqrt{c^4d^3} = \sqrt{c^4d^2 \cdot d} = c^2d\sqrt{d}$

103. $\dfrac{8\sqrt{45a^5}}{2\sqrt{5a}} = 4\sqrt{\dfrac{45a^5}{5a}}$
$= 4\sqrt{9a^4}$
$= 4 \cdot 3a^2$
$= 12a^2$

105. $\dfrac{12\sqrt{72c^5}}{4\sqrt{6c^2}} = 3\sqrt{\dfrac{72c^5}{6c^2}}$
$= 3\sqrt{12c^3}$
$= 3\sqrt{4c^2 \cdot 3c}$
$= 3 \cdot 2c\sqrt{3c}$
$= 6c\sqrt{3c}$

107. $\dfrac{36\sqrt{96x^6y^{11}}}{4\sqrt{3x^2y^4}} = 9\sqrt{\dfrac{96x^6y^{11}}{3x^2y^4}}$
$= 9\sqrt{32x^4y^7}$
$= 9\sqrt{16x^4y^6 \cdot 2y}$
$= 9 \cdot 4x^2y^3\sqrt{2y}$
$= 36x^2y^3\sqrt{2y}$

109. $\sqrt{\dfrac{3}{7}} \cdot \sqrt{\dfrac{8}{7}} = \sqrt{\dfrac{24}{49}} = \dfrac{\sqrt{24}}{7} = \dfrac{\sqrt{4 \cdot 6}}{7} = \dfrac{2\sqrt{6}}{7}$

111. $\sqrt{\dfrac{a^3}{2}} \cdot \sqrt{\dfrac{a^5}{2}} = \sqrt{\dfrac{a^8}{4}} = \dfrac{a^4}{2}$

113. $\sqrt{\dfrac{3x^5}{2}} \cdot \sqrt{\dfrac{15x^5}{8}} = \sqrt{\dfrac{45x^{10}}{16}}$
$= \dfrac{\sqrt{45x^{10}}}{\sqrt{16}}$
$= \dfrac{\sqrt{9x^{10} \cdot 5}}{4}$
$= \dfrac{3x^5\sqrt{5}}{4}$

115. $\frac{1}{2}\cdot\sqrt{\frac{3}{8}}\cdot\sqrt{\frac{15}{2}}=\frac{1}{2}\sqrt{\frac{45}{16}}$
$=\frac{1}{2}\cdot\frac{\sqrt{45}}{\sqrt{16}}$
$=\frac{1}{2}\cdot\frac{\sqrt{9\cdot 5}}{4}$
$=\frac{1}{2}\cdot\frac{3\sqrt{5}}{4}$
$=\frac{3}{8}\sqrt{5}$

Review Exercises

1. No 2. Yes

3. $6x^2-4x-3x+2x^2=8x^2-7x$

4. $(2a-3b)(4a+3b)$
$=2a\cdot 4a+2a\cdot 3b-3b\cdot 4a-3b\cdot 3b$
$=8a^2+6ab-12ab-9b^2$
$=8a^2-6ab-9b^2$

5. $(3m+5n)(3m-5n)=(3m)^2-(5n)^2$
$=9m^2-25n^2$

6. $(2x-3y)^2=(2x)^2-2(2x)(3y)+(3y)^2$
$=4x^2-12xy+9y^2$

Exercise Set 8.4

1. Like radicals have the same index and the same radicand, but their coefficients may be different.

3. $(x+3)(x+2)=x^2+2x+3x+6=x^2+5x+6,$
$(\sqrt{5}+3)(\sqrt{5}+2)=\sqrt{5^2}+2\sqrt{5}+3\sqrt{5}+6$
$=5+5\sqrt{5}+6=11+5\sqrt{5}$

In both cases, the FOIL method was used and then the expansion was simplified.

5. $9\sqrt{6}-15\sqrt{6}=(9-15)\sqrt{6}=-6\sqrt{6}$

7. $7\sqrt{a}+2\sqrt{a}=(7+2)\sqrt{a}=9\sqrt{a}$

9. $4\sqrt{5}-2\sqrt{6}+8\sqrt{5}-6\sqrt{6}$
$=(4+8)\sqrt{5}+(-2-6)\sqrt{6}$
$=12\sqrt{5}-8\sqrt{6}$

11. $3a\sqrt{5a}-4b\sqrt{7b}+8a\sqrt{5a}+2b\sqrt{7b}$
$=(3a+8a)\sqrt{5a}+(-4b+2b)\sqrt{7b}$
$=11a\sqrt{5a}-2b\sqrt{7b}$

13. $6x\sqrt[3]{9}-3x\sqrt[3]{9}=(6x-3x)\sqrt[3]{9}=3x\sqrt[3]{9}$

15. $6x^2\sqrt[4]{5x}-12x^2\sqrt[4]{5x}=(6x^2-12x^2)\sqrt[4]{5x}$
$=-6x^2\sqrt[4]{5x}$

17. $3x\sqrt{5x}+4x\sqrt[3]{5x}$
Cannot combine because the radicals are not like.

19. $\sqrt{48}-\sqrt{75}=\sqrt{16\cdot 3}-\sqrt{25\cdot 3}$
$=4\sqrt{3}-5\sqrt{3}$
$=-\sqrt{3}$

21. $\sqrt{80y}-\sqrt{125y}=\sqrt{16\cdot 5y}-\sqrt{25\cdot 5y}$
$=4\sqrt{5y}-5\sqrt{5y}$
$=-\sqrt{5y}$

23. $\sqrt{80}-4\sqrt{45}=\sqrt{16\cdot 5}-4\sqrt{9\cdot 5}$
$=4\sqrt{5}-4\cdot 3\sqrt{5}$
$=4\sqrt{5}-12\sqrt{5}$
$=-8\sqrt{5}$

25. $3\sqrt{96}-2\sqrt{54}=3\sqrt{16\cdot 6}-2\sqrt{9\cdot 6}$
$=3\cdot 4\sqrt{6}-2\cdot 3\sqrt{6}$
$=12\sqrt{6}-6\sqrt{6}$
$=6\sqrt{6}$

27. $6\sqrt{48a^3}-2\sqrt{75a^3}$
$=6\sqrt{16a^2\cdot 3a}-2\sqrt{25a^2\cdot 3a}$
$=6\cdot 4a\sqrt{3a}-2\cdot 5a\sqrt{3a}$
$=24a\sqrt{3a}-10a\sqrt{3a}$
$=14a\sqrt{3a}$

29. $\sqrt{150}-\sqrt{54}+\sqrt{24}=\sqrt{25\cdot 6}-\sqrt{9\cdot 6}+\sqrt{4\cdot 6}$
$=5\sqrt{6}-3\sqrt{6}+2\sqrt{6}$
$=4\sqrt{6}$

31. $2\sqrt{8}-3\sqrt{48}+2\sqrt{98}-\sqrt{75}$
$=2\sqrt{4\cdot 2}-3\sqrt{16\cdot 3}+2\sqrt{49\cdot 2}-\sqrt{25\cdot 3}$
$=2\cdot 2\sqrt{2}-3\cdot 4\sqrt{3}+2\cdot 7\sqrt{2}-5\sqrt{3}$
$=4\sqrt{2}-12\sqrt{3}+14\sqrt{2}-5\sqrt{3}$
$=18\sqrt{2}-17\sqrt{3}$

33. $\sqrt[3]{128}+\sqrt[3]{54}=\sqrt[3]{64\cdot 2}+\sqrt[3]{27\cdot 2}$
$=4\sqrt[3]{2}+3\sqrt[3]{2}$
$=7\sqrt[3]{2}$

35. $4\sqrt[3]{135x^5}-6x\sqrt[3]{320x^2}$
$=4\sqrt[3]{27x^3\cdot 5x^2}-6x\sqrt[3]{64\cdot 5x^2}$
$=4\cdot 3x\sqrt[3]{5x^2}-6x\cdot 4\sqrt[3]{5x^2}$
$=12x\sqrt[3]{5x^2}-24x\sqrt[3]{5x^2}$
$=-12x\sqrt[3]{5x^2}$

37. $-4\sqrt[4]{32x^9}+2x\sqrt[4]{162x^5}$
$=-4\sqrt[4]{16x^8\cdot 2x}+2x\sqrt[4]{81x^4\cdot 2x}$
$=-4\cdot 2x^2\sqrt[4]{2x}+2x\cdot 3x\sqrt[4]{2x}$
$=-8x^2\sqrt[4]{2x}+6x^2\sqrt[4]{2x}$
$=-2x^2\sqrt[4]{2x}$

39. $\sqrt{2}\left(3+\sqrt{2}\right)=\sqrt{2}\cdot 3+\sqrt{2}\cdot\sqrt{2}$
$=3\sqrt{2}+\sqrt{4}$
$=3\sqrt{2}+2$

41. $\sqrt{3}\left(\sqrt{3}-\sqrt{15}\right)=\sqrt{3}\cdot\sqrt{3}-\sqrt{3}\cdot\sqrt{15}$
$=\sqrt{9}-\sqrt{45}$
$=3-\sqrt{9\cdot 5}$
$=3-3\sqrt{5}$

43. $\sqrt{5}\left(\sqrt{6}+2\sqrt{10}\right)=\sqrt{5}\cdot\sqrt{6}+\sqrt{5}\cdot 2\sqrt{10}$
$=\sqrt{30}+2\sqrt{50}$
$=\sqrt{30}+2\sqrt{25\cdot 2}$
$=\sqrt{30}+2\cdot 5\sqrt{2}$
$=\sqrt{30}+10\sqrt{2}$

45. $4\sqrt{3x}\left(2\sqrt{3x}-4\sqrt{6x}\right)$
$=4\sqrt{3x}\cdot 2\sqrt{3x}-4\sqrt{3x}\cdot 4\sqrt{6x}$
$=8\sqrt{9x^2}-16\sqrt{18x^2}$
$=8\cdot 3x-16\sqrt{9x^2\cdot 2}$
$=24x-16\cdot 3x\sqrt{2}$
$=24x-48x\sqrt{2}$

47. $\left(3+\sqrt{5}\right)\left(4-\sqrt{2}\right)$
$=3\cdot 4-3\cdot\sqrt{2}+\sqrt{5}\cdot 4-\sqrt{5}\cdot\sqrt{2}$
$=12-3\sqrt{2}+4\sqrt{5}-\sqrt{10}$

49. $\left(3+\sqrt{x}\right)\left(2+\sqrt{x}\right)$
$=3\cdot 2+3\cdot\sqrt{x}+\sqrt{x}\cdot 2+\sqrt{x}\cdot\sqrt{x}$
$=6+3\sqrt{x}+2\sqrt{x}+\sqrt{x^2}$
$=6+5\sqrt{x}+x$

51. $\left(2+3\sqrt{3}\right)\left(3+5\sqrt{2}\right)$
$=2\cdot 3+2\cdot 5\sqrt{2}+3\sqrt{3}\cdot 3+3\sqrt{3}\cdot 5\sqrt{2}$
$=6+10\sqrt{2}+9\sqrt{3}+15\sqrt{6}$

53. $\left(\sqrt{3}+\sqrt{5}\right)\left(\sqrt{5}+\sqrt{7}\right)$
$=\sqrt{3}\cdot\sqrt{5}+\sqrt{3}\cdot\sqrt{7}+\sqrt{5}\cdot\sqrt{5}+\sqrt{5}\cdot\sqrt{7}$
$=\sqrt{15}+\sqrt{21}+\sqrt{25}+\sqrt{35}$
$=\sqrt{15}+\sqrt{21}+5+\sqrt{35}$

55. $\left(\sqrt{x}+3\sqrt{y}\right)\left(\sqrt{x}-2\sqrt{y}\right)$
$=\sqrt{x}\cdot\sqrt{x}-\sqrt{x}\cdot 2\sqrt{y}+3\sqrt{y}\cdot\sqrt{x}-3\sqrt{y}\cdot 2\sqrt{y}$
$=\sqrt{x^2}-2\sqrt{xy}+3\sqrt{xy}-6\sqrt{y^2}$
$=x+\sqrt{xy}-6y$

57. $\left(4\sqrt{2}+2\sqrt{5}\right)\left(3\sqrt{7}-3\sqrt{3}\right)$
$=4\sqrt{2}\cdot 3\sqrt{7}-4\sqrt{2}\cdot 3\sqrt{3}+2\sqrt{5}\cdot 3\sqrt{7}-2\sqrt{5}\cdot 3\sqrt{3}$
$=12\sqrt{14}-12\sqrt{6}+6\sqrt{35}-6\sqrt{15}$

59. $\left(2\sqrt{a}+3\sqrt{b}\right)\left(4\sqrt{a}-\sqrt{b}\right)$
$=2\sqrt{a}\cdot 4\sqrt{a}-2\sqrt{a}\cdot\sqrt{b}+3\sqrt{b}\cdot 4\sqrt{a}-3\sqrt{b}\cdot\sqrt{b}$
$=8\sqrt{a^2}-2\sqrt{ab}+12\sqrt{ab}-3\sqrt{b^2}$
$=8a+10\sqrt{ab}-3b$

61. $\left(\sqrt[3]{4}+5\right)\left(\sqrt[3]{4}-8\right)$
$= \sqrt[3]{4}\cdot\sqrt[3]{4}-\sqrt[3]{4}\cdot 8+5\sqrt[3]{4}-5\cdot 8$
$= \sqrt[3]{16}-8\sqrt[3]{4}+5\sqrt[3]{4}-40$
$= \sqrt[3]{8\cdot 2}-3\sqrt[3]{4}-40$
$= 2\sqrt[3]{2}-3\sqrt[3]{4}-40$

63. $\left(\sqrt[3]{9}+\sqrt[3]{4}\right)\left(\sqrt[3]{3}-\sqrt[3]{2}\right)$
$= \sqrt[3]{9}\cdot\sqrt[3]{3}-\sqrt[3]{9}\cdot\sqrt[3]{2}+\sqrt[3]{4}\cdot\sqrt[3]{3}-\sqrt[3]{4}\cdot\sqrt[3]{2}$
$= \sqrt[3]{27}-\sqrt[3]{18}+\sqrt[3]{12}-\sqrt[3]{8}$
$= 3-\sqrt[3]{18}+\sqrt[3]{12}-2$
$= 1-\sqrt[3]{18}+\sqrt[3]{12}$

65. $\left(\sqrt[3]{x}+2\right)\left(\sqrt[3]{x^2}-2\sqrt[3]{x}+4\right)$
$= \sqrt[3]{x}\cdot\sqrt[3]{x^2}-\sqrt[3]{x}\cdot 2\sqrt[3]{x}+\sqrt[3]{x}\cdot 4+2\sqrt[3]{x^2}$
$-2\cdot 2\sqrt[3]{x}+2\cdot 4$
$= \sqrt[3]{x^3}-2\sqrt[3]{x^2}+4\sqrt[3]{x}+2\sqrt[3]{x^2}$
$-4\sqrt[3]{x}+8$
$= x-2\sqrt[3]{x^2}+4\sqrt[3]{x}+2\sqrt[3]{x^2}-4\sqrt[3]{x}+8$
$= x+8$

67. $\left(4+\sqrt{6}\right)^2 = 4^2+2\cdot 4\sqrt{6}+\left(\sqrt{6}\right)^2$
$= 16+8\sqrt{6}+6$
$= 22+8\sqrt{6}$

69. $\left(4-\sqrt{2}\right)^2 = 4^2-2\cdot 4\sqrt{2}+\left(\sqrt{2}\right)^2$
$= 16-8\sqrt{2}+2$
$= 18-8\sqrt{2}$

71. $\left(2+2\sqrt{3}\right)^2 = 2^2+2\cdot 4\sqrt{3}+\left(2\sqrt{3}\right)^2$
$= 4+8\sqrt{3}+4\cdot 3$
$= 4+8\sqrt{3}+12$
$= 16+8\sqrt{3}$

73. $\left(2\sqrt{3}+3\sqrt{2}\right)^2 = \left(2\sqrt{3}\right)^2+2\cdot 6\sqrt{6}+\left(3\sqrt{2}\right)^2$
$= 4\cdot 3+12\sqrt{6}+9\cdot 2$
$= 12+12\sqrt{6}+18$
$= 30+12\sqrt{6}$

75. $\left(4+\sqrt{3}\right)\left(4-\sqrt{3}\right) = 4^2-\left(\sqrt{3}\right)^2 = 16-3 = 13$

77. $\left(\sqrt{2}+4\right)\left(\sqrt{2}-4\right) = \left(\sqrt{2}\right)^2-4^2 = 2-16 = -14$

79. $\left(6+\sqrt{x}\right)\left(6-\sqrt{x}\right) = 6^2-\left(\sqrt{x}\right)^2 = 36-x$

81. $\left(\sqrt{3}+\sqrt{2}\right)\left(\sqrt{3}-\sqrt{2}\right) = \left(\sqrt{3}\right)^2-\left(\sqrt{2}\right)^2 = 3-2 = 1$

83. $\left(\sqrt{x}+\sqrt{y}\right)\left(\sqrt{x}-\sqrt{y}\right) = \left(\sqrt{x}\right)^2-\left(\sqrt{y}\right)^2 = x-y$

85. $\left(4+2\sqrt{3}\right)\left(4-2\sqrt{3}\right) = 4^2-\left(2\sqrt{3}\right)^2$
$= 16-4\cdot 3$
$= 16-12$
$= 4$

87. $\left(3\sqrt{7}+\sqrt{13}\right)\left(3\sqrt{7}-\sqrt{13}\right) = \left(3\sqrt{7}\right)^2-\left(\sqrt{13}\right)$
$= 9\cdot 7-13$
$= 63-13$
$= 50$

89. $\sqrt{3}\cdot\sqrt{15}+\sqrt{8}\cdot\sqrt{10} = \sqrt{3\cdot 15}+\sqrt{8\cdot 10}$
$= \sqrt{45}+\sqrt{80}$
$= \sqrt{9\cdot 5}+\sqrt{16\cdot 5}$
$= 3\sqrt{5}+4\sqrt{5}$
$= 7\sqrt{5}$

91. $3\sqrt{3}\cdot\sqrt{18}-4\sqrt{18}\cdot\sqrt{12} = 3\sqrt{3\cdot 18}-4\sqrt{18\cdot 12}$
$= 3\sqrt{54}-4\sqrt{216}$
$= 3\sqrt{9\cdot 6}-4\sqrt{36\cdot 6}$
$= 3\cdot 3\sqrt{6}-4\cdot 6\sqrt{6}$
$= 9\sqrt{6}-24\sqrt{6}$
$= -15\sqrt{6}$

93. $\dfrac{\sqrt{54}}{\sqrt{3}}+\sqrt{72} = \sqrt{\dfrac{54}{3}}+\sqrt{72}$
$= \sqrt{18}+\sqrt{36\cdot 2}$
$= \sqrt{9\cdot 2}+6\sqrt{2}$
$= 3\sqrt{2}+6\sqrt{2}$
$= 9\sqrt{2}$

95. $\dfrac{\sqrt{540}}{\sqrt{3}}-4\sqrt{125}=\sqrt{\dfrac{540}{3}}-4\sqrt{125}$
$=\sqrt{180}-4\sqrt{125}$
$=\sqrt{36\cdot5}-4\sqrt{25\cdot5}$
$=6\sqrt{5}-4\cdot5\sqrt{5}$
$=6\sqrt{5}-20\sqrt{5}$
$=-14\sqrt{5}$

97. $5\sqrt{3}+4\sqrt{12}+4\sqrt{27}$
$=5\sqrt{3}+4\sqrt{4\cdot3}+4\sqrt{9\cdot3}$
$=5\sqrt{3}+4\cdot2\sqrt{3}+4\cdot3\sqrt{3}$
$=5\sqrt{3}+8\sqrt{3}+12\sqrt{3}$
$=25\sqrt{3}$

99. a) $13+10+\sqrt{5}+9+\sqrt{5}+10=\left(42+2\sqrt{5}\right)$ ft.
b) 46.5 ft. c) $46.5(\$1.89)=\$87.89$

Review Exercises

1. $2x+5$

2. $(4x+3)(4x-3)=(4x)^2-3^2=16x^2-9$

3. $\sqrt{2}$ because $\sqrt{8}\cdot\sqrt{2}=\sqrt{8\cdot2}=\sqrt{16}$

4. $\sqrt[3]{4}$ because $\sqrt[3]{2}\cdot\sqrt[3]{4}=\sqrt[3]{2\cdot4}=\sqrt[3]{8}$

5. $5y-2x=10$ slope is $\dfrac{2}{5}$, y-intercept is (0,2)
$5y=2x+10$
$\dfrac{5y}{5}=\dfrac{2x+10}{5}$
$y=\dfrac{2}{5}x+2$

6. $y=-\dfrac{1}{3}x+4$

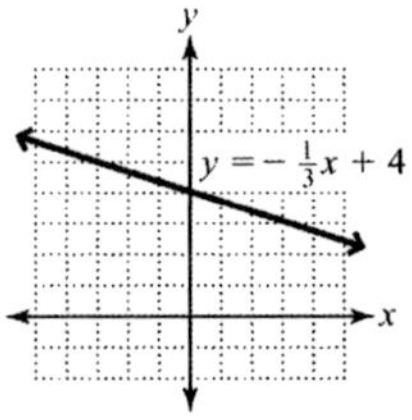

Exercise Set 8.5

1. a) $\sqrt{16}$ is rational. b) There is a radical, $\sqrt{3}$, in the denominator.

3. Multiply the fraction by a 1 so that the product's denominator has a radicand that is a perfect square.

5. $\dfrac{1}{\sqrt{3}}=\dfrac{1}{\sqrt{3}}\cdot\dfrac{\sqrt{3}}{\sqrt{3}}=\dfrac{\sqrt{3}}{\sqrt{9}}=\dfrac{\sqrt{3}}{3}$

7. $\dfrac{3}{\sqrt{8}}=\dfrac{3}{\sqrt{8}}\cdot\dfrac{\sqrt{2}}{\sqrt{2}}=\dfrac{3\sqrt{2}}{\sqrt{16}}=\dfrac{3\sqrt{2}}{4}$

9. $\sqrt{\dfrac{36}{7}}=\dfrac{\sqrt{36}}{\sqrt{7}}=\dfrac{6}{\sqrt{7}}\cdot\dfrac{\sqrt{7}}{\sqrt{7}}=\dfrac{6\sqrt{7}}{\sqrt{49}}=\dfrac{6\sqrt{7}}{7}$

11. $\sqrt{\dfrac{5}{12}}=\dfrac{\sqrt{5}}{\sqrt{12}}=\dfrac{\sqrt{5}}{\sqrt{12}}\cdot\dfrac{\sqrt{3}}{\sqrt{3}}=\dfrac{\sqrt{15}}{\sqrt{36}}=\dfrac{\sqrt{15}}{6}$

13. $\dfrac{\sqrt{7x^2}}{\sqrt{50}}=\dfrac{\sqrt{x^2\cdot7}}{\sqrt{25\cdot2}}=\dfrac{x\sqrt{7}}{5\sqrt{2}}\cdot\dfrac{\sqrt{2}}{\sqrt{2}}$
$=\dfrac{x\sqrt{14}}{5\sqrt{4}}=\dfrac{x\sqrt{14}}{5\cdot2}=\dfrac{x\sqrt{14}}{10}$

15. $\dfrac{\sqrt{8}}{\sqrt{56}}=\sqrt{\dfrac{8}{56}}=\sqrt{\dfrac{1}{7}}=\dfrac{\sqrt{1}}{\sqrt{7}}=\dfrac{1}{\sqrt{7}}\cdot\dfrac{\sqrt{7}}{\sqrt{7}}$
$=\dfrac{\sqrt{7}}{\sqrt{49}}=\dfrac{\sqrt{7}}{7}$

17. $\dfrac{5}{\sqrt{3a}}=\dfrac{5}{\sqrt{3a}}\cdot\dfrac{\sqrt{3a}}{\sqrt{3a}}=\dfrac{5\sqrt{3a}}{\sqrt{9a^2}}=\dfrac{5\sqrt{3a}}{3a}$

19. $\sqrt{\dfrac{3m}{11n}}=\dfrac{\sqrt{3m}}{\sqrt{11n}}\cdot\dfrac{\sqrt{11n}}{\sqrt{11n}}=\dfrac{\sqrt{33mn}}{\sqrt{121n^2}}=\dfrac{\sqrt{33mn}}{11n}$

21. $\dfrac{10}{\sqrt{5x}}=\dfrac{10}{\sqrt{5x}}\cdot\dfrac{\sqrt{5x}}{\sqrt{5x}}=\dfrac{10\sqrt{5x}}{\sqrt{25x^2}}=\dfrac{10\sqrt{5x}}{5x}=\dfrac{2\sqrt{5x}}{x}$

23. $\dfrac{\sqrt{6x}}{\sqrt{32x}}=\sqrt{\dfrac{6x}{32x}}=\sqrt{\dfrac{3}{16}}=\dfrac{\sqrt{3}}{\sqrt{16}}=\dfrac{\sqrt{3}}{4}$

25. $\dfrac{3}{\sqrt{x^3}}=\dfrac{3}{\sqrt{x^2\cdot x}}$
$=\dfrac{3}{x\sqrt{x}}\cdot\dfrac{\sqrt{x}}{\sqrt{x}}$
$=\dfrac{3\sqrt{x}}{x\sqrt{x^2}}$
$=\dfrac{3\sqrt{x}}{x\cdot x}$
$=\dfrac{3\sqrt{x}}{x^2}$

27. $\frac{8x^2}{\sqrt{2x}} = \frac{8x^2}{\sqrt{2x}} \cdot \frac{\sqrt{2x}}{\sqrt{2x}}$
$= \frac{8x^2\sqrt{2x}}{\sqrt{4x^2}}$
$= \frac{8x^2\sqrt{2x}}{2x}$
$= 4x\sqrt{2x}$

29. Mistake: The product of $\sqrt{2}$ and 2 is not 2.
Correct: $\frac{\sqrt{3}}{\sqrt{2}} \cdot \frac{\sqrt{2}}{\sqrt{2}} = \frac{\sqrt{6}}{\sqrt{4}} = \frac{\sqrt{6}}{2}$

31. $\frac{5}{\sqrt[3]{3}} = \frac{5}{\sqrt[3]{3}} \cdot \frac{\sqrt[3]{9}}{\sqrt[3]{9}} = \frac{5\sqrt[3]{9}}{\sqrt[3]{27}} = \frac{5\sqrt[3]{9}}{3}$

33. $\sqrt[3]{\frac{5}{2}} = \frac{\sqrt[3]{5}}{\sqrt[3]{2}} \cdot \frac{\sqrt[3]{4}}{\sqrt[3]{4}} = \frac{\sqrt[3]{20}}{\sqrt[3]{8}} = \frac{\sqrt[3]{20}}{2}$

35. $\frac{6}{\sqrt[3]{4}} = \frac{6}{\sqrt[3]{4}} \cdot \frac{\sqrt[3]{2}}{\sqrt[3]{2}} = \frac{6\sqrt[3]{2}}{\sqrt[3]{8}} = \frac{6\sqrt[3]{2}}{2} = 3\sqrt[3]{2}$

37. $\frac{m}{\sqrt[3]{n}} = \frac{m}{\sqrt[3]{n}} \cdot \frac{\sqrt[3]{n^2}}{\sqrt[3]{n^2}} = \frac{m\sqrt[3]{n^2}}{\sqrt[3]{n^3}} = \frac{m\sqrt[3]{n^2}}{n}$

39. $\sqrt[3]{\frac{a}{b^2}} = \frac{\sqrt[3]{a}}{\sqrt[3]{b^2}} \cdot \frac{\sqrt[3]{b}}{\sqrt[3]{b}} = \frac{\sqrt[3]{ab}}{\sqrt[3]{b^3}} = \frac{\sqrt[3]{ab}}{b}$

41. $\frac{4}{\sqrt[3]{2x}} = \frac{4}{\sqrt[3]{2x}} \cdot \frac{\sqrt[3]{4x^2}}{\sqrt[3]{4x^2}}$
$= \frac{4\sqrt[3]{4x^2}}{\sqrt[3]{8x^3}}$
$= \frac{4\sqrt[3]{4x^2}}{2x}$
$= \frac{2\sqrt[3]{4x^2}}{x}$

43. $\sqrt[3]{\frac{6}{25a^2}} = \frac{\sqrt[3]{6}}{\sqrt[3]{25a^2}} \cdot \frac{\sqrt[3]{5a}}{\sqrt[3]{5a}} = \frac{\sqrt[3]{30a}}{\sqrt[3]{125a^3}} = \frac{\sqrt[3]{30a}}{5a}$

45. $\frac{5}{\sqrt[4]{4}} = \frac{5}{\sqrt[4]{4}} \cdot \frac{\sqrt[4]{4}}{\sqrt[4]{4}} = \frac{5\sqrt[4]{4}}{\sqrt[4]{16}} = \frac{5\sqrt[4]{4}}{2} = \frac{5\sqrt[4]{2^2}}{2} = \frac{5\sqrt{2}}{2}$

47. $\sqrt[4]{\frac{3}{x^2}} = \frac{\sqrt[4]{3}}{\sqrt[4]{x^2}} \cdot \frac{\sqrt[4]{x^2}}{\sqrt[4]{x^2}} = \frac{\sqrt[4]{3x^2}}{\sqrt[4]{x^4}} = \frac{\sqrt[4]{3x^2}}{x}$

49. $\frac{9}{\sqrt[4]{3x^3}} = \frac{9}{\sqrt[4]{3x^3}} \cdot \frac{\sqrt[4]{27x}}{\sqrt[4]{27x}}$
$= \frac{9\sqrt[4]{27x}}{\sqrt[4]{81x^4}}$
$= \frac{9\sqrt[4]{27x}}{3x}$
$= \frac{3\sqrt[4]{27x}}{x}$

51. $\frac{3}{\sqrt{2}+1} \cdot \frac{\sqrt{2}-1}{\sqrt{2}-1} = \frac{3(\sqrt{2}-1)}{(\sqrt{2})^2 - 1^2}$
$= \frac{3\sqrt{2}-3}{2-1}$
$= \frac{3\sqrt{2}-3}{1}$
$= 3\sqrt{2}-3$

53. $\frac{4}{2-\sqrt{3}} \cdot \frac{2+\sqrt{3}}{2+\sqrt{3}} = \frac{4(2+\sqrt{3})}{2^2-(\sqrt{3})^2}$
$= \frac{8+4\sqrt{3}}{4-3}$
$= \frac{8+4\sqrt{3}}{1}$
$= 8+4\sqrt{3}$

55. $\frac{5}{\sqrt{2}+\sqrt{3}} \cdot \frac{\sqrt{2}-\sqrt{3}}{\sqrt{2}-\sqrt{3}} = \frac{5(\sqrt{2}-\sqrt{3})}{(\sqrt{2})^2-(\sqrt{3})^2}$
$= \frac{5\sqrt{2}-5\sqrt{3}}{2-3}$
$= \frac{5\sqrt{2}-5\sqrt{3}}{-1}$
$= 5\sqrt{3}-5\sqrt{2}$

57. $\frac{4}{1-\sqrt{5}} \cdot \frac{1+\sqrt{5}}{1+\sqrt{5}} = \frac{4(1+\sqrt{5})}{1^2-(\sqrt{5})^2}$
$= \frac{4(1+\sqrt{5})}{1-5}$
$= \frac{4(1+\sqrt{5})}{-4}$
$= -1(1+\sqrt{5})$
$= -1-\sqrt{5}$

59. $$\frac{\sqrt{3}}{\sqrt{3}-1}\cdot\frac{\sqrt{3}+1}{\sqrt{3}+1}=\frac{\sqrt{3}\left(\sqrt{3}+1\right)}{\left(\sqrt{3}\right)^2-1^2}=\frac{3+\sqrt{3}}{3-1}=\frac{3+\sqrt{3}}{2}$$

61. $$\frac{2\sqrt{3}}{\sqrt{3}-4}\cdot\frac{\sqrt{3}+4}{\sqrt{3}+4}=\frac{2\sqrt{3}\left(\sqrt{3}+4\right)}{\left(\sqrt{3}\right)^2-4^2}=\frac{2\cdot 3+8\sqrt{3}}{3-16}=\frac{6+8\sqrt{3}}{-13}=\frac{-6-8\sqrt{3}}{13}$$

63. $$\frac{4\sqrt{3}}{\sqrt{7}+\sqrt{2}}\cdot\frac{\sqrt{7}-\sqrt{2}}{\sqrt{7}-\sqrt{2}}=\frac{4\sqrt{3}\left(\sqrt{7}-\sqrt{2}\right)}{\left(\sqrt{7}\right)^2-\left(\sqrt{2}\right)^2}=\frac{4\sqrt{21}-4\sqrt{6}}{7-2}=\frac{4\sqrt{21}-4\sqrt{6}}{5}$$

65. $$\frac{8\sqrt{2}}{4\sqrt{2}-\sqrt{6}}\cdot\frac{4\sqrt{2}+\sqrt{6}}{4\sqrt{2}+\sqrt{6}}=\frac{32\sqrt{4}+8\sqrt{12}}{16\cdot 2-6}=\frac{32\cdot 2+8\sqrt{4\cdot 3}}{32-6}=\frac{64+8\cdot 2\sqrt{3}}{26}=\frac{64+16\sqrt{3}}{26}=\frac{\not{2}\left(32+8\sqrt{3}\right)}{\not{2}\cdot 13}=\frac{32+8\sqrt{3}}{13}$$

67. $$\frac{6\sqrt{y}}{\sqrt{y}+1}\cdot\frac{\sqrt{y}-1}{\sqrt{y}-1}=\frac{6\sqrt{y}\left(\sqrt{y}-1\right)}{\left(\sqrt{y}\right)^2-1^2}=\frac{6y-6\sqrt{y}}{y-1}$$

69. $$\frac{3\sqrt{t}}{\sqrt{t}+2\sqrt{u}}\cdot\frac{\sqrt{t}-2\sqrt{u}}{\sqrt{t}-2\sqrt{u}}=\frac{3\sqrt{t}\left(\sqrt{t}-2\sqrt{u}\right)}{\left(\sqrt{t}\right)^2-\left(2\sqrt{u}\right)^2}=\frac{3t-6\sqrt{tu}}{t-4u}$$

71. $$\frac{\sqrt{2y}}{\sqrt{x}-\sqrt{6y}}\cdot\frac{\sqrt{x}+\sqrt{6y}}{\sqrt{x}+\sqrt{6y}}=\frac{\sqrt{2y}\left(\sqrt{x}+\sqrt{6y}\right)}{\left(\sqrt{x}\right)^2-\left(\sqrt{6y}\right)^2}=\frac{\sqrt{2xy}+\sqrt{12y^2}}{x-6y}=\frac{\sqrt{2xy}+\sqrt{4y^2\cdot 3}}{x-6y}=\frac{\sqrt{2xy}+2y\sqrt{3}}{x-6y}$$

73. $$\frac{\sqrt{3}}{2}\cdot\frac{\sqrt{3}}{\sqrt{3}}=\frac{\sqrt{9}}{2\sqrt{3}}=\frac{3}{2\sqrt{3}}$$

75. $$\frac{\sqrt{2x}}{5}\cdot\frac{\sqrt{2x}}{\sqrt{2x}}=\frac{\sqrt{4x^2}}{5\sqrt{2x}}=\frac{2x}{5\sqrt{2x}}$$

77. $$\frac{\sqrt{8n}}{6}=\frac{\sqrt{4\cdot 2n}}{6}=\frac{2\sqrt{2n}}{6}=\frac{\sqrt{2n}}{3}\cdot\frac{\sqrt{2n}}{\sqrt{2n}}=\frac{\sqrt{4n^2}}{3\sqrt{2n}}=\frac{2n}{3\sqrt{2n}}$$

79. $$\frac{2+\sqrt{3}}{5}\cdot\frac{2-\sqrt{3}}{2-\sqrt{3}}=\frac{2^2-\left(\sqrt{3}\right)^2}{5\left(2-\sqrt{3}\right)}=\frac{4-3}{10-5\sqrt{3}}=\frac{1}{10-5\sqrt{3}}$$

81. $$\frac{\sqrt{5x}-6}{9}\cdot\frac{\sqrt{5x}+6}{\sqrt{5x}+6}=\frac{\left(\sqrt{5x}\right)^2-6^2}{9\left(\sqrt{5x}+6\right)}=\frac{5x-36}{9\sqrt{5x}+54}$$

83. $\dfrac{5\sqrt{n}+\sqrt{6n}}{2n}\cdot\dfrac{5\sqrt{n}-\sqrt{6n}}{5\sqrt{n}-\sqrt{6n}}=\dfrac{\left(5\sqrt{n}\right)^2-\left(\sqrt{6n}\right)^2}{2n\left(5\sqrt{n}-\sqrt{6n}\right)}$

$=\dfrac{25n-6n}{10n\sqrt{n}-2n\sqrt{6n}}$

$=\dfrac{19n}{10n\sqrt{n}-2n\sqrt{6n}}$

$=\dfrac{19\cancel{n}}{\cancel{n}\left(10\sqrt{n}-2\sqrt{6n}\right)}$

$=\dfrac{19}{10\sqrt{n}-2\sqrt{6n}}$

85. $f(x)=\dfrac{5\sqrt{2}}{x}$

a) $f\left(\sqrt{6}\right)=\dfrac{5\sqrt{2}}{\sqrt{6}}=5\sqrt{\dfrac{2}{6}}=5\sqrt{\dfrac{1}{3}}$

$=\dfrac{5}{\sqrt{3}}\cdot\dfrac{\sqrt{3}}{\sqrt{3}}=\dfrac{5\sqrt{3}}{3}$

b) $f\left(\sqrt{10}\right)=\dfrac{5\sqrt{2}}{\sqrt{10}}=5\sqrt{\dfrac{2}{10}}=5\sqrt{\dfrac{1}{5}}$

$=\dfrac{5}{\sqrt{5}}\cdot\dfrac{\sqrt{5}}{\sqrt{5}}=\dfrac{5\sqrt{5}}{\sqrt{25}}=\dfrac{5\sqrt{5}}{5}$

$=\sqrt{5}$

c) $f\left(\sqrt{22}\right)=\dfrac{5\sqrt{2}}{\sqrt{22}}=5\sqrt{\dfrac{2}{22}}=5\sqrt{\dfrac{1}{11}}$

$=\dfrac{5}{\sqrt{11}}\cdot\dfrac{\sqrt{11}}{\sqrt{11}}=\dfrac{5\sqrt{11}}{\sqrt{121}}$

$=\dfrac{5\sqrt{11}}{11}$

87. a) The graphs are identical. The functions are identical.

b) $f(x)=g(x)$

89. a) $T=\dfrac{2\pi\sqrt{L}}{\sqrt{9.8}}\cdot\dfrac{\sqrt{9.8}}{\sqrt{9.8}}$

$=\dfrac{2\pi\sqrt{9.8L}}{9.8}$

$=\dfrac{\pi\sqrt{9.8L}}{4.9}$

b) $T=\dfrac{2\pi\sqrt{L}}{\sqrt{9.8}}\cdot\dfrac{\sqrt{L}}{\sqrt{L}}$

$=\dfrac{2\pi L}{\sqrt{9.8L}}$

91. a) $s=\dfrac{\sqrt{3V}}{\sqrt{h}}\cdot\dfrac{\sqrt{h}}{\sqrt{h}}=\dfrac{\sqrt{3Vh}}{\sqrt{h^2}}=\dfrac{\sqrt{3Vh}}{h}$

b) $s=\dfrac{\sqrt{3(83{,}068{,}742)449}}{449}\approx 745$ ft.

93. a) $V_{rms}=\dfrac{V_m}{\sqrt{2}}\cdot\dfrac{\sqrt{2}}{\sqrt{2}}=\dfrac{\sqrt{2}V_m}{\sqrt{4}}=\dfrac{\sqrt{2}V_m}{2}$

b) $V_{rms}=\dfrac{163\sqrt{2}}{2}$ c) $V_{rms}\approx 115.3$

95. $\dfrac{5\sqrt{2}}{3+\sqrt{6}}\cdot\dfrac{3-\sqrt{6}}{3-\sqrt{6}}=\dfrac{5\sqrt{2}\left(3-\sqrt{6}\right)}{3^2-\left(\sqrt{6}\right)^2}$

$=\dfrac{15\sqrt{2}-5\sqrt{12}}{9-6}$

$=\dfrac{15\sqrt{2}-5\cdot 2\sqrt{3}}{3}$

$=\dfrac{15\sqrt{2}-10\sqrt{3}}{3}\Omega$

Review Exercises

1. $\pm\sqrt{28}=\pm\sqrt{4\cdot 7}=\pm 2\sqrt{7}$

2. $x^2-6x+9=(x-3)(x-3)=(x-3)^2$

3. $2x-3=5$
$2x=8$
$x=4$

4. $2x-3=-5$
$2x=-2$
$x=-1$

5. $x^2-36=0$
$(x+6)(x-6)=0$
$x+6=0$ or $x-6=0$
$x=-6$ $x=6$

6. $x^2-5x+6=0$

$(x-2)(x-3)=0$

$x-2=0$ or $x-3=0$

$x=2 \qquad x=3$

Exercise Set 8.6

1. Some of the solutions may be extraneous.

3. The principal square root of a number cannot equal a negative.

5. Subtract $3x$ from both sides to isolate the radical. This allows use of the power rule to eliminate the radical.

7. $\sqrt{x}=2$

$(\sqrt{x})^2=2^2$

$x=4$

9. $\sqrt{k}=-4$ has no real-number solution

11. $\sqrt[3]{y}=3$

$(\sqrt[3]{y})^3=3^3$

$y=27$

13. $\sqrt[3]{z}=-2$

$(\sqrt[3]{z})^3=(-2)^3$

$z=-8$

15. $\sqrt{n-1}=4$

$(\sqrt{n-1})^2=4^2$

$n-1=16$

$n=17$

17. $\sqrt{t+5}=4$

$(\sqrt{t+5})^2=4^2$

$t+5=16$

$t=11$

19. $\sqrt{3x-2}=4$

$(\sqrt{3x-2})^2=4^2$

$3x-2=16$

$3x=18$

$x=6$

21. $\sqrt{2x+24}=4$

$(\sqrt{2x+24})^2=4^2$

$2x+24=16$

$2x=-8$

$x=-4$

23. $\sqrt{2n-8}=-3$ has no real-number solution.

25. $\sqrt[3]{x-3}=2$

$(\sqrt[3]{x-3})^3=2^3$

$x-3=8$

$x=11$

27. $\sqrt[3]{3y-2}=-2$

$(\sqrt[3]{3y-2})^3=(-2)^3$

$3y-2=-8$

$3y=-6$

$y=-2$

29. $\sqrt{u-3}-10=1$

$\sqrt{u-3}=11$

$(\sqrt{u-3})^2=11^2$

$u-3=121$

$u=124$

31. $\sqrt{y-6}+2=9$

$\sqrt{y-6}=7$

$(\sqrt{y-6})^2=7^2$

$y-6=49$

$y=55$

33. $\sqrt{6x-5}-2=3$

$\sqrt{6x-5}=5$

$(\sqrt{6x-5})^2=5^2$

$6x-5=25$

$6x=30$

$x=5$

35. $\sqrt[3]{n+3}-2=-4$

$\sqrt[3]{n+3}=-2$

$(\sqrt[3]{n+3})^3=(-2)^3$

$n+3=-8$

$n=-11$

37. $\sqrt[4]{x-2}-2=-4$

$\sqrt[4]{x-2}=-2$

This equation has no real-number solution.

39. $\sqrt{3x-2}=\sqrt{8-2x}$

$(\sqrt{3x-2})^2=(\sqrt{8-2x})^2$

$3x-2=8-2x$

$5x=10$

$x=2$

Check:

$\sqrt{3(2)-2}\stackrel{?}{=}\sqrt{8-2(2)}$

$\sqrt{6-2}\stackrel{?}{=}\sqrt{8-4}$

$\sqrt{4}\stackrel{?}{=}\sqrt{4}$

$2=2$

41. $\sqrt{4x-5}=\sqrt{6x+5}$

$(\sqrt{4x-5})^2=(\sqrt{6x+5})^2$

$4x-5=6x+5$

$-10=2x$

$-5=x$

Check:

$\sqrt{4(-5)-5}\stackrel{?}{=}\sqrt{6(-5)+5}$

$\sqrt{-20-5}\stackrel{?}{=}\sqrt{-30+5}$

$\sqrt{-25}\stackrel{?}{=}\sqrt{-25}$

$\sqrt{-25}$ is not a real number

No real-number solution (-5 is an extraneous solution.)

43. $\sqrt[3]{2r+2}=\sqrt[3]{3r-1}$

$\left(\sqrt[3]{2r+2}\right)^3=\left(\sqrt[3]{3r-1}\right)^3$

$2r+2=3r-1$

$3=r$

Check:

$\sqrt[3]{2(3)+2}\stackrel{?}{=}\sqrt[3]{3(3)-1}$

$\sqrt[3]{6+2}\stackrel{?}{=}\sqrt[3]{9-1}$

$\sqrt[3]{8}\stackrel{?}{=}\sqrt[3]{8}$

$2=2$

45. $\sqrt[4]{4x+4}=\sqrt[4]{5x+1}$

$\left(\sqrt[4]{4x+4}\right)^4=\left(\sqrt[4]{5x+1}\right)^4$

$4x+4=5x+1$

$3=x$

Check:

$\sqrt[4]{4(3)+4}\stackrel{?}{=}\sqrt[4]{5(3)+1}$

$\sqrt[4]{12+4}\stackrel{?}{=}\sqrt[4]{15+1}$

$\sqrt[4]{16}\stackrel{?}{=}\sqrt[4]{16}$

$2=2$

47. $\sqrt{2x+24}=x+8$

$\left(\sqrt{2x+24}\right)^2=(x+8)^2$

$2x+24=x^2+16x+64$

$0=x^2+14x+40$

$0=(x+4)(x+10)$

$x+4=0$ or $x+10=0$

$x=-4$ $\quad$ $x=-10$

Check $x=-4$:

$\sqrt{2(-4)+24}\stackrel{?}{=}(-4)+8$

$\sqrt{-8+24}\stackrel{?}{=}4$

$\sqrt{16}\stackrel{?}{=}4$

$4=4$

Check $x=-10$:

$\sqrt{2(-10)+24}\stackrel{?}{=}(-10)+8$

$\sqrt{-20+24}\stackrel{?}{=}-2$

$\sqrt{4}\stackrel{?}{=}-2$

$2\neq-2$

-4 is the only solution. (-10 is an extraneous solution.)

49. $y-1=\sqrt{2y-2}$

$(y-1)^2=\left(\sqrt{2y-2}\right)^2$

$y^2-2y+1=2y-2$

$y^2-4y+3=0$

$(y-3)(y-1)=0$

$y-3=0$ or $y-1=0$

$y=3$ $\quad$ $y=1$

Check $y=3$:

$3-1\stackrel{?}{=}\sqrt{2(3)-2}$

$2\stackrel{?}{=}\sqrt{6-2}$

$2\stackrel{?}{=}\sqrt{4}$

$2=2$

Check $y=1$:

$1-1\stackrel{?}{=}\sqrt{2(1)-2}$

$0\stackrel{?}{=}\sqrt{2-2}$

$0\stackrel{?}{=}\sqrt{0}$

$0=0$

51. $\sqrt{3x+10}-4=x$

$$\sqrt{3x+10}=x+4$$
$$\left(\sqrt{3x+10}\right)^2=(x+4)^2$$
$$3x+10=x^2+8x+16$$
$$0=x^2+5x+6$$
$$0=(x+2)(x+3)$$

$x+2=0$ or $x+3=0$

$x=-2$ $\quad$ $x=-3$

Check $x=-2$:

$$\sqrt{3(-2)+10}-4\stackrel{?}{=}-2$$
$$\sqrt{-6+10}-4\stackrel{?}{=}-2$$
$$\sqrt{4}-4\stackrel{?}{=}-2$$
$$2-4\stackrel{?}{=}-2$$
$$-2=-2$$

Check $x=-3$:

$$\sqrt{3(-3)+10}-4\stackrel{?}{=}-3$$
$$\sqrt{-9+10}-4\stackrel{?}{=}-3$$
$$\sqrt{1}-4\stackrel{?}{=}-3$$
$$1-4\stackrel{?}{=}-3$$
$$-3=-3$$

53. $\sqrt{10n+4}-3n=n+1$

$$\sqrt{10n+4}=4n+1$$
$$\left(\sqrt{10n+4}\right)^2=(4n+1)^2$$
$$10n+4=16n^2+8n+1$$
$$0=16n^2-2n-3$$
$$0=(2n-1)(8n+3)$$

$2n-1=0$ or $8n+3=0$

$2n=1$ $\quad$ $8n=-3$

$n=\frac{1}{2}$ $\quad$ $n=-\frac{3}{8}$

Check $n=\frac{1}{2}$:

$$\sqrt{10\left(\frac{1}{2}\right)+4}-3\left(\frac{1}{2}\right)\stackrel{?}{=}\left(\frac{1}{2}\right)+1$$
$$\sqrt{5+4}-\frac{3}{2}\stackrel{?}{=}\left(\frac{1}{2}\right)+\frac{2}{2}$$
$$\sqrt{9}-\frac{3}{2}\stackrel{?}{=}\frac{3}{2}$$
$$3-\frac{3}{2}\stackrel{?}{=}\frac{3}{2}$$
$$\frac{3}{2}=\frac{3}{2}$$

Check $n=-\frac{3}{8}$:

$$\sqrt{10\left(-\frac{3}{8}\right)+4}-3\left(-\frac{3}{8}\right)\stackrel{?}{=}\left(-\frac{3}{8}\right)+1$$
$$\sqrt{-\frac{30}{8}+\frac{32}{8}}+\frac{9}{8}\stackrel{?}{=}\left(-\frac{3}{8}\right)+\frac{8}{8}$$
$$\sqrt{\frac{2}{8}}+\frac{9}{8}\stackrel{?}{=}\frac{5}{8}$$
$$\sqrt{\frac{1}{4}}+\frac{9}{8}\stackrel{?}{=}\frac{5}{8}$$
$$\frac{1}{2}+\frac{9}{8}\stackrel{?}{=}\frac{5}{8}$$
$$\frac{13}{8}\neq\frac{5}{8}$$

$\frac{1}{2}$ is the only solution. ($-\frac{3}{8}$ is an extraneous solution.)

55. $\sqrt[3]{5x+2}+2=5$

$$\sqrt[3]{5x+2}=3$$
$$\left(\sqrt[3]{5x+2}\right)^3=3^3$$
$$5x+2=27$$
$$5x=25$$
$$x=5$$

Check:

$$\sqrt[3]{5(5)+2}+2\stackrel{?}{=}5$$
$$\sqrt[3]{25+2}+2\stackrel{?}{=}5$$
$$\sqrt[3]{27}+2\stackrel{?}{=}5$$
$$3+2\stackrel{?}{=}5$$
$$5=5$$

57. $\sqrt[3]{n^2-2n+5}=2$

$$\left(\sqrt[3]{n^2-2n+5}\right)^3=(2)^3$$
$$n^2-2n+5=8$$
$$n^2-2n-3=0$$
$$(n-3)(n+1)=0$$
$$n-3=0 \quad \text{or} \quad n+1=0$$
$$n=3 \qquad\qquad n=-1$$

Check $n=3$:

$$\sqrt[3]{3^2-2(3)+5}\stackrel{?}{=}2$$
$$\sqrt[3]{9-6+5}\stackrel{?}{=}2$$
$$\sqrt[3]{8}\stackrel{?}{=}2$$
$$2=2$$

Check $n=-1$:

$$\sqrt[3]{(-1)^2-2(-1)+5}\stackrel{?}{=}2$$
$$\sqrt[3]{1+2+5}\stackrel{?}{=}2$$
$$\sqrt[3]{8}\stackrel{?}{=}2$$
$$2=2$$

59. $1+\sqrt{x}=\sqrt{2x+1}$

$$\left(1+\sqrt{x}\right)^2=\left(\sqrt{2x+1}\right)^2$$
$$1+2\sqrt{x}+x=2x+1$$
$$2\sqrt{x}=x$$
$$\left(2\sqrt{x}\right)^2=x^2$$
$$4x=x^2$$
$$0=x^2-4x$$
$$0=x(x-4)$$
$$x=0 \quad \text{or} \quad x-4=0$$
$$x=4$$

Check $x=0$:

$$1+\sqrt{0}\stackrel{?}{=}\sqrt{2(0)+1}$$
$$1+0\stackrel{?}{=}\sqrt{0+1}$$
$$1\stackrel{?}{=}\sqrt{1}$$
$$1=1$$

Check $x=4$:

$$1+\sqrt{4}\stackrel{?}{=}\sqrt{2(4)+1}$$
$$1+2\stackrel{?}{=}\sqrt{8+1}$$
$$3\stackrel{?}{=}\sqrt{9}$$
$$3=3$$

61. $\sqrt{3x+1}+\sqrt{3x}=2$

$$\left(\sqrt{3x+1}\right)^2=\left(2-\sqrt{3x}\right)^2$$
$$3x+1=4-4\sqrt{3x}+3x$$
$$(-3)^2=\left(-4\sqrt{3x}\right)^2$$
$$9=16\cdot 3x$$
$$9=48x$$
$$\frac{3}{16}=x$$

Check:

$$\sqrt{3\left(\frac{3}{16}\right)+1}+\sqrt{3\left(\frac{3}{16}\right)}\stackrel{?}{=}2$$
$$\sqrt{\frac{9}{16}+1}+\sqrt{\frac{9}{16}}\stackrel{?}{=}2$$
$$\sqrt{\frac{9}{16}+\frac{16}{16}}+\sqrt{\frac{9}{16}}\stackrel{?}{=}2$$
$$\sqrt{\frac{25}{16}}+\sqrt{\frac{9}{16}}\stackrel{?}{=}2$$
$$\frac{5}{4}+\frac{3}{4}\stackrel{?}{=}2$$
$$\frac{8}{4}\stackrel{?}{=}2$$
$$2=2$$

63. $$\sqrt{6x+7}-2=\sqrt{2x+3}$$
$$\left(\sqrt{6x+7}-2\right)^2=\left(\sqrt{2x+3}\right)^2$$
$$6x+7-4\sqrt{6x+7}+4=2x+3$$
$$6x+11-4\sqrt{6x+7}=2x+3$$
$$4x+8=4\sqrt{6x+7}$$
$$x+2=\sqrt{6x+7}$$
$$(x+2)^2=\left(\sqrt{6x+7}\right)^2$$
$$x^2+4x+4=6x+7$$
$$x^2-2x-3=0$$
$$(x+1)(x-3)=0$$
$x+1=0$ or $x-3=0$

$x=-1$ $\quad$ $x=3$

Check $x=-1$:
$$\sqrt{6(-1)+7}-2\stackrel{?}{=}\sqrt{2(-1)+3}$$
$$\sqrt{-6+7}-2\stackrel{?}{=}\sqrt{-2+3}$$
$$\sqrt{1}-2\stackrel{?}{=}\sqrt{1}$$
$$1-2\stackrel{?}{=}1$$
$$-1\neq 1$$
Check $x=3$:
$$\sqrt{6(3)+7}-2\stackrel{?}{=}\sqrt{2(3)+3}$$
$$\sqrt{18+7}-2\stackrel{?}{=}\sqrt{6+3}$$
$$\sqrt{25}-2\stackrel{?}{=}\sqrt{9}$$
$$5-2\stackrel{?}{=}3$$
$$3=3$$
3 is the only solution. (-1 is an extraneous solution.)

65. Mistake: You cannot take the principal square root of a number and get a negative.
Correct: No real-number solution.

67. Mistake: The binomial $x-3$ was not squared correctly. Correct:
$$\sqrt{x+3}=x-3$$
$$\left(\sqrt{x+3}\right)^2=(x-3)^2$$
$$x+3=x^2-6x+9$$
$$0=x^2-7x+6$$
$$0=(x-6)(x-1)$$
$x-6=0$ or $x-1=0$

$x=6$ $\quad$ $x=1$ is extraneous

69. Substitute and solve.
$$T=2\pi\sqrt{\frac{L}{9.8}}$$
$$2\pi=2\pi\sqrt{\frac{L}{9.8}}$$
$$1=\sqrt{\frac{L}{9.8}}$$
$$1^2=\left(\sqrt{\frac{L}{9.8}}\right)^2$$
$$1=\frac{L}{9.8}$$
$$9.8=L$$
The length of the pendulum is 9.8 m.

71. Substitute and solve.
$$T=2\pi\sqrt{\frac{L}{9.8}}$$
$$\frac{\pi}{2}=2\pi\sqrt{\frac{L}{9.8}}$$
$$\frac{1}{4}=\sqrt{\frac{L}{9.8}}$$
$$\left(\frac{1}{4}\right)^2=\left(\sqrt{\frac{L}{9.8}}\right)^2$$
$$\frac{1}{16}=\frac{L}{9.8}$$
$$\frac{9.8}{16}=L$$
$$0.6125=L$$
The length of the pendulum is 0.6125 m.

73. $$0.3=\sqrt{\frac{h}{16}}$$
$$(0.3)^2=\left(\sqrt{\frac{h}{16}}\right)^2$$
$$0.09=\frac{h}{16}$$
$$1.44=h$$
The distance is 1.44 ft.

75. $\frac{1}{4}=\sqrt{\frac{h}{16}}$

$\left(\frac{1}{4}\right)^2=\left(\sqrt{\frac{h}{16}}\right)^2$

$\frac{1}{16}=\frac{h}{16}$

$1=h$

The distance is 1 foot.

77. $30=\frac{7}{2}\sqrt{2D}$

$60=7\sqrt{2D}$

$\frac{60}{7}=\sqrt{2D}$

$\left(\frac{60}{7}\right)^2=\left(\sqrt{2D}\right)^2$

$73.469\approx 2D$

$36.73\approx D$

The skid distance is approximately 36.73 ft.

79. $45=\frac{7}{2}\sqrt{2D}$

$90=7\sqrt{2D}$

$\frac{90}{7}=\sqrt{2D}$

$\left(\frac{90}{7}\right)^2=\left(\sqrt{2D}\right)^2$

$165.306\approx 2D$

$82.65 \text{ ft.}\approx D$

The skid distance is approximately 82.65 ft.

81. $5=\sqrt{F_1^2+3^2}$

$5^2=\left(\sqrt{F_1^2+9}\right)^2$

$25=F_1^2+9$

$16=F_1^2$

$\sqrt{16}=F_1$

$4=F_1$

Force 1 has a value of 4 N.

83. $3\sqrt{5}=\sqrt{3^2+F_2^2}$

$\left(3\sqrt{5}\right)^2=\left(\sqrt{9+F_2^2}\right)^2$

$9\cdot 5=9+F_2^2$

$45-9=F_2^2$

$36=F_2^2$

$\sqrt{36}=F_2$

$6=F_2$

Force 2 has a value of 6 N.

85. a) $x=9$ | $x=16$ | $y=5$

$y=\sqrt{9}$ | $y=\sqrt{16}$ | $5=\sqrt{x}$

$y=3$ | $y=4$ | $5^2=\left(\sqrt{x}\right)^2$

$25=x$

b)

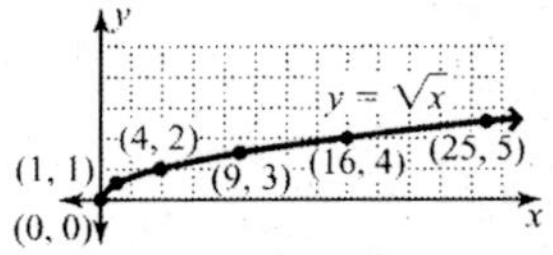

c) No. The x-values must be 0 or positive because real square roots exist only when $x\ge 0$. The y-values must be 0 or positive because, by definition, the principal square root is 0 or positive.

d) Yes, because it passes the vertical line test.

87. The graph becomes steeper from left to right.

89. The graph rises or lowers according to the value of the constant.

Review Exercises

1. $3^4=81$

2. $(-0.2)^3=-0.008$

3. $\left(\frac{2}{5}\right)^{-4}=\left(\frac{5}{2}\right)^4=\frac{625}{16}$ or 39.0625

4. $x^3\cdot x^5=x^{3+5}=x^8$

5. $\left(n^4\right)^6=n^{4\cdot 6}=n^{24}$

6. $\frac{y^7}{y^3}=y^{7-3}=y^4$

Exercise Set 8.7

1. $\sqrt{-1}$

3. No, the set of complex numbers contains both real and imaginary numbers.

5. We subtract complex numbers just like we subtract polynomials—by writing an equivalent addition and changing the signs in the second complex number.

7. $\sqrt{-36} = \sqrt{-1 \cdot 36} = \sqrt{-1} \cdot \sqrt{36} = i\sqrt{36} = 6i$

9. $\sqrt{-5} = \sqrt{-1 \cdot 5} = \sqrt{-1} \cdot \sqrt{5} = i\sqrt{5}$

11. $\sqrt{-8} = \sqrt{-1 \cdot 8} = \sqrt{-1} \cdot \sqrt{8} = i\sqrt{4 \cdot 2} = 2i\sqrt{2}$

13. $\sqrt{-18} = \sqrt{-1 \cdot 18} = \sqrt{-1} \cdot \sqrt{18} = i\sqrt{9 \cdot 2} = 3i\sqrt{2}$

15. $\sqrt{-27} = \sqrt{-1 \cdot 27} = \sqrt{-1} \cdot \sqrt{27} = i\sqrt{9 \cdot 3} = 3i\sqrt{3}$

17. $\sqrt{-125} = \sqrt{-1 \cdot 125}$
$= \sqrt{-1} \cdot \sqrt{125}$
$= i\sqrt{25 \cdot 5}$
$= 5i\sqrt{5}$

19. $\sqrt{-63} = \sqrt{-1 \cdot 63} = \sqrt{-1} \cdot \sqrt{63} = i\sqrt{9 \cdot 7} = 3i\sqrt{7}$

21. $\sqrt{-245} = \sqrt{-1 \cdot 245}$
$= \sqrt{-1} \cdot \sqrt{245}$
$= i\sqrt{49 \cdot 5}$
$= 7i\sqrt{5}$

23. $(9+3i)+(-3+4i) = 6+7i$

25. $(6+2i)+(5-8i) = 11-6i$

27. $(-4+6i)-(3+5i) = (-4+6i)+(-3-5i)$
$= -7+i$

29. $(8-5i)-(-3i) = (8-5i)+(3i)$
$= 8-2i$

31. $(12+3i)+(-15-13i) = -3-10i$

33. $(-5-9i)-(-5-9i) = (-5-9i)+(5+9i) = 0$

35. $(10+i)-(2-13i)+(6-5i)$
$= (10+i)+(-2+13i)+(6-5i)$
$= 14+9i$

37. $(5-2i)-(9-14i)+(16i)$
$= (5-2i)+(-9+14i)+(16i)$
$= -4+28i$

39. $(8i)(3i) = 24i^2 = 24(-1) = -24$

41. $(-8i)(5i) = -40i^2 = -40(-1) = 40$

43. $2i(6-7i) = 12i - 14i^2 = 12i - 14(-1) = 14+12i$

45. $-8i(4-9i) = -32i + 72i^2$
$= -32i + 72(-1)$
$= -72 - 32i$

47. $(6+i)(3-i) = 18 - 6i + 3i - i^2$
$= 18 - 3i - (-1)$
$= 18 - 3i + 1$
$= 19 - 3i$

49. $(8+5i)(5-2i) = 40 - 16i + 25i - 10i^2$
$= 40 + 9i - 10(-1)$
$= 40 + 9i + 10$
$= 50 + 9i$

51. $(8+i)(8-i) = 64 - i^2$
$= 64 - (-1)$
$= 64 + 1$
$= 65$

53. $(3-4i)^2 = 3^2 - 2 \cdot 3 \cdot 4i + (4i)^2$
$= 9 - 24i + 16i^2$
$= 9 - 24i + 16(-1)$
$= 9 - 24i - 16$
$= -7 - 24i$

55. $\frac{2}{i} = \frac{2}{i} \cdot \frac{i}{i} = \frac{2i}{i^2} = \frac{2i}{-1} = -2i$

57. $\frac{4}{5i} = \frac{4}{5i} \cdot \frac{i}{i} = \frac{4i}{5i^2} = \frac{4i}{5(-1)} = -\frac{4i}{5}$

59. $\frac{6}{2i} = \frac{3}{i} \cdot \frac{i}{i} = \frac{3i}{i^2} = \frac{3i}{-1} = -3i$

61. $\dfrac{2+i}{2i}=\dfrac{2+i}{2i}\cdot\dfrac{i}{i}$

$=\dfrac{2i+i^2}{2i^2}$

$=\dfrac{2i+(-1)}{2(-1)}$

$=\dfrac{-1+2i}{-2}$

$=\dfrac{1}{2}-i$

63. $\dfrac{4+2i}{4i}=\dfrac{2(2+i)}{4i}$

$=\dfrac{2+i}{2i}\cdot\dfrac{i}{i}$

$=\dfrac{2i+i^2}{2i^2}$

$=\dfrac{-1+2i}{-2}$

$=\dfrac{1}{2}-i$

65. $\dfrac{7}{2+i}=\dfrac{7}{2+i}\cdot\dfrac{2-i}{2-i}$

$=\dfrac{14-7i}{4-i^2}$

$=\dfrac{14-7i}{4-(-1)}$

$=\dfrac{14-7i}{4+1}$

$=\dfrac{14-7i}{5}$

$=\dfrac{14}{5}-\dfrac{7}{5}i$

67. $\dfrac{2i}{3-7i}=\dfrac{2i}{3-7i}\cdot\dfrac{3+7i}{3+7i}$

$=\dfrac{6i+14i^2}{9-49i^2}$

$=\dfrac{6i+14(-1)}{9-49(-1)}$

$=\dfrac{6i-14}{9+49}$

$=\dfrac{-14+6i}{58}$

$=-\dfrac{14}{58}+\dfrac{6}{58}i$

$=-\dfrac{7}{29}+\dfrac{3}{29}i$

69. $\dfrac{5-9i}{1-i}=\dfrac{5-9i}{1-i}\cdot\dfrac{1+i}{1+i}$

$=\dfrac{5+5i-9i-9i^2}{1-i^2}$

$=\dfrac{5-4i-9(-1)}{1-(-1)}$

$=\dfrac{5-4i+9}{1+1}$

$=\dfrac{14--4i}{2}$

$=7-2i$

71. $\dfrac{3+i}{2+3i}=\dfrac{3+i}{2+3i}\cdot\dfrac{2-3i}{2-3i}$

$=\dfrac{6-9i+2i-3i^2}{4-9i^2}$

$=\dfrac{6-7i-3(-1)}{4-9(-1)}$

$=\dfrac{6-7i+3}{4+9}$

$=\dfrac{9-7i}{13}$

$=\dfrac{9}{13}-\dfrac{7}{13}i$

73. $\frac{1+6i}{4+5i}=\frac{1+6i}{4+5i}\cdot\frac{4-5i}{4-5i}$
$=\frac{4-5i+24i-30i^2}{16-25i^2}$
$=\frac{4+19i-30(-1)}{16-25(-1)}$
$=\frac{4+19i+30}{16+25}$
$=\frac{34+19i}{41}$
$=\frac{34}{41}+\frac{19}{41}i$

75. $i^{19}=i^{16}\cdot i^3=(i^4)^4\cdot i^3=(1)^4\cdot(-i)=1\cdot(-i)=-i$

77. $i^{42}=i^{40}\cdot i^2=(i^4)^{10}\cdot i^2=1^{10}\cdot(-1)=1(-1)=-1$

79. $i^{38}=i^{36}\cdot i^2=(i^4)^9\cdot i^2=1^9\cdot(-1)=1(-1)=-1$

81. $i^{60}=(i^4)^{15}=1^{15}=1$

83. $i^{-20}=\frac{1}{i^{20}}=\frac{1}{(i^4)^5}=\frac{1}{1^5}=\frac{1}{1}=1$

85. $i^{-30}=\frac{1}{i^{30}}=\frac{1}{i^{28}\cdot i^2}=\frac{1}{(i^4)^7\cdot i^2}=\frac{1}{1^7\cdot(-1)}$
$=\frac{1}{1(-1)}=\frac{1}{-1}=-1$

87. $i^{-21}=\frac{1}{i^{21}}=\frac{1}{i^{20}\cdot i}=\frac{1}{(i^4)^5\cdot i}=\frac{1}{1^5\cdot i}=\frac{1}{1\cdot i}$
$=\frac{1}{i}\cdot\frac{i}{i}=\frac{i}{i^2}=\frac{i}{-1}=-i$

89. $i^{-35}=\frac{1}{i^{35}}=\frac{1}{i^{32}\cdot i^3}=\frac{1}{(i^4)^8\cdot i^3}=\frac{1}{1^8\cdot(-i)}=\frac{1}{1(-i)}$
$=\frac{1}{-i}\cdot\frac{i}{i}=\frac{i}{-i^2}=\frac{i}{-(-1)}=\frac{i}{1}=i$

Review Exercises

1. $4x^2-12x+9=(2x-3)(2x-3)=(2x-3)^2$

2. $\pm\sqrt{48}=\pm\sqrt{16\cdot 3}=\pm 4\sqrt{3}$

3. $3x-2=4$
$3x=6$
$x=2$

4. $2x^2-x-6=0$
$(2x+3)(x-2)=0$
$2x+3=0$ or $x-2=0$
$2x=-3$ $\quad x=2$
$x=-\frac{3}{2}$

5. $2x+3y=6$
$y=0: 2x+3\cdot 0=6$
$2x=6$
$x=3$
$(3,0)$

$x=0: 2\cdot 0+3y=6$
$3y=6$
$y=2$
$(0,2)$

6. $y=x^2-3$

x	y	(x, y)
−2	1	(−2,1)
−1	−2	(−1,−2)
0	−3	(0,−3)
1	−2	(1,−2)
2	1	(2,1)

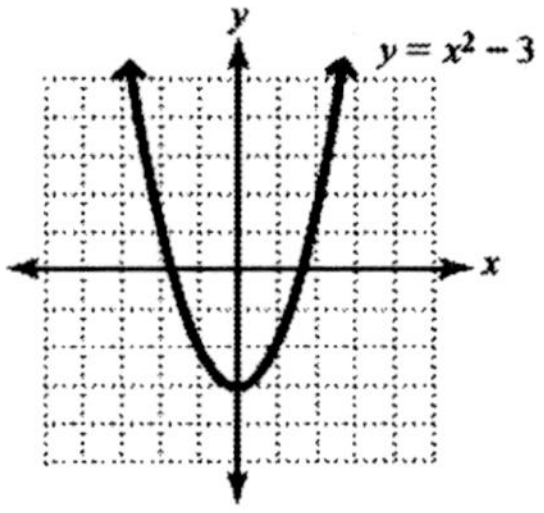

Chapter 8 Review Exercises

1. True
2. False; this statement is true only if x represents nonnegative values.
3. False; an expression is not in simplest form if it contains a radical in the denominator.
4. False; a radical equation may also have more than one solution or no solutions.
5. True
6. radicands
7. coefficients

8. Multiply
9. conjugate
10. extraneous
11. The square roots of 121 are ± 11.
12. The square roots of 49 are ± 7.
13. $\sqrt{169} = 13$
14. $-\sqrt{49} = -7$
15. $\sqrt{-36}$ is not a real number.
16. $\sqrt{\frac{1}{25}} = \frac{1}{5}$
17. $\sqrt{7} \approx 2.646$
18. $\sqrt{90} \approx 9.487$
19. $\sqrt{49x^8} = 7x^4$
20. $\sqrt{144a^6b^{12}} = 12a^3b^6$
21. $\sqrt{0.16m^2n^{10}} = 0.4mn^5$
22. $\sqrt[3]{x^{15}} = x^5$
23. $\sqrt[3]{-64r^9s^3} = -4r^3s$
24. $\sqrt[4]{81x^{12}} = 3x^3$
25. $\sqrt[5]{32x^{15}y^{20}} = 2x^3y^4$
26. $\sqrt[7]{x^{14}y^7} = x^2y$
27. $\sqrt{81x^2} = 9|x|$
28. $\sqrt[4]{(x-1)^8} = (x-1)^2$
29. $(-64)^{1/3} = \sqrt[3]{-64} = -4$
30. $(24a^4)^{1/2} = \sqrt{24a^4} = \sqrt{4a^4 \cdot 6} = 2a^2\sqrt{6}$
31. $\left(\frac{1}{32}\right)^{3/5} = \left(\sqrt[5]{\frac{1}{32}}\right)^3 = \left(\frac{1}{2}\right)^3 = \frac{1}{8}$
32. $(5r-2)^{5/7} = \sqrt[7]{(5r-2)^5}$
33. $\sqrt[8]{33} = 33^{1/8}$
34. $\frac{8}{\sqrt[7]{n^3}} = \frac{8}{n^{3/7}} = 8n^{-3/7}$
35. $\left(\sqrt[5]{8}\right)^3 = 8^{3/5}$
36. $\left(\sqrt[3]{m}\right)^8 = m^{8/3}$
37. $\left(\sqrt[4]{3xw}\right)^3 = (3xw)^{3/4}$
38. $\sqrt[3]{(a+b)^4} = (a+b)^{4/3}$
39. $x^{2/3} \cdot x^{4/3} = x^{2/3+4/3} = x^{6/3} = x^2$
40. $(4m^{1/4})(8m^{5/4}) = 32m^{1/4+5/4} = 32m^{6/4} = 32m^{3/2}$
41. $\frac{y^{3/5}}{y^{4/5}} = y^{3/5-4/5} = y^{-1/5} = \frac{1}{y^{1/5}}$
42. $\frac{b^{2/5}}{b^{-3/5}} = b^{2/5-(-3/5)} = b^{2/5+3/5} = b^{5/5} = b$
43. $(k^{2/3})^{3/4} = k^{(2/3)\cdot(3/4)} = k^{6/12} = k^{1/2}$
44. $(2xy^{1/5})^{3/4} = 2^{3/4}x^{3/4}y^{(1/5)\cdot(3/4)} = 2^{3/4}x^{3/4}y^{3/20}$
45. $\sqrt{3} \cdot \sqrt{27} = \sqrt{81} = 9$
46. $\sqrt{5x^5} \cdot \sqrt{20x^3} = \sqrt{100x^8} = 10x^4$
47. $\sqrt[3]{2} \cdot \sqrt[3]{4} = \sqrt[3]{8} = 2$
48. $\sqrt[4]{7} \cdot \sqrt[4]{6} = \sqrt[4]{42}$
49. $\sqrt[5]{3x^2y^3} \cdot \sqrt[5]{5x^2y} = \sqrt[5]{15x^4y^4}$
50. $\sqrt{\frac{49}{121}} = \frac{7}{11}$
51. $\sqrt[3]{-\frac{27}{8}} = -\frac{3}{2}$
52. $\sqrt{72} = \sqrt{36 \cdot 2} = 6\sqrt{2}$
53. $4b\sqrt{27b^7} = 4b\sqrt{9b^6 \cdot 3b}$
$= 4b \cdot 3b^3\sqrt{3b}$
$= 12b^4\sqrt{3b}$
54. $5\sqrt[3]{108} = 5\sqrt[3]{27 \cdot 4} = 5 \cdot 3\sqrt[3]{4} = 15\sqrt[3]{4}$
55. $2\sqrt[3]{40x^{10}} = 2\sqrt[3]{8x^9 \cdot 5x} = 2 \cdot 2x^3\sqrt[3]{5x} = 4x^3\sqrt[3]{5x}$

56. $2x^4\sqrt[4]{162x^7} = 2x^4\sqrt[4]{81x^4\cdot 2x^3}$
$= 2x^4\cdot 3x\sqrt[4]{2x^3}$
$= 6x^5\sqrt[4]{2x^3}$

57. $4\sqrt{6}\cdot 7\sqrt{15} = 28\sqrt{90}$
$= 28\sqrt{9\cdot 10}$
$= 28\cdot 3\sqrt{10}$
$= 84\sqrt{10}$

58. $\sqrt{x^9}\cdot\sqrt{x^6} = \sqrt{x^{15}} = \sqrt{x^{14}\cdot x} = x^7\sqrt{x}$

59. $4\sqrt{10c}\cdot 2\sqrt{6c^4} = 8\sqrt{60c^5}$
$= 8\sqrt{4c^4\cdot 15c}$
$= 8\cdot 2c^2\sqrt{15c}$
$= 16c^2\sqrt{15c}$

60. $a\sqrt{a^3b^2}\cdot b^2\sqrt{a^5b^3} = ab^2\sqrt{a^8b^5}$
$= ab^2\sqrt{a^8b^4\cdot b}$
$= ab^2\cdot a^4b^2\sqrt{b}$
$= a^5b^4\sqrt{b}$

61. $\dfrac{\sqrt{48}}{\sqrt{6}} = \sqrt{\dfrac{48}{6}} = \sqrt{8} = \sqrt{4\cdot 2} = 2\sqrt{2}$

62. $\dfrac{9\sqrt{160}}{3\sqrt{8}} = 3\sqrt{\dfrac{160}{8}}$
$= 3\sqrt{20}$
$= 3\sqrt{4\cdot 5}$
$= 3\cdot 2\sqrt{5}$
$= 6\sqrt{5}$

63. $\dfrac{8\sqrt{45a^5}}{2\sqrt{5a}} = 4\sqrt{\dfrac{45a^5}{5a}} = 4\sqrt{9a^4} = 4\cdot 3a^2 = 12a^2$

64. $\dfrac{36\sqrt{96x^6y^{11}}}{4\sqrt{3x^2y^4}} = 9\sqrt{\dfrac{96x^6y^{11}}{3x^2y^4}}$
$= 9\sqrt{32x^4y^7}$
$= 9\sqrt{16x^4y^6\cdot 2y}$
$= 9\cdot 4x^2y^3\sqrt{2y}$
$= 36x^2y^3\sqrt{2y}$

65. $-5\sqrt{n} + 2\sqrt{n} = -3\sqrt{n}$

66. $3y^3\sqrt[4]{8y} - 9y^3\sqrt[4]{8y} = -6y^3\sqrt[4]{8y}$

67. $\sqrt{45} + \sqrt{20} = \sqrt{9\cdot 5} + \sqrt{4\cdot 5} = 3\sqrt{5} + 2\sqrt{5} = 5\sqrt{5}$

68. $4\sqrt{24} - 6\sqrt{54} = 4\sqrt{4\cdot 6} - 6\sqrt{9\cdot 6}$
$= 4\cdot 2\sqrt{6} - 6\cdot 3\sqrt{6}$
$= 8\sqrt{6} - 18\sqrt{6}$
$= -10\sqrt{6}$

69. $\sqrt{150} - \sqrt{54} + \sqrt{24} = \sqrt{25\cdot 6} - \sqrt{9\cdot 6} + \sqrt{4\cdot 6}$
$= 5\sqrt{6} - 3\sqrt{6} + 2\sqrt{6}$
$= 4\sqrt{6}$

70. $4\sqrt{72x^2y} - 2x\sqrt{128y} + 5\sqrt{32x^2y}$
$= 4\sqrt{36x^2\cdot 2y} - 2x\sqrt{64\cdot 2y} + 5\sqrt{16x^2\cdot 2y}$
$= 24x\sqrt{2y} - 16x\sqrt{2y} + 20x\sqrt{2y}$
$= 28x\sqrt{2y}$

71. $\sqrt[3]{250x^4y^5} + \sqrt[3]{128x^4y^5}$
$= \sqrt[3]{125x^3y^3\cdot 2xy^2} + \sqrt[3]{64x^3y^3\cdot 2xy^2}$
$= 5xy\sqrt[3]{2xy^2} + 4xy\sqrt[3]{2xy^2}$
$= 9xy\sqrt[3]{2xy^2}$

72. $\sqrt[4]{48} + \sqrt[4]{243} = \sqrt[4]{16\cdot 3} + \sqrt[4]{81\cdot 3}$
$= 2\sqrt[4]{3} + 3\sqrt[4]{3}$
$= 5\sqrt[4]{3}$

73. $\sqrt{5}\left(\sqrt{3} + \sqrt{2}\right) = \sqrt{5}\cdot\sqrt{3} + \sqrt{5}\cdot\sqrt{2} = \sqrt{15} + \sqrt{10}$

74. $\sqrt[3]{7}\left(\sqrt[3]{3} + 2\sqrt[3]{7}\right) = \sqrt[3]{7}\cdot\sqrt[3]{3} + \sqrt[3]{7}\cdot 2\sqrt[3]{7}$
$= \sqrt[3]{21} + 2\sqrt[3]{49}$

75. $3\sqrt{6}\left(2 - 3\sqrt{6}\right) = 3\sqrt{6}\cdot 2 + 3\sqrt{6}\cdot\left(-3\sqrt{6}\right)$
$= 6\sqrt{6} - 9\sqrt{36}$
$= 6\sqrt{6} - 9\cdot 6$
$= 6\sqrt{6} - 54$

76. $\left(\sqrt{2} - \sqrt{3}\right)\left(\sqrt{5} + \sqrt{7}\right)$
$= \sqrt{2}\cdot\sqrt{5} + \sqrt{2}\cdot\sqrt{7} - \sqrt{3}\cdot\sqrt{5} - \sqrt{3}\cdot\sqrt{7}$
$= \sqrt{10} + \sqrt{14} - \sqrt{15} - \sqrt{21}$

77. $\left(\sqrt[3]{2}-4\right)\left(\sqrt[3]{4}+2\right)$
$= \sqrt[3]{2}\cdot\sqrt[3]{4}+\sqrt[3]{2}\cdot 2-4\cdot\sqrt[3]{4}-4\cdot 2$
$= \sqrt[3]{8}+2\sqrt[3]{2}-4\sqrt[3]{4}-8$
$= 2+2\sqrt[3]{2}-4\sqrt[3]{4}--8$
$= -6+2\sqrt[3]{2}-4\sqrt[3]{4}$

78. $\left(\sqrt[4]{6x}+2\right)\left(\sqrt[4]{2x}-1\right)$
$= \sqrt[4]{6x}\cdot\sqrt[4]{2x}-\sqrt[4]{6x}\cdot 1+2\cdot\sqrt[4]{2x}-2\cdot 1$
$= \sqrt[4]{12x^2}-\sqrt[4]{6x}+2\sqrt[4]{2x}-2$

79. $\left(\sqrt{5a}+\sqrt{3b}\right)\left(\sqrt{5a}-\sqrt{3b}\right)=\left(\sqrt{5a}\right)^2-\left(\sqrt{3b}\right)^2$
$= 5a-3b$

80. $\left(2\sqrt{3}-\sqrt{5}\right)\left(2\sqrt{3}+\sqrt{5}\right)=\left(2\sqrt{3}\right)^2-\left(\sqrt{5}\right)^2$
$= 4\cdot 3-5$
$= 12-5$
$= 7$

81. $\left(\sqrt{2}+1\right)^2=\left(\sqrt{2}\right)^2+2\cdot\sqrt{2}\cdot 1+1^2$
$= 2+2\sqrt{2}+1$
$= 3+2\sqrt{2}$

82. $\left(\sqrt{2}-\sqrt{5}\right)^2=\left(\sqrt{2}\right)^2-2\cdot\sqrt{2}\cdot\sqrt{5}+\left(\sqrt{5}\right)^2$
$= 2-2\sqrt{10}+5$
$= 7-2\sqrt{10}$

83. $\frac{1}{\sqrt{2}}=\frac{1}{\sqrt{2}}\cdot\frac{\sqrt{2}}{\sqrt{2}}=\frac{\sqrt{2}}{\sqrt{4}}=\frac{\sqrt{2}}{2}$

84. $\frac{3}{\sqrt[3]{3}}=\frac{3}{\sqrt[3]{3}}\cdot\frac{\sqrt[3]{9}}{\sqrt[3]{9}}=\frac{3\sqrt[3]{9}}{\sqrt[3]{27}}=\frac{3\sqrt[3]{9}}{3}=\sqrt[3]{9}$

85. $\sqrt{\frac{4}{7}}=\frac{\sqrt{4}}{\sqrt{7}}=\frac{2}{\sqrt{7}}\cdot\frac{\sqrt{7}}{\sqrt{7}}=\frac{2\sqrt{7}}{\sqrt{49}}=\frac{2\sqrt{7}}{7}$

86. $\frac{\sqrt[4]{5x^2}}{\sqrt[4]{2}}=\frac{\sqrt[4]{5x^2}}{\sqrt[4]{2}}\cdot\frac{\sqrt[4]{8}}{\sqrt[4]{8}}=\frac{\sqrt[4]{40x^2}}{\sqrt[4]{16}}=\frac{\sqrt[4]{40x^2}}{2}$

87. $\sqrt[3]{\frac{17}{3y^2}}=\frac{\sqrt[3]{17}}{\sqrt[3]{3y^2}}\cdot\frac{\sqrt[3]{9y}}{\sqrt[3]{9y}}=\frac{\sqrt[3]{153y}}{\sqrt[3]{27y^3}}=\frac{\sqrt[3]{153y}}{3y}$

88. $\frac{\sqrt[4]{9}}{\sqrt[4]{27}}=\frac{\sqrt[4]{9}}{\sqrt[4]{27}}\cdot\frac{\sqrt[4]{3}}{\sqrt[4]{3}}=\frac{\sqrt[4]{27}}{\sqrt[4]{81}}=\frac{\sqrt[4]{27}}{3}$

89. $\frac{4}{\sqrt{2}-\sqrt{3}}=\frac{4}{\sqrt{2}-\sqrt{3}}\cdot\frac{\sqrt{2}+\sqrt{3}}{\sqrt{2}+\sqrt{3}}$
$=\frac{4\sqrt{2}+4\sqrt{3}}{\left(\sqrt{2}\right)^2-\left(\sqrt{3}\right)^2}$
$=\frac{4\sqrt{2}+4\sqrt{3}}{2-3}$
$=\frac{4\sqrt{2}+4\sqrt{3}}{-1}$
$=-4\sqrt{2}-4\sqrt{3}$

90. $\frac{1}{4+\sqrt{3}}=\frac{1}{4+\sqrt{3}}\cdot\frac{4-\sqrt{3}}{4-\sqrt{3}}$
$=\frac{4-\sqrt{3}}{(4)^2-\left(\sqrt{3}\right)^2}$
$=\frac{4-\sqrt{3}}{16-3}$
$=\frac{4-\sqrt{3}}{13}$

91. $\frac{1}{2-\sqrt{n}}=\frac{1}{2-\sqrt{n}}\cdot\frac{2+\sqrt{n}}{2+\sqrt{n}}$
$=\frac{2+\sqrt{n}}{2^2-\left(\sqrt{n}\right)^2}$
$=\frac{2+\sqrt{n}}{4-n}$

92. $\frac{2\sqrt{3}}{3\sqrt{2}-2\sqrt{3}}=\frac{2\sqrt{3}}{3\sqrt{2}-2\sqrt{3}}\cdot\frac{3\sqrt{2}+2\sqrt{3}}{3\sqrt{2}+2\sqrt{3}}$
$=\frac{6\sqrt{6}+4\sqrt{9}}{\left(3\sqrt{2}\right)^2-\left(2\sqrt{3}\right)^2}$
$=\frac{6\sqrt{6}+4\cdot 3}{9\cdot 2-4\cdot 3}$
$=\frac{6\sqrt{6}+12}{18-12}$
$=\frac{6\sqrt{6}+12}{6}$
$=\sqrt{6}+2$

93. $\frac{\sqrt{10}}{6}=\frac{\sqrt{10}}{6}\cdot\frac{\sqrt{10}}{\sqrt{10}}=\frac{\sqrt{100}}{6\sqrt{10}}=\frac{10}{6\sqrt{10}}=\frac{5}{3\sqrt{10}}$

94. $\frac{\sqrt{3x}}{5}=\frac{\sqrt{3x}}{5}\cdot\frac{\sqrt{3x}}{\sqrt{3x}}=\frac{\sqrt{9x^2}}{5\sqrt{3x}}=\frac{3x}{5\sqrt{3x}}$

95. $\frac{2-\sqrt{3}}{8}=\frac{2-\sqrt{3}}{8}\cdot\frac{2+\sqrt{3}}{2+\sqrt{3}}$
$=\frac{2^2-\left(\sqrt{3}\right)^2}{8\left(2+\sqrt{3}\right)}$
$=\frac{4-3}{16+8\sqrt{3}}$
$=\frac{1}{16+8\sqrt{3}}$

96. $\frac{2\sqrt{t}+\sqrt{3t}}{5t}=\frac{2\sqrt{t}+\sqrt{3t}}{5t}\cdot\frac{2\sqrt{t}-\sqrt{3t}}{2\sqrt{t}-\sqrt{3t}}$
$=\frac{\left(2\sqrt{t}\right)^2-\left(\sqrt{3t}\right)^2}{5t\left(2\sqrt{t}-\sqrt{3t}\right)}$
$=\frac{4t-3t}{10t\sqrt{t}-5t\sqrt{3t}}$
$=\frac{t}{10t\sqrt{t}-5t\sqrt{3t}}$
$=\frac{\not{t}}{\not{t}\left(10\sqrt{t}-5\sqrt{3t}\right)}$
$=\frac{1}{10\sqrt{t}-5\sqrt{3t}}$

97. $\sqrt{x}=9$
$\left(\sqrt{x}\right)^2=9^2$
$x=81$
Check:
$\sqrt{81}\stackrel{?}{=}9$
$9=9$

98. $\sqrt{y}=-3$ has no real-number solution.

99. $\sqrt{w-1}=3$
$\left(\sqrt{w-1}\right)^2=3^2$
$w-1=9$
$w=10$
Check:
$\sqrt{10-1}\stackrel{?}{=}3$
$\sqrt{9}\stackrel{?}{=}3$
$3=3$

100. $\sqrt[3]{3x-2}=-2$
$\left(\sqrt[3]{3x-2}\right)^3=(-2)^3$
$3x-2=-8$
$3x=-6$
$x=-2$
Check:
$\sqrt[3]{3(-2)-2}\stackrel{?}{=}-2$
$\sqrt[3]{-6-2}\stackrel{?}{=}-2$
$\sqrt[3]{-8}\stackrel{?}{=}-2$
$-2=-2$

101. $\sqrt[4]{x-2}-3=-1$
$\sqrt[4]{x-2}=2$
$\left(\sqrt[4]{x-2}\right)^4=2^4$
$x-2=16$
$x=18$
Check:
$\sqrt[4]{18-2}-3\stackrel{?}{=}-1$
$\sqrt[4]{16}-3\stackrel{?}{=}-1$
$2-3\stackrel{?}{=}-1$
$-1=-1$

102. $\sqrt{y+1}=\sqrt{2y-4}$
$\left(\sqrt{y+1}\right)^2=\left(\sqrt{2y-4}\right)^2$
$y+1=2y-4$
$5=y$
Check:
$\sqrt{5+1}\stackrel{?}{=}\sqrt{2(5)-4}$
$\sqrt{6}\stackrel{?}{=}\sqrt{10-4}$
$\sqrt{6}=\sqrt{6}$

103. $\sqrt{x-6}=x+2$
$\left(\sqrt{x-6}\right)^2=(x+2)^2$
$x-6=x^2+4x+4$
$0=x^2+3x+10$
This polynomial cannot be factored. The equation has no real-number solution.

104. $\sqrt[3]{3x+10}-4=5$

$$\sqrt[3]{3x+10}=9$$
$$\left(\sqrt[3]{3x+10}\right)^3=9^3$$
$$3x+10=729$$
$$3x=719$$
$$x=\frac{719}{3}$$

Check:

$$\sqrt[3]{3\left(\frac{719}{3}\right)+10}-4\stackrel{?}{=}5$$
$$\sqrt[3]{719+10}-4\stackrel{?}{=}5$$
$$\sqrt[3]{729}-4\stackrel{?}{=}5$$
$$9-4\stackrel{?}{=}5$$
$$5=5$$

105. $\sqrt[4]{x+8}=\sqrt[4]{2x+1}$

$$\left(\sqrt[4]{x+8}\right)^4=\left(\sqrt[4]{2x+1}\right)^4$$
$$x+8=2x+1$$
$$7=x$$

Check:

$$\sqrt[4]{7+8}\stackrel{?}{=}\sqrt[4]{2(7)+1}$$
$$\sqrt[4]{15}\stackrel{?}{=}\sqrt[4]{14+1}$$
$$\sqrt[4]{15}=\sqrt[4]{15}$$

106. $\sqrt{5n-1}=4-2n$

$$\left(\sqrt{5n-1}\right)^2=(4-2n)^2$$
$$5n-1=16-16n+4n^2$$
$$0=4n^2-21n+17$$
$$0=(4n-17)(n-1)$$
$$4n-17=0 \quad \text{or} \quad n-1=0$$
$$4n=17 \qquad n=1$$
$$n=\frac{17}{4}$$

Check $n=\frac{17}{4}$:

$$\sqrt{5\left(\frac{17}{4}\right)-1}\stackrel{?}{=}4-2\left(\frac{17}{4}\right)$$
$$\sqrt{\frac{85}{4}-\frac{4}{4}}\stackrel{?}{=}4-\frac{17}{2}$$
$$\sqrt{\frac{81}{4}}\stackrel{?}{=}\frac{8}{2}-\frac{17}{2}$$
$$\frac{9}{2}\neq-\frac{9}{2}$$

Check $n=1$:

$$\sqrt{5(1)-1}\stackrel{?}{=}4-2(1)$$
$$\sqrt{5-1}\stackrel{?}{=}4-2$$
$$\sqrt{4}\stackrel{?}{=}2$$
$$2=2$$

The only solution is $n=1$ ($n=\frac{17}{4}$ is an extraneous solution).

107. $1+\sqrt{x}=\sqrt{2x+1}$

$$\left(1+\sqrt{x}\right)^2=\left(\sqrt{2x+1}\right)^2$$
$$1+2\sqrt{x}+x=2x+1$$
$$2\sqrt{x}=x$$
$$\left(2\sqrt{x}\right)^2=x^2$$
$$4x=x^2$$
$$0=x^2-4x$$
$$0=x(x-4)$$
$$x=0 \quad \text{or} \quad x-4=0$$
$$x=4$$

Check $x=0$:

$$1+\sqrt{0}\stackrel{?}{=}\sqrt{2(0)+1}$$
$$1+0\stackrel{?}{=}\sqrt{0+1}$$
$$1\stackrel{?}{=}\sqrt{1}$$
$$1=1$$

Check $x=4$:

$$1+\sqrt{4}\stackrel{?}{=}\sqrt{2(4)+1}$$
$$1+2\stackrel{?}{=}\sqrt{8+1}$$
$$3\stackrel{?}{=}\sqrt{9}$$
$$3=3$$

108. $\sqrt{3x+1}=2-\sqrt{3x}$

$$\left(\sqrt{3x+1}\right)^2=\left(2-\sqrt{3x}\right)^2$$
$$3x+1=4-4\sqrt{3x}+3x$$
$$4\sqrt{3x}=3$$
$$\left(4\sqrt{3x}\right)^2=3^2$$
$$16\cdot 3x=9$$
$$48x=9$$
$$x=\frac{9}{48}$$
$$x=\frac{3}{16}$$

Check:

$$\sqrt{3\left(\frac{3}{16}\right)+1}\stackrel{?}{=}2-\sqrt{3\left(\frac{3}{16}\right)}$$
$$\sqrt{\frac{9}{16}+\frac{16}{16}}\stackrel{?}{=}2-\sqrt{\frac{9}{16}}$$
$$\sqrt{\frac{25}{16}}\stackrel{?}{=}2-\frac{3}{4}$$
$$\frac{5}{4}\stackrel{?}{=}2-\frac{3}{4}$$
$$\frac{5}{4}\stackrel{?}{=}\frac{8}{4}-\frac{3}{4}$$
$$\frac{5}{4}=\frac{5}{4}$$

109. $\sqrt{-9}=\sqrt{-1\cdot 9}=\sqrt{-1}\cdot\sqrt{9}=3i$

110. $\sqrt{-20}=\sqrt{-1\cdot 20}$

$$=\sqrt{-1}\cdot\sqrt{20}$$
$$=i\sqrt{20}$$
$$=i\sqrt{4\cdot 5}$$
$$=2i\sqrt{5}$$

111. $(3+2i)+(5-8i)=8-6i$

112. $(7-3i)-(-2+4i)=(7-3i)+(2-4i)$

$$=9-7i$$

113. $(3i)(4i)=12i^2=12(-1)=-12$

114. $2i(4-i)=8i-2i^2=8i-2(-1)=2+8i$

115. $(6+2i)(4-i)=6\cdot 4-6\cdot i+2i\cdot 4-2i\cdot i$

$$=24-6i+8i-2i^2$$
$$=24+2i-2(-1)$$
$$=24+2i+2$$
$$=26+2i$$

116. $(5-i)^2=5^2-2(5)(i)+(-i)^2$

$$=25-10i+i^2$$
$$=25-10i-1$$
$$=24-10i$$

117. $\frac{5}{i}=\frac{5}{i}\cdot\frac{i}{i}=\frac{5i}{i^2}=\frac{5i}{-1}=-5i$

118. $\frac{3}{-i}=\frac{3}{-i}\cdot\frac{i}{i}=\frac{3i}{-i^2}=\frac{3i}{-(-1)}=\frac{3i}{1}=3i$

119. $\frac{4}{3i}=\frac{4}{3i}\cdot\frac{i}{i}=\frac{4i}{3i^2}=\frac{4i}{3(-1)}=-\frac{4i}{3}$

120. $\frac{7+i}{5i}\cdot\frac{i}{i}=\frac{7i+i^2}{5i^2}=\frac{7i-1}{5(-1)}=\frac{-1+7i}{-5}=\frac{1}{5}-\frac{7}{5}i$

121. $\frac{3}{2+i}\cdot\frac{2-i}{2-i}=\frac{3(2-i)}{2^2-i^2}$
$=\frac{6-3i}{4-(-1)}$
$=\frac{6-3i}{5}$
$=\frac{6}{5}-\frac{3}{5}i$

122. $\frac{5+i}{2-3i}=\frac{5+i}{2-3i}\cdot\frac{2+3i}{2+3i}=\frac{(5+i)(2+3i)}{2^2-(3i)^2}$
$=\frac{10+15i+2i+3i^2}{4-9i^2}=\frac{10+17i+3(-1)}{4-9(-1)}$
$=\frac{10+17i-3}{4+9}=\frac{7+17i}{13}$
$=\frac{7}{13}+\frac{17}{13}i$

123. $i^{20}=\left(i^4\right)^5=1^5=1$

124. $i^{15}=i^{12}\cdot i^3=\left(i^4\right)^3\cdot i^3=1^3\cdot(-i)=1(-i)=-i$

125. $A=\frac{1}{2}\cdot\frac{2\sqrt{5}}{5}\cdot\frac{\sqrt{15}}{\sqrt{8}}=\frac{1}{2}\cdot\frac{2\sqrt{5}}{5}\cdot\frac{\sqrt{15}}{2\sqrt{2}}$
$=\frac{2\sqrt{75}}{20\sqrt{2}}=\frac{\sqrt{25\cdot3}}{10\sqrt{2}}=\frac{5\sqrt{3}}{10\sqrt{2}}=\frac{\sqrt{3}}{2\sqrt{2}}\cdot\frac{\sqrt{2}}{\sqrt{2}}$
$=\frac{\sqrt{6}}{2\sqrt{4}}=\frac{\sqrt{6}}{2\cdot2}=\frac{\sqrt{6}}{4}$

126. a) $S=2\sqrt{2\cdot40}$
$=2\sqrt{80}$
$=2\sqrt{16\cdot5}$
$=2\cdot4\sqrt{5}$
$=8\sqrt{5}$
The speed is $8\sqrt{5}$ mph.
b) 17.9 mph

127. $50=2\sqrt{2L}$
$25=\sqrt{2L}$
$25^2=\left(\sqrt{2L}\right)^2$
$625=2L$
$312.5=L$
The skid mark length is 312.5 ft.

128. $T=2\pi\sqrt{\frac{2.45}{9.8}}$
$T=2\pi\sqrt{0.25}$
$T=\pi$
The pendulum's period is π sec.

129. Substitute and solve.
$T=2\pi\sqrt{\frac{L}{9.8}}$
$\frac{\pi}{3}=2\pi\sqrt{\frac{L}{9.8}}$
$\frac{1}{6}=\sqrt{\frac{L}{9.8}}$
$\left(\frac{1}{6}\right)^2=\left(\sqrt{\frac{L}{9.8}}\right)^2$
$\frac{1}{36}=\frac{L}{9.8}$
$\frac{9.8}{36}=L$
$0.27\overline{2}=L$
The pendulum length is approximately 0.272 m.

130. a) $t=\sqrt{\frac{40}{16}}=\sqrt{\frac{10}{4}}=\frac{\sqrt{10}}{\sqrt{4}}=\frac{\sqrt{10}}{2}$
The time is $\frac{\sqrt{10}}{2}$ sec.
b) 1.58 sec.

131. $0.3=\sqrt{\frac{h}{16}}$
$(0.3)^2=\left(\sqrt{\frac{h}{16}}\right)^2$
$0.09=\frac{h}{16}$
$1.44=h$
The distance is 1.44 ft.

132. $c=\sqrt{4^2+3^2}$
$c=\sqrt{16+9}$
$c=\sqrt{25}$
$c=5$
The connecting piece should be 5 ft.

Chapter 8 Practice Test

1. $\sqrt{36}=6$

2. $\sqrt{-49}=\sqrt{-1\cdot 49}$
$=\sqrt{-1}\cdot\sqrt{49}$
$=i\cdot 7$
$=7i$

3. $\sqrt{81x^2y^5}=\sqrt{81x^2y^4\cdot y}$
$=9xy^2\sqrt{y}$

4. $\sqrt[3]{54}=\sqrt[3]{27\cdot 2}=3\sqrt[3]{2}$

5. $\sqrt[4]{4x}\cdot\sqrt[4]{4x^5}=\sqrt[4]{16x^6}$
$=\sqrt[4]{16x^4\cdot x^2}$
$=2x\sqrt[4]{x^2}$ or $2x\sqrt{x}$

6. $-\sqrt[3]{-27r^{15}}=-\left(-3r^5\right)=3r^5$

7. $\dfrac{\sqrt{5}}{\sqrt{45}}=\sqrt{\dfrac{5}{45}}=\sqrt{\dfrac{1}{9}}=\dfrac{\sqrt{1}}{\sqrt{9}}=\dfrac{1}{3}$

8. $\dfrac{\sqrt[4]{1}}{\sqrt[4]{81}}=\dfrac{1}{3}$

9. $6\sqrt{7}-\sqrt{7}=(6-1)\sqrt{7}$
$=5\sqrt{7}$

10. $\left(\sqrt{3}-1\right)^2=\left(\sqrt{3}\right)^2-2\left(\sqrt{3}\right)(1)+(-1)^2$
$=3-2\sqrt{3}+1$
$=4-2\sqrt{3}$

11. $x^{2/3}\cdot x^{-4/3}=x^{2/3+(-4/3)}$
$=x^{-2/3}$
$=\dfrac{1}{x^{2/3}}$

12. $\left(\sqrt[3]{2}-4\right)\left(\sqrt[3]{4}+2\right)$
$=\sqrt[3]{2}\cdot\sqrt[3]{4}+\sqrt[3]{2}\cdot 2-4\cdot\sqrt[3]{4}-4\cdot 2$
$=\sqrt[3]{8}+2\sqrt[3]{2}-4\sqrt[3]{4}-8$
$=2+2\sqrt[3]{2}-4\sqrt[3]{4}-8$
$=-6+2\sqrt[3]{2}-4\sqrt[3]{4}$

13. $\sqrt[5]{8x^3}=\left(8x^3\right)^{1/5}$
$=(8)^{1/5}\left(x^3\right)^{1/5}$
$=8^{1/5}x^{3\cdot 1/5}$
$=8^{1/5}x^{3/5}$

14. $\sqrt[3]{(2x+5)^2}=(2x+5)^{2/3}$

15. $\dfrac{1}{\sqrt[3]{4}}=\dfrac{1}{\sqrt[3]{4}}\cdot\dfrac{\sqrt[3]{2}}{\sqrt[3]{2}}=\dfrac{\sqrt[3]{2}}{\sqrt[3]{8}}=\dfrac{\sqrt[3]{2}}{2}$

16. $\dfrac{\sqrt{x}}{\sqrt{x}+\sqrt{y}}=\dfrac{\sqrt{x}}{\sqrt{x}+\sqrt{y}}\cdot\dfrac{\sqrt{x}-\sqrt{y}}{\sqrt{x}-\sqrt{y}}$
$=\dfrac{\sqrt{x}\left(\sqrt{x}-\sqrt{y}\right)}{\left(\sqrt{x}\right)^2-\left(\sqrt{y}\right)^2}$
$=\dfrac{\sqrt{x^2}-\sqrt{xy}}{x-y}$
$=\dfrac{x-\sqrt{xy}}{x-y}$

17. $\sqrt{3x-2}=8$
$\left(\sqrt{3x-2}\right)^2=8^2$
$3x-2=64$
$3x=66$
$x=22$

18. $\sqrt[4]{x+8}=\sqrt[4]{2x+1}$
$\left(\sqrt[4]{x+8}\right)^4=\left(\sqrt[4]{2x+1}\right)^4$
$x+8=2x+1$
$8=x+1$
$7=x$

19. $\sqrt{2x+1}-\sqrt{x+1}=2$

$$\sqrt{2x+1}=2+\sqrt{x+1}$$
$$\left(\sqrt{2x+1}\right)^2=\left(2+\sqrt{x+1}\right)^2$$
$$2x+1=4+4\sqrt{x+1}+x+1$$
$$2x+1=5+x+4\sqrt{x+1}$$
$$x-4=4\sqrt{x+1}$$
$$(x-4)^2=\left(4\sqrt{x+1}\right)^2$$
$$x^2-8x+16=16(x+1)$$
$$x^2-8x+16=16x+16$$
$$x^2-24x=0$$
$$x(x-24)=0$$
$$x=\not{0} \quad\text{or}\quad x-24=0$$
$$x=24$$

0 is an extraneous solution.

20. $(2-i)-(4+3i)=(2-i)+(-4-3i)$
$$=-2-4i$$

21. $(4-i)(4+i)=4^2-i^2$
$$=16-(-1)$$
$$=16+1$$
$$=17 \text{ or } 17+0i$$

22. $\dfrac{2}{4-3i}=\dfrac{2}{4-3i}\cdot\dfrac{4+3i}{4+3i}$
$$=\frac{2(4+3i)}{4^2-(3i)^2}$$
$$=\frac{8+6i}{16-9i^2}$$
$$=\frac{8+6i}{16-9(-1)}$$
$$=\frac{8+6i}{16+9}$$
$$=\frac{8+6i}{25}$$
$$=\frac{8}{25}+\frac{6}{25}i$$

23. Area of a parallelogram is $\text{base}\times\text{height}$.

Area $=5\sqrt{12}\cdot2\sqrt{3}=10\sqrt{36}=10\cdot6=60\text{ m}^2$

24. a) Let $h = 12$.

$$t=\sqrt{\frac{h}{16}}=\sqrt{\frac{12}{16}}=\sqrt{\frac{3}{4}}=\frac{\sqrt{3}}{\sqrt{4}}=\frac{\sqrt{3}}{2}\text{ seconds.}$$

b) Let $t = 2$.
$$2=\sqrt{\frac{h}{16}}$$
$$2=\frac{\sqrt{h}}{\sqrt{16}}$$
$$2=\frac{\sqrt{h}}{4}$$
$$8=\sqrt{h}$$
$$8^2=\left(\sqrt{h}\right)^2$$
$$64\text{ ft.}=h$$

25. a) Let $D = 40$ feet.
$$S=\frac{7}{4}\sqrt{D}$$
$$=\frac{7}{4}\sqrt{40}$$
$$=\frac{7}{4}\sqrt{4\cdot10}$$
$$=\frac{7}{4}\cdot2\sqrt{10}$$
$$=\frac{7}{2}\sqrt{10}$$
$$=3.5\sqrt{10}$$

The speed is $3.5\sqrt{10}$ mph.

b) Let $S = 30$ mph.
$$S=\frac{7}{4}\sqrt{D}$$
$$30=\frac{7}{4}\sqrt{D}$$
$$\frac{120}{7}=\sqrt{D}$$
$$\left(\frac{120}{7}\right)^2=\left(\sqrt{D}\right)^2$$
$$\frac{14,400}{49}=D$$

The length of the skid marks is approximately 293.88 ft.

Chapters 1-8 Cumulative Review

1. False; if $|x|>3$, then $x>3$ or $x<-3$.
2. True
3. True
4. True

5. x-intercept

6. inconsistent

7. $24x^5y^2$

8. $4-6\left(5-3^2\right)+12\div 2-6$
$$=4-6(5-9)+12\div 2-6$$
$$=4-6(-4)+12\div 2-6$$
$$=4+24+12\div 2-6$$
$$=4+24+6-6$$
$$=28$$

9. $\dfrac{7.44\times 10^{-2}}{3.1\times 10^{4}}=2.4\times 10^{-2-4}$
$$=2.4\times 10^{-6}$$
$$=0.0000024$$

10. $\dfrac{\left(x^{-2}\right)^3\left(x^3\right)^4}{\left(x^{-3}\right)^3}=\dfrac{x^{-6}x^{12}}{x^{-9}}$
$$=\frac{x^{-6+12}}{x^{-9}}$$
$$=\frac{x^6}{x^{-9}}$$
$$=x^{6-(-9)}$$
$$=x^{6+9}$$
$$=x^{15}$$

11. $\left(3x^2-4y\right)^2$
$$=\left(3x^2\right)^2-2\cdot 3x^2\cdot 4y+(4y)^2$$
$$=9x^4-24x^2y+16y^2$$

12.
$$\begin{array}{r}
4x^2+6x-2 \\
2x-3\,\overline{)\,8x^3+0x^2-22x+8} \\
\underline{8x^3-12x^2} \\
12x^2-22x \\
\underline{12x^2-18x} \\
-4x+8 \\
\underline{-4x+6} \\
2
\end{array}$$

Solution: $\dfrac{8x^3-22x+8}{2x-3}=4x^2+6x-2+\dfrac{2}{2x-3}$

13. $\dfrac{2x^2+7x+3}{x^2-9}\div\dfrac{2x^2+11x+5}{x^2-3x}$
$$=\frac{2x^2+7x+3}{x^2-9}\cdot\frac{x^2-3x}{2x^2+11x+5}$$
$$=\frac{\cancel{(2x+1)}\cancel{(x+3)}}{\cancel{(x+3)}\cancel{(x-3)}}\cdot\frac{x\cancel{(x-3)}}{\cancel{(2x+1)}(x+5)}$$
$$=\frac{x}{x+5}$$

14. $2\sqrt{5a^2}\cdot 3\sqrt{10a^5}=6\sqrt{50a^7}$
$$=6\sqrt{25a^6\cdot 2a}$$
$$=6\cdot 5a^3\sqrt{2a}$$
$$=30a^3\sqrt{2a}$$

15. $3x\sqrt[3]{24x^4}-4x^2\sqrt[3]{81x}$
$$=3x\sqrt[3]{8x^3\cdot 3x}-4x^2\sqrt[3]{27\cdot 3x}$$
$$=3x\cdot 2x\sqrt[3]{3x}-4x^2\cdot 3\sqrt[3]{3x}$$
$$=6x^2\sqrt[3]{3x}-12x^2\sqrt[3]{3x}$$
$$=-6x^2\sqrt[3]{3x}$$

16. $32^{-3/5}=\dfrac{1}{32^{3/5}}=\dfrac{1}{\left(\sqrt[5]{32}\right)^3}=\dfrac{1}{(2)^3}=\dfrac{1}{8}$

17. $(4+5i)(2-3i)=8-12i+10i-15i^2$
$$=8-2i-15(-1)$$
$$=8-2i+15$$
$$=23-2i$$

18. $\dfrac{2+\sqrt{2}}{4-\sqrt{2}}=\dfrac{2+\sqrt{2}}{4-\sqrt{2}}\cdot\dfrac{4+\sqrt{2}}{4+\sqrt{2}}$
$$=\frac{8+2\sqrt{2}+4\sqrt{2}+\left(\sqrt{2}\right)^2}{4^2-\left(\sqrt{2}\right)^2}$$
$$=\frac{8+6\sqrt{2}+2}{16-2}$$
$$=\frac{10+6\sqrt{2}}{14}$$
$$=\frac{\cancel{2}\left(5+3\sqrt{2}\right)}{\cancel{2}\cdot 7}$$
$$=\frac{5+3\sqrt{2}}{7}$$

19.
$$A = \frac{1}{2}h(B+b)$$
$$2A = h(B+b)$$
$$\frac{2A}{h} = B+b$$
$$\frac{2A}{h} - B = b$$
$$\frac{2A}{h} - \frac{Bh}{h} = b$$
$$\frac{2A - Bh}{h} = b$$

20. $(x+2)(x+4) = 63$
$$x^2 + 6x + 8 = 63$$
$$x^2 + 6x - 55 = 0$$
$$(x-5)(x+11) = 0$$
$x - 5 = 0$ or $x + 11 = 0$

$x = 5$ $\quad x = -11$

21.
$$\frac{3y}{y-4} - \frac{12}{y-4} = 6$$
$$(y-4)\left(\frac{3y}{y-4} - \frac{12}{y-4}\right) = (y-4)(6)$$
$$3y - 12 = 6y - 24$$
$$-12 = 3y - 24$$
$$12 = 3y$$
$$4 = y$$

Notice that if $y = 4$, then $\frac{3y}{y-4}$ and $\frac{12}{y-4}$ are undefined. Therefore, $y = 4$ is extraneous and the equation has no solution.

22.
$$\sqrt{3x+10} = x+2$$
$$\left(\sqrt{3x+10}\right)^2 = (x+2)^2$$
$$3x + 10 = x^2 + 4x + 4$$
$$0 = x^2 + x - 6$$
$$0 = (x+3)(x-2)$$
$x + 3 = 0$ or $x - 2 = 0$

$x = -3$ $\quad x = 2$

Check $x = -3$:
$$\sqrt{3(-3)+10} \stackrel{?}{=} -3+2$$
$$\sqrt{-9+10} \stackrel{?}{=} -1$$
$$\sqrt{1} \stackrel{?}{=} -1$$
$$1 \neq -1$$

Check $x = 2$:
$$\sqrt{3(2)+10} \stackrel{?}{=} 2+2$$
$$\sqrt{6+10} \stackrel{?}{=} 4$$
$$\sqrt{16} \stackrel{?}{=} 4$$
$$4 = 4$$

The only solution is $x = 2$ ($x = -3$ is an extraneous solution).

23. $5x - 2y = -15$
$$-2y = -5x - 15$$
$$y = \frac{5}{2}x + \frac{15}{2}$$

The slope is $m = \frac{5}{2}$ and the y-intercept is $\frac{15}{2}$.

24. The slope of the given line is $m = 3$. Use $m = 3$ as the slope.
$$y - y_1 = m(x - x_1)$$
$$y - 2 = 3(x - (-4))$$
$$y - 2 = 3(x+4)$$
$$y - 2 = 3x + 12$$
$$y = 3x + 14$$

25. $y = -\frac{5}{3}x + 2$

$m = -\frac{5}{3}$; y-intercept: $(0, 2)$

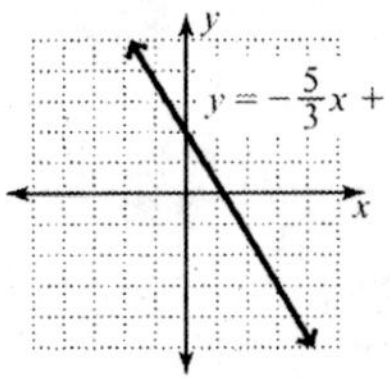

26. $\begin{cases} 2x - y > -4 \\ y > -\dfrac{2}{3}x - 2 \end{cases}$

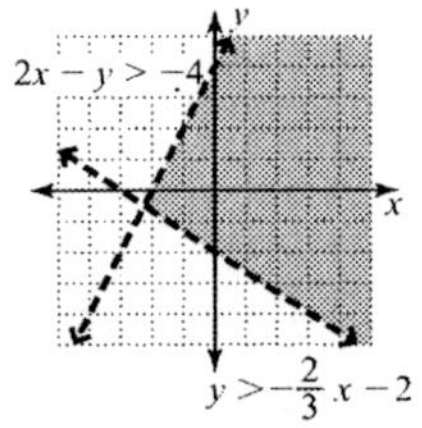

27. Create a table.

Categories	hourly wage	hours worked	pay
0-40 hours worked	w	40	$40w$
hours worked in excess of 40	$1.25w$	6	$6(1.25w)$

Translate the information in the table to an equation and solve.

$$40w + 6(1.25w) = 782.80$$
$$40w + 7.5w = 782.80$$
$$47.5w = 782.80$$
$$w = 16.48$$

Hernando's normal hourly wage is \$16.48 per hour.

28. Complete a table.

Categories	Rate	Time	Distance
against current	$r-5$	$\dfrac{20}{r-5}$	20
with current	$r+5$	$\dfrac{30}{r+5}$	30

$$\frac{20}{r-5} = \frac{30}{r+5}$$
$$30(r-5) = 20(r+5)$$
$$30r - 150 = 20r + 100$$
$$10r = 250$$
$$r = 25$$

The speed of the boat in still water is 25 mph.

29. Translate "y varies jointly as x and the square of z."

$$y = k \cdot x \cdot z^2$$
$$72 = k \cdot 4 \cdot (3)^2$$
$$72 = k \cdot 4 \cdot 9$$
$$72 = 36k$$
$$2 = k$$

Now, $y = 2xz^2$

$$y = 2(3)(4)^2$$
$$y = 2(3)(16)$$
$$y = 96$$

30. Substitute and solve.

$$T = 2\pi\sqrt{\frac{L}{9.8}}$$
$$4\pi = 2\pi\sqrt{\frac{L}{9.8}}$$
$$2 = \sqrt{\frac{L}{9.8}}$$
$$2^2 = \left(\sqrt{\frac{L}{9.8}}\right)^2$$
$$4 = \frac{L}{9.8}$$
$$39.2 = L$$

The length of the pendulum is 39.2 m.

Chapter 9

Quadratic Equations and Functions

Exercise Set 9.1

1. $x^2 = a$ has two solutions because squaring $\sqrt{a}$ and $-\sqrt{a}$ gives a for every real number a.

3. $(ax-b)^2 = c$

$$ax - b = \pm\sqrt{c}$$

$$x = \frac{b \pm \sqrt{c}}{a}$$

5. Because $x^2 - 7x + 12$ is easy to factor, factoring is a better method as it requires fewer steps than completing the square.

7. $x^2 = 49$

$$x = \pm\sqrt{49}$$

$$x = \pm 7$$

9. $y^2 = \frac{4}{25}$

$$y = \pm\sqrt{\frac{4}{25}}$$

$$y = \pm\frac{2}{5}$$

11. $n^2 = 1.44$

$$n = \pm\sqrt{1.44}$$

$$n = \pm 1.2$$

13. $z^2 = 45$

$$z = \pm\sqrt{45}$$

$$z = \pm\sqrt{9 \cdot 5}$$

$$z = \pm 3\sqrt{5}$$

15. $w^2 = -25$

$$w = \pm\sqrt{-25}$$

$$w = \pm 5i$$

17. $n^2 - 7 = 42$

$$n^2 = 49$$

$$n = \pm\sqrt{49}$$

$$n = \pm 7$$

19. $y^2 - 16 = 65$

$$y^2 = 81$$

$$y = \pm\sqrt{81}$$

$$y = \pm 9$$

21. $4n^2 = 36$

$$n^2 = 9$$

$$n = \pm\sqrt{9}$$

$$n = \pm 3$$

23. $25t^2 = 9$

$$t^2 = \frac{9}{25}$$

$$t = \pm\sqrt{\frac{9}{25}}$$

$$t = \pm\frac{3}{5}$$

25. $4h^2 = -16$

$$h^2 = -4$$

$$h = \pm\sqrt{-4}$$

$$h = \pm 2i$$

27. $\frac{5}{6}x^2 = \frac{24}{5}$

$$x^2 = \frac{6}{5} \cdot \frac{24}{5}$$

$$x^2 = \frac{144}{25}$$

$$x = \pm\sqrt{\frac{144}{25}}$$

$$x = \pm\frac{12}{5}$$

29. $2x^2 + 5 = 21$

$$2x^2 = 16$$

$$x^2 = 8$$

$$x = \pm\sqrt{8}$$

$$x = \pm\sqrt{4 \cdot 2}$$

$$x = \pm 2\sqrt{2}$$

31. $5y^2 - 7 = -97$
$$5y^2 = -90$$
$$y^2 = -18$$
$$y = \pm\sqrt{-18}$$
$$y = \pm\sqrt{-9 \cdot 2}$$
$$y = \pm 3i\sqrt{2}$$

33. $\frac{3}{4}y^2 - 5 = 3$
$$\frac{3}{4}y^2 = 8$$
$$y^2 = 8 \cdot \frac{4}{3}$$
$$y^2 = \frac{32}{3}$$
$$y = \pm\sqrt{\frac{32}{3}} = \pm\frac{\sqrt{32}}{\sqrt{3}} \cdot \frac{\sqrt{3}}{\sqrt{3}}$$
$$= \pm\frac{\sqrt{96}}{3} = \pm\frac{\sqrt{16 \cdot 6}}{3} = \pm\frac{4\sqrt{6}}{3}$$

35. $0.2t^2 - 0.5 = 0.012$
$$0.2t^2 = 0.512$$
$$t^2 = 2.56$$
$$t = \pm\sqrt{2.56}$$
$$t = \pm 1.6$$

37. $(x+8)^2 = 49$
$$x + 8 = \pm\sqrt{49}$$
$$x + 8 = \pm 7$$
$$x = -8 \pm 7$$
$$x = -8 + 7 = -1$$
$$x = -8 - 7 = -15$$

39. $(5n-3)^2 = 16$
$$5n - 3 = \pm\sqrt{16}$$
$$5n - 3 = \pm 4$$
$$5n = 3 \pm 4$$
$$n = \frac{3 \pm 4}{5}$$
$$n = \frac{3+4}{5} = \frac{7}{5}$$
$$n = \frac{3-4}{5} = -\frac{1}{5}$$

41. $(m-8)^2 = -16$
$$m - 8 = \pm\sqrt{-16}$$
$$m - 8 = \pm 4i$$
$$m = 8 \pm 4i$$

43. $(4k-1)^2 = 40$
$$4k - 1 = \pm\sqrt{40}$$
$$4k - 1 = \pm\sqrt{4 \cdot 10}$$
$$4k - 1 = \pm 2\sqrt{10}$$
$$4k = 1 \pm 2\sqrt{10}$$
$$k = \frac{1 \pm 2\sqrt{10}}{4}$$

45. $(m-7)^2 = -12$
$$m - 7 = \pm\sqrt{-12}$$
$$m - 7 = \pm\sqrt{-4 \cdot 3}$$
$$m - 7 = \pm 2i\sqrt{3}$$
$$m = 7 \pm 2i\sqrt{3}$$

47. $\left(y - \frac{3}{4}\right)^2 = \frac{9}{16}$
$$y - \frac{3}{4} = \pm\sqrt{\frac{9}{16}}$$
$$y - \frac{3}{4} = \pm\frac{3}{4}$$
$$y = \frac{3}{4} \pm \frac{3}{4}$$
$$y = \frac{3}{4} + \frac{3}{4} = \frac{3}{2}$$
$$y = \frac{3}{4} - \frac{3}{4} = 0$$

49. $\left(\frac{5}{9}d-\frac{1}{2}\right)^2=\frac{1}{36}$

$$\frac{5}{9}d-\frac{1}{2}=\pm\sqrt{\frac{1}{36}}$$
$$\frac{5}{9}d-\frac{1}{2}=\pm\frac{1}{6}$$
$$\frac{5}{9}d=\frac{1}{2}\pm\frac{1}{6}$$
$$d=\frac{9}{5}\cdot\left(\frac{1}{2}\pm\frac{1}{6}\right)$$
$$d=\frac{9}{5}\cdot\left(\frac{1}{2}+\frac{1}{6}\right)=\frac{6}{5}$$
$$d=\frac{9}{5}\cdot\left(\frac{1}{2}-\frac{1}{6}\right)=\frac{3}{5}$$

51. $(0.4x+3.8)^2=2.56$

$$0.4x+3.8=\pm\sqrt{2.56}$$
$$0.4x+3.8=\pm1.6$$
$$0.4x=-3.8\pm1.6$$
$$x=\frac{-3.8\pm1.6}{0.4}$$
$$x=\frac{-3.8+1.6}{0.4}=-5.5$$
$$x=\frac{-3.8-1.6}{0.4}=-13.5$$

53. Mistake: Gave only the positive solution.
Correct: ±7

55. Mistake: Changed –6 to 6.
Correct: $5\pm i\sqrt{6}$

57. $x^2=96$

$$x=\pm\sqrt{96}$$
$$x=\pm\sqrt{16\cdot6}$$
$$x=\pm4\sqrt{6}\approx\pm9.798$$

59. $y^2-15=5$

$$y^2=20$$
$$y=\pm\sqrt{20}$$
$$y=\pm\sqrt{4\cdot5}$$
$$y=\pm2\sqrt{5}\approx\pm4.472$$

61. $(n-6)^2=15$

$$n-6=\pm\sqrt{15}$$
$$n=6\pm\sqrt{15}\approx9.873,\ 2.127$$

63. a) $\left(\frac{14}{2}\right)^2=7^2=49$

$x^2+14x+49$

b) $(x+7)^2$

65. a) $\left(\frac{-10}{2}\right)^2=(-5)^2=25$

$n^2-10n+25$

b) $(n-5)^2$

67. a) $\left(\frac{-9}{2}\right)^2=\frac{81}{4}$

$y^2-9y+\frac{81}{4}$

b) $\left(y-\frac{9}{2}\right)^2$

69. a) $\left(\frac{-\frac{2}{3}}{2}\right)^2=\left(-\frac{1}{3}\right)^2=\frac{1}{9}$

$s^2-\frac{2}{3}s+\frac{1}{9}$

b) $\left(s-\frac{1}{3}\right)^2$

71. $\left(\frac{2}{2}\right)^2=1^2=1$

$$w^2+2w+1=15+1$$
$$(w+1)^2=16$$
$$w+1=\pm4$$
$$w=-1\pm4$$
$$w=-1+4=3$$
$$=-1-4=-5$$

73. $\left(\frac{-2}{2}\right)^2=(-1)^2=1$

$r^2-2r+1=-50+1$

$(r-1)^2=-49$

$r-1=\pm 7i$

$r=1\pm 7i$

75. $\left(\frac{-9}{2}\right)^2=\frac{81}{4}$

$k^2-9k+\frac{81}{4}=-18+\frac{81}{4}$

$\left(k-\frac{9}{2}\right)^2=\frac{9}{4}$

$k-\frac{9}{2}=\pm\frac{3}{2}$

$k=\frac{9}{2}\pm\frac{3}{2}$

$k=\frac{9}{2}+\frac{3}{2}=\frac{12}{2}=6$

$k=\frac{9}{2}-\frac{3}{2}=\frac{6}{2}=3$

77. $\left(\frac{-2}{2}\right)^2=(-1)^2=1$

$b^2-2b+1=16+1$

$(b-1)^2=17$

$b-1=\pm\sqrt{17}$

$b=1\pm\sqrt{17}$

79. $h^2-6h=-29$

$\left(-\frac{6}{2}\right)^2=(-3)^2=9$

$h^2-6h+9=-29+9$

$(h-3)^2=-20$

$h-3=\pm\sqrt{-20}$

$h-3=\pm 2i\sqrt{5}$

$h=3\pm 2i\sqrt{5}$

81. $\left(\frac{\frac{1}{2}}{2}\right)^2=\left(\frac{1}{4}\right)^2=\frac{1}{16}$

$u^2+\frac{1}{2}u+\frac{1}{16}=\frac{3}{2}+\frac{1}{16}$

$\left(u+\frac{1}{4}\right)^2=\frac{25}{16}$

$u+\frac{1}{4}=\pm\frac{5}{4}$

$u=-\frac{1}{4}\pm\frac{5}{4}$

$u=-\frac{1}{4}+\frac{5}{4}=1$

$u=-\frac{1}{4}-\frac{5}{4}=-\frac{3}{2}$

83. $\frac{4x^2}{4}+\frac{16x}{4}=\frac{9}{4}$

$x^2+4x=\frac{9}{4}$

$\left(\frac{b}{2}\right)^2=\left(\frac{4}{2}\right)^2=2^2=4$

$x^2+4x+4=\frac{9}{4}+4$

$(x+2)^2=\frac{25}{4}$

$x+2=\pm\sqrt{\frac{25}{4}}$

$x+2=\pm\frac{5}{2}$

$x=-2+\frac{5}{2}=\frac{1}{2}$

$=-2-\frac{5}{2}=-\frac{9}{2}$

85. $$\frac{9x^2}{9}-\frac{18x}{9}=-\frac{5}{9}$$
$$x^2-2x=-\frac{5}{9}$$
$$x^2-2x+1=-\frac{5}{9}+1$$
$$(x-1)^2=\frac{4}{9}$$
$$x-1=\pm\sqrt{\frac{4}{9}}$$
$$x=1\pm\frac{2}{3}$$
$$x=1+\frac{2}{3}=\frac{5}{3}$$
$$x=1-\frac{2}{3}=\frac{1}{3}$$

87. $$n^2-\frac{1}{2}n=\frac{3}{2}$$
$$\left(\frac{\frac{-1}{2}}{2}\right)^2=\left(-\frac{1}{4}\right)^2=\frac{1}{16}$$
$$n^2-\frac{1}{2}n+\frac{1}{16}=\frac{3}{2}+\frac{1}{16}$$
$$\left(n-\frac{1}{4}\right)^2=\frac{25}{16}$$
$$n-\frac{1}{4}=\pm\frac{5}{4}$$
$$n=\frac{1}{4}\pm\frac{5}{4}$$
$$n=\frac{1}{4}+\frac{5}{4}=\frac{3}{2}$$
$$n=\frac{1}{4}-\frac{5}{4}=-1$$

89. $$\frac{4x^2}{4}+\frac{16x}{4}=\frac{7}{4}$$
$$x^2+4x=\frac{7}{4}$$
$$x^2+4x+4=\frac{7}{4}+4$$
$$(x+2)^2=\frac{23}{4}$$
$$x+2=\pm\sqrt{\frac{23}{4}}$$
$$x+2=\pm\frac{\sqrt{23}}{2}$$
$$x=-2\pm\frac{\sqrt{23}}{2}$$

91. $$k^2+\frac{1}{5}k=\frac{2}{5}$$
$$\left(\frac{\frac{1}{5}}{2}\right)^2=\left(\frac{1}{10}\right)^2=\frac{1}{100}$$
$$k^2+\frac{1}{5}k+\frac{1}{100}=\frac{2}{5}+\frac{1}{100}$$
$$\left(k+\frac{1}{10}\right)^2=\frac{41}{100}$$
$$k+\frac{1}{10}=\pm\sqrt{\frac{41}{100}}$$
$$k+\frac{1}{10}=\pm\frac{\sqrt{41}}{10}$$
$$k=\frac{-1\pm\sqrt{41}}{10}$$

93. $$3a^2 + 4a = 8$$
$$a^2 + \frac{4}{3}a = \frac{8}{3}$$
$$\left(\frac{4}{3}\cdot\frac{1}{2}\right)^2 = \left(\frac{2}{3}\right)^2 = \frac{4}{9}$$
$$a^2 + \frac{4}{3}a + \frac{4}{9} = \frac{8}{3} + \frac{4}{9}$$
$$\left(a + \frac{2}{3}\right)^2 = \frac{28}{9}$$
$$a + \frac{2}{3} = \pm\sqrt{\frac{28}{9}}$$
$$a + \frac{2}{3} = \pm\frac{2\sqrt{7}}{3}$$
$$a = -\frac{2}{3} \pm \frac{2\sqrt{7}}{3}$$
$$a = \frac{-2 \pm 2\sqrt{7}}{3}$$

95. Mistake: Did not divide by 3 so that x^2 has a coefficient of 1. Then wrote an incorrect factored form.
Correct:
$$3x^2 + 4x = 2$$
$$x^2 + \frac{4}{3}x = \frac{2}{3}$$
$$x^2 + \frac{4}{3}x + \frac{4}{9} = \frac{2}{3} + \frac{4}{9}$$
$$\left(x + \frac{2}{3}\right)^2 = \frac{10}{9}$$
$$x + \frac{2}{3} = \pm\sqrt{\frac{10}{9}}$$
$$x = -\frac{2}{3} \pm \frac{\sqrt{10}}{3}$$
$$x = \frac{-2 \pm \sqrt{10}}{3}$$

97. $$A = s^2$$
$$\sqrt{A} = \sqrt{s^2}$$
$$\sqrt{A} = s$$
$$\sqrt{196} = s$$
$$14 = s$$
The length of each side is 14 inches.

99. $$V = lwh$$
$$144 = 8 \cdot h \cdot 2h$$
$$144 = 16h^2$$
$$9 = h^2$$
$$\sqrt{9} = \sqrt{h^2}$$
$$3 = h$$
height: 3 ft.
width: $2(3)$=6 ft.

101. $$A = \pi r^2$$
$$23,256.25\pi = \pi r^2$$
$$23,256.25 = r^2$$
$$\sqrt{23,256.25} = \sqrt{r^2}$$
$$152.5 = r$$
So, the diameter is $2(152.5) = 305$ m

103. Translate to the equation $(x-2)^2 = 256$ and solve.
$$(x-2)^2 = 256$$
$$\sqrt{(x-2)^2} = \sqrt{256}$$
$$x - 2 = 16$$
$$x = 18$$
The length x is 18 inches.

105. Let the length be x and the width be $x - 4$.
$$x^2 + (x-4)^2 = 20^2$$
$$x^2 + x^2 - 8x + 16 = 400$$
$$2x^2 - 8x = 384$$
$$x^2 - 4x = 192$$
$$x^2 - 4x + 4 = 192 + 4$$
$$(x-2)^2 = 196$$
$$x - 2 = \pm\sqrt{196}$$
$$x - 2 = \pm 14$$
$$x = 2 \pm 14$$
$$x = 2 - 14 = -12$$
$$x = 2 + 14 = 16$$
Length must be positive, so disregard the negative solution. The length is 16 ft. and the width is 16 – 4 = 12 ft.

107. a) Subtract the area of the interior rectangle from the area of the exterior rectangle.

$$3l \cdot 2l - 8l = 1230$$
$$6l^2 - 8l = 1230$$
$$l^2 - \frac{4}{3}l = 205$$
$$l^2 - \frac{4}{3}l + \frac{4}{9} = 205 + \frac{4}{9}$$
$$\left(l - \frac{2}{3}\right)^2 = \frac{1849}{9}$$
$$l - \frac{2}{3} = \pm\sqrt{\frac{1849}{9}}$$
$$l - \frac{2}{3} = \pm\frac{43}{3}$$
$$l = \frac{2}{3} \pm \frac{43}{3}$$
$$l = \frac{2}{3} \pm \frac{43}{3} = \frac{45}{3} = 15$$
$$l = \frac{2}{3} \pm \frac{43}{3} = -\frac{41}{3}$$

Because lengths cannot be negative, disregard the negative solution. The value of l is 15 cm.

b) length: $3 \cdot 15 = 45$ cm

width: $2 \cdot 15 = 30$ cm

109. $$9 = 16t^2$$
$$\frac{9}{16} = t^2$$
$$\sqrt{\frac{9}{16}} = t$$
$$\frac{3}{4} = t$$

The cover takes $\frac{3}{4}$ sec. to hit the floor.

111. $$400 = \frac{1}{2} \cdot 50 \cdot v^2$$
$$400 = 25v^2$$
$$16 = v^2$$
$$\sqrt{16} = \sqrt{v^2}$$
$$4 = v$$

The velocity is 4 m/sec.

Review Exercises

1. $x^2 - 10x + 25 = (x-5)^2$

2. $x^2 + 6x + 9 = (x+3)^2$

3. $\sqrt{36x^2} = 6x$

4. $\sqrt{(2x+3)^2} = 2x+3$

5. $$\frac{-4+\sqrt{8^2-4(2)(5)}}{2(2)} = \frac{-4+\sqrt{64-40}}{4}$$
$$= \frac{-4+\sqrt{24}}{4}$$
$$= \frac{-4+\sqrt{4 \cdot 6}}{4}$$
$$= \frac{-4+2\sqrt{6}}{4}$$
$$= \frac{-2+\sqrt{6}}{2} \text{ or } -1+\frac{\sqrt{6}}{2}$$

6. $$\frac{-6-\sqrt{6^2-4(5)(2)}}{2(5)} = \frac{-6-\sqrt{36-40}}{10}$$
$$= \frac{-6-\sqrt{-4}}{10}$$
$$= \frac{-6-2i}{10}$$
$$= \frac{-3-i}{5} \text{ or } -\frac{3}{5}-\frac{1}{5}i$$

Exercise Set 9.2

1. Follow the procedure for solving a quadratic equation by completing the square on $ax^2 + bx + c = 0$

3. Yes, if they have irrational or nonreal complex solutions.

5. When the discriminant is zero. The quadratic formula becomes $x = -\frac{b}{2a}$.

7. $x^2 - 3x + 7 = 0$

   $a = 1$

   $b = -3$

   $c = 7$

9. $3x^2 - 9x - 4 = 0$

   $a = 3$

   $b = -9$

   $c = -4$

11. $1.5x^2 - x + 0.2 = 0$

$a = 1.5$

$b = -1$

$c = 0.2$

13. $\frac{1}{2}x^2 + \frac{3}{4}x - 6 = 0$

$a = \frac{1}{2}$

$b = \frac{3}{4}$

$c = -6$

15. $x^2 + 9x + 20 = 0$

$a = 1, b = 9, c = 20$

$$x = \frac{-(9) \pm \sqrt{(9)^2 - 4(1)(20)}}{2(1)}$$
$$= \frac{-9 \pm \sqrt{81 - 80}}{2}$$
$$= \frac{-9 \pm \sqrt{1}}{2}$$
$$= \frac{-9 \pm 1}{2}$$
$$= -4, -5$$

17. $4x^2 + 5x - 6 = 0$

$a = 4, b = 5, c = -6$

$$x = \frac{-(5) \pm \sqrt{(5)^2 - 4(4)(-6)}}{2(4)}$$
$$= \frac{-5 \pm \sqrt{25 + 96}}{8}$$
$$= \frac{-5 \pm \sqrt{121}}{8}$$
$$= \frac{-5 \pm 11}{8}$$
$$= \frac{-5 + 11}{8} = \frac{3}{4}$$
$$= \frac{-5 - 11}{8} = -2$$

19. $x^2 - 9x = 0$

$a = 1, b = -9, c = 0$

$$x = \frac{-(-9) \pm \sqrt{(-9)^2 - 4(1)(0)}}{2(1)}$$
$$= \frac{9 \pm \sqrt{81}}{2}$$
$$= \frac{9 \pm 9}{2}$$
$$= \frac{9 + 9}{2} = 9$$
$$= \frac{9 - 9}{2} = 0$$

21. $x^2 - 8x + 16 = 0$

$a = 1, b = -8, c = 16$

$$x = \frac{-(-8) \pm \sqrt{(-8)^2 - 4(1)(16)}}{2(1)}$$
$$= \frac{8 \pm \sqrt{64 - 64}}{2}$$
$$= \frac{8 \pm \sqrt{0}}{2}$$
$$= \frac{8}{2}$$
$$= 4$$

23. $3x^2 + 4x - 4 = 0$

$a = 3, b = 4, c = -4$

$$x = \frac{-(4) \pm \sqrt{(4)^2 - 4(3)(-4)}}{2(3)}$$
$$= \frac{-4 \pm \sqrt{16 + 48}}{6}$$
$$= \frac{-4 \pm \sqrt{64}}{6}$$
$$= \frac{-4 \pm 8}{6}$$
$$= \frac{-4 + 8}{6} = \frac{2}{3}$$
$$= \frac{-4 - 8}{6} = -2$$

25. $x^2-2x+2=0$
$a=1, b=-2, c=2$
$$x=\frac{-(-2)\pm\sqrt{(-2)^2-4(1)(2)}}{2(1)}$$
$$=\frac{2\pm\sqrt{4-8}}{2}$$
$$=\frac{2\pm\sqrt{-4}}{2}$$
$$=\frac{2\pm 2i}{2}$$
$$=1\pm i$$

27. $x^2-x-1=0$
$a=1, b=-1, c=-1$
$$x=\frac{-(-1)\pm\sqrt{(-1)^2-4(1)(-1)}}{2(1)}$$
$$=\frac{1\pm\sqrt{1+4}}{2}$$
$$=\frac{1\pm\sqrt{5}}{2}$$

29. $3x^2+10x+5=0$
$a=3, b=10, c=5$
$$x=\frac{-(10)\pm\sqrt{(10)^2-4(3)(5)}}{2(3)}$$
$$=\frac{-10\pm\sqrt{100-60}}{6}$$
$$=\frac{-10\pm\sqrt{40}}{6}$$
$$=\frac{-10\pm 2\sqrt{10}}{6}$$
$$=\frac{-5\pm\sqrt{10}}{3}$$

31. $-4x^2+6x-5=0$
$a=-4, b=6, c=-5$
$$x=\frac{-(6)\pm\sqrt{(6)^2-4(-4)(-5)}}{2(-4)}$$
$$=\frac{-6\pm\sqrt{36-80}}{-8}$$
$$=\frac{-6\pm\sqrt{-44}}{-8}$$
$$=\frac{-6\pm 2i\sqrt{11}}{-8}$$
$$=\frac{3\pm i\sqrt{11}}{4}$$

33. $18x^2+15x+2=0$
$a=18, b=15, c=2$
$$x=\frac{-(15)\pm\sqrt{(15)^2-4(18)(2)}}{2(18)}$$
$$=\frac{-15\pm\sqrt{225-144}}{36}$$
$$=\frac{-15\pm\sqrt{81}}{36}$$
$$=\frac{-15\pm 9}{36}$$
$$=\frac{-15+9}{36}=-\frac{1}{6}$$
$$=\frac{-15-9}{36}=-\frac{2}{3}$$

35. $3x^2-4x+3=0$
$a=3, b=-4, c=3$
$$x=\frac{-(-4)\pm\sqrt{(-4)^2-4(3)(3)}}{2(3)}$$
$$=\frac{4\pm\sqrt{16-36}}{6}$$
$$=\frac{4\pm\sqrt{-20}}{6}$$
$$=\frac{4\pm 2i\sqrt{5}}{6}$$
$$=\frac{2\pm i\sqrt{5}}{3}$$

37. $6x^2 - 3x - 4 = 0$

$a = 6, b = -3, c = -4$

$$x = \frac{-(-3) \pm \sqrt{(-3)^2 - 4(6)(-4)}}{2(6)}$$
$$= \frac{3 \pm \sqrt{9 + 96}}{12}$$
$$= \frac{3 \pm \sqrt{105}}{12}$$

39. $2x^2 + 0.1x - 0.03 = 0$

$a = 2, b = 0.1, c = -0.03$

$$x = \frac{-(0.1) \pm \sqrt{(0.1)^2 - 4(2)(-0.03)}}{2(2)}$$
$$= \frac{-0.1 \pm \sqrt{0.01 + 0.24}}{4}$$
$$= \frac{-0.1 \pm \sqrt{0.25}}{4}$$
$$= \frac{-0.1 \pm 0.5}{4}$$
$$= \frac{-0.1 + 0.5}{4} = 0.1$$
$$= \frac{-0.1 - 0.5}{4} = -0.15$$

41. $x^2 + \frac{1}{2}x - 3 = 0$

$a = 1, b = \frac{1}{2}, c = -3$

$$x = \frac{-\left(\frac{1}{2}\right) \pm \sqrt{\left(\frac{1}{2}\right)^2 - 4(1)(-3)}}{2(1)}$$
$$= \frac{-\frac{1}{2} \pm \sqrt{\frac{1}{4} + 12}}{2}$$
$$= \frac{-\frac{1}{2} \pm \sqrt{\frac{49}{4}}}{2}$$
$$= \frac{-\frac{1}{2} \pm \frac{7}{2}}{2}$$
$$= \frac{-\frac{1}{2} + \frac{7}{2}}{2} = \frac{3}{2}$$
$$= \frac{-\frac{1}{2} - \frac{7}{2}}{2} = -2$$

43. Use: $36x^2 - 49 = 0$

$a = 36, b = 0, c = -49$

$$x = \frac{-(0) \pm \sqrt{(0)^2 - 4(36)(-49)}}{2(36)}$$
$$= \frac{0 \pm \sqrt{0 + 7056}}{72}$$
$$x = \frac{0 \pm 84}{72}$$
$$= \frac{\pm 84}{72}$$
$$= \pm \frac{7}{6}$$

45. Use: $x^2 - 2x + 3 = 0$

$a = 1, b = -2, c = 3$

$$x = \frac{-(-2) \pm \sqrt{(-2)^2 - 4(1)(3)}}{2(1)}$$
$$= \frac{2 \pm \sqrt{4 - 12}}{2}$$
$$= \frac{2 \pm \sqrt{-8}}{2}$$
$$= \frac{2 \pm 2i\sqrt{2}}{2} = 1 \pm i\sqrt{2}$$

47. Use: $100x^2+6x-50=0$
$a=100, b=6, c=-50$

$$x=\frac{-(6)\pm\sqrt{(6)^2-4(100)(-50)}}{2(100)}$$
$$=\frac{-6\pm\sqrt{36+20,000}}{200}$$
$$=\frac{-6\pm\sqrt{20,036}}{200}$$
$$=\frac{-6\pm2\sqrt{5009}}{200}$$
$$=\frac{-3\pm\sqrt{5009}}{100}$$
$$\approx -0.738, 0.678$$

49. Use: $12x^2-6x+5=0$
$a=12, b=-6, c=5$

$$x=\frac{-(-6)\pm\sqrt{(-6)^2-4(12)(5)}}{2(12)}$$
$$=\frac{6\pm\sqrt{36-240}}{24}$$
$$=\frac{6\pm\sqrt{-204}}{24}$$
$$=\frac{6\pm2i\sqrt{51}}{24}$$
$$=\frac{3\pm i\sqrt{51}}{12}$$

51. Mistake: Did not evaluate $-b$.
Correct:

$$\frac{-(-7)\pm\sqrt{(-7)^2-(4)(3)(1)}}{2(3)}=\frac{7\pm\sqrt{49-12}}{6}$$
$$=\frac{7\pm\sqrt{37}}{6}$$

53. Mistake: The result was not completely simplified.

Correct: $$\frac{2\pm\sqrt{-8}}{2}=\frac{2\pm2i\sqrt{2}}{2}=1+i\sqrt{2}$$

55. $x^2+10x+25=0$
$b^2-4ac=(10)^2-4(1)(25)=100-100=0$
1 rational solution

57. $\frac{1}{4}x^2-4x+4=0$
$b^2-4ac=(-4)^2-4\left(\frac{1}{4}\right)(4)=16-4=12$
$12>0$ and 12 is not a perfect square
2 irrational solutions

59. $3x^2+10x-8=0$
$b^2-4ac=(10)^2-4(3)(-8)=100+96=196$
$196>0$ and 196 is a perfect square $\left(196=14^2\right)$
2 rational solutions

61. $x^2-x+3=0$
$b^2-4ac=(-1)^2-4(1)(3)=1-12=-11$
$-11<0$
2 nonreal complex solutions

63. $x^2-6x+6=0$
$b^2-4ac=(-6)^2-4(1)(6)=36-24=12$
$12>0$ and 12 is not a perfect square
2 irrational solutions

65. Square root principle or factoring
$x^2-81=0$
$x^2=81$
$x=\pm\sqrt{81}$
$x=\pm9$

67. Quadratic formula
$2x^2-4x-3=0$

$$x=\frac{-(-4)\pm\sqrt{(-4)^2-4(2)(-3)}}{2(2)}$$
$$=\frac{4\pm\sqrt{16+24}}{4}$$
$$=\frac{4\pm\sqrt{40}}{4}$$
$$=\frac{4\pm2\sqrt{10}}{4}$$
$$=\frac{2\pm\sqrt{10}}{2}$$

69. Factoring
$x^2+6x=0$
$x(x+6)=0$
$x=0$ or $x+6=0$
$x=-6$

71. Square root principle

$$(x+7)^2 = 40$$
$$x+7 = \pm\sqrt{40}$$
$$x+7 = \pm 2\sqrt{10}$$
$$x = -7 \pm 2\sqrt{10}$$

73. Quadratic formula

$$x^2 - 8x + 19 = 0$$
$$x = \frac{-(-8) \pm \sqrt{(-8)^2 - 4(1)(19)}}{2(1)}$$
$$x = \frac{8 \pm \sqrt{64-76}}{2}$$
$$x = \frac{8 \pm \sqrt{-12}}{2}$$
$$x = \frac{8 \pm 2i\sqrt{3}}{2}$$
$$x = 4 \pm i\sqrt{3}$$

75. x-intercepts: $x^2 - x - 2 = 0$

$$(x-2)(x+1) = 0$$
$$x-2 = 0 \text{ or } x+1 = 0$$
$$x = 2 \qquad x = -1$$
$$(2,0), (-1,0)$$

y-intercept: $y = 0^2 - 0 - 2$
$$y = -2$$
$$(0,-2)$$

77. x-intercepts: $4x^2 - 12x + 9 = 0$

$$(2x-3)^2 = 0$$
$$2x - 3 = \pm\sqrt{0}$$
$$2x - 3 = 0$$
$$2x = 3$$
$$x = \frac{3}{2}$$
$$\left(\frac{3}{2}, 0\right)$$

y-intercept: $y = 4(0)^2 - 12(0) + 9$
$$y = 9$$
$$(0,9)$$

79. x-intercepts: $2x^2 + 15x - 8 = 0$

$$(2x-1)(x+8) = 0$$
$$2x - 1 = 0 \text{ or } x + 8 = 0$$
$$x = \frac{1}{2} \qquad x = -8$$
$$\left(\frac{1}{2}, 0\right), (-8,0)$$

y-intercept: $y = 2 \cdot 0^2 + 15 \cdot 0 - 8$
$$y = -8$$
$$(0,-8)$$

81. x-intercepts:

$$-2x^2 + 3x - 6 = 0$$
$$x = \frac{-3 \pm \sqrt{3^2 - 4(-2)(-6)}}{2(-2)}$$
$$= \frac{-3 \pm \sqrt{9-48}}{-4}$$
$$= \frac{-3 \pm \sqrt{-39}}{-4}$$

not real: no x-intercepts

y-intercept: $y = -2 \cdot 0^2 + 3 \cdot 0 - 6$
$$y = -6$$
$$(0,-6)$$

83. Let x represent the first positive integer and $x + 1$ represent the next consecutive integer.

$$x^2 + 5(x+1) = 71$$
$$x^2 + 5x + 5 = 71$$
$$x^2 + 5x - 66 = 0$$
$$(x+11)(x-6) = 0$$
$$x + 11 = 0 \quad \text{or} \quad x - 6 = 0$$

$x =$ ~~-11~~ $\qquad x = 6$

The integers are 6 and 6 + 1 = 7.

85. Let x be the first integer, $x + 1$ be the next consecutive integer, and $x + 2$ be the third consecutive integer.

$$x^2+(x+1)^2=(x+2)^2$$
$$x^2+x^2+2x+1=x^2+4x+4$$
$$2x^2+2x+1=x^2+4x+4$$
$$x^2-2x-3=0$$
$$(x-3)(x+1)=0$$
$$x-3=0 \quad \text{or} \quad x+1=0$$
$$x=3 \qquad x=\cancel{-1}$$

The sides are 3, 3 + 1 = 4, and 3 + 2 = 5.

87. If the width is w, then the length is $(3w - 3.5)$.

$$w(3w-3.5)=34$$
$$3w^2-3.5w-34=0$$
$$w=\frac{-(-3.5)\pm\sqrt{(-3.5)^2-4(3)(-34)}}{2(3)}$$
$$w=\frac{3.5\pm\sqrt{12.25+408}}{6}$$
$$w=\frac{3.5\pm\sqrt{420.25}}{6}$$
$$w=\frac{3.5\pm20.5}{6}$$
$$w=\frac{3.5+20.5}{6}=4$$
$$w=\frac{3.5-20.5}{6}=\cancel{-2.83}$$

The width is 4 ft. and the length is 3(4) – 3.5 = 8.5 ft.

89. a) $22x=1.5x\cdot x+0.5x\cdot 8$

$$22x=1.5x^2+4x$$
$$0=1.5x^2-18x$$
$$0=x(1.5x-18)$$
$$0=1.5x-18 \quad \text{or} \quad \cancel{0}=x$$
$$18=1.5x$$
$$12\text{ ft.}=x$$

b)

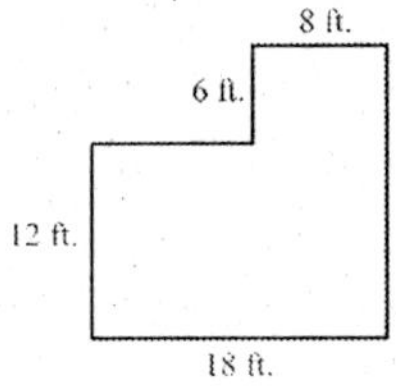

91. $460=\frac{1}{2}(-32.2)t^2+48t+485$

$$0=16.1t^2-48t-25$$
$$t=\frac{-(-48)\pm\sqrt{(-48)^2-4(16.1)(-25)}}{2(16.1)}$$
$$=\frac{48\pm\sqrt{2304+1610}}{32.2}$$
$$=\frac{48+\sqrt{3914}}{32.2}\approx 3.43$$
$$=\frac{48-\sqrt{3914}}{32.2}\approx \cancel{-0.452}$$

He was in flight for approximately 3.43 seconds.

93. a) $h=\frac{1}{2}(-9.8)t^2+0t+10$

$$h=-4.9t^2+10$$

b) $5=-4.9t^2+10$

$$4.9t^2=5$$
$$t^2=\frac{5}{4.9}$$
$$t\approx\pm\sqrt{1.02}$$
$$t\approx 1.01,\ \cancel{-1.01}$$

It takes approximately 1.01 second.

c) $0=-4.9t^2+10$

$$4.9t^2=10$$
$$t^2=\frac{10}{4.9}$$
$$t\approx\pm\sqrt{2.04}$$
$$t\approx 1.43,\ \cancel{-1.43}$$

It takes approximately 1.43 seconds.

95. $0.5n^2+2.5n=4.5n+16$

$$0.5n^2-2n-16=0$$
$$5n^2-20n-160=0$$
$$n^2-4n-32=0$$
$$(n-8)(n+4)=0$$
$$n=8,\ \cancel{-4}$$

The break-even point is 8000 units.

97. a)
$$b^2-4ac=0$$
$$(-5)^2-4(2)c=0$$
$$25-8c=0$$
$$-8c=-25$$
$$c=\frac{25}{8}$$

b)
$$b^2-4ac>0$$
$$(-5)^2-4(2)c>0$$
$$25-8c>0$$
$$-8c>-25$$
$$c<\frac{25}{8}$$

c)
$$b^2-4ac<0$$
$$(-5)^2-4(2)c<0$$
$$25-8c<0$$
$$-8c<-25$$
$$c>\frac{25}{8}$$

99. a)
$$b^2-4ac=0$$
$$12^2-4a(8)=0$$
$$144-32a=0$$
$$-32a=-144$$
$$a=\frac{-144}{-32}=\frac{9}{2}$$

b)
$$b^2-4ac>0$$
$$12^2-4a(8)>0$$
$$144-32a>0$$
$$-32a>-144$$
$$a<\frac{-144}{-32}$$
$$a<\frac{9}{2}$$

c)
$$b^2-4ac<0$$
$$12^2-4a(8)<0$$
$$144-32a<0$$
$$-32a<-144$$
$$a>\frac{-144}{-32}$$
$$a>\frac{9}{2}$$

Review Exercises

1. $u^2-9u+14=(u-7)(u-2)$

2. $3u^2-2u-16=(3u-8)(u+2)$

3. $\left(x^2\right)^2=x^{2\cdot 2}=x^4$

4. $\left(x^{1/3}\right)^2=x^{(1/3)\cdot 2}=x^{2/3}$

5.
$$5u^2+13u-6=0$$
$$(5u-2)(u+3)=0$$
$$5u-2=0 \quad \text{or} \quad u+3=0$$
$$5u=2 \qquad u=-3$$
$$u=\frac{2}{5}$$

6.
$$\frac{7}{3u}=\frac{5}{u}-\frac{1}{u-5}$$
$$3u(u-5)\cdot\frac{7}{3u}=3u(u-5)\cdot\left[\frac{5}{u}-\frac{1}{u-5}\right]$$
$$(u-5)7=3(u-5)\cdot 5-3u\cdot 1$$
$$7u-35=15u-75-3u$$
$$7u-35=12u-75$$
$$-5u=-40$$
$$u=8$$

Exercise Set 9.3

1. It can be rewritten as a quadratic equation.

3. $u=\frac{x+2}{3}$

5. $\frac{1}{x}$

7.
$$\frac{60}{x+2}=\frac{60}{x}-5$$
$$x(x+2)\left(\frac{60}{x+2}\right)=x(x+2)\left(\frac{60}{x}-5\right)$$
$$60x=60(x+2)-5x(x+2)$$
$$60x=60x+120-5x^2-10x$$
$$5x^2+10x-120=0$$
$$x^2+2x-24=0$$
$$(x+6)(x-4)=0$$
$$x+6=0 \quad \text{or} \quad x-4=0$$
$$x=-6 \qquad x=4$$

9. $$\frac{1}{x}+\frac{1}{x+2}=\frac{3}{4}$$
$$4x(x+2)\left(\frac{1}{x}+\frac{1}{x+2}\right)=4x(x+2)\left(\frac{3}{4}\right)$$
$$4(x+2)+4x=x(x+2)\cdot 3$$
$$4x+8+4x=3x^2+6x$$
$$-3x^2+2x+8=0$$
$$3x^2-2x-8=0$$
$$(3x+4)(x-2)=0$$
$$3x+4=0 \quad \text{or} \quad x-2=0$$
$$x=-\frac{4}{3} \qquad x=2$$

11. $$\frac{1}{p-4}+\frac{1}{4}=\frac{8}{p^2-16}$$
$$4(p-4)(p+4)\left(\frac{1}{p-4}+\frac{1}{4}\right)$$
$$=4(p-4)(p+4)\left(\frac{8}{(p+4)(p-4)}\right)$$
$$4(p+4)+(p-4)(p+4)=4\cdot 8$$
$$4p+16+p^2-16=32$$
$$p^2+4p-32=0$$
$$(p+8)(p-4)=0$$
$$p+8=0 \quad \text{or} \quad p-4=0$$
$$p=-8 \qquad p=4$$
Note that $p=4$ makes expressions in the original equation undefined. Therefore, $p=4$ is an extraneous solution.

13. $$\frac{6}{2y+5}=\frac{2}{y+5}+\frac{1}{5}$$
$$5(2y+5)(y+5)\left(\frac{6}{2y+5}\right)$$
$$=5(2y+5)(y+5)\left(\frac{2}{y+5}+\frac{1}{5}\right)$$
$$5(y+5)\cdot 6=5(2y+5)\cdot 2+(2y+5)(y+5)$$
$$30y+150=20y+50+2y^2+15y+25$$
$$0=2y^2+5y-75$$
$$0=(y-5)(2y+15)$$
$$y-5=0 \quad \text{or} \quad 2y+15=0$$
$$y=5 \qquad y=-7.5$$

15. $$1+2x^{-1}-8x^{-2}=0$$
$$1+\frac{2}{x}-\frac{8}{x^2}=0$$
$$x^2\left(1+\frac{2}{x}-\frac{8}{x^2}\right)=x^2\cdot 0$$
$$x^2+2x-8=0$$
$$(x+4)(x-2)=0$$
$$x+4=0 \quad \text{or} \quad x-2=0$$
$$x=-4 \qquad x=2$$

17. $$3+13x^{-1}-10x^{-2}=0$$
$$3+\frac{13}{x}-\frac{10}{x^2}=0$$
$$x^2\left(3+\frac{13}{x}-\frac{10}{x^2}\right)=x^2\cdot 0$$
$$3x^2+13x-10=0$$
$$(3x-2)(x+5)=0$$
$$3x-2=0 \quad \text{or} \quad x+5=0$$
$$x=\frac{2}{3} \qquad x=-5$$

19. $$x-8\sqrt{x}+15=0$$
$$-8\sqrt{x}=-x-15$$
$$8\sqrt{x}=x+15$$
$$\left(8\sqrt{x}\right)^2=(x+15)^2$$
$$64x=x^2+30x+225$$
$$0=x^2-34x+225$$
$$0=(x-9)(x-25)$$
$$x-9=0 \quad \text{or} \quad x-25=0$$
$$x=9 \qquad x=25$$

21. $$2x-5\sqrt{x}-7=0$$
$$2x-7=5\sqrt{x}$$
$$(2x-7)^2=\left(5\sqrt{x}\right)^2$$
$$4x^2-28x+49=25x$$
$$4x^2-53x+49=0$$
$$x=\frac{53\pm\sqrt{53^2-4(4)(49)}}{2(4)}$$
$$x=\frac{53\pm\sqrt{2025}}{8}=\frac{53\pm 45}{8}=\frac{49}{4}, \cancel{1}$$

23. $\sqrt{2a+5} = 3a-3$

$\left(\sqrt{2a+5}\right)^2 = (3a-3)^2$

$2a+5 = 9a^2 - 18a + 9$

$0 = 9a^2 - 20a + 4$

$0 = (a-2)(9a-2)$

$a-2=0$ or $9a-2=0$

$a = 2$ $\quad a = \cancel{\frac{2}{9}}$

25. $\sqrt{2m-8} - m - 1 = 0$

$\sqrt{2m-8} = m+1$

$\left(\sqrt{2m-8}\right)^2 = (m+1)^2$

$2m-8 = m^2 + 2m + 1$

$0 = m^2 + 9$

$-9 = m^2$

$\pm 3i = m$

27. $\sqrt{4x+1} = \sqrt{x+2} + 1$

$\left(\sqrt{4x+1}\right)^2 = \left(\sqrt{x+2}+1\right)^2$

$4x+1 = x+2+2\sqrt{x+2}+1$

$3x-2 = 2\sqrt{x+2}$

$(3x-2)^2 = \left(2\sqrt{x+2}\right)^2$

$9x^2 - 12x + 4 = 4(x+2)$

$9x^2 - 12x + 4 = 4x + 8$

$9x^2 - 16x - 4 = 0$

$(x-2)(9x+2) = 0$

$x-2=0$ or $9x+2=0$

$x = 2$ $\quad x = \cancel{-\frac{2}{9}}$

29. $\sqrt{2x+1} - \sqrt{3x+4} = -1$

$\left(\sqrt{2x+1}\right)^2 = \left(\sqrt{3x+4}-1\right)^2$

$2x+1 = 3x+4-2\sqrt{3x+4}+1$

$2\sqrt{3x+4} = x+4$

$\left(2\sqrt{3x+4}\right)^2 = (x+4)^2$

$4(3x+4) = x^2 + 8x + 16$

$12x + 16 = x^2 + 8x + 16$

$0 = x^2 - 4x$

$0 = x(x-4)$

$x-4=0$ or $x=0$

$x = 4$

31. Let $u = x^2$

$u^2 - 10u + 9 = 0$

$(u-9)(u-1) = 0$

$u-9=0$ or $u-1=0$

$u = 9$ $\quad u = 1$

$x^2 = 9$ $\quad x^2 = 1$

$x = \pm 3$ $\quad x = \pm 1$

33. Let $u = x^2$

$4u^2 - 13u + 9 = 0$

$(4u-9)(u-1) = 0$

$4u-9=0$ or $u-1=0$

$u = \frac{9}{4}$ $\quad u = 1$

$x^2 = \frac{9}{4}$ $\quad x^2 = 1$

$x = \pm\frac{3}{2}$ $\quad x = \pm 1$

35. Let $u = x^2$

$u^2 + 5u - 36 = 0$

$(u+9)(u-4) = 0$

$u+9=0$ or $u-4=0$

$u = -9$ $\quad u = 4$

$x^2 = -9$ $\quad x^2 = 4$

$x = \pm 3i$ $\quad x = \pm 2$

37. Let $u = x+2$

$u^2 + 6u + 8 = 0$

$(u+4)(u+2) = 0$

$u+4=0$ or $u+2=0$

$u = -4$ $\quad u = -2$

$x+2 = -4$ $\quad x+2 = -2$

$x = -6$ $\quad x = -4$

39. Let $u = x+3$

$$2u^2 - 9u - 5 = 0$$
$$(2u+1)(u-5) = 0$$

$2u+1=0$ or $u-5=0$

$u = -\frac{1}{2}$ $\quad$ $u = 5$

$x+3 = -\frac{1}{2}$ $\quad$ $x+3=5$

$x = -3\frac{1}{2}$ $\quad$ $x = 2$

41. Let $u = \frac{x-1}{2}$

$$u^2 + 8u + 15 = 0$$
$$(u+3)(u+5) = 0$$

$u+3=0$ or $u+5=0$

$u=-3$ $\quad$ $u=-5$

$\frac{x-1}{2} = -3$ $\quad$ $\frac{x-1}{2} = -5$

$x-1=-6$ $\quad$ $x-1=-10$

$x=-5$ $\quad$ $x=-9$

43. Let $u = \frac{x+2}{2}$

$$2u^2 + u - 3 = 0$$
$$(2u+3)(u-1) = 0$$

$2u+3=0$ or $u-1=0$

$u = \frac{-3}{2}$ $\quad$ $u=1$

$\frac{x+2}{2} = \frac{-3}{2}$ $\quad$ $\frac{x+2}{2} = 1$

$x+2=-3$ $\quad$ $x+2=2$

$x=-5$ $\quad$ $x=0$

45. Let $u = x^{1/3}$

$$u^2 - 5u + 6 = 0$$
$$(u-2)(u-3) = 0$$

$u-2=0$ or $u-3=0$

$u=2$ $\quad$ $u=3$

$x^{1/3}=2$ $\quad$ $x^{1/3}=3$

$\left(x^{1/3}\right)^3 = 2^3$ $\quad$ $\left(x^{1/3}\right)^3 = 3^3$

$x=8$ $\quad$ $x=27$

47. Let $u = x^{1/3}$

$$2u^2 - 3u - 2 = 0$$
$$(2u+1)(u-2) = 0$$

$2u+1=0$ or $u-2=0$

$u = -\frac{1}{2}$ $\quad$ $u=2$

$x^{1/3} = -\frac{1}{2}$ $\quad$ $x^{1/3}=2$

$\left(x^{1/3}\right)^3 = \left(-\frac{1}{2}\right)^3$ $\quad$ $\left(x^{1/3}\right)^3 = 2^3$

$x = -\frac{1}{8}$ $\quad$ $x=8$

49. Let $u = x^{1/4}$

$$u^2 - 5u + 6 = 0$$
$$(u-3)(u-2) = 0$$

$u-3=0$ or $u-2=0$

$u=3$ $\quad$ $u=2$

$x^{1/4}=3$ $\quad$ $x^{1/4}=2$

$\left(x^{1/4}\right)^4 = 3^4$ $\quad$ $\left(x^{1/4}\right)^4 = 2^4$

$x=81$ $\quad$ $x=16$

51. Let $u = x^{1/4}$

$$5u^2 + 8u - 4 = 0$$
$$(5u-2)(u+2) = 0$$

$5u-2=0$ or $u+2=0$

$u = \frac{2}{5}$ $\quad$ $u=-2$

$x^{1/4} = \frac{2}{5}$ $\quad$ $x^{1/4}=-2$

$\left(x^{1/4}\right)^4 = \left(\frac{2}{5}\right)^4$ $\quad$ $\left(x^{1/4}\right)^4 = (-2)^4$

$x = \frac{16}{625}$ $\quad$ $x =$ ~~16~~

53. Let x represent the time it takes the bus to travel 180 miles.

Vehicle	d	t	r
bus	180	x	$\frac{180}{x}$
truck	180	$x+1$	$\frac{180}{x+1}$

Rate of bus = Rate of truck + 15

$$\frac{180}{x}=\frac{180}{x+1}+15$$
$$180(x+1)=180x+15x(x+1)$$
$$180x+180=180x+15x^2+15x$$
$$0=15x^2+15x-180$$
$$0=x^2+x-12$$
$$0=(x+4)(x-3)$$
$$x=\cancel{-4},3$$

Because time cannot be negative, it takes the bus 3 hours to travel 180 miles.

55. Let x represent the time it takes the winner to run 26 miles.

Person	d	t	r
winner	26	x	$\frac{26}{x}$
other	26	$x+1$	$\frac{26}{x+1}$

Rate of winner = Rate of other + 3.79

$$\frac{26}{x}=\frac{26}{x+1}+3.79$$
$$26(x+1)=26x+3.79x(x+1)$$
$$26x+26=26x+3.79x^2+3.79x$$
$$0=3.79x^2+3.79x-26$$
$$x=\frac{-3.79\pm\sqrt{(3.79)^2-4(3.79)(-26)}}{2(3.79)}$$
$$=\frac{-3.79\pm\sqrt{408.5241}}{7.58}$$
$$\approx\frac{-3.79\pm 20.212}{7.58}\approx 2.17,-3.17$$

Because time cannot be negative, it takes the winner approximately 2.17 hours to run 26 miles.

57. Let x represent the time it takes the motorcycle to travel 300 miles.

Vehicle	d	t	r
motorcycle	300	x	$\frac{300}{x}$
bus	360	$x+3$	$\frac{360}{x+3}$

Rate of motorcycle = Rate of bus + 15

$$\frac{300}{x}=\frac{360}{x+3}+15$$
$$300(x+3)=360x+15x(x+3)$$
$$300x+900=360x+15x^2+45x$$
$$0=15x^2+105x-900$$
$$0=x^2+7x-60$$
$$0=(x+12)(x-5)$$
$$x=\cancel{-12},5$$

Because time cannot be negative, it takes the motorcycle approximately 5 hours to travel 300 miles. However, we are asked for the rate of the bus. Making the necessary substitution, we find that the bus is traveling at a rate of 45 mph.

$$\text{Rate of bus}=\frac{360}{5+3}=\frac{360}{8}=45\text{ mph}$$

59. Let x represent the time it used to take the cyclist to travel 36 miles.

	d	t	r
Old	36	x	$\frac{36}{x}$
New	36	$x-1$	$\frac{36}{x-1}$

New rate = Old rate + 6

$$\frac{36}{x-1}=\frac{36}{x}+6$$
$$36x=36(x-1)+6x(x-1)$$
$$36x=36x-36+6x^2-6x$$
$$0=6x^2-6x-36$$
$$0=x^2-x-6$$
$$0=(x-3)(x+2)$$
$$x=3,\cancel{-2}$$

Because time cannot be negative, it used to take the cyclist approximately 3 hours to travel 36 miles. a) Old rate $= \frac{36}{3} = 12$ mph

New rate $= \frac{36}{2} = 18$ mph

b) Old time is 3 hours, new time is 2 hours

61. Let x represent the number of hours for Jody to run the hoop nets.

Worker	Time to complete alone	Rate of work	Time at work	Portion of job completed
Billy	$x+2$	$\frac{1}{x+2}$	$\frac{12}{5}$	$\frac{12}{5(x+2)}$
Jody	x	$\frac{1}{x}$	$\frac{12}{5}$	$\frac{12}{5x}$

Billy's portion + Jody's portion = 1 (entire job)

$$\frac{12}{5(x+2)}+\frac{12}{5x}=1$$
$$12x+12(x+2)=5x(x+2)$$
$$12x+12x+24=5x^2+10x$$
$$0=5x^2-14x-24$$
$$0=(5x+6)(x-4)$$
$$x=\cancel{-\frac{6}{5}},4$$

Since a negative time makes no sense in the context of this problem, it takes Jody 4 hours to run the hoop nets alone and Billy 4 + 2 = 6 hours working alone.

63. Let x represent the number of hours for the new press to print the copies.

Press	Time to complete alone	Rate of work	Time at work	Portion of job completed
New	x	$\frac{1}{x}$	2	$\frac{2}{x}$
Old	$x+\frac{1}{2}$	$\frac{1}{x+\frac{1}{2}}$	2	$\frac{2}{x+\frac{1}{2}}=\frac{4}{2x+1}$

New press portion + old press portion = 1 (entire job)

$$\frac{2}{x}+\frac{4}{2x+1}=1$$
$$2(2x+1)+4x=x(2x+1)$$
$$4x+2+4x=2x^2+x$$
$$0=2x^2-7x-2$$
$$x=\frac{-(-7)\pm\sqrt{(-7)^2-4(2)(-2)}}{2(2)}$$
$$=\frac{7\pm\sqrt{49+16}}{4}=\frac{7\pm\sqrt{65}}{4}$$
$$\approx 3.77,-0.27$$

Since a negative time makes no sense in the context of this problem, it takes the new press about 3.77 hours to run the copies alone and the old press about 3.77 + 0.5 = 4.27 hours working alone.

65. String 1: tension = 50 lbs., frequency = x

String 2: tension = 60 lbs., frequency = $x + 40$ vps

$$\frac{x^2}{(40+x)^2}=\frac{50}{60}$$
$$60x^2=50(40+x)^2$$
$$60x^2=50\left(1600+80x+x^2\right)$$
$$60x^2=80,000+4000x+50x^2$$
$$10x^2-4000x-80,000=0$$
$$x^2-400x-8000=0$$

$$x = \frac{-(-400) \pm \sqrt{(-400)^2 - 4(1)(-8000)}}{2(1)}$$
$$= \frac{400 \pm \sqrt{192,000}}{2} \approx 420, -19$$

Since string frequency must be positive, string 1's frequency is 420 vps, and string 2's frequency is 420 + 40 = 460 vps.

Review Exercises

1. $-\frac{b}{2a} = -\frac{8}{2 \cdot 2} = -\frac{8}{4} = -2$

2. $2x + 3y = 6$

   x-intercept: $2x + 3 \cdot 0 = 6$
   $$2x = 6$$
   $$x = 3$$
   $$(3, 0)$$

   y-intercept: $2 \cdot 0 + 3y = 6$
   $$3y = 6$$
   $$y = 2$$
   $$(0, 2)$$

3. $f(x) = 3x^2 - 2x - 8$

   x-intercept: $3x^2 - 2x - 8 = 0$
   $$(3x + 4)(x - 2) = 0$$
   $$3x + 4 = 0 \quad \text{or} \quad x - 2 = 0$$
   $$x = -\frac{4}{3} \qquad x = 2$$
   $$\left(-\frac{4}{3}, 0\right) \qquad (2, 0)$$

   y-intercept: $3(0^2) - 2(0) - 8 = -8$
   $$(0, -8)$$

4. $f(x) = 2x^2 - 3x + 1$
   $$f(-1) = 2(-1)^2 - 3(-1) + 1$$
   $$= 2 \cdot 1 + 3 + 1$$
   $$= 2 + 3 + 1$$
   $$= 6$$

5. $f(x) = x^2 - 3$

x	$f(x)$	(x, y)
-2	1	$(-2, 1)$
-1	-2	$(-1, -2)$
0	-3	$(0, -3)$
1	-2	$(1, -2)$
2	1	$(2, 1)$

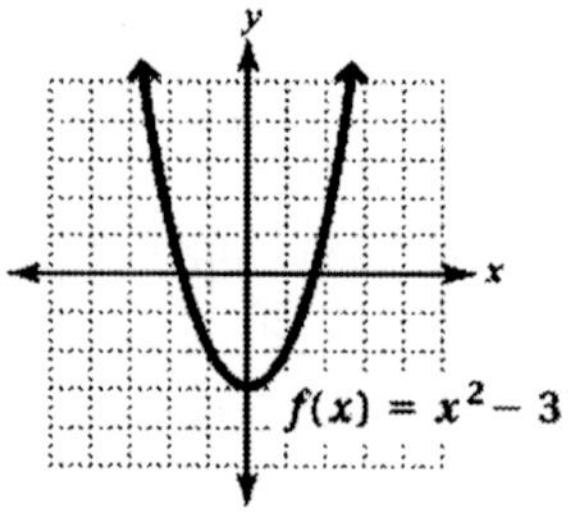

6. $f(x) = 3x^2$

x	$f(x)$	(x, y)
-2	12	$(-2, 12)$
-1	3	$(-1, 3)$
0	0	$(0, 0)$
1	3	$(1, 3)$
2	12	$(2, 12)$

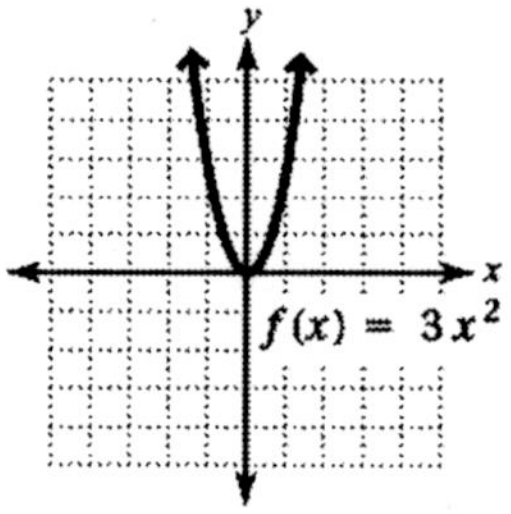

Exercise Set 9.4

1. The sign of a.

3. The axis of symmetry is the vertical line through the vertex. If $y = ax^2 + bx + c$, then the equation of the axis of symmetry is $x = -\frac{b}{2a}$.

5. There are no x-intercepts.

7. $f(x) = 5(x - 2)^2 - 3$

   Vertex: $(2, -3)$; axis of symmetry $x = 2$

9. $g(x) = -2(x + 1)^2 - 5$

   Vertex: $(-1, -5)$; axis of symmetry $x = -1$

11. $k(x)=x^2+2$

Vertex: $(0,2)$; axis of symmetry $x=0$

13. Can be rewritten as $h(x)=3(x+5)^2+0$

Vertex: $(-5,0)$; axis of symmetry $x=-5$

15. Can be rewritten as $f(x)=-0.5(x-0)^2+0$

Vertex: $(0,0)$; axis of symmetry $x=0$

17. $f(x)=x^2-4x+8$

x-coordinate: $\dfrac{-(-4)}{2(1)}=\dfrac{4}{2}=2$

y-coordinate: $y=(2)^2-4(2)+8$

$=4-8+8$

$=4$

vertex: (2, 4)
axis of symmetry: $x=2$

19. $k(x)=2x^2+16x+27$

x-coordinate: $\dfrac{-(16)}{2(2)}=\dfrac{-16}{4}=-4$

y-coordinate: $y=2(-4)^2+16(-4)+27$

$=32-64+27$

$=-5$

vertex: (–4, –5)
axis of symmetry: $x=-4$

21. $g(x)=-3x^2+2x+1$

x-coordinate: $\dfrac{-(2)}{2(-3)}=\dfrac{-2}{-6}=\dfrac{1}{3}$

y-coordinate: $y=-3\left(\dfrac{1}{3}\right)^2+2\left(\dfrac{1}{3}\right)+1$

$=-3\cdot\dfrac{1}{9}+\dfrac{2}{3}+1$

$=-\dfrac{1}{3}+\dfrac{2}{3}+1$

$=\dfrac{4}{3}$

vertex: $\left(\dfrac{1}{3},\dfrac{4}{3}\right)$

axis of symmetry: $x=\dfrac{1}{3}$

23. $h(x)=-3x^2$

a) downward

b) $(0,0)$

c) $x=0$

d)

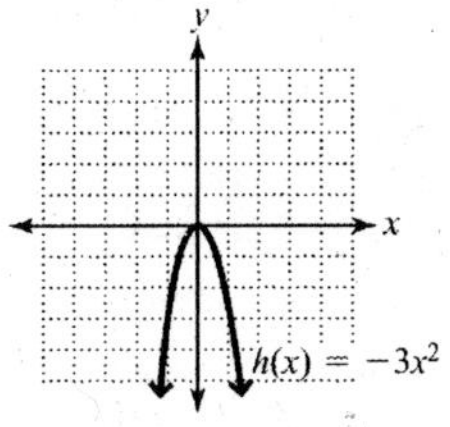

25. $k(x)=\dfrac{1}{4}x^2$

a) upward

b) $(0,0)$

c) $x=0$

d)

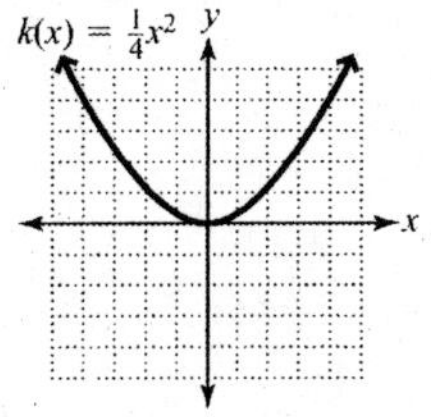

27. $f(x)=4x^2-3$

a) upward

b) $(0,-3)$

c) $x=0$

d)

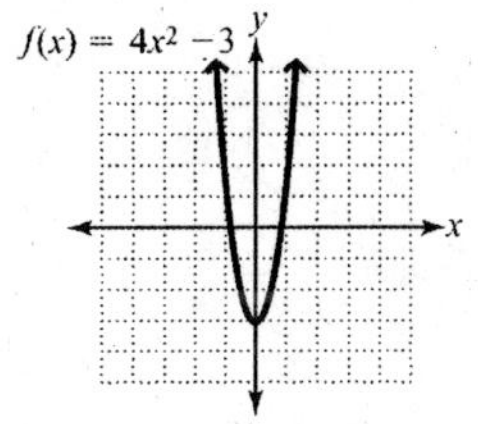

29. $g(x)=-0.5x^2+2$

a) downward

b) $(0,2)$

c) $x=0$

d)

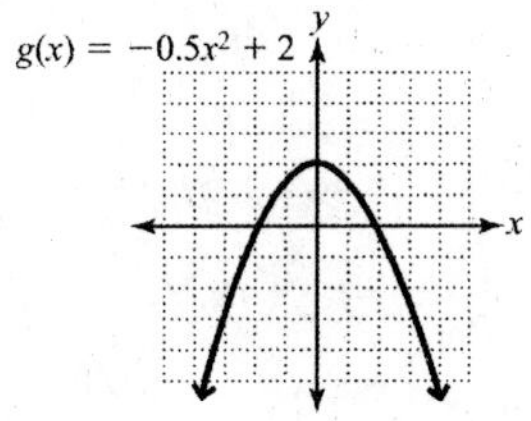

31. $f(x)=(x-3)^2+2$

a) upward

b) $(3,2)$

c) $x=3$

d)

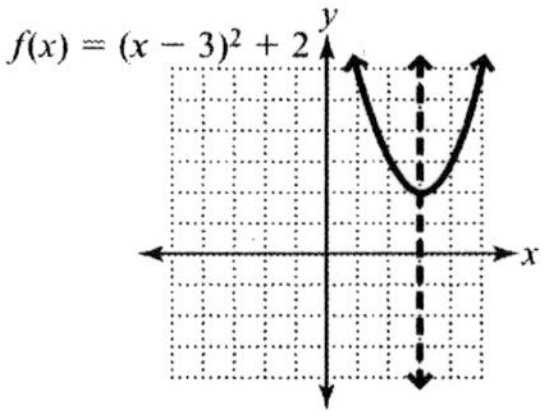

33. $k(x)=-2(x+1)^2-3$

a) downward

b) $(-1,-3)$

c) $x=-1$

d)

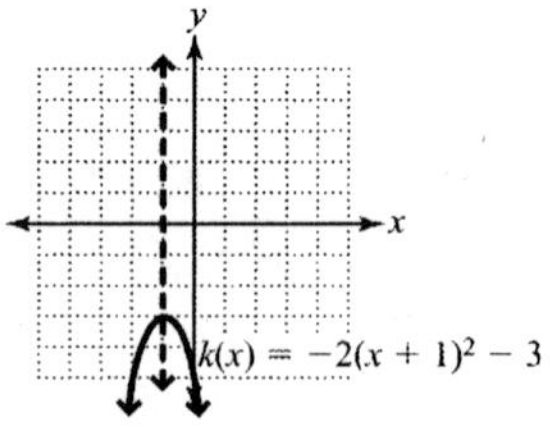

35. $h(x)=\frac{1}{3}(x-2)^2-1$

a) upward

b) $(2,-1)$

c) $x=2$

d)

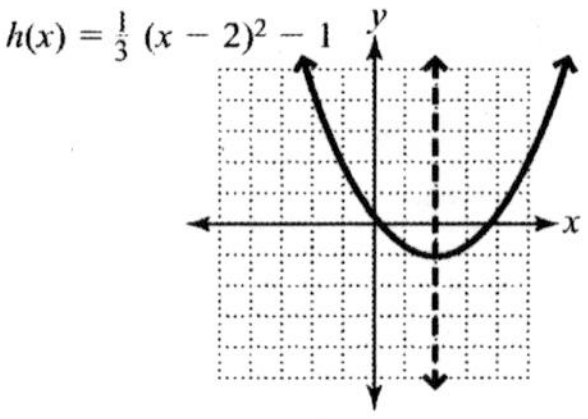

37. $h(x)=x^2+6x+9$

a) x-intercept: $x^2+6x+9=0$

$$(x+3)^2=0$$
$$x+3=\pm\sqrt{0}$$
$$x+3=0$$
$$x=-3$$
$$(-3,0)$$

y-intercept: $y=0^2+6(0)+9=9$

(0, 9)

b)
$$y=x^2+6x+9$$
$$y=(x+3)^2$$
$$h(x)=(x+3)^2$$

c) upward

d) $(-3,0)$

e) $x=-3$

f)

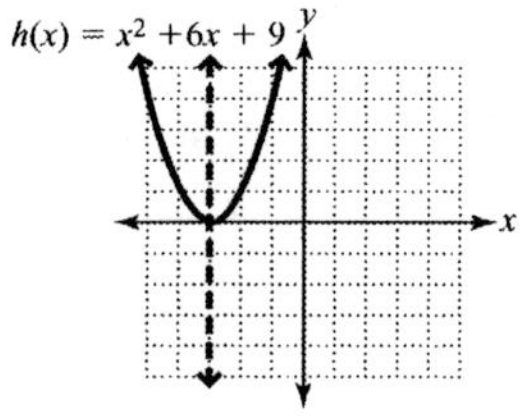

g) Domain: $\{x\,|\,x \text{ is a real number}\}$ or $(-\infty,\infty)$

Range: $\{y\,|\,y\ge 0\}$ or $[0,\infty)$

39. $g(x)=-3x^2+6x-5$

a) x-intercepts:

$$-3x^2+6x-5=0$$
$$x=\frac{-6\pm\sqrt{6^2-4(-3)(-5)}}{2(-3)}$$
$$=\frac{-6\pm\sqrt{36-60}}{-6}$$
$$=\frac{-6\pm\sqrt{-24}}{-6}$$

Since the discriminant is negative, there are no x-intercepts.

y-intercept: $y=-3(0)^2+6(0)-5=-5$

(0, −5)

b)
$$y=-3x^2+6x-5$$
$$y+5=-3(x^2-2x)$$
$$y+5-3=-3(x^2-2x+1)$$
$$y+2=-3(x-1)^2$$
$$y=-3(x-1)^2-2$$
$$g(x)=-3(x-1)^2-2$$

c) downward

d) $(1,-2)$

e) $x=1$

f)

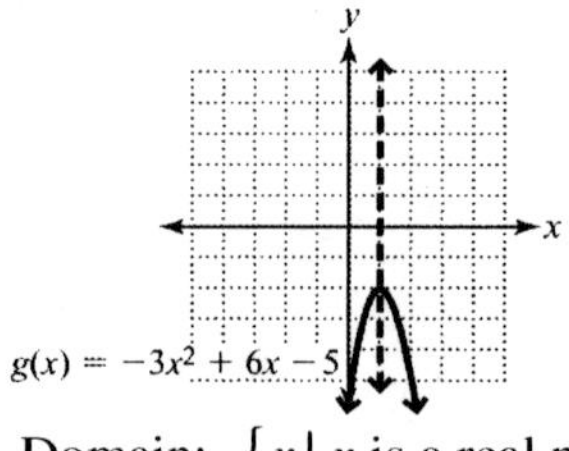

g) Domain: $\{x\,|\,x \text{ is a real number}\}$ or $(-\infty,\infty)$

Range: $\{y\,|\,y\le -2\}$ or $(-\infty,-2]$

41. $f(x) = 2x^2 + 6x + 3$

a) x-intercepts:

$2x^2 + 6x + 3 = 0$

$$x = \frac{-6 \pm \sqrt{6^2 - 4(2)(3)}}{2(2)}$$
$$= \frac{-6 \pm \sqrt{36 - 24}}{4}$$
$$= \frac{-6 \pm \sqrt{12}}{4}$$
$$= \frac{-6 \pm 2\sqrt{3}}{4}$$
$$= \frac{-3 \pm \sqrt{3}}{2}$$
$$\left(\frac{-3+\sqrt{3}}{2}, 0\right), \left(\frac{-3-\sqrt{3}}{2}, 0\right)$$

y-intercept: $y = 2(0)^2 + 6(0) + 3 = 3$

(0, 3)

b)
$$y = 2x^2 + 6x + 3$$
$$y - 3 = 2x^2 + 6x$$
$$y - 3 = 2(x^2 + 3x)$$
$$y - 3 + \frac{9}{2} = 2\left(x^2 + 3x + \frac{9}{4}\right)$$
$$y + \frac{3}{2} = 2\left(x + \frac{3}{2}\right)^2$$
$$y = 2\left(x + \frac{3}{2}\right)^2 - \frac{3}{2}$$
$$f(x) = 2\left(x + \frac{3}{2}\right)^2 - \frac{3}{2}$$

c) upward

d) $\left(-\frac{3}{2}, -\frac{3}{2}\right)$

e) $x = -\frac{3}{2}$

f)

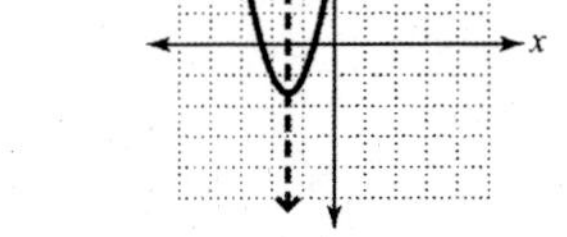

g) Domain: $\{x \mid x \text{ is a real number}\}$ or $(-\infty, \infty)$

Range: $\left\{y \mid y \geq -\frac{3}{2}\right\}$ or $\left[-\frac{3}{2}, \infty\right)$

43. Finding the vertex will allow us to know the highest point and how many seconds it takes to reach the highest point.

a) t-coordinate: $t = \frac{-45}{2(-4.9)} \approx 4.59$ sec. is the time it takes to reach the highest point.

b) h-coordinate:

$h = -4.9(4.59)^2 + 45(4.59) \approx 103.32$ meters is the maximum height that the rocket reaches.

c) The t-intercepts describe the times that the rocket is on the ground.

$h = 0: 0 = -4.9t^2 + 45t$

$$t = \frac{-(45) \pm \sqrt{(45)^2 - 4(-4.9)(0)}}{2(-4.9)}$$
$$= \frac{-45 \pm \sqrt{2025 + 0}}{-9.8} = \frac{-45 \pm 45}{-9.8}$$
$$= \frac{-45 + 45}{-9.8} = 0 \text{ (starting point)}$$

or $= \frac{-45 - 45}{-9.8} = 9.18$ sec.

45. a) x-coordinate: $-\frac{3.2}{2(-0.8)} = 2$

The maximum height will be the y-coordinate of the vertex.

y-coordinate: $y = -0.8(2)^2 + 3.2(2) + 6 = 9.2$ feet is the maximum height that the ball reaches.

b) To find how far the ball travels, we need to find the x-value when it hits the ground, which is where $y = 0$.

$y = 0: 0 = -0.8x^2 + 3.2x + 6$

$$x = \frac{-(3.2) \pm \sqrt{(3.2)^2 - 4(-0.8)(6)}}{2(-0.8)}$$
$$\approx -1.39, 5.39$$

The negative values does not make sense in this context, so the ball travels approximately 5.39 feet.

c)

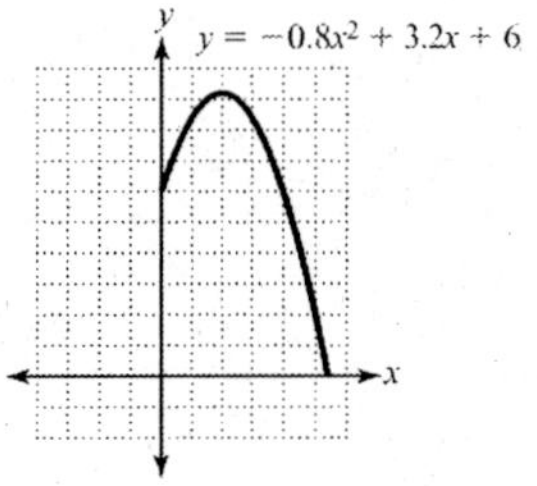

47. Finding the vertex will allow us to know the highest number of CDs sold and how many weeks it takes to reach that point.

$$n(t) = -200t^2 + 4000t$$
$$n(t) = -200(t^2 - 20t)$$
$$n(t) - 20,000 = -200(t^2 - 20t + 100)$$
$$n(t) = -200(t-10)^2 + 20,000$$

Vertex: $(10, 20,000)$

a) The tenth week b) 20,000

49. Because the enclosed space is to be rectangular, the sum of the length and width is $\frac{1}{2}(400) = 200$ feet. Let x represent the width and $200 - x$ represent the length.

$$y = x(200 - x)$$
$$y = 200x - x^2$$
$$y = -x^2 + 200x$$
$$y - 10,000 = -(x^2 - 200x + 10,000)$$
$$y - 10,000 = -(x-100)^2$$
$$y = -(x-100)^2 + 10,000$$

The vertex, $(100, 10,000)$ shows the highest point on the graph, or the maximum area to be achieved.

The area is maximized if the width is 100 feet and the length is $200 - 100 = 100$ feet.

51. The vertex would be the lowest point on the graph.

$$C(n) = n^2 - 110n + 5000$$
$$m = n^2 - 110n + 5000$$
$$m - 5000 = n^2 - 110n$$
$$m - 5000 + 3025 = n^2 - 110n + 3025$$
$$m + 1075 = (n-55)^2$$
$$m = (n-55)^2 + 1075$$
$$c(n) = (n-55)^2 + 1075$$

55 units would minimize the cost.

53. Let one of the integers be x and the other be $x + 12$.

$$y = x(x+12)$$
$$y = x^2 + 12x$$
$$y + 36 = x^2 + 12x + 36$$
$$y + 36 = (x+6)^2$$
$$y = (x+6)^2 - 36$$

The vertex $(-6, -36)$ shows one of the integers is –6 the other is –6 + 12 = 6, and the product is –36.

Review Exercises

1. $-|-2+3\cdot 4| - 4^0 = -|-2+12| - 1$
$$= -|10| - 1$$
$$= -10 - 1$$
$$= -11$$

2. $x^2 + 2x = 15$
$$x^2 + 2x - 15 = 0$$
$$(x+5)(x-3) = 0$$
$$x + 5 = 0 \quad \text{or} \quad x - 3 = 0$$
$$x = -5 \qquad x = 3$$

3. $x^2 = -20$
$$x = \pm\sqrt{-20}$$
$$x = \pm 2i\sqrt{5}$$

4. $4x - 8 \le 2x + 1$
$$2x \le 9$$
$$x \le 4.5$$

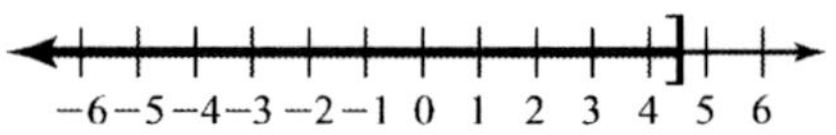

5. $|x-3| \ge 8$
$$x - 3 \le -8 \quad \text{or} \quad x - 3 \ge 8$$
$$x \le -5 \qquad x \ge 11$$

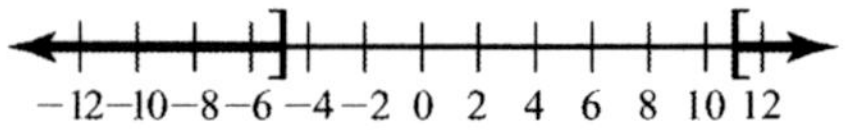

6. $\begin{cases} x+y>2 \\ 2x-3y\le 6 \end{cases}$

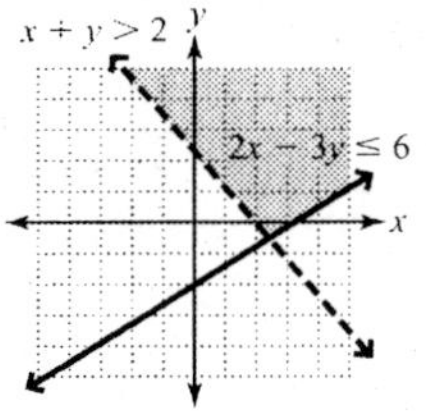

Exercise Set 9.5

1. Solve the related equation and then determine which of the three intervals satisfy the inequality by examining one value in each interval. If this one value satisfies the inequality, then all points in that interval satisfy the inequality.

3. Yes, for example, $x^2+2\le 0$ has an empty solution set since x^2+2 is always positive.

5. In both cases, the intervals to check are $(-\infty,-5),(-5,2),(2,\infty)$ and in both cases only x-values in $(-5,2)$ satisfy the original inequalities.

7. a) $x=-5,-1$ b) $(-5,-1)$
   c) $(-\infty,-5)\cup(-1,\infty)$

9. a) $x=-1,3$ b) $(-\infty,-1]\cup[3,\infty)$
   c) $[-1,3]$

11. a) $x=-2$ b) $(-\infty,-2)\cup(-2,\infty)$
   c) $\varnothing$

13. a) $\varnothing$ b) $\varnothing$
   c) $\mathbb{R}$ or $(-\infty,\infty)$

15. $(x+4)(x+2)=0$
   $x+4=0$ or $x+2=0$
   $x=-4$ $x=-2$

   For $(-\infty,-4)$, we choose –5.

   $$(-5+4)(-5+2)<0 \quad \text{False}$$
   $$-1(-3)<0$$
   $$3<0$$

   For $(-4,-2)$, we choose –3.

   $$(-3+4)(-3+2)<0 \quad \text{True}$$
   $$1(-1)<0$$
   $$-1<0$$

   For $(-2,\infty)$, we choose 0.

   $$(0+4)(0+2)<0 \quad \text{False}$$
   $$4\cdot 2<0$$
   $$8<0$$

   Solution set: $(-4,-2)$

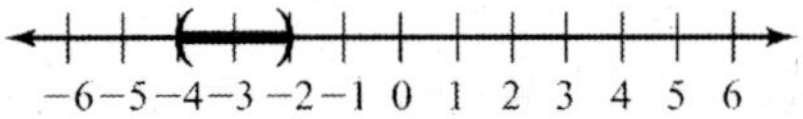

17. $(x-2)(x-5)=0$
   $x-2=0$ or $x-5=0$
   $x=2$ $x=5$

   For $(-\infty,2)$, we choose 0.

   $$(0-2)(0-5)>0 \quad \text{True}$$
   $$-2(-5)>0$$
   $$10>0$$

   For $(2,5)$, we choose 3.

   $$(3-2)(3-5)>0 \quad \text{False}$$
   $$1(-2)>0$$
   $$-2>0$$

   For $(5,\infty)$, we choose 6.

   $$(6-2)(6-5)>0 \quad \text{True}$$
   $$4(1)>0$$
   $$4>0$$

   Solution set: $(-\infty,2)\cup(5,\infty)$

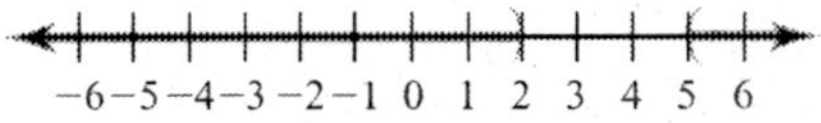

19. $x^2+5x+4=0$
   $(x+4)(x+1)=0$
   $x+4=0$ or $x+1=0$
   $x=-4$ $x=-1$

   For $(-\infty,-4)$, we choose –5.

   $$(-5)^2+5(-5)+4<0 \quad \text{False}$$
   $$25-25+4<0$$
   $$4<0$$

For $(-4,-1)$, we choose –3.

$$(-3)^2+5(-3)+4<0 \quad \text{True}$$
$$9-15+4<0$$
$$-2<0$$

For $(-1,\infty)$, we choose 0.

$$(0)^2+5(0)+4<0 \quad \text{False}$$
$$4<0$$

Solution set: $(-4,-1)$

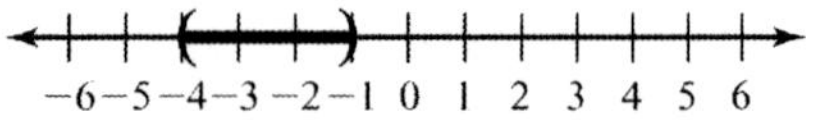

21. $x^2-4x+3=0$

$(x-3)(x-1)=0$

$x-3=0$ or $x-1=0$

$x=3$ $\quad$ $x=1$

For $(-\infty,1)$, we choose 0.

$$(0)^2-4(0)+3>0 \quad \text{True}$$
$$3>0$$

For $(1,3)$, we choose 2.

$$(2)^2-4(2)+3>0 \quad \text{False}$$
$$4-8+3>0$$
$$-1>0$$

For $(3,\infty)$, we choose 4.

$$(4)^2-4(4)+3>0 \quad \text{True}$$
$$16-16+3>0$$
$$3>0$$

Solution set: $(-\infty,1)\cup(3,\infty)$

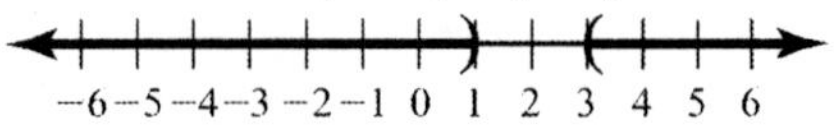

23. $b^2-6b+8=0$

$(b-4)(b-2)=0$

$b-4=0$ or $b-2=0$

$b=4$ $\quad$ $b=2$

For $(-\infty,2)$, we choose 0.

$$(0)^2-6(0)+8\le 0 \quad \text{False}$$
$$8\le 0$$

For $(2,4)$, we choose 3.

$$(3)^2-6(3)+8\le 0 \quad \text{True}$$
$$9-18+8\le 0$$
$$-1\le 0$$

For $(4,\infty)$, we choose 5.

$$(5)^2-6(5)+8\le 0 \quad \text{False}$$
$$25-30+8\le 0$$
$$3\le 0$$

Solution set: $[2,4]$

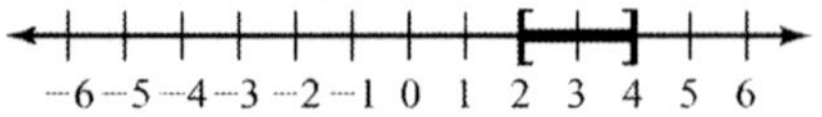

25. $y^2-4y-5=0$

$(y+1)(y-5)=0$

$y+1=0$ or $y-5=0$

$y=-1$ $\quad$ $y=5$

For $(-\infty,-1)$, we choose –2.

$$(-2)^2-5\ge 4(-2) \quad \text{True}$$
$$4-5\ge -8$$
$$-1\ge -8$$

For $(-1,5)$, we choose 3.

$$(3)^2-5\ge 4(3) \quad \text{False}$$
$$9-5\ge 12$$
$$4\ge 12$$

For $(5,\infty)$, we choose 6.

$$(6)^2-5\ge 4(6) \quad \text{True}$$
$$36-5\ge 24$$
$$31\ge 24$$

Solution set: $(-\infty,-1]\cup[5,\infty)$

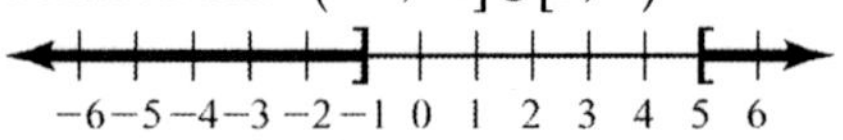

27. $a^2-3a-10=0$

$(a-5)(a+2)=0$

$a-5=0$ or $a+2=0$

$a=5$ $\quad a=-2$

For $(-\infty,-2)$, we choose -3.

$(-3)^2-3(-3)<10$ False

$9+9<10$

$18<10$

For $(-2,5)$, we choose 0.

$(0)^2-3(0)<10$ True

$0<10$

For $(5,\infty)$, we choose 6.

$(6)^2-3(6)<10$ False

$36-18<10$

$18<10$

Solution set: $(-2,5)$

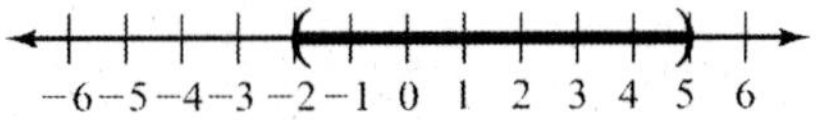

29. $y^2+6y+9=0$

$(y+3)^2=0$

Since $(y+3)^2\ge 0$ is always nonnegative, the solution set is $(-\infty,\infty)$ or $\mathbb{R}$.

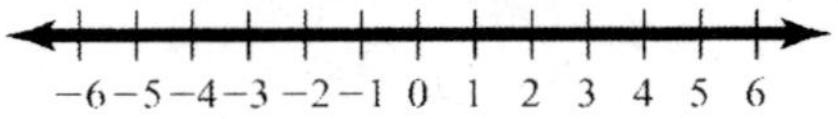

31. $2c^2-4c+7=0$

$$c=\frac{-(-4)\pm\sqrt{(-4)^2-4(2)(7)}}{2\cdot 2}$$

$$=\frac{4\pm\sqrt{16-56}}{4}$$

$$=\frac{4\pm\sqrt{-40}}{4}$$

$$=\frac{4\pm 2i\sqrt{10}}{4}=\frac{2\pm i\sqrt{10}}{2}$$

There are no c-intercepts, therefore there is no solution. The solution set is $\varnothing$.

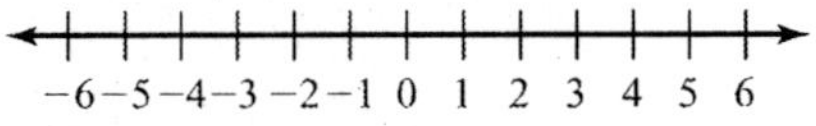

33. $4r^2+21r+5=0$

$(4r+1)(r+5)=0$

$4r+1=0$ or $r+5=0$

$r=-0.25$ $\quad r=-5$

For $(-\infty,-5)$, we choose -6.

$4(-6)^2+21(-6)+5>0$ True

$144-126+5>0$

$23>0$

For $(-5,-0.25)$, we choose -1.

$4(-1)^2+21(-1)+5>0$ False

$4-21+5>0$

$-12>0$

For $(-0.25,\infty)$, we choose 0.

$4(0)^2+21(0)+5>0$ True

$5>0$

Solution set: $(-\infty,-5)\cup(-0.25,\infty)$

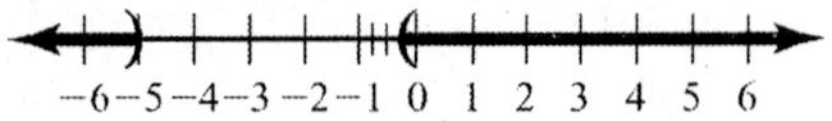

35. $x^2-5x=0$

$x(x-5)=0$

$x=0$ or $x-5=0$

$x=5$

For $(-\infty,0)$, we choose -1.

$(-1)^2-5(-1)>0$ True

$1+5>0$

$6>0$

For $(0,5)$, we choose 1.

$(1)^2-5(1)>0$ False

$1-5>0$

$-4>0$

For $(5,\infty)$, we choose 6.

$(6)^2-5(6)>0$ True

$36-30>0$

$6>0$

Solution set: $(-\infty,0)\cup(5,\infty)$

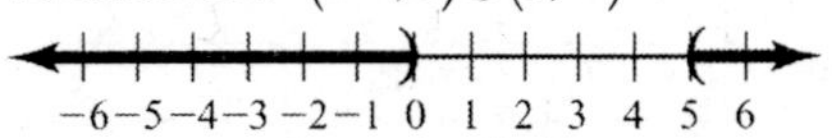

37. $x^2 - 6x = 0$

$x(x-6) = 0$

$x = 0 \quad \text{or} \quad x - 6 = 0$

$x = 6$

For $(-\infty, 0)$, we choose -1.

$(-1)^2 \le 6(-1)$ False

$1 \le -6$

For $(0, 6)$, we choose 1.

$(1)^2 \le 6(1)$ True

$1 \le 6$

For $(6, \infty)$, we choose 7.

$(7)^2 \le 6(7)$ False

$49 \le 42$

Solution set: $[0, 6]$

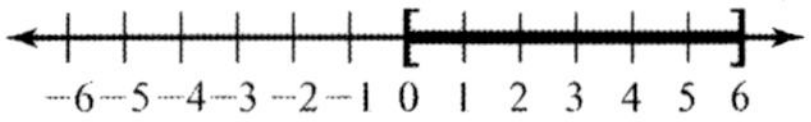

39. Since $(x+1)^2$ is always 0 or positive, every real number is a solution.

Solution set: $(-\infty, \infty)$ or $\mathbb{R}$

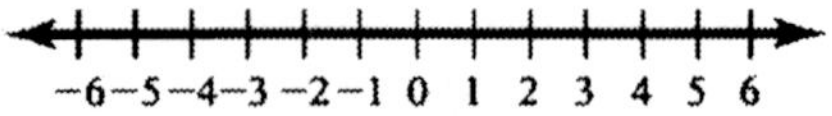

41. $(x-4)(x+2)(x+4) = 0$

$x - 4 = 0 \quad \text{or} \quad x + 2 = 0 \quad \text{or} \quad x + 4 = 0$

$x = 4 \qquad x = -2 \qquad x = -4$

Interval	$(-\infty, -4)$	$(-4, -2)$	$(-2, 4)$	$(4, \infty)$
Test No.	−5	−3	0	5
Results	−27	7	−32	63
T/F	F	T	F	T

Solution set: $[-4, -2] \cup [4, \infty)$

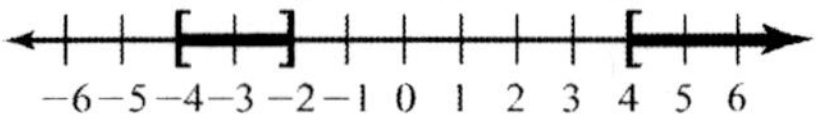

43. $(x+2)(x+6)(x-1) = 0$

$x + 2 = 0 \quad \text{or} \quad x + 6 = 0 \quad \text{or} \quad x - 1 = 0$

$x = -2 \qquad x = -6 \qquad x = 1$

Interval	$(-\infty, -6)$	$(-6, -2)$	$(-2, 1)$	$(1, \infty)$
Test No.	−7	−3	0	2
Results	−40	12	−12	32
T/F	T	F	T	F

Solution set: $(-\infty, -6) \cup (-2, 1)$

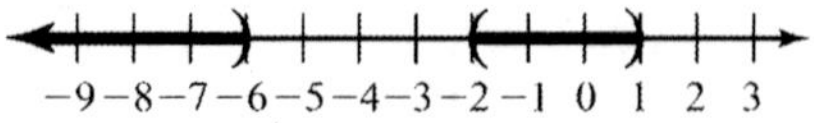

45. $3c^2 + 4c - 1 = 0$

$$c = \frac{-4 \pm \sqrt{4^2 - 4(3)(-1)}}{2(3)} = \frac{-2 \pm \sqrt{7}}{3}$$

Interval	$\left(-\infty, \frac{-2-\sqrt{7}}{3}\right)$	$\left(\frac{-2-\sqrt{7}}{3}, \frac{-2+\sqrt{7}}{3}\right)$
Test No.	−2	0
Results	3	−1
T/F	F	T

Interval	$\left(\frac{-2+\sqrt{7}}{3}, \infty\right)$
Test No.	1
Results	6
T/F	F

Solution set: $\left(\frac{-2-\sqrt{7}}{3}, \frac{-2+\sqrt{7}}{3}\right)$

47. $4r^2 + 8r - 3 = 0$

$$r = \frac{-8 \pm \sqrt{8^2 - 4(4)(-3)}}{2(4)} = \frac{-2 \pm \sqrt{7}}{2}$$

Interval	$\left(-\infty, \frac{-2-\sqrt{7}}{2}\right)$	$\left(\frac{-2-\sqrt{7}}{2}, \frac{-2+\sqrt{7}}{2}\right)$
Test No.	−3	0
Results	9	−3
T/F	T	F

Interval	$\left(\frac{-2+\sqrt{7}}{2}, \infty\right)$
Test No.	1
Results	9
T/F	T

Solution set: $\left(-\infty, \frac{-2-\sqrt{7}}{2}\right] \cup \left[\frac{-2+\sqrt{7}}{2}, \infty\right)$

49. $-0.2a^2 - 1.6a - 2 = 0$

$$a = \frac{-(-1.6) \pm \sqrt{(-1.6)^2 - 4(-0.2)(-2)}}{2(-0.2)}$$

$$= -4 \pm \sqrt{6}$$

Interval	$(-\infty, -4-\sqrt{6})$	$(-4-\sqrt{6}, -4+\sqrt{6})$
Test No.	−7	−2
Results	−0.6	0.4
T/F	T	F

Interval	$(-4+\sqrt{6}, \infty)$
Test No.	0
Results	−2
T/F	T

Solution set: $(-\infty, -4-\sqrt{6}] \cup [-4+\sqrt{6}, \infty)$

51. $\dfrac{a+4}{a-1} = 0 \qquad a - 1 = 0$

$a = 1$

$(a-1)\dfrac{a+4}{a-1} = 0(a-1)$

$a + 4 = 0$

$a = -4$

Interval	$(-\infty, -4)$	$(-4, 1)$	$(1, \infty)$
Test No.	−5	0	2
Results	1/6	−4	6
T/F	T	F	T

Solution set: $(-\infty, -4) \cup (1, \infty)$

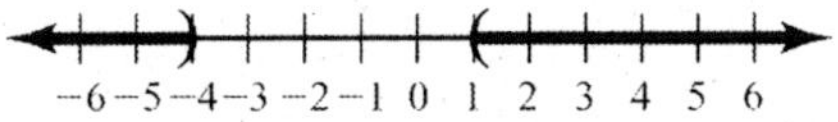

53. $\dfrac{n+1}{n+5} = 0 \qquad n + 5 = 0$

$n = -5$

$(n+5)\dfrac{n+1}{n+5} = 0(n+5)$

$n + 1 = 0$

$n = -1$

Interval	$(-\infty, -5)$	$(-5, -1)$	$(-1, \infty)$
Test No.	−6	−2	0
Results	5	−1/3	1/5
T/F	F	T	F

Solution set: $(-5, -1]$

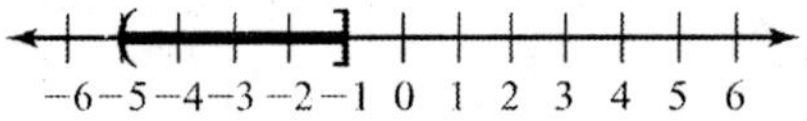

55. $\dfrac{6}{x+4} = 0 \qquad x + 4 = 0$

$x = -4$

$(x+4)\dfrac{6}{x+4} = 0(x+4)$

$6 \neq 0$

Interval	$(-\infty, -4)$	$(-4, \infty)$
Test No.	−6	0
Results	−3	3/2
T/F	F	T

Solution set: $(-4, \infty)$

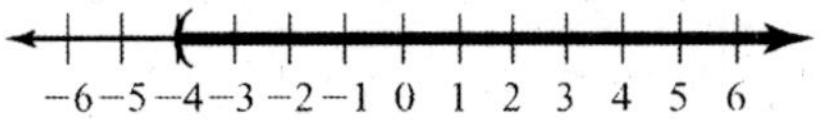

57. $\dfrac{c}{c+3} = 3 \qquad c + 3 = 0$

$c = -3$

$(c+3)\dfrac{c}{c+3} = 3(c+3)$

$c = 3c + 9$

$-2c = 9$

$c = -\dfrac{9}{2}$

Interval	$(-\infty, -9/2)$	$(-9/2, -3)$
Test No.	−6	−4
Results	$2 < 3$	$4 < 3$
T/F	T	F

Interval	$(-3, \infty)$
Test No.	0
Results	$0 < 3$
T/F	T

Solution set: $\left(-\infty, -\dfrac{9}{2}\right) \cup (-3, \infty)$

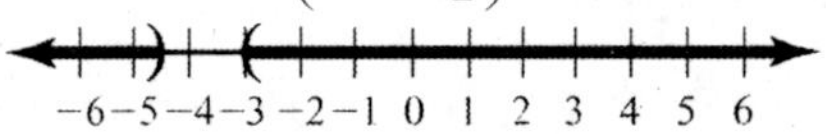

59. $$\frac{a+5}{a-4} = 4 \qquad a-4=0$$
$$a = 4$$
$$(a-4)\frac{a+5}{a-4} = 4(a-4)$$
$$a+5 = 4a-16$$
$$-3a = -21$$
$$a = 7$$

Interval	$(-\infty,4)$	$(4,7)$
Test No.	0	5
Results	$-5/4>4$	$10>4$
T/F	F	T

Interval	$(7,\infty)$
Test No.	8
Results	$13/4>4$
T/F	F

Solution set: $(4,7)$

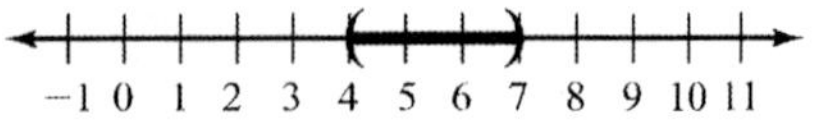

61. $$\frac{p+3}{p-3} = 4 \qquad p-3=0$$
$$p = 3$$
$$(p-3)\frac{p+3}{p-3} = 4(p-3)$$
$$p+3 = 4p-12$$
$$-3p = -15$$
$$p = 5$$

Interval	$(-\infty,3)$	$(3,5)$	$(5,\infty)$
Test No.	0	4	6
Results	$-1\ge 4$	$7\ge 4$	$3\ge 4$
T/F	F	T	F

Solution set: $(3,5]$

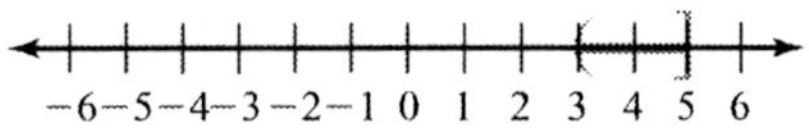

63. $$\frac{(k+3)(k-2)}{k-5} = 0$$
$$(k-5)\frac{(k+3)(k-2)}{k-5} = 0(k-5)$$
$$(k+3)(k-2) = 0$$

$$k+3=0 \qquad k-2=0 \qquad k-5=0$$
$$k=-3 \qquad k=2 \qquad k=5$$

Interval	$(-\infty,-3)$	$(-3,2)$	$(2,5)$	$(5,\infty)$
Test No.	−4	0	3	6
Results	−2/3	6/5	−3	36
T/F	T	F	T	F

Solution set: $(-\infty,-3]\cup[2,5)$

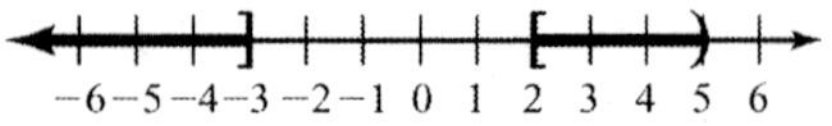

65. Because $(2x-1)^2$ is always nonnegative, we only set the denominator equal to 0; $x=0$.

Interval	$(-\infty,0)$	$(0,\infty)$
Test No.	−1	1
Results	−9	1
T/F	F	T

Solution set: $(0,\infty)$

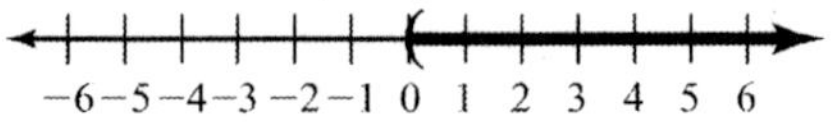

67. $$\frac{(x-5)(x-2)}{x+1} = 0$$
$$(x+1)\frac{(x-5)(x-2)}{x+1} = 0(x+1)$$
$$(x-5)(x-2) = 0$$

$$x-5=0 \qquad x-2=0 \qquad x+1=0$$
$$x=5 \qquad x=2 \qquad x=-1$$

Interval	$(-\infty,-1)$	$(-1,2)$	$(2,5)$	$(5,\infty)$
Test No.	−2	0	3	6
Results	−28	10	−0.5	4/7
T/F	T	F	T	F

Solution set: $(-\infty,-1)\cup(2,5)$

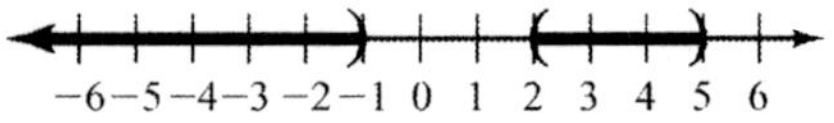

69. When the ball is on the ground, $h = 0$.

$$-16t^2 + 80t + 96 = 0$$
$$-16\left(t^2 - 5t - 6\right) = 0$$
$$t^2 - 5t - 6 = 0$$
$$(t-6)(t+1) = 0$$
$$t = 6, \cancel{-1}$$

a) 6 sec.

b) $-16t^2 + 80t + 96 = 192$

$$-16t^2 + 80t - 96 = 0$$
$$-16\left(t^2 - 5t + 6\right) = 0$$
$$t^2 - 5t + 6 = 0$$
$$(t-2)(t-3) = 0$$
$$t = 2, 3$$

The ball is 192 ft. above the ground at 2 sec. and then again at 3 sec.

c) $(2,3)$ sec.

d) $[0,2) \cup (3,6]$ sec.

71. Let $h = x - 2$ and $b = x + 6$

$$(x-2)(x+6) \ge 20$$
$$x^2 + 4x - 12 \ge 20$$
$$x^2 + 4x - 32 \ge 0$$
$$(x+8)(x-4) \ge 0$$
$$x \le -8 \text{ or } x \ge 4$$

a) $x \ge 4$ in.

b) $b \ge 4 + 6$

$b \ge 10$ in.

$h \ge 4 - 2$

$h \ge 2$ in.

73. $\dfrac{x_2 - 5}{x_2 - 2} \le \dfrac{1}{2}$

$$\frac{x_2 - 5}{x_2 - 2} = \frac{1}{2}$$
$$x_2 - 2 = 2x_2 - 10$$
$$-x_2 = -8$$
$$x_2 = 8$$

$$x_2 - 2 = 0$$
$$x_2 = 2$$

Interval	$(-\infty, 2)$	$(2,8)$	$(8,\infty)$
Test No.	0	4	10
Results	$\frac{5}{2} \le \frac{1}{2}$	$-\frac{1}{2} \le \frac{1}{2}$	$\frac{5}{8} \le \frac{1}{2}$
T/F	F	T	F

So, $2 < x \le 8$

Review Exercises

1. Add 9.

$$x^2 - 6x + 9 = (x-3)^2$$

2. Add $\dfrac{25}{4}$.

$$x^2 + 5x + \frac{25}{4} = \left(x + \frac{5}{2}\right)^2$$

3. $\sqrt{x-2} = 4$

$$\left(\sqrt{x-2}\right)^2 = 4^2$$
$$x - 2 = 16$$
$$x = 18$$

4. Downward, because the coefficient of x^2 is negative.

5. $x^2 + 4x - 12 = 0$

$$(x+6)(x-2) = 0$$
$$x + 6 = 0 \quad \text{or} \quad x - 2 = 0$$
$$x = -6 \qquad x = 2$$
$$(-6,0),\ (2,0)$$

6. $y = x^2 - 6x + 5$

$$y - 5 = x^2 - 6x$$
$$y - 5 + 9 = x^2 - 6x + 9$$
$$y + 4 = (x-3)^2$$
$$y = (x-3)^2 - 4$$

The vertex is (3, –4)

Chapter 9 Review Exercises

1. False; $\sqrt{a}$ refers only to the positive square root, whereas $\pm\sqrt{a}$ refers to both the positive and negative square roots.

2. False; it is quite possible for an equation in quadratic form to have extraneous solutions.

3. False; a quadratic equation may have a pair of complex solutions.

4. False; the discriminant can only be used to identify the type of solutions, not the actual solutions themselves.

5. True

6. $+\sqrt{a}, -\sqrt{a}$

7. $\left(\frac{b}{2}\right)^2$

8. $\frac{-b \pm \sqrt{b^2 - 4ac}}{2a}$

9. $b^2 - 4ac$

10. $(h, k); x = h$

11. $x^2 = 16$
$$x = \pm\sqrt{16}$$
$$x = \pm 4$$

12. $y^2 = \frac{1}{36}$
$$y = \pm\sqrt{\frac{1}{36}}$$
$$y = \pm\frac{1}{6}$$

13. $k^2 + 2 = 30$
$$k^2 = 28$$
$$k = \pm\sqrt{28}$$
$$k = \pm 2\sqrt{7}$$

14. $3x^2 = 42$
$$x^2 = \frac{42}{3}$$
$$x^2 = 14$$
$$x = \pm\sqrt{14}$$

15. $5h^2 + 24 = 9$
$$5h^2 = -15$$
$$h^2 = -3$$
$$h = \pm\sqrt{-3}$$
$$h = \pm i\sqrt{3}$$

16. $(x+7)^2 = 25$
$$(x+7)^2 = \pm\sqrt{25}$$
$$x + 7 = \pm 5$$
$$x = -7 \pm 5$$
$$= -7 + 5 = -2$$
$$= -7 - 5 = -12$$

17. $(x-9)^2 = -16$
$$x - 9 = \pm\sqrt{-16}$$
$$x - 9 = \pm 4i$$
$$x = 9 \pm 4i$$

18. $\left(m + \frac{3}{5}\right)^2 = \frac{16}{25}$
$$m + \frac{3}{5} = \pm\sqrt{\frac{16}{25}}$$
$$m + \frac{3}{5} = \pm\frac{4}{5}$$
$$m = -\frac{3}{5} \pm \frac{4}{5}$$
$$= -\frac{3}{5} + \frac{4}{5} = \frac{1}{5}$$
$$= -\frac{3}{5} - \frac{4}{5} = -\frac{7}{5}$$

19. $m^2 + 8m = -7$
$$\left(\frac{8}{2}\right)^2 = (4)^2 = 16$$
$$m^2 + 8m + 16 = -7 + 16$$
$$(m+4)^2 = 9$$
$$m + 4 = \pm\sqrt{9}$$
$$m + 4 = \pm 3$$
$$m = -4 \pm 3$$
$$m = -4 - 3 = -7$$
$$m = -4 + 3 = -1$$

20. $u^2-6u-12=100$

$$u^2-6u=112$$
$$\left(-\frac{6}{2}\right)^2=(-3)^2=9$$
$$u^2-6u+9=112+9$$
$$(u-3)^2=121$$
$$u-3=\pm\sqrt{121}$$
$$u-3=\pm 11$$
$$u=3\pm 11$$
$$u=3+11=14$$
$$u=3-11=-8$$

21. $2b^2-6b+7=0$

$$b^2-3b=-\frac{7}{2}$$
$$\left(\frac{-3}{2}\right)^2=\frac{9}{4}$$
$$b^2-3b+\frac{9}{4}=-\frac{7}{2}+\frac{9}{4}$$
$$\left(b-\frac{3}{2}\right)^2=-\frac{5}{4}$$
$$b-\frac{3}{2}=\pm\sqrt{-\frac{5}{4}}$$
$$b-\frac{3}{2}=\pm\frac{i\sqrt{5}}{2}$$
$$b=\frac{3\pm i\sqrt{5}}{2}$$

22. $\left(\frac{\frac{1}{4}}{2}\right)^2=\left(\frac{1}{8}\right)^2=\frac{1}{64}$

$$u^2+\frac{1}{4}u+\frac{1}{64}=\frac{3}{4}+\frac{1}{64}$$
$$\left(u+\frac{1}{8}\right)^2=\frac{49}{64}$$
$$u+\frac{1}{8}=\pm\sqrt{\frac{49}{64}}$$
$$u+\frac{1}{8}=\pm\frac{7}{8}$$
$$u=-\frac{1}{8}\pm\frac{7}{8}$$
$$u=-\frac{1}{8}-\frac{7}{8}=-1$$
$$u=-\frac{1}{8}+\frac{7}{8}=\frac{3}{4}$$

23. $p^2+2p-5=0$

$$p=\frac{-(2)\pm\sqrt{(2)^2-4(1)(-5)}}{2(1)}$$
$$=\frac{-2\pm\sqrt{4+20}}{2}$$
$$=\frac{-2\pm\sqrt{24}}{2}$$
$$=\frac{-2\pm 2\sqrt{6}}{2}$$
$$=-1\pm\sqrt{6}$$

24. $3x^2-2x+1=0$

$$x=\frac{-(-2)\pm\sqrt{(-2)^2-4(3)(1)}}{2(3)}$$
$$=\frac{2\pm\sqrt{4-12}}{6}$$
$$=\frac{2\pm\sqrt{-8}}{6}$$
$$=\frac{2\pm 2i\sqrt{2}}{6}$$
$$=\frac{1\pm i\sqrt{2}}{3}$$

25. $2t^2+t-5=0$

$$t=\frac{-(1)\pm\sqrt{(1)^2-4(2)(-5)}}{2(2)}$$
$$=\frac{-1\pm\sqrt{1+40}}{4}$$
$$=\frac{-1\pm\sqrt{41}}{4}$$

26. Use $200x^2+10x-3=0$.

$$x=\frac{-(10)\pm\sqrt{(10)^2-4(200)(-3)}}{2(200)}$$
$$=\frac{-10\pm\sqrt{100+2400}}{400}$$
$$=\frac{-10\pm\sqrt{2500}}{400}$$
$$=\frac{-10\pm 50}{400}$$
$$=\frac{-10-50}{400}=-\frac{3}{20}$$
$$=\frac{-10+50}{400}=\frac{1}{10}$$

27. $b^2-4b-12=0$

$D=(-4)^2-4(1)(-12)=64$; 2 rational

28. $6z^2-7z+5=0$

$D=(-7)^2-4(6)(5)=-71$; 2 nonreal complex

29. $k^2+6k+9=0$

$D=(6)^2-4(1)(9)=0$; 1 rational

30. $0.8x^2+1.2x+0.3=0$

$D=(1.2)^2-4(0.8)(0.3)=0.48$; 2 irrational

31. $$\frac{1}{y}+\frac{1}{y+3}=\frac{2}{3}$$
$$3y(y+3)\left(\frac{1}{y}+\frac{1}{y+3}\right)=3y(y+3)\cdot\frac{2}{3}$$
$$3(y+3)+3y=2y(y+3)$$
$$3y+9+3y=2y^2+6y$$
$$0=2y^2-9$$
$$2y^2=9$$
$$y^2=\frac{9}{2}$$
$$y=\pm\sqrt{\frac{9}{2}}$$
$$y=\pm\frac{3}{\sqrt{2}}\cdot\frac{\sqrt{2}}{\sqrt{2}}$$
$$y=\pm\frac{3\sqrt{2}}{2}$$

32. $$\frac{1}{u}+\frac{1}{u-5}=\frac{10}{u^2-25}$$
$$u(u-5)(u+5)\left(\frac{1}{u}+\frac{1}{u-5}=\frac{10}{(u-5)(u+5)}\right)$$
$$(u-5)(u+5)+u(u+5)=10u$$
$$u^2-25+u^2+5u=10u$$
$$2u^2-5u-25=0$$
$$(2u+5)(u-5)=0$$
$2u+5=0$ or $u-5=0$

$u=-\frac{5}{2}$ $\quad$ $u=\not{5}$

33. $$6-5x^{-1}+x^{-2}=0$$
$$x^2\left(6-5x^{-1}+x^{-2}\right)=x^2\cdot 0$$
$$6x^2-5x+1=0$$
$$(3x-1)(2x-1)=0$$
$3x-1=0$ or $2x-1=0$

$x=\frac{1}{3}$ $\quad$ $x=\frac{1}{2}$

34. $$2-3x^{-1}-x^{-2}=0$$
$$x^2\left(2-3x^{-1}-x^{-2}\right)=x^2\cdot 0$$
$$2x^2-3x-1=0$$
$$x=\frac{-(-3)\pm\sqrt{(-3)^2-4(2)(-1)}}{2(2)}$$
$$=\frac{3\pm\sqrt{9+8}}{4}$$
$$=\frac{3\pm\sqrt{17}}{4}$$

35. $14\sqrt{x}+45=0$
$$14\sqrt{x}=-45$$
$$\sqrt{x}=-\frac{45}{14}$$
no solution

36. $\sqrt{4m} = 3m - 1$

$$\left(\sqrt{4m}\right)^2 = (3m-1)^2$$
$$4m = 9m^2 - 6m + 1$$
$$0 = 9m^2 - 10m + 1$$
$$0 = (9m-1)(m-1)$$

$9m - 1 = 0$ or $m - 1 = 0$

~~$m = \frac{1}{9}$~~ $\qquad m = 1$

37. $\sqrt{6r+13} = 2r + 1$

$$\left(\sqrt{6r+13}\right)^2 = (2r+1)^2$$
$$6r + 13 = 4r^2 + 4r + 1$$
$$0 = 4r^2 - 2r - 12$$
$$0 = 2r^2 - r - 6$$
$$0 = (2r+3)(r-2)$$

$2r + 3 = 0$ or $r - 2 = 0$

~~$r = -\frac{3}{2}$~~ $\qquad r = 2$

38. $\sqrt{21t+2} + t = 2 + 4t$

$$\sqrt{21t+2} = 2 + 3t$$
$$\left(\sqrt{21t+2}\right)^2 = (2+3t)^2$$
$$21t + 2 = 4 + 12t + 9t^2$$
$$0 = 2 - 9t + 9t^2$$
$$0 = (2-3t)(1-3t)$$

$2 - 3t = 0$ or $1 - 3t = 0$

$t = \frac{2}{3} \qquad t = \frac{1}{3}$

39. Let $u = x^2$

$$u^2 - 5u + 6 = 0$$
$$(u-2)(u-3) = 0$$

$u - 2 = 0$ or $u - 3 = 0$

$u = 2 \qquad u = 3$

$x^2 = 2 \qquad x^2 = 3$

$x = \pm\sqrt{2} \qquad x = \pm\sqrt{3}$

40. Let $u = m^2$

$$2u^2 - 3u + 1 = 0$$
$$(2u-1)(u-1) = 0$$

$2u - 1 = 0 \qquad u - 1 = 0$

$u = \frac{1}{2} \qquad u = 1$

$m^2 = \frac{1}{2} \qquad m^2 = 1$

$m = \pm\sqrt{\frac{1}{2}\cdot\frac{2}{2}} = \pm\frac{\sqrt{2}}{2} \qquad m = \pm 1$

41. Let $u = x + 5$

$$6u^2 - 5u + 1 = 0$$
$$(2u-1)(3u-1) = 0$$

$2u - 1 = 0$ or $3u - 1 = 0$

$u = \frac{1}{2} \qquad u = \frac{1}{3}$

$x + 5 = \frac{1}{2} \qquad x + 5 = \frac{1}{3}$

$x = -\frac{9}{2} \qquad x = -\frac{14}{3}$

42. Let $u = \frac{x-1}{3}$

$$u^2 + 10u + 9 = 0$$
$$(u+1)(u+9) = 0$$

$u + 1 = 0$ or $u + 9 = 0$

$u = -1 \qquad u = -9$

$\frac{x-1}{3} = -1 \qquad \frac{x-1}{3} = -9$

$x - 1 = -3 \qquad x - 1 = -27$

$x = -2 \qquad x = -26$

43. Let $u = p^{1/3}$

$$u^2 - 11u + 24 = 0$$
$$(u-8)(u-3) = 0$$

$u - 8 = 0$ or $u - 3 = 0$

$u = 8 \qquad u = 3$

$p^{1/3} = 8 \qquad p^{1/3} = 3$

$\left(p^{1/3}\right)^3 = 8^3 \qquad \left(p^{1/3}\right)^3 = 3^3$

$p = 512 \qquad p = 27$

44. Let $u = a^{1/4}$

$$5u^2 + 13u - 6 = 0$$
$$(5u-2)(u+3) = 0$$
$$5u - 2 = 0 \quad \text{or} \quad u + 3 = 0$$
$$u = \frac{2}{5} \qquad u = -3$$
$$a^{1/4} = \frac{2}{5} \qquad a^{1/4} = -3$$
$$\left(a^{1/4}\right)^4 = \left(\frac{2}{5}\right)^4 \qquad \left(a^{1/4}\right)^4 = (-3)^4$$
$$a = \frac{16}{625} \qquad a = \cancel{81}$$

45. $f(x) = -2x^2$

a) x-intercept: $-2x^2 = 0$

$$x^2 = 0$$
$$x = \pm\sqrt{0}$$
$$x = 0$$
$$(0,0)$$

y-intercept: $y = -2(0)^2 = 0$

(0, 0)

b) downward
c) (0, 0)
d) $x = 0$
e)

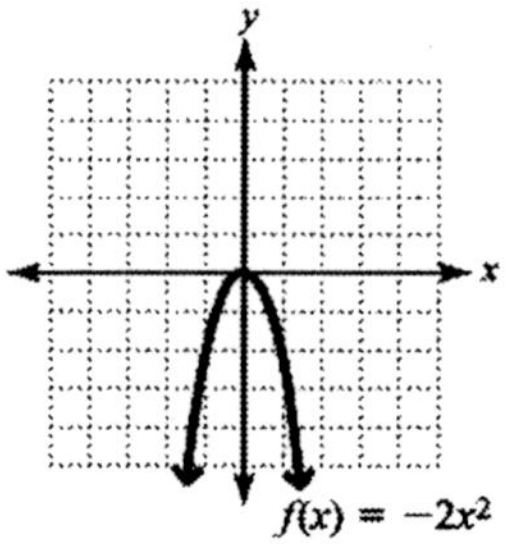

46. $g(x) = \frac{1}{2}x^2 + 1$

a) x-intercept: $\frac{1}{2}x^2 + 1 = 0$

$$\frac{1}{2}x^2 = -1$$
$$x^2 = -2$$
$$x = \pm\sqrt{-2}$$
$$x = \pm i\sqrt{2}$$

Because these solutions are not real, there are no x-intercepts.

y-intercept: $y = \frac{1}{2}(0)^2 + 1 = 1$

(0, 1)

b) upward
c) (0, 1)
d) $x = 0$
e)

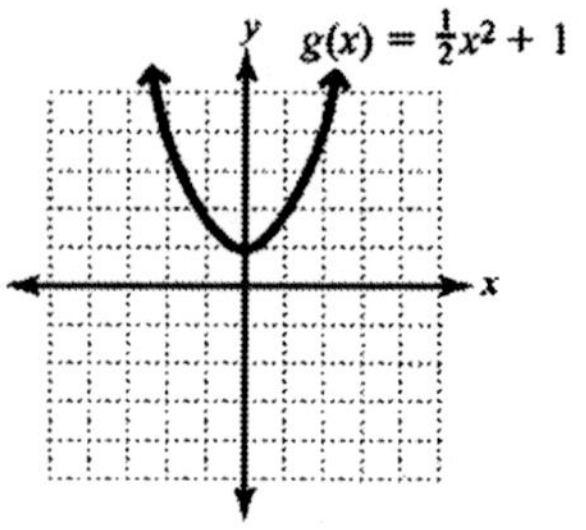

47. $h(x) = -\frac{1}{3}(x-2)^2$

a) x-intercept: $-\frac{1}{3}(x-2)^2 = 0$

$$(x-2)^2 = 0$$
$$x - 2 = \pm\sqrt{0}$$
$$x = 2 \pm \sqrt{0}$$
$$x = 2$$
$$(2,0)$$

y-intercept: $y = -\frac{1}{3}(0-2)^2 = -\frac{1}{3} \cdot 4 = -\frac{4}{3}$

$$\left(0, -\frac{4}{3}\right)$$

b) downward
c) (2, 0)
d) $x = 2$
e)

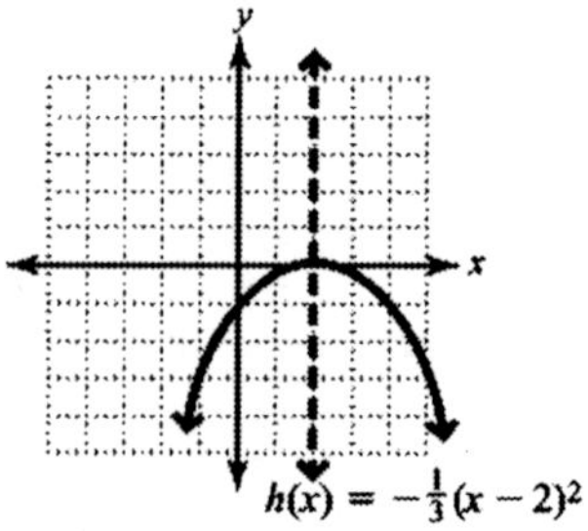

48. $k(x) = 4(x+3)^2 - 2$

a) x-intercepts:
$$0 = 4(x+3)^2 - 2$$
$$\frac{2}{4} = (x+3)^2$$
$$\pm\frac{\sqrt{2}}{2} = x+3$$
$$-3 \pm \frac{\sqrt{2}}{2} = x$$
$$\left(-3+\frac{\sqrt{2}}{2}, 0\right), \left(-3-\frac{\sqrt{2}}{2}, 0\right)$$

y-intercept:
$$y = 4(0+3)^2 - 2$$
$$y = 4 \cdot 9 - 2$$
$$y = 34$$
$$(0, 34)$$

b) upward

c) $(-3, -2)$

d) $x = -3$

e)

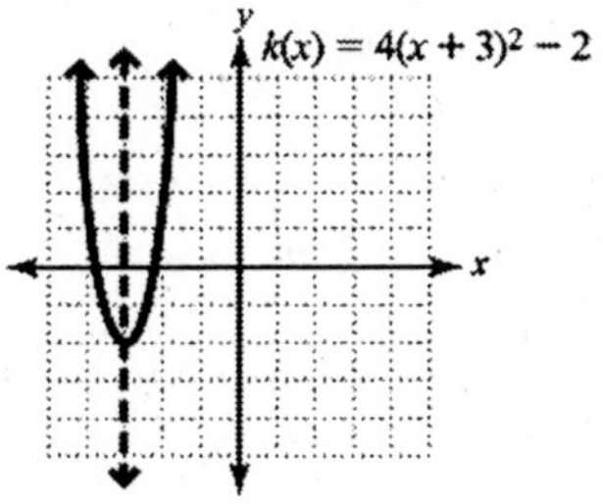

49. a)
$$y = x^2 + 2x - 1$$
$$y + 1 = x^2 + 2x$$
$$y + 1 + 1 = x^2 + 2x + 1$$
$$y + 2 = (x+1)^2$$
$$y = (x+1)^2 - 2$$
$$m(x) = (x+1)^2 - 2$$

b) x-intercepts:
$$x = \frac{-2 \pm \sqrt{2^2 - 4(1)(-1)}}{2(1)}$$
$$= \frac{-2 \pm 2\sqrt{2}}{2}$$
$$= -1 \pm \sqrt{2}$$
$$\left(-1+\sqrt{2}, 0\right), \left(-1-\sqrt{2}, 0\right)$$

y-intercept: $y = 0 + 2 \cdot 0 - 1 = -1$
$$(0, -1)$$

c) upward

d) using part (a): $(-1, -2)$

e) $x = -1$

f)

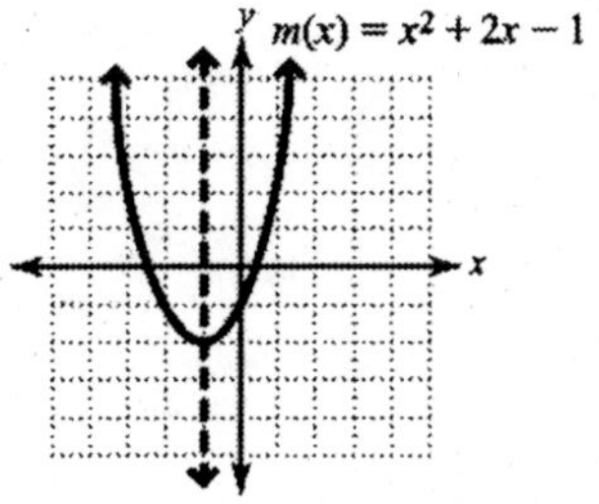

g) Domain: $\{x \mid x \text{ is a real number}\}$ or $(-\infty, \infty)$

Range: $\{y \mid y \geq -2\}$ or $[-2, \infty)$

50. a)
$$y = -0.5x^2 + 4x - 6$$
$$y + 6 = -0.5x^2 + 4x$$
$$y + 6 = -0.5(x^2 - 8x)$$
$$y + 6 - 8 = -0.5(x^2 - 8x + 16)$$
$$y - 2 = -0.5(x-4)^2$$
$$y = -0.5(x-4)^2 + 2$$
$$p(x) = -0.5(x-4)^2 + 2$$

b) x-intercepts:
$$0 = -0.5x^2 + 4x - 6$$
$$0 = 5x^2 - 40x + 60$$
$$0 = x^2 - 8x + 12$$
$$0 = (x-2)(x-6)$$
$$x - 2 = 0 \qquad x - 6 = 0$$
$$x = 2 \qquad x = 6$$
$$(2, 0), (6, 0)$$

y-intercept: $y = -0.5 \cdot 0 + 4 \cdot 0 - 6 = -6$
$$(0, -6)$$

c) downward

d) from part (a): (4, 2)

e) $x = 4$

f)

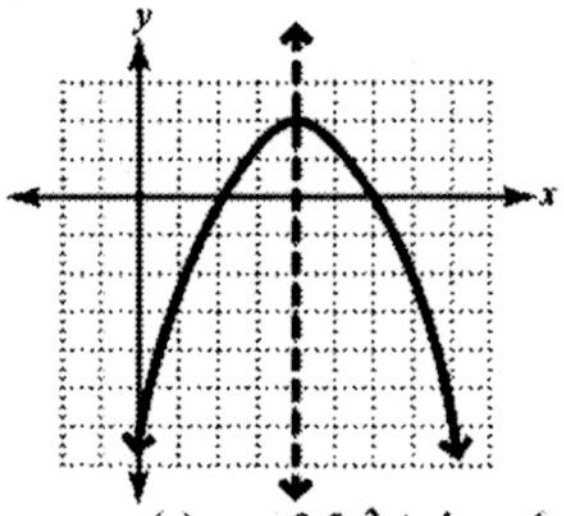

g) Domain: $\{x \mid x \text{ is a real number}\}$ or $(-\infty, \infty)$

Range: $\{y \mid y \le 2\}$ or $(-\infty, 2]$

51. $(x+5)(x-3)=0$

$x+5=0$ or $x-3=0$

$x=-5$ $\quad$ $x=3$

Interval	$(-\infty,-5)$	$(-5,3)$	$(3,\infty)$
Test No.	−6	0	4
Results	$9>0$	$-15>0$	$9>0$
T/F	T	F	T

Solution set: $(-\infty,-5)\cup(3,\infty)$

52. $n^2-6n+8=0$

$(n-4)(n-2)=0$

$n-4=0$ or $n-2=0$

$n=4$ $\quad$ $n=2$

Interval	$(-\infty,2)$	$(2,4)$	$(4,\infty)$
Test No.	0	3	5
Results	$0\le -8$	$-9\le -8$	$-5\le -8$
T/F	F	T	F

Solution set: [2, 4]

53. $x^2+9x+14=0$

$(x+7)(x+2)=0$

$x+7=0$ or $x+2=0$

$x=-7$ $\quad$ $x=-2$

Interval	$(-\infty,-7)$	$(-7,-2)$	$(-2,\infty)$
Test No.	−8	−3	0
Results	$6<0$	$-4<0$	$14<0$
T/F	F	T	F

Solution set: $(-7,-2)$

54. $(x+3)(x-1)(x-2)=0$

$x+3=0$ or $x-1=0$ or $x-2=0$

$x=-3$ $\quad$ $x=1$ $\quad$ $x=2$

Interval	$(-\infty,-3)$	$(-3,1)$	$(1,2)$	$(2,\infty)$
Test No.	−4	0	1.5	3
Results	−30	6	−1.125	12
T/F	F	T	F	T

Solution set: $[-3,1]\cup[2,\infty)$

55. $\dfrac{a+3}{a-1}=0$ $\qquad$ $a-1=0$

$a=1$

$(a-1)\dfrac{a+3}{a-1}=0(a-1)$

$a+3=0$

$a=-3$

Interval	$(-\infty,-3)$	$(-3,1)$	$(1,\infty)$
Test No.	−5	0	2
Results	$1/3\ge 0$	$-3\ge 0$	$5\ge 0$
T/F	T	F	T

Solution set: $(-\infty,-3]\cup(1,\infty)$

56. $\dfrac{r}{r+2}=2$ $\qquad$ $r+2=0$

$r=-2$

$(r+2)\dfrac{r}{r+2}=2(r+2)$

$r=2r+4$

$-4=r$

Interval	$(-\infty,-4)$	$(-4,-2)$
Test No.	−5	−3
Results	$5/3<2$	$3<2$
T/F	T	F

Interval	$(-2,\infty)$
Test No.	0
Results	$0<2$
T/F	T

Solution set: $(-\infty,-4)\cup(-2,\infty)$

57. $\frac{n-3}{n-4}=5$ $\quad n-4=0$

$n=4$

$(n-4)\frac{n-3}{n-4}=5(n-4)$

$n-3=5n-20$

$-4n=-17$

$n=\frac{17}{4}$

Interval	$(-\infty,4)$	$\left(4,\frac{17}{4}\right)$
Test No.	0	4.1
Results	$3/4\le 5$	$11\le 5$
T/F	T	F

Interval	$\left(\frac{17}{4},\infty\right)$
Test No.	5
Results	$2\le 5$
T/F	T

Solution set: $(-\infty,4)\cup\left[\frac{17}{4},\infty\right)$

58. $\frac{(k+2)(k-3)}{k-5}=0$

$(k-5)\frac{(k+2)(k-3)}{k-5}=0(k-5)$

$(k+2)(k-3)=0$

$k+2=0 \quad k-3=0 \quad k-5=0$

$k=-2 \quad k=3 \quad k=5$

Interval	$(-\infty,-2)$	$(-2,3)$	$(3,5)$	$(5,\infty)$
Test No.	−3	0	4	6
Results	$-3/4<0$	$\frac{6}{5}<0$	$-6<0$	$24<0$
T/F	T	F	T	F

Solution set: $(-\infty,-2)\cup(3,5)$

59. $A=\pi r^2$

$22,500\pi=\pi r^2$

$0=\pi r^2-22,500\pi$

$0=\pi\left(r^2-22,500\right)$

$0=r^2-22,500$

$0=(r+150)(r-150)$

$r=\cancel{-150},150$

The radius of the circle was 150 feet.

60. $400=\frac{1}{2}50v^2$

$400=25v^2$

$16=v^2$

$4=v$

The velocity was 4 m/sec.

61. $x(x+4)=285$

$x^2+4x-285=0$

$(x+19)(x-15)=0$

$x=\cancel{-19},15$

The width is 15 ft. and the length is 15 + 4 = 19 ft.

62. $\pi r^2(4)=\frac{4}{3}\pi\left(9^3\right)$

$4\pi r^2=972\pi$

$4\pi r^2-972\pi=0$

$4\pi\left(r^2-243\right)=0$

$r^2-243=0$

$r^2=243$

$r=\sqrt{243}$

$r=9\sqrt{3}$

The radius is $9\sqrt{3}$ inches.

63. $x^2+(x+9)^2=17^2$

$x^2+x^2+18x+81=289$

$2x^2+18x-208=0$

$x^2+9x-104=0$

$$x = \frac{-9 \pm \sqrt{9^2 - 4(1)(-104)}}{2(1)}$$
$$= \frac{-9 \pm \sqrt{81 + 416}}{2}$$
$$= \frac{-9 \pm \sqrt{497}}{2} \approx \cancel{-15.65}, 6.65$$

The height is 6.65 ft. and the base is 6.65 + 9 = 15.65 ft.

64. Finding the vertex will give the highest point that the acrobat will reach.

$$t = \frac{-24}{2(-16)} = \frac{3}{4} = 0.75$$

$$h = -16(0.75)^2 + 24(0.75)$$
$$h = -9 + 18$$
$$h = 9$$

The vertex is at $(0.75, 9)$

a) 0.75 sec.
b) 9 ft.
c) $-16t^2 + 24t = 0$

$$-8t(2t - 3) = 0$$
$$t = \cancel{0}, 1.5 \text{ sec.}$$

d)

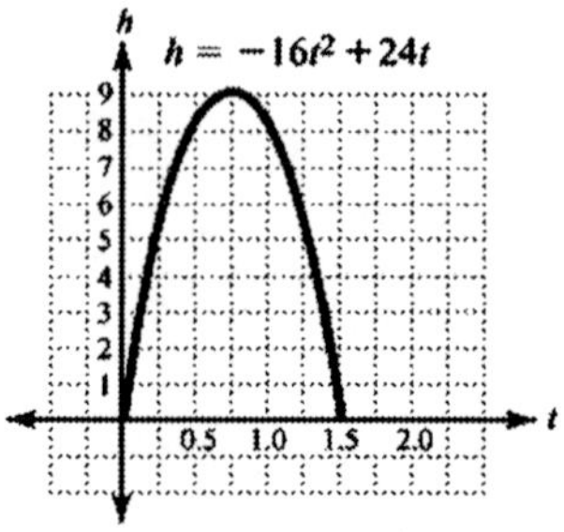

65. a) Find the vertex. $x = \dfrac{-2.16}{2(-0.03)} = 36$

$$y = -0.03(36)^2 + 2.16(36)$$
$$y = 38.88$$

The maximum height is 38.88 yd.

b) The distance of the punt is the x-coordinate of the vertex times 2, or 72 yd.

66. a) $56 \le \frac{1}{2}(x+4)(x-2)$

$$112 \le x^2 + 2x - 8$$
$$0 \le x^2 + 2x - 120$$
$$0 \le (x+12)(x-10)$$
$$x \le -12 \text{ or } x \ge 10$$

Since x must be at least 2, $x \ge 10$ in.

b) height ≥ 8 in., base ≥ 14 in.

Chapter 9 Practice Test

1. $x^2 = 81$

$$x = \pm\sqrt{81}$$
$$x = \pm 9$$

2. $(x-3)^2 = 20$

$$x - 3 = \pm\sqrt{20}$$
$$x - 3 = \pm\sqrt{4 \cdot 5}$$
$$x - 3 = \pm 2\sqrt{5}$$
$$x = 3 \pm 2\sqrt{5}$$

3. $x^2 - 8x = -4$

$$x^2 - 8x + 16 = -4 + 16$$
$$(x-4)^2 = 12$$
$$x - 4 = \pm\sqrt{12}$$
$$x - 4 = \pm\sqrt{4 \cdot 3}$$
$$x - 4 = \pm 2\sqrt{3}$$
$$x = 4 \pm 2\sqrt{3}$$

4. $3m^2 - 6m = 5$

$$\frac{3m^2 - 6m}{3} = \frac{5}{3}$$
$$m^2 - 2m = \frac{5}{3}$$
$$m^2 - 2m + 1 = \frac{5}{3} + 1$$
$$(m-1)^2 = \frac{8}{3}$$
$$m - 1 = \pm\sqrt{\frac{8}{3}}$$
$$m - 1 = \pm\frac{2\sqrt{2}}{\sqrt{3}}$$
$$m - 1 = \pm\frac{2\sqrt{2}}{\sqrt{3}} \cdot \frac{\sqrt{3}}{\sqrt{3}}$$
$$m - 1 = \frac{2\sqrt{6}}{3}$$
$$m = 1 \pm \frac{2\sqrt{6}}{3}$$
$$m = \frac{3 \pm 2\sqrt{6}}{3}$$

5. $2x^2 + x - 6 = 0$

Let $a = 2$, $b = 1$, and $c = -6$.

$$x = \frac{-(1) \pm \sqrt{(1)^2 - 4(2)(-6)}}{2(2)}$$
$$= \frac{-1 \pm \sqrt{1+48}}{4}$$
$$= \frac{-1 \pm \sqrt{49}}{4}$$
$$= \frac{-1 \pm 7}{4}$$
$$= -2, \frac{3}{2}$$

6. $x^2 - 8x + 15 = 0$

Let $a = 1$, $b = -8$, and $c = 15$.

$$x = \frac{-(-8) \pm \sqrt{(-8)^2 - 4(1)(15)}}{2(1)}$$
$$= \frac{8 \pm \sqrt{64-60}}{2}$$
$$= \frac{8 \pm \sqrt{4}}{2}$$
$$= \frac{8 \pm 2}{2}$$
$$= 3, 5$$

7. $u^2 - 16 = -6u$

$$u^2 + 6u - 16 = 0$$
$$(u+8)(u-2) = 0$$
$$u + 8 = 0 \quad \text{or} \quad u - 2 = 0$$
$$u = -8 \qquad u = 2$$

8. $x^2 = 81$

$$x = \pm\sqrt{81}$$
$$x = \pm 9$$

9. $4w^2 + 6w + 3 = 0$

Let $a = 4$, $b = 6$, and $c = 3$.

$$x = \frac{-(6) \pm \sqrt{(6)^2 - 4(4)(3)}}{2(4)}$$
$$= \frac{-6 \pm \sqrt{36-48}}{8}$$
$$= \frac{-6 \pm \sqrt{-12}}{8}$$
$$= \frac{-6 \pm 2i\sqrt{3}}{8} = \frac{-3 \pm i\sqrt{3}}{4}$$

10. $2x^2 + 4x = 0$

$$2x(x+2) = 0$$
$$2x = 0 \quad \text{or} \quad x + 2 = 0$$
$$x = 0 \qquad x = -2$$

11. $x^2 + 16 = 0$

$$x^2 = -16$$
$$x = \pm\sqrt{-16}$$
$$x = \pm 4i$$

12.
$$3k^2 = -5k$$
$$3k^2 + 5k = 0$$
$$k(3k+5) = 0$$
$$k = 0 \quad \text{or} \quad 3k+5 = 0$$
$$k = -\frac{5}{3}$$

13.
$$\frac{1}{x+2} + \frac{1}{x} = \frac{5}{12}$$
$$12x(x+2)\left(\frac{1}{x+2} + \frac{1}{x}\right) = 12x(x+2)\cdot\frac{5}{12}$$
$$12x + 12(x+2) = 5x(x+2)$$
$$12x + 12x + 24 = 5x^2 + 10x$$
$$0 = 5x^2 - 14x - 24$$
$$0 = (5x+6)(x-4)$$
$$5x+6 = 0 \quad \text{or} \quad x-4 = 0$$
$$x = -\frac{6}{5} \qquad x = 4$$

14.
$$3 - x^{-1} - 2x^{-2} = 0$$
$$x^2\left(3 - x^{-1} - 2x^{-2}\right) = x^2 \cdot 0$$
$$3x^2 - x - 2 = 0$$
$$(3x+2)(x-1) = 0$$
$$3x+2 = 0 \quad \text{or} \quad x-1 = 0$$
$$x = -\frac{2}{3} \qquad x = 1$$

15. $9\sqrt{x} + 8 = 0$ No solution
$$9\sqrt{x} = -8$$
$$\sqrt{x} = -\frac{8}{9}$$

16. $\sqrt{x+8} - x = 2$
$$\sqrt{x+8} = x+2$$
$$\left(\sqrt{x+8}\right)^2 = (x+2)^2$$
$$x+8 = x^2 + 4x + 4$$
$$0 = x^2 + 3x - 4$$
$$0 = (x+4)(x-1)$$
$$x+4 = 0 \quad \text{or} \quad x-1 = 0$$
$$x = \cancel{-4} \qquad x = 1$$

The solution -4 is extraneous. The only solution is 1.

17. Let $u = a^2$.
$$9u^2 + 26u - 3 = 0$$
$$(9u-1)(u+3) = 0$$
$$9u - 1 = 0 \quad \text{or} \quad u+3 = 0$$
$$u = \frac{1}{9} \qquad u = -3$$
$$x^2 = \frac{1}{9} \qquad x^2 = -3$$
$$x = \pm\frac{1}{3} \qquad x = \pm i\sqrt{3}$$

18. Let $u = x+1$.
$$u^2 + 3u - 4 = 0$$
$$(u+4)(u-1) = 0$$
$$u+4 = 0 \quad \text{or} \quad u-1 = 0$$
$$u = -4 \qquad u = 1$$
$$x+1 = -4 \qquad x+1 = 1$$
$$x = -5 \qquad x = 0$$

19. a) x-intercepts:
$$-x^2 + 6x - 4 = 0$$
$$x^2 - 6x + 4 = 0$$
$$x = \frac{-(-6) \pm \sqrt{(-6)^2 - 4(1)(4)}}{2(1)}$$
$$= \frac{6 \pm \sqrt{36-16}}{2}$$
$$= \frac{6 \pm \sqrt{20}}{2}$$
$$= \frac{6 \pm 2\sqrt{5}}{2}$$
$$= 3 \pm \sqrt{5}$$

The x-intercepts are $\left(3+\sqrt{5}, 0\right), \left(3-\sqrt{5}, 0\right)$.

y-intercept: Let $x = 0$.
$$y = -x^2 + 6x - 4$$
$$y = -(0)^2 + 6(0) - 4$$
$$y = -4$$

The y-intercept is (0, –4).

b)
$$y = -x^2 + 6x - 4$$
$$y + 4 = -\left(x^2 - 6x\right)$$
$$y + 4 - 9 = -\left(x^2 - 6x + 9\right)$$
$$y - 5 = -(x-3)^2$$
$$f(x) = -(x-3)^2 + 5$$

c) Because $a < 0$, the parabola opens downward.

d) From part (b) we see that $h = 3$ and $k = 5$. Therefore, the vertex is $(h, k) = (3, 5)$.

e) The axis of symmetry is given by $x = h$. Therefore, the axis of symmetry for this parabola is $x = 3$.

f)

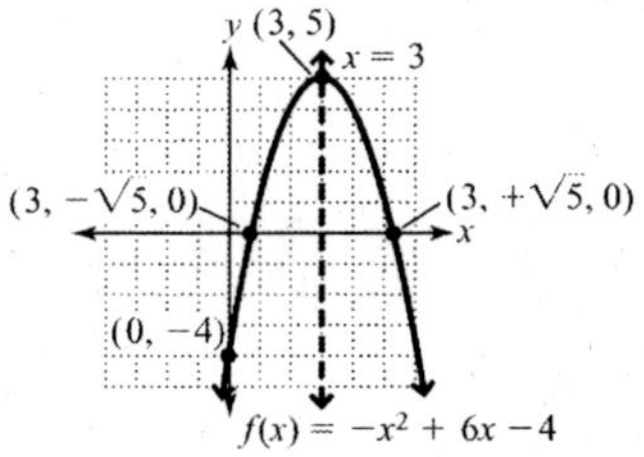

g) The domain is $\{x \mid x \text{ is a real number}\}$ or $(-\infty, \infty)$. The range is $\{y \mid y \le 5\}$ or $(-\infty, 5]$.

20. $(x+1)(x-4) = 0$

$x + 1 = 0$ or $x - 4 = 0$

$x = -1$ $\qquad$ $x = 4$

Interval	$(-\infty, -1]$	$[-1, 4]$	$[4, \infty)$
Test No.	-2	0	5
Results	$6 \le 0$	$-4 \le 0$	$6 \le 0$
T/F	F	T	F

Solution set: $[-1, 4]$

21. $\dfrac{x+2}{x-1} = 0$ $\qquad$ $x - 1 = 0$

$x = 1$

$(x-1)\dfrac{x+2}{x-1} = 0(x-1)$

$x + 2 = 0$

$x = -2$

Interval	$(-\infty, -2)$	$(-2, 1)$	$(1, \infty)$
Test No.	-5	0	2
Results	$1/2 > 0$	$-2 > 0$	$4 > 0$
T/F	T	F	T

Solution set: $(-\infty, -2) \cup (1, \infty)$

22. Let $d = 180$.

$$16t^2 + 32t = d$$

$$16t^2 + 32t = 180$$

$$16t^2 + 32t - 180 = 0$$

$$4(4t^2 + 8t - 45) = 0$$

$$4t^2 + 8t - 45 = 0$$

$$(2t - 5)(2t + 9) = 0$$

$2t - 5 = 0$ or $2t + 9 = 0$

$2t = 5$ $\qquad$ $2t = -9$

$t = \dfrac{5}{2}$ $\qquad$ $t = -\dfrac{9}{2}$

The solution $-\dfrac{9}{2}$ does not make sense in the context of the problem. Therefore, the ball takes 2.5 seconds to fall 180 feet.

23. $x(x + 20) = 400$

$$x^2 + 20x - 400 = 0$$

$$x = \frac{-20 \pm \sqrt{20^2 - 4(1)(-400)}}{2(1)}$$

$$= \frac{-20 \pm \sqrt{400 + 1600}}{2}$$

$$= \frac{-20 \pm \sqrt{2000}}{2}$$

$$\approx 12.36, -32.36$$

Since width is positive, the dimensions are 12.36 ft. by 12.36 + 20 = 32.36 ft.

24. Finding the vertex will show the maximum height that the arrow will reach.

$$x = \frac{-1.3}{2(-0.02)} = 32.5$$

$$y = -0.02(32.5)^2 + 1.3(32.5) + 8$$

$$y = 29.125$$

The vertex is (32.5, 29.125)

a) The maximum height is 29.125 m.

b) $-0.02x^2 + 1.3x + 8 = 0$

$$x = \frac{-1.3 \pm \sqrt{1.3^2 - 4(-0.02)(8)}}{2(-0.02)}$$

$$x = \frac{-1.3 \pm \sqrt{2.33}}{-0.04}$$

$$x \approx 70.66, -5.66$$

The distance can't be negative, so the arrow travels 70.66 m.

c) 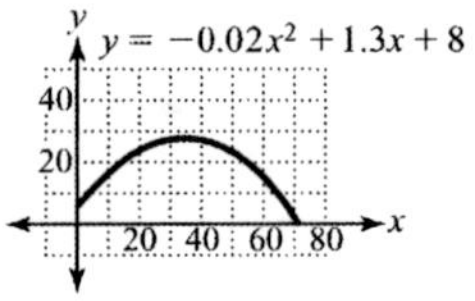

25\. a)

$$\begin{aligned} 12w(w+15) &\le 12{,}000 \\ 12w^2 + 180w - 12{,}000 &\le 0 \\ w^2 + 15w - 1000 &\le 0 \\ (w+40)(w-25) &\le 0 \\ -40 \le w &\le 25 \end{aligned}$$

Since $w > 0$, $0 < w \le 25$ ft.

b) $15 < l \le 40$ ft.

Chapters 1 – 9 Cumulative Review

1. False; $(45, -12)$ is in quadrant IV.

2. False; a system of equations that has no solution is said to be inconsistent.

3. True

4. parabola, $a > 0$

5. $\sqrt[n]{ab}$

6. $\sqrt{a}, -\sqrt{a}$

7. $(-3x-4)-(3x+2) = (-3x-4)+(-3x-2)$
$= -6x - 6$

8. $(8n^2)(-7mn^3) = -56mn^5$

9. $(x-3)(4x^2-2x+1)$
$= 4x^3 - 2x^2 + x - 12x^2 + 6x - 3$
$= 4x^3 - 14x^2 + 7x - 3$

10. $m^3 + 8 = (m+2)(m^2 - 2m + 4)$

11. $x^2 + 10x + 25 = (x+5)^2$

12. $\dfrac{4u^2+4u+1}{u+2u^2} \cdot \dfrac{u}{2u^2-u-1}$
$= \dfrac{\cancel{(2u+1)}\cancel{(2u+1)}}{\cancel{u}\cancel{(1+2u)}} \cdot \dfrac{\cancel{u}}{\cancel{(2u+1)}(u-1)}$
$= \dfrac{1}{u-1}$

13. $\dfrac{3}{x-2} - \dfrac{2}{x+2}$
$= \dfrac{3(x+2)}{(x-2)(x+2)} - \dfrac{2(x-2)}{(x+2)(x-2)}$
$= \dfrac{3x+6-2x+4}{(x-2)(x+2)}$
$= \dfrac{x+10}{(x-2)(x+2)}$

14. $x^{3/4} \cdot x^{-1/4} = x^{3/4+(-1/4)} = x^{1/2}$

15. $\dfrac{\sqrt{n}}{\sqrt{m}-\sqrt{n}} = \dfrac{\sqrt{n}}{\sqrt{m}-\sqrt{n}} \cdot \dfrac{\sqrt{m}+\sqrt{n}}{\sqrt{m}+\sqrt{n}}$
$= \dfrac{\sqrt{mn}+\sqrt{n^2}}{(\sqrt{m})^2 - (\sqrt{n})^2}$
$= \dfrac{\sqrt{mn}+n}{m-n}$

16. $3n - 5 = 7n + 9$
$-4n - 5 = 9$
$-4n = 14$
$n = -\dfrac{7}{2}$

17. $|3x+4| + 3 = 7$
$|3x+4| = 4$
$3x + 4 = -4$ or $3x + 4 = 4$
$3x = -8$ $\quad$ $3x = 0$
$x = -\dfrac{8}{3}$ $\quad$ $x = 0$

18. $2x^2 + 7x = 15$
$2x^2 + 7x - 15 = 0$
$(2x-3)(x+5) = 0$
$2x - 3 = 0$ or $x + 5 = 0$
$x = \dfrac{3}{2}$ $\quad$ $x = -5$

19.
$$\frac{5}{x-2}-\frac{3}{x}=\frac{11}{3x}$$
$$3x(x-2)\left(\frac{5}{x-2}-\frac{3}{x}\right)=3x(x-2)\cdot\frac{11}{3x}$$
$$3x\cdot 5-3(x-2)\cdot 3=11(x-2)$$
$$15x-9x+18=11x-22$$
$$6x+18=11x-22$$
$$40=5x$$
$$8=x$$

20.
$$\sqrt{5x-4}=9$$
$$\left(\sqrt{5x-4}\right)^2=9^2$$
$$5x-4=81$$
$$5x=85$$
$$x=17$$

21. $x^2-6x+11=0$
Let a = 1, b = –6, and c = 11.
$$x=\frac{-(-6)\pm\sqrt{(-6)^2-4(1)(11)}}{2(1)}$$
$$=\frac{6\pm\sqrt{36-44}}{2}$$
$$=\frac{6\pm\sqrt{-8}}{2}$$
$$=\frac{6\pm 2i\sqrt{2}}{2}$$
$$=3\pm i\sqrt{2}$$

22. a) $-2<x+4<5$
$-6<x<1$
$(-6,1)$

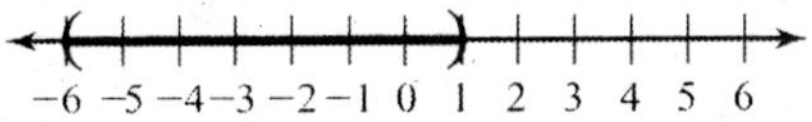

b) $\{x|-6<x<1\}$ c) $(-6,1)$

23.
$$x^2+x=12$$
$$x^2+x-12=0$$
$$(x+4)(x-3)=0$$
$x+4=0$ or $x-3=0$
$x=-4$ $x=3$

Interval	$(-\infty,-4)$	$(-4,3)$
Test No.	–5	0
Results	$20\le 12$	$0\le 12$
T/F	F	T

Interval	$(3,\infty)$
Test No.	4
Results	$20\le 12$
T/F	F

a)

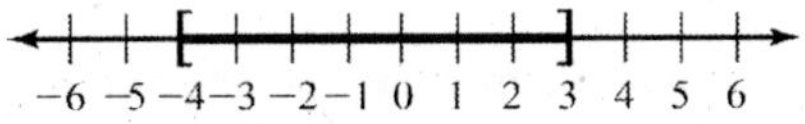

b) $\{x|-4\le x\le 3\}$ c) $[-4,3]$

24. $f(x)=-2x$

x	$f(x)$	(x,y)
–1	2	$(-1,2)$
0	0	$(0,0)$
1	–2	$(1,-2)$

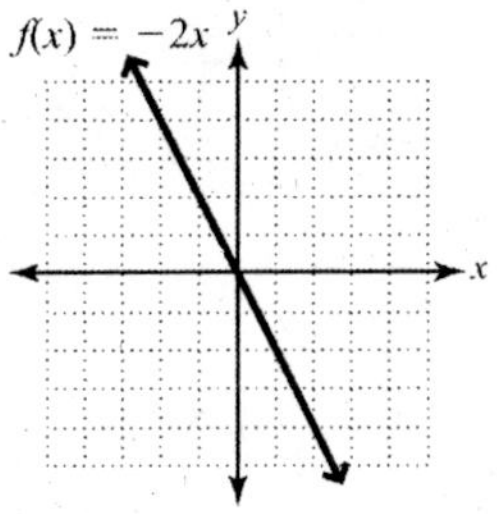

25. Solve for y. $3x-2y=6$
$$-2y=-3x+6$$
$$y=\frac{3}{2}x-3$$

So $m=\frac{3}{2}$. The perpendicular slope is $-\frac{2}{3}$.

Use $m=-\frac{2}{3}$ and $(-2,4)$.
$$y-4=-\frac{2}{3}(x-(-2))$$
$$y-4=-\frac{2}{3}x-\frac{4}{3}$$
$$3y-12=-2x-4$$
$$2x+3y=8$$

26. $\begin{cases} x+y+z=5 & \text{Eqtn. 1} \\ 2x+y-2z=-5 & \text{Eqtn. 2} \\ x-2y+z=8 & \text{Eqtn. 3} \end{cases}$

Multiply equation 1 by –1 and add to equation 3. This makes equation 4.

$$\begin{aligned} -x-y-z&=-5 \\ x-2y+z&=8 \\ \hline -3y&=3 \\ y&=-1 \quad \text{Eqtn. 4} \end{aligned}$$

Multiply equation 1 by –2 and add to equation 2. This makes equation 5.

$$\begin{aligned} -2x-2y-2z&=-10 \\ 2x+y-2z&=-5 \\ \hline -y-4z&=-15 \quad \text{Eqtn. 5} \end{aligned}$$

Substitute equation 4 into equation 5 and solve for z.

$$\begin{aligned} -(-1)-4z&=-15 \\ 1-4z&=-15 \\ -4z&=-16 \\ z&=4 \end{aligned}$$

Substitute the values for y and z to solve for x.

$$\begin{aligned} x-1+4&=5 \\ x+3&=5 \\ x&=2 \end{aligned}$$

Solution: $(2,-1,4)$

27. Complete a table.

	rate	time	distance
not experienced	$\frac{300}{t}-10$	$t+1$	300
experienced	$\frac{300}{t}$	t	300

$$\left(\frac{300}{t}-10\right)(t+1)=300$$

$$300+\frac{300}{t}-10t-10=300$$

$$\frac{300}{t}-10t-10=0$$

$$300-10t^2-10t=0$$

$$t^2+t-30=0$$

$$(t+6)(t-5)=0$$

$t+6=0 \qquad t-5=0$

$t=\cancel{-6} \qquad t=5$

It takes the experienced driver 5 hours to drive 300 miles.

28. Create a table.

	concentrate	vol. of solution	vol. of saline
15%	0.15	50	$0.15(50)$
40%	0.40	x	$0.40x$
30%	0.30	$x+50$	$0.30(x+50)$

Translate the information in the table to an equation and solve.

$$\begin{aligned} 0.15(50)+0.40x&=0.3(x+50) \\ 7.5+0.4x&=0.3x+15 \\ 0.1x&=7.5 \\ x&=75 \end{aligned}$$

75 ml of 40% solution must be added.

29. $v=\frac{k}{T}$ So, $v=\frac{48}{T}$

$6=\frac{k}{8} \qquad v=\frac{48}{15}$

$48=k \qquad v=3.2$ cm/sec.

30. a) $t=\sqrt{\frac{24}{16}}=\frac{\sqrt{24}}{4}=\frac{\sqrt{4\cdot 6}}{4}=\frac{2\sqrt{6}}{4}=\frac{\sqrt{6}}{2}$ sec.

b) $$\begin{aligned} t&=\sqrt{\frac{h}{16}} \\ 3&=\frac{\sqrt{h}}{4} \\ 12&=\sqrt{h} \\ 12^2&=\left(\sqrt{h}\right)^2 \\ 144\text{ ft.}&=h \end{aligned}$$

Chapter 10

Exponential and Logarithmic Functions

Exercise Set 10.1

1. No, $f \circ g = g \circ f$ if f and g are inverses of each other and for a few other functions.

3. The domain of a function is the range of its inverse; the range of a function is the domain of its inverse. This occurs because the ordered pairs of f and f^{-1} have each x and y interchanged.

5. If any horizontal line intersects the graph in more than one point, it is not one-to-one.

7. $$\begin{aligned}(f \circ g)(0) &= f[g(0)]\\ &= f(3)\\ &= 3 \cdot 3 + 5\\ &= 14\end{aligned}$$

9. $$\begin{aligned}(h \circ f)(1) &= h[f(1)]\\ &= h(8)\\ &= \sqrt{8+1}\\ &= 3\end{aligned}$$

11. $$\begin{aligned}(f \circ g)(-2) &= f[g(-2)]\\ &= f(7)\\ &= 3 \cdot 7 + 5\\ &= 26\end{aligned}$$

13. $$\begin{aligned}(f \circ g)(x) &= f[g(x)]\\ &= f(x^2+3)\\ &= 3(x^2+3)+5\\ &= 3x^2+9+5\\ &= 3x^2+14\end{aligned}$$

15. $$\begin{aligned}(f \circ h)(x) &= f[h(x)]\\ &= f(\sqrt{x+1})\\ &= 3\sqrt{x+1}+5\end{aligned}$$

17. $$\begin{aligned}(h \circ f)(0) &= h[f(0)]\\ &= h(5)\\ &= \sqrt{5+1}\\ &= \sqrt{6}\end{aligned}$$

19. $$\begin{aligned}(f \circ g)(x) &= f[g(x)]\\ &= f(3x+4)\\ &= 2(3x+4)-2\\ &= 6x+8-2\\ &= 6x+6\end{aligned}$$

$$\begin{aligned}(g \circ f)(x) &= g[f(x)]\\ &= g(2x-2)\\ &= 3(2x-2)+4\\ &= 6x-6+4\\ &= 6x-2\end{aligned}$$

21. $$\begin{aligned}(f \circ g)(x) &= f[g(x)]\\ &= f(x^2+1)\\ &= x^2+1+2\\ &= x^2+3\end{aligned}$$

$$\begin{aligned}(g \circ f)(x) &= g[f(x)]\\ &= g(x+2)\\ &= (x+2)^2+1\\ &= x^2+4x+4+1\\ &= x^2+4x+5\end{aligned}$$

23. $$\begin{aligned}(f \circ g)(x) &= f[g(x)]\\ &= f(3x)\\ &= (3x)^2+3(3x)-4\\ &= 9x^2+9x-4\end{aligned}$$

$$\begin{aligned}(g \circ f)(x) &= g[f(x)]\\ &= g(x^2+3x-4)\\ &= 3(x^2+3x-4)\\ &= 3x^2+9x-12\end{aligned}$$

25. $(f \circ g)(x) = f[g(x)]$
$= f(2x-5)$
$= \sqrt{2x-5+2}$
$= \sqrt{2x-3}$

$(g \circ f)(x) = g[f(x)]$
$= g(\sqrt{x+2})$
$= 2\sqrt{x+2}-5$

27. $(f \circ g)(x) = f[g(x)]$
$= f\left(\frac{x-3}{x}\right)$
$= \frac{\frac{x-3}{x}+1}{\frac{x-3}{x}}$
$= \frac{2x-3}{x-3}$

$(g \circ f)(x) = g[f(x)]$
$= g\left(\frac{x+1}{x}\right)$
$= \frac{\frac{x+1}{x}-3}{\frac{x+1}{x}}$
$= \frac{1-2x}{x+1}$

29. The domain is $[0, \infty)$ and the range is $[3, \infty)$

31. Yes

33. No

35. $(f \circ g)(x) = f[g(x)]$
$= f(x-5)$
$= x-5+5$
$= x$

$(g \circ f)(x) = g[f(x)]$
$= g(x+5)$
$= x+5-5$
$= x$

Yes

37. $(f \circ g)(x) = f[g(x)]$
$= f\left(\frac{x}{6}\right)$
$= 6 \cdot \frac{x}{6}$
$= x$

$(g \circ f)(x) = g[f(x)]$
$= g(6x)$
$= \frac{6x}{6}$
$= x$

Yes

39. $(f \circ g)(x) = f[g(x)]$
$= f\left(\frac{x+3}{2}\right)$
$= 2 \cdot \frac{x+3}{2} - 3$
$= x+3-3$
$= x$

$(g \circ f)(x) = g[f(x)]$
$= g(2x-3)$
$= \frac{2x-3+3}{2}$
$= \frac{2x}{2}$
$= x$

Yes

41. $(f \circ g)(x) = f[g(x)]$
$= f(\sqrt[3]{x+4})$
$= (\sqrt[3]{x+4})^3 - 4$
$= x+4-4$
$= x$

$(g \circ f)(x) = g[f(x)]$
$= g(x^3-4)$
$= \sqrt[3]{x^3-4+4}$
$= \sqrt[3]{x^3}$
$= x$

Yes

43. $(f \circ g)(x) = f[g(x)]$
$= f(\sqrt{x})$
$= (\sqrt{x})^2$
$= x$

$(g \circ f)(x) = g[f(x)]$
$= g(x^2)$
$= \sqrt{x^2}$
$= |x|$

No

45. $(f \circ g)(x) = f[g(x)]$
$= f(\sqrt{x})$
$= (\sqrt{x})^2$
$= x$

$(g \circ f)(x) = g[f(x)]$
$= g(x^2)$
$= \sqrt{x^2}$
$= |x|$
$= x$, since $x \geq 0$

Yes

47. $(f \circ g)(x) = f[g(x)]$
$= f\left(\dfrac{3-5x}{x}\right)$
$= \dfrac{3}{\dfrac{3-5x}{x}+5} = \dfrac{3x}{3-5x+5x}$
$= \dfrac{3x}{3} = x$

$(g \circ f)(x) = g[f(x)]$
$= g\left(\dfrac{3}{x+5}\right)$
$= \dfrac{3-5\left(\dfrac{3}{x+5}\right)}{\dfrac{3}{x+5}} = \dfrac{3-\dfrac{15}{x+5}}{\dfrac{3}{x+5}}$
$= \dfrac{3(x+5)-15}{3} = \dfrac{3x+15-15}{3}$
$= \dfrac{3x}{3} = x$

Yes

49. Yes, because $(f \circ g)(x) = (g \circ f)(x) = x$:
$(f \circ f)(x) = f[f(x)] = f\left(\dfrac{1}{x}\right) = \dfrac{1}{\dfrac{1}{x}} = x$

51. Yes. Every horizontal line that can intersect this graph does so at one and only one point, so the function is one to one.

53. No. A horizontal line can intersect this graph in more than one point, so the function is not one to one.

55. $f^{-1} = \{(2,-3),(-3,-1),(4,0),(6,4)\}$

57. $f^{-1} = \{(-2,7),(2,9),(1,-4),(3,3)\}$

59. $y = x+6$
$x = y+6$
$x-6 = y$
$f^{-1}(x) = x-6$

61. $y = 2x+3$
$x = 2y+3$
$x-3 = 2y$
$\dfrac{x-3}{2} = y$
$f^{-1}(x) = \dfrac{x-3}{2}$

63. $y = x^3 - 1$
$x = y^3 - 1$
$x+1 = y^3$
$\sqrt[3]{x+1} = y$
$f^{-1}(x) = \sqrt[3]{x+1}$

65.
$$\begin{aligned} y &= \frac{2}{x+2} \\ x &= \frac{2}{y+2} \\ x(y+2) &= 2 \\ xy + 2x &= 2 \\ xy &= 2 - 2x \\ y &= \frac{2-2x}{x} \\ f^{-1}(x) &= \frac{2-2x}{x} \end{aligned}$$

67.
$$\begin{aligned} y &= \frac{x+2}{x-3} \\ x &= \frac{y+2}{y-3} \\ x(y-3) &= y+2 \\ xy - 3x &= y+2 \\ xy - y &= 3x+2 \\ y(x-1) &= 3x+2 \\ y &= \frac{3x+2}{x-1} \\ f^{-1}(x) &= \frac{3x+2}{x-1} \end{aligned}$$

69.
$$\begin{aligned} y &= \sqrt{x-2} \\ x &= \sqrt{y-2} \\ x^2 &= y-2 \\ x^2 + 2 &= y \\ f^{-1}(x) &= x^2 + 2, x \ge 0 \end{aligned}$$

71.
$$\begin{aligned} y &= 2x^3 + 4 \\ x &= 2y^3 + 4 \\ x - 4 &= 2y^3 \\ \frac{x-4}{2} &= y^3 \\ \sqrt[3]{\frac{x-4}{2}} &= y \\ f^{-1}(x) &= \sqrt[3]{\frac{x-4}{2}} \end{aligned}$$

73.
$$\begin{aligned} y &= \sqrt[3]{x+2} \\ x &= \sqrt[3]{y+2} \\ x^3 &= y+2 \\ x^3 - 2 &= y \\ f^{-1}(x) &= x^3 - 2 \end{aligned}$$

75.
$$\begin{aligned} y &= 2\sqrt[3]{2x+4} \\ x &= 2\sqrt[3]{2y+4} \\ \frac{x}{2} &= \sqrt[3]{2y+4} \\ \frac{x^3}{8} &= 2y+4 \\ \frac{x^3}{8} - 4 &= 2y \\ \frac{1}{2}\left(\frac{x^3}{8} - 4\right) &= \frac{1}{2} \cdot 2y \\ \frac{x^3}{16} - 2 &= y \\ f^{-1}(x) &= \frac{x^3}{16} - 2 \end{aligned}$$

77.

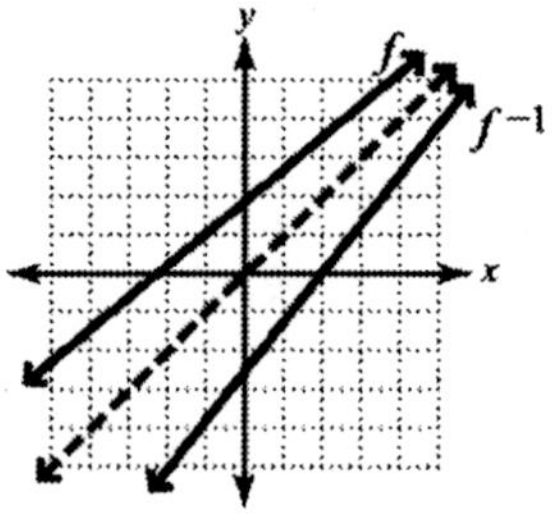

79.

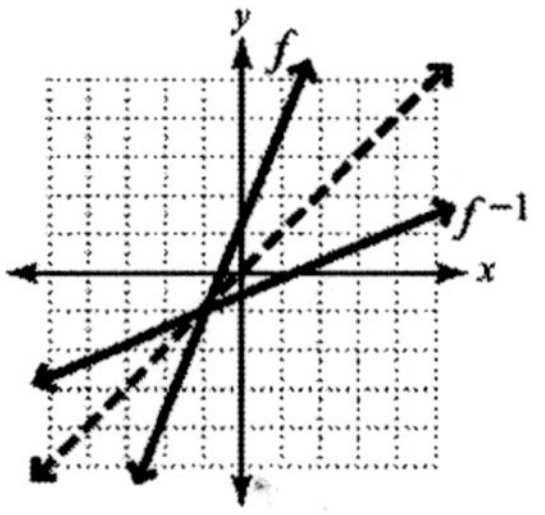

81.

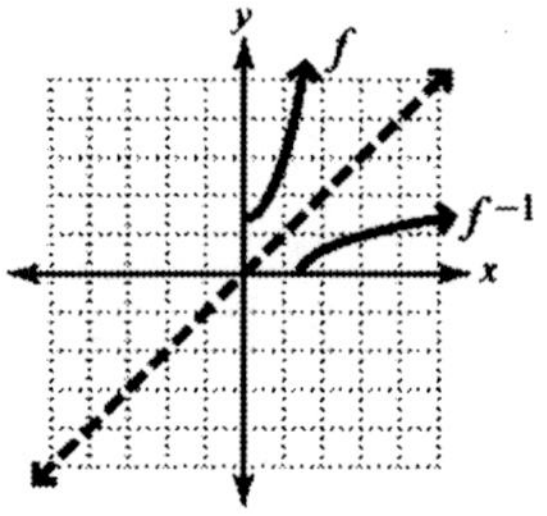

83.

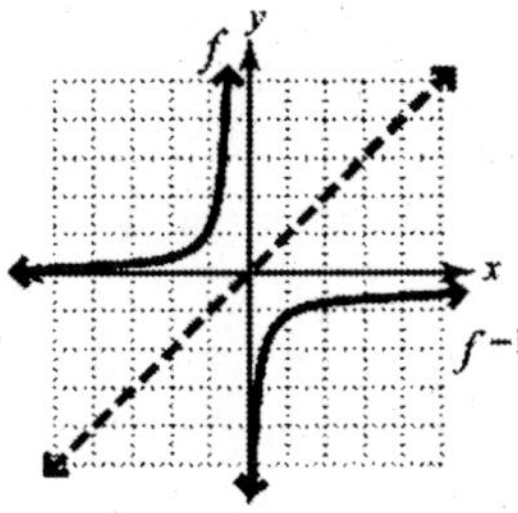

85.

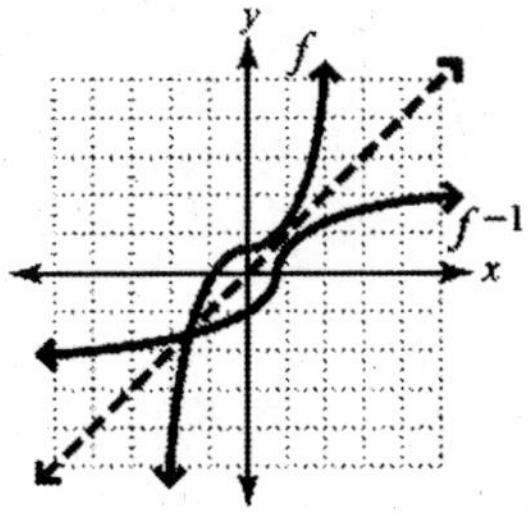

87.

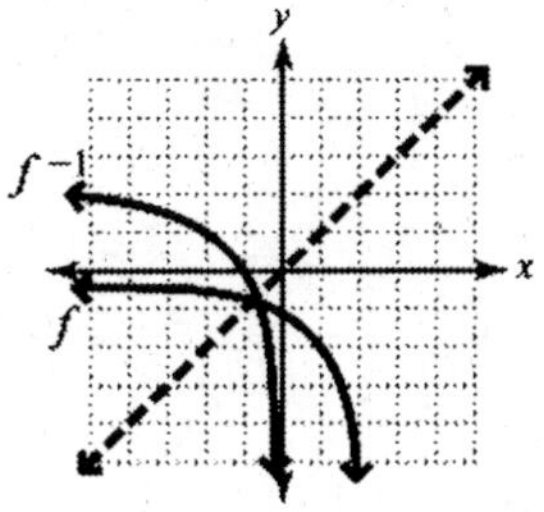

89. a)
$$y = 0.05x + 100$$
$$x = 0.05y + 100$$
$$x - 100 = 0.05y$$
$$\frac{x-100}{0.05} = y$$

b) x represents the salary, y represents the sales

c) $y = \frac{x-100}{0.05}$
$$= \frac{350-100}{0.05}$$
$$= \frac{250}{0.05}$$
$$= 5000$$
If the salary was \$350, sales were \$5000.

91. a)
$$y = 45x + 65(105 - x)$$
$$y = 45x + 6825 - 65x$$
$$y = -20x + 6825$$
$$x = -20y + 6825$$
$$x - 6825 = -20y$$
$$\frac{x-6825}{-20} = y$$
$$y = \frac{6825 - x}{20}$$

b) x represents the cost, y represents the number of 36-in. fans.

c) $f^{-1}(x) = \frac{6825 - x}{20}$
$$f^{-1}(5225) = \frac{6825-5225}{20}$$
$$= 80$$
The number of 36-inch fans was 80.

93. $(5, -4)$

95.
$$y = ax + b$$
$$x = ay + b$$
$$x - b = ay$$
$$\frac{x-b}{a} = y$$
$$f^{-1}(x) = \frac{x-b}{a}$$

Review Exercises

1. $32 = 2^5$

2. $2^3 = 8$

3. $\left(-\frac{1}{3}\right)^3 = -\frac{1}{27}$

4. $4^{-2} = \left(\frac{1}{4}\right)^2 = \frac{1}{16}$

5. $4^{3/2} = \left(\sqrt{4}\right)^3 = 2^3 = 8$

6. Vertex: (3, 1)

Exercise Set 10.2

1. Yes, the graphs are symmetric. The point $(1, 2)$ on $f(x)$ corresponds to the point $(-1, 2)$ on $g(x)$, and so on.

3. The domain is all real numbers; the range is the interval $(0, \infty)$.

5. Because the exponential function is one-to-one.

7. $f(x) = 3^x$

x	$y = f(x)$
-1	$\frac{1}{3}$
0	1
1	3
2	9

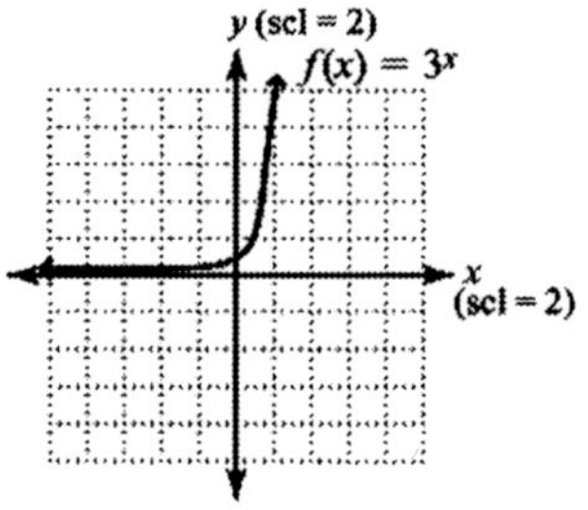

9. $f(x) = 4^x - 3$

x	$y = f(x)$
-1	$4^{-1} - 3 = -\frac{11}{4}$
0	$4^0 - 3 = -2$
1	$4^1 - 3 = 1$
2	$4^2 - 3 = 13$

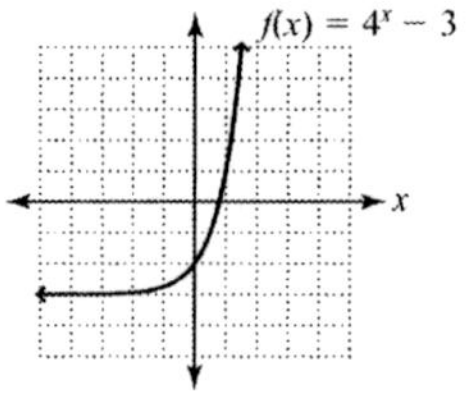

11. $f(x) = \left(\frac{1}{3}\right)^x$

x	$y = f(x)$
-1	$\left(\frac{1}{3}\right)^{-1} = 3$
0	$\left(\frac{1}{3}\right)^{0} = 1$
1	$\left(\frac{1}{3}\right)^{1} = \frac{1}{3}$
2	$\left(\frac{1}{3}\right)^{2} = \frac{1}{9}$

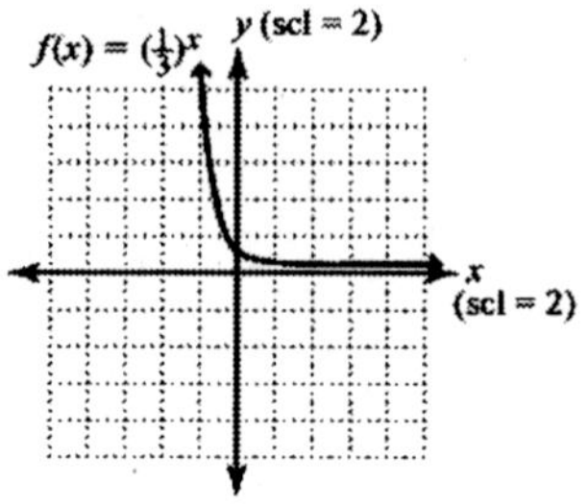

13. $f(x) = \left(\frac{2}{3}\right)^x + 2$

x	$y = f(x)$
-1	$\left(\frac{2}{3}\right)^{-1} + 2 = \frac{7}{2}$
0	$\left(\frac{2}{3}\right)^{0} + 2 = 3$
1	$\left(\frac{2}{3}\right)^{1} + 2 = \frac{8}{3}$
2	$\left(\frac{2}{3}\right)^{2} + 2 = \frac{22}{9}$

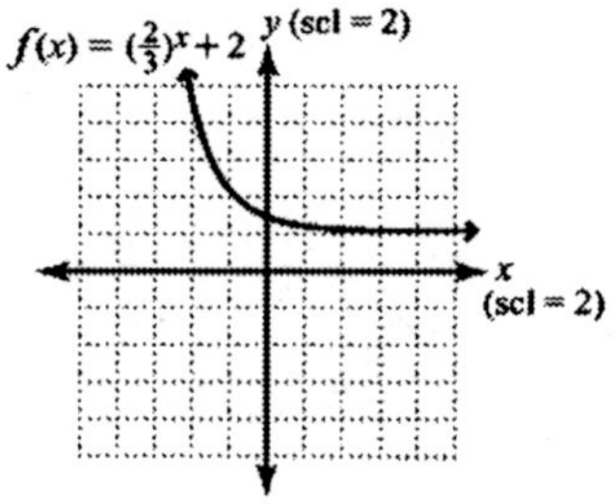

15. $f(x) = -3^x$

x	$y = f(x)$
-1	$-\frac{1}{3}$
0	-1
1	-3
2	-9

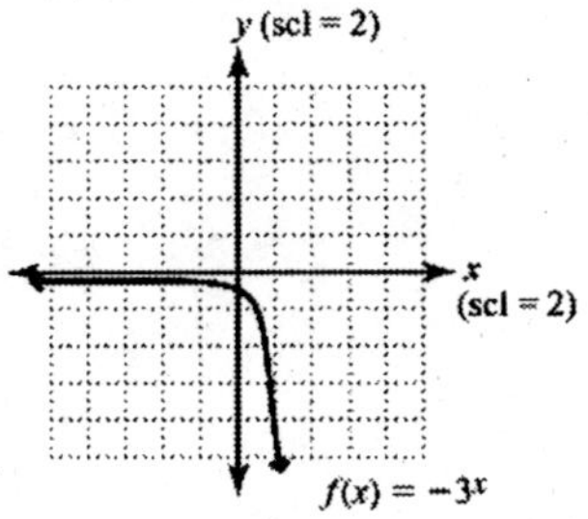

17. $f(x) = 2^{x-2}$

x	$y = f(x)$
-1	$2^{-1-2} = \frac{1}{8}$
0	$2^{0-2} = \frac{1}{4}$
1	$2^{1-2} = \frac{1}{2}$
2	$2^{2-2} = 1$

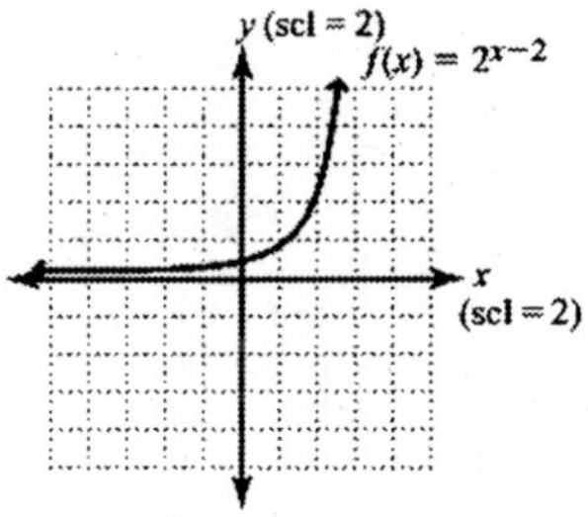

19. $f(x) = 2^{-x}$

x	$y = f(x)$
-1	$2^{-(-1)} = 2$
0	$2^{-(0)} = 1$
1	$2^{-(1)} = \frac{1}{2}$
2	$2^{-(2)} = \frac{1}{4}$

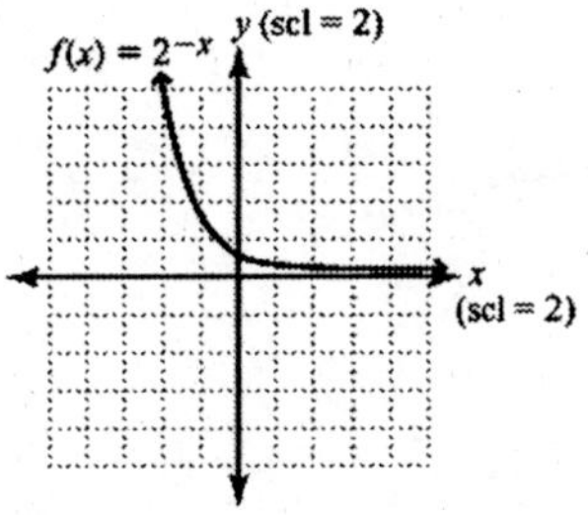

21. $f(x) = 2^{2x-3}$

x	$y = f(x)$
-1	$2^{2(-1)-3} = \frac{1}{32}$
0	$2^{2(0)-3} = \frac{1}{8}$
1	$2^{2(1)-3} = \frac{1}{2}$
2	$2^{2(2)-3} = 2$

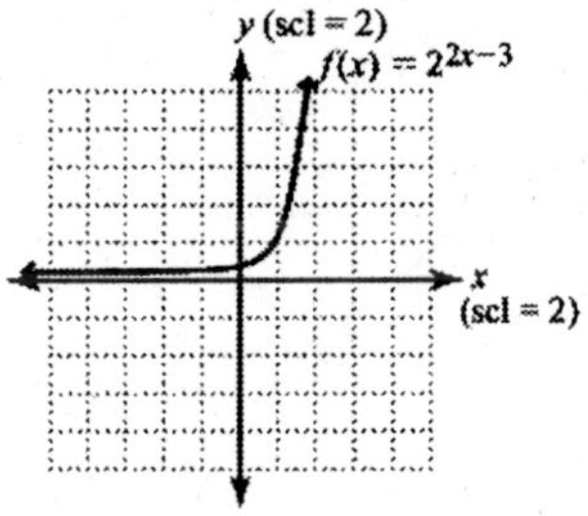

23. $f(x) = 3^{-x+2}$

x	$y = f(x)$
-1	$3^{-(-1)+2} = 27$
0	$3^{-(0)+2} = 9$
1	$3^{-(1)+2} = 3$
2	$3^{-(2)+2} = 1$

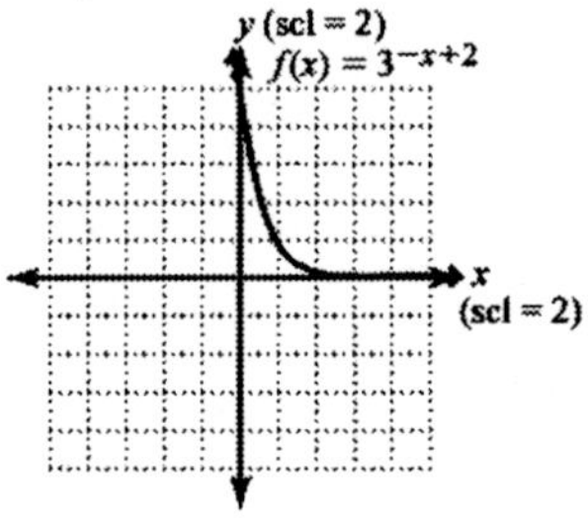

25. $2^x = 8$

$2^x = 2^3$

$x = 3$

27. $8^x = 32$

$(2^3)^x = 2^5$

$2^{3x} = 2^5$

$3x = 5$

$x = \frac{5}{3}$

29. $16^x = 4$

$(2^4)^x = 2^2$

$2^{4x} = 2^2$

$4x = 2$

$x = \frac{1}{2}$

31. $6^x = \frac{1}{36}$

$6^x = 6^{-2}$

$x = -2$

33. $\left(\frac{1}{3}\right)^x = 9$

$(3^{-1})^x = 3^2$

$3^{-x} = 3^2$

$-x = 2$

$x = -2$

35. $\left(\frac{2}{3}\right)^x = \frac{8}{27}$

$\left(\frac{2}{3}\right)^x = \left(\frac{2}{3}\right)^3$

$x = 3$

37. $\left(\frac{1}{2}\right)^x = 16$

$(2^{-1})^x = 2^4$

$2^{-x} = 4$

$-x = 4$

$x = -4$

39. $25^{x+1} = 125$

$(5^2)^{x+1} = 5^3$

$5^{2x+2} = 5^3$

$2x + 2 = 3$

$2x = 1$

$x = \frac{1}{2}$

41. $8^{2x-1} = 32^{x-3}$

$(2^3)^{2x-1} = (2^5)^{x-3}$

$2^{6x-3} = 2^{5x-15}$

$6x - 3 = 5x - 15$

$x = -12$

43. a)

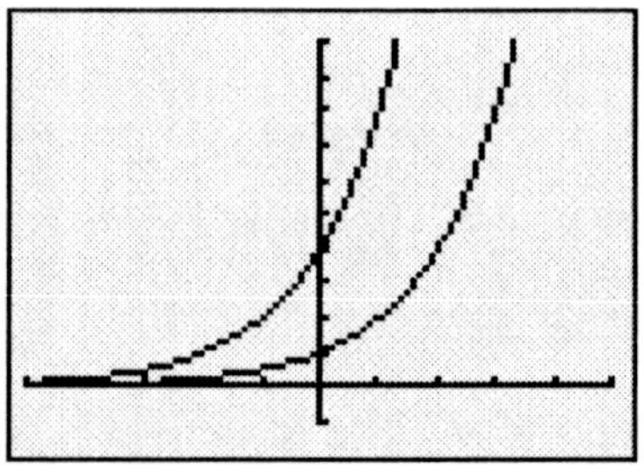

b) The graph of g is the graph of f shifted 2 units to the left.

45. a) 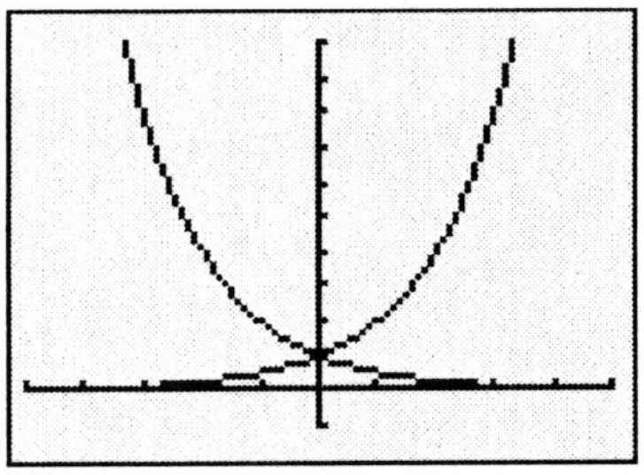

b) The graph of g is the graph of f reflected about the y-axis.

47. $A = A_0(2)^{t/20}$

$A = 100(2)^{120/20}$

$A = 6400$

After 120 minutes, there will be 6400 cells present.

49. $A = A_0(2)^{t/50}$

$A = 6(2)^{500/50}$

$A = 6144$

There will be 6144 billion people in 2500 if their growth is uncontrolled.

51. $A = P\left(1+\dfrac{r}{n}\right)^{nt}$

$A = 10{,}000\left(1+\dfrac{0.08}{4}\right)^{4\cdot 12}$

$A = 25{,}870.70$

The account value will be \$25,870.70 after 12 years.

53. $A = A_0\left(\dfrac{1}{2}\right)^{t/h}$

$A = 5\left(\dfrac{1}{2}\right)^{2160/270}$

$A = 0.020$

After 2160 days, 0.020 gram would remain.

55. $y = 2632.31(1.033)^x$

$y = 2632.31(1.033)^{56}$

$y \approx 16{,}216.42$

57. $f(x) = 2.5(0.7)^x$

$f(2) = 2.5(0.7)^2$

$= 1.225$

The chlorine level after 2 days is 1.225 parts per million.

59. $T(x) = 1.41(1.54)^x$

$T(22) = 1.41(1.54)^{22}$

$\approx 18{,}822$

If the trend continues, a total of about 18,822 million transistors could be put on a single chip in 2015.

Review Exercises

1. $5^3 = 125$

2. $\left(\dfrac{1}{3}\right)^{-2} = \left(\dfrac{3}{1}\right)^2 = 9$

3. $4^{-3} = \dfrac{1}{4^3} = \dfrac{1}{64}$

4. $5^{2/3} = \sqrt[3]{5^2} = \left(\sqrt[3]{5}\right)^2$

5. $f(x) = 3x+4$

$y = 3x+4$

$x = 3y+4$

$x-4 = 3y$

$\dfrac{x-4}{3} = y$

$f^{-1}(x) = \dfrac{x-4}{3}$

6. 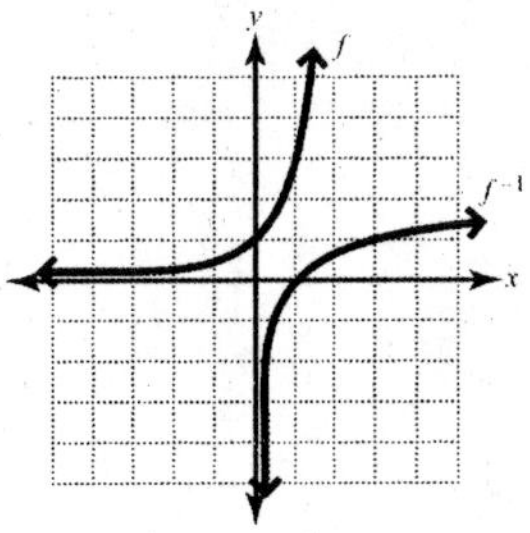

Exercise Set 10.3

1. $f^{-1}(x) = \log_3 x$, because logarithmic and exponential functions are inverses.

3. $x > 0$

5. Logarithms are exponents because logarithms are inverses of exponential functions.

7. $\log_2 32 = 5$

9. $\log_{10} 1000 = 3$

11. $\log_e x = 4$

13. $\log_5 \dfrac{1}{125} = -3$

15. $\log_{10} \dfrac{1}{100} = -2$

17. $\log_{625} 5 = \dfrac{1}{4}$

19. $\log_{1/4} \frac{1}{16} = 2$

21. $\log_7 \sqrt{7} = \frac{1}{2}$

23. $3^4 = 81$

25. $4^{-2} = \frac{1}{16}$

27. $10^2 = 100$

29. $e^5 = a$

31. $e^{-4} = \frac{1}{e^4}$

33. $\left(\frac{1}{8}\right)^2 = \frac{1}{64}$

35. $\left(\frac{1}{5}\right)^{-2} = 25$

37. $7^{1/2} = \sqrt{7}$

39. $2^5 = x$
$32 = x$

41. $5^{-2} = x$
$\frac{1}{5^2} = x$
$\frac{1}{25} = x$

43. $3^y = 81$
$3^y = 3^4$
$y = 4$

45. $5^y = \frac{1}{25}$
$5^y = 5^{-2}$
$y = -2$

47. $b^3 = 1000$
$b^3 = 10^3$
$b = 10$

49. $m^{-4} = \frac{1}{16}$
$m^{-4} = 16^{-1}$
$m^{-4} = \left(2^4\right)^{-1}$
$m^{-4} = 2^{-4}$
$m = 2$

51. $\left(\frac{1}{2}\right)^2 = x$
$\frac{1}{4} = x$

53. $\left(\frac{1}{3}\right)^{-5} = h$
$3^5 = h$
$243 = h$

55. $\left(\frac{1}{3}\right)^y = \frac{1}{9}$
$\left(\frac{1}{3}\right)^y = \left(\frac{1}{3}\right)^2$
$y = 2$

57. $\left(\frac{1}{2}\right)^t = 64$
$\left(2^{-1}\right)^t = 2^6$
$2^{-t} = 2^6$
$-t = 6$
$t = -6$

59. $\log_2 x = 4$
$x = 2^4$
$x = 16$

61. $\log_{1/4} h = 3$
$h = \left(\frac{1}{4}\right)^3$
$h = \frac{1}{64}$

63. $\log_b 16 = 4$
$b^4 = 16$
$b^4 = 2^4$
$b = 2$

65. $\frac{1}{3}\log_5 c = 1$
$\log_5 c = 3$
$c = 5^3$
$c = 125$

67. $3\log_t 9 = 6$
$\log_t 9 = 2$
$t^2 = 9$
$t^2 = 3^2$
$t = 3$

69. $\frac{1}{2}\log_4 m = -1$
$\log_4 m = -2$
$m = 4^{-2}$
$m = \frac{1}{4^2}$
$m = \frac{1}{16}$

71. $f(x) = \log_4 x$
$y = \log_4 x$
$4^y = x$

y	-1	0	1	2
x	$\frac{1}{4}$	1	4	16

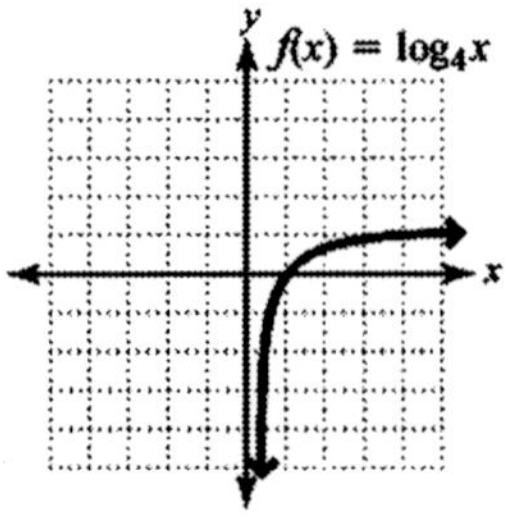

73. $f(x) = \log_{1/3} x$

$y = \log_{1/3} x$

$\left(\frac{1}{3}\right)^y = x$

y	-2	-1	0	1
x	9	3	1	$\frac{1}{3}$

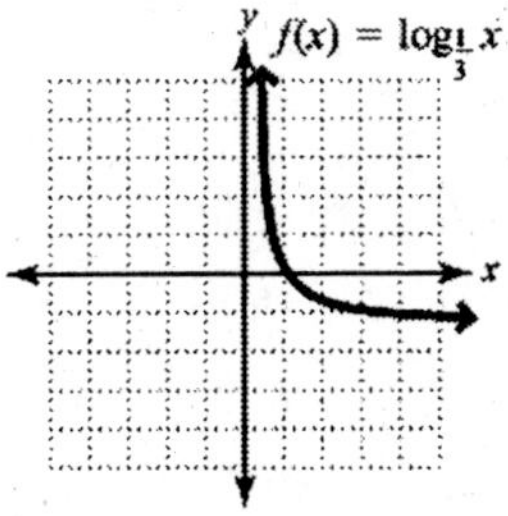

75. $f(x) = 62 + 35\log_{10}(x-4)$

$f(14) = 62 + 35\log_{10}(14-4)$

$= 62 + 35\log_{10} 10$

$= 62 + 35(1)$

$= 97$

At age 14, a girl has reached 97% of her adult height.

77. $35 = 95 - 30\log_2 x$

$-60 = -30\log_2 x$

$2 = \log_2 x$

$x = 2^2$

$x = 4$

After 4 days, 35% of the students recall important features of the lecture.

79. $\log_b b = 1$ because $b^1 = b$.

81. $\log_b 1 = 0$ because $b^0 = 1$.

Review Exercises

1. x^{-6}

2. $x^4 \cdot x^2 = x^{4+2} = x^6$

3. $\left(x^3\right)^5 = x^{3\cdot 5} = x^{15}$

4. $\frac{x^6}{x^3} = x^{6-3} = x^3$

5. $\frac{\left(x^3\right)^2 \cdot x^4}{x^5} = \frac{x^6 \cdot x^4}{x^5} = \frac{x^{10}}{x^5} = x^{10-5} = x^5$

6. $\sqrt[4]{x^3} = x^{3/4}$

Exercise Set 10.4

1. Because the exponential and logarithmic functions are inverses.

3. $\log_b(x+y)$; there is no rule for the logarithm of a sum.

5. $8^{\log_8 2} = 2$

7. $a^{\log_a r} = r$

9. $a^{\log_a 4x} = 4x$

11. $\log_3 3^5 = 5$

13. $\log_e e^y = y$

15. $\log_a a^{7x} = 7x$

17. $\log_2 5y = \log_2 5 + \log_2 y$

19. $\log_a pq = \log_a p + \log_a q$

21. $\log_4 mnp = \log_4 m + \log_4 n + \log_4 p$

23. $\log_a x(x-5) = \log_a x + \log_a (x-5)$

25. $\log_3 5 + \log_3 8 = \log_3 (5\cdot 8) = \log_3 40$

27. $\log_4 3 + \log_4 9 = \log_4 (3\cdot 9) = \log_4 27$

29. $\log_a 7 + \log_a m = \log_a 7m$

31. $\log_4 a + \log_4 b = \log_4 ab$

33. $\log_a 2 + \log_a x + \log_a (x+5) = \log_a 2x(x+5)$

$= \log_a \left(2x^2 + 10x\right)$

35. $\log_4 (x+1) + \log_4 (x+3) = \log_4 (x+1)(x+3)$

$= \log_4 \left(x^2 + 4x + 3\right)$

37. $\log_2 \frac{7}{9} = \log_2 7 - \log_2 9$

39. $\log_a \frac{x}{5} = \log_a x - \log_a 5$

41. $\log_a \frac{a}{b} = \log_a a - \log_a b = 1 - \log_a b$

43. $\log_a \frac{x}{x-3} = \log_a x - \log_a (x-3)$

45. $\log_4 \frac{2x-3}{4x+5} = \log_4 (2x-3) - \log_4 (4x+5)$

47. $\log_6 24 - \log_6 3 = \log_6 \frac{24}{3} = \log_6 8$

49. $\log_2 24 - \log_2 12 = \log_2 \frac{24}{12} = \log_2 2 = 1$

51. $\log_a x - \log_a 3 = \log_a \frac{x}{3}$

53. $\log_4 p - \log_4 q = \log_4 \frac{p}{q}$

55. $\log_b x - \log_b (x-4) = \log_b \frac{x}{x-4}$

57. $\log_x (x^2 - x) - \log_x (x-1) = \log_x \frac{x^2 - x}{x-1}$
$= \log_x \frac{x(x-1)}{x-1}$
$= \log_x x$
$= 1$

59. $\log_4 3^6 = 6\log_4 3$

61. $\log_a x^7 = 7\log_a x$

63. $\log_a \sqrt{3} = \log_a 3^{1/2} = \frac{1}{2}\log_a 3$

65. $\log_3 \sqrt[3]{x^2} = \log_3 x^{2/3} = \frac{2}{3}\log_3 x$

67. $\log_a \frac{1}{6^2} = \log_a 6^{-2} = -2\log_a 6$

69. $\log_a \frac{1}{y^2} = \log_a y^{-2} = -2\log_a y$

71. $4\log_3 5 = \log_3 5^4$

73. $-3\log_2 x = \log_2 x^{-3} = \log_2 \frac{1}{x^3}$

75. $\frac{1}{2}\log_7 64 = \log_7 64^{1/2} = \log_7 \sqrt{64} = \log_7 8$

77. $\frac{3}{4}\log_a x = \log_a x^{3/4} = \log_a \sqrt[4]{x^3}$

79. $\frac{2}{3}\log_a 8 = \log_a 8^{2/3}$
$= \log_a \sqrt[3]{8^2}$
$= \log_a \sqrt[3]{64}$
$= \log_a 4$

81. $-\frac{1}{2}\log_3 x = \log_3 x^{-1/2} = \log_3 \frac{1}{x^{1/2}} = \log_3 \frac{1}{\sqrt{x}}$

83. $\log_a \frac{x^3}{y^4} = \log_a x^3 - \log_a y^4 = 3\log_a x - 4\log_a y$

85. $\log_3 a^4 b^2 = \log_3 a^4 + \log_3 b^2 = 4\log_3 a + 2\log_3 b$

87. $\log_a \frac{xy}{z} = \log_a xy - \log_a z$
$= \log_a x + \log_a y - \log_a z$

89. $\log_x \frac{a^2}{bc^3} = \log_x a^2 - \log_x bc^3$
$= \log_x a^2 - (\log_x b + \log_x c^3)$
$= 2\log_x a - (\log_x b + 3\log_x c)$
$= 2\log_x a - \log_x b - 3\log_x c$

91. $\log_4 \sqrt[4]{\frac{x^3}{y}} = \log_4 \left(\frac{x^3}{y}\right)^{1/4}$
$= \log_4 \left(\frac{x^{3/4}}{y^{1/4}}\right)$
$= \log_4 x^{3/4} - \log_4 y^{1/4}$
$= \frac{3}{4}\log_4 x - \frac{1}{4}\log_4 y$

93. $\log_a \sqrt[3]{\frac{x^2 y}{z^3}} = \log_a \left(\frac{x^2 y}{z^3}\right)^{1/3}$
$= \log_a \left(\frac{x^{2/3} y^{1/3}}{z^{3/3}}\right)$
$= \log_a x^{2/3} y^{1/3} - \log_a z$
$= \log_a x^{2/3} + \log_a y^{1/3} - \log_a z$
$= \frac{2}{3}\log_a x + \frac{1}{3}\log_a y - \log_a z$

95. $3\log_3 2 - 2\log_3 4 = \log_3 2^3 - \log_3 4^2$
$= \log_3 8 - \log_3 16$
$= \log_3 \frac{8}{16}$
$= \log_3 \frac{1}{2}$

97. $4\log_b x + 3\log_b y = \log_b x^4 + \log_b y^3$
$= \log_b x^4 y^3$

99. $\frac{1}{2}(\log_a 5 - \log_a 7) = \frac{1}{2}\log_a\left(\frac{5}{7}\right)$

$= \log_a\left(\frac{5}{7}\right)^{1/2}$

$= \log_a\sqrt{\frac{5}{7}}$

101. $\frac{2}{3}(\log_a x^2 + \log_a y^3) = \frac{2}{3}\log_a(x^2y^3)$

$= \log_a(x^2y^3)^{2/3}$

$= \log_a\sqrt[3]{(x^2y^3)^2}$

103. $\log_b x + \log_b(3x-2) = \log_b x(3x-2)$

$= \log_b(3x^2 - 2x)$

105. $3\log_a(x-2) - 4\log_a(x+1)$

$= \log_a(x-2)^3 - \log_a(x+1)^4$

$= \log_a\frac{(x-2)^3}{(x+1)^4}$

107. $2\log_a x + 4\log_a z - 3\log_a w - 6\log_a u$

$= \log_a x^2 + \log_a z^4 - \log_a w^3 - \log_a u^6$

$= \log_a x^2z^4 - \log_a w^3 - \log_a u^6$

$= \log_a\frac{x^2z^4}{w^3} - \log_a u^6$

$= \log_a\frac{x^2z^4}{w^3u^6}$

Review Exercises

1. $(10^{9.5})(10^{-12}) = 10^{9.5+(-12)} = 10^{-2.5}$

2. $\frac{10^{-3}}{10^{-12}} = 10^{-3-(-12)} = 10^9$

3. $A = 10,000\left(1+\frac{0.06}{4}\right)^{5(4)}$

$A = \$13,468.55$

4. $\log_{10} 50 = 1.6990$

5. $10^{1.6532} = 45$

6. $e^{-1.3863} = 0.25$

Exercise Set 10.5

1. $b > 0, b \neq 1$. If $y = \log_b x$ then $b^y = x$. If $b < 0$ and $y = \frac{1}{2}$ or any other fraction whose denominator is even, then x is imaginary. If $b = 1$, then $x = 1$ for all values of y and the graph is a vertical line.

3. Positive. If $y = \ln x$, then $e^y = x$. If $x > 1$, then $e^y > 1$, which is true if $y > 0$.

5. 0.91629

7. $\log 64 = 1.8062$

9. $\log 0.0067 = -2.1739$

11. $\log 435.6 = 2.6391$

13. $\log(1.5\times10^4) = 4.1761$

15. $\log(1.6\times10^{-6}) = -5.7959$

17. $\ln 9.34 = 2.2343$

19. $\ln 79.2 = 4.3720$

21. $\ln 0.034 = -3.3814$

23. $\ln(5.4\times e^4) = 5.6864$

25. $\log e = 0.4343$

27. Error results because the domain of $\log_a x$ is $(0,\infty)$; so log 0 is undefined.

29. $\log 100 = \log_{10} 10^2 = 2$

31. $\log\frac{1}{100} = \log_{10} 10^{-2} = -2$

33. $\log\sqrt[3]{10} = \log_{10} 10^{1/3} = \frac{1}{3}$

35. $\log 0.001 = \log_{10} 10^{-3} = -3$

37. $\ln e^3 = \log_e e^3 = 3$

39. $\ln\sqrt{e} = \log_e e^{1/2} = \frac{1}{2}$

41. $d = 10\log\frac{I}{I_0}$

$d = 10\log\frac{10^{-3}}{10^{-12}}$

$d = 10\log 10^9$

$d = 10\cdot 9$

$d = 90$

The reading for a firecracker is 90dB.

43. $d = 10\log\frac{I}{I_0}$

$60 = 10\log\frac{I}{10^{-12}}$

$6 = \log\frac{I}{10^{-12}}$

$6 = \log I - \log 10^{-12}$

$6 = \log I - (-12)$

$6 = \log I + 12$

$-6 = \log_{10} I$

$I = 10^{-6}$

The sound intensity of a noisy office is 10^{-6} watts/m^2.

45. $\text{pH} = -\log\left[H_3O^+\right]$

$\text{pH} = -\log\left[1.6\times 10^{-3}\right]$

$\text{pH} = 2.796$

47. $3.5 = -\log\left[H_3O^+\right]$

$\log\left[H_3O^+\right] \doteq -3.5$

$10^{-3.5} = \left[H_3O^+\right]$

The hydronium ion concentration is $10^{-3.5}$ moles/L.

49. $t = \frac{1}{r}\ln\frac{A}{P}$

$t = \frac{1}{0.04}\ln\frac{5000}{2000}$

≈ 22.9

It will take approximately 23 years.

51. San Francisco: $7.8 = \log\frac{I}{I_0}$

$7.8 = \log I - \log I_0$

$7.8 + \log I_0 = \log I$

$10^{7.8+\log I_0} = I$

Alaska: $8.4 = \log\frac{I}{I_0}$

$8.4 = \log I - \log I_0$

$8.4 + \log I_0 = \log I$

$10^{8.4+\log I_0} = I$

$\frac{10^{8.4+\log I_0}}{10^{7.8+\log I_0}} = 10^{(8.4+\log I_0)-(7.8+\log I_0)}$

$= 10^{0.6}$

≈ 4

The 1964 Alaska earthquake was about four times as severe as the 1906 San Francisco earthquake.

53. $\log E_s = 11.8 + 1.5(7.3)$

$\log_{10} E_s = 22.75$

$E_s = 10^{22.75}$

The energy released was $10^{22.75}$ ergs.

55. $y = -69.45 + 22.7\ln x$

$y = -69.45 + 22.7\ln(40)$

$y \approx -69.45 + 84$

$y \approx 14.55$

The purchasing power of \$1.00 was approximately 40 cents in 1985.

57. $y = 3\log 110$

$y \approx 6.124$

and

$6.124(1000) = 6124$ ft.

Review Exercises

1. $\log_3(2x+1) - \log_3(x-1) = \log_3\frac{2x+1}{x-1}$

2. $3^2 = 2x+5$

3. $$\begin{aligned} 3x-(7x+2) &= 12-2(x-4) \\ 3x-7x-2 &= 12-2x+8 \\ -4x-2 &= 20-2x \\ -2x-2 &= 20 \\ -2x &= 22 \\ x &= -11 \end{aligned}$$

4. $$\begin{aligned} x^2+2x &= 15 \\ x^2+2x-15 &= 0 \\ (x+5)(x-3) &= 0 \\ x+5=0 \quad &\text{or} \quad x-3=0 \\ x=-5 \qquad & \qquad x=3 \end{aligned}$$

5. $$\begin{aligned} \frac{5}{x}+\frac{3}{x+1} &= \frac{23}{3x} \\ 3x(x+1)\left[\frac{5}{x}+\frac{3}{x+1}\right] &= 3x(x+1)\left[\frac{23}{3x}\right] \\ 3(x+1)(5)+3x(3) &= (x+1)23 \\ 15x+15+9x &= 23x+23 \\ 24x+15 &= 23x+23 \\ x &= 8 \end{aligned}$$

6. $$\begin{aligned} \sqrt{5x-1} &= 7 \\ \left(\sqrt{5x-1}\right)^2 &= 7^2 \\ 5x-1 &= 49 \\ 5x &= 50 \\ x &= 10 \end{aligned}$$

Exercise Set 10.6

1. $m=n$

3. If $\log_b x = \log_b y$, then $x=y$.

5. The solution $x=-5$ is rejected because substituting into the equation gives $\log(-5)$ and $\log(-3)$, which are both undefined.

7. $$\begin{aligned} 2^x &= 9 \\ \log 2^x &= \log 9 \\ x\log 2 &= \log 9 \\ x &= \frac{\log 9}{\log 2} \approx 3.1699 \end{aligned}$$

9. $$\begin{aligned} 5^{2x} &= 32 \\ \log 5^{2x} &= \log 32 \\ 2x\log 5 &= \log 32 \\ 2x &= \frac{\log 32}{\log 5} \\ x &= \frac{\log 32}{\log 5} \div 2 \approx 1.0767 \end{aligned}$$

11. $$\begin{aligned} 5^{x+3} &= 10 \\ \log 5^{x+3} &= \log 10 \\ (x+3)\log 5 &= 1 \\ x\log 5+3\log 5 &= 1 \\ x\log 5 &= 1-3\log 5 \\ x &= \frac{1-3\log 5}{\log 5} \approx -1.5693 \end{aligned}$$

13. $$\begin{aligned} 8^{x-2} &= 6 \\ \log 8^{x-2} &= \log 6 \\ (x-2)\log 8 &= \log 6 \\ x\log 8-2\log 8 &= \log 6 \\ x\log 8 &= \log 6+2\log 8 \\ x &= \frac{\log 6+2\log 8}{\log 8} \approx 2.8617 \end{aligned}$$

15. $$\begin{aligned} 4^{x+2} &= 5^x \\ \log 4^{x+2} &= \log 5^x \\ (x+2)\log 4 &= x\log 5 \\ x\log 4+2\log 4 &= x\log 5 \\ x\log 4-x\log 5 &= -2\log 4 \\ x(\log 4-\log 5) &= -2\log 4 \\ x &= \frac{-2\log 4}{\log 4-\log 5} \approx 12.4251 \end{aligned}$$

17. $$\begin{aligned} 2^{x+1} &= 3^{x-2} \\ \log 2^{x+1} &= \log 3^{x-2} \\ (x+1)\log 2 &= (x-2)\log 3 \\ x\log 2+\log 2 &= x\log 3-2\log 3 \\ x\log 2-x\log 3 &= -2\log 3-\log 2 \\ x(\log 2-\log 3) &= -2\log 3-\log 2 \\ x &= \frac{-2\log 3-\log 2}{\log 2-\log 3} \approx 7.1285 \end{aligned}$$

19. $$e^{3x} = 5$$
$$\ln e^{3x} = \ln 5$$
$$3x = \ln 5$$
$$x = \frac{\ln 5}{3} \approx 0.5365$$

21. $$e^{0.03x} = 25$$
$$\ln e^{0.03x} = \ln 25$$
$$0.03x = \ln 25$$
$$x = \frac{\ln 25}{0.03} \approx 107.2959$$

23. $$e^{-0.022x} = 5$$
$$\ln e^{-0.022x} = \ln 5$$
$$-0.022x = \ln 5$$
$$x = -\frac{\ln 5}{0.022} \approx -73.1563$$

25. $$\ln e^{4x} = 24$$
$$4x = 24$$
$$x = 6$$

27. $$\log_4(x+5) = 2$$
$$4^2 = x+5$$
$$16 = x+5$$
$$11 = x$$

29. $$\log_4(4x-8) = 2$$
$$4^2 = 4x-8$$
$$16 = 4x-8$$
$$24 = 4x$$
$$6 = x$$

31. $$\log_4 x^2 = 2$$
$$4^2 = x^2$$
$$16 = x^2$$
$$\pm\sqrt{16} = x$$
$$\pm 4 = x$$

33. $$\log_6\left(x^2+5x\right) = 2$$
$$6^2 = x^2+5x$$
$$36 = x^2+5x$$
$$0 = x^2+5x-36$$
$$0 = (x+9)(x-4)$$
$$x+9=0 \quad \text{or} \quad x-4=0$$
$$x=-9 \qquad\qquad x=4$$

35. $$\log(4x-3) = \log(3x+4)$$
$$4x-3 = 3x+4$$
$$x = 7$$

37. $$\ln(3x+4) = \ln(x-6)$$
$$3x+4 = x-6$$
$$2x = -10$$
$$x = \cancel{-5}$$
No solution

39. $$\log_9\left(x^2+4x\right) = \log_9 12$$
$$x^2+4x = 12$$
$$x^2+4x-12 = 0$$
$$(x+6)(x-2) = 0$$
$$x+6=0 \quad \text{or} \quad x-2=0$$
$$x=-6 \qquad\qquad x=2$$

41. $$\log_4 x + \log_4 8 = 2$$
$$\log_4 8x = 2$$
$$4^2 = 8x$$
$$16 = 8x$$
$$2 = x$$

43. $$\log_2 x - \log_2 5 = 1$$
$$\log_2 \frac{x}{5} = 1$$
$$2^1 = \frac{x}{5}$$
$$10 = x$$

45. $\log_3 x + \log_3 (x+6) = 3$

$$\begin{aligned} \log_3 x(x+6) &= 3 \\ \log_3 \left(x^2+6x\right) &= 3 \\ 3^3 &= x^2+6x \\ 27 &= x^2+6x \\ 0 &= x^2+6x-27 \\ 0 &= (x+9)(x-3) \end{aligned}$$

$$x+9=0 \quad \text{or} \quad x-3=0$$
$$x = \cancel{-9} \qquad x = 3$$

47. $\log_3 (2x+15) + \log_3 x = 3$

$$\begin{aligned} \log_3 x(2x+15) &= 3 \\ \log_3 \left(2x^2+15x\right) &= 3 \\ 3^3 &= 2x^2+15x \\ 27 &= 2x^2+15x \\ 0 &= 2x^2+15x-27 \\ 0 &= (2x-3)(x+9) \end{aligned}$$

$$2x-3=0 \quad \text{or} \quad x+9=0$$
$$x = \frac{3}{2} \qquad x = \cancel{-9}$$

49. $\log_2 (7x+3) - \log_2 (2x-3) = 3$

$$\begin{aligned} \log_2 \frac{7x+3}{2x-3} &= 3 \\ 2^3 &= \frac{7x+3}{2x-3} \\ 8(2x-3) &= 7x+3 \\ 16x-24 &= 7x+3 \\ 9x &= 27 \\ x &= 3 \end{aligned}$$

51. $\log_2 (3x+8) - \log_2 (x+1) = 2$

$$\begin{aligned} \log_2 \frac{3x+8}{x+1} &= 2 \\ 2^2 &= \frac{3x+8}{x+1} \\ 4 &= \frac{3x+8}{x+1} \\ 4x+4 &= 3x+8 \\ x &= 4 \end{aligned}$$

53. $\log_8 2x + \log_8 6 = \log_8 10$

$$\begin{aligned} \log_8 12x &= \log_8 10 \\ 12x &= 10 \\ x &= \frac{5}{6} \end{aligned}$$

55. $\ln x + \ln (2x-1) = \ln 10$

$$\begin{aligned} \ln x(2x-1) &= \ln 10 \\ \ln \left(2x^2-x\right) &= \ln 10 \\ 2x^2-x &= 10 \\ 2x^2-x-10 &= 0 \\ (2x-5)(x+2) &= 0 \end{aligned}$$

$$2x-5=0 \quad \text{or} \quad x+2=0$$
$$x = \frac{5}{2} \qquad x = \cancel{-2}$$

57. $\log x - \log (x-5) = \log 6$

$$\begin{aligned} \log \frac{x}{x-5} &= \log 6 \\ \frac{x}{x-5} &= 6 \\ x &= 6x-30 \\ -5x &= -30 \\ x &= 6 \end{aligned}$$

59. $\log_6 (3x+4) - \log_6 (x-2) = \log_6 8$

$$\begin{aligned} \log_6 \frac{3x+4}{x-2} &= \log_6 8 \\ \frac{3x+4}{x-2} &= 8 \\ 3x+4 &= 8x-16 \\ 20 &= 5x \\ 4 &= x \end{aligned}$$

61. $$8000 = 5000\left(1+\frac{.05}{4}\right)^{4t}$$

$$\begin{aligned} \frac{8}{5} &= (1.0125)^{4t} \\ \log\frac{8}{5} &= (4t)\log 1.0125 \\ \frac{\log\frac{8}{5}}{4\log 1.0125} &= t \\ 9.5 &\approx t \end{aligned}$$

It will take about 9.5 years.

63. a) $A = 8000e^{0.06(15)}$

$A = 19{,}676.82$

The value of the account will be \$19,676.82.

b) $14000 = 8000e^{0.06t}$

$\frac{14}{8} = e^{0.06t}$

$\ln\frac{7}{4} = \ln e^{0.06t}$

$\ln\frac{7}{4} = 0.06t$

$\frac{\ln\frac{7}{4}}{0.06} = t$

$9.3 \approx t$

It will take about 9.3 years.

65. a) $P(0.08) = e^{21.5(0.08)}$

$P(0.08) = 5.58\%$

b) $50 = e^{21.5b}$

$\ln 50 = \ln e^{21.5b}$

$\ln 50 = 21.5b$

$\frac{\ln 50}{21.5} = b$

$0.18 = b$

67. a) $A = 500e^{0.04(14)}$

$A \approx 875$

There will be 875 mosquitoes.

b) $10{,}000 = 500e^{0.04t}$

$20 = e^{0.04t}$

$\ln 20 = \ln e^{0.04t}$

$\ln 20 = 0.04t$

$\frac{\ln 20}{0.04} = t$

$75 \approx t$

It will take about 75 days.

69. a) $A = 6.4e^{0.01(15)}$

$A \approx 7.44$

The world population will be about 7.44 billion in 2020.

b) $7.0 = 6.4e^{0.01t}$

$\frac{7.0}{6.4} = e^{0.01t}$

$\ln\frac{7.0}{6.4} = \ln e^{0.01t}$

$\ln\frac{7.0}{6.4} = 0.01t$

$9 \approx t$

Year: $2005 + 9 = 2014$

71. a) $P = 0.48\ln(50+1)$

$P = 0.48\ln 51$

$P \approx 1.89$

The barometric pressure is 1.89 inches of mercury.

b) $1.5 = 0.48\ln(x+1)$

$3.125 = \ln(x+1)$

$e^{3.125} = x+1$

$e^{3.125} - 1 = x$

$21.8 \approx x$

The distance is about 21.8 miles.

73. a) $f(10) = 29 + 48.8\log(10+1)$

$f(10) = 29 + 48.8\log(11)$

$f(10) \approx 79.8$

At age 10, a boy has reached 79.8% of his adult height.

b) $75 = 29 + 48.8\log(x+1)$

$46 = 48.8\log(x+1)$

$0.9426 = \log(x+1)$

$10^{0.9426} = x+1$

$10^{0.9426} - 1 = x$

$7.76 = x$

A boy will attain 75% of his adult height at about age 8.

75. $y = 1.222(1.044)^x$

$y = 1.222(1.044)^{44}$

$y \approx 8.13$

In 2014, it will take \$8.13 to have the same purchasing power as \$1.00 in 1970.

77. $y = 78.62(1.092)^x$

$y = 78.62(1.092)^{29}$

$y \approx 1009.2$

Approximately \$1009.2 billion will be spent on recreation in 2014.

79. $\log_4 12 = \frac{\log 12}{\log 4} \approx 1.7925$

81. $\log_8 3 = \frac{\log 3}{\log 8} \approx 0.5283$

83. $\frac{\log 5}{\log 0.5} \approx -2.3219$

85. $\log_{1/4} \frac{3}{5} = \frac{\log \frac{3}{5}}{\log \frac{1}{4}} \approx 0.3685$

87. $P = 95 - 30\log_2 x$

$P = 95 - 30\log_2 3$

$P = 95 - 30\frac{\log 3}{\log 2}$

$P \approx 47$

After 3 days, about 47% of the lecture is retained.

Review Exercises

1. $(3x+2)-(2x-1) = (3x+2)+(-2x+1)$
$= x+3$

2. $\left(3\sqrt{5}+2\right)\left(2\sqrt{5}-1\right)$
$= 3\sqrt{5}\cdot 2\sqrt{5} + 3\sqrt{5}\cdot(-1) + 2\cdot 2\sqrt{5} + 2\cdot(-1)$
$= 30 - 3\sqrt{5} + 4\sqrt{5} - 2$
$= 28 + \sqrt{5}$

3. $f(-5) = (-5)^2 + 4 = 25 + 4 = 29$

4. $(f+g)(x) = x^2 + 4 + 2x - 1 = x^2 + 2x + 3$

5. $f \circ g = f\left[g(x)\right]$
$= (2x-1)^2 + 4$
$= 4x^2 - 4x + 1 + 4$
$= 4x^2 - 4x + 5$

6. Using the answer from #5:
$(f \circ g)(2) = 4(2)^2 - 4(2) + 5$
$= 16 - 8 + 5$
$= 13$

Chapter 10 Review Exercises

1. True

2. False; this statement is true only if $b > 0$ and $b \neq 1$.

3. False; by definition, $y = \log_a (x-4)$ is the same as $a^y = x - 4$. If $x = -3$, then we would have $a^y = -7$, which is impossible because a is restricted to be $a > 0$.

4. False; if $\log_a b = c$, then we have $a^c = b$.

5. True

6. False; base-e logarithms are called natural logarithms.

7. $(f \circ g)(3) = f\left[g(3)\right]$
$= f(7)$
$= 3(7) + 4$
$= 21 + 4$
$= 25$

8. $(g \circ f)(3) = g\left[f(3)\right]$
$= g(13)$
$= 13^2 - 2$
$= 169 - 2$
$= 167$

9. $f\left[g(0)\right] = f(-2)$
$= 3(-2) + 4$
$= -6 + 4$
$= -2$

10. $g\left[f(0)\right] = g(4)$
$= 4^2 - 2$
$= 16 - 2$
$= 14$

11. $(f \circ g)(x) = f[g(x)]$
$= f(2x+3)$
$= 3(2x+3)-6$
$= 6x+9-6$
$= 6x+3$

$(g \circ f)(x) = g[f(x)]$
$= g(3x-6)$
$= 2(3x-6)+3$
$= 6x-12+3$
$= 6x-9$

12. $(f \circ g)(x) = f[g(x)]$
$= f(3x-7)$
$= (3x-7)^2+4$
$= 9x^2-42x+49+4$
$= 9x^2-42x+53$

$(g \circ f)(x) = g[f(x)]$
$= g(x^2+4)$
$= 3(x^2+4)-7$
$= 3x^2+12-7$
$= 3x^2+5$

13. $(f \circ g)(x) = f[g(x)]$
$= f(2x-1)$
$= \sqrt{2x-1-3}$
$= \sqrt{2x-4}$

$(g \circ f)(x) = g[f(x)]$
$= g(\sqrt{x-3})$
$= 2\sqrt{x-3}-1$

14. $(f \circ g)(x) = f[g(x)]$
$= f\left(\frac{x-4}{x}\right)$
$= \frac{\frac{x-4}{x}+3}{\frac{x-4}{x}}$
$= \frac{x-4+3x}{x-4} = \frac{4x-4}{x-4}$

$(g \circ f)(x) = g[f(x)]$
$= g\left(\frac{x+3}{x}\right)$
$= \frac{\frac{x+3}{x}-4}{\frac{x+3}{x}}$
$= \frac{x+3-4x}{x+3}$
$= \frac{3-3x}{x+3}$

15. No, not a one-to-one function

16. Yes, it is a one-to-one function.

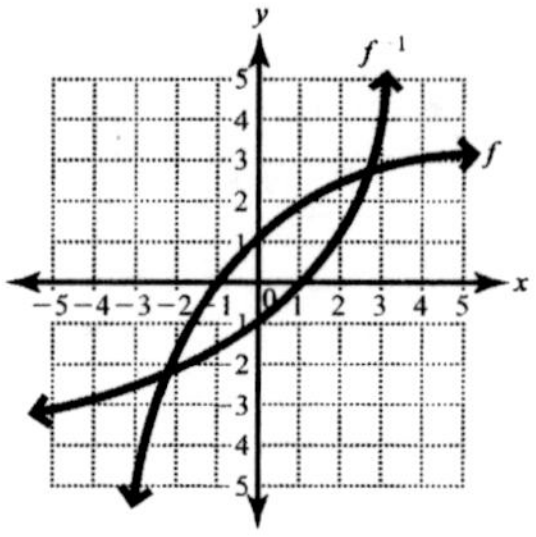

17. $(f \circ g)(x) = f[g(x)]$
$= f\left(\frac{x-2}{3}\right)$
$= 3\left(\frac{x-2}{3}\right)+2$
$= x-2+2$
$= x$

$(g \circ f)(x) = g[f(x)]$
$= g(3x+2)$
$= \frac{3x+2-2}{3}$
$= \frac{3x}{3}$
$= x$

Yes, *f* and *g* are inverse functions.

18. $(f \circ g)(x) = f[g(x)]$
$= f\left(\sqrt[3]{x-6}\right)$
$= \left(\sqrt[3]{x-6}\right)^3 + 6$
$= x - 6 + 6$
$= x$

$(g \circ f)(x) = g[f(x)]$
$= g\left(x^3 + 6\right)$
$= \sqrt[3]{x^3 + 6 - 6}$
$= x$

Yes, *f* and *g* are inverse functions.

19. $(f \circ g)(x) = f[g(x)]$
$= f\left(\sqrt{x+2}\right)$
$= \left(\sqrt{x+2}\right)^2 - 3$
$= x + 2 - 3$
$= x - 1$

No, *f* and *g* are not inverse functions.

20. $(f \circ g)(x) = f[g(x)]$
$= f\left(\frac{-4x}{x+1}\right)$
$= \frac{\frac{-4x}{x+1}}{\frac{-4x}{x+1} + 4}$
$= \frac{(x+1)\left(\frac{-4x}{x+1}\right)}{(x+1)\left(\frac{-4x}{x+1} + 4\right)}$
$= \frac{-4x}{-4x + 4x + 4}$
$= \frac{-4x}{4}$
$= -x$

No, *f* and *g* are not inverse functions.

21. $y = 5x + 4$
$x = 5y + 4$
$x - 4 = 5y$
$\frac{x-4}{5} = y$
$f^{-1}(x) = \frac{x-4}{5}$

22. $y = x^3 + 6$
$x = y^3 + 6$
$x - 6 = y^3$
$\sqrt[3]{x-6} = y$
$f^{-1}(x) = \sqrt[3]{x-6}$

23. $y = \frac{4}{x+5}$
$x = \frac{4}{y+5}$
$xy + 5x = 4$
$xy = 4 - 5x$
$y = \frac{4-5x}{x}$
$f^{-1}(x) = \frac{4-5x}{x}$

24. $y = \sqrt[3]{3x+2}$
$x = \sqrt[3]{3y+2}$
$x^3 = 3y + 2$
$x^3 - 2 = 3y$
$\frac{x^3-2}{3} = y$
$f^{-1}(x) = \frac{x^3-2}{3}$

25. (–6, 4)

26. $f^{-1}(b) = a$

27. $f(x) = 3^x$

x	$y = f(x)$
–1	$\frac{1}{3}$
0	1
1	3
2	9

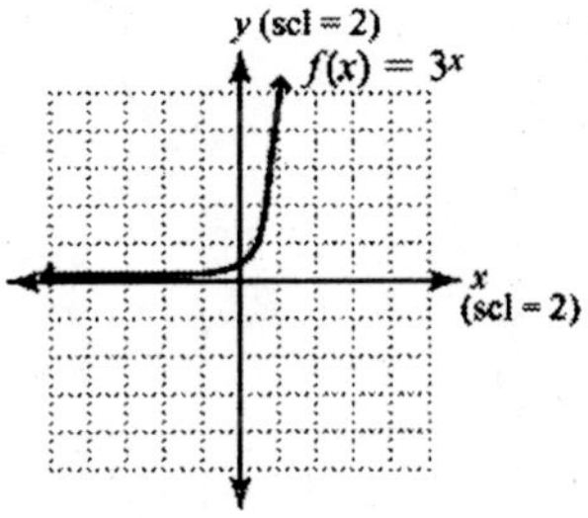

28. $f(x)=\left(\frac{1}{2}\right)^x$

x	$y=f(x)$
-1	$\left(\frac{1}{2}\right)^{(-1)}=2$
0	$\left(\frac{1}{2}\right)^{(0)}=1$
1	$\left(\frac{1}{2}\right)^{(1)}=\frac{1}{2}$
2	$\left(\frac{1}{2}\right)^{(2)}=\frac{1}{4}$

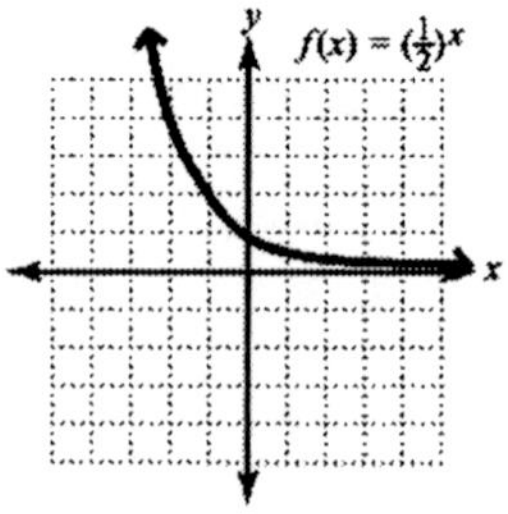

29. $f(x)=2^{x-3}$

x	$y=f(x)$
-1	$2^{-1-3}=\frac{1}{16}$
0	$2^{0-3}=\frac{1}{8}$
1	$2^{1-3}=\frac{1}{4}$
2	$2^{2-3}=\frac{1}{2}$

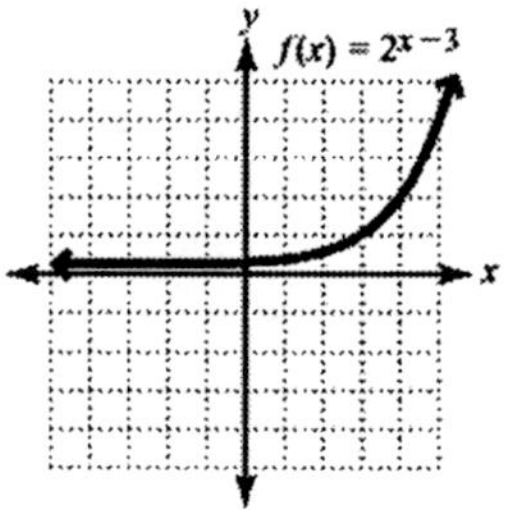

30. $f(x)=3^{-x+2}$

x	$y=f(x)$
-1	$3^{-(-1)+2}=27$
0	$3^{-(0)+2}=9$
1	$3^{-(1)+2}=3$
2	$3^{-(2)+2}=1$

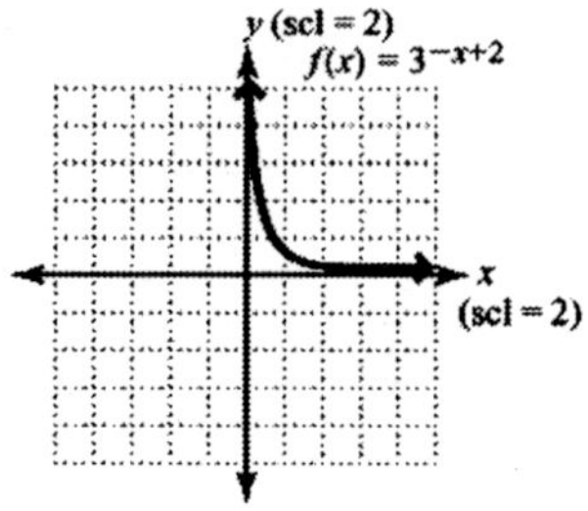

31. $5^x=625$
$5^x=5^4$
$x=4$

32. $16^x=64$
$\left(4^2\right)^x=4^3$
$4^{2x}=4^3$
$2x=3$
$x=\frac{3}{2}$

33. $6^x=\frac{1}{36}$
$6^x=\frac{1}{6^2}$
$6^x=6^{-2}$
$x=-2$

34. $\left(\frac{3}{4}\right)^x=\frac{16}{9}$
$\left(\frac{3}{4}\right)^x=\left(\frac{4}{3}\right)^2$
$\left(\frac{3}{4}\right)^x=\left(\frac{3}{4}\right)^{-2}$
$x=-2$

35. $4^{x-1}=64$
$4^{x-1}=4^3$
$x-1=3$
$x=4$

36. $5^{x+2}=25^x$
$5^{x+2}=\left(5^2\right)^x$
$5^{x+2}=5^{2x}$
$x+2=2x$
$2=x$

37. $\left(\frac{1}{3}\right)^{-x} = 27$

$\left(3^{-1}\right)^{-x} = 3^3$

$3^x = 3^3$

$x = 3$

38. $8^{3x-2} = 16^{4x}$

$\left(2^3\right)^{3x-2} = \left(2^4\right)^{4x}$

$2^{9x-6} = 2^{16x}$

$9x - 6 = 16x$

$-6 = 7x$

$-\frac{6}{7} = x$

39. $A = A_0 \cdot 2^{t/100}$

$A = 500 \cdot 2^{365/100}$

$A \approx 6277$

There are 6277 cells after one year.

40. $A = 25000\left(1 + \frac{0.06}{12}\right)^{12 \cdot 8}$

$A = 40,353.57$

The account is worth \$40,353.57 after eight years.

41. $A = A_0\left(\frac{1}{2}\right)^{t/h}$

$A = 50\left(\frac{1}{2}\right)^{300/107}$

$A \approx 7.16$

After 300 days, 7.16 grams remain.

42. $y = 4.0144(1.028)^x$

$y = 4.0144(1.028)^{25}$

$y \approx 8.01$

In 2015, the minimum wage will be approximately \$8.01.

43. $\log_7 343 = 3$

44. $\log_4 \frac{1}{64} = -3$

45. $\log_{3/2} \frac{81}{16} = 4$

46. $\log_{11} \sqrt[3]{11} = \frac{1}{3}$

47. $9^2 = 81$

48. $\left(\frac{1}{5}\right)^{-3} = 125$

49. $a^4 = 16$

50. $e^b = c$

51. $3^{-4} = x$

$\frac{1}{3^4} = x$

$\frac{1}{81} = x$

52. $\left(\frac{1}{2}\right)^{-2} = x$

$2^2 = x$

$4 = x$

53. $2^x = 32$

$2^x = 2^5$

$x = 5$

54. $\left(\frac{1}{4}\right)^x = 16$

$\left(4^{-1}\right)^x = 4^2$

$4^{-x} = 4^2$

$-x = 2$

$x = -2$

55. $x^4 = 81$

$x^4 = 3^4$

$x = 3$

56. $x^3 = \frac{1}{1000}$

$x = \sqrt[3]{\frac{1}{1000}}$

$x = \frac{1}{10}$

57. $\left(\frac{3}{4}\right)^x = \frac{9}{16}$

$\left(\frac{3}{4}\right)^x = \left(\frac{3}{4}\right)^2$

$x = 2$

58. $81^{1/4} = x$

$\sqrt[4]{81} = x$

$3 = x$

59. $f(x) = \log_4 x$

$y = \log_4 x$

$4^y = x$

y	−1	0	1	2
x	$\frac{1}{4}$	1	4	16

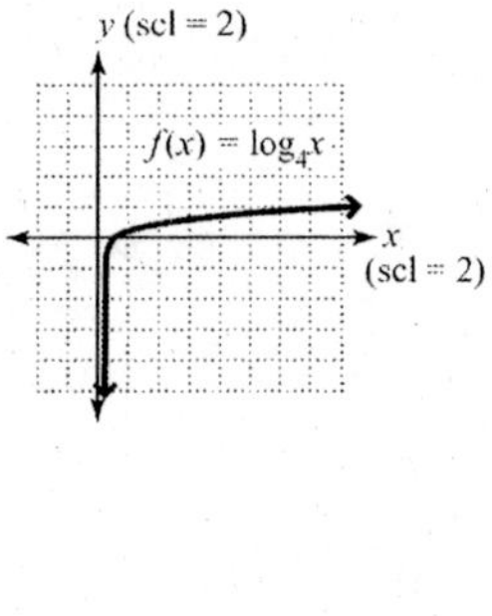

60. $f(x) = \log_{1/3} x$

$y = \log_{1/3} x$

$\left(\frac{1}{3}\right)^y = x$

y	-2	-1	0	1
x	9	3	1	$\frac{1}{3}$

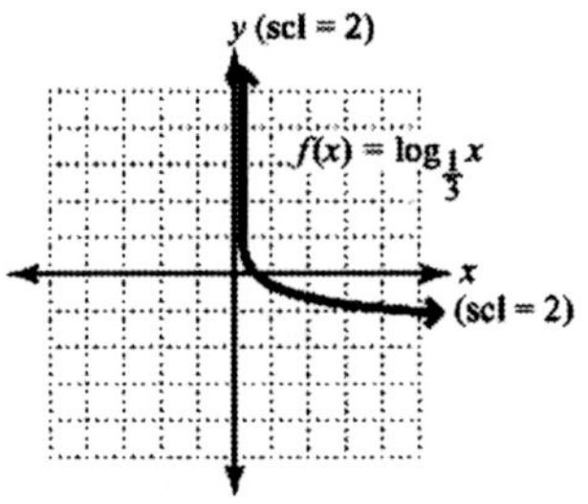

61. Domain is $(-\infty, \infty)$ and range is $(0, \infty)$ because $f(x) = \log_b x$ and $g(x) = b^x$ are inverses.

62. $d = 10\log\frac{I}{I_0}$

$d = 10\log\frac{1000I_0}{I_0}$

$d = 10\log 1000$

$d = 10\log 10^3$

$d = 10 \cdot 3$

$d = 30$

The decibel reading is 30 dB.

63. $3^{\log_3 8} = 8$

64. $\log_4 4^6 = 6$

65. $\log_6 6x = \log_6 6 + \log_6 x = 1 + \log_6 x$

66. $\log_4 x(2x-5) = \log_4 x + \log_4 (2x-5)$

67. $\log_3 4 + \log_3 8 = \log_3 (4 \cdot 8) = \log_3 32$

68. $\log_5 3 + \log_5 x + \log_5 (x-2) = \log_5 3x(x-2)$

$= \log_5 (3x^2 - 6x)$

69. $\log_b \frac{x}{5} = \log_b x - \log_b 5$

70. $\log_a \frac{3x-2}{4x+3} = \log_a (3x-2) - \log_a (4x+3)$

71. $\log_8 32 - \log_8 16 = \log_8 \frac{32}{16} = \log_8 2$

72. $\log_2 (x+5) - \log_2 (2x-3) = \log_2 \frac{x+5}{2x-3}$

73. $\log_3 7^4 = 4\log_3 7$

74. $\log_a \sqrt[3]{x} = \log_a x^{1/3} = \frac{1}{3}\log_a x$

75. $\log_4 \frac{1}{a^4} = \log_4 a^{-4} = -4\log_4 a$

76. $\log_a \sqrt[5]{a^4} = \log_a a^{4/5} = \frac{4}{5}\log_a a = \frac{4}{5}$

77. $4\log_a x = \log_a x^4$

78. $\frac{3}{5}\log_a y = \log_a y^{3/5} = \log_a \sqrt[5]{y^3}$

79. $\log_a x^2 y^3 = \log_a x^2 + \log_a y^3$

$= 2\log_a x + 3\log_a y$

80. $\log_a \frac{c^4}{d^3} = \log_a c^4 - \log_a d^3$

$= 4\log_a c - 3\log_a d$

81. $\log_a \frac{x^2 y^3}{z^4} = \log_a x^2 + \log_a y^3 - \log_a z^4$

$= 2\log_a x + 3\log_a y - 4\log_a z$

82. $\log_a \sqrt{\frac{a^3}{b^4}} = \log_a \left(\frac{a^3}{b^4}\right)^{1/2}$

$= \log_a \left(\frac{a^{3/2}}{b^{4/2}}\right)$

$= \log_a a^{3/2} - \log_a b^2$

$= \frac{3}{2}\log_a a - 2\log_a b$

$= \frac{3}{2} - 2\log_a b$

83. $3\log_x y + 5\log_x z = \log_x y^3 + \log_x z^5$

$= \log_x y^3 z^5$

84. $3\log_a 4 - 2\log_a 3 = \log_a 4^3 - \log_a 3^2$

$= \log_a \frac{4^3}{3^2}$

85. $\frac{1}{4}(2\log_a x + 3\log_a y) = \frac{2}{4}\log_a x + \frac{3}{4}\log_a y$
$= \log_a x^{2/4} + \log_a y^{3/4}$
$= \log_a x^{2/4} y^{3/4}$
$= \log_a (x^2 y^3)^{1/4}$
$= \log_a \sqrt[4]{x^2 y^3}$

86. $4\log_a (x+5) + 2\log_a (x-3)$
$= \log_a (x+5)^4 + \log_a (x-3)^2$
$= \log_a (x+5)^4 (x-3)^2$

87. $\log 326 = 2.5132$

88. $\log 0.0035 = -2.4559$

89. $\ln 0.043 = -3.1466$

90. $\ln 92 = 4.5218$

91. $\log 0.00001 = \log_{10} 10^{-5} = -5$

92. $\ln \sqrt[4]{e} = \ln e^{1/4} = \frac{1}{4}$

93. $d = 10\log \frac{I}{I_0}$
$d = 10\log \frac{10^{-3.5}}{10^{-12}}$
$d = 10\log 10^{8.5}$
$d = 10(8.5)$
$d = 85$
The decibel reading of the thunder was 85 dB.

94. $\text{pH} = -\log[H_3O^+]$
$\text{pH} = -\log 0.001259$
$\text{pH} \approx 2.9$

95. $A = Pe^{rt}$
$10{,}000 = 6000e^{0.03t}$
$\frac{10}{6} = e^{0.03t}$
$\ln\frac{5}{3} = \ln e^{0.03t}$
$\ln\frac{5}{3} = 0.03t$
$\frac{\ln\frac{5}{3}}{0.03} = t$
$17 \approx t$
It will take approximately 17 years.

96. Earthquake 1: $7.8 = \log\frac{I}{I_0}$
$7.8 = \log I - \log I_0$
$7.8 + \log I_0 = \log I$
$10^{7.8+\log I_0} = I$

Earthquake 2: $6.8 = \log\frac{I}{I_0}$
$6.8 = \log I - \log I_0$
$6.8 + \log I_0 = \log I$
$10^{6.8+\log I_0} = I$

$\frac{10^{7.8+\log I_0}}{10^{6.8+\log + I_0}} = 10^{(7.8+\log I_0)-(6.8+\log + I_0)}$
$= 10^{7.8-6.8}$
$= 10^1$
$= 10$
Earthquake 1 was ten times as severe as earthquake 2.

97. $9^x = 32$
$\ln 9^x = \ln 32$
$x\ln 9 = \ln 32$
$x = \frac{\ln 32}{\ln 9} \approx 1.5773$

98. $3^{5x} = 19$
$\ln 3^{5x} = \ln 19$
$5x\ln 3 = \ln 19$
$5x = \frac{\ln 19}{\ln 3}$
$x = \frac{\frac{\ln 19}{\ln 3}}{5} \approx 0.5360$

99. $$6^{2x-1} = 22$$
$$\ln 6^{2x-1} = \ln 22$$
$$(2x-1)\ln 6 = \ln 22$$
$$2x\ln 6 - \ln 6 = \ln 22$$
$$2x\ln 6 = \ln 22 + \ln 6$$
$$x = \frac{\ln 22 + \ln 6}{2\ln 6} \approx 1.3626$$

100. $$4^{2x-3} = 5^{x+1}$$
$$\ln 4^{2x-3} = \ln 5^{x+1}$$
$$(2x-3)\ln 4 = (x+1)\ln 5$$
$$2x\ln 4 - 3\ln 4 = x\ln 5 + \ln 5$$
$$2x\ln 4 - x\ln 5 = \ln 5 + 3\ln 4$$
$$x(2\ln 4 - \ln 5) = \ln 5 + 3\ln 4$$
$$x = \frac{\ln 5 + 3\ln 4}{2\ln 4 - \ln 5} \approx 4.9592$$

101. $$e^{4x} = 11$$
$$\ln e^{4x} = \ln 11$$
$$4x = \ln 11$$
$$x = \frac{\ln 11}{4} \approx 0.5995$$

102. $$e^{-0.003x} = 5$$
$$\ln e^{-0.003x} = \ln 5$$
$$-0.003x = \ln 5$$
$$x = -\frac{\ln 5}{0.003} \approx -536.4793$$

103. $$\log_4(x+8) = 2$$
$$4^2 = x+8$$
$$16 = x+8$$
$$8 = x$$

104. $$\log_2(x^2+2x) = 3$$
$$2^3 = x^2 + 2x$$
$$0 = x^2 + 2x - 8$$
$$0 = (x+4)(x-2)$$
$$x+4 = 0 \quad \text{or} \quad x-2 = 0$$
$$x = -4 \qquad x = 2$$

105. $$\log(3x-8) = \log(x-2)$$
$$3x-8 = x-2$$
$$2x = 6$$
$$x = 3$$

106. $$\log 5 + \log x = 2$$
$$\log 5x = 2$$
$$10^2 = 5x$$
$$100 = 5x$$
$$20 = x$$

107. $$\log_3 x - \log_3 4 = 2$$
$$\log_3 \frac{x}{4} = 2$$
$$3^2 = \frac{x}{4}$$
$$9 = \frac{x}{4}$$
$$36 = x$$

108. $$\log_2 x + \log_2(x-6) = 4$$
$$\log_2 x(x-6) = 4$$
$$\log_2(x^2-6x) = 4$$
$$2^4 = x^2 - 6x$$
$$0 = x^2 - 6x - 16$$
$$0 = (x-8)(x+2)$$
$$x-8 = 0 \quad \text{or} \quad x+2 = 0$$
$$x = 8 \qquad x = \cancel{-2}$$

109. $$\log_4 x + \log_4(x+2) = \log_4 8$$
$$\log_4 x(x+2) = \log_4 8$$
$$x(x+2) = 8$$
$$x^2 + 2x = 8$$
$$x^2 + 2x - 8 = 0$$
$$(x+4)(x-2) = 0$$
$$x+4 = 0 \quad \text{or} \quad x-2 = 0$$
$$x = \cancel{-4} \qquad x = 2$$

110. $$\log_3(5x+2) - \log_3(x-2) = 2$$
$$\log_3 \frac{5x+2}{x-2} = 2$$
$$3^2 = \frac{5x+2}{x-2}$$
$$9(x-2) = 5x+2$$
$$9x - 18 = 5x + 2$$
$$4x = 20$$
$$x = 5$$

111.
$$A = P\left(1+\frac{r}{n}\right)^{nt}$$
$$18000 = 15000\left(1+\frac{0.03}{12}\right)^{12\cdot t}$$
$$1.2 = (1.0025)^{12t}$$
$$\ln 1.2 = \ln(1.0025)^{12t}$$
$$\ln 1.2 = 12t\ln(1.0025)$$
$$\frac{\ln 1.2}{12\ln(1.0025)} = t$$
$$6.1 \approx t$$
It will take approximately 6.1 years.

112. a) $A = 7000e^{0.07(8)}$
$$A = \$12,254.71$$
b) $12,000 = 7000e^{0.07t}$
$$\frac{12}{7} = e^{0.07t}$$
$$\ln\frac{12}{7} = \ln e^{0.07t}$$
$$\ln\frac{12}{7} = 0.07t$$
$$\frac{\ln\frac{12}{7}}{0.07} = t \approx 7.7$$
It will take about 7.7 years.

113. a) $A = 400e^{0.03(12)}$
$$A \approx 573$$
The population of the colony will be about 573 ants after one year.

b) $800 = 400e^{0.03t}$
$$2 = e^{0.03t}$$
$$\ln 2 = \ln e^{0.03t}$$
$$\ln 2 = 0.03t$$
$$\frac{\ln 2}{0.03} = t$$
$$23 \approx t$$
The population will be 800 ants after about 23 months.

114. $G = \dfrac{t}{3.3\log\frac{b}{B}}$
$$G = \frac{240}{3.3\log\frac{1,000,000}{1000}}$$
$$G = \frac{240}{3.3\log 1000}$$
$$G = \frac{240}{3.3\log 10^3}$$
$$G = \frac{240}{3.3(3)} \approx 24$$
The generation time is about 24 minutes.

115. $\log_4 15 = \dfrac{\log 15}{\log 4} \approx 1.9534$

116. $\log_{1/2} 6 = \dfrac{\log 6}{\log 0.5} \approx -2.5850$

Chapter 10 Practice Test

1. Let $f(x) = x^2 - 6$ and $g(x) = 3x - 5$.
$$f\left[g(x)\right] = f(3x-5)$$
$$= (3x-5)^2 - 6$$
$$= 9x^2 - 30x + 25 - 6$$
$$= 9x^2 - 30x + 19$$

2. $f(x) = 4x - 3$

a)
$$y = 4x - 3$$
$$x = 4y - 3$$
$$x + 3 = 4y$$
$$\frac{x+3}{4} = y$$
$$f^{-1}(x) = \frac{x+3}{4}$$

b) $f\left[f^{-1}(x)\right] = f\left(\dfrac{x+3}{4}\right)$
$$= 4\left(\frac{x+3}{4}\right) - 3$$
$$= x + 3 - 3$$
$$= x$$

$$f^{-1}[f(x)] = f^{-1}(4x-3)$$
$$= \frac{4x-3+3}{4}$$
$$= \frac{4x}{4}$$
$$= x$$

c) The graphs are symmetric about the graph of $y = x$.

3.

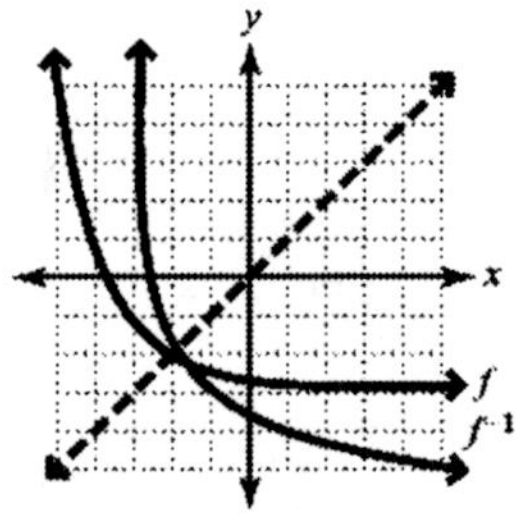

4. $f(x) = 2^{x-1}$

x	$y = f(x)$
-1	$2^{-1-1} = \frac{1}{4}$
0	$2^{0-1} = \frac{1}{2}$
1	$2^{1-1} = 1$
2	$2^{2-1} = 2$

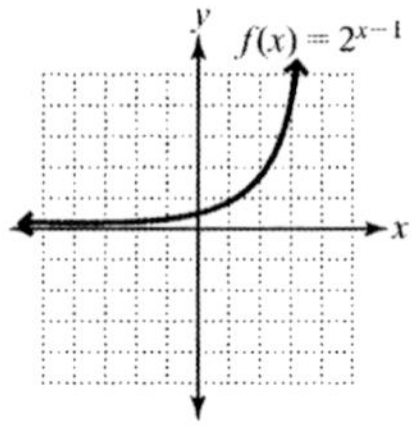

5. $$32^{x-2} = 8^{2x}$$
$$(2^5)^{x-2} = (2^3)^{2x}$$
$$2^{5x-10} = 2^{6x}$$
$$5x - 10 = 6x$$
$$-10 = x$$

6. a) $A = 20,000\left(1+\frac{0.05}{4}\right)^{4(15)}$

$A = \$42,143.63$

b) $$30,000 = 20,000\left(1+\frac{0.05}{4}\right)^{4 \cdot t}$$
$$1.5 = (1.0125)^{4t}$$
$$\ln 1.5 = \ln(1.0125)^{4t}$$
$$\ln 1.5 = 4t\ln(1.0125)$$
$$\frac{\ln 1.5}{4\ln(1.0125)} = t$$
$$8.2 \approx t$$

It will take about 8.2 years.

7. a) $A = 100\left(\frac{1}{2}\right)^{216/30}$

$A = 0.68$

About 0.68 gram remains.

b) $$20 = 100\left(\frac{1}{2}\right)^{t/30}$$
$$0.2 = (0.5)^{t/30}$$
$$\ln 0.2 = \ln(0.5)^{t/30}$$
$$\ln 0.2 = \frac{t}{30}\ln 0.5$$
$$\frac{30\ln 0.2}{\ln 0.5} = t$$
$$69.66 \approx t$$

It will take about 69.66 minutes, or 1.16 hours.

8. a) In 1960: $y = 3.17(1.026)^{60}$

$y \approx 14.8$

There were about 14.8 million people.

In 2012: $y = 3.17(1.026)^{112}$

$y \approx 56.2$

There were about 56.2 million people.

b) $$31.4 = 3.17(1.026)^x$$
$$\frac{31.4}{3.17} = 1.026^x$$
$$\ln\left(\frac{31.4}{3.17}\right) = \ln 1.026^x$$
$$\ln\left(\frac{31.4}{3.17}\right) = x\ln 1.026$$
$$\frac{\ln\left(\frac{31.4}{3.17}\right)}{\ln 1.026} = x$$
$$89 \approx x$$

89 years after 1900, in 1989.

9. $\left(\frac{1}{3}\right)^{-4} = 81$

10. $\log_6 \frac{1}{216} = x$

$$6^x = \frac{1}{216}$$
$$6^x = 216^{-1}$$
$$6^x = \left(6^3\right)^{-1}$$
$$6^x = 6^{-3}$$
$$x = -3$$

11. $\log_x 625 = 4$

$$x^4 = 625$$
$$x^4 = 5^4$$
$$x = 5$$

12. $f(x) = \log_2 x$

$$y = \log_2 x$$
$$2^y = x$$

y	-1	0	1	2
x	$\frac{1}{2}$	1	2	4

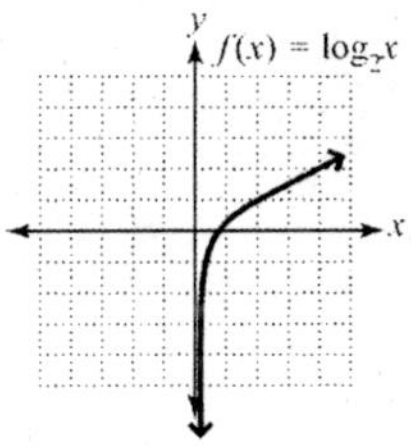

13. $\log_b \frac{x^4 y^2}{z} = \log_b x^4 y^2 - \log_b z$

$$= \log_b x^4 + \log_b y^2 - \log_b z$$
$$= 4\log_b x + 2\log_b y - \log_b z$$

14. $\log_b \sqrt[4]{\frac{x^5}{y^7}} = \log_b \left(\frac{x^5}{y^7}\right)^{1/4}$

$$= \log_b \frac{x^{5/4}}{y^{7/4}}$$
$$= \log_b x^{5/4} - \log_b y^{7/4}$$
$$= \frac{5}{4}\log_b x - \frac{7}{4}\log_b y$$

15. $\frac{3}{4}\left(2\log_b x + 3\log_b y\right) = \frac{6}{4}\log_b x + \frac{9}{4}\log_b y$

$$= \log_b x^{6/4} + \log_b y^{9/4}$$
$$= \log_b x^{6/4} y^{9/4}$$
$$= \log_b \left(x^6 y^9\right)^{1/4}$$
$$= \log_b \sqrt[4]{x^6 y^9}$$

16. a) $A = 300e^{-0.0028(500)}$

$A \approx 74$

About 74 grams remain.

b) $A = 300e^{-0.0028t}$

$$150 = 300e^{-0.0028t}$$
$$0.5 = e^{-0.0028t}$$
$$\ln 0.5 = \ln e^{-0.0028t}$$
$$\ln 0.5 = -0.0028t$$
$$\frac{\ln 0.5}{-0.0028} = t$$
$$247.6 \approx t$$

The half-life is about 247.6 days.

17. $6^{x-3} = 19$

$$\ln 6^{x-3} = \ln 19$$
$$(x-3)\ln 6 = \ln 19$$
$$x\ln 6 - 3\ln 6 = \ln 19$$
$$x\ln 6 = \ln 19 + 3\ln 6$$
$$x = \frac{\ln 19 + 3\ln 6}{\ln 6}$$
$$x \approx 4.6433$$

18. $\log_3 x + \log_3 (x+6) = 3$

$$\log_3 x(x+6) = 3$$
$$\log_3 \left(x^2 + 6x\right) = 3$$
$$3^3 = x^2 + 6x$$
$$0 = x^2 + 6x - 27$$
$$0 = (x+9)(x-3)$$

$x + 9 = 0$ or $x - 3 = 0$

$x = -9$ (crossed out) $\quad x = 3$

19. $\log(4x+2)-\log(3x-2)=\log 2$

$$\log\frac{4x+2}{3x-2}=\log 2$$
$$\frac{4x+2}{3x-2}=2$$
$$4x+2=6x-4$$
$$6=2x$$
$$3=x$$

20. a) $n=3.3\log\frac{b}{B}$

$$n=3.3\log\frac{10,000,000}{100}$$
$$n=3.3\log(100,000)$$
$$n=16.5$$

There were 16.5 generations.

b) $n=3.3\log\frac{b}{B}$

$$9.9=3.3\log\frac{b}{100}$$
$$3=\log\frac{b}{100}$$
$$10^3=\frac{b}{100}$$
$$1000=\frac{b}{100}$$
$$100,000=b$$

There will be 100,000 bacteria.

Chapters 1-10 Cumulative Review

1. True
2. True
3. False; the graph of $f(x)=\frac{x-3}{x+4}$ has a vertical asymptote at $x=-4$.
4. -8
5. x-intercepts
6. $x-3=4$; $x-3=-4$
7. $6-8\div 2\cdot 3^2-4\left(3-4\cdot 2^3\right)$

$$=6-8\div 2\cdot 3^2-4(3-4\cdot 8)$$
$$=6-8\div 2\cdot 3^2-4(3-32)$$
$$=6-8\div 2\cdot 3^2-4(-29)$$
$$=6-8\div 2\cdot 9-4(-29)$$
$$=6-4\cdot 9-(-116)$$
$$=6-36+116$$
$$=86$$

8. $\frac{\left(4a^{-4}\right)^3\left(2a^2\right)^{-3}}{\left(2a^{-2}\right)^3}=\frac{64a^{-12}\cdot 2^{-3}a^{-6}}{2^3a^{-6}}$

$$=\frac{2^6a^{-12}\cdot 2^{-3}a^{-6}}{2^3a^{-6}}$$
$$=\frac{2^3a^{-18}}{2^3a^{-6}}$$
$$=a^{-18-(-6)}$$
$$=a^{-12}$$
$$=\frac{1}{a^{12}}$$

9. $\frac{21p^2q^4-14p^5q^3+6p^3q^7}{7p^3q^5}$

$$=\frac{21p^2q^4}{7p^3q^5}-\frac{14p^5q^3}{7p^3q^5}+\frac{6p^3q^7}{7p^3q^5}$$
$$=3p^{-1}q^{-1}-2p^2q^{-2}+\frac{6}{7}q^2$$
$$=\frac{3}{pq}-\frac{2p^2}{q^2}+\frac{6q^2}{7}$$

10. $\frac{2x+3}{x^2+6x+9}-\frac{x+4}{x+3}=\frac{2x+3}{(x+3)^2}-\frac{(x+4)(x+3)}{(x+3)^2}$

$$=\frac{2x+3}{(x+3)^2}-\frac{x^2+7x+12}{(x+3)^2}$$
$$=\frac{2x+3-x^2-7x-12}{(x+3)^2}$$
$$=\frac{-x^2-5x-9}{(x+3)^2}$$

11. $4\sqrt{6c^3}\cdot 3\sqrt{10c^5} = 12\sqrt{6c^3\cdot 10c^5}$
$= 12\sqrt{60c^8}$
$= 12\sqrt{4c^8\cdot 15}$
$= 12\cdot 2c^4\sqrt{15}$
$= 24c^4\sqrt{15}$

12. $(4+3i)(2-4i) = 8-16i+6i-12i^2$
$= 8-10i-12(-1)$
$= 8-10i+12$
$= 20-10i$

13. $\log_b b^{2x} = 2x$

14. $\dfrac{d^2-d-12}{2d^2} \div \dfrac{3d^2+13d+12}{d}$
$= \dfrac{d^2-d-12}{2d^2}\cdot\dfrac{d}{3d^2+13d+12}$
$= \dfrac{(d-4)\cancel{(d+3)}}{2\cdot d\cdot \cancel{d}}\cdot\dfrac{\cancel{d}}{(3d+4)\cancel{(d+3)}}$
$= \dfrac{d-4}{2d(3d+4)}$

15. $x^4-9x^2-4x^2y^2+36y^2$
$= (x^4-9x^2)-(4x^2y^2-36y^2)$
$= x^2(x^2-9)-4y^2(x^2-9)$
$= (x^2-9)(x^2-4y^2)$
$= (x+3)(x-3)(x+2y)(x-2y)$

16. $24x^4y-30x^3y^2+9x^2y^3$
$= 3x^2y(8x^2-10xy+3y^2)$
$= 3x^2y(4x-3y)(2x-y)$

17. $|2x-3|-5=-4$
$|2x-3|=1$
$2x-3=1$ or $2x-3=-1$
$2x=4$ $\quad 2x=2$
$x=2$ $\quad x=1$

18. $2|6x-10|>-8$
$|6x-10|>-4$
Because an absolute value is always nonnegative, this statement is true for all real numbers.

19. $\begin{cases} 4x-5y=-22 \\ 3x+2y=-5 \end{cases}$

Solve the first equation for x: $x=\frac{5}{4}y-\frac{22}{4}$

Substitute $\frac{5}{4}y-\frac{22}{4}$ for x in the second equation and solve for y.
$3x+2y=-5$
$3\left(\frac{5}{4}y-\frac{22}{4}\right)+2y=-5$
$\frac{15}{4}y-\frac{66}{4}+\frac{8}{4}y=-5$
$\frac{23}{4}y-\frac{66}{4}=-5$
$4\left(\frac{23}{4}y-\frac{66}{4}\right)=4(-5)$
$23y-66=-20$
$23y=46$
$y=2$
$x=\frac{5}{4}y-\frac{22}{4}$
$x=\frac{5}{4}(2)-\frac{22}{4}$
$x=\frac{10}{4}-\frac{22}{4}$
$x=-\frac{12}{4}$
$x=-3$
Solution: $(-3,2)$

20. Let $u=2x-3$.
$u^2-2u=8$
$u^2-2u-8=0$
$(u-4)(u+2)=0$
$u-4=0$ or $u+2=0$
$u=4$ $\quad u=-2$
$2x-3=4$ $\quad 2x-3=-2$
$2x=7$ $\quad 2x=1$
$x=\frac{7}{2}$ $\quad x=\frac{1}{2}$

21. $\dfrac{x}{x-2}-\dfrac{4}{x-1}=\dfrac{2}{x^2-3x+2}$

$$\frac{x}{x-2}-\frac{4}{x-1}=\frac{2}{(x-2)(x-1)}$$

$$(x-2)(x-1)\left(\frac{x}{x-2}-\frac{4}{x-1}\right)$$

$$=(x-2)(x-1)\left(\frac{2}{(x-2)(x-1)}\right)$$

$$x(x-1)-4(x-2)=2$$

$$x^2-x-4x+8=2$$

$$x^2-5x+6=0$$

$$(x-3)(x-2)=0$$

$$x-3=0 \quad \text{or} \quad x-2=0$$

$$x=3 \qquad\qquad x=2$$

Notice that if $x=2$, then $\dfrac{x}{x-2}$ and $\dfrac{2}{x^2-3x+2}$ are undefined. Therefore, $x=2$ is extraneous. The only solution is $x=3$.

22. $\sqrt{x+4}-\sqrt{x-4}=4$

$$\sqrt{x+4}=\sqrt{x-4}+4$$

$$\left(\sqrt{x+4}\right)^2=\left(\sqrt{x-4}+4\right)^2$$

$$x+4=x-4+8\sqrt{x-4}+16$$

$$x+4=x+12+8\sqrt{x-4}$$

$$-8=8\sqrt{x-4}$$

$$-1=\sqrt{x-4}$$

$$(-1)^2=\left(\sqrt{x-4}\right)^2$$

$$1=x-4$$

$$5=x$$

Check:

$$\sqrt{5+4}-\sqrt{5-4}\stackrel{?}{=}4$$

$$\sqrt{9}-\sqrt{1}\stackrel{?}{=}4$$

$$3-1\stackrel{?}{=}4$$

$$2\neq 4$$

Because 5 is an extraneous solution, this equation has no solution.

23. $4^{x+2}=8$

$$\left(2^2\right)^{x+2}=2^3$$

$$2^{2x+4}=2^3$$

$$2x+4=3$$

$$2x=-1$$

$$x=-\frac{1}{2}$$

24. $\dfrac{1}{2}\log_2 x=2$

$$\log_2 x^{1/2}=2$$

$$2^2=x^{1/2}$$

$$4=\sqrt{x}$$

$$4^2=\left(\sqrt{x}\right)^2$$

$$16=x$$

25. $\log(3x-5)+\log x=\log 12$

$$\log\left[(3x-5)x\right]=\log 12$$

$$\log\left(3x^2-5x\right)=\log 12$$

$$3x^2-5x=12$$

$$3x^2-5x-12=0$$

$$(3x+4)(x-3)=0$$

$$3x+4=0 \quad \text{or} \quad x-3=0$$

$$x=\cancel{-\frac{4}{3}} \qquad\qquad x=3$$

26. $2x-5y=10$

$$-5y=-2x+10$$

$$y=\frac{2}{5}x-2$$

a) $m=\dfrac{2}{5}$

b) y-intercept: $(0,-2)$

x-intercept:

$$2x-5(0)=10$$

$$2x=10$$

$$x=5$$

$$(5,0)$$

c)

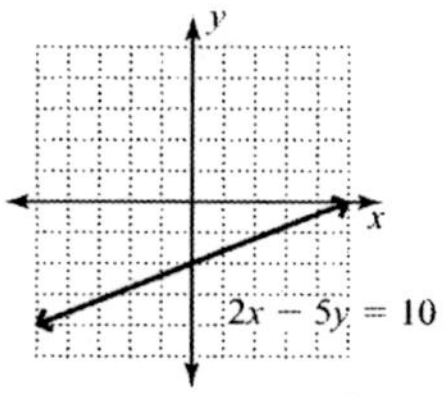

27.
$$a^2 + b^2 = c^2$$
$$n^2 + (n-1)^2 = (n+8)^2$$
$$n^2 + n^2 - 2n + 1 = n^2 + 16n + 64$$
$$n^2 - 18n - 63 = 0$$
$$(n-21)(n+3) = 0$$
$$n - 21 = 0 \quad \text{or} \quad n + 3 = 0$$
$$n = 21 \qquad n = -3$$
Lengths cannot be negative, so disregard the negative solution. One leg has a length of 21. The other leg has length $21 - 1 = 20$, and the hypotenuse has length $21 + 8 = 29$.

28. $A = P\left(1 + \frac{r}{n}\right)^{nt}$
$$A = 15{,}000\left(1 + \frac{0.04}{12}\right)^{12\cdot 3}$$
$$A = 15{,}000\left(1.00\overline{3}\right)^{36}$$
$$A \approx 16{,}909.08$$
After three years, there will be \$16,909.08 in the account.

29. Complete a table.

Categories	Rate of Work	Time at Work	Amt of Task Completed
Ethan	$\frac{1}{2}$	t	$\frac{t}{2}$
Pam	$\frac{1}{1.5} = \frac{1}{\frac{3}{2}} = \frac{2}{3}$	t	$\frac{2t}{3}$

$$\frac{t}{2} + \frac{2t}{3} = 1$$
$$6\cdot\left(\frac{t}{2} + \frac{2t}{3}\right) = 6\cdot(1)$$
$$3t + 4t = 6$$
$$7t = 6$$
$$t = \frac{6}{7} \approx 0.857$$
Working together, it takes Ethan and Pam 0.857 hour, or about 51 minutes, to complete the job.

30. Create a table.

	rate	time	distance
car	$x+10$	$\frac{300}{x+10}$	300
bus	x	$\frac{300}{x}$	300

Translate the information in the table to an equation and solve.
$$\frac{300}{x+10} + 1 = \frac{300}{x}$$
$$x(x+10)\left(\frac{300}{x+10} + 1\right) = x(x+10)\left(\frac{300}{x}\right)$$
$$300x + x(x+10) = 300(x+10)$$
$$300x + x^2 + 10x = 300x + 3000$$
$$x^2 + 10x - 3000 = 0$$
$$(x+60)(x-50) = 0$$
$$x + 60 = 0 \quad \text{or} \quad x - 50 = 0$$
$$x = -60 \qquad x = 50$$
Disregard the negative rate. If the bus travels at a rate of 50 miles per hour, then it takes $\frac{300}{50+10} = \frac{300}{60} = 5$ hours for the car to travel 300 miles.

Chapter 11

Conic Sections

Exercise Set 11.1

1. It is a parabola because only one variable is squared.

3. The parabola opens to the right because the y-variable is squared and $a = 3$, which is positive.

5. circle; $(-3, 2)$; 4

7. $y = (x-1)^2 + 2$

   opens upward

   vertex is $(1, 2)$

   axis of symmetry is $x = 1$

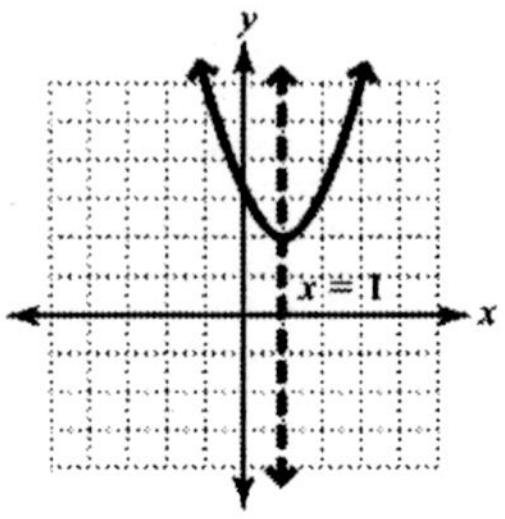

9.
$$\begin{aligned} y &= -x^2 - 2x + 3 \\ y - 3 &= -1(x^2 + 2x) \\ y - 3 - 1 &= -1(x^2 + 2x + 1) \\ y - 4 &= -1(x+1)^2 \\ y &= -1(x+1)^2 + 4 \end{aligned}$$

   opens downward

   vertex is $(-1, 4)$

   axis of symmetry is $x = -1$

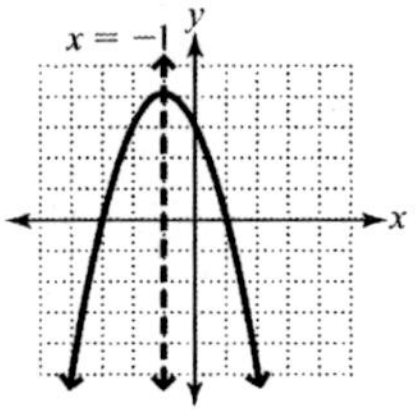

11. $x = (y+2)^2 - 2$

   opens right

   vertex is $(-2, -2)$

   axis of symmetry is $y = -2$

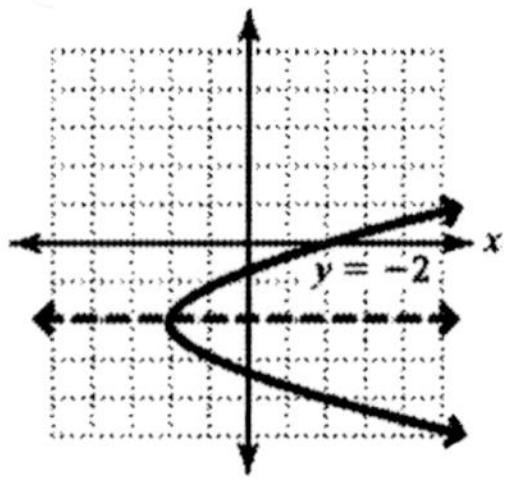

13. $x = -(y-1)^2 + 3$

   opens left

   vertex is $(3, 1)$

   axis of symmetry is $y = 1$

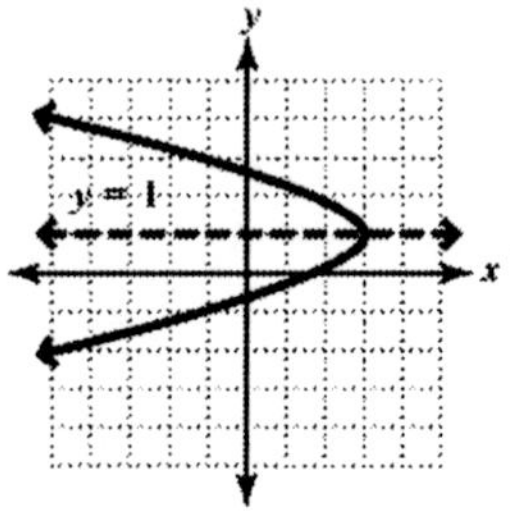

15. $x = 2(y+2)^2 - 4$

   opens right

   vertex is $(-4, -2)$

   axis of symmetry is $y = -2$

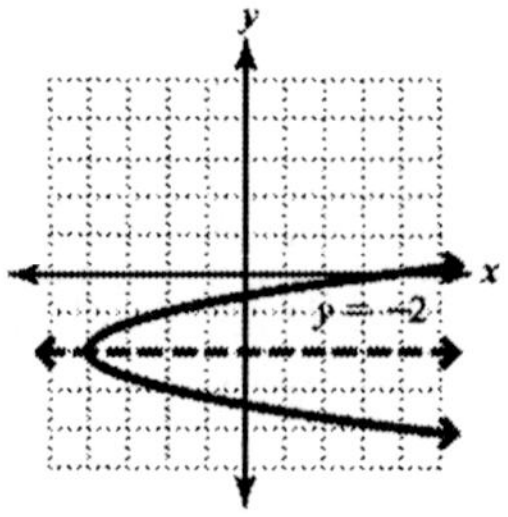

17. $x = -3(y+2)^2 - 5$
opens left
vertex is $(-5, -2)$
axis of symmetry is $y = -2$

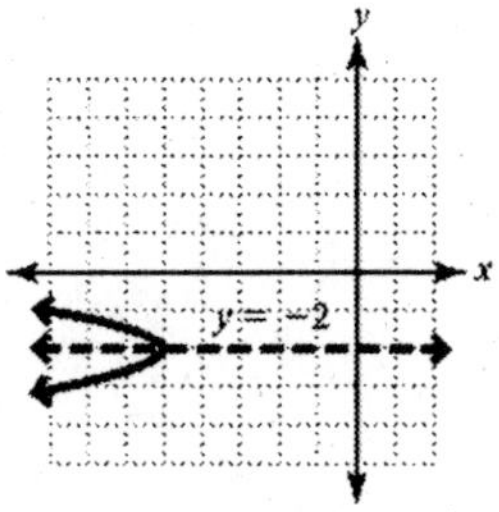

19.
$$x = y^2 + 4y + 3$$
$$x - 3 = y^2 + 4y$$
$$x - 3 + 4 = y^2 + 4y + 4$$
$$x + 1 = (y+2)^2$$
$$x = (y+2)^2 - 1$$
opens right
vertex is $(-1, -2)$
axis of symmetry is $y = -2$

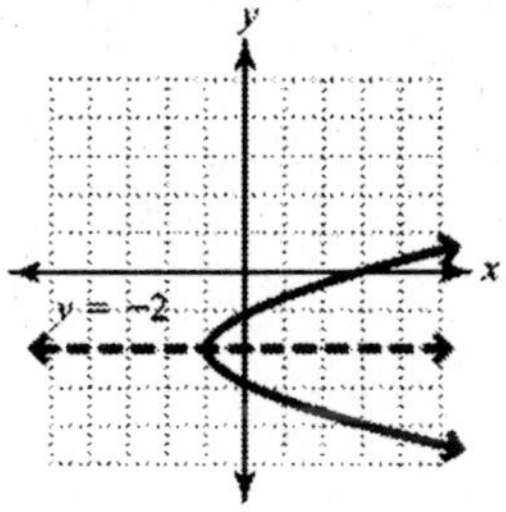

21.
$$x = -y^2 + 6y - 5$$
$$x + 5 = -(y^2 - 6y)$$
$$x + 5 - 9 = -(y^2 - 6y + 9)$$
$$x - 4 = -(y-3)^2$$
$$x = -(y-3)^2 + 4$$
opens left
vertex is $(4, 3)$
axis of symmetry is $y = 3$

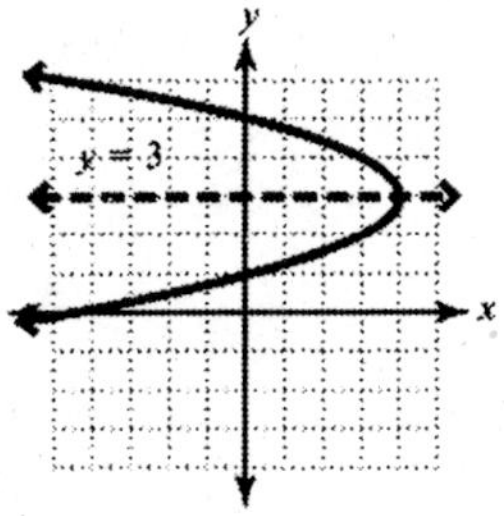

23.
$$x = 2y^2 + 8y + 3$$
$$x - 3 = 2(y^2 + 4y)$$
$$x - 3 + 8 = 2(y^2 + 4y + 4)$$
$$x + 5 = 2(y+2)^2$$
$$x = 2(y+2)^2 - 5$$
opens right
vertex is $(-5, -2)$
axis of symmetry is $y = -2$

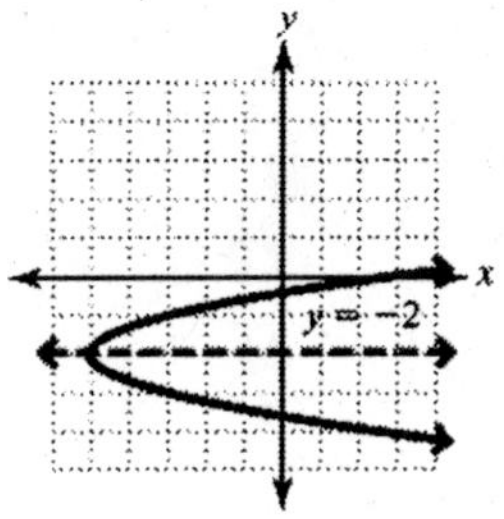

25.
$$x = 3y^2 - 6y + 1$$
$$x - 1 = 3(y^2 - 2y)$$
$$x - 1 + 3 = 3(y^2 - 2y + 1)$$
$$x = (y-1)^2 - 2$$
opens right
vertex is $(-2, 1)$
axis of symmetry $y = 1$

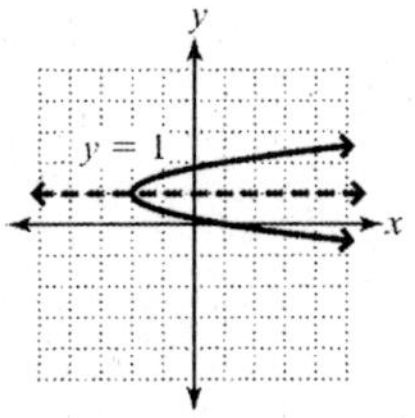

27. $$x = -2y^2 + 4y + 5$$
$$x - 5 = -2(y^2 - 2y)$$
$$x - 5 - 2 = -2(y^2 - 2y + 1)$$
$$x - 7 = -2(y-1)^2$$
$$x = -2(y-1)^2 + 7$$
opens left
vertex is $(7,1)$
axis of symmetry is $y = 1$

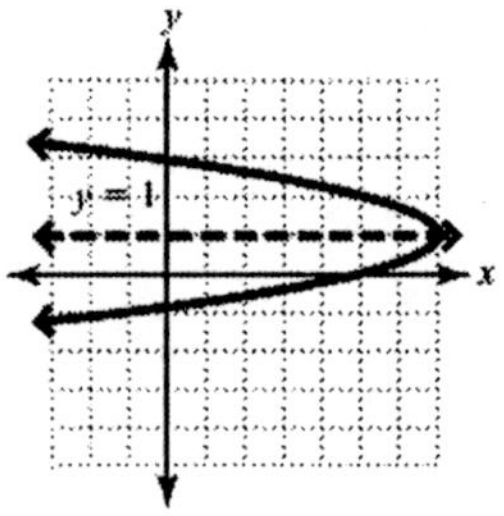

29. c

31. a

33. $(-4,2)$ and $(-1,6)$
$$d = \sqrt{(-4-(-1))^2 + (2-6)^2}$$
$$= \sqrt{(-3)^2 + (-4)^2}$$
$$= \sqrt{9+16}$$
$$= \sqrt{25}$$
$$= 5$$
$$\text{Midpoint} = \left(\frac{-4+(-1)}{2}, \frac{2+6}{2}\right)$$
$$= \left(\frac{-5}{2}, \frac{8}{2}\right)$$
$$= \left(-\frac{5}{2}, 4\right)$$

35. $(-8,-4)$ and $(-3,8)$
$$d = \sqrt{(-8-(-3))^2 + (-4-8)^2}$$
$$= \sqrt{(-5)^2 + (-12)^2}$$
$$= \sqrt{25+144}$$
$$= \sqrt{169}$$
$$= 13$$
$$\text{Midpoint} = \left(\frac{-8+(-3)}{2}, \frac{-4+8}{2}\right)$$
$$= \left(\frac{-11}{2}, \frac{4}{2}\right)$$
$$= \left(-\frac{11}{2}, 2\right)$$

37. $(-8,-10)$ and $(4,-5)$
$$d = \sqrt{(-8-4)^2 + (-10-(-5))^2}$$
$$= \sqrt{(-12)^2 + (-5)^2}$$
$$= \sqrt{144+25}$$
$$= \sqrt{169}$$
$$= 13$$
$$\text{Midpoint} = \left(\frac{-8+4}{2}, \frac{-10+(-5)}{2}\right)$$
$$= \left(\frac{-4}{2}, \frac{-15}{2}\right)$$
$$= \left(-2, -\frac{15}{2}\right)$$

39. $(2,4)$ and $(4,8)$
$$d = \sqrt{(2-4)^2 + (4-8)^2}$$
$$= \sqrt{(-2)^2 + (-4)^2}$$
$$= \sqrt{4+16}$$
$$= \sqrt{20}$$
$$= 2\sqrt{5}$$
$$\text{Midpoint} = \left(\frac{2+4}{2}, \frac{4+8}{2}\right)$$
$$= \left(\frac{6}{2}, \frac{12}{2}\right)$$
$$= (3,6)$$

41. $(-5,2)$ and $(3,-2)$

$$d=\sqrt{(-5-3)^2+(2-(-2))^2}$$
$$=\sqrt{(-8)^2+(4)^2}$$
$$=\sqrt{64+16}$$
$$=\sqrt{80}$$
$$=4\sqrt{5}$$

$$\text{Midpoint}=\left(\frac{-5+3}{2},\frac{2+(-2)}{2}\right)$$
$$=\left(\frac{-2}{2},\frac{0}{2}\right)$$
$$=(-1,0)$$

43. $(6,-2)$ and $(1,-5)$

$$d=\sqrt{(6-1)^2+(-2-(-5))^2}$$
$$=\sqrt{(5)^2+(3)^2}$$
$$=\sqrt{25+9}$$
$$=\sqrt{34}$$

$$\text{Midpoint}=\left(\frac{6+1}{2},\frac{-2+(-5)}{2}\right)$$
$$=\left(\frac{7}{2},\frac{-7}{2}\right)$$
$$=\left(\frac{7}{2},-\frac{7}{2}\right)$$

45. Find the distance from the center to a point on the circle. This is the radius.
$(4,2)$ and $(8,-1)$

$$r=\sqrt{(8-4)^2+(-1-2)^2}$$
$$=\sqrt{(4)^2+(-3)^2}$$
$$=\sqrt{16+9}$$
$$=\sqrt{25}$$
$$=5$$

47. Find the distance from the center to a point on the circle. This is the radius.
$(2,-6)$ and $(10,-1)$

$$r=\sqrt{(10-2)^2+(-1-(-6))^2}$$
$$=\sqrt{(8)^2+(5)^2}$$
$$=\sqrt{64+25}$$
$$=\sqrt{89}$$

49. $(x-2)^2+(y-1)^2=4$
Center $(2,1)$; radius 2

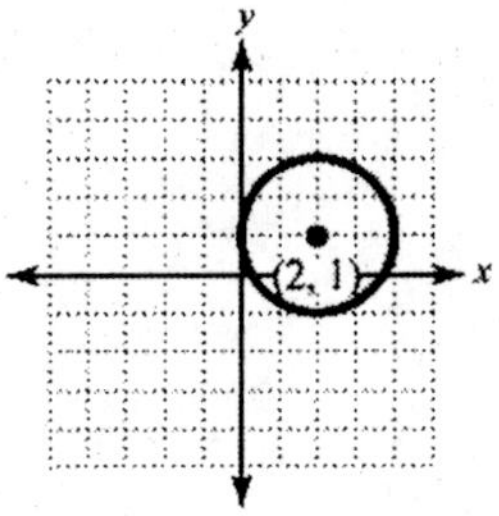

51. $(x+3)^2+(y+2)^2=81$
Center $(-3,-2)$; radius 9

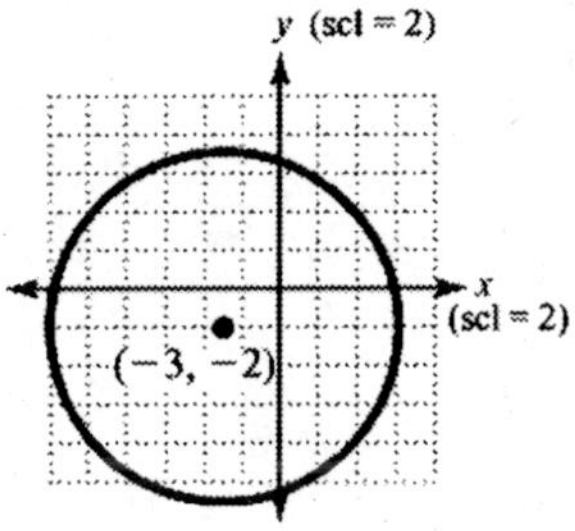

53. $(x-5)^2+(y+3)^2=49$
Center $(5,-3)$; radius 7

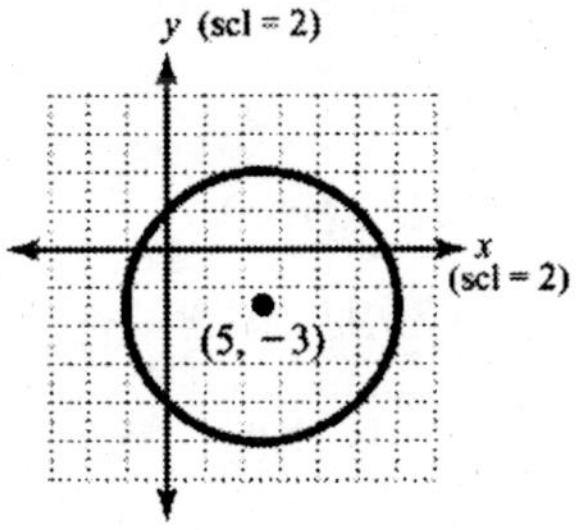

55. $(x-1)^2+(y+1)^2=12$

Center $(1,-1)$; radius $\sqrt{12}=2\sqrt{3}$

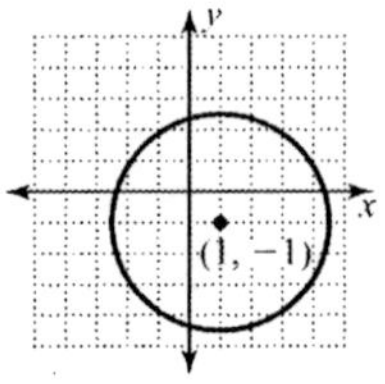

57. $(x+4)^2+(y+2)^2=32$

Center $(-4,-2)$; radius $\sqrt{32}=4\sqrt{2}$

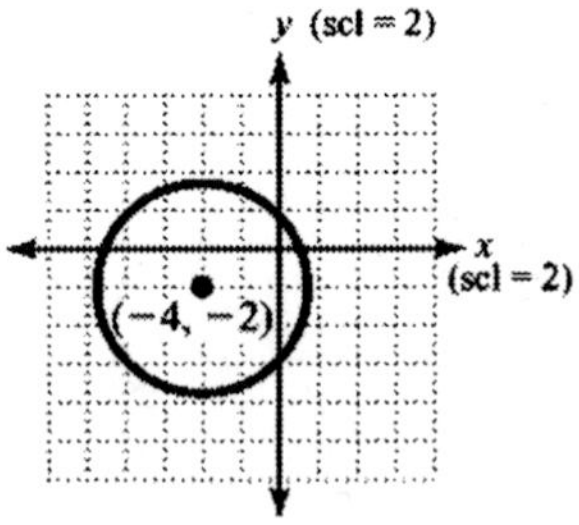

59.
$$x^2+y^2+8x-6y+16=0$$
$$\left(x^2+8x\right)+\left(y^2-6y\right)=-16$$
$$\left(x^2+8x+16\right)+\left(y^2-6y+9\right)=-16+16+9$$
$$(x+4)^2+(y-3)^2=9$$

Center $(-4,3)$; radius 3

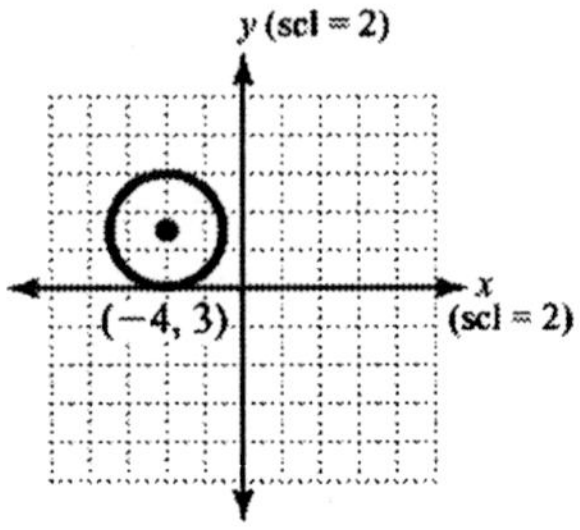

61.
$$x^2+y^2+10x-4y-35=0$$
$$\left(x^2+10x\right)+\left(y^2-4y\right)=35$$
$$\left(x^2+10x+25\right)+\left(y^2-4y+4\right)=35+25+4$$
$$(x+5)^2+(y-2)^2=64$$

Center $(-5,2)$; radius 8

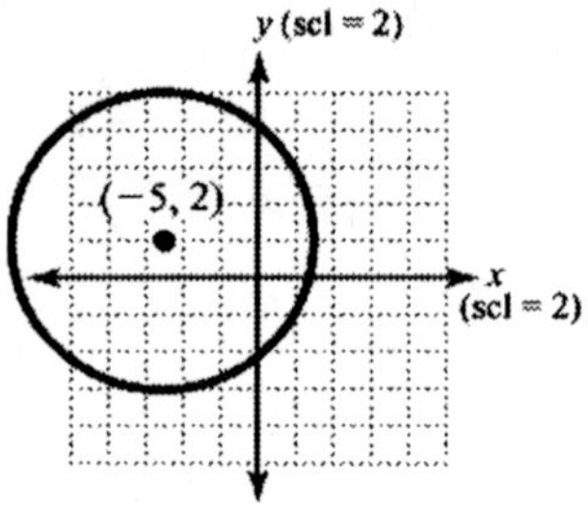

63.
$$x^2+y^2+14x-4y+49=0$$
$$\left(x^2+14x\right)+\left(y^2-4y\right)=-49$$
$$\left(x^2+14x+49\right)+\left(y^2-4y+4\right)=-49+49+4$$
$$(x+7)^2+(y-2)^2=4$$

Center $(-7,2)$; radius 2

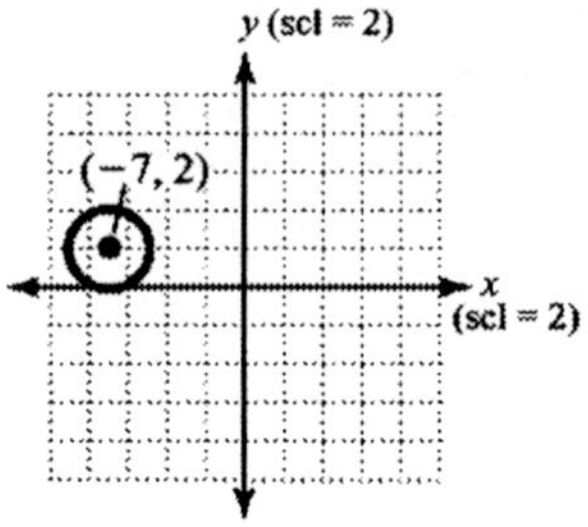

65. b

67. c

69. $x^2+y^2=49$

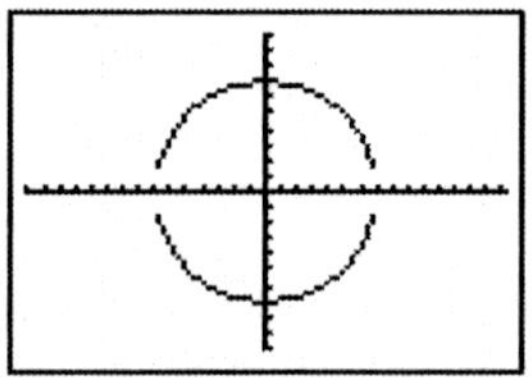

71. $(x-2)^2+(y+3)^2=25$

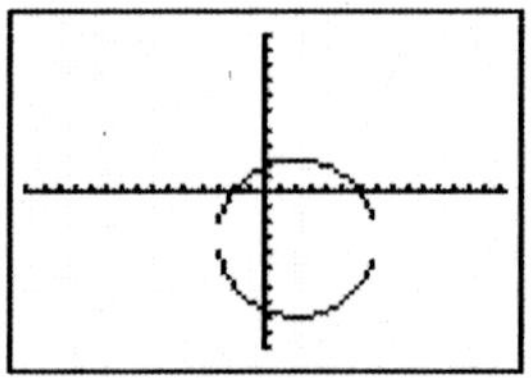

73. Center: $(4,2)=(h,k)$; radius: 4
$$(x-h)^2+(y-k)^2=r^2$$
$$(x-4)^2+(y-2)^2=4^2$$
$$(x-4)^2+(y-2)^2=16$$

75. Center: $(-4,-3)=(h,k)$; radius: 5

$$(x-h)^2+(y-k)^2=r^2$$
$$(x-(-4))^2+(y-(-3))^2=5^2$$
$$(x+4)^2+(y+3)^2=25$$

77. Center: $(6,-2)=(h,k)$; radius: $\sqrt{14}$

$$(x-h)^2+(y-k)^2=r^2$$
$$(x-6)^2+(y-(-2))^2=\left(\sqrt{14}\right)^2$$
$$(x-6)^2+(y+2)^2=14$$

79. Center: $(-5,2)=(h,k)$; radius: $3\sqrt{5}$

$$(x-h)^2+(y-k)^2=r^2$$
$$(x-(-5))^2+(y-2)^2=\left(3\sqrt{5}\right)^2$$
$$(x+5)^2+(y-2)^2=45$$

81. Find the distance from the center to a point on the circle. This is the radius.
$(2,4)$ and $(5,8)$

$$\begin{aligned} r &= \sqrt{(2-5)^2+(4-8)^2} \\ &= \sqrt{(-3)^2+(-4)^2} \\ &= \sqrt{9+16} \\ &= \sqrt{25} \\ &= 5 \end{aligned}$$

The equation in standard form is:
$(x-2)^2+(y-4)^2=25$

83. Find the distance from the center to a point on the circle. This is the radius.
$(2,4)$ and $(7,16)$

$$\begin{aligned} r &= \sqrt{(2-7)^2+(4-16)^2} \\ &= \sqrt{(-5)^2+(-12)^2} \\ &= \sqrt{25+144} \\ &= \sqrt{169} \\ &= 13 \end{aligned}$$

The equation in standard form is:
$(x-2)^2+(y-4)^2=169$

85. The center is $(2,-5)$ since all points are an equal distance from that point. Also, since all points are 8 units from the center, the radius is 8.

The equation in standard form is:
$(x-2)^2+(y+5)^2=64$

87. $(4,-2)$ and $(-2,6)$

$$\begin{aligned} \text{diameter} &= \sqrt{(4-(-2))^2+(-2-6)^2} \\ &= \sqrt{(4+2)^2+(-8)^2} \\ &= \sqrt{36+64} \\ &= \sqrt{100} \\ &= 10 \end{aligned}$$

The radius is $10 \div 2 = 5$.

$$\begin{aligned} \text{Midpoint} &= \left(\frac{4+(-2)}{2}, \frac{-2+6}{2}\right) \\ &= \left(\frac{2}{2}, \frac{4}{2}\right) \\ &= (1,2) \end{aligned}$$

The center of the circle is $(1,2)$.
The equation in standard form is:
$(x-1)^2+(y-2)^2=25$

89. $(6,-2)$ and $(-2,8)$

$$\begin{aligned} \text{diameter} &= \sqrt{(6-(-2))^2+(-2-8)^2} \\ &= \sqrt{(6+2)^2+(-10)^2} \\ &= \sqrt{64+100} \\ &= \sqrt{164} \\ &= 2\sqrt{41} \end{aligned}$$

The radius is $2\sqrt{41} \div 2 = \sqrt{41}$.

$$\begin{aligned} \text{Midpoint} &= \left(\frac{6+(-2)}{2}, \frac{-2+8}{2}\right) \\ &= \left(\frac{4}{2}, \frac{6}{2}\right) \\ &= (2,3) \end{aligned}$$

The center of the circle is $(2,3)$.
The equation in standard form is:
$(x-2)^2+(y-3)^2=41$

91. a)
$$h = -16t^2 + 96t + 112$$
$$h - 112 = -16(t^2 - 6t)$$
$$h - 112 - 144 = -16(t^2 - 6t + 9)$$
$$h - 256 = -16(t-3)^2$$
$$h = -16(t-3)^2 + 256$$
The vertex for the parabola is (3, 256). The maximum height for the rock will be 256 ft.

b) It will take 3 seconds to reach the maximum height.

c)
$$0 = -16t^2 + 96t + 112$$
$$0 = -16(t^2 - 6t - 7)$$
$$0 = t^2 - 6t - 7$$
$$0 = (t-7)(t+1)$$
$$t - 7 = 0 \qquad t + 1 = 0$$
$$t = 7 \qquad t = \cancel{-1}$$
The rock will hit the ground after 7 seconds.

93. We must find a to use the standard form $y = a(x-h)^2 + k$. The parabola has vertex (0, 18). When $y = 0$, $x = \frac{30}{2} = 15$. Then
$$y = a(x-h)^2 + k$$
$$0 = a(15-0)^2 + 18$$
$$-18 = a \cdot 225$$
$$\frac{-18}{225} = a$$
$$a = -\frac{2}{25} = -0.08$$
Substitute into the general equation.
$$y = a(x-h)^2 + k$$
$$y = -0.08(x-0)^2 + 18$$
$$y = -0.08x^2 + 18$$
or
$$y = -\frac{2}{25}x^2 + 18$$

95.
$$y = 0.0038x^2 - 0.3475x + 8.316$$
$$y = 0.0038(17)^2 - 0.3475(17) + 8.316$$
$$y \approx 3.5\%$$

97.
$$x^2 + y^2 = 20^2$$
$$x^2 + y^2 = 400$$

99.
$$x^2 + (y-110)^2 = 100^2$$
$$x^2 + (y-110)^2 = 10{,}000$$

101.
$$x^2 + y^2 = (6370 + 230)^2$$
$$x^2 + y^2 = 6600^2$$
$$x^2 + y^2 = 43{,}560{,}000$$

Review Exercises

1.

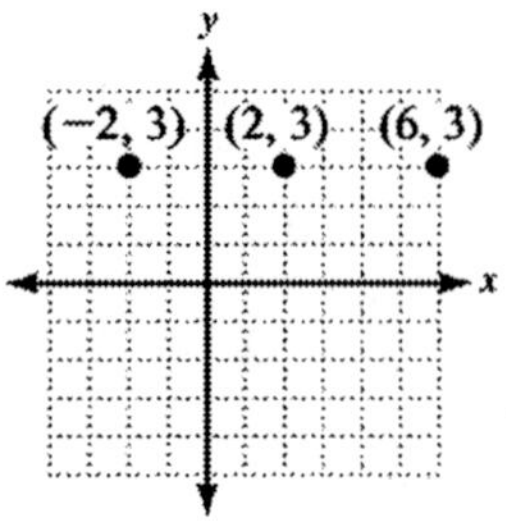

2. $25x^2 - 9y^2 = (5x+3y)(5x-3y)$

3.

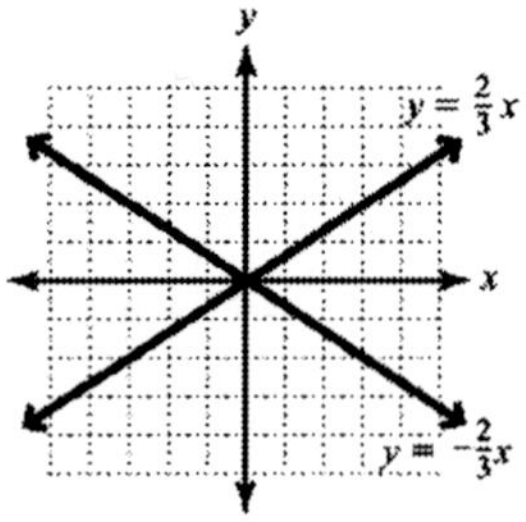

4.
$$\frac{x^2}{16} = 1$$
$$x^2 = 16$$
$$x = \pm\sqrt{16}$$
$$x = \pm 4$$

5. $9x^2 + 16 \cdot 0 = 144$

$$9x^2 = 144$$
$$x^2 = 16$$
$$x = \pm 4$$
$$(-4, 0), (4, 0)$$

$$9 \cdot 0 + 16y^2 = 144$$
$$16y^2 = 144$$
$$y^2 = \frac{144}{16}$$
$$y^2 = 9$$
$$y = \pm 3$$
$$(0, -3), (0, 3)$$

6. $25x^2 + 9 \cdot 0 = 225$

$$25x^2 = 225$$
$$x^2 = 9$$
$$x = \pm 3$$
$$(-3, 0), (3, 0)$$

$$25 \cdot 0 + 9y^2 = 225$$
$$9y^2 = 225$$
$$y^2 = \frac{225}{9}$$
$$y^2 = 25$$
$$y = \pm 5$$
$$(0, -5), (0, 5)$$

Exercise Set 11.2

1. An ellipse is the set of all points the sum of whose distances from two fixed points is constant.

3. The equation of an ellipse is a sum; the equation of a hyperbola is a difference.

5. The parabola has only one squared term.

7. $\frac{x^2}{81} + \frac{y^2}{64} = 1$

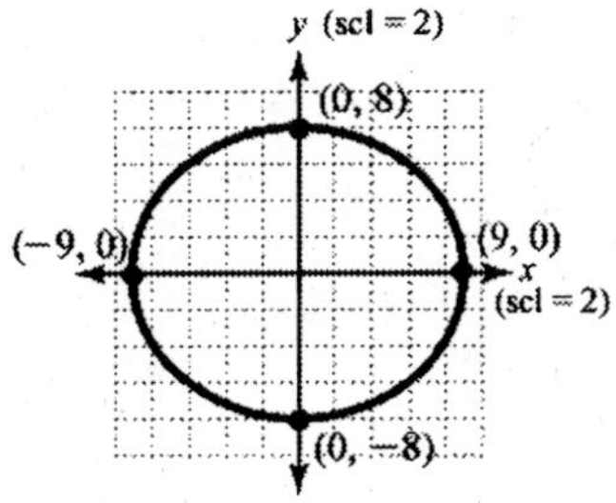

9. $\frac{x^2}{36} + y^2 = 1$

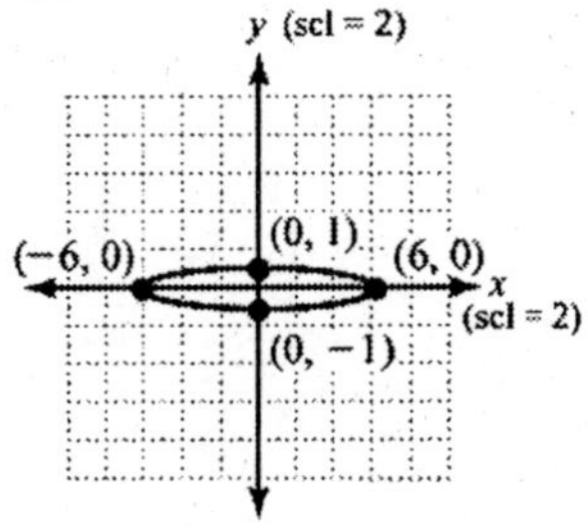

11. $4x^2 + 9y^2 = 36$

$$\frac{4x^2}{36} + \frac{9y^2}{36} = \frac{36}{36}$$
$$\frac{x^2}{9} + \frac{y^2}{4} = 1$$

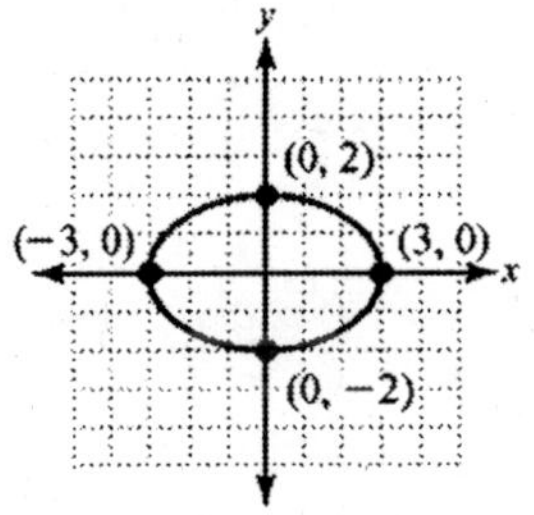

13. $36x^2 + 4y^2 = 144$

$$\frac{36x^2}{144} + \frac{4y^2}{144} = \frac{144}{144}$$
$$\frac{x^2}{4} + \frac{y^2}{36} = 1$$

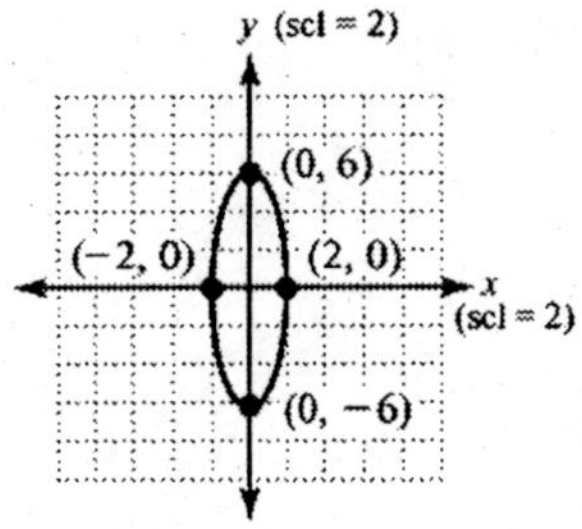

15. $\dfrac{(x-1)^2}{49}+\dfrac{(y+3)^2}{25}=1$

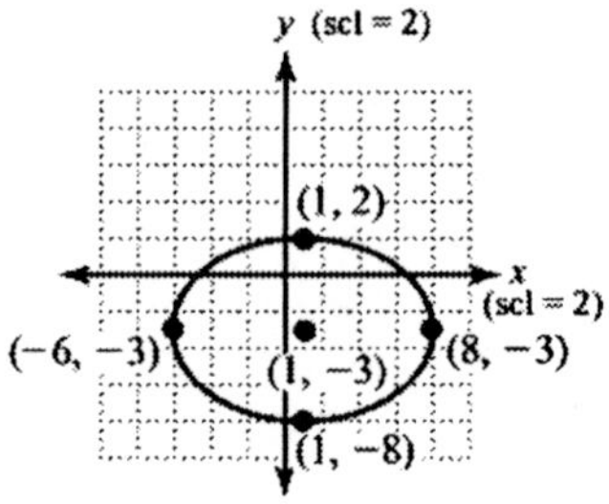

17. $\dfrac{(x-4)^2}{4}+\dfrac{(y+3)^2}{36}=1$

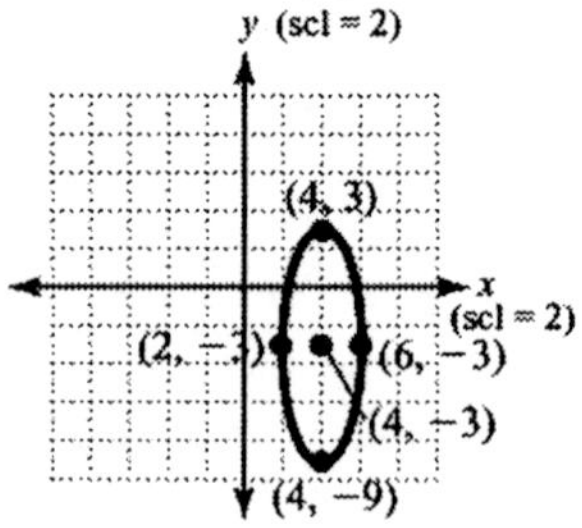

19. $\dfrac{x^2}{25}+\dfrac{y^2}{4}=1$

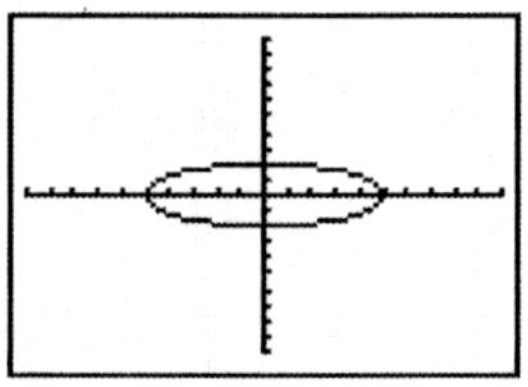

21. $\dfrac{(y-2)^2}{36}+\dfrac{(x+4)^2}{25}=1$

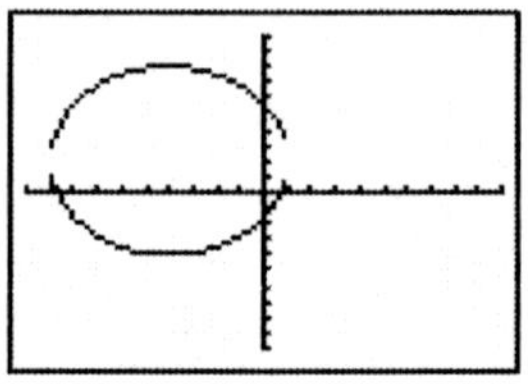

23. $\dfrac{x^2}{9}+\dfrac{y^2}{16}=1$

25. $\dfrac{x^2}{9}-\dfrac{y^2}{4}=1$

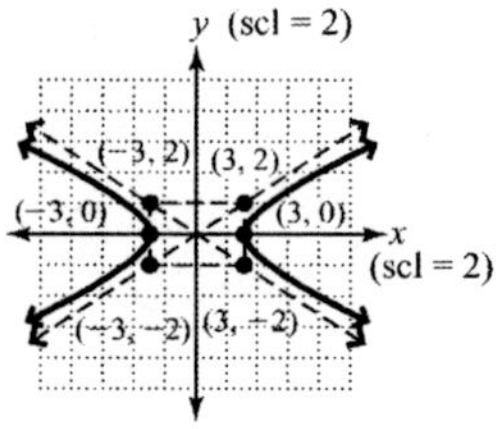

27. $\dfrac{y^2}{36}-\dfrac{x^2}{9}=1$

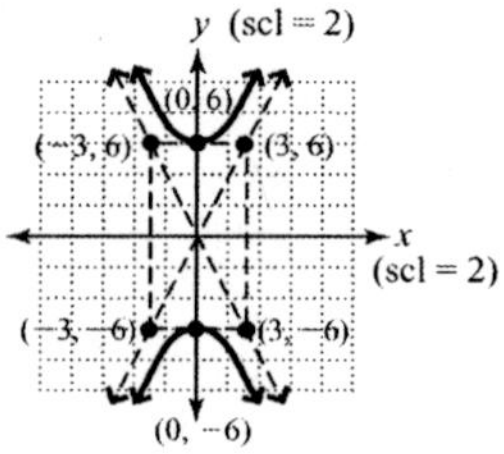

29. $9x^2-y^2=36$

$$\frac{9x^2}{36}-\frac{y^2}{36}=\frac{36}{36}$$

$$\frac{x^2}{4}-\frac{y^2}{36}=1$$

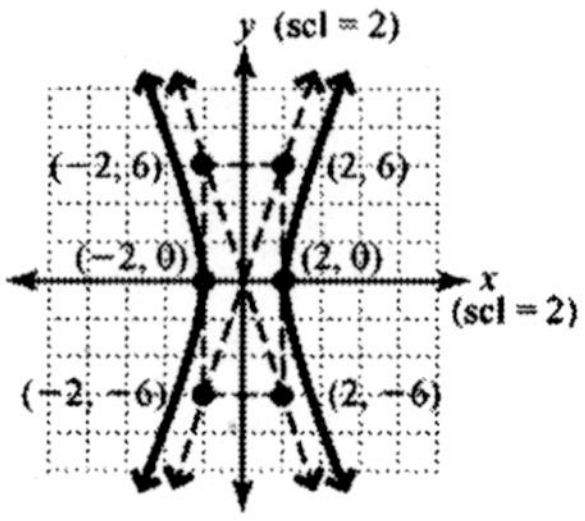

31. $9y^2-25x^2=225$

$$\frac{9y^2}{225}-\frac{25x^2}{225}=\frac{225}{225}$$

$$\frac{y^2}{25}-\frac{x^2}{9}=1$$

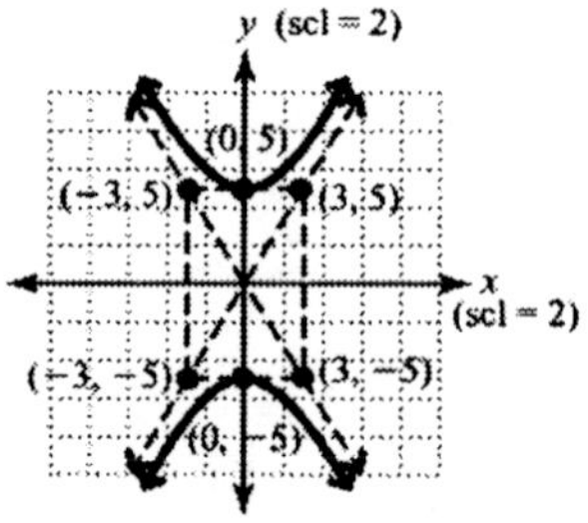

33. $\dfrac{x^2}{36}-\dfrac{y^2}{4}=1$

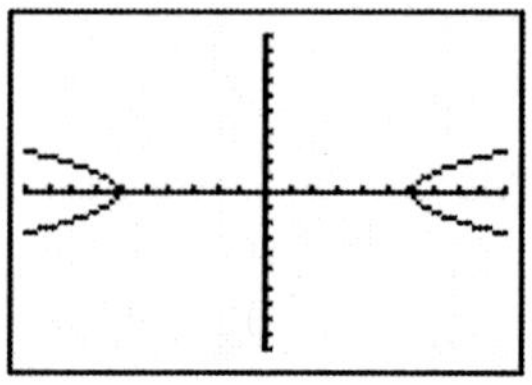

35. $\dfrac{x^2}{9}-\dfrac{y^2}{4}=1$

37. c

39. b

41. $9x^2 + 16y^2 = 144$

$$\frac{9x^2}{144} + \frac{16y^2}{144} = \frac{144}{144}$$

$$\frac{x^2}{16} + \frac{y^2}{9} = 1$$

ellipse

43. $x^2 + y^2 - 6x + 8y - 75 = 0$

$$x^2 - 6x + y^2 + 8y = 75$$

$$x^2 - 6x + 9 + y^2 + 8y + 16 = 75 + 9 + 16$$

$$(x-3)^2 + (y+4)^2 = 100$$

circle

45. $(x-2)^2 + (y+2)^2 = 49$

circle

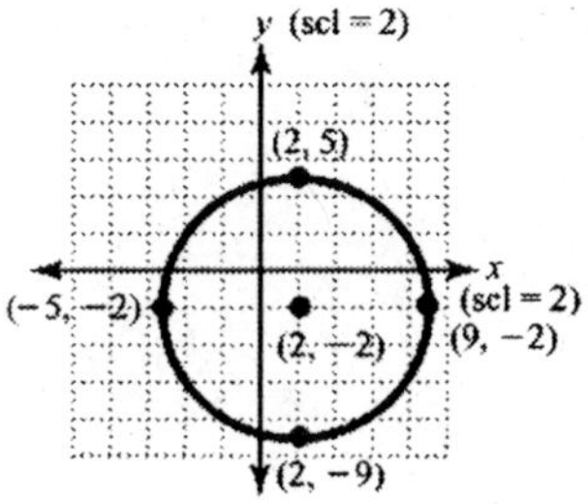

47. $y = 2(x+1)^2 + 3$

parabola

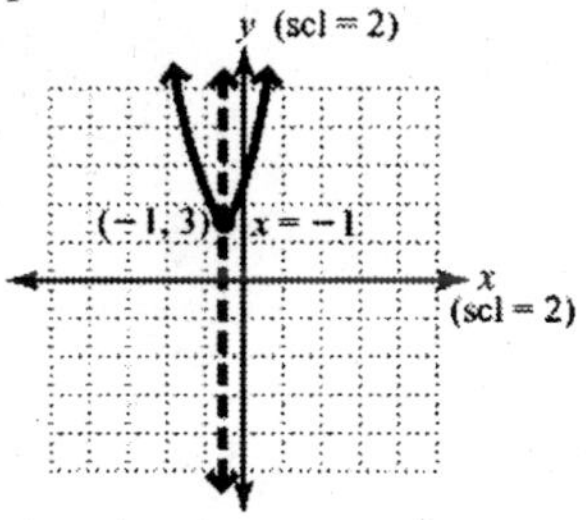

49. $\frac{x^2}{36} + \frac{y^2}{81} = 1$

ellipse

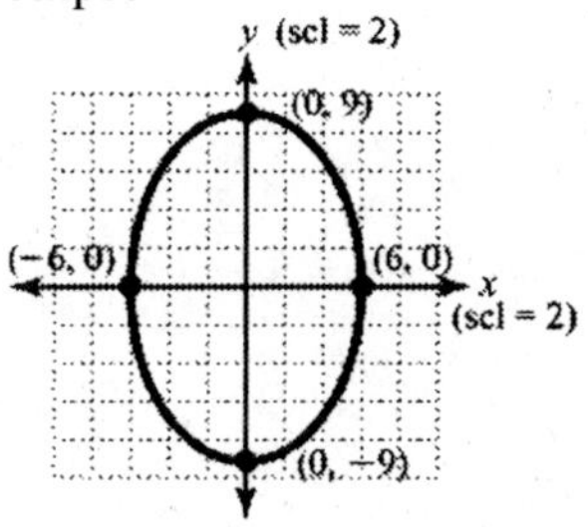

51. $\frac{x^2}{4} - \frac{y^2}{25} = 1$

hyperbola

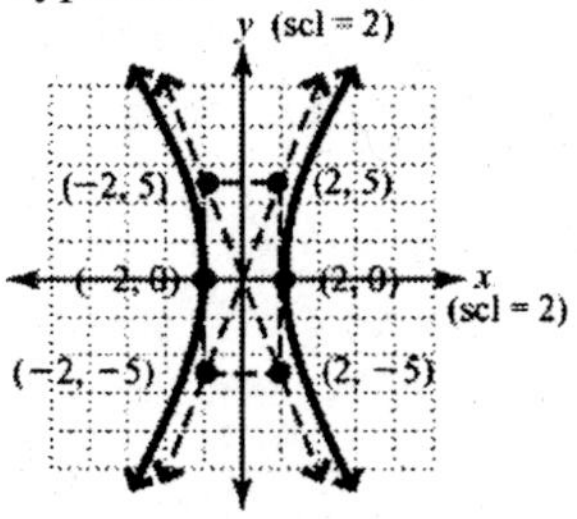

53. $y = 2x^2 + 8x + 6$

$$y - 6 = 2(x^2 + 4x)$$

$$y - 6 + 8 = 2(x^2 + 4x + 4)$$

$$y = 2(x+2)^2 - 2$$

parabola

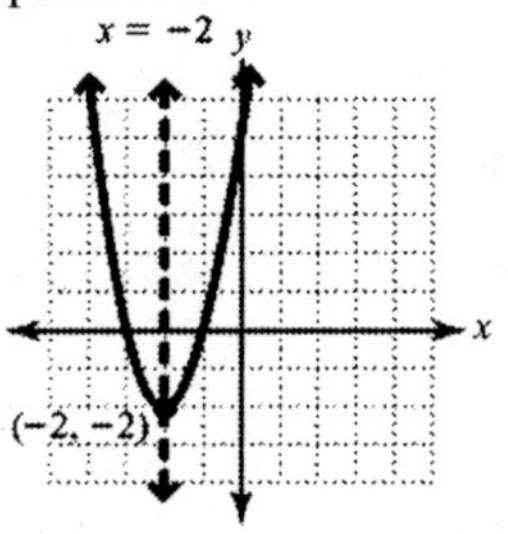

55. $\frac{x^2}{16} - \frac{y^2}{16} = 1$

hyperbola

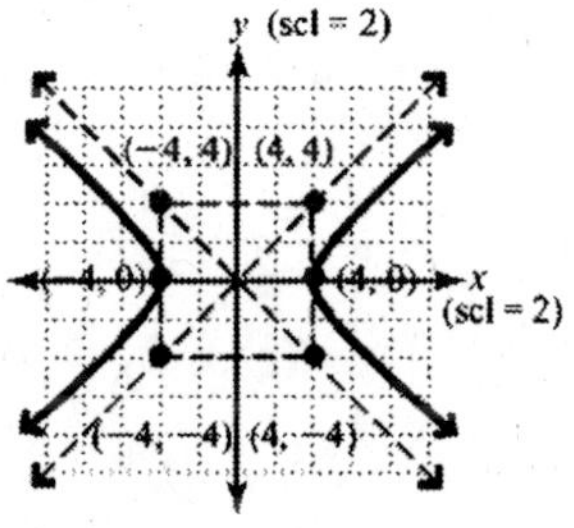

57. $\frac{(x+3)^2}{9} + \frac{(y+2)^2}{36} = 1$

ellipse

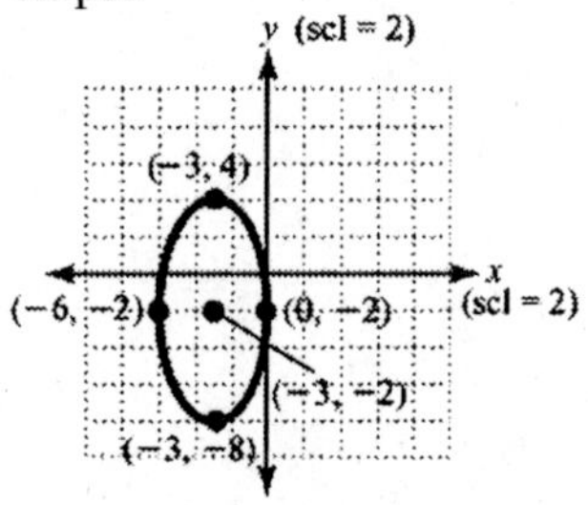

59. a) $$400x^2 + 256y^2 = 102,400$$
$$\frac{400}{102,400}x^2 + \frac{256}{102,400}y^2 = \frac{102,400}{102,400}$$
$$\frac{x^2}{256} + \frac{y^2}{400} = 1$$

Yes, the sailboat will clear the bridge. The height of the bridge at the center is $\sqrt{400} = 20$ ft., and the boat's mast is only 18 ft. above the water.

b) $\sqrt{256} = 16$

The bridge is $2(16) = 32$ ft. at the base of the arch.

61. $\frac{x^2}{3.6^2} + \frac{y^2}{2.88^2} = 1$

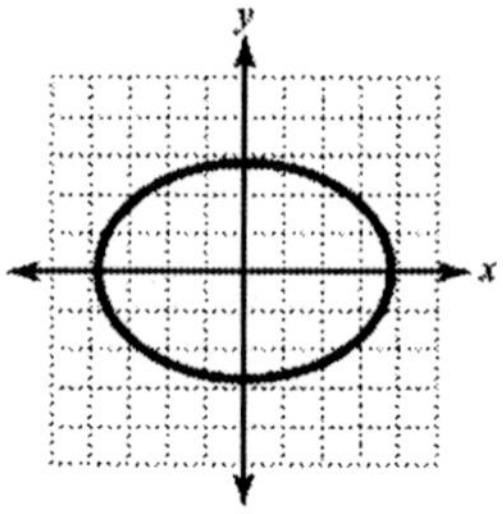

63. $\frac{x^2}{1.5^2} + \frac{y^2}{2^2} = 1$

$\frac{x^2}{2.25} + \frac{y^2}{4} = 1$

65. Rewrite as: $\frac{x^2}{5^2} + \frac{y^2}{4^2} = 1$

Now, use substitution: $c^2 = 5^2 - 4^2$

$c^2 = 25 - 16$

$c^2 = 9$

$c = \sqrt{9}$

$c = 3$ units

Review Exercises

1. $3x^2 + y = 6$

$y = -3x^2 + 6$

2. $\begin{cases} x + y = 3 \\ 2x + y = 4 \end{cases}$

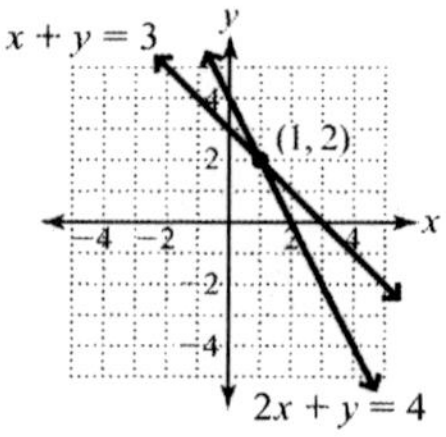

Solution: $(1, 2)$

3. $\begin{cases} 2x + y = 1 \\ 3x + 4y = -6 \end{cases}$

Solve the first equation for y: $y = 1 - 2x$

$3x + 4(1 - 2x) = -6$

$3x + 4 - 8x = -6$

$-5x = -10$

$x = 2$

$y = 1 - 2 \cdot 2$

$y = -3$

Solution: $(2, -3)$

4. $4(2x + 3y = 6)$

$3(3x - 4y = -25)$

$8x + 12y = 24$

$9x - 12y = -75$

$17x = -51$

$x = -3$

$3(-3) - 4y = -25$

$-9 - 4y = -25$

$-4y = -16$

$y = 4$

Solution: $(-3, 4)$

5. $3x^2 + 10x - 8 = 0$

$(3x - 2)(x + 4) = 0$

$3x - 2 = 0$ or $x + 4 = 0$

$x = \frac{2}{3}$ $\quad x = -4$

6. $4x^2 = 36$

$x^2 = 9$

$x = \pm 3$

Exercise Set 11.3

1. a) 0, 1, 2, 3, or 4 solutions are possible.

   b) Answers may vary.

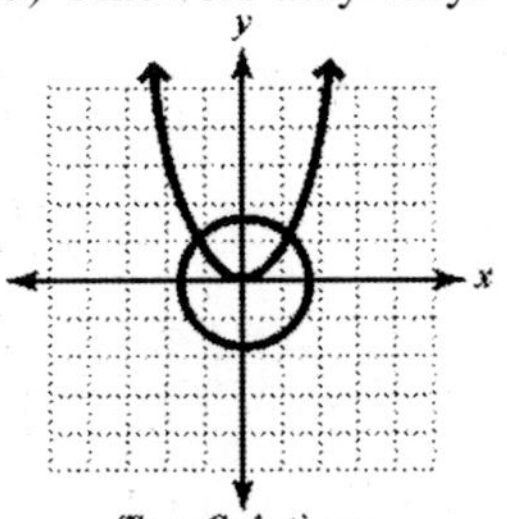

Two Solutions

3. Substitution because the first equation has a linear term and is easy to solve for y.

5. The graph of the first equation is a line, and the second is an ellipse; so the system could have 0, 1, or 2 solutions.

7. $\begin{cases} x^2+2y=1 \\ 2x+y=2 \end{cases}$

Solve the second equation for y: $y=-2x+2$

$$x^2+2y=1$$
$$x^2+2(-2x+2)=1$$
$$x^2-4x+4=1$$
$$x^2-4x+3=0$$
$$(x-3)(x-1)=0$$
$$x-3=0 \quad \text{or} \quad x-1=0$$
$$x=3 \qquad x=1$$

$x=3 \quad y=-2(3)+2=-6+2=-4 \quad (3,-4)$

$x=1 \quad y=-2(1)+2=-2+2=0 \quad (1,0)$

9. $\begin{cases} y=x^2+4x+4 \\ 3x-y=-6 \end{cases}$

The first equation is solved for y.

$$3x-(x^2+4x+4)=-6$$
$$3x-x^2-4x-4=-6$$
$$0=x^2+x-2$$
$$0=(x+2)(x-1)$$
$$x+2=0 \quad \text{or} \quad x-1=0$$
$$x=-2 \qquad x=1$$

$x=-2 \quad y=(-2)^2+4(-2)+4$
$$y=4-8+4$$
$$y=0$$
$$(-2,0)$$

$x=1 \quad y=(1)^2+4(1)+4$
$$y=1+4+4$$
$$y=9$$
$$(1,9)$$

11. $\begin{cases} x^2+y^2=25 \\ x-y=-1 \end{cases}$

Solve the second equation for x. $x=y-1$

$$(y-1)^2+y^2=25$$
$$y^2-2y+1+y^2=25$$
$$2y^2-2y-24=0$$
$$y^2-y-12=0$$
$$(y-4)(y+3)=0$$
$$y-4=0 \quad \text{or} \quad y+3=0$$
$$y=4 \qquad y=-3$$

$y=4 \quad x=4-1$
$$x=3$$
$$(3,4)$$

$y=-3 \quad x=-3-1$
$$x=-4$$
$$(-4,-3)$$

13. $\begin{cases} x^2+y^2=10 \\ 3x+y=6 \end{cases}$

Solve the second equation for y. $y=6-3x$

$$x^2+y^2=10$$
$$x^2+(6-3x)^2=10$$
$$x^2+36-36x+9x^2=10$$
$$10x^2-36x+26=0$$
$$2(5x^2-18x+13)=0$$
$$2(5x-13)(x-1)=0$$
$$5x-13=0 \quad \text{or} \quad x-1=0$$
$$x=\frac{13}{5}=2.6 \qquad x=1$$

$x = 2.6 \quad y = 6 - 3(2.6)$
$y = 6 - 7.8$
$y = -1.8$
$(2.6, -1.8)$

$x = 1 \quad y = 6 - 3(1)$
$y = 6 - 3$
$y = 3$
$(1, 3)$

15. $\begin{cases} x^2 + 2y^2 = 4 \\ x + y = 5 \end{cases}$

Solve the second equation for x. $x = 5 - y$

$(5 - y)^2 + 2y^2 = 4$
$25 - 10y + y^2 + 2y^2 - 4 = 0$
$3y^2 - 10y + 21 = 0$

$$y = \frac{10 \pm \sqrt{10^2 - 4(3)(21)}}{2(3)}$$

$$y = \frac{10 \pm \sqrt{-152}}{6}$$

Because the discriminant of the quadratic equation is negative, we know that the equation cannot be solved using real numbers. Therefore there is no real solution.

17. $\begin{cases} y = 2x^2 + 3 \\ 2x - y = -3 \end{cases}$

The first equation is solved for y.

$2x - (2x^2 + 3) = -3$
$2x - 2x^2 - 3 = -3$
$0 = 2x^2 - 2x$
$0 = 2x(x - 1)$
$2x = 0 \quad \text{or} \quad x - 1 = 0$
$x = 0 \qquad x = 1$

$x = 0 \quad y = 2 \cdot 0 + 3$
$y = 3$
$(0, 3)$

$x = 1 \quad y = 2 \cdot 1 + 3$
$y = 2 + 3$
$y = 5$
$(1, 5)$

19. $\begin{cases} y = (x - 3)^2 + 2 \\ y = -(x - 2)^2 + 3 \end{cases}$

$(x - 3)^2 + 2 = -(x - 2)^2 + 3$
$x^2 - 6x + 9 + 2 = -x^2 + 4x - 4 + 3$
$2x^2 - 10x + 12 = 0$
$x^2 - 5x + 6 = 0$
$(x - 2)(x - 3) = 0$
$x - 2 = 0 \quad \text{or} \quad x - 3 = 0$
$x = 2 \qquad x = 3$

$x = 2: \quad y = -(2 - 2)^2 + 3 = 0 + 3 = 3 \quad (2, 3)$

$x = 3: \quad y = (3 - 3)^2 + 2 = 0 + 2 = 2 \quad (3, 2)$

21. $\begin{cases} x^2 + y^2 = 20 \\ x^2 - y^2 = 12 \end{cases}$

$x^2 + y^2 = 20$
$\underline{x^2 - y^2 = 12}$
$2x^2 \quad = 32$
$x^2 \quad = 16$
$x \quad = \pm 4$

$x = -4: \quad (-4)^2 + y^2 = 20$
$16 + y^2 = 20$
$y^2 = 4$
$y = \pm 2$
$(-4, -2), (-4, 2)$

$x = 4: \quad (4)^2 + y^2 = 20$
$16 + y^2 = 20$
$y^2 = 4$
$y = \pm 2$
$(4, -2), (4, 2)$

23. $\begin{cases} 4x^2+9y^2=72 \\ x^2+y^2=13 \end{cases}$

$$\begin{aligned} 4x^2+9y^2 &= 72 \\ -4\left(x^2+y^2 = 13\right) & \end{aligned}$$

$$\begin{aligned} 4x^2+9y^2 &= 72 \\ -4x^2-4y^2 &= -52 \\ 5y^2 &= 20 \\ y^2 &= 4 \\ y &= \pm 2 \end{aligned}$$

$y=-2$ $\quad x^2+(-2)^2=13$

$$\begin{aligned} x^2+4 &= 13 \\ x^2 &= 9 \\ x &= \pm 3 \end{aligned}$$

$(-3,-2),(3,-2)$

$y=2$ $\quad x^2+(2)^2=13$

$$\begin{aligned} x^2+4 &= 13 \\ x^2 &= 9 \\ x &= \pm 3 \end{aligned}$$

$(-3,2),(3,2)$

25. $\begin{cases} 4x^2-y^2=15 \\ x^2+y^2=5 \end{cases}$

$$\begin{aligned} 4x^2-y^2 &= 15 \\ x^2+y^2 &= 5 \\ 5x^2 &= 20 \\ x^2 &= 4 \\ x &= \pm 2 \end{aligned}$$

$x=-2$ $\quad (-2)^2+y^2=5$

$$\begin{aligned} 4+y^2 &= 5 \\ y^2 &= 1 \\ y &= \pm 1 \end{aligned}$$

$(-2,-1),(-2,1)$

$x=2$ $\quad (2)^2+y^2=5$

$$\begin{aligned} 4+y^2 &= 5 \\ y^2 &= 1 \\ y &= \pm 1 \end{aligned}$$

$(2,-1),(2,1)$

27. $\begin{cases} x^2+3y^2=36 \\ x=y^2-6 \end{cases}$

The second equation is solved for x.

$$\begin{aligned} \left(y^2-6\right)^2+3y^2 &= 36 \\ y^4-12y^2+36+3y^2-36 &= 0 \\ y^4-9y^2 &= 0 \\ y^2\left(y^2-9\right) &= 0 \\ y^2(y+3)(y-3) &= 0 \\ y &= 0,-3,3 \end{aligned}$$

$y=0$ $\quad x=0^2-6$

$x=-6$

$(-6,0)$

$y=-3$ $\quad x=(-3)^2-6$

$x=9-6$

$x=3$

$(3,-3)$

$y=3$ $\quad x=(3)^2-6$

$x=9-6$

$x=3$

$(3,3)$

29. $\begin{cases} 9x^2+4y^2=36 \\ 4x^2-9y^2=36 \end{cases}$

$$\begin{aligned} &4\left(9x^2+4y^2=36\right) \\ &-9\left(4x^2-9y^2=36\right) \end{aligned}$$

$$\begin{aligned} 36x^2+16y^2 &= 144 \\ -36x^2+81y^2 &= -324 \\ 97y^2 &= -180 \\ y^2 &= -1.856 \end{aligned}$$

y is imaginary, no solution

31. $\begin{cases} y = x^2 \\ x^2 + y^2 = 20 \end{cases}$

The first equation is already solved for y.

$$x^2 + \left(x^2\right)^2 = 20$$
$$x^2 + x^4 = 20$$
$$x^4 + x^2 - 20 = 0$$
$$\left(x^2 - 4\right)\left(x^2 + 5\right) = 0$$
$$x^2 = 4$$
$$x = \pm 2$$

$x^2 + 5 = 0$ would have yielded an imaginary answer.

$x = -2 \quad y = (-2)^2$
$y = 4$
$(-2, 4)$

$x = 2 \quad y = (2)^2$
$y = 4$
$(2, 4)$

33. $\begin{cases} 25x^2 - 16y^2 = 400 \\ x^2 + 4y^2 = 16 \end{cases}$

Use elimination.

$$25x^2 - 16y^2 = 400$$
$$\underline{4\left(x^2 + 4y^2 = 16\right)}$$

$$25x^2 - 16y^2 = 400$$
$$\underline{4x^2 + 16y^2 = 64}$$
$$29x^2 = 464$$
$$x^2 = 16$$
$$x = \pm 4$$

$x = -4 \quad (-4)^2 + 4y^2 = 16$
$16 + 4y^2 = 16$
$4y^2 = 0$
$y^2 = 0$
$y = 0$
$(-4, 0)$

$x = 4 \quad 4^2 + 4y^2 = 16$
$16 + 4y^2 = 16$
$4y^2 = 0$
$y^2 = 0$
$y = 0$
$(4, 0)$

35. $\begin{cases} xy = 4 \\ 2x^2 - y^2 = 4 \end{cases}$

Solve the first equation for y. $y = \dfrac{4}{x}$

$$2x^2 - \left(\frac{4}{x}\right)^2 = 4$$
$$2x^2 - \frac{16}{x^2} = 4$$
$$2x^4 - 16 = 4x^2$$
$$x^4 - 2x^2 - 8 = 0$$
$$\left(x^2 - 4\right)\left(x^2 + 2\right) = 0$$

$x^2 - 4 = 0$ or $x^2 + 2 = 0$

$x^2 = 4 \qquad x^2 = -2$

$x = \pm 2 \qquad$ Leads to imaginary solutions

$x = -2: \quad y = \dfrac{4}{-2}$
$y = -2$
$(-2, -2)$

$x = 2: \quad y = \dfrac{4}{2}$
$y = 2$
$(2, 2)$

37. $\begin{cases} y = x^2 - 2x - 3 \\ y = -x^2 + 6x + 7 \end{cases}$

They are both solved for y, so use substitution.

$$\begin{aligned} -x^2 + 6x + 7 &= x^2 - 2x - 3 \\ 0 &= 2x^2 - 8x - 10 \\ 0 &= x^2 - 4x - 5 \\ 0 &= (x-5)(x+1) \\ x &= 5, -1 \end{aligned}$$

$x = 5 \quad y = (5)^2 - 2(5) - 3$
$= 25 - 10 - 3$
$= 12$
$(5, 12)$

$x = -1 \quad y = (-1)^2 - 2(-1) - 3$
$= 1 + 2 - 3$
$= 0$
$(-1, 0)$

39. $\begin{cases} 4x^2 + 5y^2 = 36 \\ 4x^2 - 3y^2 = 4 \end{cases}$

$$\begin{aligned} &4x^2 + 5y^2 = 36 \\ &\underline{-(4x^2 - 3y^2 = 4)} \end{aligned}$$

$$\begin{aligned} 4x^2 + 5y^2 &= 36 \\ \underline{-4x^2 + 3y^2} &\underline{= -4} \\ 8y^2 &= 32 \\ y^2 &= 4 \\ y &= \pm 2 \end{aligned}$$

$y = -2: \; 4x^2 + 5(-2)^2 = 36$
$4x^2 + 5 \cdot 4 = 36$
$4x^2 + 20 = 36$
$4x^2 = 16$
$x^2 = 4$
$x = \pm 2$
$(-2, -2), (2, -2)$

$y = 2: \; 4x^2 + 5(2)^2 = 36$
$4x^2 + 5 \cdot 4 = 36$
$4x^2 + 20 = 36$
$4x^2 = 16$
$x^2 = 4$
$x = \pm 2$
$(-2, 2), (2, 2)$

41. $\begin{cases} x = -y^2 + 2 \\ x^2 - 5y^2 = 4 \end{cases}$

The first equation is already solved for x.

$$\begin{aligned} (-y^2 + 2)^2 - 5y^2 &= 4 \\ y^4 - 4y^2 + 4 - 5y^2 &= 4 \\ y^4 - 9y^2 &= 0 \\ y^2(y^2 - 9) &= 0 \\ y &= 0, \pm 3 \end{aligned}$$

$y = 0: \; x = -0^2 + 2$
$x = 2$
$(2, 0)$

$y = -3: \; x = -(-3)^2 + 2$
$x = -9 + 2$
$x = -7$
$(-7, -3)$

$y = 3: \; x = -(3)^2 + 2$
$x = -9 + 2$
$x = -7$
$(-7, 3)$

43. Answers may vary, but one possible system is
$\begin{cases} x + y = 4 \\ x^2 + y^2 = 1 \end{cases}$.

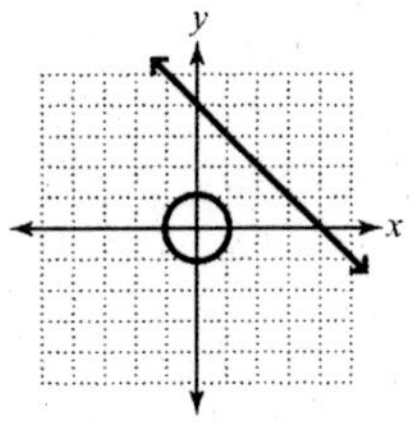

45. Let x be one of the integers and y be the other integer.

$$x^2 + y^2 = 34$$
$$\underline{x^2 - y^2 = 16}$$
$$2x^2 = 50$$
$$x^2 = 25$$
$$x = \pm 5$$

$$x = -5: \ (-5)^2 + y^2 = 34$$
$$25 + y^2 = 34$$
$$y^2 = 9$$
$$y = \pm 3$$
$$(-5,-3),(-5,3)$$

$$x = 5: \ (5)^2 + y^2 = 34$$
$$25 + y^2 = 34$$
$$y^2 = 9$$
$$y = \pm 3$$
$$(5,-3),(5,3)$$

47. Let x be one dimension and let y be the other.

$$xy = 144$$
$$2x + 2y = 52$$

Solve the second equation for x. $x = 26 - y$

$$(26 - y)y = 144$$
$$26y - y^2 = 144$$
$$0 = y^2 - 26y + 144$$
$$0 = (y-8)(y-18)$$
$$y = 8, 18$$

If $y = 8$: $xy = 144$ and $x = 18$.
If $y = 18$: $xy = 144$ and $x = 8$.

The computer keyboard is 8 in. by 18 in.

49. $-3x^2 + 120 = 11x + 28$

$$0 = 3x^2 + 11x - 92$$
$$0 = (3x+23)(x-4)$$
$$x = 4$$

$3x + 23 = 0$ would have yielded a negative answer.

So, because $x = 4$ and x is in hundreds of units, at equilibrium there would be 400 chairs.

The price per chair would be $p = 11(4) + 28 = 44 + 28 = \$72$.

51.

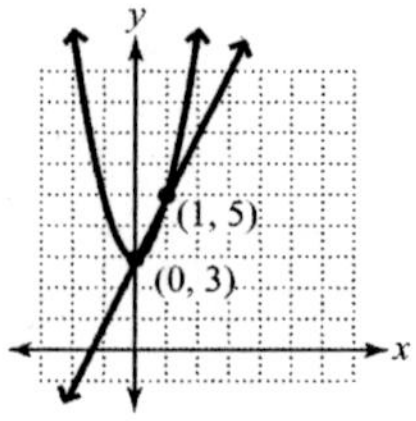

53.

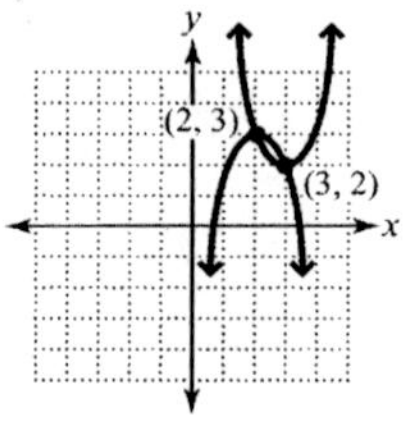

Review Exercises

1. $x + 3y \le 6$ Yes

$$0 + 3 \cdot 0 \le 6$$
$$0 \le 6$$

2. $x + 3y \le 6$ No

$$3 + 3 \cdot 4 \le 6$$
$$3 + 12 \le 6$$
$$15 \le 6$$

3. $y \ge 3$

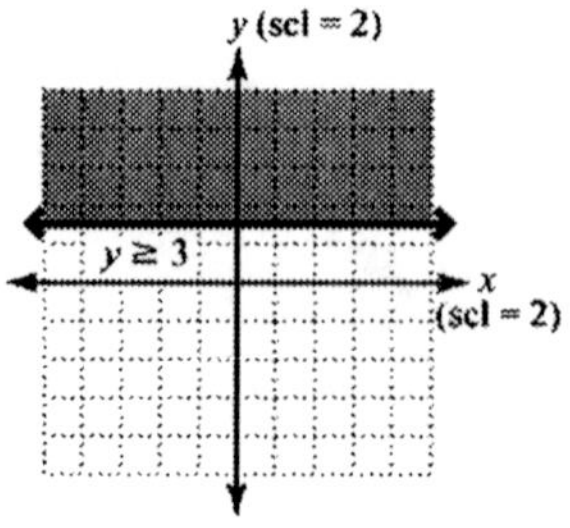

4. $y < 2x + 3$

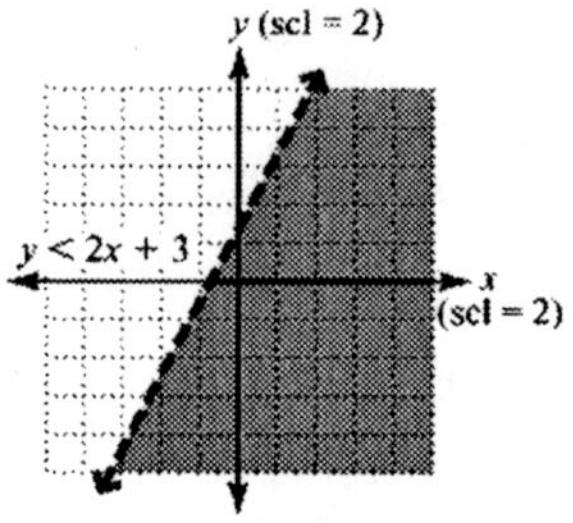

5. $2x+3y<-6$

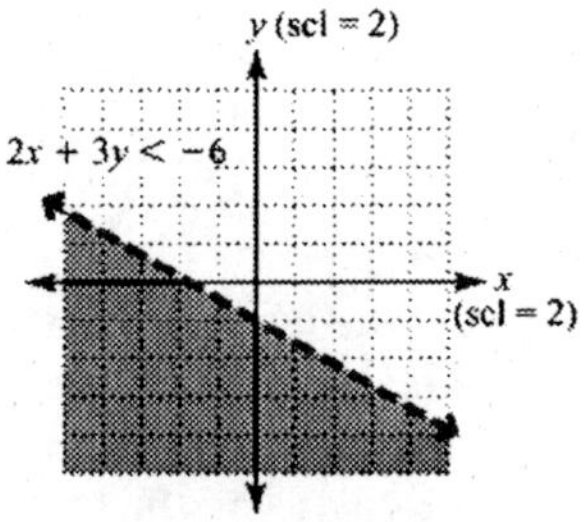

6. $x \ge -2$

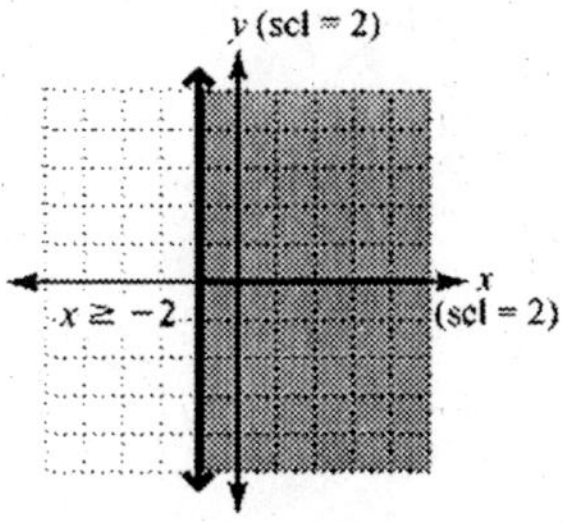

Exercise Set 11.4

1. No, the coordinates of the points on the curve make $9x^2+4y^2=36$. The curve would be a dashed curve because the points are not included in the solution.

3. One possible example is the system $\begin{cases} x^2+y^2<1 \\ x^2+y^2>4 \end{cases}$. The graph of $x^2+y^2<1$ is the region inside the circle with radius 1. The graph of $x^2+y^2>4$ is the region outside the circle with radius 2. Thus, there is no common region of solution.

5. $x^2+y^2 \le 9$

First graph the related equation $x^2+y^2=9$ with a solid line.
Let region 1 be the area inside the circle, and let region 2 be the area outside the circle.
Region 1: Choose (0, 0).

$0^2+0^2 \le 9$

$0+0 \le 9$

$0 \le 9$

True, so region 1 is in the solution set.
Region 2: Choose (4, 0).

$4^2+0^2 \le 9$

$16+0 \le 9$

$16 \le 9$

False, so region 2 is not in the solution set.

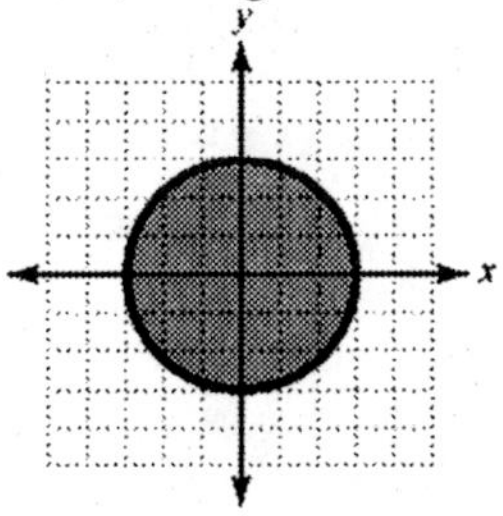

7. $y>x^2$

First graph the related equation $y=x^2$ with a dotted line.
Let region 1 be the area above the boundary, and let region 2 be the area below the boundary.
Region 1: Choose (0, 1).

$1>0^2$

$1>0$

True, so region 1 is in the solution set.
Region 2: Choose (2, 0).

$0>2^2$

$0>4$

False, so region 2 is not in the solution set.

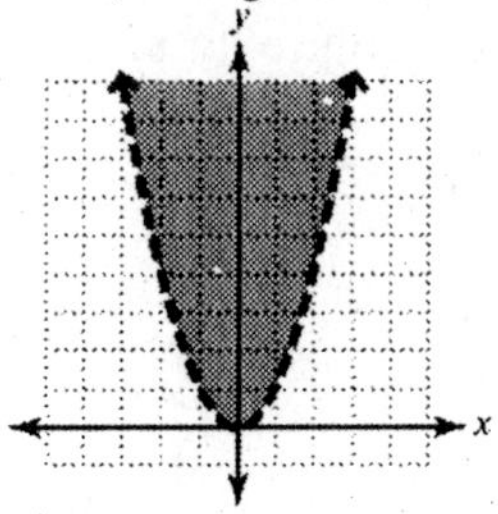

9. $y>2(x+2)^2-3$

First graph the related equation $y=2(x+2)^2-3$ with a dotted line.
Let region 1 be the area above the boundary, and let region 2 be the area below the boundary.
Region 1: Choose $(-2,0)$.

$0>2(-2+2)^2-3$

$0>2(0)^2-3$

$0>0-3$

$0>-3$

True, so region 1 is in the solution set.

Region 2: Choose (0, 0).

$0 > 2(0+2)^2 - 3$

$0 > 2(2)^2 - 3$

$0 > 2(4) - 3$

$0 > 8 - 3$

$0 > 5$

False, so region 2 is not in the solution set.

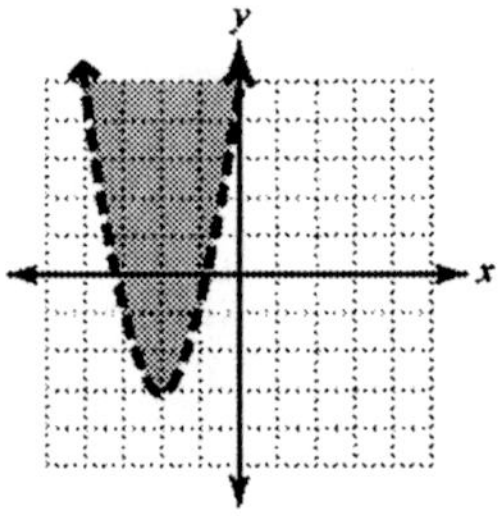

11. $\dfrac{x^2}{16} + \dfrac{y^2}{9} \le 1$

First graph the related equation $\dfrac{x^2}{16} + \dfrac{y^2}{9} = 1$ with a solid line.

Let region 1 be the area inside the ellipse, and let region 2 be the area outside the ellipse.

Region 1: Choose (0, 0).

$\dfrac{0^2}{16} + \dfrac{0^2}{9} \le 1$

$0 + 0 \le 1$

$0 \le 1$

True, so region 1 is in the solution set.

Region 2: Choose (5, 0).

$\dfrac{5^2}{16} + \dfrac{0^2}{9} \le 1$

$\dfrac{25}{16} + 0 \le 1$

$\dfrac{25}{16} \le 1$

False, so region 2 is not in the solution set.

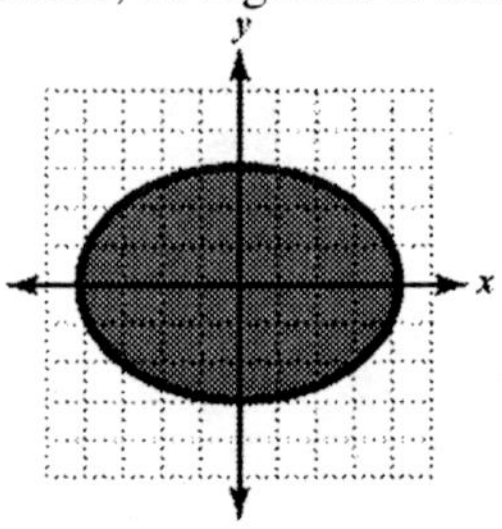

13. $\dfrac{x^2}{25} - \dfrac{y^2}{4} > 1$

First graph the related equation $\dfrac{x^2}{25} - \dfrac{y^2}{4} = 1$ with a dotted line.

Let region 1 be the area to the left of the left branch of the hyperbola, let region 3 be the area to the right of the right branch of the hyperbola, and let region 2 be the area between the two branches.

Region 1: Choose $(-8, 0)$.

$\dfrac{(-8)^2}{25} - \dfrac{0^2}{4} > 1$

$\dfrac{64}{25} - \dfrac{0}{4} > 1$

$\dfrac{64}{25} > 1$

True, so region 1 is in the solution set.

Region 2: Choose $(0, 0)$.

$\dfrac{0^2}{25} - \dfrac{0^2}{4} > 1$

$\dfrac{0}{25} - \dfrac{0}{4} > 1$

$0 > 1$

False, so region 2 is not in the solution set.

Region 3: Choose $(8, 0)$.

$\dfrac{8^2}{25} - \dfrac{0^2}{4} > 1$

$\dfrac{64}{25} - \dfrac{0}{4} > 1$

$\dfrac{64}{25} > 1$

True, so region 3 is in the solution set.

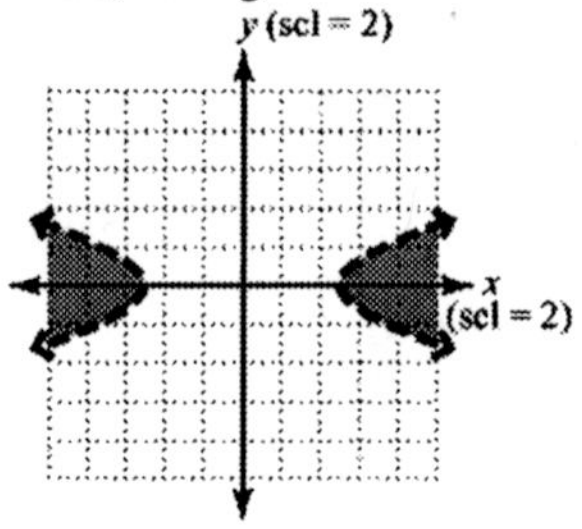

15. $x^2 + y^2 < 16$

First graph the related equation $x^2 + y^2 = 16$ with a dotted line.
Let region 1 be the area inside the circle, and let region 2 be the area outside the circle.
Region 1: Choose (0, 0).

$$0^2 + 0^2 < 16$$
$$0 < 16$$

True, so region 1 is in the solution set.
Region 2: Choose (5, 0).

$$5^2 + 0^2 < 16$$
$$25 + 0 < 16$$
$$25 < 16$$

False, so region 2 is not in the solution set.

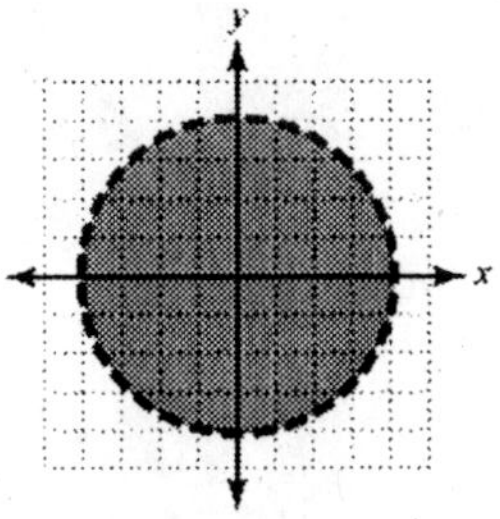

17. $y \geq -2(x-3)^2 + 3$

First graph the related equation $y = -2(x-3)^2 + 3$ with a solid line.
Let region 1 be the area above the boundary, and let region 2 be the area below the boundary.
Region 1: Choose $(0,0)$.

$$0 \geq -2(0-3)^2 + 3$$
$$0 \geq -2(-3)^2 + 3$$
$$0 \geq -2(9) + 3$$
$$0 \geq -18 + 3$$
$$0 \geq -15$$

True, so region 1 is in the solution set.
Region 2: Choose $(4,-2)$.

$$-2 \geq -2(4-3)^2 + 3$$
$$-2 \geq -2(1)^2 + 3$$
$$-2 \geq -2(1) + 3$$
$$-2 \geq -2 + 3$$
$$-2 \geq 1$$

False, so region 2 is not in the solution set.

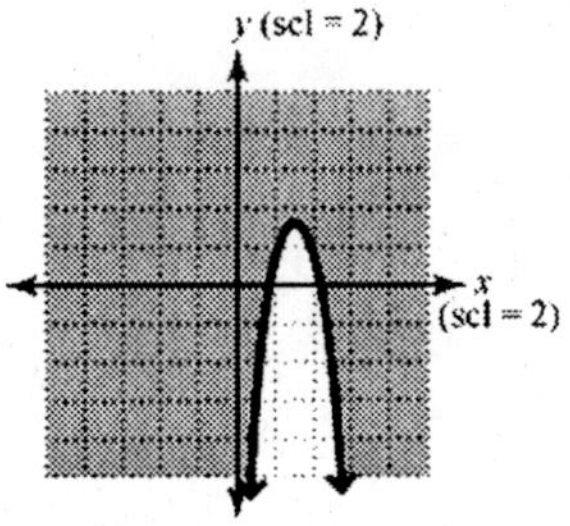

19. $\dfrac{x^2}{16} + \dfrac{y^2}{4} \leq 1$

First graph the related equation $\dfrac{x^2}{16} + \dfrac{y^2}{4} = 1$ with a solid line.
Let region 1 be the area inside the ellipse, and let region 2 be the area outside the ellipse.
Region 1: Choose (0, 0).

$$\frac{0^2}{16} + \frac{0^2}{4} \leq 1$$
$$0 + 0 \leq 1$$
$$0 \leq 1$$

True, so region 1 is in the solution set.
Region 2: Choose (5, 0).

$$\frac{5^2}{16} + \frac{0^2}{4} \leq 1$$
$$\frac{25}{16} + 0 \leq 1$$
$$\frac{25}{16} \leq 1$$

False, so region 2 is not in the solution set.

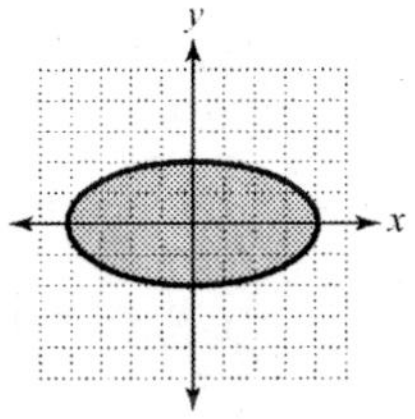

21. $\dfrac{y^2}{25} - \dfrac{x^2}{4} \leq 1$

First graph the related equation $\dfrac{y^2}{25} - \dfrac{x^2}{4} = 1$ with a solid line.
Let region 1 be the area above the top branch of the hyperbola, let region 3 be the area below the bottom branch of the hyperbola, and let region 2 be the area between the two branches.

Region 1: Choose $(0,8)$.

$$\frac{8^2}{25}-\frac{0^2}{4}\le 1$$
$$\frac{64}{25}-\frac{0}{4}\le 1$$
$$\frac{64}{25}\le 1$$

False, so region 1 is not in the solution set.

Region 2: Choose $(0,0)$.

$$\frac{0^2}{25}-\frac{0^2}{4}\le 1$$
$$\frac{0}{25}-\frac{0}{4}\le 1$$
$$0-0\le 1$$
$$0\le 1$$

True, so region 2 is in the solution set.

Region 3: Choose $(0,-8)$.

$$\frac{(-8)^2}{25}-\frac{0^2}{4}\le 1$$
$$\frac{64}{25}-\frac{0}{4}\le 1$$
$$\frac{64}{25}\le 1$$

False, so region 3 is not in the solution set.

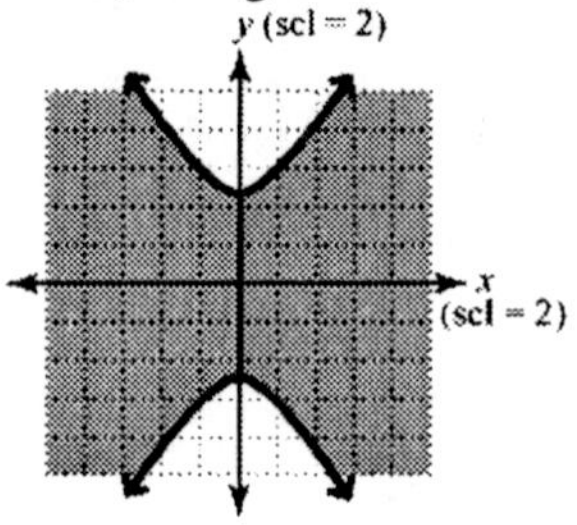

23. $y<x^2+4x-5$

First graph the related equation $y=x^2+4x-5$ with a dotted line.
Let region 1 be the area above the boundary, and let region 2 be the area below the boundary.

Region 1: Choose $(0,0)$.

$$0<0^2+4(0)-5$$
$$0<0+0-5$$
$$0<-5$$

False, so region 1 is not in the solution set.

Region 2: Choose (4, 0).

$$0<4^2+4(4)-5$$
$$0<16+16-5$$
$$0<27$$

True, so region 2 is in the solution set.

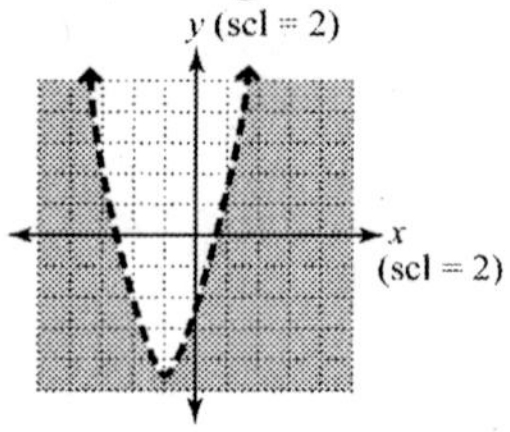

25. $\begin{cases} y<-x^2 \\ 2x-y<4 \end{cases}$

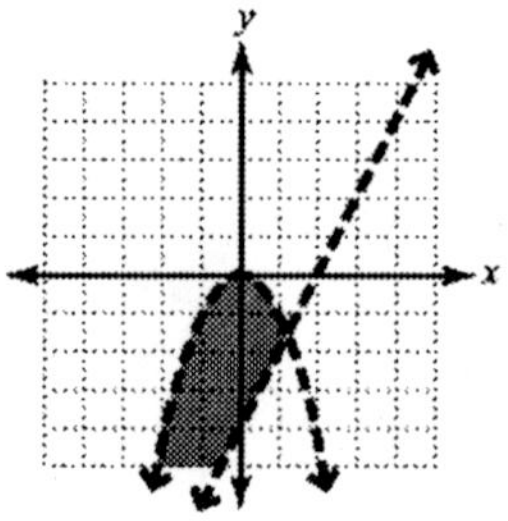

27. $\begin{cases} 3x+2y\ge -6 \\ x^2+y^2\le 25 \end{cases}$

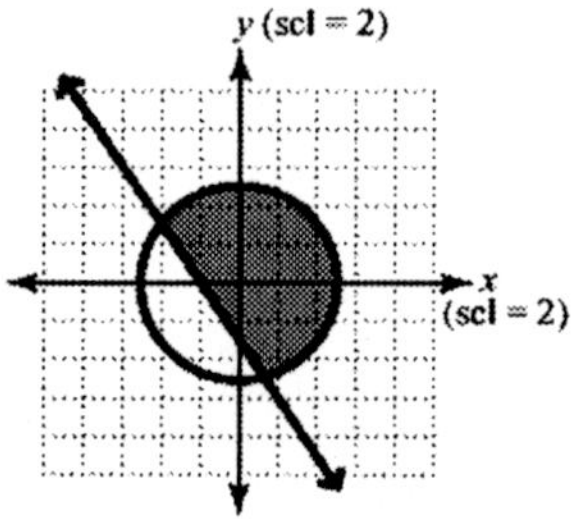

29. $\begin{cases} x^2+y^2>4 \\ x^2+y^2>9 \end{cases}$

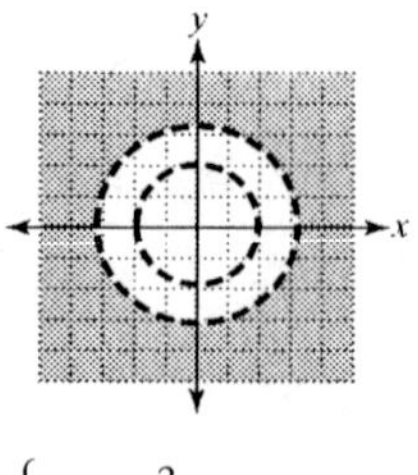

31. $\begin{cases} y>x^2+1 \\ 2x+y<3 \end{cases}$

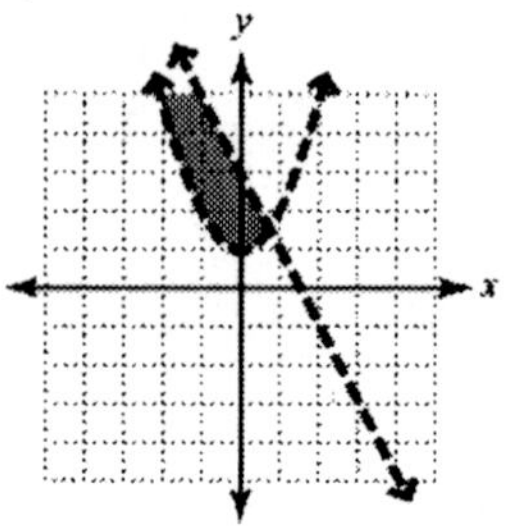

33. $\begin{cases} y < -x^2 + 3 \\ y > x^2 - 2 \end{cases}$

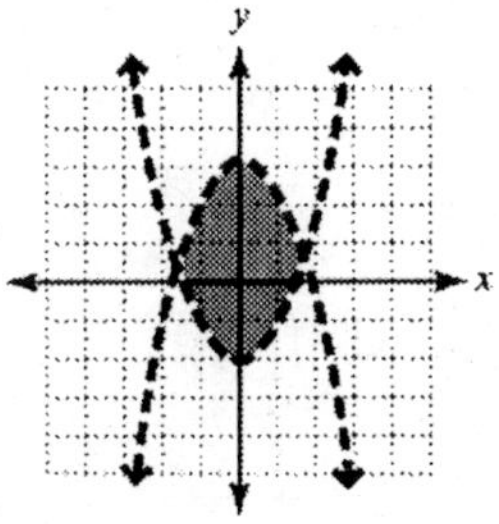

35. $\begin{cases} \dfrac{x^2}{25} + \dfrac{y^2}{9} \le 1 \\ x^2 + y^2 \ge 4 \end{cases}$

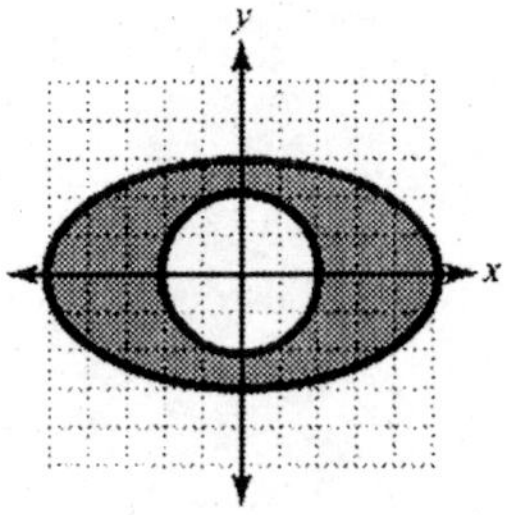

37. $\begin{cases} \dfrac{x^2}{25} + \dfrac{y^2}{9} < 1 \\ y > x^2 + 1 \end{cases}$

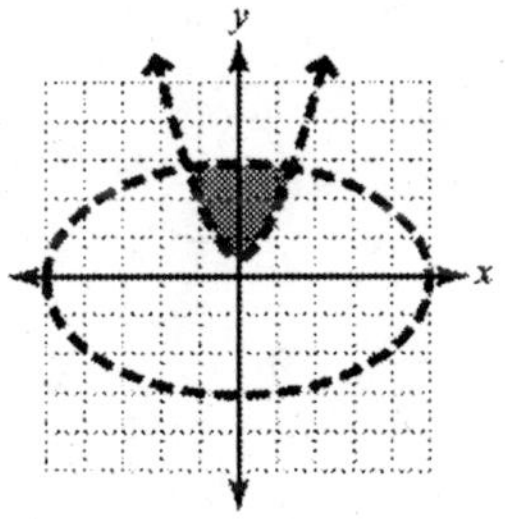

39. $\begin{cases} \dfrac{x^2}{9} - \dfrac{y^2}{4} \le 1 \\ \dfrac{x^2}{25} + \dfrac{y^2}{9} \le 1 \end{cases}$

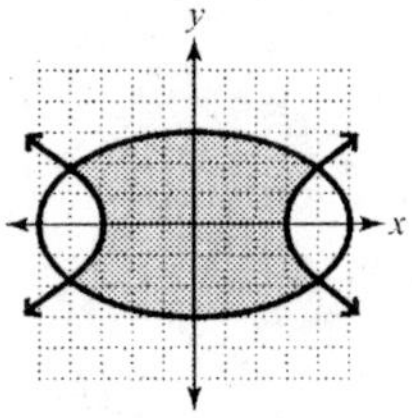

41. $\begin{cases} \dfrac{x^2}{4} - \dfrac{y^2}{4} > 1 \\ y > 2 \end{cases}$

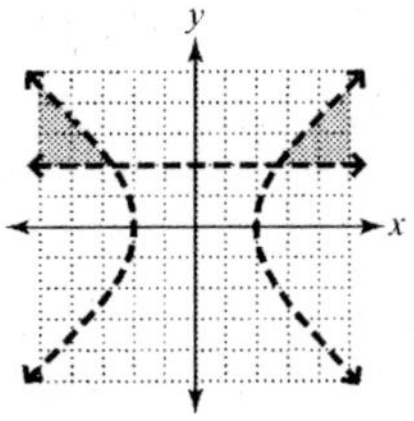

43. $\begin{cases} 3x + 2y \le 6 \\ x - y > -3 \\ x + 6y \ge 2 \end{cases}$

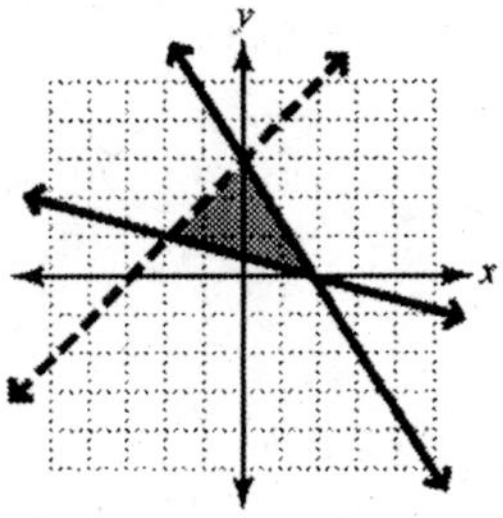

45. $\begin{cases} \dfrac{x^2}{16} + \dfrac{y^2}{4} \le 1 \\ x^2 + y^2 \le 9 \\ y \le x \end{cases}$

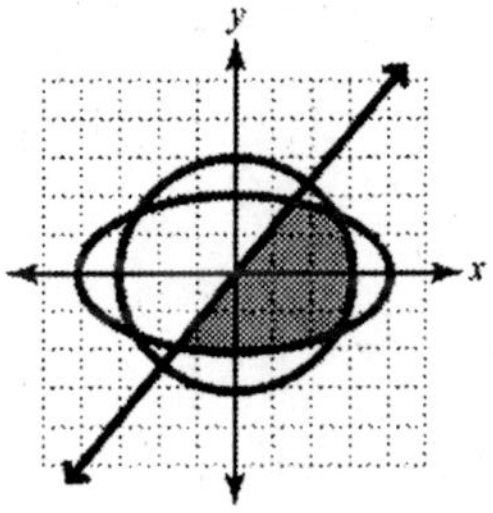

47. $\begin{cases} \dfrac{x^2}{49} - \dfrac{y^2}{16} \le 1 \\ \dfrac{x^2}{64} + \dfrac{y^2}{36} \le 1 \\ 2x - y \le -3 \end{cases}$

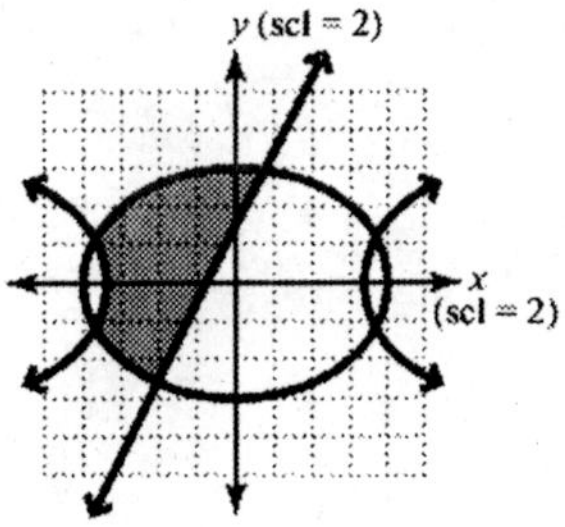

49. $\begin{cases} y < 2x^2 + 8 \\ 2x + y > 3 \\ x - y < 4 \end{cases}$

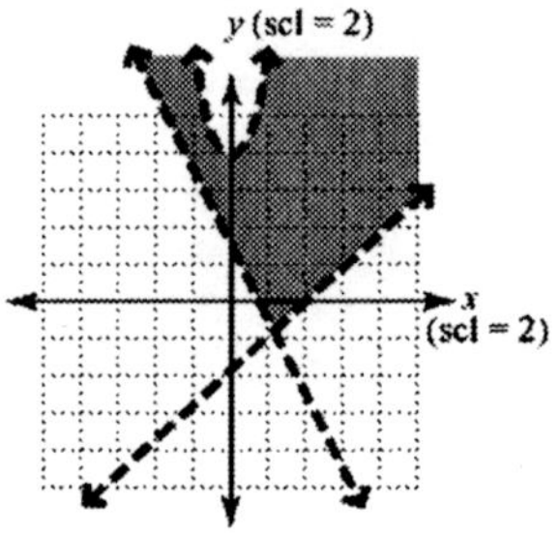

Review Exercises

1. $(3\cdot1+2)+(3\cdot2+2)+(3\cdot3+2)+(3\cdot4+2)$
$=(3+2)+(6+2)+(9+2)+(12+2)$
$=5+8+11+14$
$=38$

2. $\frac{n}{2}(a_1+a_n)=\frac{12}{2}(-8+60)$
$=6(52)$
$=312$

3. $a_1+(n-1)d=-12+(25-1)(-3)$
$=-12+(-72)$
$=-84$

4. $n=1 \quad 2(1^2)-3=2-3=-1$
$n=2 \quad 2(2^2)-3=8-3=5$
$n=3 \quad 2(3^2)-3=18-3=15$
$n=4 \quad 2(4^2)-3=32-3=29$

5. $n=1 \quad \frac{(-1)^1}{3(1)-2}=\frac{-1}{3-2}=-1$
$n=2 \quad \frac{(-1)^2}{3(2)-2}=\frac{1}{6-2}=\frac{1}{4}$
$n=3 \quad \frac{(-1)^3}{3(3)-2}=\frac{-1}{9-2}=-\frac{1}{7}$
$n=4 \quad \frac{(-1)^4}{3(4)-2}=\frac{1}{12-2}=\frac{1}{10}$

6. Let n, $(n + 2)$, and $(n + 4)$ be the three consecutive odd integers.
$n+(n+2)+(n+4)=207$
$3n+6=207$
$3n=201$
$n=67$

The three integers are 67, 69, and 71.

Chapter 11 Review Exercises

1. False; the graph of this parabola opens to the left.
2. True
3. False; a hyperbola has asymptotes.
4. True
5. True
6. False; the graph of the system lies outside the circle and above the line.
7. 10
8. $(-4,1)$, 2
9. ellipse, $(4,0)$ and $(-4,0)$
10. $(5,4),(5,-4),(-5,-4),(-5,4)$
11. $y=2(x-3)^2-5$
opens upward
vertex is $(3,-5)$
axis of symmetry is $x=3$

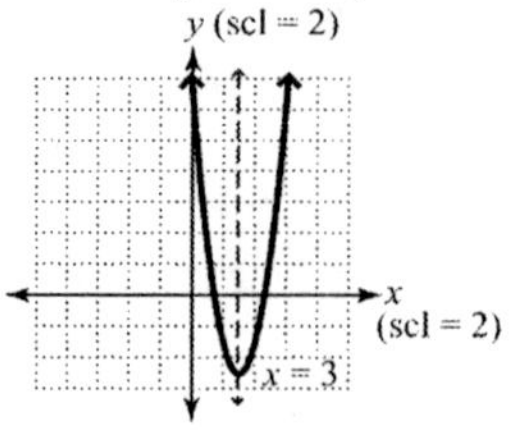

12. $y=-2(x+2)^2+3$

opens downward

vertex is $(-2,3)$

axis of symmetry $x=-2$

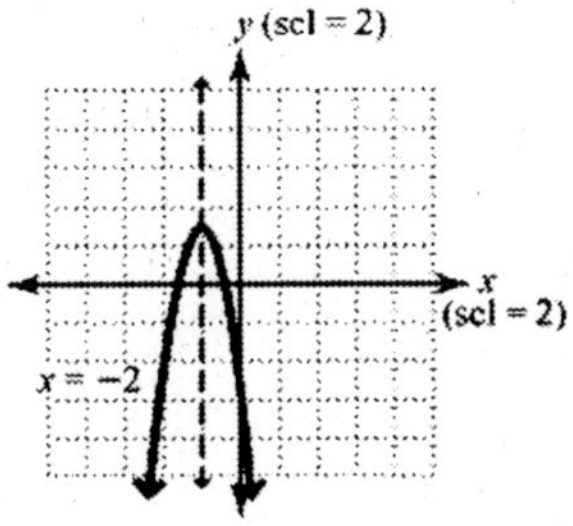

13. $x=-(y-2)^2+4$

opens left

vertex is $(4,2)$

axis of symmetry is $y=2$

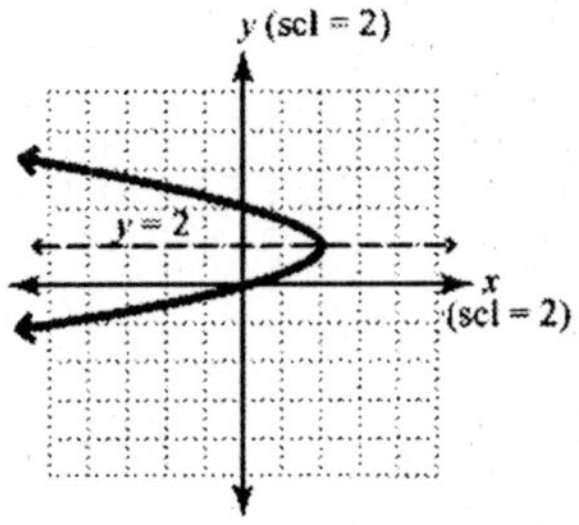

14. $x=2(y+3)^2-2$

opens right

vertex is $(-2,-3)$

axis of symmetry is $y=-3$

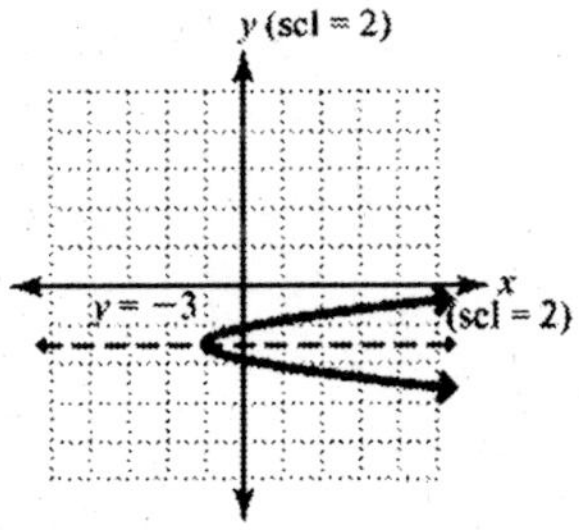

15.
$$\begin{aligned}x&=y^2+6y+8\\x-8&=y^2+6y\\x-8+9&=y^2+6y+9\\x+1&=(y+3)^2\\x&=(y+3)^2-1\end{aligned}$$

opens right

vertex is $(-1,-3)$

axis of symmetry is $y=-3$

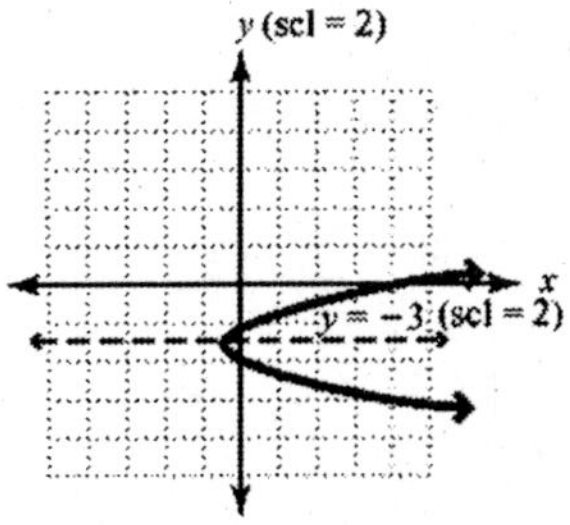

16.
$$\begin{aligned}x&=2y^2-8y-6\\x+6&=2\left(y^2-4y\right)\\x+6+8&=2\left(y^2-4y+4\right)\\x+14&=2(y-2)^2\\x&=2(y-2)^2-14\end{aligned}$$

opens right

vertex is $(-14,2)$

axis of symmetry $y=2$

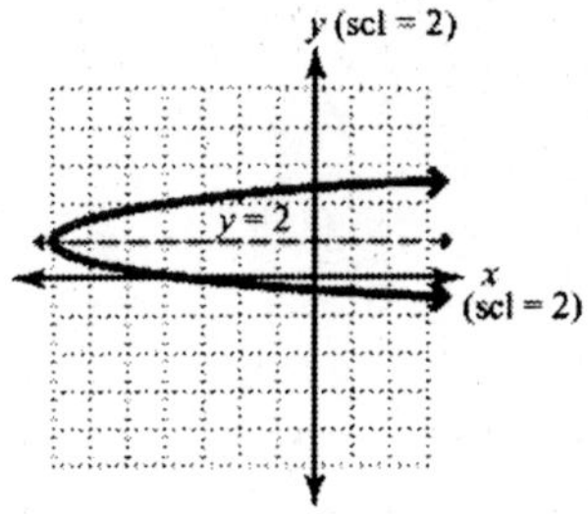

17.
$$\begin{aligned}x&=-y^2-2y+3\\x-3&=-\left(y^2+2y\right)\\x-3-1&=-\left(y^2+2y+1\right)\\x-4&=-(y+1)^2\\x&=-(y+1)^2+4\end{aligned}$$

opens left

vertex is $(4,-1)$

axis of symmetry is $y=-1$

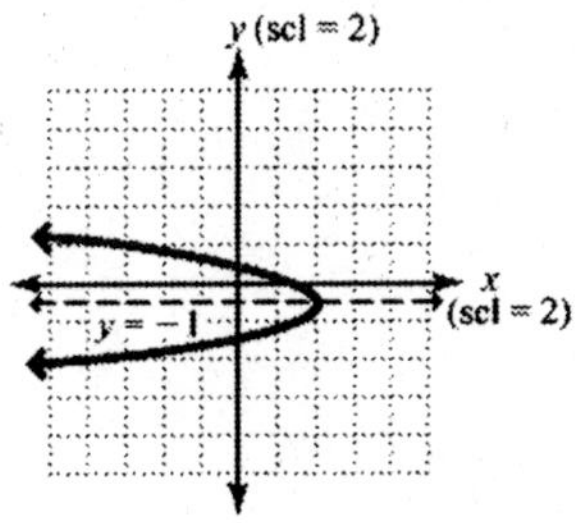

18. $$x = -3y^2 - 12y - 9$$
$$x + 9 = -3\left(y^2 + 4y\right)$$
$$x + 9 - 12 = -3\left(y^2 + 4y + 4\right)$$
$$x - 3 = -3\left(y + 2\right)^2$$
$$x = -3\left(y + 2\right)^2 + 3$$
opens left
vertex is $(3, -2)$
axis of symmetry is $y = -2$

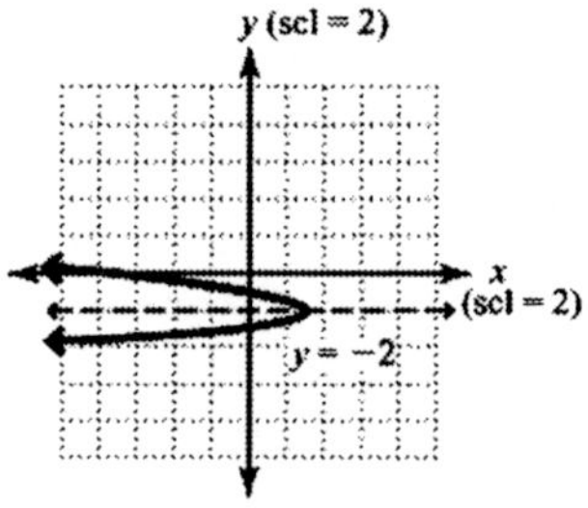

19. $(-1, -2)$ and $(-5, 1)$
$$d = \sqrt{\left(-1-(-5)\right)^2 + \left(-2-1\right)^2}$$
$$= \sqrt{(4)^2 + (-3)^2}$$
$$= \sqrt{16+9}$$
$$= \sqrt{25}$$
$$= 5$$
$$\text{Midpoint} = \left(\frac{-1+(-5)}{2}, \frac{-2+1}{2}\right)$$
$$= \left(\frac{-6}{2}, \frac{-1}{2}\right)$$
$$= \left(-3, -\frac{1}{2}\right)$$

20. $(2, -5)$ and $(-2, 3)$
$$d = \sqrt{\left(2-(-2)\right)^2 + \left(-5-3\right)^2}$$
$$= \sqrt{(4)^2 + (-8)^2}$$
$$= \sqrt{16+64}$$
$$= \sqrt{80}$$
$$= 4\sqrt{5}$$
$$\text{Midpoint} = \left(\frac{2+(-2)}{2}, \frac{-5+3}{2}\right)$$
$$= \left(\frac{0}{2}, \frac{-2}{2}\right)$$
$$= (0, -1)$$

21. Find the distance from the center to a point on the circle. This is the radius.
$$r = \sqrt{\left(6-(-2)\right)^2 + \left(-4-2\right)^2}$$
$$= \sqrt{(8)^2 + (-6)^2}$$
$$= \sqrt{64+36}$$
$$= \sqrt{100}$$
$$= 10$$

22. Find the distance from the center to a point on the circle. This is the radius.
$$r = \sqrt{\left(-6-3\right)^2 + \left(8-(-4)\right)^2}$$
$$= \sqrt{(-9)^2 + (12)^2}$$
$$= \sqrt{81+144}$$
$$= \sqrt{225}$$
$$= 15$$
The equation in standard form is:
$$(x+6)^2 + (y-8)^2 = 225$$

23. $(x-3)^2 + (y+2)^2 = 25$
Center $(3, -2)$; radius 5

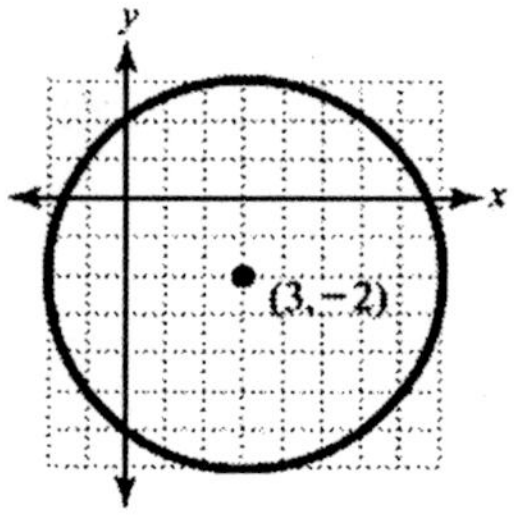

24. $(x+5)^2+(y-1)^2=4$

Center $(-5,1)$; radius 2

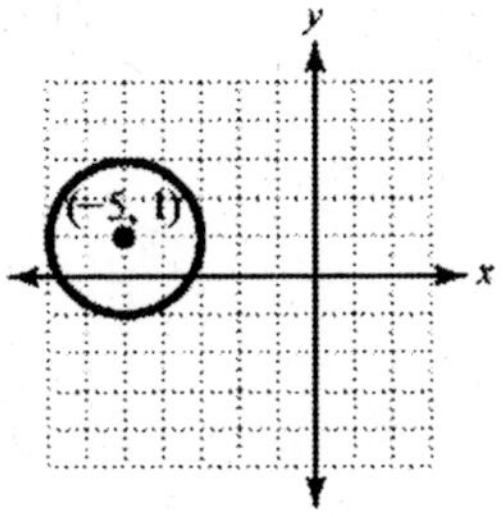

25.
$$x^2+y^2-4x+8y+11=0$$
$$\left(x^2-4x\right)+\left(y^2+8y\right)=-11$$
$$\left(x^2-4x+4\right)+\left(y^2+8y+16\right)=-11+4+16$$
$$(x-2)^2+(y+4)^2=9$$

Center $(2,-4)$; radius 3

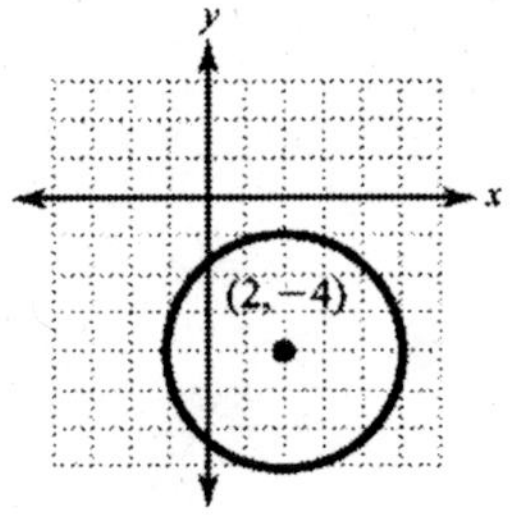

26.
$$x^2+y^2+10x+2y+22=0$$
$$\left(x^2+10x\right)+\left(y^2+2y\right)=-22$$
$$\left(x^2+10x+25\right)+\left(y^2+2y+1\right)=-22+25+1$$
$$(x+5)^2+(y+1)^2=4$$

Center $(-5,-1)$; radius 2

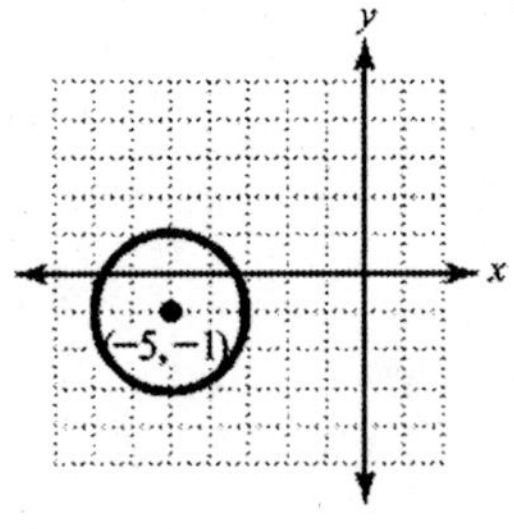

27. $(x-6)^2+(y+8)^2=9^2$

$(x-6)^2+(y+8)^2=81$

28. $(x-(-3))^2+(y-(-5))^2=10^2$

$(x+3)^2+(y+5)^2=100$

29. a)
$$h=-16t^2+32t+128$$
$$h-128=-16\left(t^2-2t\right)$$
$$h-128-16=-16\left(t^2-2t+1\right)$$
$$h-144=-16(t-1)^2$$
$$h=-16(t-1)^2+144$$

The vertex for the parabola is (1, 144). The maximum height for the rock will be 144 ft.

b) It will take 1 second to reach the maximum height.

c)
$$0=-16t^2+32t+128$$
$$0=-16\left(t^2-2t-8\right)$$
$$0=t^2-2t-8$$
$$0=(t-4)(t+2)$$

$t-4=0$ or $t+2=0$

$t=4$ $\quad t=\cancel{-2}$

The object will strike the ground after 4 seconds.

30. Since the rope is 20 ft. long and that is the distance from the center to any point on the edge of the circle, the radius is 20.

The equation in standard form is:

$x^2+y^2=20^2$

$x^2+y^2=400$

31. $\dfrac{x^2}{49}+\dfrac{y^2}{25}=1$

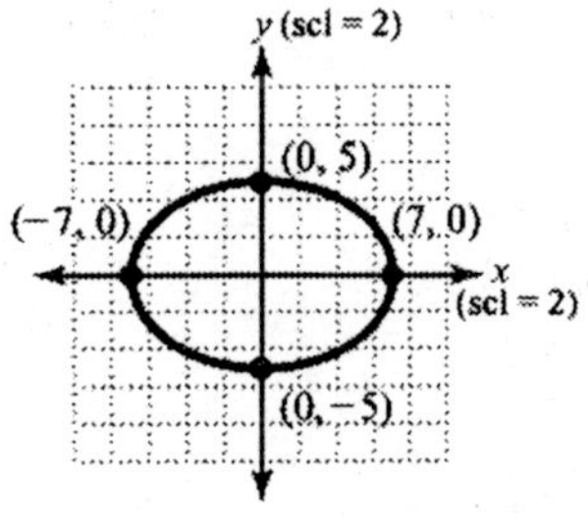

32. $\frac{x^2}{9}+\frac{y^2}{25}=1$

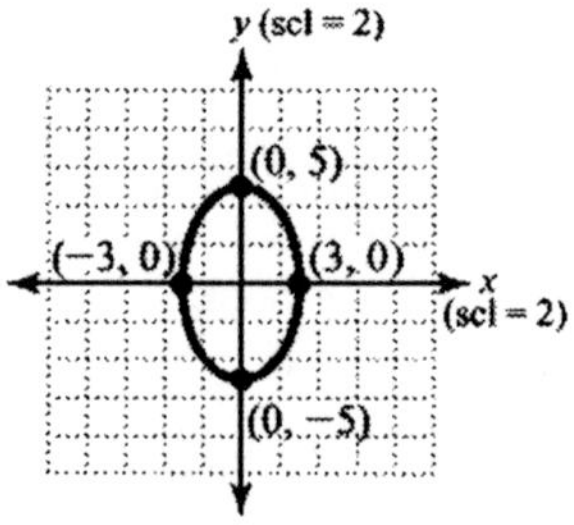

33. $4x^2+9y^2=36$

$$\frac{4x^2}{36}+\frac{9y^2}{36}=\frac{36}{36}$$

$$\frac{x^2}{9}+\frac{y^2}{4}=1$$

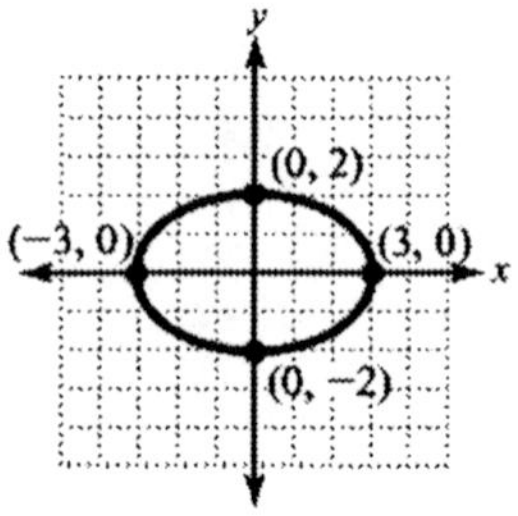

34. $25x^2+9y^2=225$

$$\frac{25x^2}{225}+\frac{9y^2}{225}=\frac{225}{225}$$

$$\frac{x^2}{9}+\frac{y^2}{25}=1$$

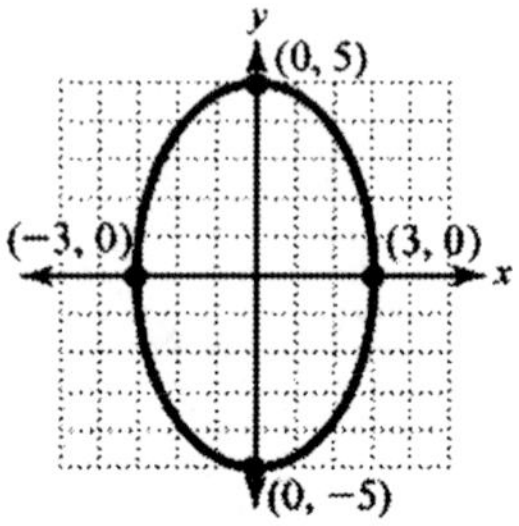

35. $\frac{(x+2)^2}{16}+\frac{(y-3)^2}{4}=1$

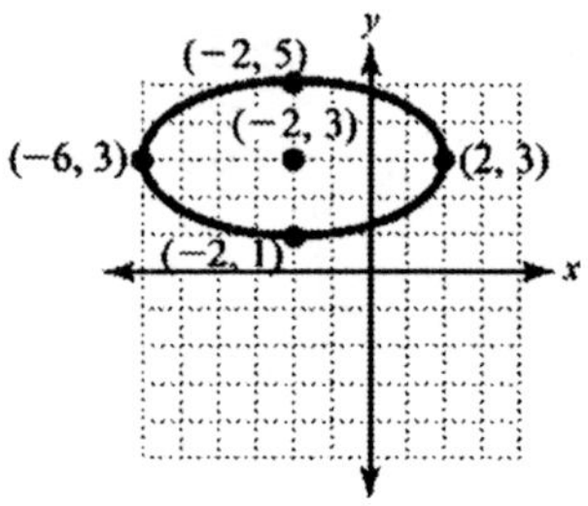

36. $\frac{(x-1)^2}{9}+\frac{(y+4)^2}{25}=1$

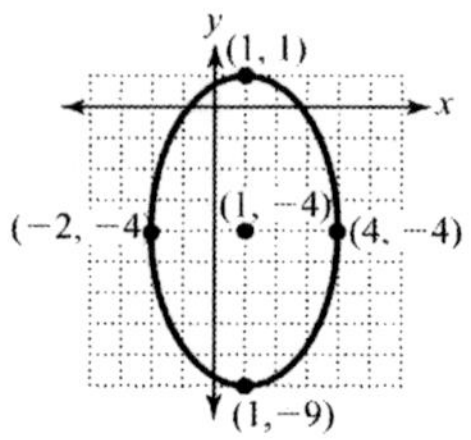

37. $\frac{x^2}{25}-\frac{y^2}{16}=1$

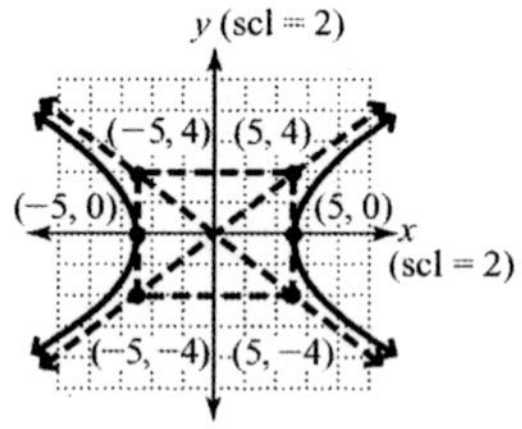

38. $\frac{x^2}{36}-\frac{y^2}{9}=1$

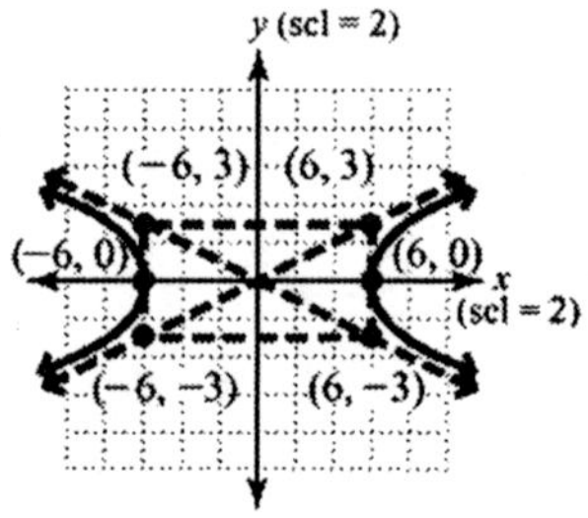

39. $y^2-9x^2=36$

$$\frac{y^2}{36}-\frac{9x^2}{36}=\frac{36}{36}$$

$$\frac{y^2}{36}-\frac{x^2}{4}=1$$

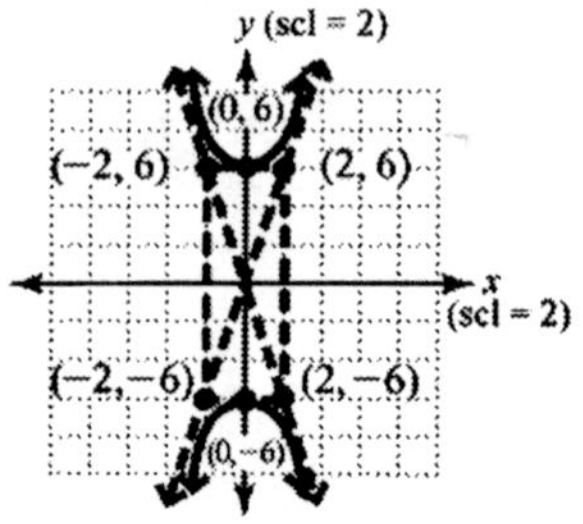

40. $25y^2 - 9x^2 = 225$

$$\frac{25y^2}{225} - \frac{9x^2}{225} = \frac{225}{225}$$

$$\frac{y^2}{9} - \frac{x^2}{25} = 1$$

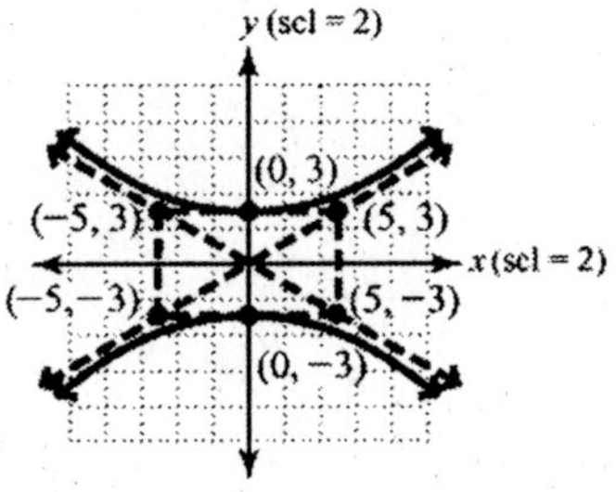

41. $\frac{x^2}{25^2} + \frac{y^2}{15^2} = 1$

$$\frac{x^2}{625} + \frac{y^2}{225} = 1$$

42. $\frac{x^2}{1.625^2} + \frac{y^2}{2^2} = 1$

$$\frac{x^2}{1.625^2} + \frac{y^2}{4} = 1$$

43. $\begin{cases} y = 2x^2 - 3 \\ 2x - y = -1 \end{cases}$

The first equation is already solved for y.

$$2x - (2x^2 - 3) = -1$$
$$2x - 2x^2 + 3 = -1$$
$$0 = 2x^2 - 2x - 4$$
$$0 = x^2 - x - 2$$
$$0 = (x-2)(x+1)$$

$x - 2 = 0$ or $x + 1 = 0$

$x = 2$ $\quad x = -1$

$x = 2$ $\quad y = 2 \cdot 2^2 - 3$

$y = 2 \cdot 4 - 3$

$y = 8 - 3$

$y = 5$

$(2, 5)$

$x = -1$ $\quad y = 2(-1)^2 - 3$

$y = 2 \cdot 1 - 3$

$y = 2 - 3$

$y = -1$

$(-1, -1)$

44. $\begin{cases} x^2 + y^2 = 17 \\ x - y = 3 \end{cases}$

Solve the second equation for x. $x = y + 3$

$$(y+3)^2 + y^2 = 17$$
$$y^2 + 6y + 9 + y^2 = 17$$
$$2y^2 + 6y - 8 = 0$$
$$y^2 + 3y - 4 = 0$$
$$(y+4)(y-1) = 0$$

$y + 4 = 0$ or $y - 1 = 0$

$y = -4$ $\quad y = 1$

$y = -4$ $\quad x = -4 + 3$

$x = -1$

$(-1, -4)$

$y = 1$ $\quad x = 1 + 3$

$x = 4$

$(4, 1)$

45. $\begin{cases} 25x^2 + 3y^2 = 100 \\ 2x - y = -3 \end{cases}$

Solve the second equation for y. $y = 2x + 3$

$$25x^2 + 3(2x+3)^2 = 100$$
$$25x^2 + 3(4x^2 + 12x + 9) = 100$$
$$25x^2 + 12x^2 + 36x + 27 = 100$$
$$37x^2 + 36x - 73 = 0$$
$$(37x + 73)(x - 1) = 0$$

$37x + 73 = 0$ or $x - 1 = 0$

$37x = -73$ $\quad x = 1$

$x = -\frac{73}{37}$

$x = -\frac{73}{37}$ $\quad y = 2\left(-\frac{73}{37}\right) + 3$

$y = -\frac{35}{37}$

$\left(-\frac{73}{37}, -\frac{35}{37}\right)$

$x = 1$ $\quad y = 2 \cdot 1 + 3$

$y = 5$

$(1, 5)$

46. $\begin{cases} x^2 + y^2 = 64 \\ x^2 - y^2 = 64 \end{cases}$

$x^2 + y^2 = 64$

$\underline{x^2 - y^2 = 64}$

$2x^2 = 128$

$x^2 = 64$

$x = \pm 8$

$x = -8$ $\quad (-8)^2 + y^2 = 64$

$64 + y^2 = 64$

$y^2 = 0$

$y = 0$

$(-8, 0)$

$x = 8$ $\quad (8)^2 + y^2 = 64$

$64 + y^2 = 64$

$y^2 = 0$

$y = 0$

$(8, 0)$

47. $\begin{cases} x^2 + y^2 = 25 \\ 25y^2 - 16x^2 = 256 \end{cases}$

$16\left(x^2 + y^2 = 25\right)$

$\underline{-16x^2 + 25y^2 = 256}$

$16x^2 + 16y^2 = 400$

$\underline{-16x^2 + 25y^2 = 256}$

$41y^2 = 656$

$y^2 = 16$

$y = \pm 4$

$y = -4$ $\quad x^2 + (-4)^2 = 25$

$x^2 + 16 = 25$

$x^2 = 9$

$x = \pm 3$

$(-3, -4), (3, -4)$

$x = 4$ $\quad x^2 + (4)^2 = 25$

$x^2 + 16 = 25$

$x^2 = 9$

$x = \pm 3$

$(-3, 4), (3, 4)$

48. $\begin{cases} x^2 + 5y^2 = 36 \\ 4x^2 - 7y^2 = 36 \end{cases}$

$-4\left(x^2 + 5y^2 = 36\right)$

$\underline{4x^2 - 7y^2 = 36}$

$-4x^2 - 20y^2 = -144$

$\underline{4x^2 - 7y^2 = 36}$

$-27y^2 = -108$

$y^2 = 4$

$y = \pm 2$

$y = 2$ $\quad x^2 + 5(2)^2 = 36$

$x^2 + 5 \cdot 4 = 36$

$x^2 + 20 = 36$

$x^2 = 16$

$x = \pm 4$

$(-4, 2), (4, 2)$

$y = -2$ $\quad x^2 + 5(-2)^2 = 36$

$x^2 + 5 \cdot 4 = 36$

$x^2 + 20 = 36$

$x^2 = 16$

$x = \pm 4$

$(-4, -2), (4, -2)$

49. $\begin{cases} y = x^2 - 2 \\ 4x^2 + 5y^2 = 36 \end{cases}$

The first equation is already solved for y.

$$4x^2 + 5(x^2 - 2)^2 = 36$$
$$4x^2 + 5(x^4 - 4x^2 + 4) = 36$$
$$4x^2 + 5x^4 - 20x^2 + 20 = 36$$
$$5x^4 - 16x^2 - 16 = 0$$
$$(5x^2 + 4)(x^2 - 4) = 0$$
$$x^2 = 4$$
$$x = \pm 2$$

$5x^2 + 4 = 0$ would have yielded an imaginary solution.

$x = -2$ $\quad y = (-2)^2 - 2$, $y = 4 - 2$, $y = 2$, $(-2, 2)$

$x = 2$ $\quad y = (2)^2 - 2$, $y = 4 - 2$, $y = 2$, $(2, 2)$

50. $\begin{cases} y = x^2 - 1 \\ 4y^2 - 5x^2 = 16 \end{cases}$

The first equation is already solved for y.

$$4(x^2 - 1)^2 - 5x^2 = 16$$
$$4(x^4 - 2x^2 + 1) - 5x^2 = 16$$
$$4x^4 - 8x^2 + 4 - 5x^2 = 16$$
$$4x^4 - 13x^2 - 12 = 0$$
$$(4x^2 + 3)(x^2 - 4) = 0$$
$$x^2 = 4$$
$$x = \pm 2$$

$4x^2 + 3 = 0$ would have yielded an imaginary solution.

$x = -2$ $\quad y = (-2)^2 - 1$, $y = 4 - 1$, $y = 3$, $(-2, 3)$

$x = 2$ $\quad y = (2)^2 - 1$, $y = 4 - 1$, $y = 3$, $(2, 3)$

51. Let x be one of the integers and y be the other integer.

$\begin{cases} x^2 + y^2 = 89 \\ x^2 - y^2 = 39 \end{cases}$

$$x^2 + y^2 = 89$$
$$x^2 - y^2 = 39$$
$$2x^2 = 128$$
$$x^2 = 64$$
$$x = \pm 8$$

$x = -8$ $\quad (-8)^2 + y^2 = 89$, $64 + y^2 = 89$, $y^2 = 25$, $y = \pm 5$, $(-8, -5), (-8, 5)$

$x = 8$ $\quad (8)^2 + y^2 = 89$, $64 + y^2 = 89$, $y^2 = 25$, $y = \pm 5$, $(8, -5), (8, 5)$

52. Let x be one dimension and let y be the other.

$\begin{cases} xy = 80 \\ 2x + 2y = 36 \end{cases}$

Solve the second equation for x. $x = 18 - y$

$$(18 - y)y = 80$$
$$18y - y^2 = 80$$
$$0 = y^2 - 18y + 80$$
$$0 = (y - 10)(y - 8)$$

$y - 10 = 0$ or $y - 8 = 0$

$y = 10$ $\quad$ $y = 8$

If $y = 10$, then $xy = 80$ and $x = 8$.
If $y = 8$, then $xy = 80$ and $x = 10$.

The rug is 8 ft. by 10 ft.

53.

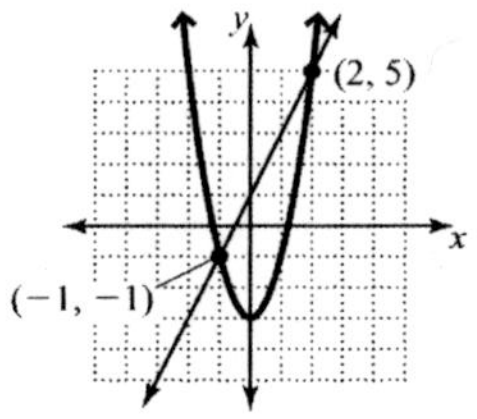

54.

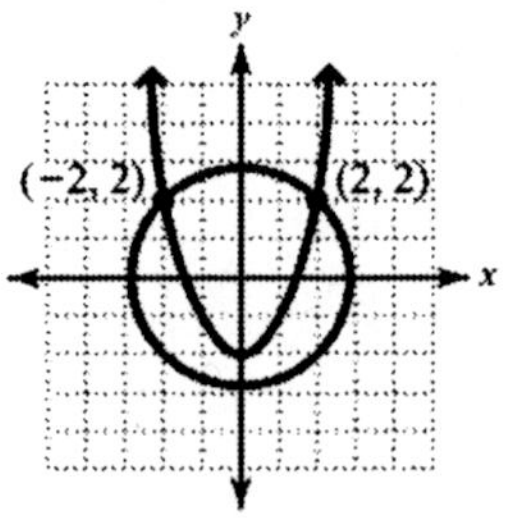

55. $x^2 + y^2 \le 64$

First graph the related equation $x^2 + y^2 = 64$ with a solid line.
Let region 1 be the area inside the circle, and let region 2 be the area outside the circle.
Region 1: Choose (0, 0).

$$0^2 + 0^2 \le 64$$
$$0 + 0 \le 64$$
$$0 \le 64$$

True, so region 1 is in the solution set.
Region 2: Choose (10, 0).

$$10^2 + 0^2 \le 64$$
$$100 + 0 \le 64$$
$$100 \le 64$$

False, so region 2 is not in the solution set.

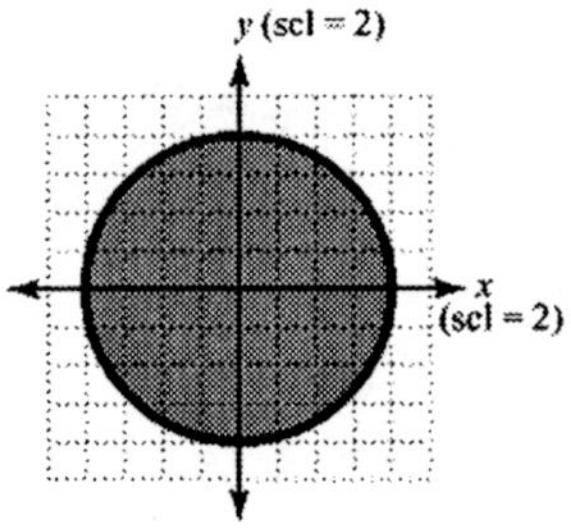

56. $y < 2(x-3)^2 + 4$

First graph the related equation $y = 2(x-3)^2 + 4$ with a dotted line.
Let region 1 be the area above the boundary, and let region 2 be the area below the boundary.
Region 1: Choose (2, 10).

$$10 < 2(2-3)^2 + 4$$
$$10 < 2(-1)^2 + 4$$
$$10 < 2(1) + 4$$
$$10 < 2 + 4$$
$$10 < 6$$

False, so region 1 is not in the solution set.
Region 2: Choose $(0, 0)$.

$$0 < 2(0-3)^2 + 4$$
$$0 < 2(-3)^2 + 4$$
$$0 < 2(9) + 4$$
$$0 < 18 + 4$$
$$0 < 22$$

True, so region 2 is in the solution set.

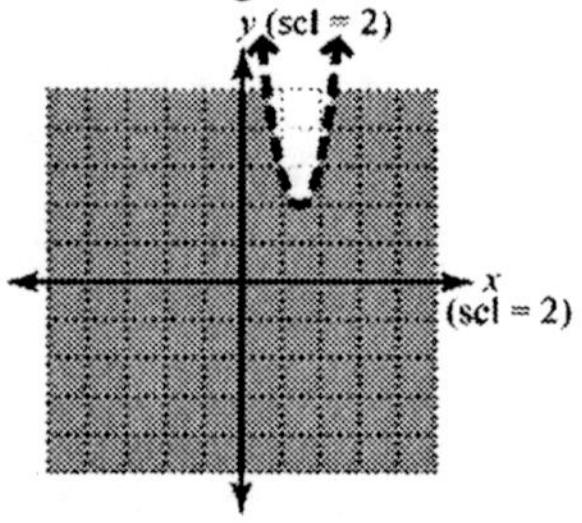

57. $\dfrac{y^2}{25}+\dfrac{x^2}{49}>1$

First graph the related equation $\dfrac{y^2}{25}+\dfrac{x^2}{49}=1$ with a dotted line.
Let region 1 be the area inside the ellipse, and let region 2 be the area outside the ellipse.
Region 1: Choose (0, 0).

$$\frac{0^2}{25}+\frac{0^2}{49}>1$$
$$0+0>1$$
$$0>1$$

False, so region 1 is not in the solution set.
Region 2: Choose (0, 6).

$$\frac{6^2}{25}+\frac{0^2}{49}>1$$
$$\frac{36}{25}+0>1$$
$$\frac{36}{25}>1$$

True, so region 2 is in the solution set.

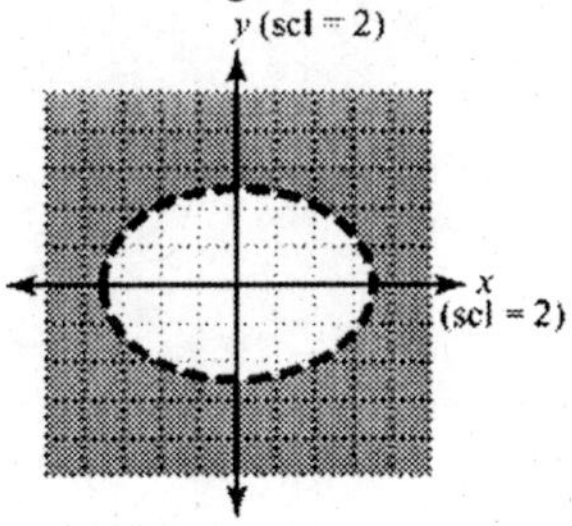

58. $\dfrac{x^2}{25}-\dfrac{y^2}{36}<1$

First graph the related equation $\dfrac{x^2}{25}-\dfrac{y^2}{36}=1$ with a dotted line.
Let region 1 be the area to the left of the left branch of the hyperbola, let region 3 be the area to the right of the right branch of the hyperbola, and let region 2 be the area between the two branches.
Region 1: Choose $(-8,0)$.

$$\frac{(-8)^2}{25}-\frac{0^2}{36}<1$$
$$\frac{64}{25}-\frac{0}{36}<1$$
$$\frac{64}{25}<1$$

False, so region 1 is not in the solution set.

Region 2: Choose $(0,0)$.

$$\frac{0^2}{25}-\frac{0^2}{36}<1$$
$$\frac{0}{25}-\frac{0}{36}<1$$
$$0<1$$

True, so region 2 is in the solution set.
Region 3: Choose $(8,0)$.

$$\frac{(8)^2}{25}-\frac{0^2}{36}<1$$
$$\frac{64}{25}-\frac{0}{36}<1$$
$$\frac{64}{25}<1$$

False, so region 3 is not in the solution set.

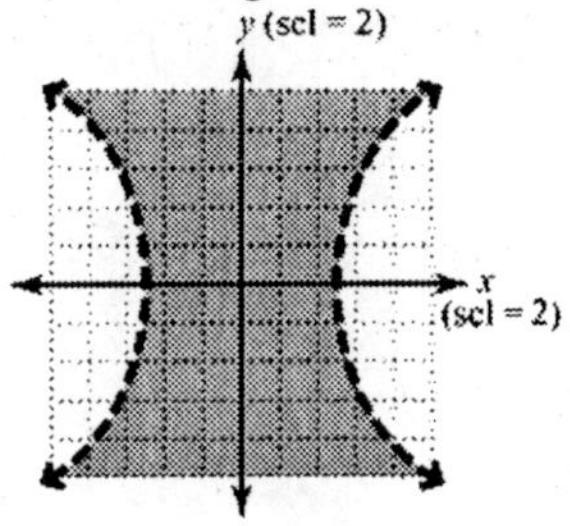

59. $\begin{cases} y \ge x^2-3 \\ 2x+y<2 \end{cases}$

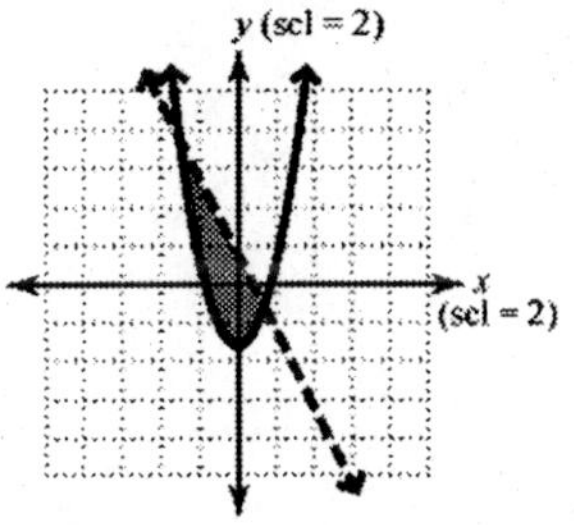

60. $\begin{cases} x+2y<4 \\ x^2+y^2 \le 25 \end{cases}$

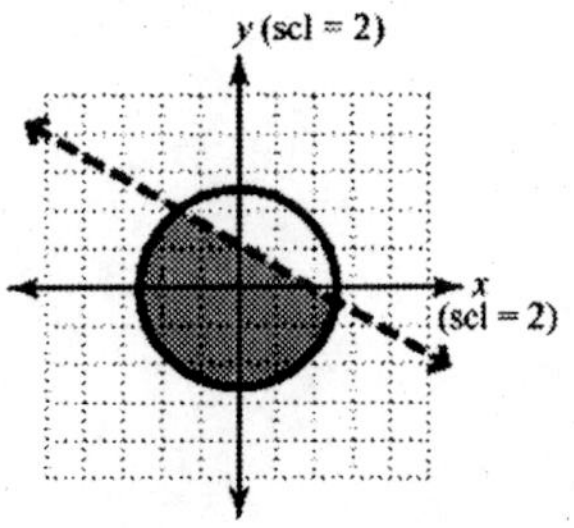

61. $\begin{cases} y > x^2 - 2 \\ y < -x^2 + 1 \end{cases}$

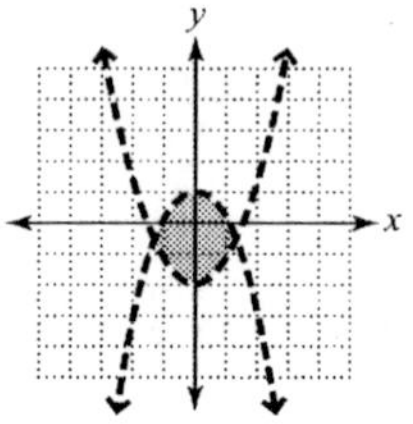

62. $\begin{cases} \dfrac{x^2}{9} + \dfrac{y^2}{25} \le 1 \\ x^2 + y^2 \ge 9 \end{cases}$

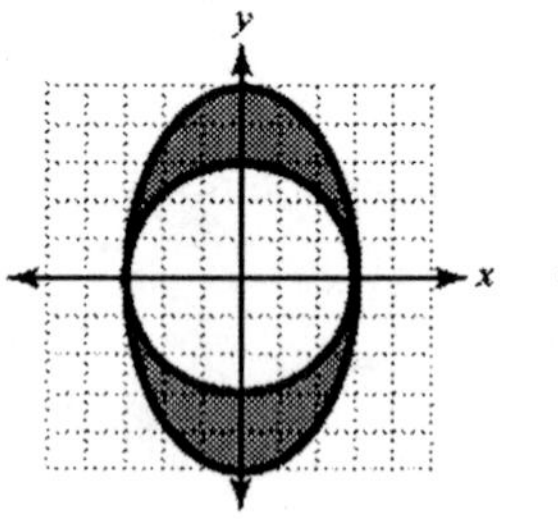

63. $\begin{cases} \dfrac{y^2}{4} - \dfrac{x^2}{9} \le 1 \\ \dfrac{x^2}{9} + \dfrac{y^2}{25} \le 1 \end{cases}$

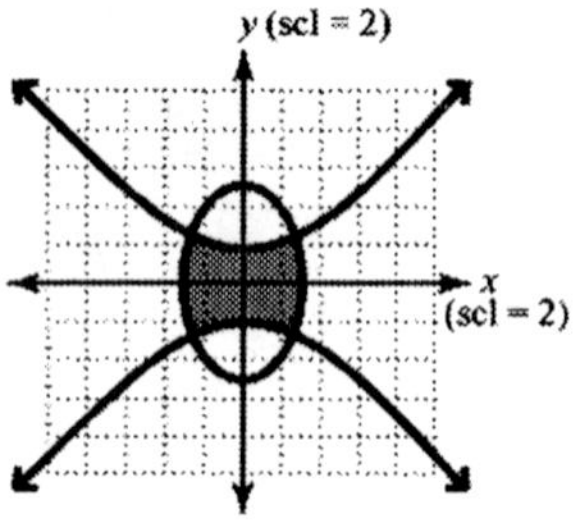

64. $\begin{cases} \dfrac{y^2}{4} + \dfrac{x^2}{16} \le 1 \\ x^2 + y^2 \le 9 \\ y \le x + 1 \end{cases}$

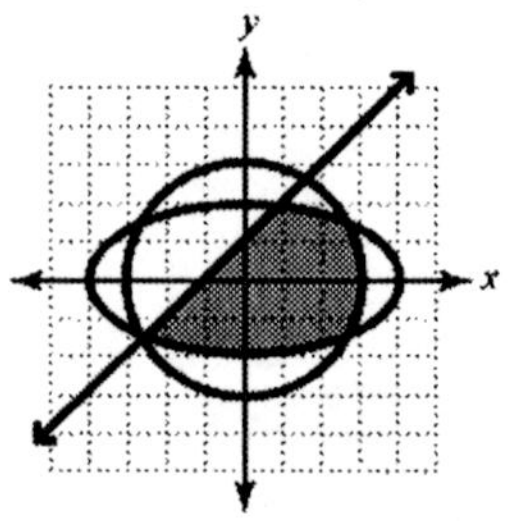

Chapter 11 Practice Test

1. $y = 2(x+1)^2 - 4$; $a = 2$, $h = -1$; and $k = -4$
The parabola opens upward, has its vertex at $(-1,-4)$, and has its axis of symmetry at $x = -1$.

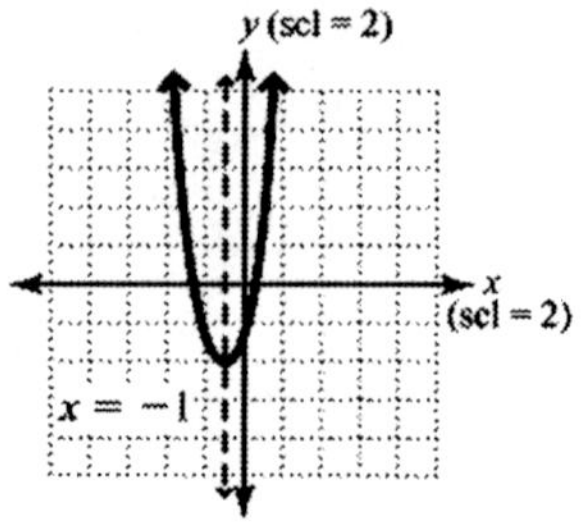

2. $x = -2(y-3)^2 + 1$; $a = -2$, $k = 3$; and $h = 1$
The parabola opens left, has its vertex at $(1,3)$, and has its axis of symmetry at $y = 3$.

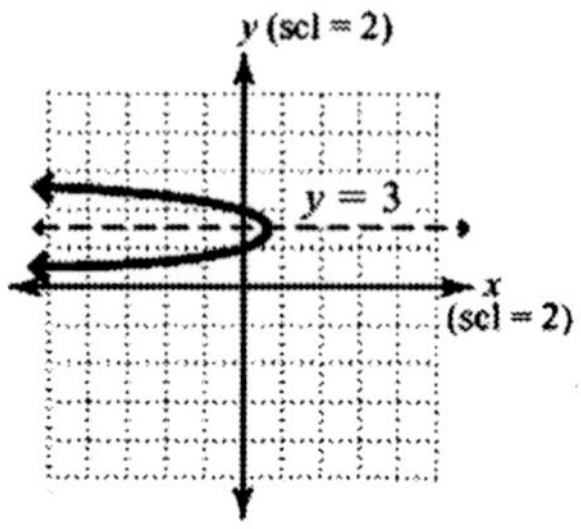

3.
$$\begin{aligned} x &= y^2 + 4y - 3 \\ x + 3 &= y^2 + 4y \\ x + 3 + 4 &= y^2 + 4y + 4 \\ x + 7 &= (y+2)^2 \\ x &= (y+2)^2 - 7 \end{aligned}$$
$a = 1$, $k = -2$; and $h = -7$
The parabola opens right, has its vertex at $(-7,-2)$, and has its axis of symmetry at $y = -2$.

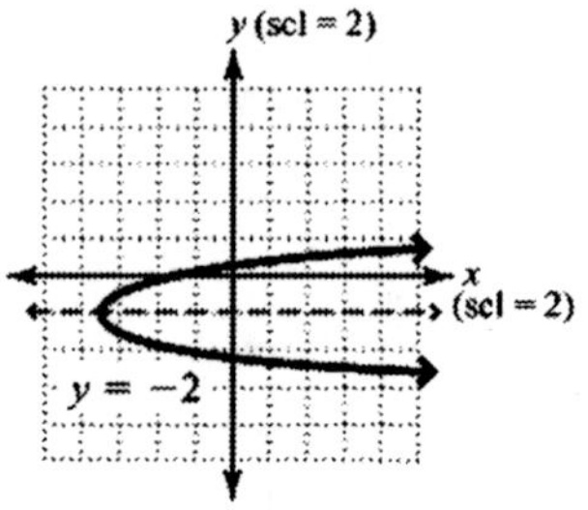

4. $(x_2, y_2) = (2, -3)$ and $(x_1, y_1) = (6, -5)$

$$\begin{aligned} d &= \sqrt{(x_2 - x_1)^2 + (y_2 - y_1)^2} \\ &= \sqrt{(2-6)^2 + (-3-(-5))^2} \\ &= \sqrt{(-4)^2 + (2)^2} \\ &= \sqrt{16+4} \\ &= \sqrt{20} \\ &= 2\sqrt{5} \end{aligned}$$

$$\begin{aligned} \text{Midpoint} &= \left(\frac{x_1 + x_2}{2}, \frac{y_1 + y_2}{2}\right) \\ &= \left(\frac{6+2}{2}, \frac{-5+(-3)}{2}\right) \\ &= \left(\frac{8}{2}, \frac{-8}{2}\right) \\ &= (4, -4) \end{aligned}$$

5. $(x+4)^2 + (y-3)^2 = 36$

Because $h = -4$, $k = 3$, and $r = 6$, the center is $(-4, 3)$ and the radius is 6.

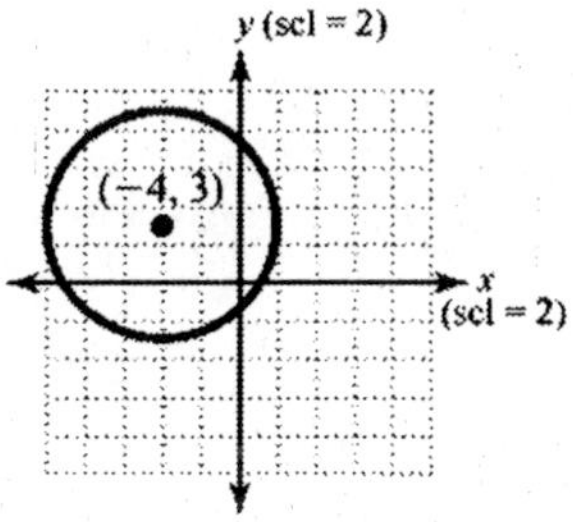

6.
$$\begin{aligned} x^2 + y^2 + 4x - 10y + 20 &= 0 \\ (x^2 + 4x) + (y^2 - 10y) &= -20 \\ (x^2 + 4x + 4) + (y^2 - 10y + 25) &= -20 + 4 + 25 \\ (x+2)^2 + (y-5)^2 &= 9 \end{aligned}$$

Because $h = -2$, $k = 5$, and $r = 3$, the center is $(-2, 5)$ and the radius is 3.

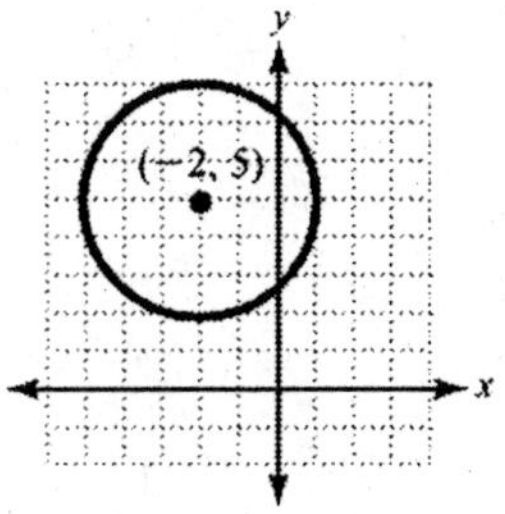

7. The center is $(2, -4)$, so $h = 2$ and $k = -4$. Find the distance from the center to a point on the circle $(-4, 4)$. This is the radius.

$$\begin{aligned} d &= \sqrt{(2-(-4))^2 + (-4-4)^2} \\ &= \sqrt{(6)^2 + (-8)^2} \\ &= \sqrt{36+64} \\ &= \sqrt{100} \\ &= 10 \end{aligned}$$

The equation in standard form is:

$(x-2)^2 + (y+4)^2 = 100$

8. $16x^2 + 36y^2 = 576$

$$\frac{16x^2}{576} + \frac{36y^2}{576} = \frac{576}{576}$$

$$\frac{x^2}{36} + \frac{y^2}{16} = 1$$

$a = 6$, $b = 4$

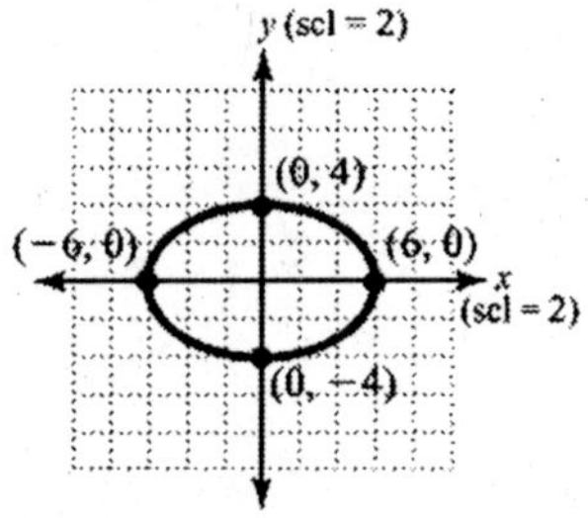

9. $\dfrac{(x+2)^2}{4} + \dfrac{(y-1)^2}{25} = 1$

$a = 2$, $b = 5$, $h = -2$, $k = 1$

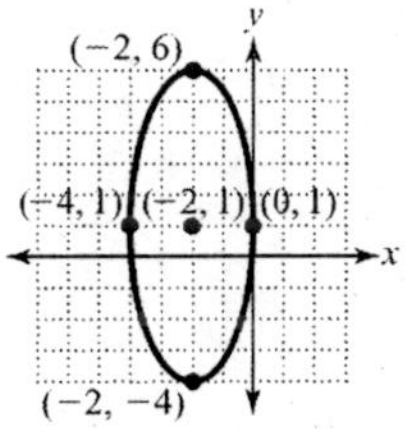

10. $\dfrac{x^2}{49} - \dfrac{y^2}{25} = 1$

$a = 7$, $b = 5$

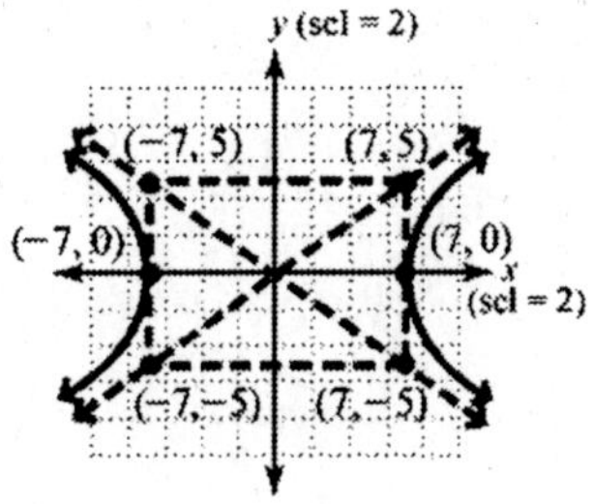

11. $\frac{y^2}{9}-\frac{x^2}{16}=1$

$a=4,\ b=3$

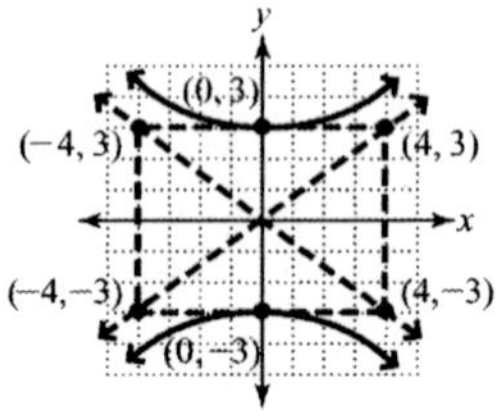

12. $\begin{cases} y=(x+1)^2+2 \\ 2x+y=8 \end{cases}$

Solve the second equation for y. $y=8-2x$.

$$y=(x+1)^2+2$$
$$8-2x=(x+1)^2+2$$
$$8-2x=x^2+2x+1+2$$
$$0=x^2+4x-5$$
$$0=(x+5)(x-1)$$
$$x=-5,1$$

$x=-5$ $\quad y=8-2(-5)$

$y=8+10$

$y=18$

$(-5,18)$

$x=1$ $\quad y=8-2(1)$

$y=8-2$

$y=6$

$(1,6)$

13. $\begin{cases} 3x-y=4 \\ x^2+y^2=34 \end{cases}$

Solve the first equation for y. $y=3x-4$.

$$x^2+y^2=34$$
$$x^2+(3x-4)^2=34$$
$$x^2+9x^2-24x+16=34$$
$$10x^2-24x-18=0$$
$$2(5x^2-12x-9)=0$$
$$5x^2-12x-9=0$$
$$(5x+3)(x-3)=0$$
$$x=-\frac{3}{5},3$$

$x=-\frac{3}{5}$ $\quad y=3\left(-\frac{3}{5}\right)-4$

$y=-\frac{29}{5}$

$\left(-\frac{3}{5},-\frac{29}{5}\right)$

$x=3$ $\quad y=3(3)-4$

$y=9-4$

$y=5$

$(3,5)$

14. $\begin{cases} x^2+y^2=13 \\ 3x^2+4y^2=48 \end{cases}$

$$-3(x^2+y^2=13)$$
$$\underline{3x^2+4y^2=48}$$
$$-3x^2-3y^2=-39$$
$$\underline{3x^2+4y^2=48}$$
$$y^2=9$$
$$y=\pm3$$

$y=-3$ $\quad x^2+(-3)^2=13$

$x^2+9=13$

$x^2=4$

$x=\pm2$

$(-2,-3),(2,-3)$

$y=3$ $\quad x^2+(3)^2=13$

$x^2+9=13$

$x^2=4$

$x=\pm2$

$(2,3),(-2,3)$

15. $\begin{cases} x^2 - 2y^2 = 1 \\ 4x^2 + 7y^2 = 64 \end{cases}$

$$-4\left(x^2 - 2y^2 = 1\right)$$
$$\underline{4x^2 + 7y^2 = 64}$$

$$-4x^2 + 8y^2 = -4$$
$$\underline{4x^2 + 7y^2 = 64}$$
$$15y^2 = 60$$
$$y^2 = 4$$
$$y = \pm 2$$

$y = -2$ $\quad x^2 - 2(-2)^2 = 1$
$$x^2 - 8 = 1$$
$$x^2 = 9$$
$$x = \pm 3$$
$$(3,-2),(-3,-2)$$

$y = 2$ $\quad x^2 - 2(2)^2 = 1$
$$x^2 - 8 = 1$$
$$x^2 = 9$$
$$x = \pm 3$$
$$(3,2),(-3,2)$$

16. $y \le -2(x+3)^2 + 2$

First graph the related equation $y = -2(x+3)^2 + 2$ with a solid line.

Let region 1 be the area above the boundary, and let region 2 be the area below the boundary.

Region 1: Choose $(0,0)$.

$$0 \le -2(0+3)^2 + 2$$
$$0 \le -2(3)^2 + 2$$
$$0 \le -2(9) + 2$$
$$0 \le -18 + 2$$
$$0 \le -16$$

False, so region 1 is not in the solution set.

Region 2: Choose $(-2,-6)$.

$$-6 < -2(-2+3)^2 + 2$$
$$-6 < -2(1)^2 + 2$$
$$-6 < -2 + 2$$
$$-6 < 0$$

True, so region 2 is in the solution set.

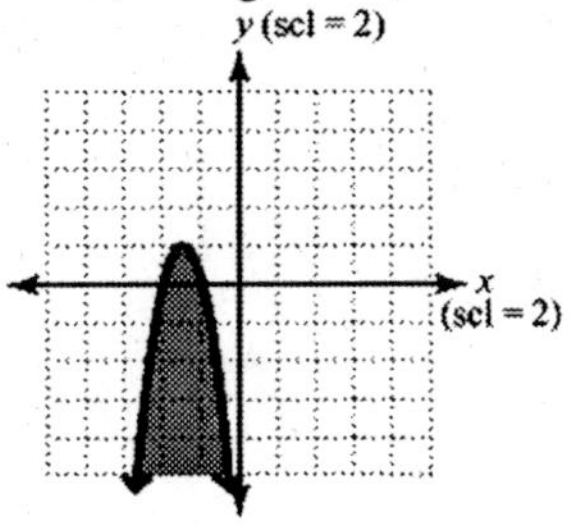

17. $\dfrac{x^2}{4} + \dfrac{y^2}{9} > 1$

First graph the related equation $\dfrac{x^2}{4} + \dfrac{y^2}{9} = 1$ with a dotted line.

Let region 1 be the area inside the ellipse, and let region 2 be the area outside the ellipse.

Region 1: Choose (0, 0).

$$\frac{0^2}{4} + \frac{0^2}{9} > 1$$
$$0 + 0 > 1$$
$$0 > 1$$

False, so region 1 is not in the solution set.

Region 2: Choose (0, 4).

$$\frac{0^2}{4} + \frac{4^2}{9} > 1$$
$$0 + \frac{16}{9} > 1$$
$$\frac{16}{9} > 1$$

True, so region 2 is in the solution set.

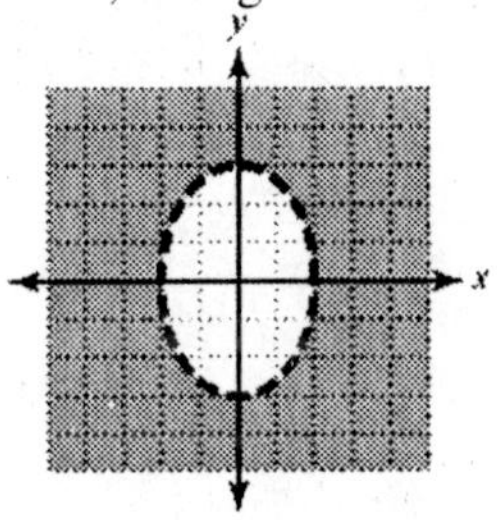

18. $\begin{cases} y \ge x^2 - 4 \\ \dfrac{x^2}{9} + \dfrac{y^2}{16} \le 1 \end{cases}$

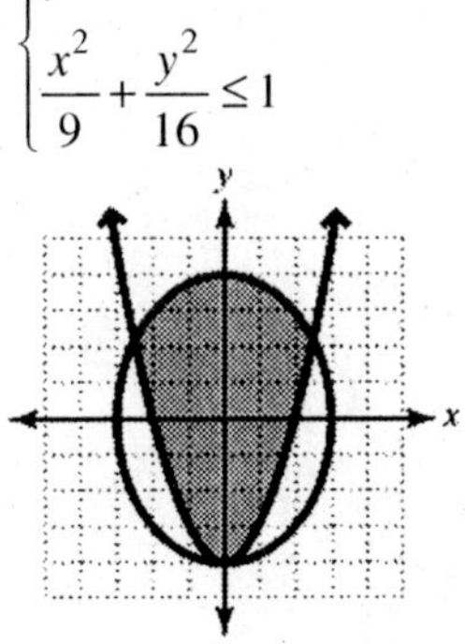

19. $x = 0$: $y = -\frac{1}{2}\cdot 0 + 18$

$y = 18$

The height is 18.

$y = 0$: $0 = -\frac{1}{2}x^2 + 18$

$-18 = -\frac{1}{2}x^2$

$36 = x^2$

$\pm 6 = x$

The distance across the base is 12.

20. $a = \frac{19}{2} = 9.5$ and $b = \frac{2}{2} = 1$

$\frac{x^2}{a^2} + \frac{y^2}{b^2} = 1$

$\frac{x^2}{9.5^2} + y^2 = 1$

Chapters 1-11 Cumulative Review

1. False; a positive base raised to a negative exponent yields a reciprocal.
2. True
3. True
4. $a_1b_2 - a_2b_1$
5. i
6. $d = \sqrt{(x_2 - x_1)^2 + (y_2 - y_1)^2}$
7. $6 - (-3) + 2^{-4} = 6 + 3 + \frac{1}{16} = 9\frac{1}{16}$
8. $(5m^3n^{-2})^{-3} = 5^{-3}m^{3(-3)}n^{-2(-3)}$

$= \frac{1}{5^3}m^{-9}n^6$

$= \frac{n^6}{125m^9}$

9. $\sqrt{-20} = \sqrt{-4\cdot 5} = 2i\sqrt{5}$
10. $ax + bx + ay + by = x(a+b) + y(a+b)$

$= (a+b)(x+y)$

11. $k^4 - 81 = (k^2+9)(k^2-9)$

$= (k^2+9)(k+3)(k-3)$

12. $|2x+1| = |x-3|$

$2x+1 = x-3$ or $2x+1 = -(x-3)$

$x = -4$ $\quad 2x+1 = -x+3$

$3x = 2$

$x = \frac{2}{3}$

13. $\sqrt{x-3} = 7$

$\left(\sqrt{x-3}\right)^2 = 7^2$

$x - 3 = 49$

$x = 52$

14. $4x^2 - 2x + 1 = 0$

$x = \frac{-(-2) \pm \sqrt{(-2)^2 - 4(4)(1)}}{2(4)} = \frac{2 \pm \sqrt{4-16}}{8}$

$= \frac{2 \pm \sqrt{-12}}{8} = \frac{2 \pm 2i\sqrt{3}}{8} = \frac{1 \pm i\sqrt{3}}{4}$

15. $9 + 24x^{-1} + 16x^{-2} = 0$

$x^2(9 + 24x^{-1} + 16x^{-2}) = x^2 \cdot 0$

$9x^2 + 24x + 16 = 0$

$(3x+4)^2 = 0$

$\sqrt{(3x+4)^2} = 0$

$3x + 4 = 0$

$x = -\frac{4}{3}$

16. $9^x = 27$

$(3^2)^x = 3^3$

$3^{2x} = 3^3$

$2x = 3$

$x = \frac{3}{2} = 1.5$

17. $\log_3(x+1) - \log_3 x = 2$

$\log_3 \frac{x+1}{x} = 2$

$3^2 = \frac{x+1}{x}$

$9x = x + 1$

$8x = 1$

$x = \frac{1}{8}$

18. $5m^2 - 3m = 0$

$m(5m-3) = 0$

$m = 0$ or $5m - 3 = 0$

$m = \frac{3}{5}$

For $(-\infty, 0)$, we choose -1.

$5(-1)^2 - 3(-1) < 0$ False

$5 + 3 < 0$

$8 < 0$

For $\left(0, \frac{3}{5}\right)$, we choose $\frac{1}{5}$.

$5\left(\frac{1}{5}\right)^2 - 3\left(\frac{1}{5}\right) < 0$ True

$\frac{1}{5} - \frac{3}{5} < 0$

$-\frac{2}{5} < 0$

For $\left(\frac{3}{5}, \infty\right)$, we choose 2.

$5(2)^2 - 3(2) < 0$ False

$20 - 6 < 0$

$14 < 0$

a) 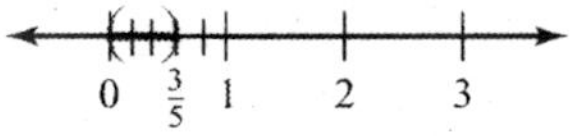

b) $\left\{ m \middle| 0 < m < \frac{3}{5} \right\}$

c) $\left(0, \frac{3}{5}\right)$

19. a) $(f+g)(x) = (2x+1) + (x^2 - 1)$

$= x^2 + 2x$

b) $(f-g)(x) = (2x+1) - (x^2 - 1)$

$= (2x+1) + (-x^2 + 1)$

$= -x^2 + 2x + 2$

c) $(f \cdot g)(x) = (2x+1)(x^2 - 1)$

$= 2x^3 - 2x + x^2 - 1$

$= 2x^3 + x^2 - 2x - 1$

d) $(f/g)(x) = \frac{2x+1}{x^2-1}; x \neq \pm 1$

e) $(f \circ g)(x) = f[g(x)]$

$= f(x^2 - 1)$

$= 2(x^2 - 1) + 1$

$= 2x^2 - 2 + 1$

$= 2x^2 - 1$

f) $(g \circ f)(x) = g[f(x)]$

$= g(2x+1)$

$= (2x+1)^2 - 1$

$= 4x^2 + 4x + 1 - 1$

$= 4x^2 + 4x$

20. $\log_5 125 = 3$

21. $\log_5 x^5 y = \log_5 x^5 + \log_5 y = 5\log_5 x + \log_5 y$

22. Find the distance from the center to a point on the circle. This is the radius.

$r = \sqrt{(2-7)^2 + (4-16)^2}$

$= \sqrt{(-5)^2 + (-12)^2}$

$= \sqrt{25 + 144}$

$= \sqrt{169}$

$= 13$

The equation in standard form is:

$(x-2)^2 + (y-4)^2 = 13^2$

$(x-2)^2 + (y-4)^2 = 169$

23. $\begin{cases} y < -x + 2 \\ y \geq x - 4 \end{cases}$

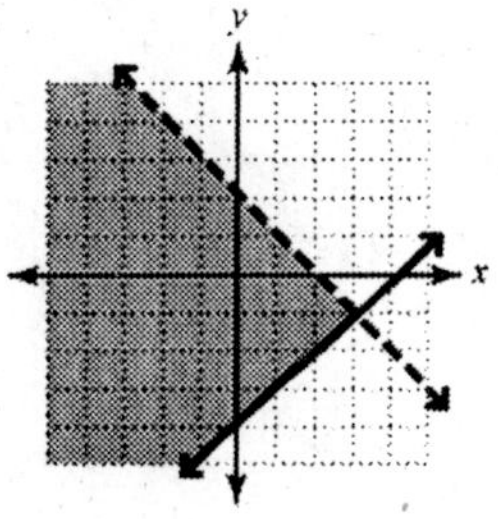

24. $f(x) = \log_3 x$

$$y = \log_3 x$$

$$3^y = x$$

y	-1	0	1	2
x	$\frac{1}{3}$	1	3	9

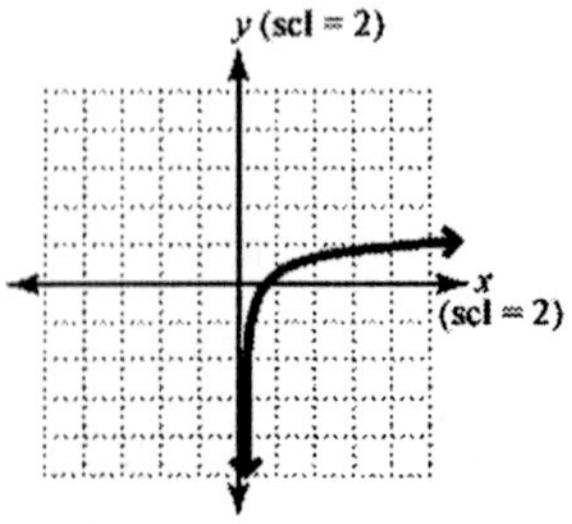

25. $\frac{x^2}{4} + \frac{y^2}{9} = 1$

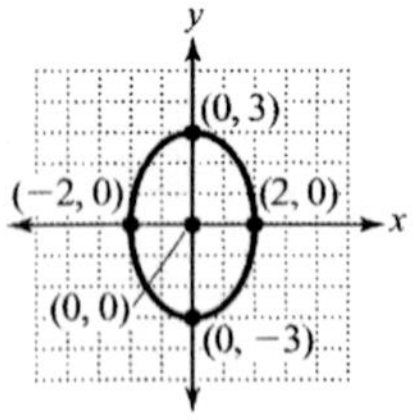

26. a) $V = \pi r^2 h$

$$\frac{V}{\pi r^2} = \frac{\pi r^2 h}{\pi r^2}$$

$$h = \frac{V}{\pi r^2}$$

b) $h = \frac{V}{\pi r^2}$

$$h = \frac{15\cancel{\pi}}{\cancel{\pi} r^2}$$

$$h = \frac{15}{r^2}$$

27. Let s be pounds of shrimp and t be pounds of tuna consumed.

Translate to a system of equations. $\begin{cases} s + t = 6.3 \\ s = 0.5 + t \end{cases}$

$0.5 + t + t = 6.3$ $\qquad s = 0.5 + 2.9$

$2t = 5.8$ $\qquad s = 3.4$

$t = 2.9$

2.9 lbs. of tuna and 3.4 lbs. of shrimp

28. Translate to a system of equations.

Let x be the money market fund, y be the income fund, and z be the growth fund.

$$\begin{cases} x + y + z = 5000 \\ 0.05x + 0.06y + 0.03z = 255 \\ z = x - 500 \end{cases}$$

$$x + y + (x - 500) = 5000$$
$$0.05x + 0.06y + 0.03(x - 500) = 255$$

$$x + y + x - 500 = 5000$$
$$0.05x + 0.06y + 0.03x - 15 = 255$$

$$-6(2x + y = 5500)$$
$$100(0.08x + 0.06y = 270)$$

$$-12x - 6y = -33,000$$
$$8x + 6y = 27,000$$

$$-4x = -6000$$
$$x = 1500$$

$z = 1500 - 500$ $\qquad 1500 + y + 1000 = 5000$

$z = 1000$ $\qquad 2500 + y = 5000$

$y = 2500$

\$1500 at 5%, \$2500 at 6%, and \$1000 at 3%

29. $M = \log \frac{I}{I_0}$

$$M = \log \frac{10^{2.9}}{10^{-4}}$$

$$M = \log 10^{6.9}$$

$$M = 6.9$$

30. $\frac{x^2}{75^2} + \frac{y^2}{20^2} = 1$

$$\frac{x^2}{5625} + \frac{y^2}{400} = 1$$